Food and Culture

THIRD EDITION

Pamela Goyan Kittler, M.S.
Cultural Nutrition Consultant

Kathryn P. Sucher, Sc.D., R.D.
Department of Nutrition and Food Science
San Jose State University

Australia | Canada | Mexico | Singapore | Spain
United Kingdom | United States

WADSWORTH
THOMSON LEARNING™

Publisher: Peter Marshall
Development Editor: Lisa Bahlinger
Assistant Editor: John Boyd
Editorial Assistant: Andrea Kesterke
Marketing Manager: Becky Tollerson
Project Editor: Sandra Craig
Print Buyer: Tandra Jorgensen
Permissions Editor: Stephanie Keough
Production: The Book Company
Text and Cover Designer: Devenish Design
Photo Researcher: Myrna Engler
Copy Editor: Laura Larson
Illustrator: Impact Publications
Cover Image: ©PhotoDisc
Compositor: Parkwood Composition
Printer: R. R. Donnelley, Willard

Printed in the United States of America
2 3 4 5 6 7 04 03 02 01

PHOTO CREDITS, Color Insert
STARCH FOODS: A. © www.comstock.com, B. © Becky Luigart-Stayner/CORBIS, C. © Danny Lehman/CORBIS, D. © Wolfgang Kaehler/CORBIS, © CORBIS; E. © 2000 PhotoDisc, Inc. PROTEIN FOODS: A. © 2000 PhotoDisc, Inc., B. © Kevin Fleming/CORBIS, C. © Pat O'Hara/CORBIS, D. © www.comstock.com, E. Courtesy of United Soybean Board. *Continued on page 496 which is a continuation of this copyright page.*

Wadsworth/Thomson Learning
10 Davis Drive
Belmont, CA 94002-3098
USA

For more information about our products, contact us:
Thomson Learning Academic Resource Center
1-800-423-0563
http://www.wadsworth.com

International Headquarters
Thomson Learning
International Division
290 Harbor Drive, 2nd Floor
Stamford, CT 06902-7477
USA

UK/Europe/Middle East/South Africa
Thomson Learning
Berkshire House
168-173 High Holborn
London WC1V 7AA
United Kingdom

Asia
Thomson Learning
60 Albert Street, #15-01
Albert Complex
Singapore 189969

Canada
Nelson Thomson Learning
1120 Birchmount Road
Toronto, Ontario M1K 5G4
Canada

Library of Congress Cataloging-in-Publication Data

Kittler, Pamela Goyan,
Food and culture / Pamela Goyan Kittler, Kathryn P. Sucher.—3rd ed.
p. cm.
Rev. ed. of: Food and culture in America. 2nd ed. c1998.
Includes bibliographical references and index.
ISBN 0-534-55164-5
1. Nutrition—United States—Cross-cultural studies. 2. Food habits—United States. I. Sucher, Kathryn. II. Kittler, Pamela Goyan, 1953– Food and culture in America. III. Title.
TX357 .K58 2001
613.2'0973—dc21

00-045588

Contents

Preface

The topic of food and culture is both compelling and complex. Anthropologists, folklorists, geographers, historians, psychologists, and sociologists have all been attracted to the subject. Research has varied from anecdotal to academic. Yet despite these efforts, information useful for practicing health and nutrition professionals has been lacking.

The need for cultural nutrition resources is undisputed. The population of the United States is increasingly heterogeneous, moving toward a plurality of uncounted ethnic, religious, and regional groups. Each of these minorities has traditional foods and food habits that may differ significantly from the so-called typical North American diet. Effective nutrition counseling, education, and food service require that these variations be acknowledged—diet is best understood within the context of culture. It is our goal to provide dietitians, nutritionists, and food service professionals with the cultural overview needed to avoid ethnocentric assumptions and the nutritional specifics concerning each group discussed. We have attempted to combine the conceptual with the technical in a way that is useful to other health professionals as well.

How the Book Is Organized

The first four chapters form an introduction to the study of food and culture. Chapter 1 discusses changing demographics, the ways in which ethnicity may affect nutrition and health status, and methods for understanding food and food habits within the context of culture. Chapter 2 focuses on the role of diet in traditional health beliefs. Some intercultural communication strategies are suggested in Chapter 3, and Chapter 4 outlines the major Eastern and Western religions and reviews their dietary practices in detail.

Chapters 5 through 14 profile North American ethnic groups and their cuisines. We have chosen breadth over depth, discussing groups with significant populations in the United States, as well as smaller, more recent, immigrant groups who have had an impact on the health care system. Other groups with low numbers of immigrants but notable influences on American cooking are briefly mentioned. Groups are considered in the approximate order of their arrival in North America.

Each chapter begins with a history of the group in the United States and current demographics. Worldview (outlook on life) is then examined, including religion, family structure, and traditional health practices. This background information illuminates the cultural context from which ethnic foods and food habits emerge and evolve. The next section of each chapter outlines the traditional diet, including ingredients, some common dishes, meal patterns, special occasions, the role of food in the society, and therapeutic uses of food. The final section explains the contemporary diet of the group, such as adaptations made by the group after arrival in the United States and influences of the group on the American diet. Reported nutritional status is reviewed, and counseling guidelines are provided.

One or more cultural food group tables are found in each of the ethnic group chapters. The emphasis is on ingredients common to the populations of the region. Important variations within regions and unique food habits are listed under the "Comments" column of the table. Known adaptations in the United States are also noted. The tables are intended as references for the reader; they do not replace either the chapter content or an in-depth interview with a client.

Chapter 15 considers the regional American fare of the Northeast, the Midwest, the South, and the West. Each section includes demographic data and an examination of the foods common in the region, followed by state-by-state descriptions of cuisine. This chapter brings the study of American cultural nutrition full circle, discussing the significant influences of different ethnic and religious groups on regional fare. Canadian regional cooking is also briefly considered.

New to This Edition

- Chapter 2 has been expanded to include an overview of botanical therapeutics in its discussion of traditional health beliefs.
- New information on nutrition education approaches has been added to Chapter 3, on intercultural communication strategies.
- A new section on South Americans and several shorter discussions of unique cuisines worldwide have been added.
- Several new tables of selected botanical remedies have been added to the chapters on various ethnic groups.
- The glossary of ethnic ingredients, designed for quick referral to foods mentioned in the chapters and cultural food group tables, has been expanded.
- The food and culture resources section has been updated to include Internet sites of interest.

Before You Begin

Food is so essential to ethnic, religious, and regional identity that dietary descriptions must be as objective as possible to prevent inadvertent criticism of the underlying culture. Yet as members of two Western ethnic and religious groups, we recognize that our own cultural assumptions are unavoidable and, in fact, serve as a starting point for our work. One would be lost without such a cultural footing. Any instances of bias are unintentional.

Any definition of a group's food and food habits implies homogeneity in the described group. In daily life, however, each member of a group has a distinctive diet, combining traditional practices with new influences. We do not want to stereotype the fare of any cultural group. Rather, we strive to generalize common U.S. food and culture trends as a basis for understanding the personal preferences of individual clients.

We have tried to be sensitive to the designations used by each cultural group, though sometimes there is no consensus among group members regarding the preferred name for the group. Also, there may be some confusion about dates in the book. Nearly all religious traditions adhere to their own calendar of events based on solar or lunar months. These calendars frequently differ from the Gregorian calendar used throughout most of the world in business and government. Religious ceremonies often move around according to Gregorian dates, yet usually they are calculated to occur in the correct season each year. Historical events in the text are listed according to the Gregorian calendar, using the terms before common era (B.C.E.) and common era (C.E.).

We believe this book will do more than introduce the concepts of food and culture. It should also encourage self-examination and individual cultural identification by the reader. We hope that it will help dietitians, nutritionists, other health care providers, and food service professionals work effectively with members of different ethnic, religious, and regional groups. If it sparks a gustatory interest in the foods of the world, we will be personally pleased. *De gustibus non est disputatum!*

Acknowledgments

The authors want to thank the many colleagues who have graciously given support and advice in the development of the third edition of this book: Jennifer Berg, New York University; Stella Cash, Michigan State University; Sharon Davis, University of Nebraska-Kearney; Jan Goodwin, University of North Dakota; Annie Hauck-Lawson, Brooklyn College; Faye Johnson, California State University-Chico; Jana Kicklighter, Georgia State University; Louise Little, University of Delaware; Audrey McCool, University of Las Vegas-Nevada; Audrey A. Spindler, San Diego State University; Christine E. Thompson, Johnson & Wales University; Josephine Umoren, Northern Illinois University.

1

Food and Culture

What do Americans eat? Meat and potatoes, according to popular myth. There's no denying that more beef is consumed than any other protein food in the United States and that franchise restaurants sell more than $5 billion worth of hamburgers and french fries each year. Yet the American diet cannot be so simply described. Just as the population of the United States contains many different ethnic and cultural groups, so are the foods and food habits of Americans equally diverse. It can no more be said that the typical U.S. citizen is white, Anglo-Saxon, and Protestant than it can be stated that meat and potatoes are what this typical citizen eats.

U.S. census and other demographic data show that one in every four Americans is of non-European heritage. These figures greatly underestimate the number and diversity of North American cultural groups, however. For instance, data often do not list members of white ethnic or religious groups, nor do they account for mixed ancestry. Census terminology, such as the category "Hispanics" (which may be defined as persons who were born in Latin America, whose parents were born in Latin America, who have a Spanish surname, or who speak Spanish), is sometimes ambiguous and confusing to census respondents. Furthermore, the census does not survey the millions of U.S. residents who are not citizens (it is estimated 200,000 arrive in America each year). Thus, the proportion of American ethnic group members is larger than statistics indicate, and, more important, it is rapidly increasing.

Asians are the fastest-growing ethnic group in America. Their population more than doubled between 1980 and 1990; four out of every ten immigrants to the United States in 1993 were born in Asia. African Americans are numerically the largest ethnic group (approximately 12 percent of the total U.S. population), although Latinos, with a growth rate of more than 50 percent between 1980 and 1990, are expected to surpass blacks by the middle of the 21st century.

Each American ethnic group has its own culturally based foods and food habits. Many of these traditions have been influenced and modified through contact with the majority culture. The foods and food habits of the majority culture have, in turn, been affected by those of the many diverse ethnic groups. Today a fast-food restaurant or street stand is just as likely to offer pizza, tacos, egg rolls, or falafel as it is to offer hamburgers. The American diet encompasses the numerous varied cuisines of the U.S. population. To understand this diet fully, one must study not only the traditional foods and food habits of the many ethnic groups but also the interactions between these traditions and those of the majority culture.

■ As suggested by their names, not even hamburgers and french fries are American in origin. Chopped beef steaks were introduced to the United States from the German city of Hamburg in the late nineteenth century and were popularized at the St. Louis World's Fair. Although the potato is a New World vegetable, it was brought to America by the Irish in 1719. The term *french fried potatoes* first appeared in the 1860s and may have come from the way the potatoes were cut or cooked. Other foods considered typically American also have foreign origins, such as hot dogs (*frankfurters*), apple pie, and ice cream.

■ Data from the 1996 Canadian census indicate that 15 percent of the population is of non-European heritage. The largest ethnic group is Native Americans (called Aboriginals in the report), followed by the Chinese and South Asians (mostly Asian Indians and Sri Lankans). Asians and Middle Easterners are the fastest-growing minorities.

■ Americans collectively consume approximately 900 billion calories each day.

Food and Culture

What Is Food?

Feeding versus Eating

Food, as defined in the dictionary, is any substance that provides the nutrients necessary to maintain life and growth when ingested. When animals feed, they repeatedly consume those foods necessary for their well-being, and they do so in a similar manner at each feeding.

Humans, however, do not feed. They eat. Eating is distinguished from feeding by the ways in which humans use food. Humans not only gather or hunt food, but they also cultivate plants and raise livestock. Food is thus regularly available to most humans, permitting the development of food habits, such as the setting of mealtimes. In addition, humans cook food, which greatly expands the number and variety of edible substances available. Choice of what to eat follows. Humans use utensils to eat food and create complex rules, commonly called manners, about how food is actually ingested. Humans share food. Standards for who may dine with whom in each eating situation are well defined.

■ Drinking rituals, such as making a toast and participating in the round of drinks, date back to ancient magical rites of communal inebriation.

■ A recent survey reports that the top three comfort foods for women are ice cream, chocolate, and cookies; men prefer ice cream, soup and pizza or pasta (Wansink & Sangerman, 2000).

Development of Food Habits

The term *food habits* refers to the ways in which humans use food, including how food is obtained and stored, how it is prepared, how it is served and to whom, and how it is consumed. A. H. Maslow's theory of human maturation as applied to food habits (Lowenberg, 1970) explains how food use progresses from eating for existence to eating for self-actualization:

1. *Physical needs for survival:* This is the most basic use of food, nearly equivalent to feeding. Daily nutrient needs must be met before more complex food use can occur.
2. *Social needs for security:* Once the immediate need for food is satisfied, future needs can be considered. The storage of food, in a granary or in a refrigerator, represents security.
3. *Belongingness:* This use of food shows that an individual belongs to a group. The need to belong is satisfied by consuming the foods that are eaten by the social group as a whole. These foods represent comfort and happiness for many people; during periods of stress or illness, people often want the foods they ate during childhood.

Humans create complex rules, commonly called manners, as to how food is actually eaten. (© Tom and DeeAnn McCarthy/PhotoEdit.)

Sometimes people adopt a special diet to demonstrate belongingness. For example, African Americans who live outside the South may choose to eat what is called *soul food* (typically southern black cuisine, such as pork ribs and greens) on certain occasions as an expression of ethnic identity.

Etiquette, the appropriate use of food, is also a way of demonstrating belonging. Entirely different manners are required when lunching with business associates at an expensive restaurant, when attending a tea, when eating in the school cafeteria, when drinking with friends at a bar, or when picnicking with a date.

4. *Status:* Food can be used to define social position. Champagne and caviar imply wealth; mesquite-grilled foods and goat cheese suggest upward mobility; beans or potatoes are traditionally associated with the poor. Status foods are used for social interaction. When a man picks up his date, he brings her chocolates, not broccoli. Wine is considered an appropriate gift to a hostess; a gallon of milk is not.

   In general, eating with someone connotes social equality with that person. Many societies regulate who can dine together as a means of establishing class relationships. Women and children may eat separately from men, or servants may eat in the kitchen, away from their employers. This separation by class was also seen in some U.S. restaurants that excluded blacks before the civil rights legislation of the 1960s.

5. *Self-realization:* This stage of food use occurs when all previous stages have been achieved to the individual's satisfaction. Personal preference takes precedence, and the individual may experiment with the foods of different ethnic or economic groups.

Food as Self-Expression

The correlation between what people eat, how others perceive them, and how they characterize themselves is striking. In one study, researchers listed foods typical of five diets: vegetarian (broccoli quiche, brown rice, avocado and bean sprout sandwich), gourmet (oysters, caviar, French roast coffee), health food (protein shake, wheat germ, yogurt), fast food (Kentucky Fried Chicken, Big Mac, pizza), and synthetic food (Carnation Instant Breakfast, Cheez Whiz). It was found that each category was associated with a certain personality type. Vegetarians were considered to be pacifists and likely to drive foreign cars. Gourmets were believed to be liberal, and sophisticated. Health food fans were described as antinuclear activists, and Democrats. Fast-food and synthetic food eaters were believed to be religious, conservative, and wearers of polyester clothing. These stereotypes were confirmed by self-description and personality tests completed by persons whose diets fell within the five categories (Sadella & Burroughs, 1981).

Another study asked college students to rate profiles of people based on their diets. The persons who ate "good" foods were judged thinner, more fit, and more active than persons with the identical physical characteristics and exercise habits who ate "bad" foods. Furthermore, the people who ate good foods were perceived by some students as being more attractive, likable, practical, methodical, quiet, and analytical than people who ate bad foods. The researchers attribute the strong morality–food effect to several factors, including the concept that "you are what you eat" and a prevailing Puritan ethic that espouses self-discipline (Stein & Nemeroff, 1995).

Food choice is, in fact, influenced by self-identity, a process whereby the food likes or dislikes of someone else are accepted and internalized as personal preferences. Research suggests that children choose foods that are eaten by admired adults (e.g., teachers), fictional characters, peers, and especially older siblings. Group approval or disapproval of a food can also condition a person's acceptance or rejection. This may explain why certain relatively unpalatable items, such as chile peppers or unsweetened coffee, are enjoyed if introduced through socially mediated events, such as family meals or workplace snack breaks. Though the mechanism for the internalization of food preference and self-identity are not well understood, it is considered a significant factor in the development of food habits (Rozin, 1996). A study on the consumption of organic vegetables,

■ The status of food can change over time. In the early years of the United States, lobster was so plentiful it was considered fit only for the poor (Root & deRochemont, 1976).

■ In a study of the morality–food effect, researchers found consistent agreement among subjects that "good" foods were healthy and unfattening, especially fruit, salad, home-made whole-wheat bread, chicken, and potatoes. "Bad" foods were identified as unhealthy and fattening, in particular, steak, hamburgers, french fries, doughnuts, and double-fudge ice cream sundaes (Stein & Nemeroff, 1995).

■ Studies suggest that contrary to popular assumption, parents have little lasting influence on the food preferences of their children.

for example, found that people who identified themselves as "green" (one who is concerned with ecology and makes consumer decisions based on this concern) predicted an intention to eat organic items independent of other attitudes, such as perceived flavor and health benefits (Shepard & Raats, 1996).

Food as self-expression is especially evident in the experience of dining out. Researchers suggest that restaurants often serve more than food; they also meet emotional needs such as belongingness, status, and self-realization. In Japan, for example, homes are private; therefore, guests are entertained in the homelike environment of a restaurant. The host chooses and pays for the meal ahead of time, the guests are all served the same dishes, and the servers are expected to be part of the conversation. Although some segments of the American restaurant business also cater to the family meal experience (i.e., those that offer playgrounds or children's entertainment), others emphasize other dining functions—the business club for financial transactions, or the candlelit neighborhood restaurant for romantic interactions (Wood, 1995). The newest restaurant, with an acclaimed chef and a need to make reservations a month in advance, represents status to some people. Ethnic restaurants appeal to those individuals seeking authenticity in the foods of their homeland or are a novelty to those interested in culinary adventure. Conversely, exposure to different foods in restaurants is sometimes the first step in adopting new food items at home (McComber & Postel, 1992).

Symbolic Use of Food

It is clear from the various uses of food that, for humans, food is more than simply nutrients. Humans use foods symbolically. A *symbol* is something that suggests something else due to relationship, association, or convention. Bread is an excellent example of food symbolism. Bread is the "staff" of life; one "breaks bread" with friends; bread represents the body of Christ in the Christian sacrament of communion. White bread was traditionally eaten by the upper classes, dark bread by the poor. A person of wealth has a lot of "bread," and whole-wheat bread is eaten by people in the United States who are concerned more with health than with status. It is the symbolic use of food that is important to each cultural group. The foods and food habits of each group are often associated with religious beliefs or ethnic behaviors. Eating, like dressing in traditional clothing or speaking in a native language, is a daily reaffirmation of cultural identity (Figure 1.1).

■ The symbolic importance of bread can be seen in some of the superstitions associated with it: Greek soldiers took a piece from home to ensure their safe, victorious return; English midwives placed a loaf at the foot of the mother's bed to prevent the woman and her baby from being stolen.

Role of Culture in Food Habits

Definition of Culture

Culture is broadly defined as the values, beliefs, attitudes, and practices accepted by a community of individuals. Cultural behavior patterns are reinforced when a group is isolated by geography or segregated by socioeconomic status. Culture is learned, not inherited; it is passed from generation to generation through a process called *enculturation* (Plawecki, 1992).

Cultural membership is defined by the term *ethnicity*. Unlike national origin or race (which may include numerous ethnic groups), it is a social identity associated with shared behavior patterns, including food habits, dress, language, family structure, and often religious affiliation. Members of the same ethnic group usually have a common heritage through locality or history and participate together with other cultural groups in a larger social system. As part of this greater community, each ethnic group may have different status or positions of power. Diversity within each cultural group, called *intraethnic variation*, is also common due to racial, regional, or economic divisions as well as differing rates of acculturation to the majority culture (Harwood, 1981).

When people from one ethnicity move to an area with different cultural norms, adaptation to the new majority society begins. This process is known as *acculturation*, and it takes place along a continuum of behavior patterns. Typically, first-generation immigrants remain emotionally connected to their culture of origin. They integrate into their new society by adopting some majority culture values and practices, but generally surround themselves with a reference group of family and friends

Figure 1.1

An edible map—food-related names of cities and towns in the United States. Food often means more than simply nutrients.

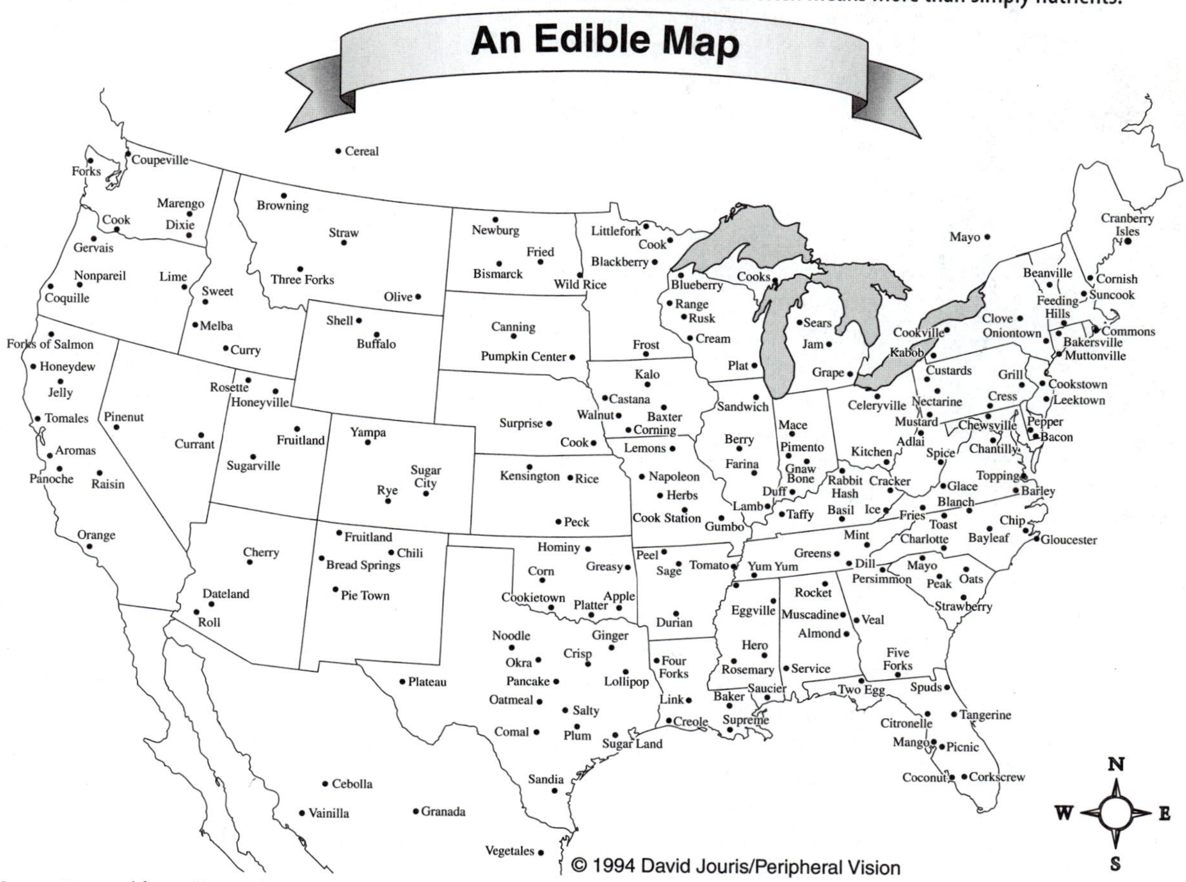

Source: Excerpted from *All Over the Map,* copyright 1994 by David Jouris, with permission from Ten Speed Press, P.O. Box 7123, Berkeley, CA 94707. Reprinted with permission.

who are from their ethnic background. For example, Asian Indians living in the United States who consider themselves to be "mostly or very Asian Indian" may encourage their children to speak English and allow them to celebrate American holidays, but they do not permit them to date non-Asian Indian peers (Sodowsky & Carey, 1988). Other immigrants become *bicultural,* which happens when the new majority culture is seen as complementing, rather than competing with, an individual's ethnicity. The positive aspects of both societies are embraced and the individual develops the skills needed to operate within either culture (Brookins, 1993). Asian Indians who call themselves Indo-Americans or Asian Indian Americans fall into this category, eating equal amounts of Indian and American foods, thinking and reading equally in an Indian language and in English (Sodowsky & Carey, 1988).

Assimilation occurs when people from one cultural group shed their ethnic identity and fully merge into the majority culture. Although some first-generation immigrants strive toward assimilation due perhaps to personal determination to survive in a foreign country or to take advantage of opportunities, most often assimilation takes place in subsequent generations. Asian Indians who identify themselves as being "mostly American" do not consider Asian Indian culture superior to American culture, and they are willing to let their children date non-Indians. It is believed that ethnic pride is reawakened in some immigrants if they

Typically, first-generation immigrants remain emotionally connected to their ethnicity, surrounding themselves with a reference group of family and friends who share their cultural background. (© Peter Mengel/Stock, Boston.)

■ The concept *conservatism of cuisine* suggests that most people are reluctant to try new foods. Acceptance occurs if, after being introduced to the item, a person determines that it is tasty, nontoxic and nutritive, and compatible with other food habits (Rozin, 1991).

become disillusioned with life in America, particularly if the disappointment is attributed to prejudice from the majority society (Sodowsky & Carey, 1988). A few immigrants exist at the edges of the acculturation process, either maintaining total ethnic identity or rejecting both their culture of origin and that of the majority culture (Meleis et al., 1992).

Culturally based food habits are often one of the last traditions people change through acculturation. Unlike speaking a foreign language or wearing traditional clothing, eating is usually done in the privacy of the home, hidden from observation by majority culture members. Adoption of new food items does not generally develop linearly as a steady progression from traditional diet to diet of the majority culture. Instead, research indicates that consumption of new items is often independent of traditional food habits (Dewey et al., 1984; Pelto et al., 1981; Szathmary et al., 1987). The lack of available native ingredients may force immediate acculturation, or convenience or cost factors may speed change. Samoans may be unable to find the coconut cream needed to prepare favorite dishes, for instance. Foods that are tasty are easily accepted, such as pastries, candies, and soft drinks; conversely, unpopular traditional foods may be the first to go. Mexican children living in America quickly reject the variety cuts of meat, such as tripe, that their parents still enjoy. It is the foods that are most associated with ethnic identity that are most resistant to acculturation. Muslims will probably never eat pork, regardless of where they live. People from China may insist on eating rice with every meal, even if it is the only Asian food on the table.

Factors That Influence Food Habits

Numerous cultural factors affect the diet of each person within a society. Experts in the field have systematically analyzed these influences to delineate the interrelationships and predict food habits. Two approaches are especially helpful in understanding individual dietary practices within the context of culture. First is the *developmental perspective of food culture* (Table 1.1), which suggests how social dynamics are paralleled by trends in food, eating, and nutrition (Sobal, 1999). Second is the *lifestyle model of dietary habits* (Figure 1.2), which outlines how specific food behaviors may result from the interaction of social factors with lifestyle factors (Pelto, 1981).

Social Dynamics

The developmental perspective of food culture is useful in conceptualizing broad trends in cultural food habits that emerge during structural changes in a society.

Table 1.1
Developmental Perspective of Food Culture

Structural Change	Food Culture Change
Globalization	Consumerization
• Local to unrestricted	• Indigenous to mass foods
Modernization	Commoditization
• Muscle to fueled power	• Homemade to manufactured
Urbanization	Delocalization
• Rural to urban residence	• Producers to only consumers
Migration	Acculturation
• Original to new settings	• Traditional to adopted foods

Source: Adapted from Sobal (1999).

As described by Sobal (1999), *globalization* is the integration of local, regional, and national phenomena into an unrestricted, worldwide organization. The parallel change in cultural food habits is *consumerization,* the transition of a society from producers of indigenous foods to consumers of mass-produced foods. Limited, seasonal ingredients such as strawberries are replaced by items grown worldwide, available any time of year. Specialty products, such as ham and other deli meats, that were at one time prepared annually or only for festive occasions can now be bought presliced, precooked, and prepackaged for immediate consumption.

The social dynamic of *modernization* encompasses new technologies and the socioeconomic shifts that result, such as during the industrial revolution when muscle power was replaced by fuel-generated engine power or during the past decade with the information age. Cultural beliefs, values, and behaviors modify in response to the dramatic structural changes that take place. *Commoditization* typifies food habits, with foods becoming processed, marketed commodities instead of home-prepared sustenance. The fresh milk from the cow in the barn becomes the plastic gallon container of pasteurized milk shipped to another part of the country and sold on-line over the Internet to a consumer limited in time (and access to dairy cows) but not money.

Urbanization occurs when a large percentage of the population abandons the low density of rural residence in favor of higher-density suburban and urban residence. Often income levels do not change in the move, but families who previously survived on subsistence farming become dependent on others for food. Delocalization occurs when the connection between growing, harvesting, cooking, and eating food is lost as meals prepared by anonymous workers are purchased from convenience markets and fast-food restaurants.

Finally, *migration* of populations from their original homes to new settlements creates significant structural change as we shift from a home-bound, culture-bound society to one in which global travel is prevalent and immigration common. Traditional food habits are in flux during *acculturation* to the diet of a new culture and as novel foods are introduced and accepted into a majority cuisine. Often wholly new traditions emerge from the contact between diverse cultural food habits.

Lifestyle Influences

The construction of a model to describe societal trends in cultural food habits is by necessity so broad that the preferences and selections of each individual are obscured. Pelto's (1981) lifestyle model of dietary habits proposes that the social factors outlined in the developmentalist model of food culture interact directly with lifestyle influences to produce specific food behaviors.

Social Factors

The food production and distribution system is responsible for the availability of foods, which differs from region to region and country to country. Individuals may have access only to homegrown food, or they may be able to purchase exotic products from around the world. Food availability influences, and in turn is influenced by, the socioeconomic and political systems. These serve to control the production and distribution of food in the culture. In the public sector, for example, farm subsidies both in the United States and in Europe promote the production of dairy and grain foods far in excess of what can be used by their populations, while deprivations during wartime required that food be strictly rationed. Government policy may also be involved with the purchasing power of consumers through programs such as food subsidies for the poor, as well as the oversight of food quality through safety standards, nutrition labeling requirements, and other production programs (Josling & Ritson, 1986).

In the private sector, advertising is another form of control, greatly influencing some food habits, such as cereal, snack, and beverage preferences. Research indicates that, in blind taste tests, people often have difficulty discriminating between different brands of the same food item. Consumer loyalty to a particular brand is believed to be more related to the sensual and emotional appeal of the name and packaging (Lannon, 1986); after all, purchase of food is a form of self-expression.

■ The developmentalist model of food culture assumes that cultures progress from underdeveloped to developed through the structural changes listed. Deliberate efforts to reverse that trend can be seen in the renewed popularity of farmer's markets in the United States and recent attacks on fast-food franchises in Europe (Sobal, 1999).

■ The governments in most developed nations often manipulate food pricing to meet agricultural objectives but are more reluctant to use food prices to meet nutrition goals (Josling & Ritson, 1986).

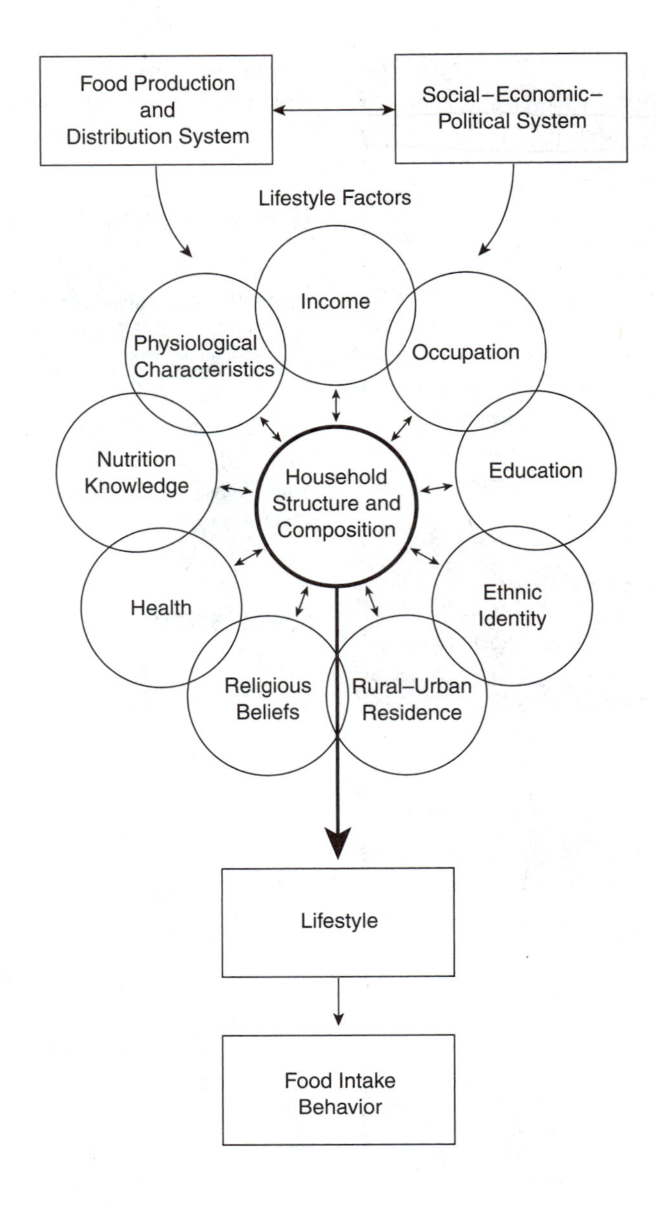

Figure 1.2
Lifestyle model of dietary habits.

Source: From G. H. Pelto (1981). "Anthropological Contributions to Nutrition Education Research," Journal of Nutrition Education, 13 *(Suppl.), S4.* Reprinted with permission.

For example, similar-tasting flake cereals such as Wheaties ("breakfast of champions"), Special K™, and Total™ target sports enthusiasts, dieters, and the health-conscious, respectively. Mass media images have increasing influence; television programming models nutrition beliefs and behaviors, as well as provides general health information. There is often substantial distortion. In content studies, snacking was found to be as prevalent as eating meals, and only 2 to 6 percent of television characters were overweight as compared to approximately 33 percent of the total American adult population (Neuendorf, 1990).

Lifestyle Factors

Both availability and control of food at the societal level affect the lifestyle factors of individuals. These influences include income, which limits what foods can be purchased. Even when nutritious food is abundantly available, the prestige of certain food items, such as lobster or truffles, is often linked to affordability. Occupation influences food habits in several ways. The amount of activity involved in a job affects the actual number of calories a person requires each day, for example. The location of the job also influences meal patterns. In some cultures everyone's job is near home and the whole fam-

Regional fare differs throughout the United States. The southwestern foods at this Texas restaurant are uncommon in many other parts of the nation. (Phil Schermeister/Corbis).

ily joins in a leisurely midday lunch. In urbanized societies, people often work far from home; therefore, lunch is eaten with fellow employees. Instead of a large, home-cooked meal, employees may eat a quick, light, fast-food meal. Furthermore, who is employed is also important. In the United States the greater the number of hours a woman spends working outside the home each day, the fewer the hours she spends in meal preparation.

Income is directly affected by occupation, and occupation is in turn affected by education. The status and self-realization phases of food use are usually, though not always, dependent on higher levels of education. Educational attainment may also influence other lifestyle factors that affect food habits, such as nutrition knowledge. One study reported that college students were more likely to try a new fruit, vegetable, or grain product if information on nutritional benefits were provided (Martins et al., 1997). Some researchers have found that attitudes about the healthfulness of certain foods is important in food preferences, and parents may purchase foods they consider health promoting for their children even if they would not select those items for themselves. It has been suggested that preferences are more often influenced by beliefs regarding nutritional quality than the actual nutritional value or health consequences of a food (Rozin, 1996; Shepard & Raats, 1996). Whether accurate or misinformed, nutrition knowledge does not always translate into knowledge-based behavior. Other lifestyle factors are often more significant in individual food preferences.

Ethnic identity may be immediate, as in persons who have recently arrived in the United States, or it may be remote, a distant heritage that has been modified or lost over the generations through acculturation. The degree of ethnic identity directly influences food habits. An individual who has just immigrated from Japan, for example, is more likely to prefer traditional Japanese cuisine than is a third- or fourth-generation Japanese American. Place of residence (rural versus urban) may also affect which foods people eat. A poor farmer, for example, may have access to more fresh ingredients than a person with the same income who lives in a city apartment. Regional differences are evident as well. The foods of New England differ from those of the Midwest, and local specialties such as Pennsylvania Dutch, Cajun, and Tex-Mex influence the cooking of all residents in those areas.

- College education is associated with fruit consumption among blacks and vegetable consumption among Hispanics and whites (Devine et al., 1999).

- In 1991 the Ford Institute discovered the following regional food preferences: People in the Northeast eat more frozen vegetables and doughnuts and drink 66 percent more tea than the national average; southerners eat more fish and fruit jellies; people in the north-central states eat more ice cream and 25 percent more candy and potato chips; and westerners eat more fresh fruits, vegetables, and cheese and eat 100 percent more whole-wheat bread than the national average.

Religious beliefs are similar to ethnic identity in that they may have a great impact on food habits or they may have no influence at all. It depends both on what religion is followed and on the degree of adherence (see chapter 4, "Food and Religion," for details on specific religions). Many Christian denominations have no food restrictions, but some, such as the Seventh-Day Adventists, have strict guidelines about what church members may eat. Judaism requires that only certain foods be consumed, in certain combinations, yet most Jews in the United States do not follow these rules strictly. Health beliefs also influence food habits in varying ways. Specific foods are often credited with health-promoting qualities, such as ginseng in Asia, chicken soup in Eastern Europe, and corn in Native American culture. A balance of hot and cold or yin and yang foods may be consumed (see the section on the therapeutic use of foods in chapter 2). Vegetarianism is another way in which health beliefs can affect food habits.

One of the lifestyle factors most influential on food habits is physiological characteristics, including age, gender, body image, and state of health. Food preferences and the ability to eat and digest foods vary in the life cycle. Pregnant and lactating women commonly eat differently than other adults. In the United States, women are urged to consume more food when they are pregnant, especially dairy products. They are also believed to crave unusual food combinations, such as pickles and ice cream. They may avoid certain foods, such as strawberries, because they are believed to cause red birthmarks. In some societies with subsistence economies, pregnant women may be allowed to eat more meat than other people; in others, pregnant women avoid beef because it is feared that the cow's cloven hoof may cause a cleft palate in the child. Most cultures also have rules regarding what foods are appropriate for infants; milk is generally considered wholesome, and frequently any liquid resembling milk, such as nut "milk," is also believed to be nourishing.

Puberty is a time for special food rites in many cultures. In America, adolescents are especially susceptible to advertising and peer pressure at this time in their lives. They tend to eat quite differently from both children and adults, rejecting those foods typically served at home and consuming more fast foods and soft drinks. A rapid rate of growth at this time also affects the amount of food that teenagers consume. The opposite is true of older adults. As their metabolism slows, their caloric needs decrease. In addition, they may find that their tolerance for fatty foods or highly spiced foods decreases. Elders often face other eating problems related to age, such as the inability to chew certain foods or a disinterest in cooking and dining alone.

Gender has also been found to influence what a person eats. In some cultures, women are prohibited from eating specific foods or are expected to serve the best pieces of food to the men. In other societies, food preference is related to gender. Some people in the United States consider steak to be a masculine food and salad to be a feminine one; men drink beer, and women drink white wine.

A person's state of health has an impact on what is eaten. A chronic disease such as diabetes requires an individual to restrict or omit certain foods. An individual who is sick may not be hungry or may find it difficult to eat. Even minor illnesses may result in dietary changes, such as drinking ginger ale for an upset stomach or hot tea for a cold. Those who are exceptionally fit, such as student or professional athletes, may practice other food habits, including carbohydrate loading or consumption of high-protein bars.

Finally, at the center of all the lifestyle influences are household structure and composition. Lifestyle is defined by and organized around the family unit, which may be a nuclear family, an extended family, a single-parent family, a couple with no children, a group of unrelated adults with or without children, or an adult living alone. In each household, food behavior develops from the complex interaction of lifestyle influences. Although the lifestyle of each family is unique, similar household composition, influenced by similar societal and lifestyle factors, results in similar food habits.

The Complexity of Cuisine

The Sobal and Pelto approaches detail the many connections between diet and culture. One example of the complexity of food habits

■ Old age is a cultural concept; among some Native Americans and Southeast Asians a person becomes an elder in their 40s (Wray, 1992).

■ In the United States, men are infrequently involved in meal planning (23 percent), shopping (36 percent), or preparation (27 percent). Men who are heads of households in small families, those whose wives work full-time, and those who are younger are more likely to be involved in these activities (Harnack et al., 1998).

■ In many homes, few meals are eaten together as a family. The term *grazing* refers to individuals who eat throughout the day, consuming a doughnut here and a hot dog there, usually on the go.

within a society can be seen in the impact that house porches had on nutrition in Mississippi during the 1930s (Camp, 1991). Researchers were puzzled as to why farm families who moved into low-income housing units reduced their consumption of vegetables during the winter, even though annual supplies were unaltered. What was discovered was the new homes did not have front porches and the residents no longer spent evenings outside shelling peas and breaking beans as they socialized with their family and neighbors. Canning activity therefore also dwindled, and those families who could not afford to buy commercially prepared vegetables simply did without.

In some ways, food can be likened to the chemical spark that bridges the gap (*synapse*) between nerve cells. In this *synaptic model,* food is the link between biological impulses for sustenance and social needs for fellowship, demonstrated in the way food is served at both festive and solemn occasions (Camp, 1989). Food habits are social constructs integral to human interaction. As stated by one sociologist,

> the system which they form cannot be understood apart from the ways in which it interrelates with other social institutions, in the process of historical development. At one and the same time, the system provides a communicative resource, a language, which both expresses the main themes and values of the society and enables individuals to pursue their individual projects and purposes. Every occasion of usage is, then, both a reaffirmation of a world view and a subtle modification of its shape as the individual interprets and restates it. (Gofton, 1986, p. 131)

The Study of Cultural Foods

Considering the numerous factors that influence food habits, learning about a particular cuisine requires more than a glossary of ingredients and recipes for traditional dishes. What a person eats must be examined within a cultural context. Researchers from the fields of anthropology, nutrition, psychology, and sociology have all contributed methods and approaches to the study of food habits.

Cultural Perspective

Everyone is so intimately involved with his or her own culture that it is nearly impossible to study another cultural group objectively.

Food habits that appear illogical to an outsider to a group usually make sense to insiders. Complex symbolic, economic, sociological, ecological, or even physiological reasons for how a culture uses food often escape an outsider's recognition. A frequently cited example of well-meaning, but misguided, outsider viewpoint was the U.S. Agency for International Development's provision of powdered milk to undernourished populations in developing nations after World War II. Complaints by recipients that drinking the milk caused stomachaches, bloating, gas, and diarrhea were attributed to eating the powder without mixing it with water, improper dilution, or impure water sources. It was assumed that if milk was a nutritious food for most Americans, it was a nutritious food for hungry people elsewhere in the world. It was not discovered until 1965 that most adults in the world cannot digest milk, a condition called *lactose intolerance.*

■ *Lactose intolerance,* the inability to digest the milk sugar lactose, develops as a person matures. It is believed that only 15 percent of the adult population in the world (those of northern European heritage) can drink milk without some digestive discomfort.

The assumption that powdered milk was a beneficial food for any malnourished person also illustrates another aspect of cultural perspective, the concept of *cultural relativity.* This refers to understanding a culture within the context of that culture, avoiding the biases of one's own culture. The bias in favor of milk was so strong that it prevented the donors from recognizing that the milk was causing illness, not alleviating hunger, even after the clinical discovery of lactose intolerance was made.

The study of culturally based food habits is not an exact science; further, there are no absolute right or wrong ways to use food. It is sometimes difficult not to apply value judgments to other peoples' food habits, especially those that are repugnant within the context of one's own culture. The use of dog meat in some Asian cultures is an example. It is tempting to label such food habits as immoral or disgusting, yet this is cultural bias. Most Asians do not share the Americans' adoration of dogs as pets. Instead, within the context of their cultures, dogs are considered an acceptable food source and may be raised for this purpose. *Ethnocentric* is the term applied to a person who does not practice cultural relativity and who believes that his or her own habits are superior in some way to those of another culture.

■ The same food practice (content) often has different implications interculturally (context): Consuming a small portion of food may mean that more food is unaffordable; it may connote characteristics of self-control or health-consciousness; it may be due to concerns about body image (dieting); it may reflect gender role within the family (saving larger amounts for the husband or children); it may be adherence to religious proscriptions regarding moderation of intake.

Methodology

Given the appropriately sensitive cultural perspective, the next step in studying culturally based food habits is to describe observations systematically for further analysis. In the United States, food has typically been classified by food group (as in the Basic Four Food Groups: protein, dairy, cereal and grain, fruit and vegetables), by percentage of important nutrients (as identified in the Recommended Dietary Allowances [RDA] for energy, protein, vitamins, and minerals), or according to recommendations for health (the U.S. Department of Agriculture [USDA] Food Guide Pyramid; see Figure 1.3). The way in which food is classified is also a matter of culture. For instance, the Basic Four Food Groups, which include a category for dairy foods, is useful only in cultures whose members consume significant amounts of milk and milk products. Such categories are also limited in the types of information they provide about food use in different cultures. Although they list what foods people eat, they reveal nothing about how, when, or why foods are consumed.

Researchers have suggested many other ways of categorizing foods and food habits. Some models have been developed based on intercultural uses of food, such as the culinary triangle (Levi-Strauss, 1969) of raw, cooked, and rotten food. In this universal approach, it is postulated that all raw food is transformed culturally by cooking or naturally by decomposition. Other categories, found in both developing and industrialized societies, include cultural superfoods, usually staples that have a dominant role in the diet; prestige foods, often protein foods, usually expensive or rare; body image foods, those believed to influence health and well-being; sympathetic magic foods, which by their form or color are associated with specific effects on the body; and physiologic group foods (those that are reserved for, or forbidden to, groups with certain physiologic status, such as gender, age, or health condition) (Jelliffe, 1967).

Foods and food habits can also be classified according to their symbols (content) or their meanings (context). Content includes the observable practices of food use, such as ingre-

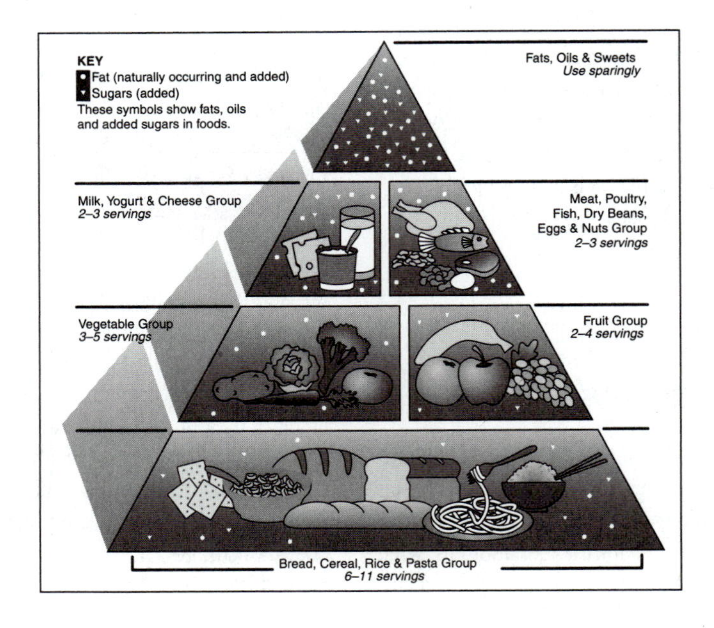

Figure 1.3
The USDA Food Guide Pyramid is one way in which food is classified.

Source: From Whitney and Rolfes *Understanding Nutrition,* 6e, West Publishing, p. 45.

dients, preparation methods, equipment, and food-handling roles. All content has meaning within context. Context can be connotative (physical and economic properties of food, such as appearance, aroma, flavor, convenience, and availability) and image-based self-expression (communicating cultural or social identity). The complexity of food habits can be ranked in both content and context categories (Fewster et al., 1973; Hertzler et al., 1982). Some classification approaches that are helpful in understanding the food habits of a culture include (1) which foods are considered edible and (2) how often those foods are consumed. The daily, weekly, and yearly use of food in a culture can be characterized through meal patterns and meal cycles.

Edible or Inedible?

One approach (National Research Council, 1945) describes the process each person uses to determine what is considered food.

1. *Inedible foods:* These foods are poisonous or are not eaten because of strong beliefs or taboos. Which foods are defined as inedible varies culturally. Examples of foods that are frequently prohibited (*taboo*) include animals that are useful to the cultural group, such as cattle in India; animals that are dangerous to catch; animals that have died of unknown reasons or of disease; animals that consume garbage or excrement; and plants or animals that resemble a human ailment (i.e., strawberries or beef during pregnancy, as described previously). The reasons behind food taboos are often unknown.
2. *Edible by animals, but not by me:* These foods are items such as locusts, ants, and termites in the United States or corn in France (where it is used only as a feed grain). Again, the foods in this category vary widely by culture.
3. *Edible by humans, but not by my kind:* These foods are recognized as acceptable in some societies, but not in one's own culture. Examples of unacceptable foods in the United States that are acceptable elsewhere include giant snails in Africa, dog meat in Asia (as discussed previously), iguana in the Caribbean, horse meat or blood sausage in Europe, and bear paw in Mongolia.
4. *Edible by humans, but not by me:* These foods include all those that are accepted by a person's cultural group but not by the individual, due to factors such as preference (e.g., tripe, raw oysters), expense, or health reasons (a low-sodium or low-cholesterol diet may eliminate many traditional American foods). Other factors, such as religious restrictions, may also influence food choices.
5. *Edible by me:* These are all foods accepted as part of an individual's diet.

There are always exceptions to the ways in which foods are categorized. It is assumed, for instance, that poisonous plants and animals would always be avoided. In Japan, however, *fugu* (blowfish or globefish) is considered a delicacy despite the deadly toxin contained in the liver, intestines, testes, and ovaries. These organs must be deftly removed by a certified chef as the last step of cleaning (if they are accidentally damaged, the poison spreads rapidly through the flesh). Eating the fish supposedly provides a tingle in the mouth prized by the Japanese. Several people die each year from fugu poisoning.

Core Foods

Those foods that are defined as edible within a culture can be grouped according to frequency of consumption. Passim and Bennett's (1943) expanded concept of *core foods* states that the staples regularly included in a person's diet, in unmodified form, are at the core of food habits. Foods that are widely but less frequently eaten are termed *secondary core foods.* Foods that are eaten only sporadically are called *peripheral foods.* These foods are characteristic of individual food preference, not cultural group habit. It is hypothesized that changes in food behaviors happen most often with peripheral foods and least often with core foods. A Mexican American who is adapting to life in the United States, for example, is much more likely to give up some peripheral food item such as *nopalitos* (cactus) than to eliminate core foods, such as tortillas or beans.

■ *Taboo* is Polynesian, from the Tongan term *tabu,* meaning "marked as Holy."

■ Children younger than the age of 2 will eat anything and everything. Children between 3 and 6 years of age begin to reject culturally unacceptable food items. By age 7, children are completely repulsed by foods that their culture categorizes as repugnant (Rozin et al., 1985).

■ Insects, such as termites and ants, provide 10 percent of the protein consumed worldwide.

■ Among the most universal of food taboos is cannibalism, although anthropologists have discovered numerous examples of prehistoric human consumption in European and new world excavations.

Meals and Meal Cycles

A more comprehensive method of categorizing food habits has been developed through the structural analysis of eating patterns (Douglas, 1972). Food can be considered a code that transmits messages about social relations and events in a culture by means of meals and meal cycles.

The first step in decoding eating patterns is to determine what constitutes a meal. In each culture a meal is made up of certain elements. In the United States, for instance, cocktails and appetizers, or coffee and dessert, are not considered meals (Figure 1.4) A meal contains a main course and side dishes, typically a meat, vegetable, and starch. In the western African nation of Cameroon, a meal is a snack unless *cassava* (manioc) paste is served. In many Asian cultures a meal is not considered a meal unless rice is included, no matter how much other food is consumed.

The elements that define a meal must also be served in their proper order. In the United States, appetizers come before soup or salad, followed by the entrée, and then by dessert; in France the salad is served after the entrée. All foods are served simultaneously in Vietnam so that each person may combine flavors and textures according to taste. Furthermore, the foods served should be appropriate for the meal or situation. Although some cultures do not differentiate among foods that can be served at different meals, eggs and bacon are considered breakfast foods in the United States, while cheese and olives are popular in the Middle East for the morning meal. Soup is commonly served at breakfast in Southeast Asia; in the United States, soup is a lunch or dinner food, and, in parts of Europe, fruit soup is sometimes served as dessert. Cake and ice cream are appropriate for a child's birthday party in the United States; wine and cheese are not.

■ A one-pot dish is considered a meal if it contains all the elements of a full meal. For example, American casserole dishes often feature meat, vegetables, and starch, such as shepherd's pie (ground beef, green beans, and tomato sauce topped with mashed potatoes) or tuna casserole (tuna, peas, and noodles).

■ The sprig of parsley added to a plate of food may have originated as a way to safeguard the meal from evil.

■ Research has found people eat a meal if it is offerred at the appropriate time and they have not eaten recently. One study conducted with patients who suffer memory loss found that after consuming a full lunch, patients would eat another meal 10 minutes later when they forgot that they had eaten. Some even started a third meal when it was served after another short break (Rozen, 1996).

Other aspects of the meal message include who prepares the meal and what culturally specific preparation rules are used. In the United States, catsup goes with french fries; in Great Britain, vinegar is sprinkled on *chips* (fried potatoes). Orthodox Jews consume meat only if it has been slaughtered by an approved butcher in an approved manner and has been prepared in a particular way (see chapter 4 for more information on Judaism). Who eats the meal is also important. A meal is frequently used to define the boundaries of interpersonal relationships. Americans are comfortable inviting friends for dinner, but they usually invite acquaintances for only drinks and hors d'oeuvres. For a family dinner, people may include only some of the elements that constitute a meal, but serving a meal to guests requires that all elements be included in their proper order.

The final element of what constitutes a meal is typical portion size. In many cultures, one meal a day is designated the main meal and usually contains the largest portions. The amount of food considered appropriate varies, however. A traditional serving of beef in China may be limited to one ounce added to a dish of rice. In France, a three- or four-ounce filet is more typical. In the United States, a six- or even eight-ounce steak is not unusual, and some restaurants specialize in twelve-ounce cuts of prime rib. American tradition is to clean

Figure 1.4
Each culture defines which foods are needed to constitute a meal.

Source: Blondie. Reprinted with special permission of King Features Syndicate.

Special dishes that include costly ingredients or are time-consuming to prepare are characteristic of feasting in many cultures. (© Robert Brenner/PhotoEdit.)

one's plate regardless of how much is served, while in other cultures, such as those in the Middle East, it is considered polite to leave some food on the plate to indicate satiety.

Beyond the individual meal is the cycle in which meals occur. These include the everyday routine, such as how many meals are usually eaten and when. In much of Europe a large main meal is traditionally consumed at noontime, for example, while in most of the United States today the main meal is eaten in the evening. In poor societies only one meal per day may be eaten, whereas in wealthy cultures three or four meals are standard. The meal cycle in most cultures also includes feasting or fasting, and often both.

Feasting celebrates special events, occurring in nearly every society where a surplus of food can be accumulated. Religious holidays such as Christmas and Passover; secular holidays such as Thanksgiving and the Vietnamese New Year's Day, known as *Tet;* and even personal events such as births, marriages, and deaths are observed with appropriate foods. In many cultures, feasting means simply more of the foods consumed daily and is considered a time of plenty when even the poor have enough to eat. Special dishes that include costly ingredients or are time consuming to prepare also are characteristic of feasting. The elements of a feast rarely differ from those of an everyday meal. There may be more of an everyday food or several main courses with additional side dishes and a selection of desserts, but the meal structure does not change. For example, Thanksgiving typically includes turkey and often another entrée, such as ham or lasagna (meat); several vegetables; bread or rolls, potatoes, sweet potatoes, and stuffing (starch); as well as pumpkin, mincemeat, and pecan pies or other dessert selections. Appetizers, soups, and salads may also be included.

Fasting is often partial, the elimination of just some items from the diet, such as the traditional Roman Catholic omission of meat on Fridays or a Hindu personal fast day, when only foods cooked in milk are eaten. Complete fasts are less common. During the holy month of *Ramadan,* Muslims are prohibited from taking food or drink from sunrise to sunset, but they may eat in the evening. *Yom Kippur,* the day of atonement observed by Jews, is a total fast from sunset to sunset (see chapter 4, for more details).

■ Feasting functions to redistribute food from rich to poor, to demonstrate status, to motivate people toward a common goal (e.g., a political fund-raising dinner), to mark the seasons and life-cycle events, and to symbolize devotion and faith (e.g., Passover, Eid-al-Fitr, and Communion).

Nutrition and Food Habits

The Need for Cultural Competency

Suppose you wish to know as much as you can about the foods a person likes and eats, and can only ask one question. What should that question be? Without doubt, the question should be a distinctly social one: what is your culture or ethnic group? There is no other question that is nearly as informative. (Rozin, 1996, p. 101)

In recent years, the significance of culturally based food habits on health and diet has been recognized and the need for intercultural competencies in the areas of nutrition research, assessment, counseling, and education has been cited. Health professionals require skills "in discovering relevant cultural patterns and cultural changes among the clients and groups they serve, and need to incorporate a knowledge of these patterns into interventions" (Terry, 1994, p. 503). Data accuracy is dependent on respect for different values and a trusting relationship between respondent and researcher (Cassidy, 1994); effective intercultural communication is a function of understanding and accepting a client's perspective and life experience (Fong, 1991). New standards of nutrition care issued by professional accreditation organizations reflect similar guidelines (Dougherty et al., 1995). Looking toward the future, it has been proposed that health care professionals should move beyond the theoretical concepts that are the foundation of cultural sensitivity and relevance to the practicalities of cultural competency. Language skills, managerial expertise and leadership are needed to guide diverse communities in healthy lifestyle changes, to serve hard-to-reach populations, and to effect change in the health care system (Boyle, 2000).

- "Respect for diverse viewpoints and individual differences" is an American Dietetic Association value.
- About one-third of the annual increase in Latinos is through immigration.
- The population of California became the first state plurality in 2000.

Diversity in the U.S. Population

The growing need for cultural competency is evident in current demographic trends (Figure 1.5). Since the 1970s, the United States has moved increasingly toward a *cultural plurality,* where no single ethnic group is a majority. Pluralities already exist in 186 U.S. counties, according to the 1990 U.S. census data, including those within the cities of New York and Los Angeles. Nationwide, demographers estimate that non-Hispanic whites will become less than 50 percent of the total population by the year 2050.

This change can be seen in the dramatic differences in ethnic group growth between 1980 and 1990 (see Figure 1.5). Although the total population increased by about 10 percent, many ethnic groups, such as Asian Indians, Chinese, Koreans, and Vietnamese, more than doubled their populations; Aleuts, Filipinos, Guamanians, Mexicans, and Samoans experienced growth rates five times the national average.

These demographic trends have continued through the 1990s. For the first time since the 18th century, the number of non-Hispanic whites added to the U.S. population was less than the number of Latinos added in 1994 (883,000 compared to 902,000). African Americans are increasing by about 400,000 annu-

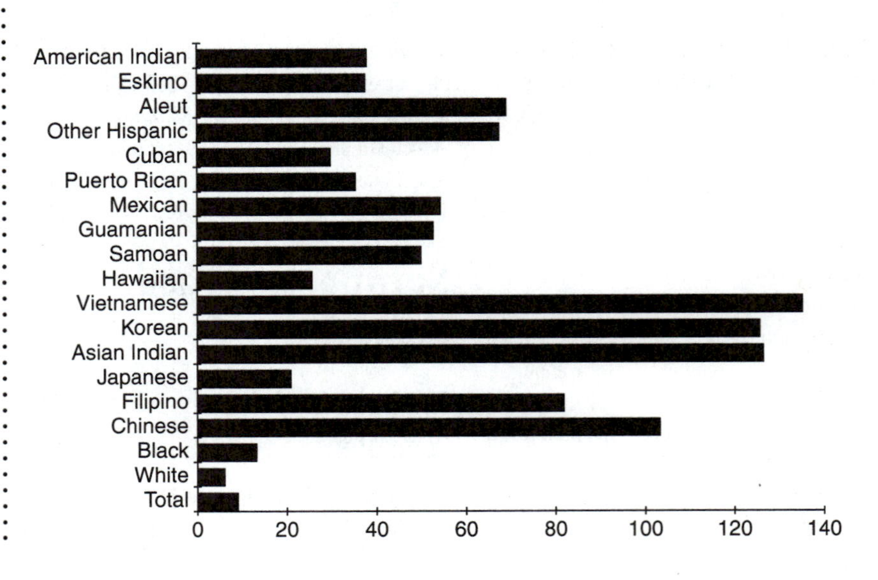

Figure 1.5
Percentage change in U.S. population groups from 1980 to 1990.

Note: The category "Other Hispanic" includes Brazilians, Chileans, Costa Ricans, El Salvadorans, Jamaicans, Nicaraguans, and so forth.

Diversity in Canada

The Canadian census is conducted differently than the U.S. count. Canadians in 1996 were asked to list their ethnicity in an open-ended question, and multiple responses listing one or more ethnicities were accepted. This has provided a broader picture of ancestry, particularly as single responses and multiple responses were reported separately. For example, of the 1.1 million Aboriginals (including Native American Indians, Métis—people of mixed Aboriginal and non-Aboriginal heritage—and Inuit), 477,630 listed this ethnicity as a single response and 624,330 listed it as part of a multiple response.

A separate question inquired if the census respondent was a member of a *visible minority,* defined by the Employment Equity Act as "persons, other than Aboriginal peoples, who are non-Caucasian in race or non-white in color." The act specifically lists Chinese, South Asians (i.e., Asian Indians, Pakistanis, Sri Lankans), blacks, Arabs and West Asians, Filipinos, Latin Americans, Japanese, Koreans, and Pacific Islanders.

Immigration growth (14.5 percent) in Canada has dramatically exceeded overall population growth (4 percent) in recent years. Immigrants, as of 1996, represent over 17 percent of the total Canadian population, the largest share in over 50 years. Of greater importance, immigration patterns have shifted during the past three decades. In 1961, 90 percent of immigrants came from Europe; by 1991, only 19 percent were of European background. Most recent immigrants are visible minorities, with over 57 percent from Asia.

Nearly all (93 percent) recent immigrants to Canada have settled in urban areas. Seven out of ten live in three metropolitan areas: Toronto, Vancouver, and Montreal. Toronto is most popular, with over 42 percent of Canada's total visible minority population, although only 15 percent of the overall national population live in the city. Almost half of all Canadian South Asians and blacks, and 40 percent of Canadian Chinese, Koreans, and Filipinos have made Toronto home. Vancouver has 18 percent of the visible minority population, 90 percent of whom are Asian. The largest concentration of Japanese Canadians lives in Vancouver. Montreal also has sizeable visible minority communities, including the largest Arab and West Asian population in the country. Over 21 percent of blacks, 22 percent of Southeast Asians, and 26 percent of Latin Americans live in the city. Other urban areas with disproportionately large recent immigrant populations include Calgary, Edmonton, and Ottawa-Hull. While Winnipeg and Halifax have lower visible minority populations than the nation as a whole, Winnipeg hosts a notable Filipino community, and significant numbers of blacks and Arabs live in Halifax.

ally, and Native Americans by about 25,000. Asians and Pacific Islanders, who are adding about 300,000 to the population each year, have the fastest annual growth rate at 3.5 percent each year, compared to Latinos at 3.4 percent, blacks at 1.3 percent, Native Americans at 1.2 percent, and whites at 0.41 percent.

Notably, many American ethnic populations have an average age significantly lower than that of the total population. In 1993, U.S. census figures indicates that the non-Hispanic white population of adults and children totaled 73 percent, while the number of non-Hispanic white youth under 5 years of age was only 64 percent. Predicted demographic changes are often seen first among children and young adults.

Ethnicity and Health

Health is not enjoyed equally by all in the United States. Disparities in mortality rates, chronic disease incidence, and access to care are prevalent among American minority ethnic groups. Poor health status among ethnic groups in the United States is frequently associated with poverty (Blane, 1995) and low educational attainment (Centers for Disease Control and Prevention, 1994; Greenlund et al., 1996). Acculturation to the majority culture is also a significant factor independent of socioeconomic status. First noted in hypertension and heart disease (Henry & Cassel, 1969; Marmot & Syme, 1976), modernization has also been linked to increased fat in the diet (Romero-Gwynn et al., 1993; Tsunehara et al., 1990; Wenkam & Wolff, 1970); increased blood cholesterol and triglyceride levels (McMurray et al., 1991); increased blood pressure levels (Dressler et al., 1987); obesity (McGarvey, 1991; Murphy et al., 1995; National Research Council, 1998); non-insulin-dependent diabetes (Ekoe, 1988; Swinburn et al., 1991); and cancer (Whittemore et al., 1990). The stress of adaptation to the pressures of a fast-paced society are believed to be significant (Hull, 1979). Hereditary

■ As of 1997, there are more Latino youths than African American children in the United States, even though blacks still outnumber Latinos over all.

■ Acculturation is not an inherent risk factor in health (Palinkas & Pickwell, 1995). Some changes in diet, such as a reduction in pickled food intake associated with stomach cancer or increased availability of fruits and vegetables, can be beneficial. Better educational opportunities and health care services can also promote health.

■ A recent study suggests that the mortality risk among African Americans is higher than for whites even after adjusting for socioeconomic factors (Sorlie et al., 1995).

predisposition toward developing certain health conditions may also play a role (Fackelman, 1991; Haffner et al., 1991; Howard et al., 1991).

The specific impact of ethnicity on health status is not well delineated through the limited research data available. If only age-adjusted mortality rates for the United States (1990) are examined, it appears that Asians/Pacific Islanders are the healthiest population group with the lowest number of deaths, followed by Native Americans/Alaska Natives, Hispanics, whites, and African Americans (Centers for Disease Control and Prevention, 1992). However, these figures obscure significant differences in the prevalence and progression of certain diseases among groups, and the diversity of health status within each broad ethnic designation.

Studies on non-insulin-dependent diabetes, for example, indicate that every other ethnicity (as grouped in the preceding paragraph) has incidence rates above those of the white population: two to three times for African Americans, two to five times for Hispanics, two to six times for Asians/Pacific Islanders, and three to more than ten times for Native Americans/Alaska Natives (Ekoe, 1988; also see individual chapters on each ethnic group for further data). Deaths from the disease are nearly equal between whites and Asians/Pacific Islanders, almost double for Hispanics compared to whites, and more than twice as high for African Americans and Native Americans/Alaska Natives than for whites (Desenclos & Hahn, 1992). Closer inspection reveals considerable variation within these broader ethnic designations as well. In Hawaii, native Hawaiians have more than twice the diabetes prevalence rate of the Korean and Filipino populations, and more than three times that of the Chinese population; Japanese, Asian Indians, and Samoans have all demonstrated rapidly increasing rates after moving to the United States compared to other Asian/Pacific Islander groups. Nowhere is the difference as great as in Native Americans/Alaska Natives, where the prevalence rate for diabetes in Alaska Inuit (older than age 20) is 1.9 percent; for Cherokee Indians (older than age 35), 31 percent; and for Pima Indians (older than age 35), who have the highest rates of diabetes in the world, 50 percent.

Research comparing both ethnic and racial factors in the development of kidney failure among Hispanic men found that in Hispanic whites, 45 percent of end-stage renal disease was attributed to diabetes; while in non-Hispanic whites only 28 percent was due to diabetes, and nearly as many cases were due to hypertension (27 percent) and other causes. Among Hispanic blacks, 40 percent of renal failure was attributed to diabetes and 26 percent attributed to hypertension. Nearly the reverse was true for non-Hispanic blacks, 40 percent of whose renal failure was due to hypertension and 25 percent due to diabetes. Hypertension rates among Hispanics approximate those in whites, which are nearly 50 percent lower than those in African Americans. Hispanics in Florida (mostly Cubans) showed substantially lower rates of kidney failure due to diabetes than did Puerto Ricans or Hispanics in the Southwest. This study was noteworthy for its investigation of ethnic and racial, as well as intraethnic differences, and indicates the complexity of ethnicity and disease (Chiapella & Feldman, 1995).

The variable role of ethnicity is also seen in U.S. infant mortality trends. Although dramatic declines have occurred since 1950, these gains are not evenly distributed throughout the population. The gap between white and black infant deaths has increased; the mortality rate for African Americans is 112 percent higher than for whites. Much of the discrepancy is attributed to disorders resulting from prematurity and low birth weight. Variation within broad ethnic designations is also demonstrated. Chinese infant deaths are 30 percent lower than for white infants. Japanese, Filipino, Cuban, Central and South American, and Mexican mortality rates are also significantly lower than for whites, while Puerto Ricans, native Hawaiians, and Native Americans have mortality rates 26 percent, 33 percent, and 55 percent higher than for whites, respectively. Overall, infant deaths in the United States remain higher than those found in other developed nations, probably due to the excess mortality found among some ethnic groups (Singh & Yu, 1995).

Furthermore, a growing number of studies have documented inequalities in health care treatment for nonwhite groups. Preventive care such as immunizations and cholesterol screenings lag behind the U.S. average, and clinical care disparities abound. For example, African Americans are much less likely than whites to have renal transplants for kidney disease or coronary artery bypass surgery for heart disease, but they are significantly more likely than whites to have lower limbs amputated due to diabetic neuropathy and gastrostomy tubes used on elder patients. Though it is believed that health care access and low health insurance rates may be a factor in these differences, even ethnic patients with comprehensive government benefits often receive unequal treatment (Gornick et al., 1996; Grant et al., 1998; Henry J. Kaiser Family Foundation, 1999; McBean & Gornick, 1994). Some researchers are concerned that managed competition in health care will exacerbate these differences because of capitated payment programs, reduced access due to consolidation of services, and protocols (usually standardized for white patients) dictated by third-party payors that disallow variability appropriate for different populations (Geiger, 1996; Salmond, 1999).

As these examples suggest, ethnicity can be a significant factor in the development of certain disease conditions, the way they are experienced, and how they are ultimately resolved (see chapter 2, "Traditional Health Beliefs and Practices," for further information). The explosive growth of ethnic groups in the U.S. population since the mid-1980s, the rapid movement toward cultural pluralism, and the undeniable connection between heritage and health evidence the urgent need for cultural competency among American health care providers.

Applications

The study of cultural foods has specific applications in determining nutritional status and implementing dietary change. Even the act of obtaining a diet record has cultural implications (see chapter 3, "Intercultural Communication"). Questions such as what was eaten at breakfast, lunch, and dinner not only ignore other daily meal patterns but also make assumptions about what constitutes a meal. Snacks, and consumption of food not considered a meal, may be overlooked. Common difficulties in data collection, such as under- or over-reporting food intake may also be culturally related to the perceived status of an item, for example, or portion size estimates may be an unknown concept, complicated by the practice of sharing food from other family members' plates. Terminology can be particularly troublesome. Words in one culture may have different meanings in another culture, or even among ethnic groups within a culture. Nutritional jargon, such as "fiber," may be unknown or vaguely defined (Cassidy, 1994).

Stereotyping is another pitfall in culturally sensitive nutrition applications, resulting from the overestimation of association between group membership and individual behavior (Gudykunst & Nishida, 1994). Stereotyping occurs when a person ascribes the collective traits associated with a specific group to every member of that group, discounting individual characteristics. A health professional knowledgeable about cultural food habits may inadvertently make stereotypical assumptions about dietary behavior if the individual preferences of the client are neglected. Cultural competency in nutrition implies not only familiarity with the food habits of a particular culture, but recognition of intraethnic variation within a culture as well.

Researchers suggest that health care providers working in intercultural nutrition become skilled in careful observation of client groups, visiting homes, neighborhoods, and markets to learn about where food is purchased, what food is available, and how it is stored, prepared, served, and consumed. Participation in community activities, such as reading local newspapers and attending neighborhood meetings or events, is another way to gather relevant information. Informant interviewing reveals the most about a group; individual members of the group, group leaders, and other health care professionals serving the group are potential sources (Terry, 1994). Combining qualitative approaches such as indepth, open-ended interviews with clients and quantitative measures through questionnaires is one of the most culturally sensitive methods of obtaining data about a group (Cassidy, 1994). Qualitative information obtained through the interviews should alert the researcher to nutrition issues within the group and guide development of the assessment tool; the quantitative results should confirm the data provided through the interview in a larger

■ Minority Americans are twice as likely as whites to be uninsured for health care.

■ In developing stereotypes, it is believed that individuals typically remember the positive traits of their own ethnic group and tend to remember the negative traits of other ethnic groups (Gudykunst & Nishida, 1994).

sample (see chapter 3, "Intercultural Communication," for more information).

Cultural perspective is particularly important when evaluating the nutritional impact of a person's food habits. Ethnocentric assumptions about dietary practices should be avoided (see "Self-Evaluation of Food Habits" at the end of this chapter to examine personal food beliefs, behaviors, and attitudes). A food behavior that on first observation is judged detrimental may actually have limited impact on a person's physical health. Sometimes other moderating food habits are unrecognized. For example, a dietitian may be concerned that an Asian patient is getting insufficient calcium because she eats few dairy products. Undetected sources of calcium in this case might be the daily use of fermented fish sauces or broth rich in minerals made from vinegar-soaked bones. Likewise, a food habit that the investigator finds repugnant may have some redeeming nutritional benefits. Examples include the consumption of raw meat and organs by the Inuits, which provided a source of vitamin C that would have otherwise been lost during cooking, or the use of mineral-rich ashes or clay in certain breads and stews in Africa and Latin America (Wilson, 1985).

■ Sometimes culturally based food habits have vital nutritional benefits. One example is the use of corn tortillas with beans in Mexico. Neither corn nor beans alone supply the essential amino acids (chemical building blocks of protein) needed to maintain optimum health. Combined, they provide complete protein.

In addition, physiological differences among populations can affect nutritional needs. The majority of the research on dietary requirements has been conducted on young, white, middle-class American men. Extrapolation of findings to other populations should be done with caution.

Thus, diet should be carefully evaluated within the context of culture. One effective method is to classify food habits according to nutritional impact: (1) food use that has positive health consequences and that should be encouraged, (2) neutral food behaviors that have neither adverse nor beneficial effects on nutritional status, (3) food habits that are unclassified due to insufficient culturally specific information, and (4) food behaviors that have demonstrable harmful affects on health and that should be repatterned (Jelliffe & Bennett, 1961).

When modification of diet is necessary, it should be attempted in partnership with the client, respectful of culturally based foods and food habits. Compliance is associated with an approach that is congruent with the client's traditional health beliefs and practices (see chapter 2, "Traditional Health Beliefs and Practices," for more information). For instance, educators developed a food guide for

■ One example of a multicultural creation is the California roll, the addition of avocado in traditional Japanese crab sushi. It is called "American sushi" in Japan.

Asian tofu is the main ingredient in this vegetarian adaptation of shepard's pie, a traditional British entrée popular in the United States. (Luigart-Stayner/Corbis)

Caribbean Islanders living in the United States that grouped cultural foods into three categories: growth, protection, and energy, reflecting client group perceptions of how food affects health (Stowers, 1992). A weight loss program for African American women also used ethnic foods and trained black women as group leaders (Kanders et al., 1994), overcoming a distrust of nonblack health care providers common in the African American community. Culturally appropriate nutrition interventions have the best chances of success in a diverse society.

The American Melting Pot

The term *melting pot* has been used to describe cultural pluralism in the United States. It suggests not just separate ethnic and religious groups living in the same country, but a physical and spiritual blending of all cultures through exposure to a diversity of values, beliefs, and practices. Foods and food habits are a part of this mix. Although many Americans of every ethnic background enjoy hamburgers and french fries, American cuisine goes far beyond meat and potatoes. Just as it is impossible to describe "the American diet," it cannot be said that members of American ethnic groups eat only their traditional foods. Exotic combinations of ingredients and preparation techniques result in new American dishes, such as tofu lasagna, tuna croissant sandwiches, and chili-topped spaghetti. It is this unique blending of the traditional and the new that makes the study of American foods and food habits so exciting and challenging.

■ **In 1995, Americans consumed 14 million hamburgers each day.**

Self-Evaluation of Food Habits

Meal Composition/Cycle

How many meals do you eat each day?

What elements (bread, rice, meat, vegetables, or other) are needed to make a meal for you?

What is a typical serving size of meat, starch, and vegetable for you?

Do you eat dessert? How often and at what meals?

Is the meal served in "courses"? If yes, what are they? What is the order of service?

How often do you snack each day? What types of foods do you consider a snack?

What are the major holidays you and your family celebrate each year?

What special foods are served for these holidays?

Do you ever fast? If yes, when? What, if any, foods are avoided and/or consumed?

Rules

What, if any, utensils do you use when you eat?

At a family meal, who would be present?

Is someone at the family meal served first? If yes, who and why?

List three rules (i.e., don't eat with your mouth open) that you follow when you eat a meal.

Do these rules change if you eat informally? If yes, how?

Attitudes

Were you aware of your own food habit norms before you completed this evaluation?

List two personal biases you discovered through this evaluation.

Are there any food habits that you find morally or ethically repugnant?

What is your opinion about people who do not share your food habit norms?

Application

How do your food habits differ from your family norms? Those of friends? Those of people you work with? Those of clients? In what significant ways do they differ?

How can you personally avoid ethnocentric judgments regarding food habits?

Are you willing to try new foods?

2

Traditional Health Beliefs and Practices

Health and illness in America are usually considered the specialty of mainstream biomedicine. Furthermore, health promotion is based on scientific findings of researchers regarding diet, exercise, and lifestyle issues such as smoking cessation and stress management; disease is treated according to the latest technologies. In reality, health care is pluralistic in the United States, as well as in most other cultures. It is estimated that 70 to 90 percent of all sickness is managed outside the biomedical system. It is also projected that 4 out of every 10 Americans utilizes alternative medical care, spending over $21 billion in 1997 for treatment (Eisenberg et al., 1998; Kleinman et al., 1978). Most people in the United States never consult a physician or allied health care provider when physical or emotional symptoms occur, relying on home remedies and popular therapies rather than professional help. When biomedical care is sought, it is often in conjunction with these other systems.

Culture determines how a person defines health, recognizes illness, and seeks treatment. Traditional health beliefs and practices can be categorized in a variety of ways: through the etiology of illness (due to personal, natural, social, or supernatural causes), or by the therapies that are employed (the use of therapeutic substances, physical forces, or magicoreligious interventions). There is no consensus, however, on these classifications. In this chapter, home remedies, popular approaches such as folk and alternative traditions, and professional systems (including U.S. biomedicine, traditional Chinese medicine, and Ayurvedic medicine) are reviewed within the cultural context of health and illness. Specific beliefs and practices are detailed in the following chapters on each American ethnic group.

Worldview

Cultural Outlook

Each culture has a unique outlook on life, based on the beliefs and values that they share with other members of their group. These standards typically express what is worthy in a life well lived within a particular society and are the measures through which one assesses personal behavior and that of others. This cultural outlook, or *worldview,* influences individual perceptions about health and illness as well as the role each plays within the structure of the society (Randall-David, 1989; Schilling & Brannon, 1986).

Majority American values, which are shared by most whites and to some degree by many other ethnic groups in the United States, emphasize individuality and control over fate (Table 2.1). Personal responsibility and self-help are considered cultural cornerstones. When compared to the British and French, for example, it was found that nearly twice as many Americans take individual responsibility for preventing illness, while more than one-third of the French respondents indicated that chance is a significant factor in health (Retchin et al., 1992). Most other cultures worldwide believe that fate—including the

■ *Biomedicine* is the term used to describe the conventional system of health care in the United States and other Western nations based on the principles of the natural sciences, including biology, physiology, and biochemistry.

■ *Ayurvedic medicine* is the ancient Asian Indian system of healing.

■ The term *majority* describes the culture of the predominant White, Anglo-Saxon, Protestant population in the United States. It is important to note, however, that many ethnic groups have contributed to majority American culture and that non-whites may also share characteristic beliefs, values and customs.

Table 2.1
Comparison of Common Values

Majority American Culture	Other Cultural Groups
Mastery over nature	Harmony with nature
Personal control over the environment	Fate
Doing—activity	Being
Time dominates	Personal interaction dominates
Human equality	Hierarchy/rank/status
Individualism/privacy	Group welfare
Youth	Elders
Self-help	Birthright inheritance
Competition	Cooperation
Future orientation	Past or present orientation
Informality	Formality
Directness/openness/honesty	Indirectness/ritual/"face"
Practicality/efficiency	Idealism
Materialism	Spiritualism/detachment

Source: Adapted from E. Randall-David, *Strategies for Working with Culturally Diverse Communities.* Association for the Care of Children's Health, 19 Mantua Rd. Mt. Royal, NJ 08061. Copyright 1989. Reprinted with permission.

■ Majority Americans say clocks *run*. The Spanish say clocks *walk*. Native Americans say clocks *tick*. (Henderson & Primeaux, 1981).

■ Some majority Americans find eating a meal a disruption of daily tasks; others adhere to strict meal schedules. In polychronistic societies, meals are usually leisure events, a chance to enjoy the blessings of food in the company of family and friends.

will of God, the actions of supernatural agents, or birthright (i.e., astrological alignment or cosmic karma)—is a primary influence in health and illness. Although most cultures have complex practices regarding the maintenance of health, the concept of preventative health care, such as annual checkups, is unknown in some cultures where fate dominates.

The significance of fate often coincides with differences in perceptions of time. Many Americans place great value on promptness and schedules; they are also future oriented, meaning that they are willing to work toward long-term goals or make sacrifices so that they or their children will reap rewards in the future. The majority members in the United States are also monochronistic, with a preference for concentrating on one issue or task at a time in a sequential manner. Many other cultural groups live in the present and are often polychronistic, or comfortable doing many things at once. A Mexican American who is talking with his grandmother while fixing an appliance and watching a baseball game on television is unlikely to cut the visit short just because he has a medical appointment. Immediate interests and responsibilities, including interpersonal relationships, are more important than being on time. A few cultures, such as certain Native American nations, are past oriented, living according to historical direction.

Most majority Americans are very task oriented and desire direct participation in their health care; they feel best when they can *do* something. Other cultures place a greater value on *being* and feel comfortable with inactivity. Self-worth is based more on personal relationships than on accomplishments. The expectation is that the health care practitioner will take responsibility for treatment. The whole idea of the provider–client partnership may be alien to Asians who often expect to be fully directed in their care. While many Americans value patient autonomy and confidentiality, other cultural groups, such as Middle Easterners, believe that the family should be involved in all health care decisions—the welfare of the group outweighs that of the individual.

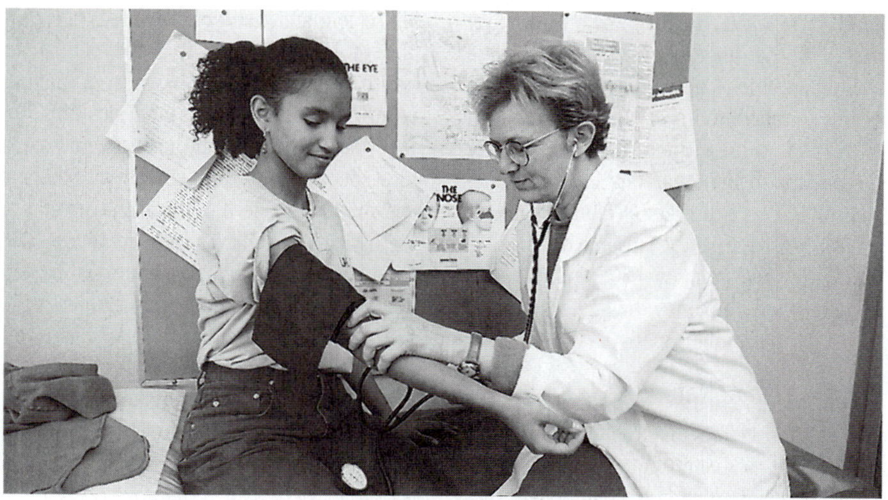

The concept of preventive health care, such as annual checkups provided by a biomedical professional, is often unknown in some cultures where fate is believed to determine health. (© Michael Newman/PhotoEdit.)

Americans consider honest, open dialogue essential to effective communication, and informality is usually a sign of friendliness. Many other cultures prefer indirect communication techniques and expect a formal relationship with everyone but intimate family members. In cultures where identity with a group is more significant than individuality, social status and hierarchy are respected, which can have an impact on the practioner–client relationship (see chapter 3, "Intercultural Communication," for more information).

Worldview is especially evident in serious, life-and-death health care decisions. Southeast Asians may appear indifferent to a terminally ill family member and have little interest in prolonging life because of a faith in reincarnation. Some African Americans distrust white American health care recommendations regarding do-not-resuscitate orders in part because they contradict the critical role of faith in African American healing. An Orthodox Jewish patient may believe that physicians are mandated to preserve life and that any person who assists death through denial of sustaining care is a murderer; a non-Orthodox Jew may believe that no one should endure unrelievable pain, and thus dying should not be prolonged. Middle Easterners traditionally demand that everything be done to keep a person alive because death is in God's hands, and one must never give up hope. A Mexican American family might view death as part of God's plan for a relative; they might be against anything that would quicken death, or they may expect the practitioner to make the decision (Blackhall, et al., 1995; Klessig, 1992; Ndidi Uche Griffin, 1994).

Most health care situations are not cases of life or death, and worldview encompasses many other, less catastrophic aspects of health and illness as well. It is useful to examine the biomedical worldview and understand the perspective of most U.S. health care providers before learning about other traditional health beliefs and practices. Comparisons between biomedicine and other medical systems can reveal areas of potential disagreement or conflict regarding how and why illness occurs and the expectations for treatment before working with a client. Compliance increases with clinical approaches that concur with the client's worldview (Kleinman et al., 1978; Leininger, 1991; Kumanyika & Morssink, 1997).

■ A U.S. survey on physician-assisted suicide for terminally ill patients in pain found that 74 percent of whites felt it should be allowed, compared to 70 percent of Hispanics, 63 percent of Asians, and 51 percent of blacks (Smith, 1998).

■ The world *health* comes from the Anglo-Saxon term *hal,* meaning "wholeness."

Biomedical Worldview

Biomedicine is a cultural subdivision of the American majority worldview. It shares many beliefs with the dominant outlook but differs in a few notable areas (Helman, 1990; Kleinman et al., 1978; Spector, 1991). There are certainly exceptions to the biomedical worldview within

■ Some researchers have noted that though the biomedical community often calls clients whose cultural background differs from the majority "hard to reach," this term is equally applicable to health professionals who refuse to provide culturally appropriate care (Kumanyika & Morssink, 1997).

certain specialties, and by some providers, yet many of the underlying assumptions are culture-specific. There is "a tendency for health personnel to impose their beliefs, practices and values upon another culture because they believe that their ideas are superior to those of another person or group" (Leininger, 1991 p. 37). This process is called *cultural imposition,* which impacts nearly all client care.

Relationship to Nature

Biomedicine adheres to the concept of mastery over nature. Practitioners are soldiers in the "war" on cancer (or other conditions). They "fight" infection, "conquer" disease, and "kill" pain. Technology is considered omnipotent; its tools are the arsenal used to battle pain and illness.

One factor in this approach is the attitude that health can be measured numerically and that there are standardized definitions of disease. Blood and urine analyses, X rays, scans, and other diagnostic tests are used to define whether a patient is within normal physical or biochemical ranges. Figures falling within designated parameters means the patient is functioning normally; if the data are too high or too low, the patient is in an abnormal state that may indicate disease. Diagnosis occurs independent of the idiosyncratic characteristics of the individual, usually without consideration of cultural factors such as ethnic background or religious faith. Symptoms occurring outside numerical confirmation are frequently determined to be of psychosomatic origins (Helman, 1990).

Personal Control or Fate?

The conventional U.S. medical system leaves little room for chance or divine intervention. Scientific rationality dictates that there is a biomedical cause for every condition, even if it is as yet undiscovered. Each individual inherits a certain physiological constitution, and there is a personal responsibility to make the choices that prevent illness. Receiving immunizations and getting regular checkups are biomedical ways in which an individual can preserve health. Being obese, smoking cigarettes, consuming immoderate amounts of alcohol, and failing to manage stress are biomedical examples of how an individual may endanger health. A person who behaves in a manner believed to cause disease is often stigmatized.

When ill, the biomedical assumption is that a person will reliably comply with therapy and that treatment, if undertaken correctly by the patient, will alleviate the condition. The onus of cure is dependent on personal behavior. From the patient's perspective, there is the presumption that health care professionals will provide mistake-free care. Malpractice suits filed when care was less than perfect have led to extensive charting and record keeping in the U.S. biomedical system.

State of Being

Congruent with the value on personal control, biomedical patients are expected to be active partners in their cure. Complacency and noncompliance are greatly disliked by biomedical practitioners (Spector, 1991). Changes in lifestyle can help preserve health; taking medications and completing therapeutic regimens can relieve symptomatic pain or cure disease. The biomedical emphasis is on doing, not being. Other worldviews may expect client passivity and acceptance of adverse conditions . Clients are the recipients of healing, not participants.

Role of the Individual

Similar to the American majority worldview, individuality is honored in U.S. biomedicine, and client confidentiality is nearly inviolate. Individuals are seen as a single, biological unit, not as members of a family or a particular cultural group. It is assumed that a person desires privacy, and clients are sometimes encouraged by providers to keep medical matters quiet, even if it means withholding information from relatives. Treatment typically is focused on each client, in keeping with the beliefs of personal responsibility and the provider–patient partnership.

Human Equality

A fundamental premise in American biomedicine is that all patients deserve equal access to care, although practically speaking, cost, location, and convenience prevent many patients from receiving adequate health services. This is a relatively unique perspective; most other societies deliberately ration health care through assessing physical status (e.g., a young person may receive services denied a terminally ill elderly person) or through socioeconomic status (e.g., the wealthy can purchase care, the poor are left to whatever society offers).

The biomedical worldview on human equality differs substantially from the mainstream American outlook in one way, however. A hierarchy of biomedical professionals is strictly observed in the United States, with physicians having the highest status and allied health professionals substantially less. Health care workers outside the professional system, such as clerical and custodial workers, and those beyond the reach of biomedicine, such as folk healers, are accorded even lower standing. Deference to those of superior rank is expected. The client is typically inferior to biomedical professionals within this hierarchy.

Aging

Biomedicine supports the majority American worldview in its value on youthfulness. Many aspects of health care practice are dedicated to postponing the aging process, from plastic surgery to the technological prolongment of life. The fear of aging is so pervasive in the U.S. culture that it influences health care outside the conventional biomedical system as well. Numerous alternative traditions promise everlasting youth through the use of certain products. The emphasis on youthfulness is in direct conflict with other cultural worldviews that honor the wisdom that comes with aging and that hold high esteem for elders.

Perceptions of Time

Biomedicine is future-oriented, with a focus on what can be done today so that the client will be better tomorrow. Often treatments are unpleasant, invasive, and even painful at the moment of their application, yet the hope is that they will benefit the client in the future. Long-term management of disease and illness prevention strategies such as diet are even more oriented toward future benefits.

Although being on time for appointments and taking medications when scheduled is valued in clients, biomedical practitioners are notorious for their disrespect of the client's time. Clients are frequently asked to arrange nonemergency consultations weeks or even months in advance and may be kept waiting on the day of their appointments.

Degree of Formality/Degree of Directness

The established biomedical hierarchy, as well as the emphasis on timeliness, is often reflected in the degree of informality observed in dialogue between provider and patient. The provider often addresses the client by his or her first name, yet expects the patient to use formal titles in return. The provider usually spends limited time on small talk and attempts to get quickly to the problem; there is the expectation that the patient will also use direct approaches. Extensive jargon without explanation is often employed.

Biomedical practitioners value honest, open communication with patients because it enhances their ability to diagnose and treat disease, and it assists in issues such as informed consent. Other cultural worldviews, however, value indirect or intuitive communication with health care practitioners (see chapter 3, "Intercultural Communication," for more information). Some cultures also believe that the family, not the patient, should be told about serious conditions (Blackhall et al., 1995).

■ The number of elders in the United States is expected to double by the year 2050; figures among some ethnic groups, such as African Americans, Asian Americans, and Latinos, show even greater growth.

Materialism or Spirituality?

Each disease from the biomedical viewpoint has its own physiological characteristics: a certain cause, specific symptoms, expected test results, and a predictable response to treatment. To many biomedical health care providers, an illness isn't "real" unless it is clinically significant; emotional or social issues are the domain of other specialists. Biomedicine differs from most traditional health care approaches in the recognition of the mind–body duality. Nearly all other cultures consider the mind and body as a unified whole. *Somatization* refers to the expression of emotions through bodily complaints. In biomedical culture, somatic symptoms are often interpreted as a maladaptive

■ The separation between physical and emotional or psychological health is so embedded in American culture that there is no English word to even express the concept of mind–body unity.

incoherent

emotional response, yet they are the most common presentation of psychological distress in patients worldwide (Ots, 1990). In folk medicine and some alternative traditions, the emotional needs of the patient are addressed through physical therapies. Spiritual intervention is frequently sought concurrently.

What Is Health?

Cultural Definitions of Health

Meaning of Health

The World Health Organization (WHO) describes health as "a state of complete physical, mental, and social well-being, not merely an absence of disease or infirmity" (Helman, 1990). Although comprehensive from a biomedical perspective, this definition does not fit the worldview of many cultural groups, because it ignores the natural, spiritual, and supernatural dimensions of health.

Most Native Americans believe that health is achieved through harmony with nature, which includes the family, the community, and the environment. Africans also emphasize a balance with nature and believe that malevolent environmental forces such as those of nature, God, the living, or the dead may disrupt a person's energy and bring illness. Many African Americans, Latinos, Middle Easterners, and some southern Europeans attribute health to living according to God's will. Gypsies maintain health through avoiding contact with non-Gypsies, who are considered inherently polluted. Most Asians believe that health is dependent on their relationship to the universe and that a balance between polar elements, such as yin and yang, must be maintained. Some Southeast Asians are concerned with pleasing their ancestor spirits, who may cause accidents or sickness when angry. Pacific Islanders believe that fulfilling social obligations is essential to health and that disharmony with family or village members can result in illness. Asian Indians consider mind, body, and soul to be interconnected and believe that spirituality is as important to health as a good diet or getting proper rest (see individual chapters on each ethnic group for more details).

- In Ayurvedic medicine, a distinction is made between general health and optimal health.

- In traditional Chinese medicine, 15 separate pulses are identified, each associated with an internal organ and each with its own characteristics.

- Health attributes not only differ interculturally but also can change within the same culture over time. Flatulence, for example, was considered an aphrodisiac in medieval Europe.

Health in other cultures is less dependent on symptoms than on the ability to accomplish daily responsibilities. Among Koreans, there is a strong desire to avoid burdening children with one's health problems. Mexican men may ignore physical complaints because it is weak and unmanly to acknowledge pain. Even within a single culture, socioeconomic differences may contribute to the definition of health; daily aches are tolerated when a weekly paycheck is essential (Helman, 1990).

Health Attributes

As health is defined culturally, so are the characteristics identified with health. Physical attributes are most commonly associated with well-being, including skin color, weight maintenance, and hair sheen. Normal functioning of the body, such as regular bowel movements, routine menstruation, and a steady pulse is expected, as is the use of arms, legs, hands and the senses. Undisturbed sleep and appropriate energy levels also suggest good health. Behavioral norms within the context of marriage, family, and community are sometimes considered a sign of well-being. It is the cultural specifics of health characteristics that tend to vary.

Healthy hair in the United States is advertised as clean, shiny, and flake-free. In many cultures, oily hair is the norm and dandruff is not a significant concern. Americans count on a single, strong pulse of about 72 beats per minute when resting, while in other medical systems there is more than one pulse of importance to health, and it is a primary diagnostic tool in illness. Pregnancy is a medical condition in the United States that warrants regular exams by biomedical professionals, whereas in many societies pregnancy is a normal aspect of a healthy woman's cycle and prenatal care is uncommon. Generally speaking, Americans expect to be content in their lives; many other cultures have no such assumptions and do not link happiness with well-being.

Body Image

One area of significant cultural variation regarding health is body image. Perceptions of weight, health, and beauty differ worldwide. In the United States, there is significant societal pressure to be thin. Although there is no scientific concurrence on the definition of ideal or even healthy weight for individuals, being overweight is usually believed to be a character flaw in the majority American culture. Even health care professionals reportedly make moral judgments about obesity, depicting overweight persons as weak-willed, ugly, and self-indulgent (Cassell, 1995). The health risks associated with being overweight cause some providers to presume ill health in their obese clients. Thinness corresponds to the biomedical worldview regarding mastery of nature, the idea that the intellect can control the appetite (deGarine & Pollack, 1995).

Historically, thinness has been associated with a poor diet and disease. In many cultures today, including those of some Africans, Caribbean Islanders, Filipinos, Mexicans, Middle Easterners, Native Americans, and Pacific Islanders, being overweight is a protective factor that is indicative of health as well as an attribute of beauty. Overweight African American women, for example, demonstrate a positive self-image and are more than twice as likely as overweight white women to be happy with their weight and to consider themselves attractive (Kumanyika et al., 1993; Stevens et al., 1994). A larger ideal body image is the norm for most black women regardless of age, education, or socioeconomic status (Becker et al., 1999). A study of Puerto Ricans in Philadelphia found that the majority of respondents defined normal weight for men as between 16 and 23 percent above the American white ideal weight for height and approximately 10 percent above white ideal weight for women (Massara & Stunkard, 1979). African American, black Caribbean Islanders, and Puerto Rican women also report a larger body size as attractive to family and peers when compared to Anglo, Eastern European American, and Italian American women (Mossavar-Rahmani et al., 1996). Although low socioeconomic status is often associated with obesity in the United States (Sobal & Stunkard, 1989), increased income has been correlated with weight gain among some Asians, Asian Indians, Latinos, and Native Americans (Furnham & Alibhai, 1983; Jeffery, 1991).

Researchers have found that attitudes about weight sometimes change when an immigrant enters a culture with different perceptions regarding health and beauty. Kenyans who relocated to Great Britain and Puerto Ricans living on the mainland United States expressed a desire for thinness that is between that of their country of origin and that of the majority culture in their new homeland (Furnham & Alibhai, 1983; Mossavar-Rahmani et al., 1996). A survey of Samoans living in Hawaii indicated that most now associate obesity with being unhealthy (Fitzpatrick & Nietschmann, 1983).

Health Maintenance

Health Habits

Just as with health attributes, there are some broad areas of intercultural agreement on health habits. Nearly all people identify a good diet, sufficient rest, and cleanliness as necessary to preserving health. It is in the definitions of these terms that cultural variations occur. For example, majority Americans typically identify three "square" meals each day as a good diet. Asians may indicate a balance of yin and yang foods is a requirement. Middle Easterners may be concerned with sufficient quantity, and Asian Indians may be concerned with religious purity of the food. To most Americans, keeping clean means showering daily; while some Filipinos bathe several times each day to maintain a proper hot-cold balance.

Beyond diet, sleep, and cleanliness other methods of preserving health are more culturally specific. An annual survey of self-reported health and safety habits in the United States (Rodale Press, 1995) lists numerous American habits considered important in avoiding illness and accidents by the biomedical culture in the United States. Approximately 90 percent of respondents drink alcohol in moderation, avoid driving after drinking, and use smoke detectors in their homes. A majority do not smoke, avoid fat in their diet and try to eat enough high-fiber foods, calcium-rich foods, and cabbage family (cruciferous) vegetables. Fewer than one-half avoid cholesterol-rich foods. Only one-third of the respondents exercise vigorously at least three times each week, and fewer than one in five said that their weight was within the range recommended for their height, build, and gender. Small increases in the numbers of persons seeking regular medical exams (i.e., cholesterol

- Health food advocate Sylvester Graham, of graham cracker fame, was among the first in the mid–19th century to associate obesity with moral weakness, calling gluttony the greatest of all causes of evil (Cassell, 1995).

- In a survey of 58 cultures, it was found that 81 percent consider plumpness to be the female ideal (Raymond, 1986).

- Despite a preoccupation with thinness, it is estimated that between 20 and 50 percent of the American adult population in the United States is obese, depending on the criteria used to define overweight.

screenings, mammograms) and decreases in the number of persons getting 7 to 8 hours of sleep each night have also been noted. Another survey, limited to dietary habits, reported similar findings regarding fat, cholesterol, and fiber consumption (American Dietetic Association, 1993). When compared to the British and French, substantially more Americans have made changes in their health behaviors in recent years, including losing weight; decreasing meat, egg, and alcohol consumption; increasing exercise; and quitting smoking (Retchin et al., 1992).

Other cultural groups would find many of these survey listings irrelevant to health maintenance. Alcohol consumption may be prohibited (as in Islam) or smoking may not be considered a factor in disease prevention. Physical labor is often a factor in preserving health, but recreational exercise is rare throughout much of the world. Eating fresh fruits and vegetables may be limited in cold climates and not identified as essential to a healthful diet. Preventative biomedical care is unusual in many cultures.

■ Smoking tobacco is a religious rite in some Native American tribes.

■ A survey of Americans reported 95 percent believe "certain foods have benefits that go beyond basic nutrition" (IFIC, 1999).

■ In ancient China, nutritionists were ranked highest among health professionals.

■ The neutral category of foods is usually expanded by those Asians who prefer biomedicine and believe that traditional health systems are unsophisticated (Anderson, 1987).

■ The humoral qualities of dry and moist are found in the traditional Italian meal, which features a wet *(minestra)* or a dry *(asciutta)* course before the entrée.

Culturally specific health practices differ particularly in those beliefs passed on within families. A small survey of U.S. college students from many backgrounds (Spector, 1991) revealed notable variation in health habits beyond general concepts regarding diet, sleep, and cleanliness. Dressing warmly (Eastern European, French, French Canadian, Iranian, Irish, Italian, Swedish) and avoiding going outdoors with wet hair (Eastern European, Italian) were listed by some. Daily doses of cod liver oil (British, French, French Canadian, German, Norwegian, Polish, Swedish) and molasses (African American, French, French Canadian, German, Irish, Swedish) as a laxative in the spring were frequently reported maintenance measures. Natural amulets were traditionally worn in some families to prevent illness, such as camphor bags (Austrian, Canadian, Irish) or garlic cloves (Italian). Faith was important to many of the students, expressed as blessing of the throat (Irish, Swedish), and wearing holy medals (Irish), as well as daily prayer (Canadian, Ethiopian).

Health-Promoting Food Habits

Food habits are often identified as the most important way in which a person can maintain health. Nearly all cultures classify certain foods as necessary to strength, vigor, and mental acuity. Some also include items that create an equilibrium within the body and soul.

General dietary guidelines for health usually include the concepts of balance and moderation. In the United States, current recommendations include a foundation of complex carbohydrates in the form of whole grains, vegetables, and fruits; supplemented by smaller amounts of protein foods such as meats, legumes, and dairy products; and limited intakes of fats and sugar. A survey of American dietary habits found that vegetables and fruits were the items selected most often by respondents as the foods needed to achieve a healthy diet (American Dietetic Association, 1993). The Chinese system of *yin-yang* encourages a balance of those foods classified as yin (items that are typically raw, soothing, cooked at low temperatures, white or light green in color) with those classified as yang (mostly high-calorie foods, cooked in high heat, spicy, red-orange-yellow in color), avoiding extremes in both. Some staple foods, such as boiled rice, are believed to be perfectly balanced and are therefore neutral (Anderson, 1987). Although which foods are considered yin or yang vary regionally in China, the concept of keeping the body in harmony through diet remains the same, usually adjusted seasonally to compensate for external changes in temperature and for physiological conditions such as age and gender (see chapter 11 on Asians for more information).

Aspects of the yin-yang diet theory are found in many other Asian nations, and a similar system of balance focused on the *hot-cold* classification of foods is practiced in the Middle East, parts of Latin America, the Philippines, and India. Hot-cold concepts developed out of ancient Greek humoral medicine that identified four characteristics in the natural world (air-cold, earth-dry, fire-hot, water-moist) associated with four body humors: hot and moist (blood), cold and moist (phlegm), hot and dry (yellow/green bile), and cold and dry (black bile). Applied to daily food habits, this system usually focuses on only the hot and cold aspects of food (defined by characteristics such as taste, preparation method or proximity to the sun) balanced to account for personal constitution and the weather. In Lebanon, it is believed that the body must have time to adjust to a hot food before a cold item can be eaten. In Mexico, the categorization of hot and cold foods is related to a congruous relationship with the natural world. Asian Indians associate a hot-cold balanced diet with spiritual harmony.

Quantity of food is often associated with health as well. African Americans, for example, traditionally eat heavy meals, reserving light foods for ill and recuperating family members. In the Middle East, ample food is necessary for good health, and a poor appetite is sometimes regarded as an illness in itself. As discussed previously, being overweight is frequently associated with well-being in some cultures.

In addition to balance and moderation, specific foods are sometimes identified with improved strength or vitality. In the United States, milk builds strong bones, carrots improve eyesight, and candy provides quick energy. Navajos consider milk to be a "weak" food, but meat and blue cornmeal are "strong" foods. Asians call strengthening items *pu* or *bo* foods, including protein-rich soups with pork liver or oxtail in China, and bone marrow or dog meat soup in Korea. Puerto Ricans drink eggnog or malt-type beverages to improve vitality.

The *sympathetic quality* of a food, meaning a characteristic that looks like a human body part or organ, accounts for many health food beliefs. It is an obvious extension of the "you are what you eat" concept: that the properties of a food entering the mouth are transformed into physical traits. American weightlifters eat meat to build muscle, Asian Indians eat walnuts to increase their brain power, Italians drink red wine to improve their blood, and Vietnamese eat gelatinized tiger bones to become strong. Throughout Asia and parts of the United States, ginseng, which is a root that resembles a human figure, is believed to increase strength and stamina. Other foods are believed to prevent specific illnesses. Americans, for instance, are urged to eat cabbage family (cruciferous) vegetables to reduce their risk of certain cancers. Oatmeal, and fish (high in omega-3 fatty acids), have both been promoted as preventing heart disease.

Some cultures believe that locally grown foods prepared at home are healthiest and, in the United States, organic foods (those produced without the use of chemical additives or pesticides) have increased in popularity. Vegetarianism, macrobiotics, customized diets that account for an individual's food sensitivities or allergies, and very low-fat diets are a few of the many other ways in which health is promoted by some people through food habits.

Disease, Illness, and Sickness

Cultural Definitions of Disease, Illness, and Sickness

When health is diminished, a person experiences difficulties in daily living. Weakness, pain and discomfort, emotional distress, or physical debilitation may prevent an individual from fulfilling responsibilities or obligations to the family or society. Researchers call this experience *illness,* referring to a person's perceptions of and reactions to a physical or psychological condition, understood within the context of worldview. In biomedical culture, illness is caused by *disease,* defined as abnormalities or malfunctioning of body organs and systems. The term *sickness* is used for the entire disease–illness process. When an individual becomes sick, questions such as how did the illness occur, how are the symptoms experienced, and how is it cured arise—answered primarily through cultural consensus on the meaning of sickness.

Becoming Sick

During the onset of a sickness, a person becomes aware that a problem exists, through physical or behavioral complaints. The development may be slow and the symptoms may take time to manifest into a disease condition, or they may occur suddenly, with little question as to the presentation of illness.

Except in emergencies, an individual usually seeks confirmation of illness first from family or friends. Symptoms are described and a diagnosis is sought. A knowledgeable relative is often the most trusted person in determining whether a condition is cause for concern and whether further care should be pursued; in

- Chicken soup, a traditional tonic among Eastern European Jews, has become a well-accepted health-promoting food throughout the United States.

- Ritualistic cannibalism, especially where the heart or liver of a brave and worthy enemy was consumed, is an extreme example of the sympathetic qualities of food.

- Macrobiotics is a Japanese diet based on brown rice, miso soup, and vegetables that was popularized in Europe as promoting health in the 1920s. Serious nutritional deficiencies have been identified in infants and toddlers on this restricted diet (Dagnelie & van Stavern, 1994).

- In his popular books and lectures, Dr. Dean Ornish has promoted the use of a very low-fat diet (less than 10 percent of calories) to reverse the buildup of cholesterol-rich plaque in the arteries and to reduce the risk of heart disease.

many cultures, a mother or grandmother is the medical expert within a family. This is a major step in social legitimization of the sickness. If there is agreement that the person is ill, then the individual can adopt a new role within the family or community—sick person. In this capacity, the sick person is excused from many daily obligations regarding work and family, as well as social and religious duties. A reprieve from personal responsibility for well-being is also given, with care provided by relatives, healers, or health professionals. The role of sick person provides a socially accepted, temporary respite from the physical and psychological burdens of everyday life, with the understanding that sickness is not a permanent condition and that recovery should occur (Helman, 1990; Spector, 1991).

■ Regarding the 1918 influenza epidemic, the *New York Post* reported that *'epidemics are the punishment which nature inflicts for the violation of her laws and ordinances'* (Garrett, 1994).

Explanatory Models

When unexpected events occur, there is a human need to explain the origins and causes of seemingly random occurrences. Explanatory models that are congruent with a culture's worldview are used to account for why good or evil happens to a person or a community and to calm individual fears of being victimized. In sickness, the explanatory model details the cause of disease, the ways in which symptoms are perceived and expressed, the ways in which the illness can be healed and prevented from reoccurring and why one person develops a sickness while another remains healthy (Clark, 1983; Kleinman et al., 1978).

The etiology of sickness is of central concern because the reason an illness occurs often determines the patient's outlook regarding the progression and cure of the sickness. In biomedical culture, three causes of disease are identified: (1) immediate causes, such as bacterial or viral infection, toxins, tumors, or physical injury; (2) underlying causes, including smoking, high cholesterol levels, glucose intolerance, or nutritional deficiencies; and (3) ultimate causes, such as hereditary predisposition, environmental stresses, obesity, or other factors (Clark, 1983). The causes of illness are generally more complex. Helman (1990) describes four theories on the etiology of sickness prevalent in most societies (Figure 2.1): those originating in the patient, those from the natural world, those from the social world, and those due to supernatural causes.

Sickness Due to the Patient. First are those sicknesses that develop within the individual patient, usually attributable to a person's constitution or lifestyle; that is, an individual has a genetic or psychological vulnerability to illness or disease, either an inherited weakness or emotional susceptibility, such as depression. Sickness may also be due to how a person lives life. A person in the United States may be blamed for a heart attack if he has become overweight, eats fatty foods, smokes tobacco, and never exercises. A person who fails to wear a seat belt, then is injured in an auto accident, may also be found at fault. Responsibility for sickness falls primarily on the patient, although in many other cultures when a person's actions are unfavorable to health, it is outside forces that are thought to actually cause an illness or accident in retaliation for the offense.

Sickness Due to the Natural World. Etiology in the natural world includes environmental

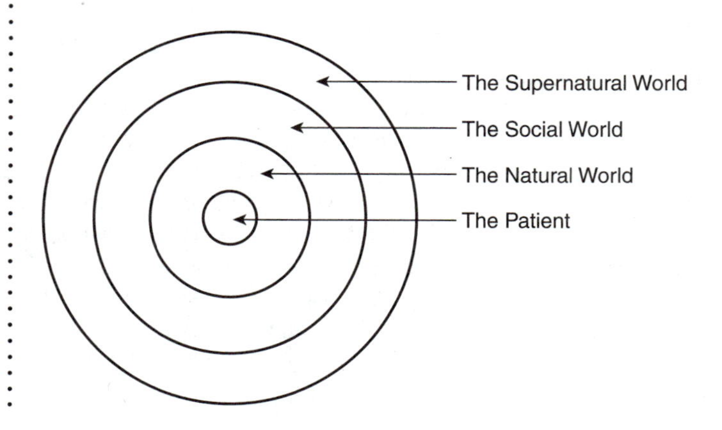

Figure 2.1
Client theories on the etiology of sickness.

Source: From C. G. Helman, (1990), *Culture, Health, and Illness* (2nd ed.). Reprinted with permission from Butterworth-Heinemann.

elements such as the weather, allergens, smoke, pollution, and toxins. *Wind* or *bad air* is of particular concern in many cultural groups, including some Arabs, Chinese, Italians, Filipinos, and Mexicans, because it can enter the body through pores, orifices, or wounds in the body, causing illness. Viruses, bacteria, and parasites are natural biological agents of sickness. Humoral systems, which associate various body humors with natural elements (as described previously), connect illness and disease with disharmony in the environment. Astrology, which determines an individual's fate (including health status) through planetary alignment at the time of birth, is another natural world phenomena. Injuries due to natural forces, such as lightning or falling rocks are sometimes categorized with this group, however, many cultures believe such accidents are actually the result of supernatural injunction.

Sickness Due to the Social World. Sickness attributed to social causes occurs around interpersonal conflict within a community. It is common to blame an enemy for pain and suffering. Inadvertent or purposeful malevolence is the source of illness and disease in many cultures. Among the most common causes is the *evil eye.* it is widely believed in parts of Africa, Asia, Europe, Greece and the Middle East, India, Latin America, and the United States that an individual who stares (especially with envy) can project harm on another person, even if the gaze is unintentional. Children are believed to be particularly vulnerable to the evil eye, resulting in colic, crying, hiccups, cramps, convulsions, and seizures. Among adults, the evil eye can cause headache, malaise, complications in pregnancy and birth, impotence and sterility in men, and insanity (Hand, 1980).

Protections against the evil eye include such practices as placing a red bag filled with herbs on an infant's crib in Guatemala; knotting black or red string around children's wrists in India; leaving children unwashed and unadmirable in Iran; wearing a charm in the form of a black hand *(mano negro)* in Puerto Rico; and painting a house white and blue to blend with the sky, thus avoiding notice, in Greece. Eastern European Jews wear a red ribbon, Sephardic Jews wear a blue ribbon. In Scotland, a fragment of the Bible is kept on the body, and in Muslim areas of Southeast Asia, a piece of the Koran is worn (Spector, 1991).

Conjury is another frequent social cause of sickness. A person with imputed powers to manipulate the natural or supernatural world directs illness or injury toward an individual, or sells the magic charms or substances necessary for a normal person to inflict harm. Conjury is practiced by witches (called *brujos* or *brujas* in Spanish), sorcerers, root doctors, herb doctors, *voodoo* or *hoodoo* doctors (see chapter 8 for more information about voodoo), underworld men, conjure men, and *goofuhdus* men, most of whom obtain their powers from the Devil or other evil spirit. For example, a conjurer might sprinkle graveyard dust under a person's feet, causing him to waste away, known as *fading* in African American tradition. A bundle of sticks placed in the kitchen will cause illness in the people who consume the food prepared there. A brujo can cause illness in Latinos through contagious magic, using bits of a person's hair or fingernails when casting a spell. Native American conjury often utilizes animals or natural phenomena (such as lightning) to attack a victim, or causes natural objects to be inserted into the body, resulting in pain. The *witched* Native American may behave in inexplicable, disruptive ways and may be abandoned by the community if the individual is considered incurable and unable to change undesirable conduct (Graham, 1976; Hand, 1980; Jackson, 1976; Spector, 1991). There is often an overlap between sickness attributed to the social world and that caused by supernatural forces.

Sickness Due to the Supernatural World. In the supernatural realm, sickness is caused by the actions of gods, spirits, or the ghosts of ancestors. The will of God is a prominent factor in illness and disease suffered by many Jews, Christians, and Muslims. Sickness is sometimes considered a punishment for the violation of religious covenants, and other times it is viewed as simply a part of God's unknowable plan for humanity. Even those persons who do not follow a specific faith may ascribe illness to "fate," "luck," or "an act of God" (Helman, 1990).

- Other names for the evil eye in the United States include the *bad eye,* the *look,* the *narrow eye,* and the *wounding eye.* Those who are victims have been *blinked, eye bitten, forelooked,* or *overlooked.*

- Fear of the evil eye is mentioned in Talmudic writings, the Bible, and the Koran.

- The term *goofuhdus man* probably originated with the graveyard dust used by some conjurers, called *gooferdust.*

- Similar to the evil eye is the Latino condition of *envidia,* when a person is jealous of another individual, sometimes resulting in serious illness, even death.

Body image is one area where the viewpoint of the client may vary from the biomedical assumptions of the health care provider. Many anorectics do not consider themselves in need of medical intervention. (© Tony Freeman/PhotoEdit.)

■ When Navajos dream frequently of death, it is usually considered a sign of serious illness.

■ The Khanty of Siberia tattoo a bird on their shoulder to prevent soul loss. Originally the tattoos were given only to elders, but in more recent times they have also been given to youth leaving home for other parts of Russia.

■ The Chinese say *gan zhu nu*—"the liver is the host of anger." The heart is associated with anxiety, the spleen with melancholia, and the kidneys with declining vitality (Ots, 1990).

Some Africans, Asians, Latinos, Middle Easterners, Native Americans, and Pacific Islanders believe that malevolent spirits can attack a person, causing illness. For example, among Cambodians, death can occur when the nightmare spirit immobilizes a victim through sitting on his or her chest and causing extreme fright (Adler, 1995). In other situations, *spirit possession* takes place. An evil spirit inhabits the body of a victim who then exhibits aberrant behavior, such as incoherent speech or extreme withdrawal. Many Southeast Asians associate caretaker spirits with body organs and life forces who may desert a person when angered or frightened, leaving that individual vulnerable to sickness. In addition, the ghosts of ancestors usually protect their living relatives from harm, but may inflict pain and illness when ignored or insulted.

One of the most common causes of sickness in many cultures is *soul loss,* when the soul detaches from a person's body, usually due to emotional distress or from spirit possession. The symptoms typically include general malaise, listlessness, depression, a feeling of suffocation, or weight loss. If left untreated, soul loss can lead to more serious illness.

Folk Illnesses

Inasmuch as sickness is culturally sanctioned and explained through culture-specific models, it follows that each culture recognizes different disorders. Certain symptoms, complaints, and behavioral changes are associated with specific conditions and are termed *folk illnesses* or *culture-bound syndromes* by researchers (Helman, 1990; Kleinman, 1980). Examples of such sicknesses are numerous, such as cases of soul loss, experienced by some Asians, Latinos (who call it *susto* or *espanto*), Native Americans, Pacific Islanders, and Southeast Asians. *Muso,* experienced by young Samoan women as mental illness, and sudden unexpected nocturnal death syndrome (SUNDS) suffered by Cambodians (see the previous section, "Sickness Due to the Supernatural World") are cases of folk illness due to evil spirits. Strong emotions, particularly fright or anger, cause many folk conditions, such as stroke precipitated in *bilis* or *colera* in some Guatemalans, the cooling of the blood and organs in *ceeb* among the Hmong, or the stomach and chest pain of *hwabyung* in some Koreans. Psychological distress is usually expressed through somatic complaints in most cultures; for instance, an Asian Indian may present symptoms of extreme stress as burning on the soles of the feet.

Diet-related folk illnesses are common. *High blood* and *low blood* among African Americans are examples. Depending on the cultural group, imbalance in the digestive system results in numbness of the extremities *(si zhi ma mu)* in some Chinese; nausea and the feeling of a wad of food stuck in the stomach *(empacho)* among Mexicans; and paralysis in some Puerto Ricans *(pasmo)*. The eating disorder *anorexia nervosa* is described by many researchers as a culture-bound syndrome in the United States and other westernized nations, associated with issues such as body image, femininity and control, and, in some cases, religious asceticism (Banks, 1992; Swartz, 1987). In the case of anorexia, it is usually the biomedical culture that identifies the symptoms as a disease state. Many anorectics do not consider themselves ill or in need of medical intervention. Such differences in the definition of sickness account for why some conditions, such as anorexia or other folk illnesses, are difficult to cure with biomedical approaches. Effective treatment of many sicknesses depends on agreement between the patient and the practitioner regarding how the illness has occurred, the meaning of the symptoms, and how the sickness is healed (Kleinman et al., 1978; Swartz, 1987).

Healing Practices

Biomedicine has increasingly focused on disease to the exclusion of illness. Biomedical health professionals attempt to diagnose and *cure* the structural and functional abnormalities found in patient organs or systems. In contrast, *healing* addresses the experience of illness, alleviating the infirmities of the sick patient even when disease is not evident. Healing responds to the personal, familial, and social issues surrounding sickness (Helman, 1990; Kleinman et al., 1978; Muecke, 1983).

Seeking Care

When sickness occurs, a person must make choices regarding healing. Professional biomedical care, if available, is usually initiated when the onset of symptoms is acute or an injury is serious. Nearly all cultural groups recognize the value of biomedicine in emergency situations.

Choice of care often depends on the patient's view of the illness in cases where the sickness is not life-threatening. In these situations, home remedies are generally the first treatment applied (Helman, 1990; Kleinman, 1980; Kleinman et al., 1978). Therapies may be determined by the patient alone, or in consultation with family members, friends, or acquaintances. If the remedies are ineffective, if other people encourage further care, or if the individual experiences continued disruption of work, social obligations, or personal relationships, professional advice may be sought. The type of healer chosen depends on factors such as availability, cost, previous care experiences, referrals by relatives or friends, and how the patient perceives the problem. If the patient suffers from a folk illness, a folk healer may be sought immediately because biomedical professionals are considered ignorant about such conditions. Otherwise, biomedical care may be undertaken, independently, or simultaneously with other approaches. A study of Taiwanese patients revealed that in acute illness, biomedical care was initially sought; if treatment was ineffective, traditional Chinese medical practitioners were employed; if there was no progress in healing, another traditional specialist would be tried; and if the patient was still afflicted, sacred healers would be sought. In chronic or recurrent sickness, biomedical, traditional Chinese medical and spiritual approaches would be attempted concurrently (Kleinman, 1980).

Research suggests that large numbers of Americans obtain health care outside the biomedical system for minor and major illnesses (Eisenberg et al., 1993; Kao & Devine, 2000; Murray & Rubel, 1992). As many as one-third to one-half of patients with intractable conditions (i.e., back pain, chronic renal failure, arthritis, insomnia, headache, depression, gastrointestinal problems); terminal illnesses such as cancer or acquired immune deficiency syndrome (AIDS); and eating disorders seek unconventional treatment. Nearly all do so without the recommendation of their biomedical doctor, integrating multiple therapies on their own (Chrisman & Kleinman, 1980; Eisenberg et al., 1993; Kao & Devine, 2000).

Biomedicine is rejected by some people because their experience with care was impersonal, costly, inconvenient, or inaccessible (Murray & Rubel, 1992). Further, conventional treatments may have been painful or harmful. Some clients believe that biomedical professionals are hostile or uninterested in ethnic health issues (Henry J. Kaiser Family Foundation, 1999; Ndidi Uche Griffin, 1994). Often, the biomedical approach is incongruent with the patient's perspective on sickness. The health care professional may express disdain for explanations of etiology that disagree with the biomedical viewpoint, or dismiss complaints with no discernible clinical diagnosis as insignificant. Folk healers and other alternative practitioners can provide an understanding of illness within the context of the patient's worldview and can offer care beyond the cure of disease, including sincere sympathy and renewed hope.

Healing Therapies

No consensus prevails among researchers on the classification of what is called unconventional, alternative, or folk medical care. Home remedies (i.e., herbal teas, megavitamins, relaxation techniques), popular therapies (i.e., chiropractic, homeopathy, hypnosis, massage), and professional practices (i.e., those that require extensive academic training in conventionally recognized medical systems, such as biomedicine, traditional Chinese medicine, and Ayurvedic medicine) include a variety of treatments that fall into three broad categories: (1) administration of therapeutic substances, (2) application of physical forces or devices, and (3) magicoreligious interventions (Spector, 1991; Murray & Rubel, 1992). Most patients use unconventional therapies without the supervision of a biomedical doctor or any other kind of health care provider. Popular and professional practitioners, when consulted, may use one or several of these treatments in healing a patient.

Administration of Therapeutic Substances. Biomedical pharmaceutical and diet prescriptions are two of the most common types of

■ Anorexia nervosa is identified by an intense fear of fat, a refusal to maintain body weight, resulting in a weight 15 percent or more below that recommended, and the absence of three consecutive menstrual cycles in women of menstruating age. A 1985 survey reported 9 percent of girls 13 to 18 years old believe they have symptoms of anorexia (Banks, 1992).

■ Rates of anorexia nervosa are increasing in cultures exposed to western values, including those in Japan, South Korea, Argentina, India, Pakistan, the Philippines, and even some Arab nations.

■ Unconventional approaches to health care are significantly more popular in the western regions of the United States than in other areas of the country (Eisenberg et al., 1993).

■ Preconquest Aztec medicine is also considered a professional system. An intellectual elite was certified through a training academy with access to a herbarium and a zoo. The Spanish killed all Aztec medical practitioners and brought humoral medicine to the region.

■ At the turn of the century, 20 to 25 percent of American urban doctors practiced homeopathy. Arizona, Connecticut, and Nevada are the only states that currently license *homeopaths* (practitioners of homeopathy).

■ Naturopathic doctors trained in the United States attend a 4-year program that includes many biomedical disciplines. Doctors of Chiropractic (DC) are the third largest category of health care practitioners in the United States, following physicians and dentists. Osteopaths are licensed as Doctors of Osteopathy (DO) in all 50 states.

■ Research on why people choose alternative therapies suggests they do so not due to dissatisfaction with biomedical approaches as much as that health care alternatives are more congruent with their own worldview (Astin, 1998).

■ *Kur* (spa) therapy is popular throughout Europe, particularly in Germany. In addition to the hydrotherapeutic qualities, the mineral content of the water is believed to be stimulating.

therapeutics in this category, which also includes over-the-counter medications, health food preparations, and prepackaged diet meals, as well as vitamins and mineral supplements. Home remedies and health practitioners other than biomedical professionals often emphasize the use of *botanical medicine*, which includes plant leaves, fruits, and roots (particularly herbs), and occasionally animal parts, such as antlers or organs, or certain powdered mineral elements. It is estimated in 1997 that over 15 million Americans took botanical remedies or high doses of vitamins concurrently with prescription drugs (Eisenberg et al., 1998). In many cultures, there are healers who specialize in the use of herbal preparations; often they are elder men or women with intimate knowledge of the natural environment. *Root doctors* and *remèdemen* in the American South, and the proprietors of *botánicas* (herbal pharmacies) found in some Latino neighborhoods are a few examples. In addition to folk healing, both traditional Chinese medicine and Ayurvedic medicine make extensive use of botanical medicine (see the chapters on each American ethnic group for more details).

Homeopathy also prescribes therapeutic substances, such as botanical medicine, diluted venom, or bacterial solutions, and biomedical drugs. Originating in Germany, homeopathy is based on the concept that symptoms in illness are evidence that the body is curing itself, and acceleration or exaggeration of the symptoms speeds healing. *Naturopathic* medicine also focuses on aiding the body to heal itself, usually through noninvasive, natural treatments (including some physical manipulations, as the following section describes), although biomedical drugs and surgery are used in certain cases.

Application of Physical Forces or Devices. Manipulations of the body operate on the shared premise that internal functioning improves with minor adjustments of physical structure. *Chiropractic* theory states that misalignments of the spine interfere with the nervous system, resulting in damage to surrounding muscles. *Osteopathic* medicine proposes that blood and lymph flow, as well as nerve function improves through manipulation of the bones, particularly the correction of posture problems, mobilization of bone joints, and spine alignment.

Several Asian healing therapies can be classified as the application of physical forces or devices and are practiced by family members in the home, by folk healers, by specialists in the therapies, or by traditional Chinese medical physicians. *Massage* therapy, *acupressure*, and *pinching* or *scratching* techniques are used to release the vital energy flow through the 14 meridians of the body identified in traditional Chinese medicine, primarily by relieving muscle tension so that oxygen and nutrients can be delivered to organs and wastes removed. *Coining* is a related practice, when a coin or spoon is rubbed across the skin instead of pressing or pinching specific points. *Acupuncture* is similar to acupressure in that it attempts to restore the balance of vital energy in the body along the 14 meridians, but it differs in that it stimulates specific junctures through the insertion of nine types of very fine needles. The needles do not cause bleeding or pain. Acupuncture is considered useful in correcting conditions where there is too much heat (yang) in the body. In conditions of too much cold (yin) another technique is preferred, called *moxibustion*, in which a small burning bundle of herbs (e.g., wormwood) or a smoldering cigarette is touched to specific locations on the meridians to restore the balance of energy. A similar method is *cupping*, the placement of a heated cup or a cup with a scrap of burning paper in it, over the meridian points (Marti, 1995; Spector, 1991).

Application of electricity is used in various *electrotherapies*, primarily to stimulate muscle or bone healing, especially in sports medicine. *Biofeedback* also uses small electric pulses to teach a person how to consciously monitor and control normally involuntary body functions, such as skin temperature and blood pressure, to alleviate health problems, which include insomnia, gastrointestinal conditions, and chronic pain. *Hydrotherapy* involves the application of baths, showers, whirlpools, saunas, steam rooms, and poultices to relieve the discomforts of back pain, muscle tension, arthritis, hypertension, cirrhosis of the liver, asthma, bronchitis, and head colds.

Botanical Remedies

More than 80 percent of the world's population uses herbal remedies to prevent illness, treat sickness, and optimize health. It is estimated that one in three Americans used an over-the-counter herbal preparation in 1997 and that the market is growing 25 to 50 percent each year (Chitwood, 1999a; Stavish, 1998).

Technically, an herbal medicine contains only leafy plants that do not have a woody stem (DeBusk, 1999). A more comprehensive term is *botanical,* including all therapeutic parts of all plants, from the root (e.g., ginseng), the bark (e.g., willow), the sap (e.g., from aloe), gum (e.g., frankincense) or oil (e.g., from nutmeg), the flowers (e.g., echinacea), the seeds (e.g., gingko biloba), to the fruit (e.g., bilberries). Botanical remedies often use the whole plant, which practitioners claim is superior to using a single active extract because other components in the plant may work together synergistically in the preparation to enhance the therapeutic value and to buffer any side effects. For this same reason, plants are often combined in formulary mixtures, particularly in traditional Chinese medicine. Many botanical remedies have proved effective in clinical trials, such as saw palmetto for enlarged prostate, St. John's wort for depression, and fenugreek seeds for diabetes (Chitwood, 1999b; O'Connell, 1999).

Most consumers select botanical remedies instead of biomedicine because they believe they are safer and more effective than prescription drugs or they are treating chronic conditions for which biomedicine has little to offer in the way of relief. One survey found that regular users of dietary supplements (including herbal medicines) are significantly more likely than nonusers to believe that supplements are safe and effective, are tested adequately, and make valid advertising claims (National Public Radio, 1999). Some proponents note that botanicals have been used for centuries and that reported deaths each year number in only the hundreds, compared to over 100,000 deaths due to prescription drug–induced conditions (Lazarou et al., 1998). The key word, is *reported.* The Dietary Supplement Health and Education Act (DSHEA) passed by Congress in 1994 defines dietary supplements as separate from food and drugs, thus outside the scope of federal monitoring. Manufacturers are exempt from regulations requiring that complaints, injuries, or deaths due to consumption of their product be reported to the Food and Drug Administration (FDA). Though the FDA retains the right to protect the public from harmful products, the burden of proof is on the government to prove that a particular botanical remedy is unsafe. Many manufacturers have voluntarily adopted good manufacturing processes, and the American Herbal Products Association has created a botanical safety rating system that classifies herbs as 1, safe when consumed appropriately; 2, restricted for certain uses; 3, use only under the supervision of an expert; and 4, insufficient data to make a safety classification.

Unfortunately, the explosive, unregulated growth of the industry has resulted in numerous problems due to inappropriate dosing, products adulterated with prescription drugs or toxins, or harmful interactions with biomedical therapies. Among more notable news media reports are 20 to 30 deaths due to overdoses of ephedra (also known as *ma huang*); the discovery by the California Department of Health Services that up to one-third of imported Asian botanicals were spiked with drugs not listed on the label or contained arsenic, lead, and/or mercury; 53 cases of renal failure and three deaths in Belgium due to a mislabeled Chinese weight loss remedy; and several cases of acute hepatitis due to the consumption of chaparral and/or bee pollen.

Of particular concern is the interaction of botanical drugs with other concurrent therapies. Adverse interactions with anticoagulants are one example (de Lemos et al., 1999). Garlic, gingko biloba, ginseng, and salvia *(danshen)* taken with Warfarin have been found to potentiate the anticoagulation activity and in some cases resulted in hemorrhaging. Other botanicals with the potential for adverse interactions include echinacea, feverfew, and certain herbal teas. Additional, possibly fatal interactions can occur between licorice root and the heart drug Lanoxin; kava-kava and antianxiety drugs such as Xanax; St. John's wort and antidepressants, including Prozac, Paxil, and Zoloft; and ephedra with any MAO inhibitor used to treat depression, such as Marplan, Nardil, or Parnate, or the heart drug Lanoxin (Graedon & Graedon, 1999). In addition, some botanical remedies can interact adversely with anesthesia, and others can increase side effects of radiation therapy (Kao et al., 2000).

Difficulties arise when patients often use these botanical remedies to alleviate symptoms for which they are also being treated biomedically, but they do not inform their health care practitioner because they assume that herbs are safe. Some pharmacies have recently added dietary supplements to their computer database and have instructed pharmacists to ask about botanical supplement use by clients in an effort to identify potentially harmful combinations (see each chapter on U.S. ethnic groups for "Selected Botanical Remedies" tables).

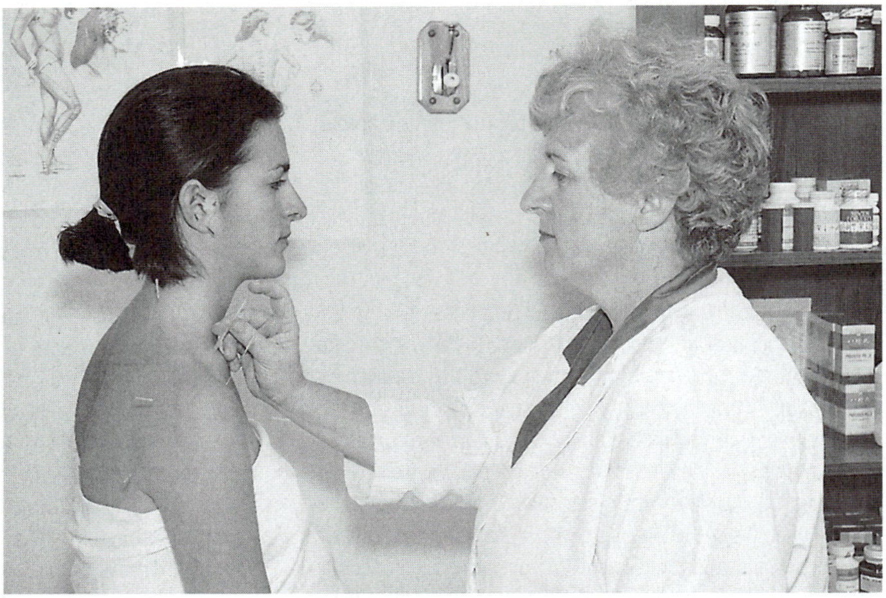

Acupuncture attempts to restore the balance of vital energy in the body through the use of special needles. (© W. B. Spunbarg/PhotoEdit.)

■ Seventy-nine percent of Americans believe faith can help recovery, 63 percent want physicians to discuss faith, 48 percent would like their doctors to pray with them, and 25 percent use prayer to help healing. However, reviews of the research on the salutary effect of religious faith have found studies that suggest church attendance is significantly related to good health are flawed (Levin & Vanderpool, 1987; Sloan et al., 1999).

■ When a faith healer, such as a *traiteur,* is unsuccessful in healing a patient, it is usually attributed to the patient's lack of faith in God or due to God's will.

Magicoreligious Interventions. Spiritual healing practices are associated with nearly all religions and cosmologies. They typically fall into two divisions: those actions taken by the individual, and those taken on behalf of the individual by a sacred healer.

In western religious traditions, God has power over life and death. Sickness represents a breach between humans and God. Healing is integrally related to salvation because both mend broken ties (Yoder, 1976). Living according to God's will is necessary to prevent illness, and prayer is the most common method of seeking God's help in healing. Roman Catholics, for example, make appeals to the saints identified with certain afflictions: St. Teresa of Avila for headaches, St. Peregrine for cancer, St. John of God for heart disease, St. Joseph for the terminally ill, and St. Bruno for those who are possessed are just a few examples. Pilgrimages to the shrines of these saints are made for special petitions. In Eastern religions, health is determined mostly by correct conduct in this and past lives, as well as in the virtuous behavior of ancestors. Religious offerings are made regularly; for instance, Hindus choose a personal deity to worship daily at a home shrine. Improper actions leading to disharmony within a person, family, community, or the supernatural realm can cause sickness. Healing occurs through restoration of balance, often including offerings to the deities or spirits of the living and dead who have been offended.

Individual healing practices that developed out of religious ritual include *meditation,* a contemplative process of focused relaxation; *yoga,* the control of breathing and use of systematic body poses to restrain the functions of the mind and promote mind–body unity; and *visualization* or *guided imagery,* induced relaxation and targeted willing away of health problems. Each concentrates the power of mind on reducing health risks, such as stress, high blood pressure, and decreased immune response, or on alleviating specific medical conditions. *Hypnotherapy* works in a similar manner; although it is generally done with the aid of a hypnotherapist, self-hypnosis can be learned for personal use.

In many cases, the spiritual skills of the individual are inadequate for the problem, and the help of a sacred healer is sought. These health practitioners generally work through interventions with the supernatural world, which may include prayers, blessings, chanting or singing, charms, and conjury, as well as the use of therapeutic substances (i.e., herbal remedies) and application of physical cures (i.e., the laying on of hands). *Faith healers,* most of whom get their healing gifts from God, are common among many Christian groups. Some are affiliated with certain denominations and rites, such as the Cajun *traiteurs* of Louisiana, who specialize in treating one or two ailments through prayers and charms associated with Catholicism (Brandon, 1976; Leistner, 1996). Others, such as the sympathy healers of the Pennsylvania Dutch who practice *powwowing* (also known by the German name *Brauche* or *Braucherei*), are considered the direct instruments of God (Hostetler, 1976).

Persons with a spiritual calling are often employed to treat illness. *Neng* among the Cambodians, Mexican *curanderos* (or *curanderas*), practitioners of voodoo in the American South, and *espiritos* or *santeros* (or *santeras*) in the Caribbean may communicate with the spirits or saints to heal their patients. Ceremonial invocation is the primary therapy, although charms and spells to counteract witchcraft and botanical preparations to ease physical complaints are used as well (Bankston, 1995; Graham, 1976; Jackson, 1976).

Shamans, called *medicine men* among many Native American groups, are sacred healers with exceptional powers. They originated in Russia an estimated 20,000 years ago and spread throughout the world to the indigenous cultures of the Americas, Southeast Asia, Indonesia, Polynesia, and Australia. Remote tribal groups found in Africa, India, and Korea have similar headers. A shaman is a composite priest, magician, and doctor; the position passed on from generation to generation, or through a calling that could include fainting spells or convulsive fits due to attacks by spirits. Shamans typically complete lengthy apprenticeships and are initiated through a series of trials simulating death and rebirth. In shamanic systems, sickness is due to spiritual crisis and healing emphasizes strengthening of the soul through redirection of the life forces or, in cases of serious illness, retrieval of the soul, which may have been stolen by evil spirits. Shamanic practices include visualization techniques to create harmony between the patient and the universe, singing, chanting, prognostication, dream analysis, and séances. Shamans are often expert herbalists (Balzer, 1987; Freedman, 1996; Sheikh et al., 1989).

Pluralistic Health Care Systems

The enduring popularity of traditional health beliefs and practices is based in cultural congruency. Healing sickness, with or without the services of an expert provider, takes place according to a patient's worldview. Humans value what validates their beliefs, and discount anything that differs, regardless of statistical data or scientific claims; they give disproportionate credence to persons they like and respect (Anderson, 1987).

Medical Pluralism

Medical pluralism is the term for the consecutive or concurrent use of multiple health care systems (Clark, 1983). Although it is believed that Americans of non-European heritage and the poor (who may have little contact with biomedicine) are most likely to rely on folk traditions (Chrisman & Kleinman, 1980), many patients using home remedies and popular therapies are well-educated, upper-class members of the white majority culture (Eisenberg et al., 1993; Murray & Rubel, 1992). Research indicates the use of folk healers among Korean Americans and Latinos also increases with wealth and advanced education; acculturation is not associated with a rise in the use of biomedical services (Graham, 1976; Marks et al., 1987; Sawyers & Eaton, 1992; Solis et al., 1990). Medical pluralism is widespread in the United States.

Biomedical Healing

Clients using traditional health practices are generally seeking to alleviate the difficulties experienced in illness through understandable, flexible, and convenient treatment from a warm and caring provider. The personal relationship with the healer is as important as the actual therapy (see chapter 3, "Intercultural Communication").

Biomedical professionals often operate in partnership, knowingly or not, with unconventional health care practitioners. Researchers suggest care is optimized when providers work

■ *Powwowing* is unrelated to Native American healing practices; the faith healer asks for direct assistance from God in curing the sick person through the use of charmed objects, such as knotted strings or nails, and prayer or incantation.

■ The word *shaman* comes from the Russian term *saman,* meaning "ascetic."

■ 64 percent of U.S. medical schools reported that they offer elective courses in alternative medicine or include such topics in required classes (Wetzel et al., 1998).

■ A small study of prostate cancer patients reported that 37 percent used alternative therapies in conjunction with radiation treatment. A separate survey found that their physicians believed only 4 percent of their patients used any other health care practices (Kao & Devine, 2000).

■ Medications using digitalis, opiates, and salicylates, common today as biomedical therapeutics, were first used by folk healers.

■ In one study, pregnant Hmong women showed dramatic increases in acceptance of invasive procedures, such as pelvic exams and blood tests, after viewing a video in the Hmong language acknowledging the value of traditional practices and explaining the rationale of the biomedical procedures (Spring et al., 1995).

■ According to the 5th-century physician Hippocrates, "Life is short, art is long, opportunity fleeting, experiment deceptive, and judgment difficult. Hence not only the physician, but also the patient, and everyone else involved in the situation, must cooperate."

together rather than at cross purposes. Patients are sometimes put in the confusing position of choosing between biomedical and traditional systems that reject each other (Pang, 1989). Studies suggest that some unconventional therapies are effective, benefiting the patient physiologically or psychologically, and should be accepted as complementary to biomedical approaches (Murray & Rubel, 1992). Cooperative monitoring by a biomedical professional can also detect those few instances when a home remedy or popular practice is harmful to the patient.

Further, biomedical health care providers can adopt certain healing strategies. Understanding the patient's perspective on illness and attending to differences in the patient-provider explanatory models (Kleinman et al., 1978) is one approach (see "Self-Evaluation of Therapeutic Food Use" at the end of this chapter). Recommending alternate, experimental biomedical programs in cases of advanced chronic disease is another (Murray & Rubel, 1992). A more comprehensive methodology is offered by Leininger, who developed the culture care theory of nursing to guide "judgments, decisions, or actions so as to provide *cultural congruent care* that is beneficial, satisfying, and meaningful" to clients (Leininger, 1991, p. 41). She identifies three modes of effective care: (1) cultural care preservation and/or maintenance, (2) cultural care accommodation and/or negotiation, and (3) cultural care repatterning or restructuring.

Cultural care preservation and/or maintenance is used when a traditional health belief or practice is known to be beneficial in its effect and is encouraged by the provider. Cultural care accommodation and/or negotiation is accomplished between the provider and the patient (or the patient's family) when there is an expectation for care that is outside biomedical convention. Cultural care repatterning or restructuring occurs when both provider and patient agree that a habit is harmful to health, and a cooperative plan is developed to introduce a new and different lifestyle. Applied to food habits, the culture care theory acknowledges that some traditional beliefs and practices regarding diet have beneficial or neutral consequences, some have unknown consequences, and some may be deleterious to the health of a client (see chapter 1 on nutrition and food habits).

Success of the culture care theory modalities is dependent on the intercultural knowledge and sensitivity of the provider, and must be undertaken in a coparticipative manner with the patient, as well as the patient's family (Leininger, 1991). The same could be said for all effective medical systems: healing is not necessarily the sole domain of home remedies and popular health care approaches. Biomedicine can help heal sickness through recognizing and understanding the therapeutic advantages of traditional health beliefs and practices.

Self-Evaluation of Therapeutic Food Use

Health Maintenance

What is your idea of a balanced diet?

List any foods you eat to stay healthy.

List any foods you eat to improve strength, endurance, or vitality.

List any foods you avoid to help prevent illness or disease.

What do you consider to be a healthy body image (thin, plump, muscular, or other)?

Disease, Illness, and Sickness/Healing Practices

List one food that your mother fed you when you were sick.

Are there any foods you desire when you are sick?

List any foods you eat to cure illness or disease when you are sick.

List several home, popular, or traditional therapies involving food, herbs, and/or vitamins and minerals. Would you ever try any of them? Why or why not?

Attitudes

Were you aware of your own therapeutic uses of food before you completed this evaluation?

Are your therapeutic uses of foods based on biomedical research? On information you obtained from a newspaper, magazine, television, or computer? On information learned from family or friends?

What is your opinion about people who use home, popular, or traditional therapies to treat illness and disease?

What is your opinion about overweight persons? About overly thin persons?

Application

How do your food habits differ from your family norms? Those of friends? Those of people you work with? Those of clients or patients? In what significant ways do they differ?

What can you do to avoid assumptions about therapeutic food habits that seem illogical or unfounded?

3

Intercultural Communication

Health providers in the United States take pride in their technical expertise and mastery of knowledge. They spend years understanding biochemical and physiological processes, laboratory assessments, diagnostic data, and therapeutic strategies, yet little of that time is devoted to how valuable information is effectively communicated to clients.

Words are the primary tool of the clinician following diagnosis. While the surgeon depends on the scalpel, most other providers rely on language to inform and guide patients in the treatment and lifestyle changes necessary to maintain or improve health (Tumulty, 1970). The surgeon has significant control within the surgical setting; in many cases the patient is not even conscious. In contrast, the clinician interacts directly with a patient who has independent, sometimes contradictory, ideas about health, illness, and treatment. The provider can control only her or his side of the conversation; if the words are ineffective, the client may reject recommended medications or therapies. Although the actions of the surgeon are generally limited to the patient, the advice of the health care provider often impacts not only the patient, but also the patient's family. Dietary modifications in particular may have long-term implications; if cultural food habits are changed, the new ways of eating may be passed on for generations.

How can a health care practitioner successfully communicate with clients from culturally different backgrounds? Intercultural communication is a specialty in itself. The field encompasses language and the context in which words are interpreted, including gestures, posture, spatial relationships, concepts of time, the status and hierarchy of persons, the role of the individual within a group, and the setting. This chapter presents a broad and limited overview of intercultural communication concepts useful in counseling individuals or nutrition education programs with groups. Later chapters on each American ethnic group provide more specific information as reported by researchers and practitioners familiar with cultural communication characteristics.

Role of Communication in Health Care

In most cultures, health care is a team effort. Team members in the United States typically include one or more physicians, associated allied health professionals, and the patient. In other societies, the extended family and traditional health practitioners (including those who provide spiritual intervention) are also significant players. Care is effective to the degree that these interdependent team members successfully communicate despite possible differences in language, ethnicity, religious affiliation, gender, age, educational background, occupation, health beliefs, or other cultural factors (Kreps & Kunimoto, 1994).

■ "Observe the nature of each country; diet; customs; the age of the patient; speech; manners; fashion; even his silence.... One has to study all these signs and analyze what they portend." Hippocratic writings, 5th century B.C.E.

Purpose of Communication in Health Care

> Health care providers and consumers depend on their abilities to communicate sensitively and effectively with one another to relieve discomfort, save lives, and promote health. Ineffective communication in health care can, and often does, result in unnecessary pain, suffering, and death. (Kreps & Kunimoto, 1994, p. 8)

Interaction between Provider and Client

In a simple model of successful communication, the health care professional relies on the client to provide accurate, detailed information about his or her medical history and current symptoms so that the appropriate diagnosis and treatment can be determined. The client depends on the practitioner to explain any medical condition in terms that are understandable and to describe treatment strategies and expectations clearly. This basic conversation is repeated between providers and their clients daily; it is the essence of clinical health care. In practice, however, the complexity of the interaction between provider and client is greatly underestimated in this model.

Numerous barriers to the sharing and understanding of knowledge can prevent successful communication in the health care setting. For example, a client may be fearful or in pain when seeking help, more focused on immediate discomfort than on conversing clearly with the provider. During times of stress, a client is also more likely to use her or his mother tongue than English if it is a second language (Haffner, 1992). In biomedicine, the provider usually assumes the role of the expert, leaving little room for participation of the client as the authority on what he or she is experiencing physically or emotionally (Brookins, 1993; Cassidy, 1994). The provider may rely on medical jargon because it is difficult to translate many terms without extensive explanations or oversimplification (Kreps, 1990). A provider may be most concerned with the technical aspects of a health problem and inadvertently ignore the interpersonal dimensions of the relationship with the patient or may be too rushed to express care and compassion (Kreps & Kunimoto, 1994). Furthermore, cultural communication customs may interfere directly with the trust and respect necessary for effective health care (Fong, 1991).

Researchers in effective communication have identified five ways in which misunderstandings occur that are applicable to the health care setting (Beck, 1988; Gudykunst & Nishida, 1994):

The client depends on the practitioner to explain any medical condition in terms that are understandable and to describe treatment strategies and expectations clearly. (© Jeff Dunn/The Picture Cube.)

1. A provider can never fully know a client's thoughts, attitudes, and emotions, especially when the client is from a different cultural background.
2. A provider must depend on verbal and nonverbal signals from the client to learn what the client believes about health and illness, and these signals may be ambiguous.
3. A provider uses his or her own cultural understanding of communication to interpret verbal and nonverbal signals from the client, which may be inadequate for accurate deciphering of meaning in another cultural context.
4. A provider's state of mind at any given time may bias interpretation of a client's behavior.
5. There is no correlation between what a provider believes are correct interpretations of a client's signals or behaviors and the accuracy of the provider's belief. Misunderstandings of meaning are common.

The results of ineffective communication in health care can be serious. Noncompliance issues are among the most important for the clinician. Patients may reject recommendations or fail to return for followup appointments because they are dissatisfied with their relationship with their health care provider (Brookins, 1993; Lynch, 1990; Ndidi Uche Griffin, 1994). Conversely, patients who received care in accordance with their desired care reported significantly better dietary management of their diabetes in one study (Anderson et al., 1995). Patients often report better outcomes with traditional healers than with biomedical practitioners because there is more time spent on explanation and understanding of the condition (Kleinman et al., 1978). Development of the interpersonal relationship with the practitioner is crucial to a patient understanding and accepting treatment strategies, particularly if recommendations conflict with cultural perceptions regarding health and illness (Fong, 1991).

Intercultural Challenges

Researchers use an iceberg analogy (Figure 3.1) to describe how a person's cultural heritage can impact communication (Kreps & Kunimoto, 1994). Ethnicity, age, and gender are the most visible personal characteristics that affect dialogue, the so-called tip of the iceberg. Beneath the surface but equally influential may be degree of acculturation or assimilation, socioeconomic status, health condition, religion, educational background, group membership, sexual orientation, or political affiliation.

Most people are comfortable conversing with persons who are culturally similar to themselves. Communication is sometimes described as an *action chain* (Hall, 1977), meaning that one phrase or action leads to the

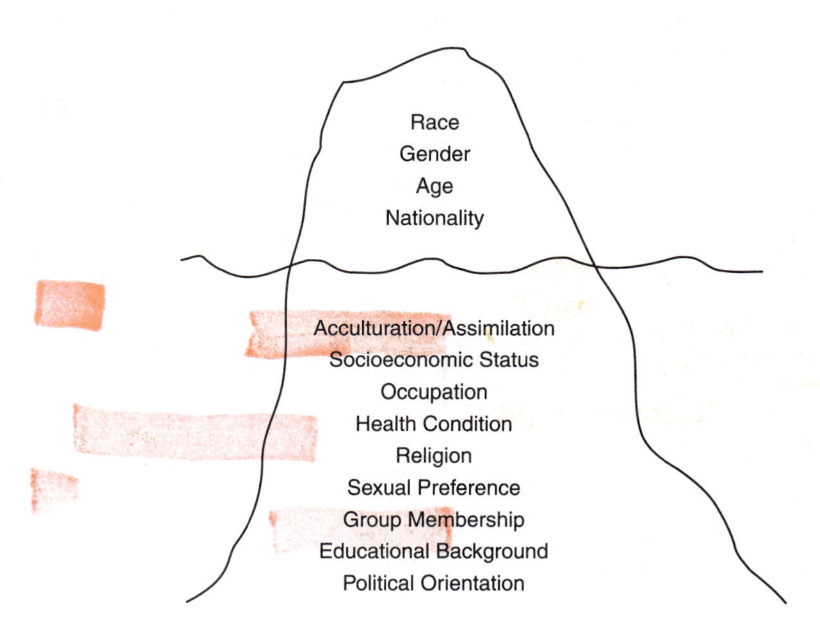

Figure 3.1
Iceberg model of multicultural influences on communication.

Source: Adapted from G. L. Kreps & E. N. Kunimoto, *Effective Communication in Multicultural Health Care Settings* (Thousand Oaks, CA: Sage, 1994), 6. Copyright 1994. © Reprinted with permission.

next: in the United States, a person who extends her hand in greeting expects the other person to take her hand and shake it, or when a person says "thank you," a "you're welcome" should follow. Communication is a whole series of unwritten expectations regarding how a person should reciprocally respond, and such expectations are largely cultural in origin. If a person understands the communication action chain, and responds as expected, a successful relationship can develop. When a person does not respond as expected, communication can break down and the relationship can deteriorate (Kreps & Kunimoto, 1994).

When meeting a person for the first time, the only information speakers usually have to work from are their own cultural norms. They use them to predict how a person will respond to their words and what conversational approaches are appropriate. They may also use social roles to determine their communication behavior. Furthermore, speakers modify their words and actions as they get to know a person individually, observing personal cues about communication idiosyncrasies that vary from cultural or social customs (Gudykunst & Nishida, 1994). An employee, for example, may make certain assumptions about a supervisor based on ethnicity, gender, or age, and especially occupational status, and then make adaptations. For example, an employee may start out calling his boss "Mrs. Smith" as a sign of respect for her position but use the more informal "Sue" when she requests that he call her by her first name.

Interpersonal relationships between two individuals are based mostly on personal communication preferences; group interactions commonly depend on cultural or social norms. There is a greater possibility for misinterpretations to occur at the cultural or social levels of communication because they are more generalized. As mentioned in chapter 1, stereotyping occurs if a person overestimates another individual's degree of association with any particular cultural or social group. It is believed that only about one-third of the people from any given group actually behave in ways typified by the group as a whole (Gudykunst & Nishida, 1994). Assumptions about how a person of a different cultural heritage should communicate can elicit certain types of reactions based on norms in that culture: Stereotyping can become a self-fulfilling prophecy. For instance, Asians have a reputation for being restrained in conversation compared to the typical American approach. A nutrition education provider teaching a prenatal care program may respond to more aggressive speakers during a group meeting, failing to actively involve a Vietnamese American participant. The Vietnamese American woman may feel the provider is disinterested in her questions or comments, becoming even less likely to offer input at the next meeting, reinforcing the provider's perception that all Asians are quiet.

The challenge is to increase familiarity with cultural communication behaviors, while remaining aware of personal cues and moving toward an interpersonal relationship as quickly as possible.

Responsibilities of Health Care Provider

Although communication requires the active participation of at least two persons, the health care provider has certain responsibilities in interactions with clients. The provider often assumes the superordinate position in the relationship because she is accorded that status by the client or because the client is distracted by pain or discomfort. In that role, it is the practitioner's obligation to understand what is said by the client and to provide the client with the information needed to participate in treatment. This may require that the clinician be familiar with cultural norms, listen carefully and seriously to the client (observing personal cues), take action based on what is said by the client, plus underpromise and overdeliver regarding therapeutic expectations. Caring and considered communication can empower the client within the relationship and improve treatment efficacy (Kreps & Kunimoto, 1994).

Intercultural communication awareness is believed to occur in four stages. First is *unconscious incompetence,* when a speaker misunderstands communication behaviors but doesn't even know misinterpretation has occurred. The second stage is *conscious incompetence,* when a speaker is aware of misunderstandings but makes no effort to correct them. Third is *conscious competence,* when a speaker considers his or her own cultural communication characteristics and makes modifications as needed to prevent misinterpretations. The final stage is *unconscious competence,* when a speaker is skilled in intercultural communication practices and no longer needs to think about them during conversation (Howell, 1982).

Effective intercultural communication begins when the speaker is mindful of his or her own communication behaviors and is sensitive to misinterpretations that may result from them (Gudykunst & Nishida, 1994). A willingness to listen carefully to a client without assumptions or bias and to recognize that the client is the expert when it comes to information about his or her experience are prerequisites to understanding the client (Cassidy, 1994). To provide successful health care, the practitioner must move past mastery of knowledge to a level of communication that promotes understanding and acceptance among clients from many cultural backgrounds (Fong, 1991).

Intercultural Communication Concepts

Communication uses symbols to represent objects, ideas, or behaviors. Thoughts, emotions, and attitudes are translated into language and nonverbal actions (i.e., gestures, posture, eye contact) to send messages from one person to another. Only the person sending the message knows the meaning of the message: the person receiving the message must use what she knows about cultural and social norms, as well as what she knows about the speaker personally, to interpret the message.

The two components of the message are the content and the relationship between the speaker and the receiver. Depending on the situation, the content or the relationship may assume greater prominence in interpretation of meaning. Messages that violate cultural expectations may be accurate in content but have a negative impact on the relationship. If the message consistently offends the receiver, the relationship will deteriorate and the message will be disregarded. Advising a Chinese client to increase calcium intake through increased milk consumption is a high content message, but does little to acknowledge the role of milk in the Chinese diet: Is the client lactose-intolerant? Does the client like milk? Does the client classify milk as health promoting or as a cause of illness? How does milk fit into the balance of the diet according to the client? Unless the provider gains an understanding of how the client conceptualizes the situation, the content of the message may be ignored because the client

Only the person sending the message knows the meaning of the message—the recipient must use what he or she knows about cultural and social norms, as well as what is known about the speaker personally, to interpret the message. (© Tom McCarthy/PhotoEdit.)

assumes disinterest or even disrespect for personal beliefs and expectations. Thus, the provider–client relationship is weakened. Messages that demonstrate respect for the individuality of the receiver are called *personal messages* and improve relationships; those that are disrespectful are termed *object messages* and often degrade relationships. Communication occurs in a continuum between personal and object messages (Kreps & Kunimoto, 1994).

Verbal messages are most useful for communicating content, while nonverbal messages usually convey information about relationships. If the nonverbal message is consistent with the verbal message it can build the relationship and help the receiver correctly interpret the meaning intended by the speaker. When the nonverbal message is inconsistent with the verbal message, both the relationship and the content are undermined. Successful communication is dependent on verbal and nonverbal skills, each of which are significantly impacted by cultural considerations.

Verbal Communication

The abstract nature of language means that it can only be correctly interpreted within context. The cultural aspects of context are so embedded that a speaker often believes they are innate—that is, that all other people must communicate according to the same expectations. Context includes issues common to cultural worldview, such as the role of the individual in a group and perceptions of power, authority, status, and time. Additionally, context in communication also encompasses the significance of affective and physical expression (termed *low-* or *high-context*) and level of tolerance for uncertainty and ambiguity (called *uncertainty avoidance*). Verbal communication occurs within these cultural premises, often operating at an unconscious level in the speaker.

Low- and High-Context Cultures

The context in verbal communication varies culturally. Conversational context can be defined as the affective and physical cues a speaker uses to indicate meaning, such as tone of voice, facial expression, posture, and gestures.

In most Western cultures, messages usually concern ideas presented in a logical, linear sequence. The speaker tries to say what is meant through precise wording, and the content of the language is more objective than personal along the continuum of personal and object messages. This communication style is termed *low-context* because the actual words are more important than who is receiving the message, how the words are said, or the nonverbal actions that accompany them. Communication in a low-context culture is so dependent on words that the underlying meaning is undecipherable if wording is chosen poorly or deliberately to mislead the recipient. Nearly every American has also experienced the obtuse professional languages of attorneys or scientists who fail to convey their message in common terminology. The Swiss, Germans, and Scandinavians are considered examples of low-context cultures (Hall, 1977).

In cultures with a high-context communication style, most of the meaning of a message is found in the context, not in the words. In fact, the wording used may be vague, circuitous, or incomplete. The content of the language is more personal than objective, dependent on the relationship between speaker and listener. Attitudes and feelings are more prominent in the conversation than thoughts. Communication in high-context cultures is analogous to the saying "reading between the lines." Misunderstandings easily occur if either participant is unfamiliar with the meaning of the nonverbal signifiers being used, such as small eye movements or sounds that are made when in agreement, or disagreement, or when upset. For example, white, majority Americans tend to squirm a bit when uncomfortable with a conversation, while the Japanese will quickly suck in air. High-context cultures are most prevalent among homogeneous populations with a common understanding of the affective and physical expression used in sending the message (see the following "Nonverbal Communication" section). Asian, Middle Eastern, and Native American cultures are very high-context. Latino societies are moderately high-context. American culture is thought to be toward the low end, but more middle-context than many European societies.

In low-context cultures, communication is usually explicit, straightforward, and unam-

■ Bureaucrats operate in a low-context setting, using technical words with limited nonverbal communication.

■ One area of conflict often found between blacks and whites is due to differences in high- and low-context communication. African Americans tend to be more high-context than white Americans, using cognitive, affective, and physical responses that may appear disruptive or overly emotional to whites (Gilbert & Gay, 1989).

biguous. The focus is on the speaker, who uses words to send messages that are often intended to persuade or convince the receiver. In high-context cultures, indirect communication is preferred. Implicit language is used, and many qualifiers are added; nonverbal cues are significant to interpreting the message. The locus of conversation is the receiver; the speaker makes adjustments in consideration of the listener's feelings (Gudykunst & Nishida, 1994). Low-context listeners are often impatient with high-context speakers, wondering when the speaker will get to the point of the conversation. Low-context listeners also frequently miss the affective and physical expression in the message.

Biomedicine tends to be extremely low-context. The conversation is focused on the provider who delivers a verbal message to the client with little consideration for the nonverbal message. The communication is high on content and low on relationship. Clients from high-context cultures are likely to be dissatisfied, even offended, by such impersonal, objective interactions. Communication problems may not be evident to a low-context clinician until the client leaves and never returns.

Interactions may range from low- to high-context depending on the situation, regardless of the overall cultural preference. In uncomfortable or embarrassing situations, a low-context communicator may be very sensitive and indirect. In high-context cultures, direct language is frequently used in intimate relationships.

As an example of low- versus high-context communication in different situations, consider a researcher presenting current nutritional data on spinach to a group of other professionals. She will likely speak in a relatively monotone voice and use scientific jargon. She will present her points in a sequential manner, supporting her thesis with examples, then restating her ideas in the conclusion. She will probably stand erect and limit the expressive use of her hands and face. The message is almost entirely in the content of the words she says. In contrast, this same woman might behave very differently when feeding her reluctant toddler spinach for dinner. She might smile and make yummy sounds as she offers him a spoonful or pretend the spinach is a plane coming in for a landing in his mouth. She might give him a spoonful of meat or potato, then try the spinach again. She might even dance around his high chair a little or hum a few bars of the old cartoon theme song about a sailor who liked spinach. She doesn't try to get him to eat spinach by explaining its nutrient content, as she did at her meeting. The message is nonlinear and not dependent on the content of the words she uses. This is not to say that a health care provider should burst out in song when working with a client from a high-context culture. But it does suggest that indirect, expressive approaches may be more effective in some intercultural clinical, educational, or counseling settings. Identification of a culture as either low- or high-context provides a general framework for communication but may be affected by other situational factors.

■ One way to patronize a person is to speak in a low-context mode, elaborating beyond what is needed for understanding.

■ Even within low-context cultures, intimate conversations are usually highly contextual; words and phrases may be significantly shortened or abbreviated—just a look may be enough for understanding.

■ It is evident that the model for food habits using Maslow's Theory of Human Maturation (see chapter 1) is, in fact, a cultural construct. Self-realization is the pinnacle of development only in cultures that value individuality over the needs of the group.

Individuals and Groups

The relationship of the individual to the group is determined in part by whether a culture is low- or high-context. In low-context cultures the individual is typically separate from the group and self-realization is an important goal. Self-esteem is dynamic, based on successful mastery or control of a situation. In high-context cultures the individual is usually defined by group association and a person desires oneness with the group, not individuation. A mutual dependency exists and self-esteem is based on how well a person can adjust to a situation (Gudykunst & Nishida, 1994). Individualism is a prominent characteristic in Australia, Canada, Great Britain, New Zealand, the Netherlands, and the United States. Collectivism is especially valued in the nations of Denmark, Ghana, Guatemala, Indonesia, Nigeria, Panama, Peru, El Salvador, Sierra Leone, Taiwan, Thailand, and Venezuela.

In societies that emphasize individuality, a person must communicate to gain acceptance by the group, whether it is the family, the workplace, or the community. Communication is used to establish the self within an individual or group relationship. When meeting someone new, the action chain in the conversation is flexible, with few expectations. The two people may focus on one or the other speaker and often delve into personal preferences, such as favorite restaurants or sports teams. When group identity is the focus of a society, there is no need for a person to seek acceptance from the group or to communicate individuality. Silence is highly valued. Interactions between strangers tend to be ritualized, and if the action chain is broken, communication cannot continue. The expectation is that each speaker will indicate group affiliation and that such identity conveys all the information that is needed to know that person.

The role of the individual within the group can have an impact on health care delivery. Cultures that are more group oriented require greater participation of their members in matters of health and illness and may expect that relatives will participate in giving patient histories, oversee physical exams, or make decisions regarding treatment (Gostin, 1995). Middle Easterners expect to go to the hospital with an ill family member to provide care. Next of kin is determined along bloodlines among Latinos, and care decisions are often the responsibility of a grandmother or mother instead of the spouse. Koreans prefer that the whole family make decisions regarding treatment for a terminally ill patient. Native Americans are so strongly associated with the group that it is difficult for them to communicate individual needs, which may be seen as narcissistic.

Uncertainty Avoidance

Related to role of the individual in a group is tolerance for uncertainty and ambiguity. Some groups exhibit great discomfort with what is unknown and different; these are defined as high uncertainty avoidance cultures. They may become anxious about behavior that deviates from the norm; there is a desire for consensus in high uncertainty avoidance cultures. Argentina, Belgium, Chile, Columbia, Costa Rica, Croatia, Egypt, France, Greece, Guatemala, Israel, Japan, Korea, Mexico, Panama, Peru, Portugal, Turkey, Serbia, and Spain are stronger in uncertainty avoidance than the United States, as are most African and other Asian nations. They typically have a history of central rule and complex laws that regulate individual action on behalf of the group (Gudykunst, 1994; Scarborough, 1998).

Cultures with low or weaker uncertainty avoidance include Canada, Denmark, Great Britain, Hong Kong, India, Indonesia, Jamaica, the Netherlands, the Philippines, Sweden, and the United States. People from these nations are usually curious about the unknown and different. They are more informal, willing to accept dissent within a group, and open to change.

It is important to distinguish the differences between risk avoidance and uncertainty avoidance. A person from a high uncertainty avoidance culture may be quite willing to take familiar risks or even new risks in order to minimize the ambiguity of a situation. But in general, risks that involve change and difference are difficult for people with strong uncertainty avoidance; this is especially a concern when changes threaten acceptance by the group. For example, researchers suggest African American women may resist certain preparations, or seasonings if family members object or if the foods might undermine ethnic identity. Furthermore, weight loss may be avoided if being thin means the potential loss of a peer group that values a larger figure (Kumanyika & Morssink, 1997).

Working with family or peers in a group setting to effect dietary change may be more successful for persons with a low tolerance for uncertainty, especially when the positive value of change is accepted and group consumption patterns are modified.

Power, Authority, and Status

The perception of power, or power "distance," can strongly influence communication patterns. In low-context cultures, where individuality is respected, power or status is usually attributed to the role or job that a person fulfills. Power distance is small. People are seen as equals, differentiated by their accomplishments. It is common for an individual to question directions or instructions; the belief is that a person must understand "why" before a task can be completed. A client may desire a full explanation of a condition and expected outcomes before undertaking a specific therapy. In many high-context cultures, where group identification is esteemed, superiors are seen as fundamentally different from subordinates. Authority is rarely questioned. For example, a health care provider counseling an Ethiopian patient with Type II diabetes may believe that a culturally sensitive approach is to ask him about his perceptions of the disease. What does he call it? How does he think it can be cured? Unknown to the provider, the Ethiopian man has a large power distance, and he assumes that the provider is the expert. Why would she ask such questions of him? Doesn't she know what she is doing? He expects her to provide all the answers, with little participation from him. He may even become uncooperative, or fail to return for a follow-up visit, because he questions her expertise.

Although there is usually some combination of both small and large power distance tendencies in a culture, one is predominant. Some countries with small power distance include Austria, Canada, Denmark, Germany, Great Britain, Ireland, Israel, the Netherlands, New Zealand, Sweden, and the United States. Those with larger power distance perception include most African, Asian, Latino, and Middle Eastern cultures, including (but not exclusively) Egypt, Ethiopia, Ghana, Guatemala, India, Malaysia, Nigeria, Panama, Saudi Arabia, and Venezuela (Gudykunst, 1994; Scarborough, 1998). Client empowerment, particularly in setting goals and objectives, may be resisted by persons from groups who come from cultures with a larger power distance; persons from groups with a smaller power distance may prefer maximum personal responsibility.

Several cultures are gender oriented as well. In masculine cultures, power is highly valued. Some, such as Germany, Hong Kong, and the United States, are considered masculine due to their aggressive, task-oriented, materialistic culture. Others, such as Italy, Japan, Mexico, and the Philippines, are characterized as masculine because sex roles are strongly differentiated. Men are accorded more authority in masculine societies. In more feminine cultures, quality of life is important; men and women share more equally in the power structure. More feminine countries can be task oriented and materialistic, such as Denmark, the Netherlands, Norway, and Sweden, but hard work and good citizenship are seen as benefiting the whole society, not a hierarchy of superiors within an organization or nation. Workplace relationships and an obligation to others are characteristically emphasized. Nearly all other nations combine masculine and feminine power qualities. Strong gender orientation can cause conflict. A health care team, for instance, may include several experts, often both women and men. An Italian American patient might show little respect for the female members and may ignore their directions unless restated by one of the male practitioners, regardless of his area of expertise. Even women from masculine societies may find it difficult to accept the authority of a female provider. Generally, masculine-oriented American health care practitioners can communicate more successfully with most ethnic clients by using a less assertive, less autocratic approach that includes compromise and consensus (Scarborough, 1998).

Time Perception

Being "on time," sticking to a schedule, and not "taking too much" of a person's time are valued concepts in America that are unimportant in societies in which the idea of time is less structured. Low-context cultures tend to be *monochronistic,* meaning that they are interested in completing one thing before progressing to the next. Monochronistic societies are well suited to industrialized accomplishments. *Polychronistic* societies are often found in high-context cultures. Many tasks may be pursued simultaneously, but not to the exclusion of personal relationships. Courtesy and kindness are more important than deadlines in polychronistic groups (Hall, 1977). Exceptions occur, however. The French have a relatively low-context culture but are polychronistic; the Japanese can become monochronistic when conducting business transactions with Americans.

People who are single-minded often find working with those performing multiple tasks frustrating. Monochronistic persons may see polychronistic behaviors, such as interrupting a face-to-face conversation for a phone call or being late for an appointment, as rude or contemptuous. Yet no disrespect is intended, nor is it believed that polychronistic persons are less productive than monochronistic people. In fact, "multitasking," the ability to do many things simultaneously, is valued in many organizations today.

■ **The Inuit conception of time is governed by the tides—one set of tasks is done when the tide is out, another when the tide comes in.**

■ **The Arabs say, *"Bukra insha Allah,"* which means "Tomorrow, if God wills."**

■ **To many Chinese the gift of a clock means the same as "I wish you were dead." Each tick is perceived as a reminder of mortality.**

Nonverbal Communication

High-context cultures place great emphasis on nonverbal communication in the belief that body language reveals more about what a person is thinking and feeling than words do (Hall,

1977). Yet customs about touching, gestures, eye contact, and spatial relationships vary tremendously among cultures, independent of low- or high-context communication style. As discussed previously, such nonverbal behavior can reinforce the content of the verbal message being sent, or it can contradict the words and confuse the receiver. Successful intercultural communication depends on consistent verbal and nonverbal messages. During personal and group interactions, persons move together in a synchronized manner. Barely detectable motions, such as the tilt of the head or the blink of an eye, are imitated when people are in synch and communicating effectively. The ways in which a person moves, however, are usually cultural and often unconscious. Although body language is closely linked to ethnicity, most people believe that the way they move through the world is universal. Misinterpretations in the subtleties of nonverbal communication are common and often inadvertent. More than 7,000 different gestures have been identified (Axtell, 1991), and meaning is easily misunderstood when awareness of differences is limited.

Touching

Touching includes handshakes, hugging, kissing, placing a hand on the arm or shoulder, and even unintentional bumping. In China, for example, touching between strangers, even handshaking if one person is male and the other female, is uncommon in public. Orthodox Jewish men and women are prohibited from touching unless they are relatives or are married. To Latinos, touching is an expected and necessary element of every relationship. The *abrazo,* a hug with mutual back patting, is a common greeting. Touching norms frequently vary according to attributes like gender, age, or even physical condition. In the United States, it is acceptable for an adult to pat the head of a child, but questionable with another adult. Women may kiss each other, men and women may kiss, but men may not publicly kiss other men. It is admirable to take the arm of an elderly person crossing the street but rude to do so for a healthy young adult.

Cultures in which touching is mostly avoided include those of the United States, Canada, Great Britain, Scandinavia, Germany, the Balkans, Japan, and Korea. Those in which touching is expected include the Middle East and Greece, Latin America, Italy, Spain, Portugal, and Russia. Cultures that fall in between are those of China, France, Ireland, and India (Axtell, 1991), as well as those in Africa, Southeast Asia, and the Pacific Islands. Health care professionals should take careful note of cultural touching behaviors. Vigorous handshaking is often considered aggressive behavior and a reassuring hand on the shoulder may be insulting. Of special mention are attitudes about the head. Many cultures consider the head sacred, and an absent-minded pat or playful cuff to the chin may be exceptionally offensive. Conversely, persons from cultures in which frequent touching is the norm may be insulted by reticence to a hug or a kiss, and may be unaware of legal issues regarding inappropriate touching in the United States.

Gesture, Facial Expression, and Posture

Gestures include overt movements like waving, or standing to indicate respect when a person enters the room, as well as more indirect motions like handing an item to a person or nodding the head in acknowledgment. Facial expression includes deliberate looks of attention or questioning and unintentional wincing or grimacing. Even smiling has specific cultural connotations.

Confusion occurs when movements have significantly different meanings to different people. Crossed arms are often interpreted as a sign of hostility in the United States, yet do not have similar negative associations in the Middle East, where it is a common stance while talking. The thumbs-up gesture is obscene in Afghanistan, Australia, Nigeria, and many Middle Eastern nations. The crooked-finger motion used in the United States to beckon someone is considered lewd in Japan; is used to call animals in Croatia, Malaysia, Serbia, and Vietnam; and summons prostitutes in Australia and Indonesia. To many Southeast Asians, it is an insolent or threatening gesture (Dresser, 1996).

Asians find it difficult to directly disagree with a speaker and may tilt their chins quickly upward to indicate no in what appears to Americans to be an affirmative nod. Some Asian Indians, Greeks, Turks, and Iranians shake

- **Anthropologists speculate that the handshake, the hug, and bow with hands pressed together all originated to demonstrate that a person was not carrying weapons.**
- **In Japan the small bow used in greetings and departures is a sign of respect and humility. The inferior person in the relationship always bows lower and longer than the person in the superior position.**
- **Many Asians completely avoid touching strangers, even in transitory interactions, such as returning change after a purchase. This can be offensive to persons who consider physical contact a sign of acceptance, such as African Americans and Latinos.**
- **Some East African tribes spit at each others' feet in greeting to prevent witchcraft.**
- **Kissing may have originated when mothers orally passed chewed solid food to their infants during weaning.**

Touching norms frequently vary according to attributes such as gender, age, or physical condition. (© Michael Newman/ PhotoEdit.)

their heads back and forth to show agreement and nod up and down to express disagreement. Puerto Ricans may smile in conjunction with other facial expressions to mean "please," "thank you," "excuse me," or other phrases. The Vietnamese may smile when displeased; in India, smiles are used only between intimates. An Asian Indian client may not know what to make of a health care provider who smiles in a friendly fashion.

Good posture is an important sign of respect in nearly all cultures. Slouching or putting one's feet up on the desk are generally recognized as impolite. In many societies the feet are considered the lowest and dirtiest part of the body, so it is rude to point the toe at a person when one's legs are crossed or to show the soles of one's shoes.

Eye Contact

The subtlest nonverbal movements involve the eyes. Rules regarding eye contact are usually complex, varying according to issues such as social status, gender, and distance apart. Majority Americans consider eye contact indicative of honesty and openness, yet staring is thought rude. To Germans, direct eye contact is an indication of attentiveness. African Americans may be uncomfortable with prolonged eye contact but may also find rapid aversion insulting. In general, blacks tend to look at a person's eyes when speaking and look away when listening. To Filipinos, direct eye contact is an expression of sexual interest or aggression. Among Native Americans, direct eye contact is considered rude, and averted eyes do not necessarily reflect disinterest. When Asians and Latinos avoid eye contact, it is a sign of respect. Middle Easterners believe that the minute motions of the eyes and pupil are the most reliable indication of how a person is reacting in any situation.

Spatial Relationships

Each person defines his or her own "space"—the surrounding area that is reserved for the individual. Acute discomfort can occur when another person stands or sits within the space identified as inviolate. Middle Easterners prefer to be no more than 2 feet from whomever they are communicating with so that they can observe their eyes. Latinos enjoy personal closeness with friends and acquaintances. African Americans are likely to be offended if a person moves back or tries to increase the distance between them. Intercultural communication is most successful when spatial preferences are flexible.

In addition to distance, the way a person is positioned affects communication in some cultures. It is considered rude in Samoan and Tongan societies, for instance, to speak to a person

- The one-finger salute is considered an insult in many cultures. This gesture dates back to Roman times, when it was called *digitus impudicus*, or the "impudent finger."
- Koreans say, "A man who smiles a lot is not a real man."
- The Japanese demonstrate attentiveness by closing their eyes and nodding.

Different cultural expectations regarding nonverbal behaviors, such as eye contact and spatial preferences, can lead to misunderstandings in intercultural communication. (Yogi, Inc./Corbis)

unless the parties are positioned at equal levels, for example, both sitting or both standing.

Successful Intercultural Communication

The biomedical system emphasizes the mastery of knowledge often at the expense of communication skill development. In the time-pressured and cost-constrained setting of health care delivery, object messages are more common than personal messages, and content is considered more relevant than the relationship. Forgotten in the process is that the empathetic connection between health care providers and patients can be therapeutic in itself (Woloshin et al., 1995). Sharing control, seeking feedback, and demonstrating sincere concern are just a few communication strategies important to patient satisfaction and compliance (Thompson, 1990).

Intercultural Communication Skills

Reading about culturally based communication differences is an intellectual undertaking. Actually applying intercultural communication concepts is much more challenging. Successful face-to-face interactions require understanding cultural communication expectations and being familiar with the idiosyncratic style of the other person. Often there is contradictory information to assimilate. The Japanese, for example, come from a culture that is male dominated, assertive, and achievement oriented, yet also high in uncertainty avoidance, emphasizing consensus and ritualistic communication practices. Should an explicit or implicit approach be used when speaking with someone of Japanese heritage? Clearly, much depends on the circumstances surrounding the conversation and the particular people involved. A few guidelines on applied intercultural communication skills are a beginning. Numerous books, articles, and courses on health care communication are available to supplement this brief overview (see also Table 3.1).

Name Traditions

Determine how clients prefer to be addressed. Majority Americans are among the most informal worldwide, frequently calling strangers and acquaintances by their given names. Nearly all other cultures expect a more respectful approach. This can include the use of title or prefixes (Mr.,

■ One researcher reports that the use of cultural food potlucks among hospital staff members has facilitated cultural understanding between employees of diverse backgrounds (Burner et al., 1990).

■ Studies suggest that the provider-client relationship improves when the client is also trained in communication skills. A meeting with coaches to guide clients in formulating appropriate questions and to practice negotiation techniques before an appointment has been successful in some health care settings (Ruben, 1990).

Mrs., Miss, Ms., Dr., Sir, Madam, etc.), use of surname, and proper pronunciation. Never use "Dear," "Honey," "Sweetie," "Fella," "Son," or other endearments or pejorative terms in place of proper names.

Name order is often different than the American pattern of title, given name, middle name, and surname. In many Latin American countries a married woman uses her given name and her maiden surname, followed by *de* (of) and her husband's family surname. Children typically use their given name, their father's surname, then their mother's family surname also. This causes confusion, because while most Latino men prefer to be addressed by their father's family surname, it is often the name that sequentially is placed on the "middle name" line of forms. Persons reading the form often assume the mother's family surname is a Latino's last name (Dresser, 1996).

Middle Easterners use their title, given name, and surname. They may also use *bin* (for men) or *bint* (for women) meaning "of" a place or "son/daughter of." It is not to be confused with the given name Ben (although it is pronounced similarly). If the grandparent of a Middle Easterner is well known, his or her last name may be added to the name order, following the surname, with another *bin.* The Vietnamese, Mein, Hmong, and Cambodians place the surname first, followed by the given name (although many make the switch to the American name order with acculturation). The Chinese and Koreans use a similar system, including a generational name following the family name and before the given name (the generational name is sometimes hyphenated with the given name or the two names are run together). Married women in China and Korea do not take their husband's surname. In Japan the surname is followed by *san,* meaning Mr. or Ms. Given names are only used among close friends and intimates. Hindus in India do not traditionally use surnames. A man goes by his given name, preceded by the initial of his father's given name. A woman follows the same pattern, using her father's initial until marriage, at which time she uses her given name, followed by her husband's given name. Muslims in India use the Middle Eastern order, and Christian Indians use the American pattern (Morrison et al., 1994). There are numerous other name traditions and preferences of address. When in doubt, it is best to ask.

Appropriate Language

Use unambiguous language when working with clients who are limited in English. Choose common terms (not necessarily simple words), avoiding those with multiple meanings, such as *to address,* which may mean to talk to someone, to give a speech, to send an item, or to consider an issue. Vague verbs, such as *get, make,* and *do,* may cause confusion. Use specific verbs, such as *purchase, complete,* or *prepare* when directing clients.

Slang and idioms may have no meaning in another culture. Many new English speakers interpret words literally. "How's it going?" makes no sense if one does not understand the meaning of *it* (Dresser, 1996). Sports analogies, including "score" and "strike out," are indecipherable if the game is unfamiliar. Phrases that suggest a mental picture, such as "run that by me" or "easy as pie" or "dodge a bullet," are barriers to comprehension (Harris & Moran, 1987). Some persons with limited English skills are embarrassed to admit that they do not understand what is being said or to ask that something be repeated. When comprehension is critical, respectfully request clients to repeat instructions in their own words or ask that they demonstrate a skill so that misunderstandings can be corrected.

■ The given name Benjamin is derived from Middle Eastern tradition. It means "son of Jamin."

Avoid asking questions that can be answered with a simple yes or no. For example, "Do you understand?" will often prompt a positive response in practitioner–client conversations, regardless of comprehension level. It is better to ask leading questions—for example, "What confuses you?" or "Tell me what you don't understand" (Dresser, 1996). In some Asian cultures, it is impossible to say no to a request. The Japanese, for instance, have developed many ways to avoid a negative response, such as answering "maybe," countering or criticizing the question, issuing an apology, remaining silent, or leaving the room (Usunier, 1993).

The direct, explicit communication style of majority Americans is predicated on the assumption that each person is saying what he or she means. This can cause difficulties when

Table 3.1
Communication Style Differences

Native Americans	Asians	Latinos
Speak slowly and softly	Speak softly	Speak softly: may perceive normal white voice as yelling
Indirect gaze when speaking and listening	Avert eyes as sign of respect	Avert eyes as sign of respect
Seldom make responses to indicate listening or offer encouragement to continue, rarely interject	Rules similar to Native Americans'	Rules similar to Native Americans'
Delayed auditory (silence valued)	Mild delay (silence valued)	Mild delay
Expression restrained	Polite, restrained; articulation of feelings considered immature	Men restrained, women expressive
Prefer direct approach	Indirect approach: may appear indifferent	Indirect approach
Rarely ask questions; yes/no responses considered complete	Rarely ask questions	Will ask questions when encouraged
Smile, handshake (*not* vigorous) customary greeting	Touching between strangers inappropriate	Shake hands, embrace, kiss cheeks, pat backs; sit and stand closer

conversing with persons for whom negotiation is standard practice. For example, a practitioner offers coffee or tea to an Iranian client. She refuses, so he sits down and begins the discussion. The client is upset, because she was being polite and had expected the health care professional to ask again, then insist that she have something to drink (Dresser, 1996).

Typically Americans not only believe what is said initially; they also consider answers absolute. In some cultures, it is acceptable to make a commitment, then decide later to change the terms of the agreement or decline altogether. People from these groups assume that one cannot predict intervening events or future needs. Further, there can be differences in what is accepted as "truth." In the United States, truth is considered objective, supported by immutable facts. In many other cultures, truth is subjective, often based on emotions (Harris & Moran, 1987; Morrison et al., 1994). A Filipino American client may report to the practitioner that she has not lost *any* weight on her low-calorie diet this week. But when she gets on the scale, she weighs 3 pounds less. Asked about it, the client explains she is frustrated because she gained one-half a pound yesterday and does not feel she is progressing. Understanding that the definitions of truth vary culturally can help explain some miscommunication.

Use of a Translator

Language can be the most difficult of all intercultural communication barriers to overcome. It is estimated that nearly 14 million people living in the United States have poor English skills (Woloshin et al., 1995). Many are recent immigrants and others view their stay in America as temporary and therefore see no need to learn English (Haffner, 1992).

According to Title VI of the Civil Rights Act of 1964, all persons in the United States are guaranteed equal access to health care services regardless of national origin, which has been interpreted to mean that there can be no discrimination based on language. Programs that do not offer access to persons with limited English skills may lose federal funding, including

Table 3.1
Communication Style Differences—continued

Middle Easterners	Whites	African Americans
Speak softly	Speak loudly, fast; control of listener	Speak fast; with affect and rhythm
Direct gaze, man to man and woman to woman, but woman may avert eyes with man	Eye contact when speaking and listening; staring rude	Direct eye contact when speaking (may avert if prolonged), look away when listening
Facial gestures responsive	Head nodding, murmuring	Interject (turn taking) often
Mild delay	Quick response	Quicker response
Very expressive, emotional	Task-oriented, focused	Very expressive, demonstrative
Indirect approach	Direct approach, minimal small talk	Respectful, direct approach
Will ask polite questions	Ask direct questions	Assertive questioning
Numerous greeting rituals, mild handshake; sit or stand very close, frequent touching (but no backslapping), use hands while speaking	Firm handshake; smile greeting; moderate touching (shoulder patting, backslapping)	Firm handshake, smile, hugging, kissing greeting; reluctance to touch may be interpreted as personal rejection

Source: Adapted from D. W. Sue, & D. Sue, *Counseling the Culturally Different: Theory and Practice* (2nd ed.) (New York, Wiley, 1990). Copyright 1990. © Reprinted with permission.

Medicare and Medicaid reimbursements. However, the regulation is vague and difficult to enforce. Compliance is mostly complaint driven. At present, few medical institutions have adequate professional translators available to meet the needs of their non-English-speaking clientele (Woloshin et al., 1995).

Unfortunately, many health care providers resort to nonprofessional translators, such as the client's family or friends, to facilitate communication. The inadequacies of such translations are numerous. Patients may be reluctant or embarrassed to discuss certain conditions in front of their relatives, or family members may decide that the information provided by the practitioner isn't really needed by the patient so they do not interpret accurately. Untrained translators are often unfamiliar with medical terminology. One study indicated that 23 to 52 percent of phrases were mistranslated by nonprofessionals; for example, *laxative* was the term used for diarrhea, and *swelling* was confused with getting fat. The translator tended to ignore questions about bodily functions altogether (Ebden et al., 1988). Even individuals who are totally bilingual may not be familiar with all dialects; the terms used in one part of a country may be very different from those used in another region. Ethical issues arise when children are used as translators. Children may be frightened of medical procedures, and dependence on a child for communication can invert family dynamics causing unnecessary intergenerational stress.

Issues regarding informed consent, patient safety, and noncompliance occur when translations are inadequate. Some health care providers attempt to use their personal skills in a foreign language because they believe that it is better to try to communicate directly than to lose some control through interpretations. Although conversing in the language of the client is often

■ A professional translator reports that unintended results due to limited language skills can be confusing, insulting, or even comic. A friendly physician meant to ask one of his clients, "Cuantos años tiene usted?" (How old are you?") He mispronounced the word as *anos,* however, meaning "How many anuses do you have?" (Haffner, 1992).

greatly appreciated, it is important for the provider not to overestimate fluency. Obtaining the services of a professional translator is warranted in all but emergency situations when a delay could be life-threatening (Haffner, 1992).

When using a translator, the practitioner should speak directly to the client, then watch the client rather than the translator during translation. If the nonverbal response doesn't fit the comment, confirmation with the translator can ensure that the meaning is clear. Sometimes a translator may appear to answer for the patient; the translator may be very familiar with the patient's history based on previous translations for other health providers. Conversely, it may take a translator considerably longer to translate a comment than it takes to say it in English, in part because certain cultural interpretations and explanations may be necessary (Muecke, 1983). The technique of back interpretation, meaning that instructions are repeated back to the clinician, can prevent miscommunication and open the conversation to any further questions by the client.

■ Health care providers note that age often affects intercultural communication because older minority members are often socially isolated and may be unwilling to communicate with health care providers from a different culture (Wood, 1989).

Providers can increase effective communication through a translator by using a positive tone of voice and avoiding a judgmental or condescending attitude. Short, direct phrases; avoiding metaphors or colloquialisms; and repeating important information more than once can improve client understanding (Randall-David, 1989).

In areas where translators are unavailable, telephone interpretation services may be an alternative. The AT&T Language Line is available throughout the United States and offers translators in most major languages trained in medical terminology (Woloshin et al., 1995).

Intercultural Counseling

Precounseling Preparation

Researchers have made many recommendations regarding effective intercultural communication. The basic competencies needed by practitioners include (1) *information transfer*—the verbal and nonverbal ability to convey object messages; (2) *relationship development and maintenance*—the ability to create rapport, establish trust, and demonstrate empathy and respect; and (3) *compliance gaining*—the ability to obtain client cooperation (Ruben, 1990). Practically speaking, a health care provider cannot be expected to become an expert in intercultural communication or to fully understand the communication modes best suited to each of the many clients from different cultural heritages. Most patients living in the United States do not expect to be treated as they would in their homeland. But familiarity with intercultural communication attitudes, knowledge, and skills can greatly enhance health care efficacy.

■ Demographic data on practitioners show a disproportionate number of whites (among dietitians, 90.5 percent are non-Hispanic whites), suggesting that intercultural counseling will become increasingly prevalent until greater diversity in the health care professions is achieved (Saracino & Michael, 1996).

The In-depth Interview

The in-depth interview is essential in intercultural counseling to determine many of the iceberg issues that may affect communication and cooperation in health care, including ethnicity, age, degree of acculturation or bicultural adaptation, socioeconomic status, health condition, religious affiliation, educational background, group membership, sexual orientation, or political affiliation. However, a client may believe that personal questions about his background are invasive or unnecessary, especially if he comes from a high-context culture. Direct inquiry may even suggest to the client that the practitioner is incompetent because she cannot determine the problem through indirect methods.

One culturally sensitive approach is the *respondent-driven interview* (Cassidy, 1994), in which simple, open-ended questions by the provider initiate conversation. The client can express her understanding and experience in her own words. The practitioner exerts little control over the flow of the response yet elicits data through careful prompting. Useful questions to ask during the conversation (Anderson et al., 1995; Kleinman et al., 1978) include these:

1. What do you call your problem? What name do you give it?
2. What do you think caused it?
3. Why did it start when it did?
4. What does your sickness do to your body? How does it work?
5. Will you get better soon, or will it take a long time?
6. What do you fear about your sickness?
7. What problems has your sickness caused for you personally? for your family? at work?

Checklist for Intercultural Nutrition Counseling

Attitudes

✓ I am open-minded and willing to be a learner instead of the expert when it comes to the client's life experiences and ways of knowing.

✓ I am sincerely interested in different cultural perspectives on reality and I can respect other cultural orientations other than my own.

✓ I can tolerate the ambiguities of intercultural communication. I can accept that some interactions will be uncomfortable or unfamiliar to me.

✓ I am patient; I attempt to understand the ideas and feelings of the client.

Knowledge

✓ I understand that although some cultural influences on communication are readily apparent, others are hidden and require development of a personal relationship so that salient factors in communication and compliance can be identified.

✓ I know that body language can provide significant information about the client's concerns and feelings; the relationship can improve or deteriorate through nonverbal communication.

✓ I understand that modification of culturally held beliefs and behaviors can have significant, long-term effects on the client and the client's family. Attempts to force change may result in ineffective communication and noncompliance by the client.

✓ I am familiar with cultural food habits of my clients.

✓ I have learned about traditional health beliefs and practices.

Skills

✓ I explain diet rationale in common terms within the context of a client's worldview, including concepts regarding the cause, prevention, and treatment of illness; I set realistic goals.

✓ I emphasize continuation of positive cultural food habits and recommend modification of only those food habits that may be detrimental to the client's health. I avoid personal bias.

✓ I attempt to send nonverbal messages consistent with my verbal messages.

✓ I engage in effective intercultural communication with all participants in the health care system to help clients through illness and improve health through supportive personal relationships, through cooperation with families and through the gathering and sharing of relevant data with other health care professionals.

Source: Adapted from Cassidy (1994); Gudykunst and Nishida (1994); Kavanagh and Kennedy (1992); Kreps and Kunimoto (1994); Sanjur (1995).

8. What kind of treatment will work for your sickness? What results do you expect from treatment?
9. What home remedies are common for this sickness? Have you used them?

Furthermore, information about traditional healers should be requested:

10. How would a healer treat your sickness? Are you using that treatment?

For nutritional assessment within the context of the client's condition, questions about food habits are appropriate:

11. Can what you eat help cure your sickness or make it worse?
12. Do you eat certain foods to keep healthy? To make you strong?
13. Do you avoid certain foods to prevent sickness?
14. Do you balance eating some foods with other foods?
15. Are there foods you will not eat? Why?

Learning about how a client understands his illness, including expectations about how the illness will progress, what the provider should do, and what he has established as therapeutic goals, allows the provider to compare her own view of the illness and to resolve any discrepancies that might interfere with care (Kleinman et al., 1978). Demonstrating sincere interest in cultural health beliefs through an open-ended conversation can elicit the information needed to begin assessment and to determine the approaches that would be most effective for each individual.

Intercultural Nutrition Assessment

Several difficulties in the collection and analysis of cultural health data have emerged in recent years. Researchers have discovered that standardized tools can introduce systematic bias into results or provide misleading information when utilized with different cultural populations.

Generalized approaches to the use of the 24-hour food recall, food frequency forms, and nutrient databases can produce large errors in assessment (Cassidy, 1994; Forsythe & Gage, 1994; Loria et al., 1991, 1994; Sanjur, 1995). Cultural unfamiliarity with concepts such as fiber; terminology differences, such as using one word for several foods or not having a name for a certain category of food (e.g., there is no Native American word for "vegetables"); lack of differentiation between meals and snacks; and translation mistakes (e.g., literal translations of food names, use of a brand name for a generic item, use of the name for a traditional food for a similar American item) are a few ways collected data can be invalidated. Frequent consumption of mixed dishes can result in omission of some nutrient sources (as when rice is prepared with dried peas or beans, yet reported as rice by some Caribbean Islanders) and over- or underestimation of intake due of complications in portion-size estimates. Tremendous variability in the amounts of food eaten has been reported between individuals and cultural groups (Hankin & Wilkins, 1994). Nutritional variety may be artificially reduced if the composition data for a cultural food prepared in significantly different ways among subgroups is used without allowances for recipe modifications. Using food composition data for similar foods when specific listings on a cultural item are unavailable can lead to intake miscalculations if numerous substitutions are necessary.

■ A study of Chinese American schoolchildren found that USDA food composition databases supplied information on only 65 percent of the 120 different food items identified in the 3-day diet records (Wang & Sabry, 1994).

Food lists derived from data on the U.S. population as a whole may miss significant dietary nutrient sources in subgroups. A study comparing a generalized food frequency questionnaire with ethnic-specific tools developed for African Americans, whites, and Mexican Americans found improvement in assessing total fat, vitamin A, and vitamin C intake with the modified forms. Calculations increased by 6 to 7 percent for fat sources, 1 to 3 percent for vitamin A, and 1 to 2 percent for vitamin C when 28 cultural food items were added to the 74 food items typical of the total population (Borrud et al, 1989). A review of food frequency questionnaires in minority populations, however, found that the number of published examples was so few, and the methodologies were so varied, that no conclusions regarding how best to develop valid, reliable instruments could be reported. Questions regarding how extensive questionnaires should be (especially number of included foods), if food groupings need to be modified, and use of portion size remain unanswered (Coates & Monteilh, 1997).

Other assessment tools may be questionable in intercultural settings as well. A model used to determine health attitudes among whites was found unreliable when used with Mexican Americans, even after operational adjustments for cultural differences (Schwab et al., 1994). Contradictions were found, for example, between self-administered and interviewer-directed questionnaire responses; participants were more likely to express disagreement about an item when completed individually than when asked about it by the interviewer. Cultural attitudes regarding pleasing authorities are believed to have influenced respondent answers, calling into question the use of the interview as a valid tool for gathering data in this ethnic group. In another study on low literacy populations the opposite was found; self-reported data on food frequency questionnaires was found unreliable when compared to comments made by respondents during follow-up interviews (Bettin, 1994).

Furthermore, anthropometric measurement tools are sometimes inappropriate for certain populations. Height and weight growth curves are particularly vulnerable to misinterpretation due to cultural variation, especially among Asian groups. Stature prediction equations for whites were inaccurate on Latinos (Chumlea et al., 1998; Freimer et al., 1983; Gardner, 1994; Netland & Brownstein, 1985) and African Americans (Hoerr et al., 1992). The

Standardized height and weight growth curves have not been validated for all ethnic groups and should be applied cautiously with cultural variation in mind. (© David Young-Wolff/PhotoEdit.)

predictive value of waist-to-hip ratios and body mass index (BMI) may vary in some populations (Croft et al., 1995; Lusky et al., 2000; Patel et al., 1999; Potvin et al., 1999). Even physiologic calculations, such as basal metabolic rate equations, may differ culturally (Case et al., 1997; Liu et al., 1995).

The development of culturally specific techniques and tools is a critical need in nutritional assessment. For an individual client, switching from a quantitative to a more qualitative approach can establish trust and cooperation in initial interviews. The 24-hour recall, for instance, can be conducted in an open-ended manner, requesting simply that all foods consumed the previous day be remembered (Cassidy, 1994). This eliminates difficulties with quantifying portions or differentiating meals. The dietitian does not need to predetermine food items or categories. In subsequent meetings, more information such as frequency and amount of given items can be requested after explanation of why this type of data is needed.

When working with many clients of a single cultural heritage, it may be useful to prepare quantitative tools based on qualitative research. Several detailed interviews with individuals can provide information on appropriate language, concepts, and formatting of the instrument. Monthly 24-hour recalls of small, representative samples are useful in determining overall consumption patterns, especially where seasonal variation occurs (Hankin & Wilkins, 1994). Focus groups have been found effective in selecting what food items to include and quantification measures in the preparation of multicultural food frequency questionnaires (Forsythe & Gage, 1994). Guidance from the targeted population is essential (Hankin & Wilkins, 1994; Jackson & Broussard, 1987; Sanjur, 1995). Two questions regarding the accuracy of data collection are suggested. First, how do cultural perceptions about food affect the

■ A comparison of food composition tables in 9 European nations exemplifies the difficulty in obtaining accurate nutrient data. Differences in definitions, analysis methods, and expression made it impossible to compare local tables with international data (Deharveng et al., 1999).

■ Kumanyika and Morssink (1997) question the biomedical paradigm that behavioral change is always best accomplished through one-on-one work with an individual. One example of cultural differences cited was research that found while white adolescent girls demonstrate poor outcomes when counseled with their mothers, black adolescent girls show significantly improved outcomes when their mothers participate in weight loss sessions (Brownell et al., 1983; Kumanyika & Morssink, 1997).

way in which clients report their intake? Second, how is the report of intake affected by the client's relationship with the interviewer, the setting, and the assessment tool? Answers to these questions can suggest culturally sensitive approaches and improve validity of collected data. (Cassidy, 1994; Sanjur, 1995)

Access to cultural foods composition data and culturally specific anthropometric and physiological measurements is more problematic. Requests for recipes can be used to expand current databases, although this technique may be too time consuming to complete with every client. Being mindful that data analysis is often approximate and that standardized measurements may be questionable, dietary modifications should be made carefully and cautiously with all clients from cultural backgrounds other than the American majority.

Intercultural Nutrition Education

Successful nutrition education strategies for groups are as dependent on intercultural communication skills as is nutrition counseling with individuals. Kumanyika and Morssink (1997) have aptly described how culture can affect program outcomes in a group weight loss setting. Negative results are possible at any phase of the process, from motivation and attendance to skill acquisition and behavior change (see Figure 3.2). At each point cultural influences may reinforce or contradict the content and context of the education messages conveyed by the health care practitioner. When communication conflict develops, an inexact period of time exists when the client is willing to negotiate toward resolution of the message. If dissatisfaction continues, a poor weight loss outcome results because the person (1) is never motivated to sign up, (2) drops out of the program before completion, (3) attends but never learns skills, or (4) learns skills but does not apply them in practice. Program designers must do more than superficially modify the program materials and the setting to communicate effectively with a different cultural group. Understanding the cultural health beliefs, attitudes, and values of a target audience; developing education programs within the context of those perceptions; and utilizing culturally appropriate, consistent verbal and nonverbal messages in an accepted medium increases communication efficacy.

Culturally Relevant Program Preparation

Although health education program models typically advise a step-by-step process of planning and execution, the reality is that some aspects of preparation and implementation occur concurrently. A health educator may begin planning with a general idea of goals for a population group, but will probably modify and refine objectives as more information

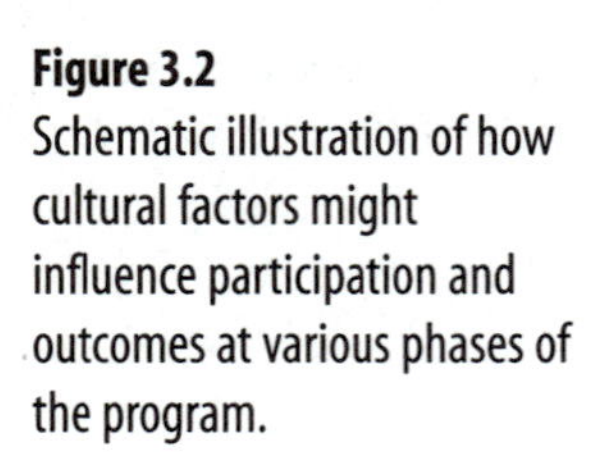

Figure 3.2
Schematic illustration of how cultural factors might influence participation and outcomes at various phases of the program.

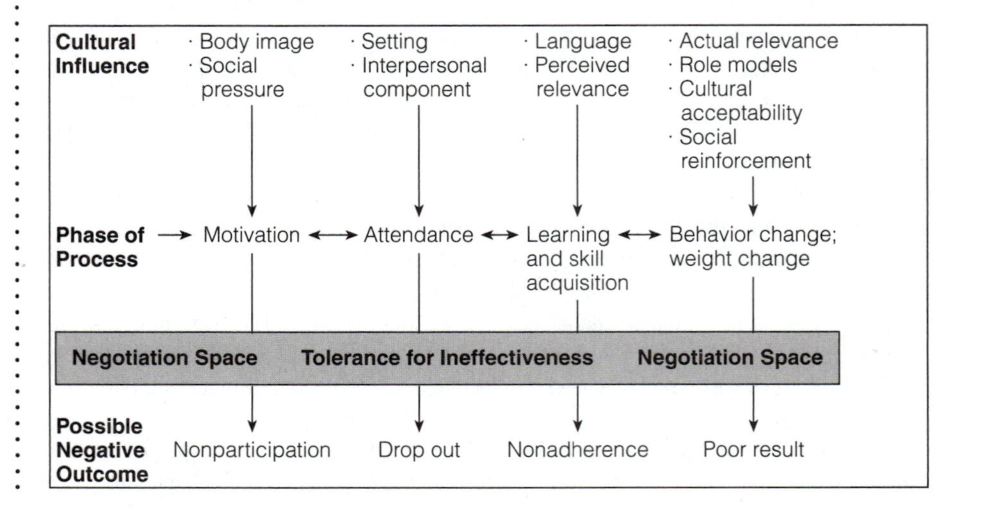

about the target audience is gathered. Ongoing evaluation may suggest better message formats or more suitable influence channels as the effort proceeds. Effective programs are often nonlinear, with each element in planning and implementation connected through feedback and assessment into a continuous improvement loop.

Targeting the Audience

Identification of the target audience in nutrition education efforts is among the most important steps in program planning. Learning about the cultural orientation of the group is imperative; campaigns to change behaviors aimed independently at the individual are usually misdirected (Brown & Einsiedel, 1990). What appear to be significant needs to the health educator may be considered unimportant or too difficult to remedy by members of the target population.

Definitions of health differ widely among cultures (see chapter 2, "Traditional Health Beliefs and Practices"). A common belief is that illness is a matter of heredity, fate, punishment by God, or due to supernatural causes. In a study of health perceptions held by hard-to-reach populations in the United States, it was found that although individuals believed that lifestyle might impact acute infection, there was almost no association between diet or exercise and chronic disease; respondents had limited motivation to improve health behaviors because they felt they had little personal control over their health (White & Maloney, 1990). The role of the individual within the group can also affect responsibility for health maintenance; in some cultures, the extended family is held accountable for the health of each member.

Demographic information about the target audience can guide program development. Primary language should be identified, as well as gender, average age, socioeconomic status, educational attainment, religious affiliation, and other iceberg factors in communication. Assessment of acculturation or bicultural adaptation is equally important. The more culturally homogeneous the target population, the more appropriate are the messages that can be created (Kreps & Kunimoto, 1994; Rabin, 1994). In many cases, the larger, heterogeneous audience may be stratified into smaller, segmented target groups that share similar cultural beliefs and attitudes.

Involving members of the targeted audience in program planning is one of the best ways to determine cultural orientation. Of special note is the role of community leaders or spokespersons in the process. Seeking the respect, trust, and endorsement of influential persons within the target audience for a particular nutrition education program can open intercultural communication channels otherwise limited to the formal interactions reserved for strangers (Kavanagh & Kennedy, 1992; Rabin, 1994; Randall-David, 1989). The educator establishes a relationship with the group through asking for permission to present the health message to its members.

Goals and Objectives

The next step in intercultural program planning is to define goals and objectives that are clear and realistic within the cultural context of the target audience. Even culturally sensitive education messages do not necessarily translate into sustained modification of food habits without follow-up support and overly ambitious expectations are a common reason for failure (Donohew, 1990; Elshaw et al., 1994). Nevertheless, strategies that emphasize continuation of positive cultural dietary patterns or portion control rather than elimination of certain foods are reportedly of interest to African Americans, Asian Americans, Latinos, and Native Americans, as well as whites (Chew, 1983; Jackson & Broussard, 1987; White & Maloney, 1990). Programs that coordinate objectives with cultural beliefs about the role of food in health, such as the balance of hot and cold or yin and yang items in a meal, can also successfully incorporate dietary change into cultural lifestyle. Consulting health care practitioners in the targeted community can provide information on local needs and concerns useful in defining achievable goals and objectives.

Triangulation

An especially useful step in the program design process has been described by Goldberg, Rudd, and Dietz (1999). Called *triangulation*, it is a

■ It has been suggested that in capturing local dietary intake patterns, the services of an ethnographer can help determine appropriate, effective assessment methods (Jerome, 1997).

■ Canadian researchers found that among reasons aboriginal community leaders supported their food intake research was documentation of the culture, assistance in land claims to assess toxicological risks, and self-determination (Wein, 1995).

■ Content of the message can be critical. For example, one study on the effects of public education efforts to reduce bulimic eating behaviors revealed that some women learned about vomiting as a weight control method from the campaign (Swartz, 1987).

■ The four barriers to healthy eating identified by African American women in the triangulation focus groups were taste, cost, time, and lack of information such as recipes, shopping tips, and a chart comparing healthful and unhealthful choices (Goldberg et al., 1999).

■ More than 20 percent of the U.S. population is considered functionally illiterate, reading below an eighth-grade level.

method of confirming congruence between data collected on the target audience and proposed program goals and objectives (Figure 3.3). Triangulation means that information gathered through one source or method is used to confirm and extend information gathered through other sources and methods. The researchers note that in addition to corroboration, triangulation can provide improved understanding of local issues and perspectives. In the pilot, community nutritionists were interviewed to help define the target population of young African American women and for direction in program development. Then target group women participated in focus groups and were asked to discuss benefits and obstacles to healthy eating. A final step surveyed community resources on the availability of quality food products. When data from the three qualitative studies were compared, the researchers found that their target audience was confirmed as appropriate, that the nutritionists had correctly identified the need for culturally relevant skill-building messages, and that there was a barrier to achieve program goals and objectives due to the lack or excessive cost of fresh and frozen foods. They also discovered that nutritionist recommendations regarding an appropriate spokesperson for the campaign were rejected by the targeted audience. The triangulation process provided concrete data on target population needs and credible communication channels, directing program planning towards culturally relevant interventions and resource development.

Developing the Message

It is believed that the more fundamental the health message is in relation to a group's survival, safety, or social needs the more effective it will be interculturally (Kreps & Kunimoto, 1994). The message must satisfy the individual's need to gain knowledge or offer a solution to a perceived problem before it is worth a person's time to process the information (Brown & Einsiedel, 1990). Messages should be as direct and explicit as allowed within cultural norms (Randall-David, 1989). The language relevant to the group should be used in development of the message and translation of existing materials should be avoided to prevent inappropriate phrasing and terminology. Common words used by the target audience are effective, although it is important that they are not used in a way that is insincere or condescending. Written materials should be brief and prepared at the reading level of the target population.

Marketing experts recognize that many cultural groups are high-context communicators and have greater abilities than the white American majority culture to send and receive messages through nonverbal modes (Rabin, 1994). Body language must be culturally congruent with the verbal message for successful

Figure 3.3
Triangulation

From J. Goldberg et al.: JADA. (c) The American Dietetic Association. Reprinted by permission from *Journal of the American Dietetic Association,* Vol. 99:717–722.

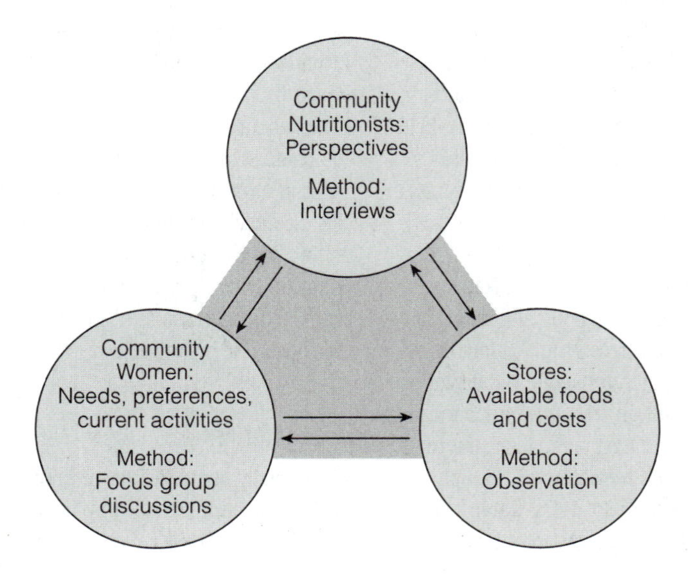

communication to occur. The use of pictures, cartoons, and photographic images can symbolically enhance content meaning of a message in a high-context culture, as well as aid target populations with mixed English language skills or reading abilities.

Educational messages are most effective when they are more personal than objective; the emotional dimension is as important as the content. Many researchers recommend the universally accepted format of storytelling to deliver the message (Esquivel & Keitel, 1990; Rabin, 1994). Actors and other celebrities are especially suited to recounting personal experiences about health issues. Stories can transcend many cultural boundaries; if a message is targeted toward one cultural group yet applicable to many audiences, a spokesperson identified with the intended target group also may have broader appeal when using a narrative approach.

A pilot test of the message with targeted audience members can improve success. Focus groups can be especially useful in assessing cultural appropriateness of education materials and in identifying any resistance triggers that may be inadvertently included (Brown & Einsiedel, 1990; White & Maloney, 1990).

Implementation Strategies

Dissemination of a nutrition education message should include analysis of cultural influence channels and media preferences, development of an effective marketing mix, and evaluation of the program. Whether a person actually hears, sees, and understands a message is dependent on frequency, timing and accessibility. Exciting, informative, culturally appropriate messages fail if they never reach the target audience (Brown & Einsiedel, 1990; Donohew, 1990).

Influence Channels

Influence channels are the ways in which message materials are transmitted to the target audience. They include television, video, computers, radio, magazines, newspapers, newsletters, direct mail, and telephones.

Each cultural group demonstrates distinct media use patterns and is best approached through those influence channels. Oral traditions are strong among some populations, while written messages are favored by others. Most African Americans (97 percent) listen to the radio for more than 30 hours each week; Latinos enjoy television programs that focus on family and relationship issues, watching on average 15 hours of Spanish programming and 10 hours of English programming each week (Mogelonsky, 1998; Rabin, 1994). Asians watch about 6 hours of native-language television weekly. In a study of Native Americans in California, a majority of respondents reported that they would prefer receiving nutrition information through newsletters (69 percent) or videotapes (67 percent); approximately one- quarter indicated they would like to receive a visit from a health professional, and only 6 percent selected a workshop with family and friends (Ikeda et al., 1993). Computer-based, interactive nutrition education programs are an emergent educational tool, particularly suitable for audiences with low literacy or limited English language skills. The Internet is another promising technology, offering 24-hour access to health education materials and easy access to group support through bulletin boards, chat rooms, and individualized therapy through e-mail (Ammerman et al., 1994; Gleason, 1995; Winzelberg et al., 2000).

According to marketing experts, the most effective presentation of a message requires a combination of pictures, sounds, and words in the broadcast and print media (Rabin, 1994). The use of multiple influence channels and frequent repetitions of the message at times when the target audience is listening or watching is also important. Beyond the mass media, health fairs, neighborhood clinics, local markets and grocery stores, traditional healers, churches, schools, food banks or social service centers, carnivals, and sporting events are a few of the locations where culturally relevant nutrition education materials can be successfully distributed on a smaller scale. Low-income, hard-to-reach whites, blacks, and Latinos express interest in nonjudgmental small-group support meetings similar to Alcoholics Anonymous and Tupperware-style home meetings with food samples and cooking demonstrations as settings for nutrition education programs (White & Maloney, 1990).

■ Mass media campaigns are believed to influence a change in health behavior by about 10 percent of the targeted audience, which can be a significant number in a large campaign (Brown & Einsiedel, 1990).

■ One California television station broadcasts programming in Cantonese, Mandarin, Korean, Japanese, Vietnamese, Tagalog, Khmer, Farsi, and Hindi. Cable now offers Telemundo, TV Asia, Filipino Channel, Native American Nations, and World African Network.

■ Television programs, particularly medical shows and soap operas, are a primary source of health care information in the United States (Neuendorf, 1990).

■ Interpersonal follow-up is believed to increase efficacy in communicating a complex health care message (Brown & Einsiedel, 1990).

■ Use of a self-reported shelf inventory was found to be an effective method in evaluating community nutrition intervention programs with Hispanics, blacks, and whites (Beto et al., 1997).

Marketing Mix

The four *P*s of the marketing mix are product, price, placement, and promotion (Kreps & Kunimoto, 1994). They refer to a well-developed message (product) that advances program goals and objectives at minimal economic or psychological cost to target audience members (price) and presents this in a method congruent with target audience media preferences (placement) in such a way that the target audience members are encouraged to become more involved in the program, through phone numbers for further information or through attendance at group meetings (promotion). Attention to all four areas of the marketing mix assures that the health care message is fully accessible to the target audience.

Evaluation

Process evaluation keeps track of progress throughout the program, especially the identification of larger community conditions that may be presenting barriers to dissemination of the message. Summative evaluation is used after completion of the effort to assess program results. Evaluation data is useful in refining intercultural nutrition education strategies both during implementation and in future programs. Publication of culturally sensitive nutrition education program results greatly benefits other health professionals and their clients through shared knowledge about intercultural communication techniques and tools.

4

Food and Religion

The function of religion is to explain the inexplicable, thus providing humans with a sense of comfort in a chaotic world. Food, because it sustains life, is an important part of religious symbols, rites, and customs, those acts of daily life that are intended to bring about an orderly relationship with the spiritual or supernatural realm.

In the Western world, Judaism, Christianity, and Islam are the most prevalent religions, whereas Hinduism and Buddhism are common in the East. The Western religions, originating in the Middle East, are equated with the worship of a single God and the belief that that God is omnipotent and omniscient. It is for God to command and for humankind to obey. This life is a time of testing and a preparation for life everlasting, when humans will be held accountable to God for their actions on Earth.

The Eastern religions Hinduism and Buddhism developed in India. Unlike the Western religions, they do not teach that God is the lord and maker of the universe who demands that humankind be righteous. Rather, the principal goal of the Indian religions is deliverance or liberation of the human soul, which is immortal, from the bondage of the body. Moreover, nearly all Indian religions teach that liberation, given the right disposition and training, can be experienced in the present life.

This chapter discusses the beliefs and food practices of the world's major religions. Other religions of importance to specific cultures are introduced in the following chapters on each ethnic group. As with any description of food habits, it is important to remember that religious dietary practices vary enormously even among members of the same faith. Many religious food practices were codified hundreds or thousands of years ago for a specific locale and, consequently, have been reinterpreted over time and to meet the needs of expanding populations. As a result, most religions have areas of questionable guidelines. For example, fish without scales are banned under kosher food laws. Are sturgeon, which are born with scales but lose them as they mature, considered fit to eat for Jews? The most Orthodox say no, whereas many Conservative and Reform Jews say yes: smoked sturgeon can be found at almost any Jewish deli. Hindus, who avoid fish with "ugly forms," identify those that are undesirable according to local tradition. In addition, religious food practices are often adapted to personal needs. Catholics, encouraged to make a sacrifice during Lent, traditionally gave up meat but today may choose pastries or candy instead. Buddhists may adopt a vegetarian diet only during the period when as an elder they become a monk or nun. Because religious food prescriptions are usually written in some form, it is tempting to see them as being black-and-white. Yet they are among the most variable of culturally based food habits.

- Nearly 148 million Americans claim membership in a religion; of those, the majority belong to a Western faith: Approximately 60 percent are Protestant Christians, 26 percent are Roman Catholics, 4 percent are Jewish, 4 percent are Muslims, and 2 percent are Eastern Orthodox. Estimates often differ in each faith due to variations in how membership is defined.

- In the 1991 Canadian census, nearly half of all people in the nation were identified as Roman Catholics, and another 36 percent were listed as Protestant, many of whom belong to the United Church or the Anglican Church. It is estimated that there are 387,000 followers of the Eastern Orthodox Church (1.5 percent of the population), 318,100 adherents of Judaism, 253,300 Muslims, 163,400 Hindus, 157,000 Buddhists, and 147,000 Sikhs.

Western Religions

Judaism

The Jewish religion, estimated to be 4,000 years old, started when Abraham received God's earliest covenant with the Jews. Judaism was originally a nation as well as a religion. However, after the destruction of its capital, Jerusalem, and its main sanctuary, the Temple of Solomon, in 70 C.E. by the Romans, it had no homeland until the birth of Israel in 1948.

During the Diaspora (the dispersion of Jews outside the homeland of Israel), Jews scattered and settled all over the ancient world. Two sects of Judaism eventually developed: the *Ashkenazi,* who prospered in Germany, northern France, and the eastern European countries; and the *Sephardim,* originally from Spain, who now inhabit most southern European and Middle Eastern countries. *Hasidic* Jews are observant Ashkenazi Jews who believe salvation is to be found in joyous communion with God as well as in the Bible. Hasidic men are evident in larger U.S. cities by their dress, which includes long black coats and black or fur-trimmed hats (worn on Saturdays and holidays only), and by their long beards with side curls.

The cornerstone of the Jewish religion is the Hebrew Bible, particularly the first five books of the Bible, the Pentateuch, also known as the books of Moses, or the *Torah.* It consists of Genesis, Exodus, Leviticus, Numbers, and Deuteronomy. The Torah chronicles the beginnings of Judaism and contains the basic laws that express the will of God to the Jews. The Torah not only sets down the Ten Commandments, but also describes the right way to prepare food, give to charity, and conduct one's life in all ways. The interpretation of the Torah and commentary on it are found in the *Talmud.*

The basic tenet of Judaism is that there is only one God, and His will must be obeyed. Jews do not believe in original sin (that humans are born sinful), but rather that all people can choose to act in a right or wrong way. Sin is attributed to human weakness. Humans can achieve, unaided, their own redemption by asking for God's absolution (if they have sinned against God) or by asking forgiveness of the person they sinned against. The existence of the hereafter is recognized, but the main concern in Judaism is with this life and adherence to the laws of the Torah. Many Jews belong to or attend a *synagogue* (temple), which is led by a *rabbi,* who is a scholar, teacher, and spiritual leader. In the United States, congregations are usually classified as Orthodox, Conservative, or Reform, although American Jews represent a spectrum of beliefs and practices. The main division among the three groups is their position on the Jewish laws. Orthodox Jews believe that all Jewish laws, as the direct commandments of God, must be observed in all details. Reform Jews do not believe that the laws are permanently binding but that the moral law is valid. They believe that the laws are still being interpreted and that some laws may be irrelevant or out of date, and they observe only certain religious practices. Conservative Jews hold the middle ground between Orthodox and Reform beliefs.

Immigration to the United States

In the early 19th century, Jews, primarily from Germany, sought economic opportunities in the New World. By 1860 there were approximately 280,000 Jews living in the United States. Peak Jewish immigration occurred around the turn of the century (1880–1920); vast numbers of Jews moved from eastern Europe because of poverty and *pogroms* (organized massacres practiced by the Russians against the Jews before World War II).

During the Great Depression, Jews continued to immigrate into the United States, primarily to escape from Nazi Germany. Their numbers were few, however, because of restrictions in the immigration quota system. Today Jews continue to come to the United States, especially from Russia. Some come from Israel as well. The Jewish population in the United States is approximately 6 million; more than one-half of these Jews live in the northeastern region of the nation. Large populations are found in New York, Chicago, and Los Angeles. Most Jews in the United States are Ashkenazi: 10 percent identify themselves as Orthodox, 34 percent as Conservative, 29 percent as Reform, and the rest are not affiliated with a specific congregation.

Kashrut, the Jewish Dietary Laws

Some people in the United States believe that Jewish food consists of dill pickles, bagels and lox (smoked salmon), and chicken soup. In actuality, the foods Jews eat reflect the region where their families originated. Because most Jews in the United States are Ashkenazi, their diet includes the foods of Germany and eastern Europe. Sephardic Jews tend to eat foods similar to those of southern Europe and the Middle Eastern countries, while Jews from India prefer curries and other Asian foods.

All Orthodox and some Conservative Jews follow the dietary laws, ***kashrut,*** that were set down in the Torah and explained in the Talmud. ***Kosher*** or ***kasher*** means "fit" and is a popular term for Jewish dietary laws and permitted food items. *Glatt kosher* means that the strictest kosher standards are used in obtaining and preparing the food.

Kashrut is one of the pillars of Jewish religious life and is concerned with the fitness of food. Many health-related explanations have been postulated about the origins of the Jewish dietary restrictions; however, it is spiritual health, not physical health or any other factor, that is the sole reason for their observance. Jews who "keep kosher" are expressing their sense of obligation to God, to their fellow Jews, and to themselves.

The dietary laws governing the use of animal foods can be classified into the following categories:

1. *Which animals are permitted for food and which are not:* Any mammal that has a completely cloven foot and also chews the cud may be eaten, and its milk may be drunk. Examples of permitted, or clean, animals are all cattle, deer, goats, oxen, and sheep. Unclean animals include swine, carnivorous animals, and rabbits. Clean birds must have a crop, gizzard, and an extra talon, as do chickens, ducks, geese, and turkeys. Their eggs are also considered clean. All birds of prey, and their eggs, are unclean and cannot be eaten. Among fish, everything that has fins and scales is permitted; everything else is unclean. Examples of unclean fish are catfish, eels, rays, sharks, and all shellfish. As discussed previously, Orthodox rabbis consider swordfish a prohibited food; Conservative authorities list it as kosher. Caviar, which comes from sturgeon is similarly disputed (Greenberg, 1989). All reptiles, amphibians, and invertebrates are also unclean.
2. *Method of slaughtering animals:* The meat of permitted animals can be eaten only if the life of the animal is taken by a special process known as ***shehitah.*** If an animal dies a natural death or is killed by any other method, it may not be eaten. The *shohet* (person who kills the animal) must be trained and licensed to perform the killing, which is done by slitting the neck with a sharp knife, cutting the jugular vein and trachea at the same time. This method, which is quick and painless, also causes most of the blood to be drained from the carcass.
3. *Examination of the slaughtered animal:* After the animal is slaughtered, it is examined by the shohet for any blemishes in the meat or the organs that would render the animal *trefah,* meaning unfit for consumption. Disease in any part of the animal makes the whole animal unfit to eat.
4. *The parts of a permitted animal that are forbidden:* Two parts of the animal body are prohibited. Blood from any animal is strictly forbidden; even an egg with a small bloodspot in the yolk must be discarded. *Heleb* (fat that is not intermingled with the flesh, forms a separate solid layer that may be encrusted with skin or membrane, and can easily be peeled off) is also proscribed. The prohibition against heleb only applies to four-footed animals.
5. *The preparation of the meat:* For meat to be kosher, the heleb, blood, blood vessels, and sciatic nerve must be removed. Much of this work is now done by the Jewish butcher, although some Jewish homemakers still choose to remove the blood. This is known as ***koshering,*** or ***kashering,*** the meat. It is accomplished by soaking the meat (within 72 hours after slaughter) in water and then covering it with kosher salt. After the blood has been drawn out, the meat is rinsed several times with water. The liver cannot be made kosher in the ordinary way because it contains too much

■ Lender's Bagels, founded in 1927 by Polish immigrant Harry Lender, was the first company to popularize bagels outside New York City, including the creation of green bagels for St. Patrick's Day.

■ Most gelatin is obtained from processed pig tissues. Kosher, gelatin-like products are available.

■ Kosher cheese must be made with rennet obtained from a calf that was killed according to the Jewish laws of slaughtering.

■ The prohibition of the sciatic nerve is based on the biblical story of Jacob's nighttime fight with a mysterious being who touched him on the thigh, causing him to limp. Because the nerve is difficult to remove, the entire hindquarter of the animal is usually avoided.

■ Because it is not known how much salt remains on the meat after rinsing, Orthodox Jews with hypertension are often advised to restrict their meat consumption.

■ *Tevilah* is the ritual purification of metal or glass pots, dishes, and utensils through immersion in the running water of a river or ocean. China and ceramic items are exempt.

■ Two breads, or one bread with a smaller one braided on top, are usually served on the Sabbath, symbolic of the double portion of *manna* (nourishment) provided by God on Fridays to help sustain the Israelites when they wandered in the desert for 40 years after their exodus from Egypt.

blood. Instead, its surface must be cut across or pierced several times, then it must be rinsed in water, and finally it must be broiled or grilled on an open flame until it turns a greyish white color.

6. *The law of meat and milk:* Meat *(fleischig)* and dairy *(milchig)* products may not be eaten together. It is generally accepted that after eating meat a person must wait 6 hours before eating any dairy products. Only 1 hour is necessary if dairy products are consumed first. Many Jews are lactose intolerant and do not drink milk. However, other dairy items such as cheese, sour cream, and yogurt are often included in the diet. Separate sets of dishes, pots, and utensils for preparing and eating meat and dairy products are usually maintained. Often there are separate linens and washing implements. Eggs, fruits, vegetables, and grains are *pareve,* neither meat nor dairy, and can be eaten with both. Olives are considered dairy foods, prohibited with meat, if they are prepared using lactic acid.
7. *Products of forbidden animals:* The only exception to the rule that products of unclean animals are also unclean is honey. Although bees are not fit for consumption, honey is kosher because it is believed that it does not contain any parts from the insect.
8. *Examination for insects and worms:* Because small insects and worms can hide on fruits, vegetables, and grains, these foods must be carefully washed twice and examined before being eaten. Kosher-certified prepackaged produce is available from some suppliers.

A processed food product (including therapeutic dietary formulas) is considered kosher only if a reliable rabbinical authority's name or insignia appears on the package. The most common insignia is a "K," permitted by the U.S. Food and Drug Administration (FDA), indicating rabbinical supervision by the processing company. Other registered symbols include those found in Figure 4.1 (Greenberg, 1989).

Religious Holidays

The Sabbath. The Jewish Sabbath, the day of rest, is observed from shortly before sundown on Friday until after nightfall on Saturday. Traditionally the Sabbath is a day devoted to prayer

Figure 4.1
Examples of kosher food symbols

The Union of Orthodox Jewish Congregations
New York, New York

O.K. (Organized Kashrut) Laboratories
Brooklyn, New York

Kosher Supervision Service
Teaneck, New Jersey

Asian-American Kashrus Services
San Rafael, California

The Heart "K" Kehila Kosher
Los Angeles, California

Chicago Rabbinical Council
Chicago, Illinois

Orthodox Vaad of Philadelphia
Philadelphia, Pennsylvania

Vaad Hakahrus of Dallas, Inc.
Dallas, Texas

Vaad Harabonim of Greater Detroit and Merkaz
Southfield, Michigan

Orthodox Rabbinical Council of S. Florida
(Vaad Harabonim De Darom Florida)
Miami Beach, Florida

Vaad Horabonim of Massachusetts
Boston, Massachusetts

Vaad Hakashrus of the Orthodox Jewish
Council of Baltimore
Baltimore, Maryland

Atlanta Kashruth Commission
Atlanta, Georgia

Vaad Hakashrus of Denver
Denver, Colorado

Vaad Harabonim of Greater Seattle
Seattle, Washington

Kashruth Council Orthodox Division
Toronto Jewish Congress
Willowdale, Ontario, Canada

Montreal Vaad Hair
Montreal, Canada

Vancouver Kashruth
British Columbia, Canada

and rest, and no work is allowed. All cooked meals must be prepared before sundown on Friday, because no fires can be kindled on the Sabbath. *Challah,* a braided bread, is commonly served with the Friday night meal. In most Ashkenazi homes the meal would traditionally contain fish or chicken or *cholent,* a bean and potato dish that can be prepared Friday afternoon and left simmering until the evening meal on Saturday. *Kugel,* a pudding, often made with noodles, is a typical side dish.

Rosh Hashanah. The Jewish religious year begins with the New Year, or Rosh Hashanah, which means "head of the year." Rosh Hashanah is also the beginning of a 10-day period of penitence that ends with the Day of Atonement, *Yom Kippur.* Rosh Hashanah occurs in September or October; as with all Jewish holidays, the actual date varies from year to year because the Jewish calendar is based on lunar months counted according to biblical custom and does not coincide with the secular calendar.

For this holiday the challah is baked in a round shape that symbolizes life without end and a year of uninterrupted health and happiness. Apples are dipped in honey, and a special prayer is said for a sweet and pleasant year. No sour or bitter foods are served on this holiday and special sweets and delicacies, such as honey cakes, are usually prepared.

Yom Kippur, the Day of Atonement. Yom Kippur falls 10 days after Rosh Hashanah and is the holiest day of the year. On this day, every Jew atones for sins committed against God and resolves to improve and once again follow all the Jewish laws.

Yom Kippur is a complete fast day (no food or water; medications are allowed) from sunset to sunset. Everyone fasts, except boys under 13 years old, girls under 12 years old, persons who are very ill, and women in childbirth. The meal before Yom Kippur is usually bland to prevent thirst during the fast. The meal that breaks the fast is typically light, including dairy foods or fish, fruit, and vegetables.

Sukkot, Feast of Tabernacles. Sukkot is a festival of thanksgiving. It occurs in September or October and lasts 1 week. On the last day, *Simchat Torah,* the reading of the Torah (a portion is read every day of the year), is completed for the year and started again. This festival is very joyous, with much singing and dancing. Orthodox families build a *sukkah* (hut) in their yards and hang fruit and flowers from the rafters, which are built far enough apart so that the sky and stars are visible. Meals are eaten in the sukkah during Sukkot.

Hanukkah, the Festival of Lights. Hanukkah is celebrated for 8 days, usually during the month of December, to commemorate the recapture of the Temple in Jerusalem in 169 C.E. Families celebrate Hanukkah by lighting one extra candle on the *menorah* (candelabra) each night, so that on the last night all eight candles are lit. Traditionally potato pancakes, called *latkes,* are eaten during Hanukkah. Other foods cooked in oil, such as doughnuts, are sometimes eaten as well.

Purim. Purim, a joyous celebration that takes place in February or March, commemorates the rescue of the Persian Jews from the villainous Haman by Queen Esther. It is a *mitzvah* (good deed) to eat an abundant meal in honor of the deliverance. The feast should include ample amounts of meat and alcoholic beverages. Customarily, people dress in disguise for the day, to hide from Haman, to add surprise to gift giving, or to "hide" from God to overindulge in anonymity. Foods associated with the holiday include *hamantaschen* (literally "Haman's pockets," but usually interpreted to mean Haman's ears clipped in the humiliation of defeat; triangular-shaped pastries, filled with sweetened poppy seeds or fruit jams made from prunes or apricots) and *kreplach* (triangular- or heart-shaped savory pastries stuffed with seasoned meat or cheese and boiled like ravioli). Purim challah (a sweet bread with raisins) and fish cooked for the holiday in vinegar, raisins, and spices are often served. Seeds, beans, and cereals are offered in remembrance of the restricted diet the pious Queen Esther ate.

Passover. Passover, called *Pesach* in Hebrew, is the 8-day festival of spring and of freedom. It occurs in March or April and celebrates the anniversary of the Jewish exodus from Egypt. The Passover *seder,* a ceremony carried out at

■ In poor Ashkenazi homes, *gefilte* (filled) fish became popular on the Sabbath. Similar to the concept of meatloaf, it is made by extending the fish through pulverizing it with eggs, bread, onion, sugar, salt, and pepper, then stewing the balls or patties with more onions.

■ In Israel it is traditional to send baskets of sweets and gifts for Purim, called *mishlo'ah manot* (baskets for friends) and *mantanat la-evyonim* (gifts for the poor).

■ It is customary to invite strangers to share the Passover meal.

■ Ashkenazi Jews traditionally avoided pepper during Passover because it was sometimes mixed with bread crumbs or flour by spice traders.

■ In some Sephardic homes, matzah is layered with vegetables and cheese or meat for the Passover meal.

Typical seder meal. (© Michael Newman/PhotoEdit.)

home, includes readings from the seder book, the *Haggadah,* recounting the story of the exodus, of the Jews' redemption from slavery, and of the God-given right of all humankind to life and liberty. A festive meal is a part of the seder; the menu usually includes chicken soup, matzo balls, and meat or chicken.

When Moses led the Jews out of Egypt, they left in such haste that there was no time for their bread to rise. Today *matzah,* a white flour cracker, is the descendant of the unleavened bread or "bread of affliction." During the 8 days of Passover, no food that is subject to a leavening process or that has come in contact with leavened foods can be eaten. The foods that are forbidden are wheat, barley, rye, and oats. Wheat flour can be eaten only in the form of matzah or matzah meal, which is used to make *matzo* balls. In addition, beans, peas, lentils, maize, millet, and mustard are also avoided. No leavening agents, malt liquors, or beers can be used.

Observant Jewish families have two sets (as milk and meat cannot be mixed at any time) of special dishes, utensils, and pots that are used only for Passover. The entire house, especially the kitchen, must be cleaned and any foods subject to leavening removed before Passover. It is customary for Orthodox Jews to sell their leavened products and flours to a non-Jew before Passover. It is very important that all processed foods, including wine, be prepared for Passover use and be marked "Kosher for Passover."

■ The Torah prohibits the drinking of wine made by non-Jews because it might have been produced for the worship of idols. Some Orthodox Jews extend the prohibition to any grape product, such as grape juice or grape jelly.

■ Cottage cheese is associated with Shavout because the Israelites were late in returning home after receiving the Ten Commandments and the milk had curdled. Many dishes served during the holiday contain cheese fillings.

The seder table is set with the best silverware and china and must include candles, kosher wine, the Haggadah, three pieces of *matzot* (the plural of *matzah*) covered separately in the folds of a napkin or special Passover cover, and a seder plate. The following items go on the seder plate:

1. *Z'roah.* Z'roah is a roasted shank bone, symbolic of the ancient paschal lamb in Egypt, which was eaten roasted.
2. *Beitzah.* Beitzah is a roasted egg, representing the required offering brought to the Temple at festivals. Although the egg itself was not sacrificed, it is used in the seder as a symbol of mourning. In this case, it is for the loss of the Temple in Jerusalem.
3. *Marror.* Marror are bitter herbs, usually horseradish (although not an herb), symbolic of the Jews' bitter suffering under slavery. The marror is usually eaten between two small pieces of matzot.
4. *Haroset.* Haroset is a mixture of chopped apple, nuts, cinnamon, and wine. Its appearance is a reminder of the mortar used by the Jews to build the palaces and pyramids of Egypt during centuries of slavery. The haroset is also eaten on a small piece of matzo.
5. *Karpas.* A green vegetable, such as lettuce or parsley, is placed to the left of the haroset, symbolic of the meager diet of the Jews in

bondage. It is dipped into salt water in remembrance of the tears shed during this time. It also symbolizes springtime, the season of Passover.

6. A special cup, usually beautifully decorated, is set on the seder table for Elijah, the prophet who strove to restore purity of divine worship and labored for social justice. (Elijah is also believed to be a messenger of God, whose task it will be to announce the coming of the Messiah and the consequent peace and divine kingdom of righteousness on Earth.)

Shavout, Season of the Giving of the Torah. The 2-day festival of Shavout occurs 7 weeks after the second day of Passover and commemorates the revelation of the Torah to Moses on Mt. Sinai. Traditional Ashkenazi foods associated with the holiday include *blintzes* (extremely thin pancakes rolled with a meat or cheese filling, then topped with sour cream), *kreplach*, and *knishes* (dough filled with a potato, meat, cheese, or fruit mixture, then baked).

Fast Days

There are several Jewish fast days in addition to Yom Kippur (see Table 4.1). On Yom Kippur and on *Tisha b'Av,* the fast lasts from sunset to sunset and no food or water can be consumed. All other fast days are observed from sunrise to sunset. Most Jews usually fast on Yom Kippur, but other fast days are observed only by Orthodox Jews. All fasts can be broken if it is dangerous to a person's health; those who are pregnant or nursing are exempt from fasting.

Additional information about Jewish dietary laws and customs associated with Jewish holidays can usually be obtained from the rabbi at a local synagogue. The Union of Orthodox Jewish Congregations of America also publishes a directory of kosher products.

Christianity

Throughout the world, more people follow Christianity than any other single religion. The three dominant Christian branches are Roman Catholicism, Eastern Orthodox Christianity, and Protestantism. Christianity is founded on recorded events surrounding the life of Jesus, believed to be the Son of God and the Messiah, that are chronicled in the New Testament of the Bible. The central convictions of the Christian faith are found in the Apostles' and the Nicene Creed. The creed explains that people are saved through God's grace, the life and death of Jesus, and his resurrection as Christ.

For most Christians the sacraments mark the key stages of worship and sustain the individual worshiper. A sacrament is an outward act derived from something Jesus did or said, through which an individual receives God's grace. The sacraments observed and the way they are observed vary among Christian groups. The seven sacraments of Roman Catholicism,

■ The commemoration of the Last Supper is called *Corpus Christi,* when Jesus instructed his disciples that bread was his body and wine his blood. In Spain and many Latin American countries, Corpus Christi is celebrated by parading the bread (called the *Host*) through streets covered with flowers.

Table 4.1
Jewish Fast Days

Fast Day	Time of Year	Purpose
Tzom Gedaliah	Day after Rosh Hashanah	In memory of Gedaliah, who ruled after the first Temple was destroyed
Yom Kippur	10 days after Rosh Hashanah	Day of Atonement
Tenth of Tevet	December	Commemorate an assortment of national calamities listed in the Talmud
Seventeenth of Tamuz	July	Commemorate an assortment of national calamities listed in the Talmud
Ta'anit Ester	Eve of Purim	In grateful memory of Queen Esther, who fasted when seeking divine guidance
Ta'anit Bechorim	Eve of Passover	Gratitude to God for having spared the first-born of Israel. Usually only the first-born son fasts.
Tisha b'Av	August	Commemorates the destruction of the Temple in Jerusalem

for example, are baptism (entering Christ's church), confirmation (the soul receiving the Holy Ghost), communion (partaking of the sacred presence by sharing bread and wine), marriage, unction (sick and dying are reassured of salvation), reconciliation (penance and confession), and ordination of the clergy.

Roman Catholicism

The largest number of persons adhering to one Christian faith in the United States are Roman Catholics (approximately 57 million people in 1990). The head of the worldwide church is the pope, considered infallible when defining faith and morals. The seven sacraments are conferred on the faithful.

Although some Roman Catholics immigrated to the United States during the colonial period, substantial numbers came from Germany, Poland, Italy, and Ireland in the 1800s and from Mexico and the Caribbean in the 20th century. There are small groups of French Catholics in New England (primarily in Maine) and in Louisiana. In addition, most Filipinos and some Vietnamese people living in the United States are also Catholics.

Feast Days. Most Americans are familiar with Christmas (the birth of Christ) and Easter (the resurrection of Christ after the crucifixion). Other feast days celebrated in the United States are New Year's Day, the Annunciation (March 25th), Palm Sunday (the Sunday before Easter), the Ascension (40 days after Easter), Pentecost Sunday (50 days after Easter), the Assumption (August 15th), All Saint's Day (November 1st), and the Immaculate Conception (December 8th).

Holiday fare depends on the family's country of origin. For example, the French traditionally serve *bûche de Noël* (a rich cake in the shape of a Yule log) on Christmas for dessert, while the Italians may serve *panettone,* a fruited sweet bread (see individual chapters on each ethnic group for specific foods associated with holidays).

■ St. Agnes Day, celebrated on January 21 in Great Britain, honors the patron saint of girls. On the eve of the day, British girls traditionally baked a cake and walked upstairs backward saying a prayer to Agnes and making a wish for the man of their dreams.

■ St. Valentine's Day traditions may date back to *Lupercalia,* a Roman festival held in mid-February at which a young man would draw the name of a young woman out of a box to be his sweetheart for a day.

■ Lent is the 40 days before Easter; the word originally meant "spring." The last day before Lent is a traditional festival of exuberant feasting and drinking in many regions where Lenten fasting is observed. In France and in Louisiana, it is known as *Mardi Gras;* in Britain, *Shrove Tuesday;* in Germany, *Fastnacht;* and *Carnival* throughout the Caribbean and in Brazil.

Italian American Catholics often serve *panettone,* a sweet bread with dried fruits, on feast days, especially Christmas. (Courtesy of Grossich and Bond, Inc.)

Fast Days. Fasting permits only one full meal per day at midday. It does not prohibit the taking of some food in the morning or evening; however, local custom as to the quantity and quality of this supplementary nourishment varies. Abstinence forbids the use of meat, but not of eggs, dairy products, or condiments made of animal fat and is practiced on certain days and in conjunction with fasting. Only Catholics older than the age of 14 and younger than the age of 60 are required to observe the dietary laws.

The fast days in the United States are all the days of Lent, the Fridays of Advent, and the Ember Days (the days which begin each season), but only the most devout fast and abstain on all of these dates. More common is fasting and abstaining only on Ash Wednesday and Good Friday. Before 1966, when the U.S. Catholic Conference abolished most dietary restrictions, abstinence from meat was observed on every Friday that did not fall on a feast day. Abstinence is now encouraged on the Fridays of Lent in remembrance of Christ's sacrificial death.

Some older Catholics and those from other nations may observe the pre-1966 dietary laws. In addition, Catholics are required to avoid all food and liquids, except water, for one hour before receiving communion.

Eastern Orthodox Christianity

The Eastern Orthodox Church is as old as the Roman Catholic branch of Christianity, although not as prevalent in the United States. In the year 300 C.E., there were two centers of Christianity, one in Rome and the other in Constantinople (now Istanbul, Turkey). Differences arose over theological interpretations of the Bible and the governing of the church, and in 1054 the fellowship between the Latin and Byzantine churches was finally broken. Some of the differences between the two churches concerned the interpretation of the Trinity (the Father, the Son, and the Holy Ghost), the use of unleavened bread for the communion, the celibacy of the clergy, and the position of the pope. In the Eastern Orthodox Church, leavened bread is used for communion, the clergy are allowed to marry before entering the priesthood, and the authority of the pope is not recognized.

The Orthodox Church consists of 14 self-governing churches, 5 of which—Constantinople, Alexandria (the Egyptian Coptic Church), Antioch, Jerusalem, and Cyprus—date back to the time of the Byzantine Empire. Six other churches represent the nations where the majority of people are Orthodox (Russia, Rumania, Serbia, Bulgaria, Greece, and the former Soviet state of Georgia). Three other churches exist independently in countries where only a minority profess the religion. The beliefs of the churches are similar; only the language of the service differs.

The first Eastern Orthodox church in America was started by Russians on the West Coast in the late 1700s. It is estimated that more than 3 million persons in the United States are members of the Eastern Orthodox religion, with the largest following (2 million) being Greek. Most Eastern Orthodox churches in the United States recognize the patriarch of Constantinople as their spiritual leader.

Feast Days. All the feast days are listed in Table 4.2. Easter is the most important holiday in the Eastern Orthodox religion and is celebrated on the first Sunday after the full moon after March 21, but not before the Jewish Passover. Lent is preceded by a pre-Lenten period lasting 10 weeks before Easter or 3 weeks before Lent. On the third Sunday before Lent (Meat Fare Sunday), all the meat in the house is eaten. On the Sunday before Lent (Cheese Fare Sunday), all the cheese, eggs, and butter in the house are eaten. On the next day, Clean Monday, the Lenten fast begins. Fish is allowed on Palm Sunday and on the Annunciation Day of the Virgin Mary. The Lenten fast is traditionally broken after the midnight services on Easter Sunday. Easter eggs in the Eastern Orthodox religion range from the highly ornate (eastern Europe and Russia) to the solid red ones used by the Greeks.

Fast Days. In the Eastern Orthodox religion there are numerous fast days (see Table 4.3). Further, those receiving Holy Communion on Sunday abstain from food and drink before the service. Fasting is considered an opportunity to prove that the soul can rule the body. On fast days no meat or animal products (milk, eggs, butter, and cheese) are consumed. Fish is also

■ In the Eastern Orthodox religion the leavened communion bread called *Phosphoron* is prepared by the women of the church.

■ The Ethiopian church is an Orthodox denomination that is similar to the Egyptian Coptic Church. *Timkat* (Feast of the Epiphany) is the most significant Christian holiday of the year, celebrating the baptism of Jesus. Beer brewing, bread baking, and eating roast lamb are traditional.

■ *Koljivo,* boiled whole wheat kernels mixed with nuts, dried fruit, and sugar, must be offered before the church altar 3, 9, and 40 days after the death of a family member. After the koljivo is blessed by the priest, it is distributed to the friends of the deceased. The boiled wheat represents everlasting life and the fruit represents sweetness and plenty.

■ The red Easter egg symbolizes the tomb of Christ (the egg) and is a sign of mourning (the red color). The breaking of the eggs on Easter represents the opening of the tomb and belief in the resurrection.

Table 4.2
Eastern Orthodox Feast Days

Feast Day	Date*
Christmas	Dec. 25, Jan. 7
Theophany	Jan. 6, Jan. 19
Presentation of Our Lord into the Temple	Feb. 2, Feb. 15
Annunciation	Mar. 25, Apr. 7
Easter	See text for date
Ascension	40 days after Easter
Pentecost (Trinity) Sunday	50 days after Easter
Transfiguration	Aug. 6, Aug. 19
Dormition of the Holy Theotokos	Aug. 15, Aug. 28
Nativity of the Holy Theotokos	Sept. 8, Sept. 21
Presentation of the Holy Theotokos	Nov. 21, Dec. 4

*Date depends on whether the Julian or Gregorian calendar is followed.

■ In Greece the 3 great saints of the Orthodox Church, St. Basil, St. Gregory, and St. John Chrysotom, are honored on January 30, the Holiday of the Three Hierarchs.

■ Welch's Grape Juice was developed by Dr. Thomas Welch, dentist, Methodist, and ardent prohibitionist, in 1869 as an alternative to communion wine.

■ When 7 millers got together in 1891 to market oatmeal, they chose the name Quaker Oats even though none were members of the Friends, because that religious group was generally regarded as pure and wholesome.

avoided, but shellfish is allowed. Older or more devout Greek Orthodox followers do not use olive oil on fast days, but will eat olives.

Protestantism

The 16th century religious movement known as the Reformation established the Protestant churches by questioning the practices of the Roman Catholic Church and eventually breaking away from its teachings. The man primarily responsible for the Reformation was Martin Luther, a German Augustinian monk who taught theology. He started the movement when, in 1517, he nailed a document containing 95 protests against certain Catholic practices on the door of the castle church in Wittenberg. He later broadened his position. A decade later, several countries and German principalities organized the Protestant Lutheran Church based on Martin Luther's teachings.

Luther placed great emphasis on the individual's direct responsibility to God. He believed that every person can reach God through direct prayer without the intercession of a priest or saint; thus, every believer is, in effect, a minister. Although everyone is prone to sin and inherently wicked, a person can be saved by faith in Christ, who by his death on the Cross atoned for the sins of all people. Consequently, to Luther, faith was all-important and good works alone could not negate evil deeds. Luther's theology removed the priest's mystical function, encouraging everyone to read the Bible and interpret the Scriptures.

The beliefs taught by Martin Luther established the foundation of most Protestant faiths. Other reformers who followed Luther are associated with specific denominations. In the mid-16th century, John Calvin developed the ideas that led to the formation of the Presbyterian, Congregationalist, and Baptist churches; John Wesley founded the Methodist movement in the 18th century. Other denominations in the United States include Episcopalians (related to the English Anglican Church started under King Henry VIII); Seventh-Day Adventists; Jehovah's Witnesses; Disciples of Christ; Church of Jesus Christ of Latter Day Saints (Mormons); Church of Christ, Scientist (Christian Scientists); and Friends (Quakers).

Christmas and Easter are the primary holidays celebrated in the Protestant tradition. Although the role of food is important in these holidays, the choice of items served is determined by family ethnicity and preference rather than religious practice. Fasting is also uncommon in most Protestant denomina-

Table 4.3
Eastern Orthodox Fast Days and Periods

Fast Days
Every Wednesday and Friday except during fast-free weeks:
Week following Christmas till Eve of Theophany (12 days after Christmas)
Bright Week, week following Easter
Trinity Week, week following Trinity Sunday
Eve of Theophany (Jan. 6 or 18)
Beheading of John the Baptist (Aug. 29 or Sept. 27)
The Elevation of the Holy Cross (Sept. 14 or 27)
Fast Periods
Nativity Fast (Advent): Nov. 15 or 28 to Dec. 24 or Jan. 6
Great Lent and Holy Week: 7 weeks before Easter
Fast of the Apostles: May 23 or June 5 to June 16 or 29
Fast of the Dormition of the Holy Theotokos (Aug. 1 or 14 to Aug. 15 or 28)

Dates depend on whether the Julian or Gregorian calendar is followed.

tions. Some churches or individuals may use occasional fasting, however, to facilitate prayer and worship. Only a few of the Protestant denominations, such as the Mormons and the Seventh-Day Adventists, have dietary practices integral to their faith.

Mormons. The Church of Jesus Christ of Latter Day Saints is a purely American institution that emerged in the early 1800s. Its founder, Joseph Smith, Jr., had a vision of the Angel Moroni, who told him of golden plates hidden in a hill and the means by which to decipher them. The resulting *Book of Mormon* was published in 1829, and in 1830 a new religious faith was born.

The Book of Mormon details the story of two bands of Israelites who settled in America and from whom certain Native Americans are descended. Christ visited them after his resurrection, and they thus preserved Christianity in its pure form. The tribes did not survive, but the last member, Moroni, hid the nation's sacred writings, compiled by his father, Mormon.

The Mormons believe that God reveals himself and his will through his apostles and prophets. The Mormon Church is organized along biblical lines. Members of the priesthood are graded upward in six degrees (deacons, teachers, priests, elders, seventies, and high priests). From the priesthood are chosen, by the church at large, a council of twelve apostles, which constitutes a group of ruling elders; from these, by seniority, a church president rules with life tenure. There is no paid clergy. Sunday services are held by groups of Mormons, and selected church members give the sermon.

To escape local persecution, Brigham Young led the people of the Mormon Church to Utah in 1847. Today Utah is more than 80 percent Mormon, and many western states have significant numbers of church members. The main branch of the church is headquartered in Salt Lake City, but a smaller branch, the Reorganized Church of Jesus Christ of Latter Day Saints, is centered in Independence, Missouri. All Mormons believe that Independence will be the capital of the world when Christ returns.

Joseph Smith, through a revelation, prescribed the Mormon laws of health, dealing particularly with dietary matters. They prohibit the use of tobacco, strong drink, and hot drinks. Strong drink is defined as alcoholic beverages; hot drinks means tea and coffee. Many Mormons do not use any product that

■ One study found that Protestants (especially Baptists) were most likely to be overweight; Jews, Moslems, and Buddhists least likely. However, differences in affiliation were inconsequential when controlled for socioeconomic status and ethnicity. Of significance, however, was degree of devotion and risk of obesity (Ferraro, 1998).

■ Loma Linda Foods began as a bakery in 1906 providing whole-wheat bread and cookies to the patients and staff of Loma Linda University Medical Center in southern California.

■ The American breakfast cereal industry is the result of the dietary and health practices of the Seventh-Day Adventists. In 1886, Dr. John Kellogg became director of the Adventists' sanitarium in Battle Creek, Michigan, and in his efforts to find a tasty substitute for meat, he invented corn flakes.

contains caffeine. Followers are advised to eat meat sparingly, and to base their diets on grains, especially wheat. In addition, all Mormons are required to store a year's supply of food and clothing for each person in the family. Many also fast 1 day per month (donating the money that would have been spent on food to the poor).

Seventh-Day Adventists. In the early 1800s, many people believed that the Second Coming of Christ was imminent. In the United States, William Miller predicted that Christ would return in 1843 or 1844. When both years passed and the prediction did not materialize, many of his followers became disillusioned. However, one group continued to believe that the prediction was not wrong but that the date was actually the beginning of the world's end preceding the coming of Christ. They became known as the Seventh-Day Adventists and were officially organized in 1863.

The spiritual guide for the new church was Ellen G. Harmon, who later became Mrs. James White. Her inspirations were the result of more than 2,000 prophetic visions and dreams she reportedly had during her life. Mrs. White claimed to be not a prophet but a conduit that relayed God's desires and admonitions to humankind.

Today, there are more than half a million Seventh-Day Adventists in the United States and more than 1 million worldwide. Besides the main belief in Christ's advent, or second coming, the Seventh-Day Adventists practice the principles of Protestantism. They believe that the advent will be preceded by a monstrous war, pestilence, and plague, resulting in the destruction of Satan and all wicked people; the Earth will be purified by holocaust. Although the hour of Christ's return is not known, they believe that dedication to his work will hasten it.

The church adheres strictly to the teachings of the Bible. The Sabbath is observed from sundown on Friday to sundown on Saturday and is wholly dedicated to the Lord. Food must be prepared on Friday and dishes washed on Sunday. Church members dress simply, avoid ostentation, and wear only functional jewelry. The church's headquarters are in Tacoma Park, Maryland, near Washington, D.C., where they were moved after a series of fires ravaged the previous center in Battle Creek, Michigan. Each congregation is led by a pastor (a teacher more than a minister), and all the churches are under the leadership of the president of the general conference of Seventh-Day Adventists.

Adventists follow Apostle Paul's teaching that the human body is the temple of the Holy Spirit. Many of Mrs. White's writings concern health and diet and have been compiled into such books as *The Ministry of Healing, Counsels on Diet and Foods, and Counsels on Health.*

Adventists believe that sickness is a result of the violation of the laws of health. One can preserve health by eating the right kinds of foods in moderation and by getting enough rest and exercise. Overeating is discouraged. Vegetarianism is widely practiced because the Bible states that the diet in Eden did not include flesh foods. Most Adventists are lacto-ovo vegetarians (eating milk products and eggs, but not meat). Some do consume meat, although they avoid pork and shellfish. Mrs. White advocated the use of nuts and beans instead of meat, substituting vegetable oil for animal fat, and using whole grains in breads. Like the Mormons, the Adventists do not consume tea, coffee, or alcohol and do not use tobacco products. Water is considered the best liquid and should be consumed only before and after the meal, not during the meal. Meals are not highly seasoned, and condiments such as mustard and pepper are avoided. Eating between meals is discouraged so that food can be properly digested. Mrs. White recommended that 5 or 6 hours elapse between meals.

Islam

Islam is the second largest religious group in the world. Although not widely practiced in the United States, Islam is the dominant religion in the Middle East, northern Africa, Pakistan, Indonesia, and Malaysia. Large numbers of people also follow the religion in parts of sub-Saharan Africa, India, Russia, and Southeast Asia.

Islam, which means "submission" (to the will of God), is not only a religion but also a way of life. One who adheres to Islam is called a Muslim, "he who submits." Islam's founder, Mohammed, was neither a savior nor a messiah but rather a prophet through whom God delivered his messages. He was born in 570 C.E. in Mecca, Saudi Arabia, a city located along the spice trade route. Early in Mohammed's life he acquired a respect for Jewish and Christian monotheism. Later the archangel Gabriel appeared to him in many visions. These revelations continued for a decade or more, and the archangel told Mohammed that he was a prophet of Allah, the one true God.

Mohammed's teachings met with hostility in Mecca, and in 622 he fled to Yathrib. The year of the flight *(hegira)* is the first year in the Muslim calendar. At Yathrib, later named Medina, Mohammed became a religious and political leader. 8 years after fleeing Mecca, he returned triumphant and declared Mecca a holy place to Allah.

The most sacred writings of Islam are found in the *Quran* (sometimes spelled *Koran* or *Qur'an*), believed to contain the words spoken by Allah through Mohammed. It includes many legends and traditions that parallel those of the Old and New Testaments, as well as Arabian folk tales. The Quran also contains the basic laws of Islam, and its analysis and interpretation by religious scholars have provided the guidelines by which Muslims lead their daily lives.

The Muslims believe that the one true God, Allah, is basically the God of Judaism and Christianity but that his word was incompletely expressed in the Old and New Testaments and was only fulfilled in the Quran. Similarly, they believe that Mohammed was the last prophet, superseding Christ, who is considered a prophet and not the Son of God. The primary doctrines of Islam are monotheism and the concept of the last judgment, the day of final resurrection when all will be deemed worthy of either the delights of heaven or the terrors of hell.

Mohammed did not institute an organized priesthood or sacraments but instead advocated the following ritualistic observances, known as the Five Pillars of Islam:

1. *Faith,* shown by the proclamation of the unity of God, and belief in that unity, as expressed in the creed, "There is no God but Allah; Mohammed is the Messenger of Allah."
2. *Prayer,* performed 5 times daily (at dawn, noon, midafternoon, sunset, and nightfall), facing Mecca, wherever one may be; and on Fridays, the day of public prayer, in the *mosque* (a building used for public worship). On Fridays, sermons are delivered in the mosque after the noon prayer.
3. *Almsgiving, zakat,* as an offering to the poor and an act of piety. In some Islamic countries, Muslims are expected to give 2.5 percent of their net savings or assets in money or goods. The money is used to help the poor or to support the religious organization in countries where Islam is not the dominant religion. In addition, zakat is given to the needy on certain feast and fast days (see the next section on dietary practices for more details).
4. *Fasting,* to fulfill a religious obligation, to earn the pleasure of Allah, to wipe out previous sins, and to appreciate the hunger of the poor and the needy
5. *Pilgrimage to Mecca, hadj,* once in a lifetime if means are available. No non-Muslim can enter Mecca. Pilgrims must wear seamless white garments; go without head covering or shoes; practice sexual continence; abstain from shaving or having their hair cut; and avoid harming any living thing, animal or vegetable.

There are no priests in Islam; every Muslim can communicate directly with God, so a mediator is not needed. The successors of the prophet Mohammed and the leaders of the Islamic community were the *caliphes* (or *kalifah*). No caliphes exist today. A *mufti,* like a lawyer, gives legal advice based on the sacred laws of the Koran. An *imam* is the person appointed to lead prayer in the mosque and deliver the Friday sermon.

The following prominent sects in Islam have their origin in conflicting theories on the office of caliph *(caliphate):* (1) The *Sunni,* who form the largest number of Muslims and hold that the caliphate is an elected office that must be occupied by a member of the tribe of Koreish, the tribe of Mohammed. (2) The *Shi'ites,* the second largest group, who believe that the caliphate was a God-given office held rightfully by Ali, Mohammed's son-in-law, and his descendants. The Shi'ites are found primarily in Iran, Iraq, Yemen, and India. (3) The *Khawarij,* who believe that the office of caliph

■ If one is unable to attend a mosque, the prayers are said on a prayer rug facing Mecca.

■ The *Kaaba,* in Mecca, is the holiest shrine of Islam and contains the Black Stone given to Abraham and Ishmael by the Archangel Gabriel. During the hadj, each pilgrim touches the stone and circles the shrine.

■ No one claiming title to the office of caliphate has been recognized by all Muslim sects since its abolition by the Turkish government in 1924 following the fall of the Ottoman Empire. The role of the caliphate in modern Islam is uncertain.

is open to any believer whom the faithful consider fit for it. Followers of this sect are found primarily in eastern Arabia and North Africa. (4) The *Sufis,* ascetic mystics who seek a close union with God now, rather than in the hereafter. Only 3 percent of present-day Muslims are Sufis.

It is estimated that nearly 6 million Muslims live in the United States; many came from the Middle East. In addition, some African Americans believe Allah is the one true God; they follow the Quran and traditional Muslim rituals in their temple services. The movement was originally known as the Nation of Islam and its adherents identified as Black Muslims. A split in the Nation of Islam resulted in one faction of Black Muslims becoming an orthodox Islamic religion called the World Community of Al-Islam in the West. It is accepted as a branch of Islam. The other Black Muslim faction has continued as the Nation of Islam under the leadership of Louis Farrakhan.

■ Some devout Muslims also avoid land animals without external ears, such as snakes and lizards.

Halal, Islamic Dietary Laws

In Islam, eating is considered to be a matter of worship. Muslims are expected to eat for survival and good health; self-indulgence is not permitted. Muslims are advised not to eat more than two-thirds of their capacity, and sharing food is recommended. Food is never to be thrown away, wasted, or treated with contempt. The hands and mouth are washed before and after meals. If eating utensils are not used, only the right hand is used for eating, as the left hand is considered unclean.

Permitted or lawful foods are called *halal.* Unless specifically prohibited, all food is edible. Unlawful or prohibited *(haram)* foods prescribed in the Quran include:

1. All swine, four-footed animals that catch their prey with their mouths, and birds of prey that seize their prey with their talons.
2. Improperly slaughtered animals. An animal must be killed in a manner similar to that described in the Jewish laws, by slitting the front of the throat; cutting the jugular vein, carotid artery, and windpipe; and allowing the blood to drain completely. In addition, the person who kills the animal must repeat at the instant of slaughter, "In the name of God, God is great." Fish and seafood are exempt from this requirement.

Some Muslims believe that a Jew or a Christian can slaughter an animal to be consumed by Muslims as long as it is done properly. Others will eat only kosher meat,

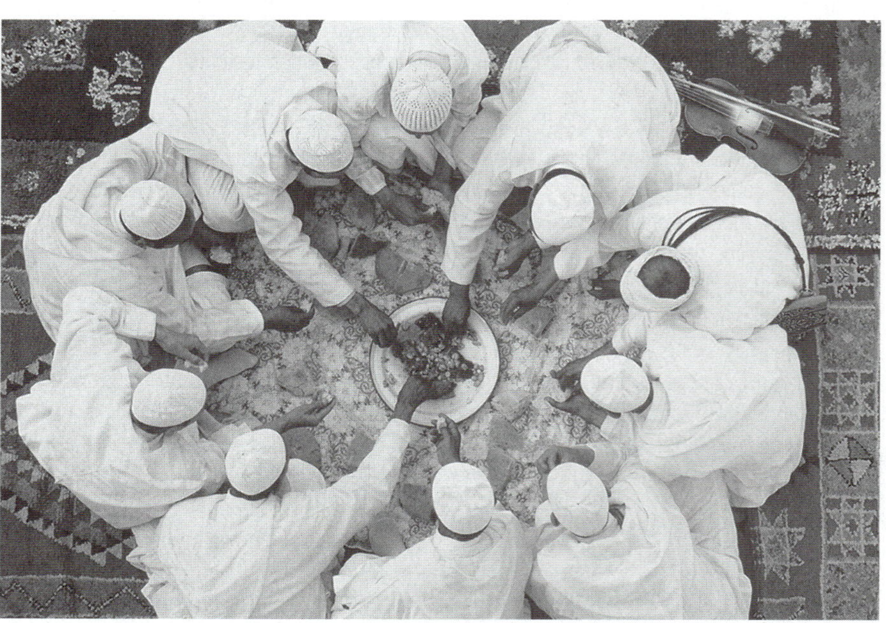

Islamic laws consider eating to be a matter of worship, and Muslims are encouraged to share meals. (J. Boisberranger/Viesti Associates.)

while some abstain from meat unless they know it is properly slaughtered by a Muslim or can arrange to kill the animal themselves. Meat of animals slaughtered by people other than Muslims, Jews, or Christians is prohibited. Meat is also prohibited if it came from an animal that was slaughtered when any name besides God's was mentioned.

3. Alcoholic beverages and intoxicating drugs, unless medically necessary. The drinking of stimulants, such as coffee and tea, is discouraged, as is smoking; however, these prohibitions are practiced only by the most devout Muslims.

A Muslim can eat or drink prohibited food under certain conditions, such as when the food is taken by mistake, is forced by others, or there is fear of dying by hunger or disease. When in doubt regarding whether a food is halal or haram, a Muslim is encouraged to avoid the item. Foods that are in compliance with Islamic dietary laws are sometimes marked with symbols that are registered with the Islamic Food and Nutrition Council of America (IFNCA) (Figure 4.2), signifying the food is fit for consumption by Muslims anywhere in the world (Hussaini, 1993).

Feast Days

The following are the feast days in the Islamic religion:

1. *Eid al-Fitr,* the Feast of Fast Breaking—the end of Ramadan is celebrated by a feast and the giving of alms.
2. *Eid al-Azha,* the Festival of Sacrifice—the commemoration of Abraham's willingness to sacrifice his son, Ishmael, for God. It is customary to sacrifice a sheep and distribute its meat to friends, relatives, and the needy.
3. *Shab-i-Barat,* the night in the middle of *Shaban*—originally a fast day, this is now a feast day, often marked with fireworks. It is believed that God determines the actions of every person for the next year on this night.
4. *Nau-Roz,* New Year's Day—primarily celebrated by the Iranians, it is the first day after the sun crosses the vernal equinox.
5. *Maulud n'Nabi*—the birthday of Mohammed

Feasting also occurs at birth, after the consummation of marriage, at *Bismillah* (when a child first starts reading the Koranic alphabet), after circumcision of boys, at the harvest, and at death.

■ The term for a food that is questionably halal or haram is *mashbooh.*

■ Iranians celebrate the new year with a display, called *haft-sinn,* 7 items beginning with the letter *s*, denoting life, rebirth, happiness, prosperity, joy, beauty, and health. These are vinegar *(serke),* apple *(seeb),* dried fruit or olives *(sanjed),* sumac seasoning *(sumagh),* sweet wheat pudding *(samanu),* garlic *(seer),* or hyacinth *(sonbul),* and sprouted seeds *(sabzi),* respectively.

■ The month of Ramadan can fall during any part of the year. The Muslim calendar is lunar but does not have a leap month; thus, the months occur at different times each year.

Fast Days

On fast days, Muslims abstain from food, drink, smoking, and coitus from dawn to sunset. Food can be eaten before the sun comes up and again after it sets. Muslims are required to fast during *Ramadan,* the 9th month of the Islamic calendar. It is believed that during Ramadan, "the gates of Heaven are open, the gates of Hell closed, and the devil put in chains." At sunset the fast is usually broken by taking a liquid, typically water, along with an odd number of dates.

All Muslims past the age of puberty (15 years old) are required to fast during Ramadan. A number of groups are exempt from fasting, but most must make up the days before the next Ramadan. They include sick individuals with a recoverable illness; people who are traveling; women during pregnancy, lactation, or menstruation; elders who are physically unable to fast; insane people; and those engaged in hard labor. During Ramadan, it is customary to invite guests to break the fast and dine in the evening; special foods are eaten, especially sweets. Food is often given to neighbors, relatives, and needy individuals or families.

Muslims are also encouraged to fast 6 days during *Shawwal,* the month following Ramadan; the 10th day of the month of *Muharram;* and the 9th day of *Zul Hijjah,* but not during the pilgrimage to Mecca. A Muslim

Figure 4.2
Examples of halal food symbols

■ Women who wish to undertake voluntary fasts must seek permission from their husbands.

may fast voluntarily, preferably on Mondays and Thursdays. Muslims are not allowed to fast on two festival days, *Eid al-Fitr* and *Eid al-Azha* (see the previous section for more information) or on the days of sacrificial slaughter (*Tashriq*—the 12th, 13th, and 14th days of *Dhu-al-Hijjah*). It is also undesirable for Muslims to fast excessively (because Allah provides food and drink to consume) or to fast on Fridays.

Eastern Religions

Hinduism

■ Some Hindu worshipers break coconuts on the temple grounds to symbolize the spiritual experience: The hard shell is a metaphor for the human ego, and once it is cracked open, the soft, sweet meat representing the inner self is open to becoming one with the Supreme Being (Dresser, 1996).

■ According to legend, Vishnu rested on a 1,000-headed cobra between the creation and destruction of the world. During the festivities of *Naga Panchami,* snakes are venerated at Hindu temples, and milk is offered to cobras to prevent snakebite.

Hinduism is considered to be the world's oldest religion, and, like Judaism, it is the basis of other religions such as Buddhism. Although Hinduism was once popular throughout much of Asia, most Hindus now live in India, its birthplace.

The common Hindu scriptures are the *Vedas,* the *Epics,* and the *Bhagavata Purana.* The Vedas form the supreme authority for Hinduism. There are four Vedas: the *Rigveda,* the *Samaveda,* the *Yajurveda,* and the *Aatharvaveda.* Each consists of four parts.

The goal of Hinduism is not to make humans perfect beings or life a heaven on earth but to make humans one with the Universal Spirit or Supreme Being. When this state is achieved, there is no cause and effect, no time and space, no good and evil; all dualities are merged into oneness. This goal cannot be reached by being a good person, but it can be obtained by transforming human consciousness or liberation, *moksha,* into a new realm of divine consciousness that sees individual parts of the universe as deriving their true significance from the central unity of spirit. The transformation of human consciousness into divine consciousness is not achieved in one lifetime, and Hindus believe that the present life is only one in a series of lives, or reincarnations. Hindus believe in the law of rebirth, which postulates that every person passes through a series of lives before obtaining liberation; the law of *karma;* and that one's present life is the result of what one thought or did in one's past life. In each new incarnation an individual's soul moves up or down the spiritual ladder; the goal for all souls is liberation.

There is one supreme being, Brahmin, and all the various gods worshiped by men are partial manifestations of him. Hindus choose the form of the supreme being that satisfies their spirit and make it an object of love and adoration. This aspect of worship makes Hinduism very tolerant of other gods and their followers; many different religions have been absorbed into Hinduism.

The three most important functions of the supreme being are the creation, protection, and destruction of the world, and these functions have become personified as three great gods: Brahma, Vishnu, and Siva (the Hindu triad or trinity). The supreme being as Vishnu is the protector of the world. Vishnu is also an *avatara,* meaning he can take on human forms whenever the world is threatened by evil. Rama and Krishna are regarded as two such embodiments and are also objects of worship.

Hindus believe that the world passes through repeating cycles; the most common version of the creation is connected to the life of Vishnu. From Vishnu's navel grows a lotus, and from its unfolding petals is born the god Brahma who creates the world. Vishnu governs the world until he sleeps; then Siva destroys it and the world is absorbed into Vishnu's body to be created once again.

The principles of Hinduism are purity, self-control, detachment, truth, and nonviolence. Purity is both a ceremonial goal and a moral ideal. All rituals for purification and the elaborate rules regarding food and drink are meant to lead to purity of mind and spirit. Self-control governs both the flesh and the mind. Hinduism does not teach its followers to suppress the flesh completely, but to regulate its appetites and cravings. The highest aspect of self-control is detachment. Complete liberation from this world and union with the divine are not possible if one clings to the good or evil of this existence. Pursuit of truth is indispensable to the progress of man, and truth is always associated with nonviolence, *ahimsa.* These principles are considered to be the highest virtues. India's greatest exponent of this ideal was Mahatma Gandhi, who taught that nonviolence must be practiced not only by individuals, but also by communities and nations.

One common belief of Hinduism is that the world evolved in successive stages, begin-

Meat Prohibitions

It is believed that only 1 percent of the world's population refuses to eat all types of meat, poultry, and fish and that only one-tenth of 1 percent are total vegans, avoiding all animal products. In nearly all societies, the consumption of animal flesh increases with affluence and availability. Yet many cultures impose some restrictions on what meats may be consumed, mostly in accordance with prevailing religious dietary laws. The devout of each faith see little reason to ask why a particular food is prohibited. In fact, it is considered presumptuous or sacrilegious for humans to question the directives of God or church.

This has not deterred researchers from speculating on the rationale of meat taboos. Some have investigated the whole field of taxonomy and how animals are classified as "different," thus abominable, due to their physical characteristics (Douglas, 1966). Others have focused on the use of the term *unclean* in relation to biblical and Quranic pork prohibitions, claiming that pork consumption is unhealthful. This idea was supported by the mid-19th century discovery that trichinosis is spread through eating undercooked pork products. Many researchers discard the trichinosis theory because it is thought that ancient populations could have made the association between eating pork and the slow development of the disease, not to mention that other animals that carry fatal illnesses (e.g., spongiform encephalitis or "mad cow" disease) are not avoided. However, many studies continue today to determine how pork consumption is incompatible with human health (Chaudry, 1992). Seventh-Day Adventists, who abstain from all animal flesh for health reasons, are well regarded for their sponsored research on vegetarian diets.

Anthropologist Marvin Harris (1988) has proposed a socioecological theory for why certain meats are avoided. He suggests that if an animal is much more valuable alive than dead or, conversely, if it does not fit well into the local ecology or economy, consumption will be prohibited. Religious dietary codes often reinforce preexisting food practices and prejudices. He reviews the examples of beef in India, pork in the Middle East, and horsemeat in medieval Europe from this perspective:

1. In India, 80 percent of the population lives in rural areas. Cattle are the primary power source in this agrarian society; tractors and the fuel to run them are far too costly for the average Indian farmer. In addition, cattle provide dung that is dried to produce a clean, slow-burning cooking fuel. India is limited in forest resources and fossil fuels, so manure energy is important to both the economy and the environment. Cows also provide milk for the dairy products vital in some Indian vegetarian fare. Even dead cows serve a purpose in India, providing the very poor with the skins to craft leather products to survive financially and the protein from the scavenged carrion to survive physically.
2. Archeological records show that pork was a part of the diet in the ancient Middle East. However, by about 1900 B.C.E., pigs had become unpopular sources of meat in Babylonia, Egypt, and Phonecia. This timing occurs with an expanding population and deforestation of the region. Though pigs are efficient producers of meat, they compete with humans for food sources. In addition, they do not thrive in hot, dry climates. Cows, goats, and sheep, on the other hand, are ruminants. They graze over large areas and survive on the cellulose in plants unavailable to human metabolism. They do not need protection from the sun. The nomadic Hebrews were unlikely to herd pigs in their early history, and by the time they settled there was a broad aversion to pigs by many Middle Easterners. The first followers of Mohammed were also pastoral people, which perhaps explains why the only explicitly prohibited animal flesh in Islam is pork.
3. Christians, who defined their faith partially through their rejection of food laws, at one time banned consumption of horsemeat through papal decree. Horse consumption was frequent in prehistoric Europe but diminished as the region became more forested and pastureland diminished. Though an uncommon meat, it was still enjoyed when available and ritually sacrificed by northern Europeans. Prohibitions against horses began with their use in cavalry. Most Middle Easterners do not consume horsemeat (banned by Jews and by custom among Moslems), and it was avoided by the Romans. Asian nomads who roamed on horseback consumed horse milk and blood but only ate the flesh in emergencies. During the 8th century when European Christian strongholds came under attack from Moslem cavalry in the south and mounted nomads from the west, Gregory III recognized the need for horses in the defense of the church. He prohibited consumption of horsemeat as "unclean and detestable." However, horse consumption was never entirely eliminated, especially during periods when other meats were limited, and gradually religious restrictions were eased. By the 19th century, horsemeat had regained favor, especially in France and Belgium, where it is a specialty item today. Despite the initial need for horsepower, the prohibition was unsustainable over time because it contradicted prevailing food traditions.

■ Gandhi's birthday is commemorated in India on October 3 with devotional songs and the spinning of cotton thread in honor of his support for textile workers.

■ *Yoga* means yoke, as in yoking together or union.

■ Ganesh got his elephant head when he angered his father, Siva, who cut off his human head. When his mother pleaded with Siva to replace his head, Siva used the head of a nearby elephant. Hindus honor Ganesh through offerings of the foods he favored.

ning with matter and going on through life, consciousness, and intelligence to spiritual bliss or perfection. Spirit first appears as life in plants, then as consciousness in animals, intelligence in humans, and finally bliss in the supreme spirit. A good person is closer to the supreme spirit than a bad person is, and a person is closer than an animal. Truth, beauty, love, and righteousness are of higher importance than intellectual values (e.g., clarity, cogency, subtlety, skill) or biological values (e.g., health, strength, vitality). Material values (e.g., riches, possessions, pleasure) are valued least.

The organization of society grows from the principle of spiritual progression. The Hindu lawgivers tried to construct an ideal society in which people are ranked by their spiritual progress and culture, not according to their wealth or power. The social system reflects this ideal, which is represented by four estates, or castes, associated originally with certain occupations. The four castes are the *Brahmins* (teachers and priests), the *Kshatriyas* (soldiers), the *Vaisyas* (merchants and farmers), and the *Sundras* (laborers). Existing outside social recognition are the untouchables (e.g., butchers, leather workers), a group of persons who do not fall into the other four categories; this designation was outlawed by the Indian government in 1950.

The four castes are represented as forming parts of the Creator's body, respectively his mouth, arms, thighs, and feet. The untouchables were created from darkness that Brahma discarded in the process of creation. The castes also conform to the law of spiritual progression, in that the most spiritual caste occupies the top and the least spiritual the bottom. The Hindus believe that nature has three fundamental qualities: purity, energy, and inertia. Those in whom purity predominates form the first caste, energy the second caste, and inertia the third and fourth caste. Each caste should perform its own duties, follow its hereditary occupation, and cooperate with the others for the common welfare. People's good actions in this life earn them promotion to a higher caste in the next life.

There are thousands of subdivisions of the four main castes. The subcastes often reflect a trade or profession, but some scholars contend that the latter was imposed on the former. In reality, the subcaste is very important to daily life, while what major caste one belongs to makes little difference to non-Brahmins (see chapter 14 for more information).

The ideal life of a Hindu is divided into four successive stages, called *asramas*. The first stage is that of the student and is devoted entirely to study and discipline. The *guru* becomes an individual's spiritual parent. After this period of preparation the student should settle down and serve his or her marriage, community, and country. When this active period of citizenship is over, he or she should retire to a quiet place in the country and meditate on the higher aspects of the spirit (become a recluse). The recluse then becomes a *sannyasin,* one who has renounced all earthly possessions and ties. This stage is the crown of human life. The goals of life are *dharma* (righteousness), *artha* (worldly prosperity), *kama* (enjoyment), and *moksha* (liberation). The ultimate aim of life is liberation, but, on their way to this final goal, people must satisfy the animal wants of the bodies, as well as the economic and other demands of their families and communities. However, all should be done within the moral law of dharma.

Common practices in Hinduism include rituals and forms of mental discipline. All Hindus are advised to choose a deity on whose form, features, and qualities they can concentrate their mind and whose image they can worship every day with flowers and incense. The deity is only a means of realizing the supreme being by means of ritualistic worship. Externally the deity is worshiped as a king or honored guest. Internal worship consists of prayer and meditation. Mental discipline is indicated by the word *yoga*. Along with mental discipline, yoga has come to mean a method of restraining the functions of the mind and their physiological consequences.

Hindus can be divided into three broad sects according to their view of the supreme being. They are the *Vaishnava,* the *Saiva,* and the *Sakta,* who maintain the supremacy of Vishnu, Siva, and the Sakti (the female and active aspects of Siva), respectively. Different sects are popular in different regions of India. Many Hindus do not worship one God exclusively. Vishnu may be worshiped in one of his full (Krishna or Rama) or partial embodiments. In addition there are hundreds of lesser deities, much like saints. One is Siva's son, the elephant-headed Ganesh, who is believed to bring good luck and remove obstacles.

There may be as many as 950,000 Hindus in the United States, although it is not known for certain how many Hindus live here, because the religion is not centered around a temple and its affairs are not conducted by priests. It is assumed that a significant percentage of the Asian Indian population in America is Hindu. A small percentage of non-Indian Americans have become followers of the Hindu religion. The International Society for Krishna Consciousness, founded in 1966 by devotees of a 16th-century Bengali ascetic, has the largest number of converts.

Hindu Dietary Practices

In general, Hindus avoid foods that are believed to hamper the development of the body or mental abilities. Bad food habits will prevent one from reaching mental purity and communion with God. Dietary restrictions and attitudes vary among the castes.

The Laws of Manu state that "no sin is attached to eating flesh or drinking wine, or gratifying the sexual urge, for these are the natural propensities of men; but abstinence from these bears greater fruits." Many Hindus are vegetarians. They adhere to the concept of *ahimsa,* avoiding the infliction of pain on an animal by not eating meat. Although the consumption of meat is allowed, the cow is considered sacred and is not to be killed or eaten. If meat is eaten, pork as well as beef is usually avoided. Fish that have ugly forms, and the heads of snakes, snails, crabs, fowl, cranes, ducks, camels, and boars are forbidden. No fish or meat should be eaten until it has been sanctified by the repetition of mantras offering it to the gods. Pious Hindus may also abstain from drinking alcohol. Garlic, turnips, onions, mushrooms, and red-colored foods such as tomatoes and red lentils may be avoided.

Intertwined in Hindu food customs is the concept of purity and pollution. Complex rules regarding food and drink are meant to lead to purity of mind and spirit. Pollution is the opposite of purity and should be avoided or ameliorated. To remain pure is to remain free from pollution; to become pure is to remove pollution.

Certain substances are considered both pure in themselves and purifying in their application. These include the products of the living cow, such as milk products, dung, and urine, and water from sources of special sanctity, such as the Ganges River. They also include materials commonly employed in rituals, such as turmeric and sandalwood paste. All body products (feces, urine, saliva, menstrual flow, and afterbirth) are polluting. Use of water is the most common method of purification, because water easily absorbs pollution and carries it away.

Feast Days

The Hindu calendar has eighteen major festivals every year. Additional important feast days are those of marriages, births, and deaths. Each region of India has its own special festivals; it has been said that there is a celebration going on somewhere in India every day of the year. All members of the community eat generously on festive occasions, and these may be the only days that very poor people eat adequately. Feasting is a way of sharing food among the population, because the wealthy are responsible for helping the poor celebrate the holidays.

One of the gayest and most colorful of the Hindu festivals is *Holi,* the spring equinox and the celebration of one of Krishna's triumphs. According to legend, Krishna had an evil uncle who sent an ogress named Holika to burn down Krishna's house. Instead, Krishna escaped and Holika burned in the blaze. It is traditional for Indians to throw colored water or powder at passersby during this holiday.

The 10-day celebration of *Dusshera* in late September or early October commemorates the victory of Prince Rama (one of Vishnu's embodiments) over the army of the demon Ravana and is a grateful tribute to the goddess Durga, who aided Rama. The first 9 days are spent in worshiping the deity, and the tenth day is spent celebrating Rama's victory.

Divali, celebrated throughout India in November, marks the darkest night of the year, when souls return to earth and must be shown

- Hindus are encouraged to practice moderation—they are advised not to eat too early, not to eat too late, and not to eat too much.

- Students who are studying the Vedas and other celibates are usually vegetarians and may restrict irritating or exciting foods, such as honey.

- Water is the beverage of choice at meals. Standing water is easily defiled if it is touched by a member of a lower caste; flowing water is so pure that even an untouchable standing in it does not pollute it.

- The Hindu calendar is lunar; thus, its religious holidays do not always fall on the same day on the Western calendar. Every 3 to 5 years the Hindu calendar adds a 13th leap month (a very auspicious period) to reconcile the months with the seasons.

- Prince Rama was aided by Hanuman, a monkey hero; thus, monkeys are revered throughout India despite the serious crop damage they cause.

The numerous religious holidays and secular events celebrated in India include feasting, which serves to distribute food throughout the community. (United Nations/J. Isaac.)

■ In southern India the rice harvest is celebrated in the festival called *Pongal*—new rice is cooked in milk and when it begins to bubble, the family shouts, "pongal!" ("It boils!").

the way by the lights in the houses. For many, Divali is also the beginning of the new year, when everyone should buy new clothes, settle old debts and quarrels, and wish everyone else good fortune.

Fast Days

In India, fasting practices vary according to one's caste, family, age, sex, and degree of orthodoxy. A devoutly religious person may fast more often and more strictly than one who is less religious. Fasting may mean eating no food at all or abstaining from only specific foods or meals. It is rare for an individual to go without any food at all, because Hinduism has numerous fast days. The fast days in the Hindu calendar include the first day of the new and full moon of each lunar month; the 10th and 11th days of each month; the feast of *Sivaratri;* the 9th day of the lunar month *Cheitra;* the 8th day of *Sravana;* days of eclipses, equinoxes, solstices, and conjunction of planets; the anniversary of the death of one's father or mother; and Sundays.

Buddhism

Siddhartha Gautama, who later became known as Buddha (the Enlightened One), founded this eastern religion in India in the 6th century B.C.E. Buddhism flourished in India until 500 C.E., when it declined and gradually became absorbed into Hinduism. Meanwhile, it had spread throughout southeastern and central Asia. Buddhism remains a vital religion in many Asian countries, where it has been adapted to local needs and traditions.

Buddhism was a protestant revolt against orthodox Hinduism, but it accepted certain Hindu concepts, such as the idea that all living beings go through countless cycles of death and rebirth, the doctrine of karma, spiritual liberation from the flesh, and that the path to wisdom includes taming the appetites and passions of the body. Buddha disagreed with the Hindus about the methods by which these objectives were to be achieved. He advocated the "Middle Way" between asceticism and self-indulgence, stating that both extremes in life should be avoided. He also disagreed with the Hindus on caste distinctions, believing that all persons were equal in spiritual potential.

The basic teachings of Buddha are found in the *Four Noble Truths* and the *Noble Eightfold Path.* The Four Noble Truths are as follows:

1. *The Noble Truth of Suffering:* Suffering is part of living. Persons suffer when they experience birth, old age, sickness, and death. They also suffer when they fail to obtain what they want.

2. *The Noble Truth of the Cause of Suffering:* Suffering is caused by a person's craving for life, which causes rebirth.
3. *The Noble Truth of the Cessation of Suffering:* A person no longer suffers if all cravings are relinquished.
4. *The Noble Truth to the Path That Leads to the Cessation of Suffering:* This is the Eightfold Path. By following this path (right view, right thought, right speech, right action, right livelihood, right effort, right mindfulness, and right concentration), craving is extinguished and deliverance from suffering ensues.

The third and fourth phases of the Eightfold Path (Figure 4.3), right speech and right action, have been extended into a practical code of conduct that is known as the Five Precepts. These are (1) to abstain from the taking of life, (2) to abstain from the taking of what is not given, (3) to abstain from all illegal sexual pleasures, (4) to abstain from lying, and (5) to abstain from consumption of intoxicants because they tend to cloud the mind.

The person who perfects Buddha's teachings achieves *nirvana,* a state of calm insight, passionlessness, and wisdom. In addition, the person is no longer subject to rebirth into the sorrows of existence. Because the ideal practice of Buddhism is impractical in the turmoil of daily life, Buddhism has encouraged a monastic lifestyle. The ideal Buddhists are monks, following a life of simplicity and spending considerable time in meditation. They own no personal property and obtain food by begging. They are usually vegetarians and are permitted to eat only before noon. The monk confers a favor or merit (good karma) on those who give him food.

There are numerous sects in Buddhism and two great schools of doctrine: *Theravada* (also known as *Hinayana*) Buddhism, which is followed in India and Southeast Asia; and *Mahayana* Buddhism, which is followed in China, Japan, Korea, Tibet, and Mongolia. Theravada Buddhism is primarily a spiritual philosophy and system of ethics. It places little or no emphasis on deities, teaching that the goal of the faithful is to achieve nirvana. In Mahayana, a later form of Buddhism, Buddha is eternal and cosmic, appearing variously in many worlds to make known his truth, called *dharma.* This has resulted in a pantheon of Buddhas who are sometimes deified and, for some sects, a hierarchy of demons. Some sects promise the worshiper a real paradise, rather than the perfected spiritual state of nirvana.

The number of Buddhists in the United States may be as high as 2 million; the majority are immigrants from Japan, China, and Southeast Asia and their descendants. A small number of non-Asians also have more recently converted to Buddhism. *Zen* Buddhism, a Chinese sect that spread to Japan around the year 1200, has gained followers in the West.

Dietary Practices

Buddhist dietary restrictions vary considerably depending on the sect and country. Buddhist doctrine forbids the taking of life; therefore,

■ In both Theravada and Mahayana temples, worshipers may offer food at the altar, such as apples, bananas, grapes, oranges, pineapples, candy, rice, dried mushrooms, and oil.

■ A Zen Buddhist monastery, Tassajara, located in central California, is famous for its vegetarian restaurant and popular cookbook.

■ Zen macrobiotics is not associated with Zen Buddhism. It is a Japanese diet created in the 20th century and popularized in Europe as a way to achieve health and spiritual enlightenment.

■ At some Buddhist temples a vegetariian lunch is offered after the service. The congregation eats after the monks have finished their meal.

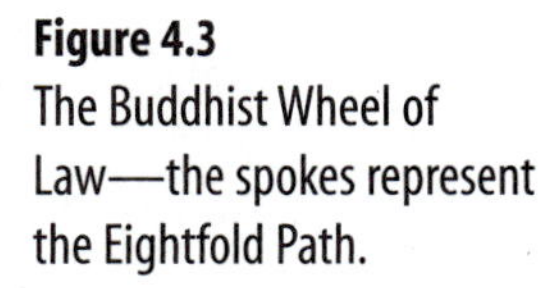

Figure 4.3
The Buddhist Wheel of Law—the spokes represent the Eightfold Path.

■ Some Buddhists celebrate *Magha Puja,* the Four Miracles Assembly, in February or March when Buddha appointed the first Buddhist brotherhood of monks at a coincidental meeting of 1,250 disciples at a shrine.

■ Buddhist monks in Tibet carve sculptures in butter (as high as 15 feet) and parade them during an evening in March, lit by lanterns, for *Chogna Choeba,* the Butter Lamp Festival. Afterward they are dismantled and thrown in the river, symbolic of the impermanence of life.

many followers are lacto-ovo vegetarians (eating dairy products and eggs, but no meat), some eat fish, and others abstain only from beef. Others believe that if they were not personally responsible for killing the animal, it is permissible to eat its flesh.

Feasts and Fasts

Buddhist festivals vary according to region. From July to October, Buddhist monks are directed to remain in retreat and meditate, coinciding with the rainy season and the sprouting of rice in the fields. The first day of retreat is a time for worshipers to bring gifts of food and articles of clothing to the monks; the retreat ends with *pravarana,* the end of the rainy season, when worshipers once again offer gifts to the monks, invite them to a meal, and organize processions. Mahayana Buddhists commemorate the birth of Buddha, his enlightenment, and his death on three separate days (which vary according to the regional calendar); Theravada Buddhists honor the Buddha on a single holiday called *Vesak,* celebrated during April or May. Buddhist monks may fast twice a month, on the days of the new and full moon. They also do not eat any solid food after noon.

Starch Foods

Starchy foods form the foundation of nearly all diets.

A. Rice is eaten by millions worldwide, and comes in many varieties, including short-grain (far right), long-grain (in scoop), and wild rice (long, black grains in lower left mix). Rice products, such as noodles and papers, are also common.

B. Wheat is popular in drier regions, typically eaten as bread, pasta, and in cereal form, such as couscous (lower bag), and bulgar (upper bag).

C. Corn is an important new world starch, traditionally prepared as flat breads (including these tortillas), as dumplings, in steamed packets, in stews, and as gruel.

D. In tropical areas, fruit and root vegetables are significant sources of starch, including breadfruit (upper basket), cassava (lower basket), yams (lower left corner), taro root (left upper corner), lotus root (cut root with hollow spaces in center), sweet potatoes (ruby orange roots on right), and burdock root (long, pencil-thin roots).

E. Acorns are a starch consumed in some Native American, European, and Middle Eastern cultures.

Protein Foods

Protein foods include a wide variety of meat products, dairy foods, fish and shellfish, as well as numerous legumes.

A. Sausages are eaten in nearly every culture. They come in hundreds of types and make use of miscellaneous cuts and leftovers, such as blood (dark red links).

C. It is estimated that nearly 30 percent of the population worldwide is dependent on fish, such as this dried salmon.

B. Dairy products, including yogurt and cheese, are available in even more varieties. Yet many cultural groups consume only limited amounts of milk or other dairy foods.

D. Legumes, such as beans, peas, and lentils, are eaten daily in many cultures.

E. Soy products, such as soy milk and bean curd (known as tofu or tobu) are especially significant in the diet of many Asians.

Vegetables

Vegetables are featured in the cuisines of almost all cultures.

A. Greens are especially common, such as this mizuna and bok choy.

B. Examples of vegetables grown on bushes or vines include eggplants, which come in many shapes (from round to oblong, to long-thin), and colors (white, striped, green, orange, purple), and chile peppers, which vary in heat from mild to mouth-searing tomatillos, which are crunchy, slightly acidic vegetables covered with a papery husk.

C. Olives, harvested from trees, are unusual because they are high in fat compared to other vegetables, and must be processed to remove bitterness before consumption.

D. Mushrooms, fungi that are eaten as vegetables, are usually edible, but can be highly toxic. Types include shiitake (large, dark brown cap); oyster mushrooms (yellow, funnel-shaped); lobster mushsooms (orange, knobby); and black cloud ears (dark, curly fungus).

Fruits

Fruit is a favorite worldwide.

A. Some regions have only a few types available, such as desert regions where prickly pear cactus fruit is a specialty.

B. Temperate regions have a broader selection of fruits, supporting applies, citrus fruits, berries, and more unusual fruit such as pomegranates.

C. Tropical areas have an enormous variety of fruits. The jelly-like flesh of immature, green coconuts is scooped out as a sweet treat.

D. Durian, another tropical specialty, has a strong odor similar to rotting onions esteemed by its fans.

E. Carambola (also known as starfruit) is juicy and one of the only tropical fruits with a crunch.

F. Other examples of tropical fruits include cocao pods (reddish-orange, ribbed fruit); nutmeg (yellow, lobed fruit); casimiroa (round, green); breadfruit; guavas, mangoes; and papaya.

5

Native Americans

The designation "Native American" includes the greatest number of ethnic groups of any minority population in the United States. Each of the approximately 400 American Indian and Alaska Native nations has its own distinct cultural heritage.

Nearly 2 million self-declared Native Americans live in the United States, just under 1 percent of the total population. The vast majority of Native Americans today live west of the Mississippi River. Roughly half live in rural areas, either on government reservations or on nearby farms. Native American ethnic identity varies tremendously, from tenacious maintenance of heritage to total adoption of the majority culture.

Traditional Native American foods have made significant contributions to today's American diet. Corn, squash, beans, cranberries, and maple syrup are just a few of the items Native Americans introduced to European settlers. Historians question whether the original British colonists would have survived their first years in America without the supplies they obtained and cooking methods they learned from the Native Americans. The diet of Native Americans has changed dramatically from its origins, yet recent renewed interest in Native American culture has prevented the complete disappearance of many traditional foods and food habits. This chapter reviews both the past and present diet of Native American ethnic groups.

Cultural Perspective

History of Native Americans

Settlement Patterns

It is hypothesized that the Native Americans came to North America approximately 20,000 to 50,000 years ago across the Bering Strait, which links Asia to Alaska, although some evidence suggests that earlier migrations may have occurred. Archaeological research provides little insight into the settlement patterns and diversification of Native American culture in the years before European contact in the 1600s. Furthermore, the Native American languages were entirely verbal, so written historical records are nonexistent. There are, consequently, enormous gaps in what is known of early Native American societies.

In contrast, the observations of the Native Americans by white settlers have been well documented. They identified three major centers of Native American culture during the seventeenth century. In the Southeast the sophisticated social organization of the Cherokees, Chickasaws, Choctaws, Creeks, and Seminoles led the Europeans to call them the "Five Civilized Tribes." The Iroquois, in what is now New York State, ruled a democratic confederacy of five nations. Religion and the arts flourished in Pueblo communities adjacent to the Rio Grande and Little Colorado rivers in the Southwest.

The introduction of horses, firearms, and metal knives changed the lifestyles of many

■ *Native American* is used as a designation for the indigenous peoples of the Americas, which includes American Indians and Alaska Natives, who are American Indians, Inuits, and Aleuts. (In Canada, these people are called *Aboriginals.*) Some Native Americans, and some researchers, prefer the term *American Indian,* but this ignores Alaska Natives. It has also been suggested that native Hawaiians be included as Native Americans; however, the history and culture of these peoples are substantially different from those of Native Americans in the U.S. mainland, Alaska, and Canada, so they are discussed in the chapter on Southeast Asians and Polynesians. Native American cultures of Mexico, Central America, and South America are considered under Latinos.

■ During the early 19th century, the Cherokee had a written language, a bilingual newspaper, a school system, a court system, and a Cherokee Nation constitution. They were a prosperous tribe; many owned black slaves.

Traditional Native American foods: Some typical foods in traditional Native American diets include beans, berries, corn, fish, *jerky,* maple syrup, squash, and tomatoes. (Photo by Laurie Macfee.)

nations, especially those that used the new tools to exploit the resources of the Great Plains. This initial interaction between white settlers and Plains Indians resulted in the development of the stereotype of the buffalo-hunting horseman with feathered headdress who came to represent all Native American ethnic groups in the white imagination.

European diseases and the massacre of whole nations reduced the number of both Native American individuals and ethnic groups. In addition, many Native Americans were forced to migrate west to accommodate white expansion. It is estimated that the hardships of involuntary relocation and the deaths caused by illness and assault caused the extinction of nearly one-quarter of all Native American ethnic groups.

■ The largest Native American populations are the Cherokee (369,000) and the Navajo (225,000).

Native American lands dwindled as white settlers moved westward. By the late 19th century, the majority of Native Americans lived on lands held in trust for them by the U.S. government called federal reservations. Still others resided in state reservation communities. Although they were not required by law to live on reservations, there were few other viable Native American communities.

The Bureau of Indian Affairs (BIA) took over the administration of the reservations near the turn of the 20th century. They established a program of cultural assimilation designed to bring the Native American residents into mainstream American society. Before the 1930s, Native American children were sent to off-reservation boarding schools where white values were encouraged. Later, public reservation schools attempted similar indoctrination. The BIA program usually failed to force Native Americans to accept white values, however. The Native Americans changed their dress, occupation, and social structure, but they did not fully assimilate. In many cases their religious beliefs, crafts, music, and dance were strengthened to support their ethnic identity.

Current Demographics

Many Native Americans left the reservations for the employment opportunities available during World War II (Figure 5.1). Some joined the armed services, where they became fluent in English and the ways of the majority society. Others took war-related industry jobs. In the 1950s and 1960s the BIA Employment Assistance Program was a major factor in the continuing outmigration of Native Americans from the reservations to the cities. By 1980 just more than one-half of all Native Americans resided in urban areas and towns outside of reservations. The cities with the largest Native American populations are (in descending order) Los Angeles, the San Francisco Bay Area, Tulsa, Minneapolis-St. Paul, Oklahoma City, Chicago, and Phoenix (see also Table 5.1). The

Figure 5.1
Native American nations in the United States

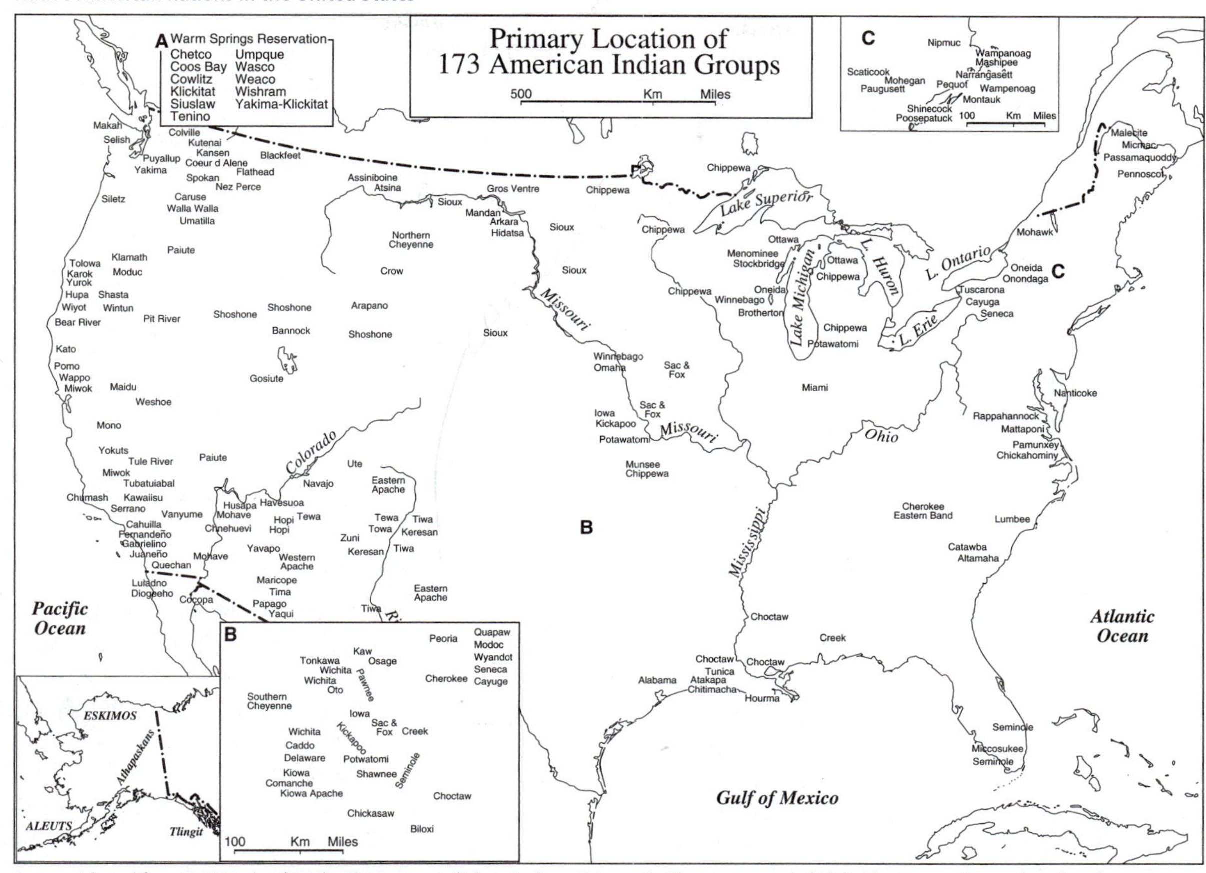

Source: Adapted from T. W. Taylor (1972), *The States and Their Indian Citizens.* In Thernstrom et al. (Eds.), *The Harvard Encyclopedia of American Ethnic Groups* (Cambridge, MA: Harvard University Press, 1980). Copyright 1980. Reprinted with permission by the President and Fellows of Harvard College.

remainder of the Native American population resides in rural areas, including reservations. Many first-generation urban Native Americans maintain close ties with the reservation of their ethnic group and travel often between the city and tribal land. Members of the second generation living in urban regions are more likely to think of the city as their permanent home.

Socioeconomic Status

The socioeconomic status of Native Americans declined drastically with the forced migrations of the 19th century. Even those Native American nations that were agriculturally self-sufficient suffered when relocated to regions with poor growing conditions. Further, there were few native occupations that were valued in the job market outside the reservations. BIA education efforts were generally unsuccessful, and most Native Americans did not begin to find employment until World War II and the development of the BIA Employment Assistance Program. More recently the Indian Self-Determination and Education Act of 1975 was enacted to promote Native American participation in government and education.

Economic improvement has been slow. Although unemployment rates among Native Americans dropped slightly in the 1980s (unemployment rates are two to three times the national average), poverty levels have increased. Nearly one-third (32 percent) of Native American families fell below the poverty level in the 1990 U.S. census. Congressional amendments to the Older Americans Act in 1987 found that 61 percent of Native American elders older than the age of 60 also live in poverty. Rates of formal education have been improving, yet 43 percent of all adult Native

- Efforts to increase prosperity on reservations include utilization of natural resources and establishment of gambling operations, which are legal on tribal lands.

- The number of U.S. businesses owned by Native Americans nearly doubled in the early 1990s, growing at a rate more than three times that of the overall increase in firms.

Table 5.1
Ten States with the Largest Native American Populations

State	Native American Population	Percentage of State Population
California	306,690	1
Oklahoma	260,029	8
Arizona	255,463	6
New Mexico	158,036	9
Washington	100,309	2
Alaska	97,098	16
North Carolina	95,398	1
Texas	93,343	0.5
New York	74,483	0.4
Michigan	59,678	1

Source: U.S. Census Bureau.

Americans have not finished high school or obtained an equivalency diploma.

Native American Organizations

Few Native American neighborhoods develop in urban areas. Native Americans who settle in the cities usually arrive as individuals or small family groups and typically do not live near others of their nation. Sometimes longtime city residents exhibit a sense of superiority over recent arrivals from the reservations. In general, the difficulties of adjusting to urban white society stimulate many young Native Americans to identify not only with Native Americans of other nations but also with people from other ethnic groups. Activists for Native American rights often come from the cities and are not always supported by Native Americans who live on tribal lands. Native American organizations have done much to maintain Indian identity. Most areas with large Native American populations have their own clubs and service associations. Organizations to promote ethnic identity have been founded by the Navajo, Pueblos, Tlingit, Haida, and Pomo. Other groups serve the social needs of the pan-Native American community, such as athletic clubs and dance groups. The United South and Eastern Tribes works to strengthen Native American unity; the All Indian Pueblo Council advocates on behalf of Pueblo tribes. The National American Indian Council was founded in 1982 to represent the needs of urban Native Americans.

■ The Navajo concept that integrates everything positive in life, including beauty, goodness, order, and harmony, is called *hozho* (Rhodes, 1995).

■ Religious ceremonies held by the Native American Church, called peyote meetings, feature singing, prayers, and traditional foods such as tripe soup. Continued success and good health for tribal members are entreated.

Worldview

Harmony best describes the Native American approach to life. Each individual strives to maintain a balance between spiritual, social, and physical needs in a holistic approach. Only what is necessary for life is taken from the natural environment; it is believed that the Earth should be cared for and treated with respect. Generosity is esteemed and competitiveness is discouraged, yet individual rights are also highly regarded. Personal autonomy is protected through the principle of noninterference. Among the Navajo, for example, an individual would never presume to speak for another, even a close family member. For most Native Americans, time is conceptualized as being without beginning or end, and the culture is present-oriented, meaning that the needs of the moment are emphasized over the possible rewards of the future.

Religion

Traditional Native American religions vary from an uncomplicated belief in the power of a self-declared evangelist to elaborate theological systems with organized hierarchies of priests. Yet they all share one characteristic: the religion permeates all aspects of life. Rather than a separate set of beliefs practiced at certain times in specific settings, religion is an integral part of the Native American holistic worldview. Religious concepts influence both the physical and emotional well-being of the individual.

Many Native American nations have rejected all attempts at Christian conversion, especially in the Southwest. The Navajo, Arizona Hopi, Rio Grande Pueblos, Potawatomi, Lakota, and Dakota have retained most of their native religious values and rituals. Other religions unique to Native Americans emerged after European contact, such as the Drum Dance cult and the Medicine Bundle religions, which combine spiritual elements from several different ethnic groups. In addition, religions that mix Christianity with traditional beliefs have been popular since the late 19th century; the Native American Church has been especially successful. Other groups claim Christian fellowship but continue to practice native religions as well. Finally, many Native Americans now adhere to Roman Catholicism or some form of Protestantism, especially in urban areas, where churches have been established to serve all Native American congregations.

■ *Sioux* was the French term for the Lakota, Dakota, and Nakota peoples. It derives from the word for snake and is considered derogatory by some Native Americans.

Family

The primary social unit of Native Americans is the extended family. Children are valued highly and there is great respect for elders. All blood kin of all generations are considered equal; there is no differentiation between close and distant relatives. Aunts and uncles are often considered to be like grandparents, and cousins are viewed as brothers or sisters. Even other tribal members are sometimes accepted as close kin. In many Native American societies an individual without relatives is considered poor. Many Native American nations are matrilineal, meaning that lineage is inherited from the mother. Even in such systems the men are the family providers and heads of the household; women are typically in charge of domestic matters. Native American children are expected to assist their parents in running the home.

Traditional Health Beliefs and Practices

In Native American culture, health reflects a person's relationship to nature, broadly defined as the family, the community, and the environment. Every illness is due to an imbalance with supernatural, spiritual, or social implications. Treatment focuses on the cause of the imbalance, not just the symptoms, and is holistic in approach. Traditional Native American medicine is concerned with physical, mental, and spiritual renewal through health maintenance, prevention of illness, and restoration of health.

The primary social unit of Native Americans is the extended family, which includes all close and distant kin. (© Lionel Delevingne/Stock, Boston.)

Many reasons account for illness. Some Navajos believe that witchcraft, through agents such as animals, lightning and whirlwinds, transgressions committed at ceremonial occasions, or evil spirits (especially ghosts), may cause fainting, hysteria, or other conditions. Witchcraft may also take the form of intrusive objects, causing pain where the object is inserted as well as emaciation. Possession by a spirit may dislodge the soul, resulting in a feeling of suffocation (or possession may be a sign of a gift for healing). Soul loss may also cause mental disorders. Violation of a taboo, whether an actual breach by an individual or contact with evil objects that have committed mythical breaches results in general seizures (Bachman-Carter et al., 1998; Kunitz & Levy, 1981). These beliefs are shared by many other Native Americans. For example, many Iroquois believe in a similar list of reasons for illness, adding that unfulfilled dreams or desires may also be a contributing factor (Wilson, 1983).

■ The Cherokee medicine man, Sequoyah, explained that "Indian medicine is a guide to health, rather than a treatment. The choice of being well instead of being ill is not taken away from an Indian" (Garrett, 1990, p. 180).

■ Navajos believe the soul (or "wind") enters the body at birth and leaves at death—a child's first laugh signals the firm attachment of the soul.

Some Native Americans reject the concept that poor nutrition, bodily malfunctions, or an infection by a virus or bacteria can cause sickness. An evil external source is often identified instead. Some Dakotas, for instance, blame diabetes mellitus on disease-transmitting foods provided by whites with the intention of eliminating all Native Americans (Lang, 1985). Small bags of herbs (called "medicine bundles" by certain Plains Indians), fetishes, or symbols may be worn to protect against malevolent forces.

Fetishes are used when an animal that has been harmed or killed causes an ailment; a fetish in the form of the animal is rubbed on the afflicted body part with appropriate chants.

Traditional healers often specialize in their practice. Navajo medicine men and women usually exert a positive influence in preventing disharmony through rituals such as the sweat bath to promote peace. They also have negative powers, which can be used to counteract witchcraft or evil acts by a person's enemies. Diagnosticians may be called on to identify the cause of an illness through stargazing or listening (if crying is heard, the patient will die). Hand motions or trembling also may be involved, sometimes including "painting" with white-, blue-, yellow-, and black-colored sand to produce a picture of magical healing power. Herbalists, often women, assist in the treatment of illness through the ceremonial collection and application of wild plant remedies (see the section "Therapeutic Uses of Foods" later in this chapter for examples.) Other traditional Navajo practitioners are singers, who cure with sacred chanting ceremonies, and healers, who have specific responsibility for care of the soul.

Among California Native Americans, illness was treated first with home remedies. If that proved ineffective, nonsacred healers such as herbalists or masseuses were contacted. If the patient still did not improve, a diviner would be consulted for a diagnosis. If spiritual or supernatural intervention was needed, a shaman (medicine man) was employed.

Traditional healers continue to hold an important role in Native American culture. In addition to administering cures, medicine men and women are often seen as "culture brokers" preserving Native American identity within rapid social change (Bean 1976).

■ An outbreak of serious respiratory infections due to the Hanta virus was explained by Navajo healers as being due to rejection of traditional ways and adoption of convenience foods. Prayer over corn pollen and herb tea was the prescribed cure (Plawecki et al., 1994).

■ The tradition of the Pacific Northwest *potlatch* reflects the abundance of the region. Commonly held for important events such as marriage or birth of a son, these ostentatious feasts with excessive amounts of food were followed by gift giving of items such as blankets, canoes, edible oils, furs, jewelry, and wood carvings. The potlatch established the wealth and socioeconomic status of the host. The nations of this region were the only Native American groups to develop a class system based on possession of material goods. Anthropologists suggest that the potlatch was also a method of food distribution among nations from different areas.

■ Although the wild turkey was found in the Plymouth area, it may not have been eaten at the first Thanksgiving. Goose was the preferred holiday bird during colonial times.

Traditional Food Habits

The traditional food habits of Native Americans were influenced primarily by geography and climate. Each Native American nation adopted a way of life that allowed it to maximize indigenous resources. Many were agriculturally based societies, others were predominantly hunters and gatherers, and some survived mainly on fish. Most of each day was spent procuring food.

Ingredients and Common Foods

Indigenous Foods

Archaeological records and early descriptions of America by European settlers indicate that Native Americans on the East Coast enjoyed an abundance of food. Fruits, including blueberries, cranberries, currants, grapes, persimmons, plums, and strawberries, as well as vegetables, such as beans, corn, and pumpkins, were mentioned by the New England colonists. They describe rivers so full of life that fish could be caught with frying pans, sturgeon so large they were called "Albany beef," and lobster so plentiful that they would pile up along the shoreline after a storm. Game included deer, moose, partridge, pigeon, rabbit, raccoon, squirrel, and turkey. Maple syrup was used to sweeten foods. Farther south, Native Americans cultivated groundnuts (Apios americana, or "Indian potatoes") and tomatoes, and collected wild Jerusalem artichokes (a starchy tuber related to the sunflower). Native Americans of the Pacific Northwest collected enough food, such as salmon and fruit, during the summer to support them for the remainder of the year. Peoples of the plains hunted buffalo and those of the northeastern woodlands gathered wild rice; nations of the Southwest cultivated chile peppers and squash amidst their corn. (See Table 5.2 for a listing of foods native to the Americas.)

Native Americans not only introduced whites to indigenous foods but also shared their methods of cultivation and food preparation as well. Legend has it that the Pilgrims nearly starved despite the plentiful food supply because they were unfamiliar with the local foods. One version of the tale is that Squanto, the sole surviving member of the Pautuxets (the other members had succumbed to smallpox following contact with earlier European explorers), saved the Pilgrims, who were mostly merchants, by teaching them to grow corn. He showed them the Native American method of planting corn kernels in mounds with a fish head for fertilizer and using the corn stalks as supports for beans. In Virginia, Native Americans are attributed with providing the Cavaliers, also from England, with enough food to ensure their survival.

Table 5.2
Indigenous Foods of the Americas*

Fruits	Berries (blackberries, blueberries, cranberries, gooseberries, huckleberries, loganberries, raspberries, strawberries), cactus fruit (*tuna*), cherimoya, cherries (acerola cherries, chokecherries, groundcherries), grapes (e.g., Concord), guava, mamey, papaya, passion fruit (*granadilla*), pawpaw, persimmon (American), pineapple, plums (American, beach), soursop (*guanabana*), zapote (*sapodilla*)
Vegetables	Avocado (alligator pear), bell peppers (sweet peppers, pimento), cactus (*nopales, nopalitos*), chayote (*christophine, chocho, huisquil, mirliton,* vegetable pear), pumpkins, squash, tomatillo, tomatoes
Tubers/roots	Arrowroot, cassava (*yuca, manioc, tapioca*), groundnut, Indian breadroot, Jerusalem artichoke, jicama, malanga (*yautia*), potatoes, sweet potatoes
Grains/cereals	Amaranth, corn (maize), quinoa, wild rice
Nuts/seeds	Brazil nuts, cashews, hickory nuts, pecans, pumpkin seeds (*pepitas*), sunflower seeds, walnuts (black)
Legumes	Beans (green beans, most dried beans), peanuts
Poultry	Turkey
Seasonings/flavorings	Allspice, chiles (e.g., hot and sweet chiles), chocolate (cocoa), maple syrup, sassafras (filé powder), spicebush, vanilla

*Foods native to North, Central, or South America. Some items not indigenous to the United States (e.g., pineapple, potatoes) were popularized only after acceptance in Europe and introduction by European settlers. Other foods (e.g., avocado, jicama, tomatillo) have become more common in the United States with the growing Latino population.

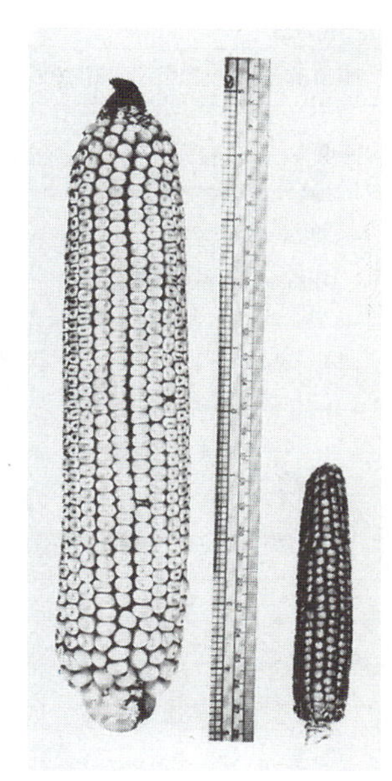

Modern Indian corn compared to Indian corn circa 500 C.E. (From LaFarge, 1956, with permission from Crown Publishers, Inc.)

Foods Introduced from Europe

Foods introduced by the Europeans, especially the French Jesuits in the North and the Spanish in the South, were well accepted by Native Americans. Apples, apricots, carrots, lentils, peaches, purslane, and turnips were some of the more successful new foods. Settler William Penn noted that he found peaches in every large Native American farm he encountered, barely 100 years after they were introduced to the Iroquois. The Europeans also brought rye and wheat. However, few Native American nations replaced corn with these new grains.

Livestock made a much greater impact on Native American life than did the new fruits and vegetables. Cattle, hogs, and sheep reduced the Native Americans' dependence on game meats. The Creeks and Cherokees of the Southeast fed their cattle on corn and fattened their suckling pigs and young lambs on apples and nuts. The Powhatans of Virginia fed their hogs peanuts, then cured the meat over hickory smoke. Lamb and mutton became staples in the Navajo diet after the introduction of sheep by the Spanish. In addition, the Europeans brought horses and firearms, which made hunting easier, and metal knives and iron pots, which simplified food preparation. They also introduced the Native Americans to distilled spirits.

■ Carrots escaped vegetable gardens in many areas, reverting to the uncultivated flowering form known as Queen Anne's lace.

Staples

The great diversity of Native American cultures has resulted in a broad variety of cuisines. The cooking of one region was as different from that of another as French food is from German food today. Native American cooking featured local ingredients and often reflected the need to preserve foods for future shortages. The only staple foods common to many, though not all, Native American nations were beans, corn, and squash. The cultural food groups are listed in Table 5.3.

■ Corn is the largest crop grown in the United States today. It is the American staple grain, although most corn is used for sweeteners or animal feed, and is ultimately consumed in the form of soft drinks, meat, and poultry. Some estimates suggest that up to 30 percent of food eaten in the United States derives from corn.

Table 5.3
Cultural Food Groups: Native American

Group	Comments
PROTEIN FOODS	
Milk/milk products	High incidence of lactose intolerance among Native Americans with a high percentage of Native American heritage
Meats/poultry/fish/eggs/legumes	Meat is highly valued, considered healthful. Meats are mostly grilled or stewed, preserved through drying and smoking. Beans are an important protein source.
CEREALS/GRAINS	Corn is primary grain; wild rice is available in some areas.
FRUITS/VEGETABLES	Indigenous plants are major source of calories in diet of some Native American nations. Fruits and vegetables are either gathered or cultivated; fruit is a popular snack food.
ADDITIONAL FOODS	
Seasonings	
Nuts/seeds	Nuts and seeds are often an important food source; acorns are sometimes a staple.
Beverages	Herbal teas often consumed for enjoyment, therapeutic value, or spiritual value.
Fats/oils	Traditional diets vary in fat content, from extremely low in the mostly vegetarian cooking of California and Nevada Indians to very high in the primarily animal-based fare of Native Alaskans.
Sweeteners	Consumption of sweets are low in traditional diets.

Table 5.3
Cultural Food Groups: Native American—continued

Common Foods	Postcontact Adaptation
No common milk products in traditional diets	Powdered milk and evaporated milk are typical commodity products, usually added to coffee, cereal, and traditional baked goods; ice cream is popular with some groups. Some reports have been made of frequent milk consumption.
Meat: bear, buffalo (including jerky, pemmican), deer, elk, moose, oppossum, otter, porcupine, rabbit, raccoon, squirrel. *Poultry and small birds:* duck, goose, grouse, lark, pheasant, quail, seagull, wild turkey *Fish, seafood, and marine mammals:* abalone, bass, catfish, clams, cod, crab, eel, flounder, frogs, halibut, herring, lobster, mussels, olechan, oysters, perch, red snapper, salmon, seal, shad, shrimp, smelts, sole, sturgeon, trout, turtle, walrus, whale *Eggs:* bird, fish *Legumes:* many varieties of the common bean (kidney, navy, pinto, etc.), *tepary* beans	Beef is well accepted; lamb and pork are also popular. Canned and cured meats (bacon, luncheon meat) may be common if income is limited. Game is rarely eaten. Meats remain a favorite food. Chicken eggs are commonly eaten.
Cornmeal breads (baked; steamed), hominy, gruels, corn tortillas, *piki,* toasted corn; wild rice	Wheat has widely replaced corn; store-bought or commodity breads and sugared cereals are common. Cakes, cookies, pastries are popular.
Fruits: blackberries, blueberries, buffalo berries, cactus fruit (tuna), chokeberries, cherries, crab apples, cranberries, currants, elderberries, grapes, groundcherries, huckleberries, persimmons, plums, raspberries, salal, salmonberries, strawberries (beach and wild), thimbleberries, wild rhubarb	Apples became common after European introduction. Apples, bananas, oranges, peaches, pineapple have been well accepted; canned fruits are popular. Wild berries are still gathered in rural areas.
Vegetables: camass root, cacti (*nopales*), chiles, fiddleheads, groundnuts, Indian breadroot, Jerusalem artichokes, lichen, moss, mushrooms, nettles, onions, potatoes, pumpkin, squash, squash blossoms, sweet potatoes, tomatoes, wild greens (cattail, clover, cow parsnip, creases, dandelion, ferns, milkweed, pigweed, pokeweed, saxifrage, sunflower leaves, watercress, winter cress), wild turnips, *yuca* (cassava)	Some traditional vegetables are eaten when available. Green peas, string beans, instant potatoes are common commodity items. Intake of vegetables is low; variety is limited. Potato chips and corn chips often are popular as snacks.
Chiles, garlic, hickory nut cream, onions, peppermint, sage, salt, sassafras, seaweed, spearmint, and other indigenous herbs and spices	
Acorn meal, black walnuts, buckeyes, chestnuts, hazelnuts, hickory nuts, mesquite tree beans, pecans, peanuts, *piñon* nuts (pine nuts), pumpkin seeds, squash seeds, sunflower seeds, seeds of wild grasses	
Teas of buffalo berries, mint, peyote, rose hip, sassafras, spicebush, sumac berries, *yerba buena;* honey and water	Coffee, tea, soft drinks are common beverages. Alcoholism is prevalent.
Fats rendered from buffalo, caribou, moose, and other land mammals; seal and whale fat	Butter, lard, margarine, vegetable oils have replaced rendered fats in most regions; seal and whale fat are still consumed by Inuits and Aleuts.
Maple syrup, other tree saps, honey	Sugar is primary sweetener; candy, cookies, jams, and jellies are popular.

Regional Variations

Native American fare has been divided by regions into five major types: northeastern, southern, plains, southwestern, and Pacific Northwest/Alaska Native. Although each area encompasses many different Native American nations, they share similarities in foods and food habits.

Northeastern. The northeastern region of the United States was heavily wooded, with numerous freshwater lakes and a long Atlantic coastline. It provided the local Native Americans, including the Iroquois and Powhatan, with abundant indigenous fruits, vegetables, fish, and game. Most nations also cultivated crops such as beans, corn, and squash. Many of the foods that are associated with the cooking of New England have their origins in northeastern Native American recipes. The clambake was created when the Narragansett and the Penobscot steamed their clams in beach pits lined with hot rocks and seaweed. Dried beans were simmered for days with maple syrup (the precursor of Boston baked beans). The dish that today is called succotash comes from a stew common in the diet of most Native Americans; it combined corn, beans, and fish or game. In the Northeast, it was usually flavored with maple syrup. Clam chowder, codfish balls, brown bread, corn pudding, pumpkin pie, and the dessert known as Indian pudding are all variations of northeastern Native American recipes.

In addition to clams, the Native Americans of the region ate lobster, oysters, mussels, eels, and many kinds of salt- and freshwater fish. Game, such as deer and rabbit, was eaten when available. Wild ducks, geese, and turkeys were roasted with stuffings featuring crab apples, grapes, cranberries, or local mushrooms. Corn was the staple food. It was prepared in many ways, such as roasting the young ears; cooking the kernels or meal in soups, gruels, and breads; steaming it in puddings; or preparing it as popcorn. Pumpkins and squash were baked almost daily, and beans were added to soups and stews. Local green leafy vegetables were served fresh. Sweets included cherries stewed with maple syrup, cranberry pudding, crab apple sauce, and hazelnut cakes.

Southern. The great variety of foods found in the northeastern region of the United States was matched in the plentiful fauna and lush flora of the South. Oysters, shrimp, and blue crabs washed up on the warm Atlantic beaches during tropical storms. The woodlands and swamplands teemed with fish, fowl, and game, including bear, deer, raccoon, and turtle, as well as ample fresh fruit, vegetables, and nuts. The Native Americans of this region, such as the Cherokee, Creek, and Seminole, were accomplished farmers, growing crops of beans, corn, and squash.

When Africans were first brought as slaves to America, they were often housed at the periphery of farms. Initially a great deal of interaction took place between blacks and local Native Americans, who taught them how to hunt the native game without guns and to use the indigenous plants. Some of the Native American cooking techniques were later introduced into white southern cuisine by African American cooks, and many of the flavors typical of modern southern cooking come from

■ Maple syrup is produced from the sap of the maple tree found in New England and eastern Canada. It takes 40 gallons of sap (the entire year's production from one mature tree) to make 1 gallon of syrup. Maple sugar is made by condensing the syrup further until it solidifies.

■ Dried, ground sassafras was called *filé* (meaning "thread") powder by the French settlers of the South due to the stringy consistency it gave to soups and stews when boiled. Filé powder, which often includes ground thyme in addition to sassafras, is still used in southern cooking today.

Sample Menu

A TRADITIONAL ALGONQUIN MEAL

Maple-sugar baked beans

Smoked trout

Indian pudding or maple popcorn balls

Wild strawberry juice

traditional Native American foods. Hominy (dried corn kernels with the hulls removed) and grits (made of coarsely ground hominy) were introduced to the settlers by the Native Americans. The chicken dish known as Brunswick stew is an adaptation of a southern Native American recipe for squirrel. The Native Americans also made sophisticated use of native plants for seasoning, and they thickened their soups and stews with sassafras. The staple foods of corn, beans, and squash were supplemented with the indigenous woodland fruits and vegetables. Blackberries, gooseberries, huckleberries, raspberries, strawberries, crab apples, grapes, groundcherries, Jerusalem artichokes, green leafy vegetables, persimmons (pounded into a paste for puddings and cakes), and plums were some of the numerous edible native plants. The Native Americans of the South also used beechnuts, hazelnuts, hickory nuts, pecans, and black walnuts in their cooking. The thick, creamlike oil extracted from hickory nuts was used to flavor corn puddings and gruels. Honey was the sweetener used most frequently, and it was mixed with water for a cooling drink. Teas were made from mint, sassafras, or spicebush (*Lindera benzoin*), and during the summer "lemonade" was made from citrus-flavored sumac berries.

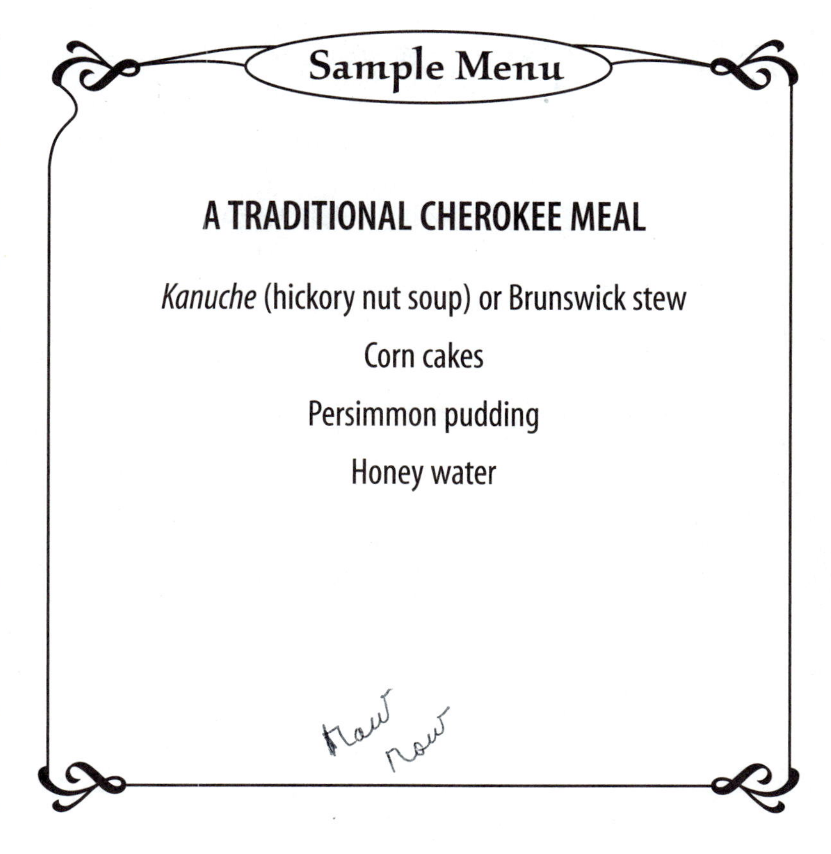

Sample Menu

A TRADITIONAL CHEROKEE MEAL

Kanuche (hickory nut soup) or Brunswick stew

Corn cakes

Persimmon pudding

Honey water

Plains. The Native Americans who lived in the area that is now the American Midwest were mostly nomadic hunters, following the great herds of bison across the flat plains for sustenance. The land was rugged and generally unsuitable for agriculture. Those nations that settled along the fertile Mississippi and Missouri River valleys, however, developed farm-based societies supported by crops of beans, corn, and squash.

Bison meat was the staple food for most plains nations such as the Arapaho, Cheyenne, Crow, Dakota, and Pawnee. The more tender cuts were roasted or broiled, while the tougher ribs, joints, and other bones with marrow were prepared in stews and soups. Pieces of meat, water, and sometimes vegetables would be placed in a hole in the ground lined with cleaned buffalo skin. The stew would then be "stone-boiled": rocks that had been heated in the fire would be added to the broth until the mixture was thoroughly cooked. All parts of the bison were eaten, including the liver and kidneys (which were consumed raw immediately after the animal was slaughtered), udder, tongue, and hump. Extra meat was preserved by cutting it into very thin strips and then dehydrating it in the sun or over the fire. This tough, dried meat would keep for several years and was known as *jerked* buffalo or *jerky*. The jerky would be pulverized and mixed with water or corn gruel, or, in emergencies, eaten dry. Most often it was shredded and mixed with buffalo fat and berries, then formed into cakes called *pemmican*.

When bison were unavailable, the plains nations would hunt deer, rabbit, and game birds. Fresh green leafy vegetables were consumed in season and root vegetables, such as wild onions, prarie turnips (*Psoralea esculenta*, also called breadroot, *tipsin*, and *timpsila*) and Jerusalem artichokes, were eaten throughout the year. Wild rice was collected in the northern parts of the Midwest, where it was served with bison or venison and used as a stuffing for grouse, partridge, and duck. Blackberries, juneberries (also called serviceberries

- A traditional Cherokee soup is *kanuche*, made from pounded hickory nuts.

- The Sioux word for Jerusalem artichokes is *topeka*, also the name of the capital of Kansas.

- The term *jerked* or *jerky* is believed to have come from the Spanish word for meat dried in the sun, *charqui*.

■ Wild rice may have provided 25 percent of the total diet of the Ojibwa Indians.

■ It was up to the Hopi woman to propose marriage to a man. She did this by preparing a blue cornmeal cake called a *piki* and placing it on her intended's doorstep. If the piki was taken into the house, her proposal was accepted.

■ Peaches were planted in the Southwest by Spanish priests. They were especially valued by the Hopi and Navajo, who preserved the fruit by drying it.

or saskatoon berries), cherries, crab apples, grapes, persimmons, and plums were available in some areas, but the most popular fruit was the scarlet buffalo berry (*Shepherdia canadensis*), so called because it was so often served in sauces for bison meat or dried for pemmican. In addition, berries were traditionally boiled with bison suet and/or blood to make a thick pudding called *wojapa*.

Southwestern. Some of the oldest Native American settlements in North America were located along the river valleys of the arid Southwest. Despite the semidesert conditions, many Native Americans such as the Hopi, Pima, Pueblo, and Zuni lived in *pueblo* (Spanish for "town" or "village") communities and were mostly farmers, cultivating beans, chiles, corn, and squash. Others, including the Apache and Navajo, were originally roving hunters and gatherers. After the Spanish introduced livestock, some of these nomadic groups began to raise sheep. Mutton has since become associated as a traditional staple food of the region.

Until the arrival of livestock, the diet of the region was predominantly plant based, providing a nourishing diet when supplemented with small game such as rabbit and turkey. Corn was the primary food and at least five different colors of corn were cultivated. Each color symbolized one of the cardinal points for the Zuni, and each had its own use in cooking. White corn (East) was ground into a fine meal and was used in gruels and breads. Yellow corn (North) was roasted and eaten in kernel form or off the ear. The rarer red (South), blue (West), and black (the nadir) corn was used more for special dishes, such as the lacy, flat Hopi bread made from blue cornmeal known as *piki.* Multicolored corn represented the zenith. The Hopi also attached importance to the color of corn and cultivated 20 different varieties. In many areas, corn was prepared in ways similar to those of the northern Mexican Native Americans—*tortillas* (flat, griddle-fried cornmeal bread), *pozole* (hominy), and the *tamale-like chukuviki* (stuffed cornmeal dough packets). Juniper ash (considered a good source of calcium and iron) was often added to cornmeal dishes for flavoring (Christensen et al., 1998).

Beans were the second most important crop of the southwestern region. Many varieties were grown, including the domesticated indigenous *tepary* beans and pinto beans from Mexico. Both squash and pumpkins were commonly consumed, and squash blossoms were fried or added to soups and salads. Squash and pumpkin seeds were also used to flavor dishes and chiles were used as vegetables and to season stews. Muskmelons were also grown after they were introduced by the Spanish.

When crops were insufficient, the southwestern Native Americans relied on wild plants and to add variety tender amaranth greens were eaten in summer. *Piñon* seeds (also called pine nuts) flavored stews and soups. Both the fruit (*tunas*) and the pads (*nopales*) of the prickly pear cactus were eaten, as were the pulp and fruit of other succulents, such as yucca (the starchy fruit known today as "Navajo bananas"). The beans of the mesquite tree were a staple in some desert regions; they were ground into a flour and used in gruels, breads, and sunbaked cakes.

Northwest Coast/Alaska Natives. This culinary region incorporates a diverse geographic area. The climate of the Pacific Northwest coast is temperate. The luxuriantly forested hills and mountain slopes abound with edible plants and game, and the sea supplies fish, shellfish, and marine mammals. Farther north, in Alaska and Canada, the growing season shortens to only a few summer months and temperatures in the winter regularly plunge to −50 degrees Fahrenheit. Two-thirds of Alaska is affected by permafrost, and the vast stretches of tundra are inhospitable to humans.

The Native American ethnic groups that inhabit this region include Indians as well as Inuits (Eskimos) and Aleuts, known as Alaska Natives. Native American nations such as the Tlingit and Kwakiutl inhabit the northwest coastal area and some interior Alaskan regions. The Aleuts live on the 1,000-mile-long chain of volcanic islands that arch into the Pacific from Alaska called the Aleutians. The Inuits, including the Yupik and Inupiat, live in the northern and western areas of Alaska, as well as in Canada, Greenland, and Siberia.

The Native Americans of the Northwest Coast had no need for agriculture. Food was plentiful, and salmon was their staple. The fish were caught annually in the summer as they swam upstream to spawn. They were roasted over the fire when fresh, and the eggs, known today as red caviar, were a favorite treat when dried in the sun into chewy strips. Extra fish were smoked to preserve them for the winter. In addition, cod, clams, crabs, halibut, herring, shrimp, sole, smelt, sturgeon, and trout were consumed. Ocean mammals such as otter, seal, and whale were also hunted. Bear, deer, elk, and mountain goats were eaten, as were numerous wildfowl and game birds.

Despite the abundant fish and game, wild plants made up more than half the diet of the Northwest Coast Native Americans. More than 100 varieties of indigenous fruits, vegetables, and even lichen were consumed, including acorns, blackberries, blueberries, camass roots (*Camassia quamash*), chokecherries (*Prunus virginiana*), desert parsley, leafy green vegetables, hazelnuts, huckleberries, mint, raspberries, salal (*Gaultheria shallon*), and strawberries. Roots were roasted or baked, and fresh greens were popular.

In contrast to the plenty of the Northwest Coast, the diet of many Alaska Natives was often marginal. The Inuits and Aleuts were usually seminomadic, traveling as necessary to fish and hunt. Fish and sea mammals such as seal, walrus, and whale were the staple foods. Arctic hare, caribou, ducks, geese, mountain goats, moose, musk oxen (hunted to extinction in Alaska by the 1870s), polar bear, and mountain sheep were consumed when available. Some items were boiled, but many were eaten raw due to the lack of wood or other fuel. The fat of animals was especially valued as food. *Muktuk* (also called *muntak* or *onattak* is still a commonly consumed item, consisting of chunks of meat with the layer of fat and skin attached. Muktuk is typically frozen before use. Walrus or whale muktuk can also be preserved by rolling it in herbs (with no salt) and fermenting it in a pit for several months, to make a treat known as

Sample Menu

A TRADITIONAL NAVAJO MEAL

Green chile stew or Pozole

Frybread

Navajo cake or Peach crisp

■ The word *Eskimo* comes from the Algonquin word *eskimantik,* meaning "eater of raw flesh." *Inuit,* meaning "the people," is the name preferred by the nation.

■ Adult salmon, which are normally ocean-dwelling fish, swim up rivers to spawn and then die. The Native Americans of the Pacific Northwest believed that salmon were immortal. It was thought that the salmon came yearly to the rivers to feed the people, then returned to the oceans, where they were reborn. The Native Americans would ceremoniously return the fish skeletons to the river so that this rebirth could occur.

■ In 1995 the Makahs of northwest Washington petitioned the International Whaling Commission for the right to resume the hunting of gray whales (with a limit of five per year). It is widely thought that other Native American coastal groups also will seek approval.

■ Acorns contain tannic acid, a bitter-tasting substance that is toxic in large quantities. To make the acorns edible, Native American women would first crack and remove the hard hull, then grind the meat into a meal, add water to make a dough, and then leach the tannic acid from the dough by repeatedly pouring hot water through it. Acorns were sometimes leached at sandy-bottomed streams as well.

Inuit women preparing a dead seal for butchering. (Corbis/Bettmann.)

kopalchen. *Akutok,* a favored dish, was a mixture of seal oil, berries, and caribou fat. Even the stomachs of certain game were examined for edible undigested foods, such as lichen in elk and clams in walruses. Wild plants were very limited, including willow shrubs, seaweed, mosses, lichen, a few blueberries, salmonberries and cranberries. Leaves from an aromatic bush (*Ledum palastre*) were brewed to make tundra tea (also called Hudson Bay tea), a beverage still popular today.

Sample Menu

A TRADITIONAL YUPIK MEAL

Duck soup or Grilled salmon

Pilot bread

Agutok (Inuit ice cream) with salmonberries

Tundra tea

Other Native American Cuisines. Many traditional Native American diets do not fit conveniently within the five major regional cuisines. Among them was the fare of the population found in what is now Nevada and parts of California, called "Digger Indians" by the first whites to encounter them because they subsisted mostly on dug-up roots, such as Indian breadroot, supplemented by small game and insects. In central California, numerous nations, such as the Miwok and Pomo, had an acorn-based diet. In the rugged northern mountains and plains lived nations such as the Blackfeet, Crow, Shoshone, and Sioux, who were nomadic hunters of game. Although many may have hunted bison at one time, they were limited to the local bear, deer, moose, rabbits, wildfowl, and freshwater fish when the expansion of other Native Americans and whites into the Midwest pushed them northward and westward. Wild plants added variety to their diet.

Meal Composition and Cycle

Daily Patterns

Traditional meal patterns varied according to ethnic group and locality. In the Northeast, one large, hearty meal was consumed before noon, and snacks were available throughout the day. Two meals per day were more common in the Southwest. The women would rise before dawn to prepare breakfast, eaten at sunrise. The afternoon was spent cooking the evening meal, which was eaten before sunset. Two meals per day were also the pattern among the Native Americans of the Pacific Northwest.

In regions with limited resources, meals were often monotonous. The two daily meals of the southwestern Native Americans, for example, regularly consisted of cornmeal gruel or bread and boiled dehydrated vegetables. No distinction was made between morning or evening menus. Other dishes such as game, fresh vegetables, or fruit were included when seasonally available. The single meal of the northeastern Native Americans often included roasted game; the Northwest Coast Native Americans frequently included some form of salmon twice a day in addition to the many local edible greens and roots.

Food was simply prepared. It was roasted over the fire or in the ashes, or cooked in soups or stews. The northeastern and Northwest Coast Native Americans steamed seafood in pits; southwestern Native Americans baked cornmeal bread in adobe ovens called *hornos.* (After the introduction of hogs, flat breads were commonly fried in lard.) Seasonal items were preserved by drying them in the sun or smoking them over a fire; for meat, fish, and oysters, special wood was often used to impart a distinctive flavor. Other foods were ground into a meal or pounded into a paste. In Alaska, meats, greens, and berries were preserved in "fermented" (aged) blubber. All nations liked sweets, but they were limited to fruits and dishes flavored with maple syrup, honey, or other indigenous sweeteners.

Special Occasions

Many Native American religious ceremonies were accompanied by feasts. Among the northeastern Iroquois, seasonal celebrations were held for the maple, planting, strawberry, green corn, harvest, and New Year's festivals. The southern nations held an elaborate Green Corn Festival in thanks for a plentiful summer harvest. No one was allowed to eat any of the new corn until the ceremony was complete. Each home was thoroughly cleaned, the fires were extinguished, and all old pieces of pottery and clothing were replaced with

Baking bread in a southwestern outdoor oven. (Corbis/Bettmann.)

newly made items. The adult men bathed and purged with an emetic. When everything and everyone was thoroughly clean in body and spirit, a central fire was traditionally lit by rubbing two sticks together and each hearth fire was relit with its flames. The feasting on new corn then began. Amnesty was granted for all offenses except murder, and the festival signified the beginning of a new year for marriages, divorces, and periods of mourning.

Role of Food in Native American Culture

Many Native American nations, especially in the inland regions, experienced frequent food shortages. Food is valued as sacred and, in the holistic worldview shared by most Native American ethnic groups, food is also considered a gift of the natural realm. In some nations, elaborate ceremonies accompanied cultivation of crops, and prayers were offered for a successful hunt.

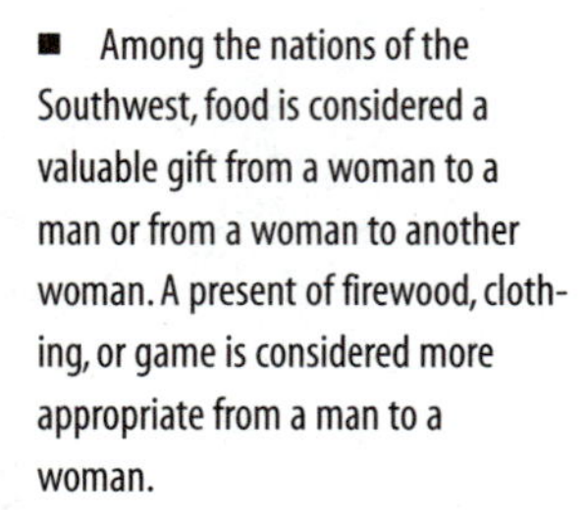

■ Among the nations of the Southwest, food is considered a valuable gift from a woman to a man or from a woman to another woman. A present of firewood, clothing, or game is considered more appropriate from a man to a woman.

■ The Winnebago nation held feasts at the onset of any epidemic in an effort to halt the spread of illness.

■ An Iroquois Thanksgiving prayer illustrates the relationship between food and health: "We return thanks for all herbs, which furnish medicines for the cure of our diseases. We return thanks to the corn, and to her sisters, the beans and squash, which give us life. We return thanks to the bushes and trees, which provide us with fruit."

■ Some Native Americans steeped willow bark as a tea for use as a pain killer. It is high in salicyclic acid, the active ingredient in aspirin.

The men in many nations were traditionally responsible for hunting or the care of livestock. The job of food gathering, preparation, and storage usually belonged to the women; they also made the cooking utensils, such as watertight baskets or clay pots. In predominantly horticultural societies, both men and women were frequently involved in cultivation of the crops. Among the nations of the Northeast the men ate first, followed by women and children. In the Southwest, men prepared the game they caught and served it to the women.

Sharing food is an important aspect of most Native American societies. Food is usually offered to guests, and extra food is often given to members of the extended family. In some nations of the Southwest, meals are prepared and eaten communally. Each woman makes a large amount of one dish and shares it with the other families, who in turn share what they have prepared. Many Native Americans find the idea of "selling" food inconceivable; it is suggested that this is one reason that there are few restaurants featuring Native American specialties.

Therapeutic Uses of Food

The role of food in spiritual and physical health is still important for many Native Americans, and many food plants provide medicine in some form (see Table 5.4). Corn is of significance in some healing ceremonies. Cornmeal may be sprinkled around the bed of a patient to protect him or her against further illness. Corn pollen may be used to ease heart palpitations, and fine cornmeal is rubbed on children's rashes. Navajo women drink blue cornmeal gruel to promote the production of milk after childbirth, and Pueblo women use a mixture of water and corn ear smut (*usti lago maydis,* a kind of fungus) to relieve diarrhea and to cure irregular menstruation. A similar drink was given to Zuni women to speed childbirth and to prevent postpartum hemorrhaging. Corn silk tea was used as a diuretic and was prescribed for bladder infections.

Numerous other indigenous plants are used by Native Americans for medicinal purposes. For example, agave leaves (from a succulent common in the Southwest) are chewed as a general tonic, and the juice is applied to fresh wounds. Pumpkin pastes soothe burns. Chile peppers are used in compresses for arthritis and applied directly to warts. Infusions are used for many remedies, such as wild strawberries for diarrhea and mint tea to ease colic, indigestion, and nausea. Traditionally, maple sugar lozenges were used for sore throats. Bitter purges and emetics are administered because they are distasteful and repugnant to any evil spirits that might cause illness (Vogel, 1981).

Food restrictions are still common during illness. Depending on the nation, many Native Americans believe that cabbage, eggs, fish, meat, milk, onions, or organ meats should be eliminated from a patient's diet. Conversely, some foods may be considered important to maintain strength during sickness, such as meat among the Seminole in Florida (Joos, 1980) and both meat and blue cornmeal among the Navajo (Schilling & Brannon, 1986). Some foods are prohibited after childbirth, such as cod, halibut, huckleberries, and spring salmon for Nootka women of the Northwest Coast.

Many plants were found by the Native Americans to have psychotherapeutic properties. They were used to relax and sedate patients, to stupefy enemies, and to induce hypnotic trances during religious ceremonies. The opiates in the roots of California poppies

Table 5.4
Selected Native American Botanical Remedies

Scientific Name	Common Name	Parts	Traditional Use
Apocynum spp. ☠	Bitter root; dogbane; Indian hemp; black hemp; flytrap	Root	Worms; weak heart; venereal disease; kidney conditions; liver problems; gall stones; gout; purgative; contraceptive
Arctium lappa	Burdock	Whole plant	Kidney disorders; gout; hypercholesterolemia; blood toxins
Arctostaphylos uva-ursi ☹	Bearberry; kinnikinnick; bear's grape; mountain cranberry	Leaves; fruit	Diabetes; urinary tract infections; kidney diseases (esp stones); diuretic; smoked as tobacco substitute
Asclepias syriaca ☠	Milkweed; emetic root; snake milk; milk ipecac	Root	Rheumatism; dyspepsia; blood conditions; contraceptive; rashes
Baptisia tinctoria ☠	Wild indigo; horesfly weed; yellow broom; milkweed	Root; leaves	Malaise; blood problems; kidney conditions
Caulophyllum thalictroices ☠	Blue cohosh; beechdrops; papoose root; squawroot; blue ginseng; yellow ginseng	Root	Menstrual cramps; childbirth; colic
Chenopodium ambrosioides ☠	Wormseed; Mexican tea	Leaves; seeds	Parasites
Chimaphila umbellata	Pipsissewa; ground holly; wintergreen	Whole plant	Diabetes; urinary tract infections; kidney problems; TB; rheumatism
Cimicifuga racemosa ☹	Black cohosh; snakeroot; squawroot; rattleroot; bugbane	Root	Menstrual problems; childbirth; cough; diarrhea; kidney problems
Datura stramonium ☠	Jimsonweed; thorn apple; stinkweed; angel's trumpet	Leaves; seeds	Snake bite; spider bite; narcotic; rheumatism; stomach problems
Echinacea angustifolia ☹	Purple coneflower; black sampson; niggerhead	Roots; flowers	Blood toxins; snake bite; rabies; toothache; sore throat
Ephedra nevadensis	Mormon tea; joint-fir	Leaves	Diuretic; headache; lung congestion; kidney conditions; venereal diseases
Eriodictyon californicum	Yerba santa	Leaves	Malaise; colds; coughs; asthma
Eupatorium purpurea ☹	Gravel root; joe-pye weed; feverwort; boneset; Indian sage; queen of the meadow; agueweed	Root	Urinary tract infections; gallstones; bladder stones; kidney problems; edema; diuretic; flu
Fagus spp.	Beech; beechnut	Bark; root	Diabetes; dysentery, stomach ulcers; liver problems; kidney conditions; anorexia; rheumatism; TB
Geranium maculatum	Cranesbill; crowfoot; dovesfoot; alum root; stork's bill	Root	Dysentery; edema; infected gums, oral thrush
Gillenia trifoliata	American ipecac; dime-a-bottle plant; Indian physic	Root	Dyspepsia; menstrual difficulties; rheumatism, edema; emetic; fever
Glecoma hederacea	Alehoof; ground ivy; cat's foot	Leaves	Diabetes; blood toxins; diuretic; kidney aliments; liver problems
Heuchera spp.	Alum root; American sanicle; coral bells	Root	Diabetes; diarrhea; malaise; fever

Table 5.4
Selected Native American Botanical Remedies—continued

Scientific Name	Common Name	Parts	Traditional Use
Hydrastis canadensis ☹	Goldenseal; yellow root; yellow puccoon; ground raspberry	Root	Stomach problems; kidney ailments; liver conditions; skin ulcers; eye wounds
Kalmia latifolia ☠	Calico bush; mountain laurel; spoonwood; lambkill; leatherleaf	Leaves	Diarrhea; dysentery; liver conditions; edema; venereal diseases; rheumatism; suicide
Lobelia inflata ☠	Indian tobacco; emetic weed; asthma weed; gagroot	Leaves	Emetic; venereal diseases; bronchial spasms; asthma; earaches; smoked as mild stimulant
Panax quinquefolium	American ginseng; five finger root; fivefinger; gift of the gods; redberry; sang	Root	Fertility; menstrual disorders; digestive problems; aphrodisiac
Phoradendron juniperinum ☠	Juniper mistletoe	Fruit	Stomach disorders; diarrhea; baldness
Phytolacca americana ☠	Poke; pokeweed; weed; poke salat; pigeonberry; osa'	Root; flowers; fruit	Swollen glands (lymph, thyroid disorders); blood toxins; liver insufficiency
Pinus strubus ☹	White pine	Bark; resin	Colds; sore throats; dyspepsia; kidney disorders; hemorrhage; coughs; TB; rheumatism
Podophyllum peltatum ☠	Mandrake; mayapple; hog apple; Indian apple; raccoonberry	Root	Purgative; emetic; parasites; stomach problems; urinary tract disorders; liver complaints
Populus spp.	Poplar; quaking ash; balm of gilead; lemon balm; bitterwood	Bark; buds	Cold; flu; parasites; chest pain; rheumatism; malaise
Psilocybe spp. ☠	Blue halo; blue-foot; liberty cap; wavy cap; San Isidro; psilocybin mushroom	Fungus	Muscle pain; hallucinogen
Rhus spp.	Sumac; sumach	Bark; fruit	Diabetes; diuretic; diarrhea; cough; colds
Salix alba ☹	White willow	Bark	Inflammation; fever; headache; malaria; rheumatism
Salix nigra ☹	Black willow	Bark	Anaphrodisiac; premature ejaculation
Sanguinaria canadensis	Bloodroot; Indian plant; red puccon	Root	Blood conditions; dyspepsia; liver ailments; cough; asthma; TB
Scutellaria lateriflora ☠	Skullcap; blue skull; hoodwort; mad dog weed	Whole plant	Anxiety; insomnia; menstrual problems; childbirth; rabies
Seronoa serrulata ☹	Saw plametto; fan palm; sabal	Fruit	Aphrodisiac; digestive disorders; weakness
Solanum dulcamara ☠	Bitter sweet; common nightshade; climbing nightshade	Bark	Swollen glands; venereal diseases; skin conditions; amenorrhea
Salanum nigrum ☠	Black nightshade; garden nightshade; bull nettle; Carolina horsenettle	Leaves	Sedative; narcotic; toothache, eye wash
Stigmata maydis; zea mays	Corn silk	Stamens	Diuretic (during pregnancy); kidney and liver weakness; kidney stones; urinary tract disorders

Table 5.4
Selected Native American Botanical Remedies—continued

Scientific Name	Common Name	Parts	Traditional Use
Ustilago maydis	Maize mushroom; corn smut; huitlacoche	Fungus	Parturient; uterine hemorrhage
Vaccinium myrtillus	Bilberry; huckleberry; whortleberry	Leaves; fruit	Stomach conditions; diarrhea
Viburnum opulus	High-bush cranberry; crampbark; snowball tree	Bark	Muscle pain; asthma; pregnancy; edema; hysteria
Zanthoxylum americanum	Prickly ash; toothache tree	Bark; fruit	Blood tonic; arthritis; rheumatism; toothache

Note: Data on some plants are very limited; adverse effects may occur even if not indicated.
Key: ☠, all or some parts reported to be harmful or toxic in large quantities or specific preparations; ☹, may be contraindicated in some medical conditions/with certain prescription drugs.

dulled the pain of toothache, for example. Lobelia was smoked as an antispasmodic for asthma and bronchitis. In the Southwest, knobs from the peyote cactus were used to produce hallucinations, sometimes in combination with other intoxicants. Jimsonweed (Datura stramonium) was used to keep boys in a semiconscious state for 20 days so that they could forget their childhood during Algonquin puberty rites (Emboden, 1976).

Contemporary Food Habits

Native American ethnic identity is changing. Traditional beliefs and values are often in direct conflict with those of the majority society, and Native Americans' self-concept has undergone tremendous changes in the process of acculturation. A 1978 study described three transitional adaptations of members of the Ojibwa (also called Chippewa), identified by three different lifestyles; these may serve as a model for the adaptations made by members of other Native American groups (Primeaux & Henderson, 1981). The first stage of adaptation is traditional in which parents and grandparents speak the Ojibwa language at home, practice the *Midewiwin* religion, and participate in Native American cultural activities such as feasts and powwows. The second stage is more acculturated. English is the primary language, although some Ojibwa also is spoken. Catholicism is the preferred religion, and the family is involved in activities of the majority society. In the third Ojibwa lifestyle, the pan-traditional stage, the family speaks either English or Ojibwa exclusively; practices a religion that is a combination of Native American and Christian beliefs, such as the Native American Church; and is actively involved in activities from both traditional Native American and white societies.

Adaptation of Food Habits

Food habits reflect changes in Native American ethnic identity. Many Native Americans eat a diet that includes few traditional foods. Others are consciously attempting to revive the foods and dishes of their ancestors.

Ingredients and Common Foods

When Native Americans were uprooted from their lands and their known food supplies, many immediately became dependent on the foods provided to the reservations. One study evaluating the diets of Havasupai Indians living on a reservation in Arizona found that 58 percent of the subjects ate only foods purchased or acquired on the reservation during

■ Native American food is becoming trendy, especially the fare of the Southwest. In 1984 the late James Beard, considered the father of American gastronomy, wrote a glowing article praising the delicious dishes served at a New York celebration of Native American cuisine.

■ A study that examined Dogrib food habits in acculturating communities in Canada found that new food items are added to traditional diets rather than substituted for traditional foods, resulting in an increase in total calorie intake compared to less acculturated communities (Szarthmay et al., 1987).

the 24-hour recall period (Vaughn et al., 1997). Commodity foods currently include items such as canned and chopped meats, poultry, fruit juices, peanut butter, eggs, evaporated and powdered milk, dried beans, instant potatoes, peas, and string beans. Researchers report that many of these foods such as kidney beans, noodles, and peanut butter are discarded by the Navajo; powdered milk may also be rejected because it is disliked, or it is considered a "weak" food suitable only for infants or the elderly (Kunitz & Levy, 1981). Other sources of food include gardening (reportedly practiced by between 43 and 91 percent of rural Native Americans), fishing, hunting, gathering indigenous plants, and raising livestock. One study of California Miwok, for instance, reports that 67 percent of respondents recall that their grandparents harvested wild greens, nuts, berries, and mushrooms, and that 47 percent of respondents continue to supplement their diet this way (Ikeda et al., 1993). Among Baffin Inuit, traditional foods (e.g., sea and land mammals, fish) make up about one-third of energy intake. Men were found to eat more traditional items than women (Kuhnlein, 1995).

Over the years, traditional foods and methods of preparation were lost and substitutions were made. For example, beef is a commonly accepted substitute for game among many Native American ethnic groups. Traditional foods compose less than 25 percent of the daily diet among the Hopi. Older Hopi women lament the fact that younger Hopi are no longer learning how to cook these dishes (Kuhnlein et al., 1979). In a study of Cherokee women (Terry & Bass, 1984) the consumption of traditional foods was found to be directly related to the amount of Native American inheritance of the woman in charge of the food supply in the home. Corn and corn products such as hominy were among the most popular traditional foods; game meat, hickory nuts, raspberries, and winter squash were the least commonly served. Among Cherokee teenagers, traditional items such as fry bread, bean bread (corn bread with pinto beans), and chestnut bread (made from chestnuts and cornmeal) were well accepted. Although more than 80 percent of the adolescents were familiar with typical Cherokee dishes, including native greens and game meat (bear, deer, groundhog, rabbit, raccoon, squirrel, and wild boar), these foods were rarely eaten (Storey et al., 1986). Pima consume traditional items, such as tepary beans and cactus stew, mostly at community get-togethers (Smith et al., 1996). California Miwok list mostly southwestern items such as beans, rice, and tortillas as those they most associate with Native American foods and recall numerous items eaten by their grandparents but not consumed now, such as squirrels, rabbit, deer, acorn mush, and certain insects. Access to wild game is limited due to hunting restrictions.

Navajo women eat traditional foods infrequently, with the exception of fry bread, mutton, and tortillas. Blue cornmeal mush (with ash), hominy, and sumac berry pudding are a few of the native dishes consumed occasionally (Wolfe & Sanjur, 1988). Dakota women of all ages take pride in traditional foods but prepare them only when it is convenient or for special occasions. Uncooked meat is consumed in some nations; it has been noted that some Native Americans do not care about the temperature of foods (Bass & Wakefield, 1974).

■ A national Canadian survey reported that 66 percent of aboriginal peoples obtained some of their meat, poultry, and fish through hunting and fishing; 10 percent obtained most; and 5 percent obtained all this way (Wein, 1995).

■ Fry bread, a flat bread made from wheat flour and typically cooked in lard, has been prepared in the Southwest for about 100 years. Although made from ingredients introduced by the Europeans, it is frequently identified as a traditional food.

■ Some Native Americans use the phrase "way back foods" to differentiate between indigenous traditional foods and those developed after European contact.

■ Meat portions among northern Plains Indians is reportedly large, typically 5 to 8 ounces per serving.

■ A Native American hotel chef, George Crum, is attributed with the invention of potato chips in 1853. Today Americans consume an average of 17 pounds of potato chips per person each year.

Meal Composition and Cycle

Little has been reported regarding current Native American meal patterns. It is assumed that three meals per day has become the norm, especially in families without income constraints. One or two main meals is reportedly the norm for northern Plains Indians (Woolf et al., 1999) and irregular meals have been described among Alaska Natives (Halderson, 1998).

One study of Dakotas noted that there was little variety in daily menus (Bass & Wakefield, 1974). Cereal with milk; fried potatoes; fried eggs, chopped meat, milk, and bread with butter; oatmeal or cornmeal mush; or just bread with butter was eaten for breakfast. Coffee was the most popular beverage. Lunch was the main meal, consisting of bologna sandwiches with potato chips and a carbonated beverage or just fried potatoes. Fried potatoes and chopped meat were eaten for dinner. Other northern Plains Indians enjoy soups or sandwiches at lunch, with a large meal of fried meats, potatoes with gravy, canned vegetables, and fruit for dinner (Woolf et al., 1999). Navajo women were found to eat fry bread or tortillas, potatoes, eggs, sugar, and coffee were most frequently consumed. Fried foods were preferred for breakfast, and lunch and dinner consisted of one boiled meal and one fried or roasted meal. Pima in Arizona prefer eggs, bacon or sausage, and fried potatoes for breakfast, while Southwest specialties, such as tacos, tamales, and chili con carne, are common at other meals.

Snacking is common in some Native American groups, including fruit, soft drinks, potato or corn chips, sunflower seeds, crackers, cookies or cake, candy, and ice cream (Halderson, 1993; Ikeda et al., 1993; Woolf et al., 1999). Eating at fast-food restaurants is popular among the Navajo and some younger Plains Indians (Wolfe & Sanjur, 1988; Woolf et al., 1999) but is infrequent among the Pima (Smith et al., 1996).

Special Occasions

Numerous traditional celebrations are maintained by Native American tribal groups (Kavasch, 1995). Among the largest is the 5-day Navajo Nation Fair held each Labor Day weekend; the Pawnee Veteran's Day Dance and Gathering where ground meat with pecans and corn with yellow squash are served; the Miccosukee Arts Festival and the Seminole Fair in Miami, where alligator meat is featured; the Iroquois Midwinter Festival held in January to mark the new year; the Upper Mattaponi Spring Festival in Virginia over Memorial Day Weekend; the 3-day Creek Nation Festival and Rodeo; the Yukon International Storytelling Festival at which wild game such as caribou and musk ox are available; and the Apache Sunrise Ceremony which features gathered foods such as amaranth leaves and the pulp and fruit from the saguaro and prickly pear cacti. More local festivities are also common. Pueblo Feast Days are observed in honor of the Catholic patron saint of each village with a soup of posole (hominy) and beef or pork ribs. Northwest Coast potlatches are common in the spring, featuring herring roe, fish or venison stews, euchalon, salmon, and other traditional foods. In addition, there are all-Indian festivals which draw attendees from throughout the country, such as O'Odham Tash Indian Days in Casa Grande Arizona, the Red Earth Festival in Oklahoma City, and the Gallup Intertribal Indian Ceremonial in New Mexico. Native Americans may also eat traditional foods on special occasions such as birthdays, but for holidays of the majority culture, other foods are considered appropriate. For example, turkey with all the trimmings is served by Dakotas for Thanksgiving and Christmas (Bass & Wakefield, 1974).

Nutritional Status

Nutritional Intake

Research on the nutritional status of Native Americans is limited. Severe malnutrition was documented in the 1950s and 1960s, including numerous cases of kwashiorkor and marasmus. today, low socioeconomic status and lack of transportation, fuel, refrigeration, and running water contribute to an inadequate diet for many Native Americans.

The few studies of what some Native Americans now eat suggest that a diets high in refined carbohydrates (starchy and sugary foods) and fat, and low in fruits, and vegetables is common. For example, the caloric and protein intake of Alaska Natives declined during the past 15 to 20 years, as the traditional diet was replaced by processed, canned, and packaged foods that are low in nutrients (Jackson, 1986). It is estimated that the carbohydrate content of the Alaska Native diet before contact with westerners was exceptionally low (3 to 5 percent of daily calories) due to a dependence on sea mammals and fish. In 1978, that figure had increased to 50 percent of total calories, much of it from low nutrient density foods (Murphy et al., 1995). Low intake of calcium, iron, phosphorus, vitamins A, C, and D, and riboflavin, as well as zinc and fiber, have been reported. Traditional Alaska Native diets have been found higher in protein, phosphorus, iron, zinc, copper, magnesium, and vitamin A (Kuhnlein et al., 1996). Physicians in Canada report a northern infant syndrome with clinical symptoms of hepatitis, respiratory distress, rickets, and hemolytic anemia that is likely due to vitamin deficiencies (Godel & Hart, 1984).

Although life expectancy for Native Americans has increased in recent years, it is still lower than that of the overall U.S. population. The age-adjusted mortality rates for Native

- The adoption of the new Cherokee Nation constitution (following their forced dislocation) is commemorated on September 6. The event honors reunification of all Cherokees and is celebrated over the Labor Day weekend with a State of the Nation address, parade, arts and crafts, and games.

- High rates of lactose intolerance have been reported for both Native Americans and Alaska Natives. The occurrence among individuals, however, is related to percentage of Native American heritage.

- A diet high in simple carbohydrates has led to an increase in dental caries. This problem was virtually unknown among Native Americans before European contact.

- Recent studies looking at specific consumption patterns have reported low intake of fruit and vegetables. Of Native Americans living in California, 60 percent said they had not eaten any fruit the previous day, and 28 percent reported they had not consumed any vegetables. Among the Lakota of South Dakota, nearly 60 percent stated they ate fruit only two to eight times per month; nearly half reported eating vegetables (including potatoes) over five times weekly. Barriers to increased consumption of fruit and vegetables included cost, availability, and quality (Harnack, 1999; Ikeda, 1998).

Americans are nearly 50 percent higher, and one-third of Native Americans die before the age of 45, compared to just 11 percent of people in the total U.S. population.

The birthrate among Native Americans is nearly 50 percent higher than for the total U.S. population and among Alaska Natives the rate is more than double the average U.S. rate. Native American mothers are more likely to be younger than in the general population and less likely to be married; yet the number of low-birth-weight infants is slightly lower than that for the total U.S. population. Infant mortality rates are 55 percent above the U.S. average; maternal death rates are more than four times that of all U.S. mothers.

Breast-feeding has traditionally been considered the proper way to feed infants among the Navajo. Elders in one study reported that breast-fed infants were better able to hear traditional teachings and were better disciplined. It also shows that the children are loved (Wright et al., 1993). Eighty-one percent of subjects initiated breast-feeding. Most added infant formula within the 1st week and used this combined feeding practice for more than 5 months. Other estimates of breast-feeding show rates of 24 to 44 percent (Story et al., 1998). Baby-bottle tooth decay, due most often to extended use of a bottle with formula, milk, juice, or soda affects more than one-half of all Native American and Alaska Native children.

■ White bread, french fries, and milk contributed almost 20 percent of calories in a 24-hour recall completed by a sample of Mohawk children. Sugar intake was also high (Trifonopoulos et al., 1998).

■ Chronic conditions complicating diabetes were found more prevalent among northern Plains Indians than among the general population including hypertension and end-stage renal disease (Woolf, et al., 1999).

Obesity is prevalent among Native Americans. Comprehensive studies of school children and adolescents show consistently higher weight for height ratios than for white, black, or Hispanic populations, with rates for obesity being as high as one-third for girls and nearly one-half for boys (Bernard et al., 1995; Jackson, 1993; Storey et al., 1986b, 1998). A high incidence of obesity among adult Native Americans has also been reported: 63 percent of Navajo sampled were more than 120 percent of their ideal body weight; 83 percent of Havasupai subjects and 60 percent of Seminoles were identified as obese; and 75 percent of Native Americans from Oklahoma nations averaged 145 percent of their ideal body weight (Jackson, 1986; Vaughen et al., 1997).

Figures on obesity contradict some nutritional intake data, which show the caloric intake of many Native Americans to be normal or less than the recommended dietary allowances. Some metabolic differences in obese Native Americans have been demonstrated (Howard et al., 1991), and researchers have suggested that a hereditary adaptation to a feast-and-famine existence in the past may account for this tendency to gain weight, exacerbated in some cases by a sedentary lifestyle (Bass & Wakefield, 1974). However, a study of Hualapai women of Arizona indicated that both nonobese and obese subjects consume adequate energy, and that the daily caloric intake of obese women was significantly higher than for the nonobese women. Sweetened beverages and alcoholic beverages accounted for the differences (Teufel & Dufour, 1990). Dieting behaviors among adult Native American women mostly involve healthy approaches such as eating more fruits and vegetables and exercising more, according to one study, although skipping meals, fasting, and disordered eating such as self-induced vomiting were also mentioned; 10 percent engaged in binge eating. A national survey of Native American youth revealed that over 40 percent reported binge-eating behavior and vomiting rates of 4 to 6 percent. Frequent dieting was also common. All disordered eating occurred more frequently among overweight respondents (Neumark-Sztainer, 1997; Sherwood et al., 2000).

The incidence of non-insulin-dependent diabetes mellitus, especially among some Native Americans of the Plains and Southwest, is estimated to be between three and four times that of the general population; the Pima exhibit the highest rates of the disease in the world, affecting nearly 50 percent of the adult population as they age. The death rate from the disease is more than twice as high for Native Americans as for non–Native Americans. Notably, diabetes was rare among Native Americans 50 years ago (Parker, 1994). Higher rates of diabetes are found among Alaska Natives who have significant increased intake of nonindigenous protein (i.e., beef, chicken), carbohydrates (i.e., white bread, potatoes or rice, soft drinks), and fat (i.e., butter, shortening), combined with a lower intake of native foods such as salmon, caribou, berries, and seal oil. One theory for the high rates of diabetes among Native Americans is the dietary change from indigenous starches (i.e., corn, beans, acorns) to the refined flours and sugars of the adapted diet. Traditional starches are harder to digest and absorb, leading to lower blood sugar levels and insulin responses that may be protective in the development of diabetes (Brand et al., 1990). Other researchers suggest the increased intake of fat in the modern Native American diet may be responsible (Swinburn et al., 1991).

The prevalence of heart disease, although still less than that of the total U.S. population, has greatly increased during the 25 years among Native Americans and is now the leading cause of death. Increases in hypertension, a major risk factor in the development of heart diseases, have also been noted. Both chronic conditions may be attributable to dietary changes approximating the diet of the majority culture and high rates of obesity. Elderly urban Native Americans are especially at risk (Kramer, 1992).

The incidence of alcoholism among Native Americans has dramatically decreased in the recent years, but it still remains a significant medical and social problem. High unemployment rates and loss of tribal integrity, ethnic identity, and self-esteem are frequently cited as reasons for substance abuse among both reservation and urban Native Americans. The rate for alcohol-related deaths is more than five times that of the general U.S. population.

Respiratory infections continue to occur among Native Americans more frequently than among non–Native Americans, probably because of factors such as malnutrition, overcrowding, and inadequate sanitation. Death rates from tuberculosis, pneumonia, and influenza are much higher than those of the total U.S. population. Gastrointestinal infections, particularly gastroenteritis and bacillary dysentery, are also prevalent.

Counseling

A survey of Native American nurses identified attitudes, skills, and knowledge needed to serve Native Americans successfully. They listed being open-minded, avoiding ethnocentrism, and using intercultural communication skills, especially the ability to listen carefully and to provide respectful silence. Learning about Native American worldview, traditional health beliefs, differences between nations, and the history of each group was considered essential to effective interaction (Weaver, 1999). Of particular importance is recognition of diversity within Native American groups and understanding of local culture.

Access to biomedical health care may be limited for some Native Americans because of low income or inadequate transportation. Others may hold beliefs that cause them to avoid biomedical treatment; pregnancy is often considered a healthy state, for example, and Native American women may not seek prenatal care for this reason. Some older Native Americans report fear of non–Native American providers, and others find biomedical physicians too "negative" because of impersonal care. Disclosure of risks may also be regarded as negative, in violation of a positive approach to life (Carrese & Rhodes, 1995, 2000). Some Native Americans may be angry about their condition, such as having diabetes, blaming it on a Western conspiracy. One social worker reports that her clients at a dialysis center in Arizona are so accustomed to friends and family members with kidney failure that they fatalistically expect a similar outcome for themselves and are sometimes relieved when it finally happens (Juarez, 1990). Research shows that many Native Americans believe that renal failure, amputations, and blindness are inevitable consequences of diabetes (Parker, 1994). Traditional attitudes about time may cause delays in seeking care—the importance of finishing a current project or commitment may outweigh that of keeping an appointment with a health provider.

In general, both verbal and nonverbal communication with Native Americans must take into account cultural traditions. Among the Navajo, the patient should be asked directly for his medical history, as even family members may believe that that have no right to speak for another. It is estimated that more than one-half of Native Americans speak their native language in addition to English. Native American languages are primarily verbal, and some Native Americans may experience difficulties with written information or instructions. Further, some English words, such as "germ," may not exist in Native American languages (Plawecki et al., 1994). It has been reported, for example, that increasing

■ Research suggests that a diet high in fish oils that contain large amounts of omega-3-fatty acids may lower serum cholesterol and triglyceride levels, thus reducing the risk of heart disease. The low rate of cardiovascular illness among Inuits is attributed to the abundant amount of these fatty acids in their traditional diet.

■ Nearly one in every four Native Americans believe their general health is "fair" or "poor" (Denny & Taylor, 1999).

■ A study of how diabetes affects daily living among Native Americans reported on the tremendous toll caused by the disease in some nations. One respondent stated his concern for the future of his people when so many were confined to wheelchairs or blinded by the condition (Parker, 1994)

■ Some Native Americans believe diabetes is caused by eating too much sugar.

■ The Indian Health Service (IHS) has a federal trust to provide health care to all Native Americans who are members of tribes recognized by the U.S. government. It began operation under the BIA in 1924 and was placed under what is now called the Department of Health and Human Services in 1955. They operate 43 hospitals and work in collaboration with state, tribal, and private health care facilities to provide comprehensive services. Under the Indian Self-Determination and Education Assistance Act of 1975, tribes are given the option of staffing and managing IHS programs in their communities. However, numbers of Native American health professionals are limited, and intercultural care is the norm rather than the exception in many facilities.

■ Alaska Natives suffer the highest incidence of botulism in the world, due mostly to the way in which indigenous foods, such as seal, are butchered on the ground, then stored for long periods in plastic bags underground (Segal, 1992).

vegetables in the diet of some Native Americans has met with resistance because the closest equivalent Native American word for vegetable is "weeds" (Jackson & Broussard, 1987). Many Native Americans are comfortable with periods of silence in a conversation, using the time to compose their thoughts or to translate responses. A yes or no response may be considered a complete answer to a question. A Native American may answer "I don't know" if she thinks that a question is inappropriate or does not wish to discuss the topic. Information may be withheld until she feels she can trust the provider. She may not ask questions during an interview because that would suggest that the health care provider was not communicating clearly. Among the Navajo, direct questioning by the provider suggests that the practitioner is unknowledgeable or incompetent. The very concept of a dietary interview may be interpreted by some Native Americans as interference with their personal autonomy. Quiet, unhurried conversations are most conducive to successful interaction.

Nonverbal communication is very sophisticated among some Native Americans. It has been reported that children may not be taught to speak until other senses are developed (Wilson, 1983). A Native American client may expect the practitioner to intuit the problem through nonverbal techniques rather than though an interview. Although a smile and a handshake are customary, a vigorous handshake may be considered a sign of aggressiveness. Direct eye contact may be considered rude, and the health professional should not interpret averted eyes as evidence of disinterest.

A client may have misconceptions about biomedicine. Because the Navajo health system classifies illness by cause rather than by symptom, clients may have difficulty understanding the necessity of a physical exam or medical history. There may also be the expectation that medication can cure illness and that an injection is needed for every disorder (Kunitz & Levy, 1981).

Very low compliance rates among some Native Americans on special diets have been reported (Jackson & Broussard, 1987). A Native American may be confused about recommendations for weight loss because slimness is associated with disease or witchcraft. Native American clients may follow traditional health practices in conjunction with biomedical therapies. One study of urban Native Americans found that 28 percent practiced traditional Native American medicine; researchers speculate that use of herbs, special food items, teas, and purification rituals may be even more prominent in rural regions and on reservations (Fuchs & Bashshur, 1975). At one Navajo reservation it was found that 90 percent of clients visiting a biomedical clinic had first sought traditional care (Wilson, 1983). Traditional Native American healers, such as herbalists, and medicine men and women, or shamans, often will decline to treat conditions unfamiliar to them and will refer a patient to a biomedical health care provider for treatment of any "white man's disease" (Jackson & Broussard, 1987). Compliance is most effective when traditional practices are accepted and encouraged as an integral part of complete care.

At in-depth interview is necessary to determine not only ethnic identity but also degree of acculturation. Traditional medical beliefs and customs, if practiced, should be acknowledged. Personal dietary preferences are of special importance due to the variety of Native American foods and food habits. Note taking may be considered exceptionally rude, so memory skills or a tape recorder may be preferable during the interview.

6

Northern and Southern Europeans

Some of the largest American ethnic groups come from northern and southern Europe (see the map in Figure 6.1). Immigrants from these regions began arriving in what is now the United States in the 16th century and are still coming, significantly influencing American majority culture. Today, many of what we consider to be American foods and food habits were introduced by these settlers. The northern European idea of a meal, consisting of a large serving of meat, poultry, or fish with smaller side dishes of starch and vegetable, was quickly adopted and expanded in the United States to include even bigger portions of protein foods. Adaptations of some southern European specialties have become commonplace American fare. Each ethnic group from northern and southern Europe has brought a unique cuisine that combined with indigenous ingredients; blended with the cooking of Native Americans, other Europeans, and Africans; and flavored with the foods of Latinos, Asians, and Middle Easterners to form the foundation of the typical American diet. This chapter discusses the traditional foods and food habits of Great Britain, Ireland, France, Italy, Spain, and Portugal and examines their contributions to the cooking of the United States.

Northern Europeans

Great Britain includes the countries of England, Scotland, Wales, and Northern Ireland. Ireland is now a sovereign country. Although quite northern, the climate is temperate due to the warming influence of the Gulf Stream, and the lowlands are arable.

Just across the English Channel is France, regarded for centuries as the center of western culture politically, as well as in the arts and sciences. Its capital, Paris, is one of the world's most beautiful and famed cities. France contains some of the best farmland in Europe, and three-fifths of its land is under cultivation. It is especially well known for premium wine production.

In 1607, people from Great Britain began immigrating to what has become the United States. They brought with them British trade practices and the English language, literature, law, and religion. By the time the United States gained independence from Britain, the British and their descendants constituted one-half of the American population. They produced a culture that remains unmistakably British-flavored, even today. The French came to the United States later, and in smaller numbers, yet have made significant regional contributions. The traditional foods of northern Europe and their influence on American cuisine are examined in this next section.

Cultural Perspective

History of Northern Europeans in the United States

Immigration Patterns

Great Britain. The British who immigrated in the 17th century settled primarily in New

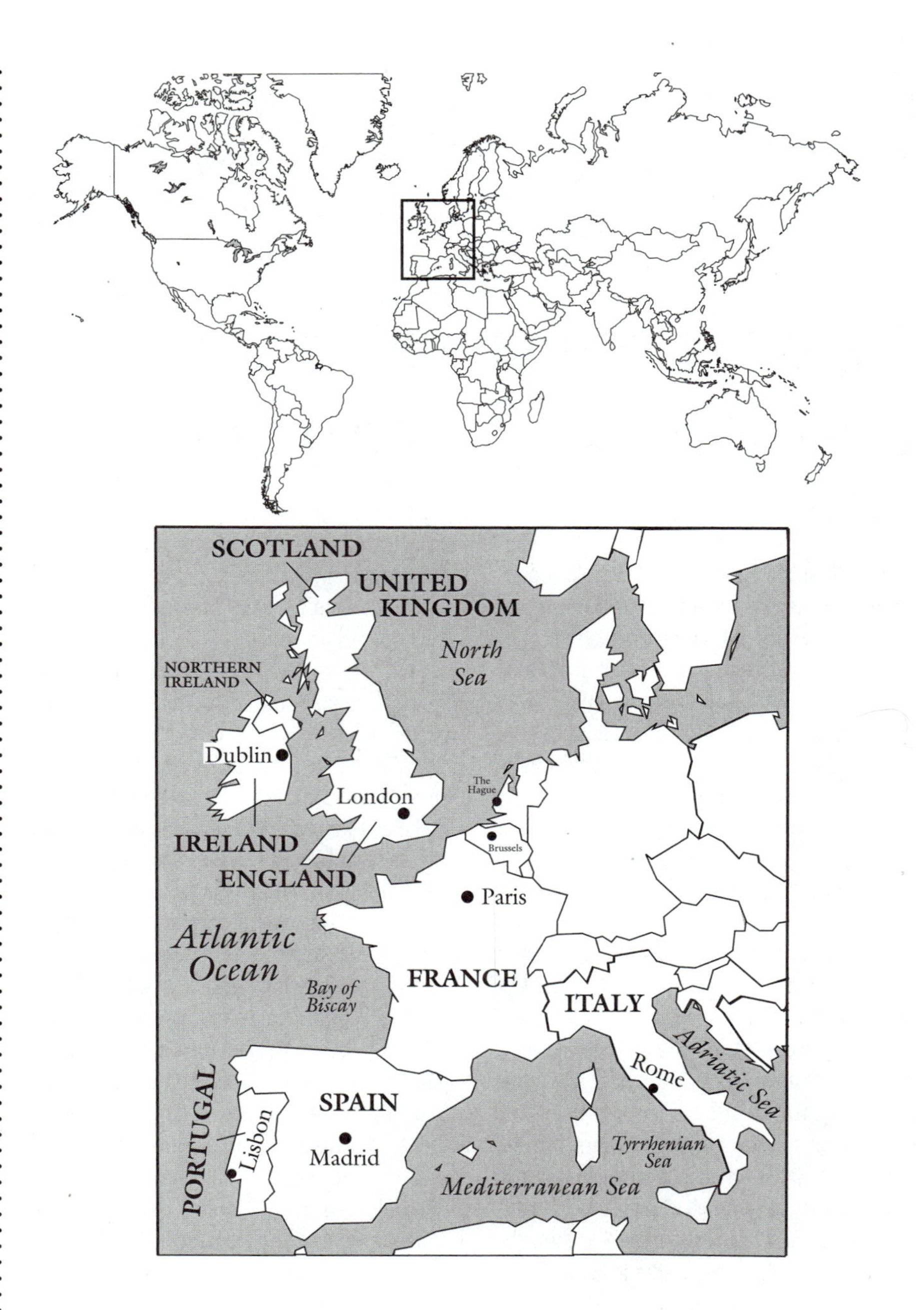

Figure 6.1
Northern and Southern Europe

England, Virginia, and Maryland. While many originally came to avoid religious persecution, such as the Puritans in New England and the Catholics in Maryland, most later immigrants earned their passage to America by signing on as indentured servants.

By the 18th century, British immigration had slowed. After independence, British immigration to the United States further declined due to American hostility and disapproval by the British government. However, reported arrivals of British in the 19th century increased substantially. Early in the century most immigrants were families from rural areas of southern and western England. In the latter half of the century the majority of immigrants were from large English towns; many were seasonal unskilled workers who repeatedly returned to Britain.

It is said that there have been Scots in America as long as there have been Europeans on the continent. More than 100 towns and cities in the United States bear Scottish names, and it has been estimated that 1.5 million

Scots immigrated to America. Although the majority of Scots came during the 18th and 19th centuries, 400,000 immigrated between 1921 and 1931, when Scotland suffered a severe economic depression. The Scottish settled over most of the United States and were often professionals or skilled laborers.

Although British immigration did not decline in the early 20th century, the United States was no longer the country of first choice for those leaving Great Britain. During the Great Depression in the 1930s, more British people returned to Britain than came to America. After World War II, an increase in immigration was attributable to British war brides returning to the United States with their American husbands. Since the 1970s, British immigration has been constant at about 10,000 to 20,000 persons per year.

Ireland. The first Irish people to immigrate in substantial numbers to the United States were the descendants of Scottish Presbyterians who settled in Northern Ireland in the 17th century. Large-scale immigration began in the 18th century, and by 1775 there were an estimated 250,000 Scotch Irish living in the American colonies. Most of the immigration was the result of an economic depression brought on by a textile slump in Ireland.

Initially the Scotch Irish settled in Pennsylvania. Before long the direction of Scotch Irish immigration was westward to the frontier, first up the Delaware River and then beyond the Susquehanna into the rich farmlands of the Cumberland Valley. The Scotch Irish played an important role in the settlement of the trans-Allegheny region and eventually clustered around the site of Pittsburgh and in other areas of southwestern Pennsylvania. They also settled in the frontier regions of western Maryland, the Shenandoah Valley of Virginia, and the backcountry of Georgia.

Irish Catholics started to arrive in the United States by 1820, and their immigration reached an apex between 1840 and 1860, when approximately 2 million people arrived. The impetus to leave Ireland was not only religious persecution but also repeated crop failures. The potato blight that destroyed their principal crop in 1845 resulted in death by starvation of 1 million Irish people.

The Irish Catholics were the first great ethnic minority in American cities, and their early history set the pattern for later minority immigrant groups. They settled in the

Traditional foods of northern Europe. Some typical foods in the northern European diet include apples, bacon, beef, cheese, cream, French bread, oatmeal, salt cod, and tripe. (Photo by Laurie Macfee.)

St. Patrick's Day in Savannah. (© Robert Brenner/PhotoEdit.)

northeastern cities and were at the bottom of the socioeconomic ladder. Whereas the Scotch Irish, who were often of relatively high economic standing and Protestant, found it fairly easy to move into mainstream American society, the Irish Catholics were often stereotyped as drunkards, brawlers, and incompetents. The Irish achieved success with painful slowness. For many, their first entry into the American mainstream came by way of city politics.

France. Immigration directly from France has been the smallest, yet most constant, of that from any European country, but the return rate has been high. Most of the estimated 1 million persons who have immigrated to the United States from France have been middle-class and skilled and have come for economic opportunity. A smaller number came because of religious persecution. More than 12,000 Huguenots, French Protestants, settled in the American colonies in the 18th century. They were considered to be excellent skilled workers.

Generally, French people who settled in the United States were eager to assimilate and able to do so because they were economically successful. Few pockets of French culture remain in the United States, with the exception of southern Louisiana, originally a French holding, and northern New England. However, the Frenchness of these areas is probably due more to the influence of French Canadian immigration than to direct French immigration.

French Canadians are the descendants of explorers and settlers who came from France, primarily Normandy and Brittany, during the seventeenth century. They established New France in what is today known as Canada. When the English gained control of Canada, many French Canadians moved to the United States; in some instances they were deported from Canada. Most settled in northern New England, especially Maine, and their descendants are known as Franco-Americans. Others from Acadia (Nova Scotia, New Brunswick, Prince Edward Island, and part of Maine) relocated, often not by choice, to central and southern Louisiana; their descendants are known as Cajuns.

Current Demographics and Socioeconomic Status

British and Irish. The British assimilated into American mainstream society easily. Distinct groups from specific regions of Great Britain can still be found, however. For example, Cornish immigrants of the nineteenth century were often miners, and their descendants are still living in certain old mining regions, such as Grass Valley, California; Butte City, Montana; and the areas around Lake Superior. The Welsh who immigrated in the 19th century were miners and millworkers. They settled in the mid-Atlantic and midwestern states, especially Ohio and Pennsylvania; many were Baptist, Calvinist, or Methodist. Remnants of Welsh communities in the United States still celebrate St. David's Day (the feast day of the patron saint of Wales) and the annual festival of the National Gymanfa Ganu Association (an assembly that sings Welsh hymns or folk songs). Forty-three million Americans listed English or British ancestry in the 1990 U.S. census.

Close to 5 million Irish Catholics have immigrated to the United States, and today there are approximately 38 million Americans (about 15 percent of the total population) who claim to be of Irish descent. Although they started out on a lower economic rung than other older immigrant groups, they are now scattered throughout the occupational structure. In the 1950s the Irish were overrepresented as clergymen, fire fighters, and police officers. Today there are disproportionately more in law, medicine, and the sciences. Numerous Irish people have also excelled in the literary arts and in sports, especially boxing and baseball. Although Irish Catholics have to some extent assimilated into mainstream American society, they still remain an identifiable ethnic group.

The Scotch Irish are widely dispersed and are not easily identifiable. It has been estimated that 1 in every 30 Americans (some 8 million people) is of Scotch Irish descent.

French. Over 10 million Americans listed French as their ancestry in the 1990 U.S. census, and it is believed that more than 2 million people of French Canadian descent live in the United States. In Louisiana, more than 800,000 are of Acadian ancestry. The Cajuns settled in rural and inaccessible areas of southern Louisiana, the bayous, and along the Mississippi River. Primarily farmers, fishermen, and herders, they were self-sufficient and kept to themselves. Today they are still rural, but their occupations reflect local economic conditions.

The French Canadians who settled in the New England states were less likely to be farmers; instead, they worked in factories that processed textiles, lumber, and bricks. Since 1950 there has been an increase in the percentage of Franco-Americans holding white-collar jobs, but they still lag behind other ethnic groups economically. Compared to the French who immigrated directly from France and assimilated rapidly into American culture, the descendants of French Canadians have clung to their heritage, including language, customs, and religious affiliation.

Worldview

Religion

British. Nearly all early British immigrants to America were Protestant. Although many came to escape persecution by the Church of England, others maintained this faith and established congregations throughout the American colonies. The Church of England in the United States became the Episcopal church during the late 18th century.

British ethnicity was often expressed through religious affiliation, particularly with the Episcopal, Methodist, Baptist, and Quaker faiths. Many immigrants established distinctively English congregations, but, within a generation, most became indistinguishable from other American churches. Today Americans of British descent participate in most U.S. faiths.

Irish. Religion is a cornerstone of Irish Catholic society, and in the United States, it is centered around the parish. Over time the Catholic Church in America came to be dominated by the Irish, often to the resentment of other Catholic immigrants. The church spared no effort to aid its members; it established schools, hospitals, and orphanages across the United States. The church helped to bridge the cultural gap for many Irish immigrants through advice, job placement, savings clubs, and temperance societies. Today religion plays a less important role in Irish Catholic life, although the role of the Irish in the church is still significant.

French. Among French Americans the Catholic Church provided the nucleus of the community, gave it stability, and helped preserve the language and traditions of the people. The church today still plays a central role.

Family

British. The immigrant English family formed the model for the typical American family. It included a father, a mother, and their

- **The Welsh honor the patron saint of Wales, St. David, by wearing a leek stalk on their lapel. St. David identified his followers by placing a leek in the brim of their caps.**

- **Canadian census data from 1996 reported 5.3 million people of Canadian ancestry; 4.9 million of British (17 percent of the total population), 4.3 million of Scottish heritage; 3.8 million of Irish background; and 2.7 million of French ancestry (9 percent of the total population).**

- **New England Puritans and English Quakers were among the first in the United States to promote free public education.**

children. This family group would sometimes reside near other relatives but more often would establish solitary households. The father was in charge of the public and business aspects of the family, while the mother controlled the domestic and social responsibilities. Traditionally the oldest children in the home were well educated and were sent to private schools, if affordable. Such an education was considered an investment in the future, and children were expected to continue the family business and to maintain the family's social position.

Due to the similarities between the British family and the emerging American family, new immigrants from England assimilated quickly. It was very common for British immigrants to marry non-English spouses.

Irish. Many of the characteristics of the Irish family in the 19th century persisted into the 20th century. Irish Catholics tended to marry at a late age, have large families, and divorce rarely. Today, however, most first- and second-generation Irish Catholics are likely to marry outside their group and, with increasing frequency, outside Catholicism. Traditionally the father was the breadwinner in the Irish Catholic family, but the mother's position was a strong one. Daughters were often as well educated as sons. The Irish people's relatively egalitarian attitudes toward sex roles may be responsible for the high concentration of Irish American women in professional jobs and white-collar work.

French. Until the 20th century, Cajuns lived in rural areas in extended family households with as many as 10 or 12 children per couple. The whole family worked as a unit, and decisions that affected the group were made jointly by all the adults. Until 1945, many Cajuns were illiterate and spoke only Cajun French. The use of Cajun French was prohibited by the public schools in 1921; as a result, many younger Cajuns today do not speak or understand French. The average family size today is smaller, and there is more marriage outside the community, but Cajuns still retain strong ties to their families.

The Franco-Americans in New England are similar to the Cajuns in that they maintained French traditions; this was not due to isolation, however, but rather to their continued contact with French relatives in Quebec. They had little desire to acculturate. During the 1930s the bond to Canada weakened because of new laws restricting reentry into the United States and the Great Depression, which diminished new French Canadian immigration. Today the descendants of the French Canadians speak French infrequently and often marry outsiders. Family ties are still strong, but, as with the Cajuns, family size has decreased. Franco-American women have traditionally had higher status and more authority than their counterparts in France.

Traditional Health Beliefs and Practices

Many of what would be characterized as American majority cultural beliefs regarding health originated in northern Europe. For example, when students were surveyed on family home remedies (Spector, 1991), those of British, Irish, and French descent shared similar health maintenance practices such as a "good" diet, plentiful sleep, and daily exercise. Fresh air, cleanliness, and keeping warm and dry were also commonly mentioned. Chicken soup, tea with honey, lemon and/or whiskey, and hot milk were used often as home cures. Practices less common today included sulfur and molasses as a laxative, regular use of cod liver oil, mustard packs and oatmeal plasters, and blessing of the throat. The Irish traditionally wear protective religious medallions and drink hot whiskey with cloves for a cough.

Among the British and the Irish there is the more generalized belief that good health is dependent on "proper" attitude (which includes religious faith) and a rigorous, regular lifestyle. Many northern Europeans associate a moderate diet with maintaining bowel regularity and laxative use is common. Stomach ailments may be explained as due to food that is too spicy, spoiled, or incompatible (causing an allergic reaction).

The traditional French lifestyle, which features leisurely meals and little structured exercise, presents a paradox to researchers. Although the French consume more total saturated fat and cholesterol than Americans, their death rate from heart disease is less than one-half of that in the United States. Scientists speculate that some other protective factor in the French diet or lifestyle may account for this discrepancy, such as the higher intake of wine, or more frequent walking. Genetic factors do not appear to be a cause. Studies to compare the French with Americans of French ancestry have not been reported.

The French Canadians who settled in Louisiana brought numerous traditional remedies to the region. In addition, the Native Americans were admired for their knowledge of indigenous plants and cures. Today, Americans of French descent in Louisiana, including the Cajuns and urban dwellers of all socioeconomic groups, often use home remedies and consult folk healers. Teas made from magnolia leaves, elderberry flowers, sassafras, or citronella are just a few of the infusions commonly prescribed for colds (Brandon, 1976). Salves of whiskey and camphor, or sheep's tallow and turpentine, also are believed to be beneficial for colds. Sore throats are treated by gargling herbal teas or hot water with dissolved honey, salt, and baking soda. Nausea is abated with an infusion of flies or chicken gizzards, sassafras tea is used to cleanse the blood, garlic is ingested for worms, and tobacco smoke blown into the ears to cure earaches. A string of garlic is tied around the neck of a baby with worms. Red flannel pouches filled with camphor or asafetida are worn to prevent illness. It should be noted that many Americans of French descent may also consult practitioners of voodoo for health problems (see "Traditional Health Beliefs and Practices" in chapter 7 for more details).

Traditional Food Habits

The influence of France on the food habits of Great Britain and Ireland and vice versa has led to many similarities in the cuisines of these countries, although the ingredients of southern French cooking differ in that they are more like those of Mediterranean countries. The influence of these northern European cuisines on American foods and food habits has been extensive.

Ingredients and Common Foods

Staples and Regional Variations

Great Britain and Ireland. Animal products are of key importance in Great Britain and Ireland. Some form of meat or fish is present at most meals, in addition to eggs, milk, and cheese. In Britain and Ireland, lamb is a commonly eaten meat, as is roast beef, which is often made for Sunday dinner with *Yorkshire pudding* (a popover cooked in meat drippings). Pork is often served as sausages *(bangers)* and bacon. Various game birds also are eaten. The cultural food groups list (Table 6.1) includes a more complete detailing of ingredients.

The Irish and British diet also contains a variety of seafood. A well-known fast-food item is *fish and chips.* The fish is battered and deep-fried, served with fried potatoes, and seasoned with salt and malt vinegar. Fish is often preserved and served as an appetizer or at breakfast. Examples are smoked Scottish salmon and *kippers,* salted and smoked fish.

■ A *pub,* or public house, is a bar that serves beer, wine, hard liquor, and light meals. The British pub is often the place where friends and family meet to socialize.

■ Oatcakes, called *bannocks,* were traditionally eaten to celebrate the pagan Celtic holiday of Beltane on May 1. One section was burnt or covered with ash; the unlucky person who received the marked portion was sacrificed (in more recent times the victim would leap through a small bonfire three times instead).

Sample Menu

AN IRISH SUNDAY SUPPER

Mustard soup

Pork with apples and ale

Colcannon or "Mash"

Soda bread

Summer pudding (with fresh berries)

Stout

Table 6.1
Cultural Food Groups: Northern European

Group	Comments
PROTEIN FOODS	
Milk/milk products	The English and Irish drink milk as a beverage; the French may not.
Meat/poultry/fish/eggs/legumes	Meat, poultry, or fish is usually the centerpiece of the meal. Meats are generally roasted or broiled in Great Britain; also prepared as stews or in pies. Smoked, salted, or dried fish is popular in England.
CEREALS/GRAINS	Wheat bread usually accompanies the meal. In Britain and Ireland, oatmeal or porridge is common for breakfast.
FRUITS/VEGETABLES	Potatoes are frequently eaten in Ireland. Arrowroot starch is used as thickener and tapioca (from cassava tubers) is eaten.
ADDITIONAL FOODS	
Seasonings	British and Irish dishes emphasize naturalness of foods with mild seasoning, then served with flavorful condiments or sauces used to taste. French dishes often prepared with complementary sauces or gravies that enhance food flavor.
Nuts/seeds	Nuts especially popular; used primarily in desserts.
Beverages	Alcoholic beverages consumed as part of the meal.
Fats/oils	Butter used extensively in cooking of northern and central France; olive oil more common in southern regions of the country.
Sweeteners	

Table 6.1
Cultural Food Groups: Northern European—continued

Common Foods	Adaptations in the United States
Cheese (cow, sheep, and goat milk), cream, milk, sour cream, yogurt	
Meat: Beef (roasts; variety cuts such as brains, kidneys, liver, sweetbreads, tongue, and tripe), horsemeat, lamb, oxtail, pork, rabbit, snails, veal, venison	The Irish consume more animal protein.
Poultry and small birds: chicken, duck, goose, partridge, pheasant, pigeon, quail, thrush, turkey	
Fish and shellfish: anchovies, bass, clams, cod, crab, crawfish, haddock, herring, lobster, mackerel, mullet, mussels, oysters, perch, pike, pompano, salmon, sardines, scallops, shad, shrimp, skate, sole, sturgeon, trout, whiting	
Eggs: poultry and fish	
Legumes: kidney beans, lentils, lima beans, split peas	
Barley, hops, oats, rice, rye, wheat	Corn and corn products are consumed more.
Fruits: apples, apricots, cherries, currants, gooseberries, grapes (many varieties), lemons, melons, oranges, peaches, pears, plums, prunes, raisins, raspberries, rhubarb, strawberries	Native and transplanted fruits and vegetables, such as bananas, blueberries, okra, and squash, were added to the diet.
Vegetables: artichokes, asparagus, beets, brussels sprouts, cabbage, carrots, cauliflower, celery, celery root, cucumbers, eggplant, fennel, green beans, green peppers, lettuce (many varieties), leeks, mushrooms (including chanterelles, cèpes), olives, onions, parsnips, peas, potatoes, radishes, salsify, scallions, sorrel, spinach, tomatoes, turnips, truffles, watercress	
Angelica (licorice-flavored plant), bay leaf, capers, chiles, chives, chocolate, chutney, cinnamon, cloves, coffee, cognac, fennel seeds, garlic, ginger, horseradish, juniper berries, mace, malt vinegar, marjoram, mint, mustard, nutmeg, oregano, paprika, parsley, pepper (black, white, green, and pink), rosemary, saffron, sage, shallots, sweet basil, Tabasco sauce (and other hot sauces), tarragon, thyme, vanilla, Worcestershire sauce	Cajun and Creole cooking are highly spiced. Stews are thickened with *filé* powder (sassafras).
Nuts: almonds (sweet and bitter), chestnuts, filberts (hazelnuts), pecans, walnuts (including black)	
Seeds: sesame	
Beer (ale, stout, bitters), black and herbal tea (mint, anise, chamomile, etc.), cider, coffee, gin, hot chocolate, liqueurs, port, sherry, whiskey, wine (red, white, champagne, and fruit/vegetable)	
Butter, goose fat, lard, margarine, olive oil, vegetable oil, salt pork	
Honey, sugar	Molasses and maple syrup are used as sweeteners. Irish Americans use more sugar than members of other groups.

Tea time in Great Britain has become an afternoon meal, with small sandwiches, scones (on the second rack of the silver tray), and an assortment of cookies and pastries. (Courtesy of Grossich and Partners.)

Dairy products and eggs also play an important role in the diet of the British and Irish. Eggs are traditionally served for breakfast, and cheese is the key ingredient in the traditional *ploughman's lunch* served in pubs. It consists of a piece of cheddar cheese, bread, pickled onions, and a pint of beer. Other cheeses produced in England are the slightly salty, nutty Cheshire and Stilton, a blue cheese. Devonshire is known for its rich cream products, such as double cream (which has twice as much butter fat as ordinary cream) and clotted cream, a slightly fermented, thickened cream. It is often spread on *scones,* biscuits made with baking powder.

Although not the main focus of the meal, breads are not overlooked. In Ireland, soda bread, a bread made with baking soda instead of yeast, was traditionally prepared every day to accompany the meal. Wheat flour is commonly used for baking, and oatmeal is eaten as a porridge for breakfast in Scotland or used in making bread and biscuits throughout Britain and Ireland. Biscuits, or "biskcake," in England can refer to bread, cake, cookies, crackers, or what are known in America as biscuits. Scottish shortbread is an example of a sweet, buttery biscuit.

Fruits and vegetables are limited to those that grow best in cool climates. Potatoes, brought to Ireland from the New World in the seventeenth century, are the mainstay of the Irish diet and are found in British fare as well. Potatoes are found in stews or pies, such as *stobhach Gaelach,* an Irish stew of lamb's neck, and *shepherd's* or *cottage pie,* a meat pie made of leftover ground meat and onions and topped with mashed potatoes. Mashed potatoes are often just referred to as "mash," as in "bangers and mash" (sausages and mashed potatoes). Some side dishes made of potatoes are *boxty,* a type of potato pancake; *bubble*

■ *Colcannon* was customarily served for the harvest dinner and on Halloween in Ireland. For Halloween, coins were wrapped and buried in the dish so the children could find them as they ate.

■ *The Guinness Book of World Records* was originally published by the family who makes Guinness stout to help settle pub arguments.

Home of french mustard, Dijon, France.

and squeak, a dish made of leftover cabbage and potatoes, chopped and fried together; and *colcannon,* mashed and seasoned boiled white vegetables with onions or leeks.

The most common beverages consumed by adults in Ireland and England are tea, beer, and whiskey. Tea, which has become synonymous with a meal or break in the afternoon, was introduced to England in 1662 by the wife of Charles II. Drunk with most meals and as a refreshment, strong black tea is preferred, served with milk and sugar. Frequently consumed alcoholic beverages include beer and whiskey. The British and the Irish do not drink the bottom-fermented style of beer common in the United States. Instead, in Ireland a favorite is *stout,* a dark, rich, heavy beer that can provide substantial calories to the diet; in Britain the pubs usually serve "bitters," a rich amber-colored beer, strongly flavored with hops. Both beers are served at cellar temperature and are naturally carbonated.

Whiskey is made in both Ireland and Scotland, but the Irish are usually credited with its invention and name. In Ireland it is distilled from mashed, fermented barley. Scotch, or Scotch whisky (spelled without the *e*), is distilled from a blend of malted whiskey (in which the barley germinated before fermentation) and unmalted whiskey. Scotch is traditionally a much stronger, smokier-tasting beverage than Irish whiskey. Other alcoholic drinks popular in Britain are gin, port (a brandy-fortified wine made in Portugal), and sherry (also a fortified wine, from Spain). A less common beverage but still popular in some regions is *mead,* a type of honey wine made from the fermentation of honey and water. The Welsh prefer a stronger, highly spiced variety called *metheglyn.*

France. The cooking of France has traditionally been divided into classic French cuisine (*haute* or *grande* cuisine) and provincial or regional cooking. Classic French cooking is elegant and formal, mostly prepared in restaurants using the best ingredients from throughout the country. Provincial cuisine is simpler fare made at home featuring fresh local ingredients.

The ancestors of most French Americans are from two of France's northern provinces, Brittany and Normandy. Brittany, known as *Bretagne,* is located in the northwest; its shores are washed by the English Channel and the Bay of Biscay. Seafood, simply prepared, is common, and delicate Belon oysters are shipped throughout France. It is said that mutton and vegetables from the region have a naturally salty taste because of the salt spray. Apples are the prevalent fruit, and cider is widely exported.

Located along the English Channel and east of Brittany is Normandy, also known for its seafood and apples. *Calvados,* an apple brandy, is thought to be the mother of applejack, an alcoholic apple drink used to clear the

■ The term *honeymoon* originated with the European custom of newlyweds drinking mead for the first lunar month following their wedding.

■ Fresh cream is called *fleurette* and is often added to sauces or whipped for dessert. Also popular is *crème fraîche,* cream that is fermented until it is thickened and slightly tangy.

■ The method for producing sparkling wine was discovered in the 17th century by the monk Dom Perignon, who reputedly shouted when he first sipped the beverage, "I think I am drinking stars!" Today the champagne bearing his name is considered one of the finest in the world.

■ *Pâté* is a spread of finely ground, cooked, seasoned meats. A *terrine* is commonly made with leftover meats cut into small pieces, mixed with spices and a jelling substance, then baked in a loaf pan.

■ In the small town of Pamiers, the city council has banned any ready-made or mass-produced food (e.g., frozen pizza) from the local school cafeteria to "ensure our kids stay healthy, teach them the taste of proper French food, and help keep our small farmers in business" (Henley, 2000).

palate during meals in Louisiana. Another alcoholic drink produced in the region is *Bénédictine,* named after the Roman Catholic monks who still make it at the monastery in Fecamp. Normandy is also renowned for its rich dairy products; its butter is considered one of the best in France. *Camembert,* a semisoft cheese with a mild flavor, and *Pont-l'Évêque,* a hearty aromatic cheese, are produced in the area. Dishes from Normandy are often prepared with rich cream sauces. *Crêpes,* very thin, unleavened pancakes, originated in this region; they are typically served topped with sweet or savory sauces or rolled with meat, poultry, fish or seafood, cheese, or fruit fillings.

Champagne, bordered by the English Channel and Belgium, has a cuisine influenced by the Germanic cultures. Beer is popular, as are sausages, such as *andouille* and *andouillette,* large and small intestinal casings stuffed with pork or lamb stomach. *Charcuterie,* cold meat dishes such as sausages, *pâtés,* and *terrines,* which often are sold in specialty stores, are especially good from this region. Throughout the world, Champagne is probably best known for its naturally carbonated wines. Only sparkling wines produced in this region can be legally called *champagne* in France.

The province that borders Germany, Alsace-Lorraine, has been alternately ruled by France and Germany. One of its principal cities is Strasbourg. Many German foods are favored in the region, such as goose, sausages, and sauerkraut. Goose fat is often used for cooking, and one of the specialties of the area is *pâté de fois gras,* pâté made from the enlarged livers of force-fed geese. Another famous dish is *quiche Lorraine,* pie pastry baked with a filling of cream, beaten eggs, and bacon. Alsace-Lorraine is a wine-producing area; its wines are similar to the German Rhine wines but are usually not as sweet. Distilled liquors produced in the region are *kirsch,* a cherry brandy, and the brandy *eau de vie de framboise,* made from raspberries.

Located south of Normandy and Brittany in the west-central part of France is Touraine, the province that includes the fertile Loire valley. Along the river one can see the beautiful chateaux or palaces built by the French nobility. Known as the "garden of France," Touraine produces some of the finest fruits and vegetables in the country. A dry white wine produced in the area is *Vouvray.* In the north-central region is the area surrounding the city of Paris called the *Ile-de-France,* the home of classic French cuisine. Some of the finest beef and veal, as well as a variety of fruits and vegetables, are produced in this fertile region. Brie, semisoft and mild flavored, is the best-known cheese of the area. Dishes of the Ile-de-France include *lobster à l'américaine,* lobster prepared with tomatoes, shallots, herbs, white wine, and brandy; *potage St. Germain,* pea soup; and *filet de bœuf béarnaise,* filet of beef with a béarnaise sauce.

Located southeast of Paris is Burgundy, one of the foremost wine-producing regions of France. Burgundy's robust dishes start to take on the flavor of southern France; they contain garlic and are often prepared with olive oil. Dijon, a principal city, is also the name of the mustards of the region prepared with white wine and herbs. Dishes of the area are *escargot,* snails (raised on grape vines)

A FRENCH LUNCH

Pork and veal terrine

Cassolet or *Poulet a la Normande* (chicken with cream and *Calvados* sauce)

Green salad

Baguette

Selection of cheeses (*Brie, Pont-l'Evêque,* etc.), Fresh fruit tart, or *Petits fours*

Wine

cooked in a garlic butter and served in the shell; *coq au vin,* rooster or chicken cooked in wine; and *bœuf bourguignon,* a hearty red wine beef stew. In Burgundy the red wines are primarily made from the pinot noir grape and the white wines from the chardonnay grape. The great wines of the area are usually named after the villages in which they are produced; for example, Gevrey-Chambertin, Vosne-Romanée, and Volnay. *Cassis,* a black currant liquor, is also produced in the region, and brandy from Cognac is a specialty. To the east, along the border with Switzerland, is the mountainous Franche-Comte region, known for its exceptionally tender and flavorful Bresse chicken.

The other major wine-producing region of France is Bordeaux, which is also the name of its principal city. Famous for its hearty dishes, the term *à la bordelaise* can mean (1) prepared in a seasoned sauce containing red or white wine, marrow, tomatoes, butter, and shallots; (2) use of *mirepoix,* a finely minced mixture of carrots, onions, and celery seasoned with bay leaves and thyme; (3) accompanied by *cèpes,* large fleshy mushrooms; or (4) accompanied by an artichoke and potato garnish.

A red Bordeaux wine is full-bodied and made primarily from the cabernet sauvignon grape. (In Great Britain, Bordeaux is called claret.) Among the wines produced are St. Julien, Margaux, Graves, St. Emilion, Pomerol, and Sauternes, a sweet white dessert wine.

In the south of France is Languedoc, famous for *cassoulet,* a complex dish containing duck or goose, pork or mutton, sausage, and white beans, among other ingredients. Provence, located on the Mediterranean Sea, is a favorite vacation spot because of its warm Riviera beaches. Provence is also known for the large old port city of Marseilles, its perfumes from the city of Grasse, and the international film festival in Cannes.

The cooking of Provence is similar to that of Italy and Spain. Staple ingredients are tomatoes, garlic, and olive oil; *à la Provençal* means that a dish contains these three items. Other common food items are seafood from the Mediterranean, eggplant, and zucchini. Popular dishes from the region are *bouillabaisse,* the famed fish stew made with tomatoes, garlic, olive oil, and several types of seafood, seasoned with saffron, and usually served with *rouille,* a hot red pepper sauce; *ratatouille,* tomatoes, eggplant, and zucchini cooked in olive oil; *salade Niçoise,* a salad originating in Nice, containing tuna, tomatoes, olives, lettuce, other raw vegetables, and sometimes hard-boiled eggs; and *pan bagna,* a French bread sandwich slathered with olive oil and containing a variety of ingredients, such as anchovies, tomatoes, green peppers, onions, olives, hard-boiled eggs, and capers. One unique specialty item in the region associated with haute cuisine is black truffles. This costly, pungent underground fungus flavors or garnishes many classic French dishes.

Cooking Styles

Although the ingredients of the countries on the opposite sides of the English Channel are not substantially different, their cooking styles vary greatly. France's cuisine is admired and imitated around the world, while Britain and Ireland's cuisine is often described as being simple and hearty.

Great Britain and Ireland. Both the British and the Irish take pride in the naturalness of their fare and their ability to cook foods so that the flavors are enhanced rather than obscured. Meat is usually roasted or broiled, depending on the cut, and lightly seasoned with herbs and spices. Strong-flavored condiments like *Worcestershire sauce* (flavored with anchovies, vinegar, soy, garlic, and assorted spices) on roast beef or mint jelly on lamb are often served. *Chutneys,* highly spiced fruit or vegetable pickles originally from India, are also popular. Leftover meat is finely chopped, then served in a stew, pie, or pudding.

While most Americans think of pies and puddings as being sweet desserts, in Britain and Ireland this is not necessarily the case. A pie is a baked pastry consisting of a mixture of meats, game, fish, and vegetables, or fruit, covered with or enclosed in a crust. A *Cornish pasty* is a small, pillow-shaped pie filled with meat, onions, potatoes, and sometimes fruit (a handy all-in-one meal of meat, side vegetable, and dessert) that miners took down in the mines to eat for lunch or dinner. Another well-known British dish is steak and kidney pie.

■ Tomatoes were introduced to Europe in 1523 from the New World. Reaction was mixed; some people thought they were poisonous, while others believed they brought luck. Tomato-like pin cushions developed from the latter superstition.

■ Mincemeat pies, which are traditionally served at Christmas and Thanksgiving in some parts of the United States, were originally prepared with seasoned, ground meats, suet, and fruit, but today they are usually made with only dried and candied fruit, nuts, and spices.

■ It is said that the Devil refused to visit Cornwall because the wives of Cornish miners had a reputation for putting anything in a pasty (Lockwood & Lockwood, 1991).

■ In Elizabethan England, prunes (dried plums) were offered free to brothel guests because of their reputation as an aphrodisiac.

■ The French do not usually use bread plates but place their portion directly on the table.

■ The first eating chocolate was introduced by the British in 1847. Europeans now consume 20–25 pounds of chocolate per person each year—twice the amount eaten by U.S. citizens.

■ *Deviled* describes a dish prepared with a spicy hot sauce or seasoning.

Pudding is a steamed, boiled, or baked dish that may be based on anything from custards and fruits to meat and vegetables. An example of a sweet pudding is *plum pudding,* which is served traditionally at Christmas. It is a steamed dish of suet, dried and candied fruit, and other ingredients. *Trifle* is a layered dessert made from custard, pound cake, raspberry jam, whipped cream, sherry, and almonds.

France. Classic French cuisine implies a carefully planned meal that balances the texture, color, and flavor of the dishes. The soul of French cooking is its sauces, often painstakingly prepared from stocks that are simmered for hours to bring out the flavor. A white stock is made from fish, chicken, or veal, and a brown stock is made from beef or veal.

Sauces are subtly flavored with natural ingredients, such as vegetables, wine, and herbs. They must never overwhelm the food, but rather complement it. The five basic sauces are *espagnole* (or brown sauce), made with brown stock, mirepoix, and *roux* (thickening agent made from flour cooked in butter or fat drippings); *velouté,* made with white stock, roux, onions, and spices; *béchamel* (or cream sauce), made with white stock, milk, and roux; tomato, made with white stock, tomatoes, onions, carrots, garlic, and roux; and, *hollandaise,* which combines egg yolks and drawn butter (béarnaise is similar, flavored with tarragon). Examples of classic cold sauces are mayonnaise and *vinaigrette,* a mixture of vinegar and oil.

Some common rules in preparing French dishes are (1) never mix sweet and sour flavors in the same dish; (2) never serve sweet sauces over fish; (3) do not under- or overcook food; (4) with the exception of salad and fruit, do not serve uncooked food; (5) always use the freshest, best-tasting ingredients; and (6) wine is an integral part of the meal and must complement the food.

French breads and pastries are particularly noteworthy. Breads are typically made with white flour, shaped into long loaves (e.g., thin *baguettes*), rounds, braids, or rings, then baked in a wood-fired oven. Sweeter breads, such as eggy *brioche* or buttery, flaky *croissants,* are between breads and pastries. Specialty doughs, such as cream puff pastry, multilayered puff pastry, and the classic sponge cake *génoise,* are used to create the numerous desserts of France, such as cakes, *petits fours* (small, bite-size pastries), and tarts. Chocolate, fresh fruits, and pastry cream thickened with egg yolks enrich these pastries.

Meal Composition and Cycle

Daily Pattern

Great Britain and Ireland. In Britain, four meals are traditionally served each day—breakfast, lunch, tea, and an evening meal (dinner). In the 19th and early 20th centuries, breakfast was a very substantial meal, consisting of oatmeal; bacon, ham, or sausage; eggs (prepared several ways); bread, fried in bacon grease; toast with jam or marmalade; grilled tomatoes or mushrooms; and possibly smoked fish or deviled kidneys. All this was washed down with tea. In Scotland, oatmeal is usually eaten for breakfast, while in England today, packaged breakfast cereals are often eaten during the week, and the more extensive breakfast is reserved for weekends and special occasions.

Lunch was originally a hearty meal and still is on Sunday, but during the week it is squeezed in between work hours. It may include a meat pie, fish and chips, or a light meal at the pub with a pint of bitters. Both Sunday lunch and the weekday dinner are much like an American dinner. They consist of meat or fish, vegetable, and starch. The starch is often potatoes or rice, and bread also accompanies the meal. Dessert (often called pudding) follows the main course.

In the late afternoon in Britain and Ireland, most people take a break and have a pot of tea and a light snack. In some areas a "high tea" is served. This can be a substantial meal that includes potted meat, fish, shrimp, ham salad, salmon cakes, fruits, and a selection of cakes and pastries. High tea is associated with working-class or rural families who have maintained the custom of a large lunch, and high tea serves as dinner. It is thought that the upper British classes add the term *high* to tea as a dinner when it is served occasionally in place of dinner as a novelty or to children as an informal substitute for dinner. Whether snack or meal, the British most often just call it tea.

France. The French eat only three meals a day—breakfast, lunch, and dinner. There is very little snacking between meals. Breakfast, in contrast to the British meal, is very light, consisting of a *croissant* or French bread with butter and jam, and strong coffee with hot milk or hot chocolate. The French breakfast is what is known in the United States as a "continental breakfast." Lunch is traditionally the largest meal of the day and, in many regions of France, businesses close at midday for 2 hours so that people can return home to eat. The meal usually starts with an appetizer (*hors d'oeuvre*) such as pâté. The main course is a meat, fish, or egg dish accompanied by a vegetable and bread. If salad is eaten, it is served after the main course. Dessert at home is usually cheese and fruit. In a restaurant, ice cream (more like a fruit sherbet or sorbet), cakes, custards, and pastries are served in addition to fruit and cheese. Wine is served with the meal and coffee after the meal.

Special Occasions

Christmas and Easter are the most important Christian holidays celebrated in England, Ireland, and France. Ireland and France are predominantly Catholic countries and tend to observe all the holy days of obligation and patron saints' days. France commemorates the beginning of the French Revolution on July 14, Bastille Day.

Great Britain and Ireland. The British celebrate Christmas by serving hot punch or mulled wine; roast goose, turkey, or ham; plum pudding and mince pies; and, afterward, port with nuts and dried fruit. The plum pudding is traditionally splashed with brandy, then flamed before serving. Boxing Day, the day after Christmas, is when friends and relatives visit one another.

Foods served at Easter include hot cross buns and *Shrewsbury simnel.* In ancient times the cross on the buns is believed to have symbolized both sun and fire; the four quarters represented the seasons. Today the cross represents Christ and the resurrection. Shrewsbury simnel is a rich spice cake topped with twelve decorative balls of marzipan originally representative of the astrological signs. It is also served on Mother's Day.

Another holiday celebrated throughout Great Britain is New Year's Day on January 1. The Scottish traditionally eat *haggis* on New Year's Eve. It is a sheep's stomach stuffed with a pudding made of sheep's innards and oatmeal. After it is served, the diners drench their portions with Scotch whisky before eating.

St. Patrick's Day began as a religious commemoration for the patron saint of Ireland. The Irish American custom of eating a corned beef and cabbage meal on March 17 is now as popular in Ireland as it is in America.

France. In France the main Christmas meal is served after mass on the night of December 24. Two traditional dishes are a black or white "pudding" (blood or pork sausage) and a goose or turkey with chestnuts. In Provence the Christmas Eve meal is meatless, usually cod, but the highlight is that it is followed by 13 desserts.

On Shrove Tuesday (*Mardi Gras*) the French feast on pancakes, fritters, waffles, and various biscuits and cakes. During Lent, no eggs, fat, or meat are eaten. Dishes served during Lent often contain cod or herring. Cod is also the traditional dish served on Good Friday; in some regions, lentils are eaten to "wash away one's sins." Easter marks the return of the normal diet, and eggs are often served hard-boiled (also colored), in omelets, or in breads and pastries. French toast (*croûtes dorée*) is a traditional Easter dish. Also common are pies filled with minced meats.

Therapeutic Uses of Food

Most northern Europeans share a belief that a "good" diet is essential to maintaining health. Traditional home remedies for minor illness include chicken soup, tea with honey or lemon or whiskey, hot milk, or hot whiskey with cloves in it. Practices less common today are taking sulfur with molasses as a laxative and regular use of cod liver oil. (See Table 7.1, "Selected European Botanical Remedies," in chapter 7 for more examples.)

The French Canadians who settled in Louisiana brought numerous traditional therapeutics to the region. In addition, the Native Americans were admired for their knowledge of indigenous plants and cures. Today Americans of French descent, including Cajuns and urban

■ *Potted* is an English term for fish, meat, poultry, or game pounded with lard or butter into a coarse or smooth pâté, then preserved in jars or pots.

■ The Associated Press reported in 1995 that the number of small neighborhood cafés in France, known as *bistros,* was declining. Nearly 500,000 existed in 1900; only 50,000 are open today. The popularity of fast foods and less time to socialize are some of the reasons responsible for the lifestyle changes.

■ In Britain a piece of mincemeat pie is eaten at midnight on New Year's Eve while making a wish for the upcoming year.

■ When the Christmas pudding is prepared, it is a British superstition for each family member to take a turn stirring it clockwise (the direction the sun was in ancient times assumed to rotate around the Earth). To stir counterclockwise is to ask for trouble.

■ In 1992, U.S. customs officials prevented a shipment of *haggis* from entering the country, claiming it unfit for human consumption.

La boucherie: French-speaking Cajuns in Louisiana maintain the hog-butchering traditions of their past. Before the days of refrigeration, everyone in the community helped prepare the meat and lard. Participants went home with fresh pork cuts and spicy sausages called *boudin.* La boucherie continues today at many Cajun festivals. (Courtesy of the Louisiana Office of Tourism.)

dwellers of all socioeconomic groups, often use home remedies. Teas made from magnolia leaves, elderberry flowers, sassafras, or citronella are just a few of the infusions commonly prescribed for colds. Sore throats are treated by gargling herbal teas or hot water with dissolved honey, salt, and baking soda. Nausea is abated with an infusion of flies or chicken gizzards, sassafras tea is used to cleanse the blood, and garlic is ingested for worms.

■ Some Cajuns believe that milk and fish should not be consumed together at the same meal.

■ Cornish pasties are still popular in parts of the country where immigrants from Cornwall came to work in the mines, such as the Upper Peninsula in Michigan, where May 24 was declared Pasty Day in 1968.

Contemporary Food Habits in the United States

Adaptations of Food Habits

Ingredients and Common Foods

British and Irish. Many American dishes have their origins in Great Britain. The Puritans, adapting Native American fare, made a pudding with cornmeal, milk, molasses, and spices. Today this is called *Indian pudding.* Pumpkin pie is just a custard pie to which the native American squash, pumpkin, is added. *Sally Lunn,* a sweet hot bread made in Virginia, is said to have been sold on the streets of Bath, England, by a young lass who gave it her name. Others claim that the name is a corruption of the French *sol et lune,* a sun and moon cake. *Syllabub,* a milk and wine punch drunk in the American South at Christmastime, is also an English recipe.

French. French cooking has had less influence on everyday American cooking (except for french fries), but there are probably few cities that don't have a French restaurant (which may or may not be owned by a French immigrant). French Americans adapted their cuisine and meal patterns to the available ingredients and other ethnic cooking styles. The best example of this is found in Louisiana, where *Creole* and *Cajun* cooking developed. Creole cooking is to Cajun cooking what French grande cuisine is to provincial cooking. Some dishes may sound typically French, such as the fish stew *bouillabaisse,* but it is made with fish from the Gulf of Mexico, not from the Mediterranean. Even the coffee is slightly different, as it is flavored with the bitter chicory root.

Ingredients for Cajun cooking reflect the environment of Louisiana: "Bayou Cajun" foods are from lake and swamp areas, while "prairie Cajun" dishes are found in inland areas. Fish and shellfish abound, notably crawfish, crabs, oysters, pompano, redfish, and shrimp, to name just a few. Shellfish is commonly eaten raw on the half shell (oysters) or boiled in a spicy mixture. *Gumbo* and *jambalaya* are often made with seafood. Gumbo is a thick, spicy soup made with a variety of seafood, meat, and vegetables. It is thickened with either okra or filé powder and then ladled over rice. Jambalaya, also a highly seasoned stew made with a combination of seafood, meats, and vegetables, was brought to New Orleans by the Spanish. Originally made only with ham (*jambon*), it was later modified. The base for these stews and gravies is roux; however, the Cajun roux is unique in that the flour and fat (usually vegetable oil) are cooked very slowly until the mixture turns brown and has a nutlike aroma and taste.

Other key ingredients in Cajun cooking are rice (which has been grown in Louisiana since the early 1700s), red beans, tomatoes, chayote squash, eggplant, spicy hot sauce, and a variety of pork products. One of the better-known hot sauces, *Tabasco,* is produced in the bayous of southern Louisiana from fermented chile peppers, vinegar, and spices. A deep-fried rice fritter, *calas,* is the Louisiana version of a doughnut. Other rice dishes are *red beans and rice* and *dirty rice.* Dirty rice derives its name from the fact that its ingredients, bits of chicken gizzards and liver, give the rice a brown appearance. Cajun *boudin* sausages are a specialty. *Boudin blanc* is made with pork and rice; *boudin rouge* has pork blood added to it. *Cochon de lait,* a suckling pig roasted over a wood fire, is prepared at Cajun festivals in central Louisiana. *Fricot* is a popular soup made with potatoes and sausage or shredded meat. Cracklings, known as *gratons,* are fried, bite-size bits of pork skin (often with meat attached) that are popular in some regions.

Pecan *pralines* are a famous New Orleans candy. Pecans are native to Louisiana; pralines are large flat patties made from brown sugar, water or cream, and butter. Another confection eaten often with coffee are *beignets,* round or square puffed French doughnuts dusted with powdered sugar. French toast, or *pain perdu,* is another French specialty that was transported to New Orleans and is now familiar to most Americans.

The cuisine of French Americans in New England tends to be traditionally French, but it is influenced by common New England foods and food habits. Franco-Americans use more herbs and spices than other New Englanders and take time to prepare the best-tasting food. Traditional French dishes are pork pâté, called *creton* by the Franco-Americans, and the traditional yule log cake (*bûche de Noël*) served at Christmas. Franco-American cuisine offers numerous soups and stews. One of the most elaborate of the stews, which is also called a pie, is *cipate,* known as *cipaille, si-pallie, six-pates,* and "sea pie" in some areas. A typical recipe calls for chicken, pork, veal, and beef plus four or five kinds of vegetables layered in a heavy kettle, covered with pie crust. It is slowly cooked after chicken stock has been added through vents in the crust.

Maple syrup is commonly used. One unique breakfast dish is eggs poached in the syrup. Maple syrup is also served over bread dumplings or just plain bread. Franco-Americans appreciate wine and distilled spirits. One unusual combination of both is *caribou,* a mixture of "white whiskey" (a distilled, colorless liquor) and red wine, which is drunk on festive occasions. (See also chapter 5, "Regional Americans.")

Meal Composition and Cycle

British and Irish. American food habits have been greatly influenced by British and Irish immigrants. Meal patterns and composition are very similar to those in Great Britain. Americans eat a hearty breakfast that often includes ham or bacon and eggs. The heart of the evening meal is usually meat, accompanied by a vegetable and a starch. Bread is served with most meals.

Festive meals also reflect the British and Irish influence. A traditional Christmas dinner includes roast turkey or ham, stuffing, and mashed vegetables. For dessert a pie is customary, often mincemeat. Two holidays that Americans think of as being typically American, Thanksgiving and Halloween, are actually of British and Irish origin. Thanksgiving combined the tradition of an old British harvest festival with the Pilgrims' celebration of surviving in their new environment. In Great Britain and Ireland, Halloween, or *All Hallow's Eve,* is believed to have originated in ancient times.

■ Crawfish are also known as crayfish (especially in New Orleans), crawdads, crawdaddy crab (in the Great Lakes area), clawfish, and mudbugs, among others. They are small crustaceans that look like miniature lobsters, found in the fresh waters of Louisiana, Lake Michigan, California, and the Pacific Northwest.

■ *Gumbo* is the African Bantu word for okra.

■ Louis Armstrong, the famous jazz trumpeter from New Orleans, always signed his letters, "Red beans and ricefully yours."

■ The name for the popular Cajun music style, *zydeco,* is derived from the French term for green bean, *haricot* ("ar-ee-cō") because it is snappy, like a bean.

Ghosts and witches were thought likely to wander abroad on Halloween night.

French. Americans of French descent have adopted the American meal cycle with the main meal in the evening. In Louisiana, the best-known celebration is Mardi Gras, culminating on Shrove Tuesday, just before the beginning of Lent. In New Orleans there are parades, masquerading, and general revelry; the festival reaches its climax at a grand ball before midnight. After this day and night of rich eating and grand merriment the 40 days of fasting and penitence of Lent begin. In the Cajun countryside, Mardi Gras is celebrated with "run": Men on horseback ride from farmhouse to farmhouse collecting chickens and sausages to add to a community gumbo. Participants enjoy beer, boudin, and *faire le maque* ("make like a monkey" or clown around) at each stop. During the rest of the year, Cajuns sponsor many local festivals, such as the crawfish, rice, and yam festivals.

Franco-Americans, like their French ancestors, serve meat pies on religious holidays. The special pie for Easter has sliced hard-boiled eggs laid down on the bottom crust, then a layer of cooked meat topped with well-seasoned pork and beef meatballs. For Christmas, *tourtière,* a pie made with simmered seasoned pork, is eaten cold after midnight Mass.

■ A study of men of Irish ancestory living in Scotland was unable to account for excess premature deaths by established risk factors (Abbotts et al., 1999).

■ Some Cajuns believe that being thin means a person is "puny" or "unattractive" (Leistner, 1996).

Nutritional Status

The influence of the British and French on American cuisine is undoubtedly one reason the U.S. diet is high in cholesterol and fat, and low in fiber and complex carbohydrates. Although few studies have been conducted on the nutritional status of Americans of French, Irish, and British descent, it is assumed that they have the same nutritional advantages and disadvantages of the general American population.

Nutritional Intake

Irish. A study to determine differences in mortality from coronary heart disease examined Irish brothers: one group in Ireland, one group living in the United States (Boston), and a third control group of first-generation Irish Americans in Boston (Kushi *et al.,* 1985). Although there was no significantly different relative risk for death from heart disease among the three groups, it wa found that their diets varied significantly. The Boston brothers and the first-generation Irish Americans had a higher intake (as a percentage of caloric intake) of animal protein, total fat (more vegetable and less animal), sugar, fiber, and cholesterol, and a lower intake of starch. The Irish brothers had a higher caloric intake than the Boston brothers and the first-generation Irish Americans, yet their relative weight was significantly lower.

Celiac sprue, an intolerance to gluten that results in malabsorption, appears to be prevalent among the Irish. It has been reported (Stadler, 1980) that a greater incidence of the disease exists in western Ireland (1/303) and Scotland (1/778) than in England (1/6,300) or the United States (1/2,000). It is not known whether this tendency is shared by Irish Americans.

It is commonly assumed that a high rate of alcoholism prevails among Irish Americans. It has been reported that they drink more frequently than any other ethnic group, but they are no more likely to be "problem drinkers" than Slavs or Native Americans. However, another comparative study found Irish American men had more physical, psychological, sexual, and/or occupational problems with alcohol misuse than did Puerto Rican men (Greeley 1972; Johnson, 1997).

French. A study on the causes of inherited chylomicronemia indicates that the frequency of lipoprotein lipase deficiency is very high among French Canadians (Ma et al., 1991). Franco-Americans may also have high rates of this genetic defect, leading to elevated triglycerides and the necessity of a very low-fat diet.

Counseling

Most Americans of British, Irish, or French descent are completely acculturated; however, studies of people in France suggest that the French are likely to undertake many activities at once, change plans frequently, ignore schedules, and communicate indirectly with enthusiastic body language. Direct eye contact is important. A quick, light handshake when meeting is customary (Hall & Hall, 1990).

The French Paradox

Researchers have been puzzling over what they call the French paradox: Why it is that French men have the lowest death rates from cardiovascular disease (CVD) of all industrialized nations despite high intake of dietary fats and cholesterol and high rates of smoking, hypertension, and diabetes? Though mortality rates overall in France are only 8 percent lower than those in the United States and 6 percent less than those in Great Britain, CVD deaths are 36 percent and 39 percent fewer than the United States and Great Britain, respectively.

Numerous hypotheses have been postulated. Most often mentioned is the relationship between drinking wine with meals, particularly red wine, which is high in cardioprotective ingredients such as the antioxidant resveratrol. It is believed these agents prevent platelet aggregation, as well as elevate high-density lipoproteins in the blood (Constant, 1997; Das et al., 1999; Renaud & Gueguen, 1998). Other researchers have suggested that the betaine in inexpensive wines (from the addition of beet sugar to increase alcohol content) could be significant because it may lower plasma homocysteine levels, an independent risk factor in arteriosclerosis (Mar & Zeisel, 1999). Folate from a high consumption of fruits and vegetables may work similarly in reducing homocysteine levels (Parodi, 1997). One study found that although 84 percent of respondents consumed more than 30 percent of their calories each day from fat, and 96 percent consumed more than 10 percent of their calories daily from saturated fats, the French sample rated very high in overall diversity of their diet (Drewnowski et al., 1996). Other factors may be the high activity levels of the French, their moderate portion sizes, and infrequent snacking.

Other researchers are less enthralled with the virtues of the French diet and lifestyle. Two British researchers have suggested that prior to the 1980s, the French actually had a significantly lower intake of fat and cholesterol than did the people of Great Britain. What seems like a paradox is actually a time lag, and they predict CVD mortality rates will increase in France when the long-term effects of the dietary changes begin to take their toll (Law & Wald, 1999). A French scientist has proposed that the failure to report all CVD death and high rates of alcohol-related deaths among young French men (including liver cirrhosis, gastrointestinal cancer, accidents, violence, and suicides) accounts partially for the paradox (de Lorgeril, 1999). Further study of the French diet and improved statistical analysis may finally solve the riddle.

Americans of British and Irish descent are often stoic in the face of illness and reserved in the communication of their symptoms. Some Irish believe that the best way to stay healthy is to avoid doctors unless very ill. The British, Irish, and French all tend to be more formal than Americans and politeness is expected. Socioeconomic status and religious practice are likely to have greater impact on foods and food habits than country or origin. The indepth personal interview should reveal any notable ethnically based preferences.

Southern Europeans

Southern European countries lie along the Mediterranean Sea and include Italy, southern France, Spain, and Portugal. Italy, shaped like a boot, sticks out into the Mediterranean and includes the island of Sicily, which lies off the boot toe. Italy is separated from the rest of Europe by the Alps, which form its northern border. Spain, located to the west of France (the Pyrenees Mountains form a natural border between the two countries), occupies the majority of the Iberian peninsula. Portugal sits on the western end of the peninsula, but also includes the Azore and Madeira Islands located in the Atlantic Ocean. Most of southern Europe enjoys a warm Mediterranean climate except in the cooler mountainous regions.

Immigration to the United States from southern Europe has been considerable, primarily from the poorer regions of southern Italy. Many Americans, even those of non-European descent, enjoy Italian cuisine in some form. The foods of Spain and Portugal are similar to those of Italy and France because of both the Greek and Roman influence in the region and the shared climate, but their preparations differ.

The following section reviews the traditional diets of Italy, Spain, and Portugal. The influence of these cuisines on American fare is also discussed.

■ Historically, *Mac* before a family name meant "son of," whereas the letter O signified "descended from."

Traditional foods of southern Europe. Some foods typical of the traditional southern European diet include almonds, artichokes, basil, cheese, eggplant, garlic, chickpeas, olive oil, olives, onions, pasta, *prosciutto,* salt cod, sweet bread, and tomatoes. (Photo by Laurie Macfee.)

■ The decline of the Roman Empire is believed due in part to the spice trade. The Roman demand for cinnamon, ginger, pepper, and other seasonings led to a trade imbalance that weakened the entire economy.

■ Rome gave the Carthaginian name of Hispania to what is known as Spain today, because of its abundance of rabbits, *sphan.*

■ The word *salary* comes from the Latin word for "salt," because Roman soldiers were paid with rations of the mineral.

■ The market for processed spaghetti sauce in the United States is more than $500 million annually.

Cultural Perspective

History of Southern Europeans in the United States

Immigration Patterns

The majority of immigrants from southern Europe were Italians, who swelled the population of U.S. cities on the eastern seaboard during the late 19th and early 20th centuries. Next in number were the Portuguese, primarily from the Azore Islands. Smaller numbers of Spanish immigrants have been documented.

Italians. According to immigration records, more than 5 million Italians have settled in the United States. The majority came from the poorer southern Italian provinces and from Sicily between 1880 and 1920. Although earlier immigrants from northern Italy settled on the west coast of the United States during the gold rush, most of the later immigrants settled in the large industrial cities on the east coast. Several cities still boast Italian neighborhoods such as the North End in Boston and North Beach in San Francisco.

Many Italians came to America for economic reasons; more than one-half of the immigrants, mostly men, returned to their homeland after accumulating money. Peasants in their native land, Italians in the United States often became laborers in skilled or semi-skilled professions, especially the building trades and the clothing industry. Immigration from Italy fell sharply after World War I, yet more than 500,000 Italians have immigrated since World War II.

Spaniards More than one-quarter of a million people from Spain have immigrated to the United States since 1820. However, the majority of the Spanish-speaking population in the United States comes from American acquisition of Spanish territories and the immigration of people from Latin American countries (see chapters 9 and 10 for more detail).

Half of the Spanish immigrants to the United States came in the late 19th and early 20th centuries, probably because of depressed economic conditions in Spain. In 1939, after the fall of the second Spanish republic, a small number of refugees immigrated for political reasons. Many of the early Spanish immigrants were from the Basque region, located in northeastern Spain on the border with France (there are also French Basques). They came to the United States in the 19th century, settling first in California, then spreading north and east throughout the West. The Basques are thought to be one of the oldest surviving ethnic groups in Europe; they lived in their homeland before the invasion of the Indo-Europeans around 2000 B.C.E. Their language, *Euskera,* is not known to be related to any other living language.

Portuguese. Beginning in the early 19th century, two waves of Portuguese immigrants arrived in the United States. Early immigrants were primarily from the Azore Islands and Cape Verde Islands, and they often located in the whaling ports of New England and Hawaii. The gold rush attracted a large number of Portuguese to the San Francisco Bay area. Many of these immigrants came for economic opportunity.

After World War II, a small number of Portuguese from Macao, a Portuguese settlement on the coast of China near Hong Kong, settled in California. They are well educated and many hold professional jobs. A much more significant number of Portuguese, more than 150,000, entered the United States after 1958, again mostly from the Azore Islands, following a series of volcanic eruptions that devastated the region. It has been estimated that the Portuguese currently have one of the higher rates of new arrivals among European groups.

Current Demographics and Socioeconomic Status

Italians. Today there are approximately 14 million Americans of Italian descent, most of whom live in or around major cities. Economically, Italian Americans shared in the general prosperity after World War II, and today most are employed in white-collar jobs or as skilled laborers.

Four generations of Italians living in the United States have been identified. The elderly living in urban Italian neighborhoods are one group, those who are middle age and living in either urban or suburban settings are the second generation, the well-educated younger Italian Americans living in mostly suburban areas are the third generation, and fourth are the very recent immigrants from Italy (Harwood, 1981).

Spaniards. More than 500,000 citizens identified themselves as Spaniards in the 1990 U.S. census; another 93,000 claimed to be Spanish Americans. Unlike the Spaniards, who settled in cities such as New York and later Tampa, Florida, the Basques settled mostly in the rural regions of California, Nevada, Idaho, Montana, Wyoming, Colorado, New Mexico, and Arizona and became ranchers. Some Basque immigrants, however, were drawn to the mining jobs of West Virginia and the rubber and steel plants of Ohio, Illinois, Michigan, and Pennsylvania. There are between 50,000 and 100,000 Basque Americans living in the United States. Today most Basque descendants are involved in some aspect of animal husbandry or small business; few have entered other professions. Newer Basque communities now exist in Connecticut and Florida because *jai alai* (a Basque sport) facilities were established there.

Portuguese. As of 1993, over 1,150,000 Americans are of Portuguese descent, 50,000 claim Cape Verdean ancestry, and 4,000 are of Azore Islands heritage. Initially the Portuguese Americans on the West Coast were farmers and ranchers, but eventually their descendants moved into professional, technical and administrative positions. On the East Coast the descendants of the Portuguese who settled in the whaling ports now make up a significant part of the fishing industry, though only 3 percent of all Portuguese Americans work in this occupation. The percentage of Portuguese familis living in poverty is half that of the U.S. average.

■ Among the Basques, it is said that the Devil once came to the region to learn their language, *Euskera,* so that he could entrap the inhabitants. He gave up after 7 years when he was only able to master two words: *bai* and *ez* ("yes" and "no").

Worldview

Religion

Italians. In Italy the Roman Catholic Church was a part of everyday life. Immigrants to America, however, found the church to be more remote and puritanical, as well as staffed by the Irish. The church responded by establishing national parishes (parishes geared toward one ethnic group with a priest from that group), which helped immigrants adjust to America. Some religious festivals, part of daily

■ There are over 1.2 million Canadians of Italian ancestry.

■ Boise, Idaho, is considered the Basque capital of the United States.

■ Immigrants from the Cape Verde and Azore islands and those from Madeira may not feel Portuguese. Instead they identify with their island or city of origin.

spiritual life in Italy, were transferred to America and are still celebrated today, such as the Feast of San Gennaro in New York's Little Italy.

Spanards. Most Spaniards are Roman Catholic. The Jesuit Order was founded in Basque country and has significantly influenced Basque devotion. Basque Americans are very involved in their parishes, and there is the expectation that religion is part of daily life and sacrifice.

Portuguese. The Roman Catholic Church also helped the Portuguese ease into the mainstream of American life. Local churches and special parishes often sponsor traditional religious *festas* that include Portuguese foods, dances, and colorful costumes.

Family

Italians. The social structure of rural villages in southern Italy was based on the family, whose interests and needs molded each individual's attitudes toward the state, church, and school. The family was self-reliant and distrusted outsiders. Each member was expected to uphold the family honor and fulfill familial responsibilities. The father was head of the household; he maintained his authority with strict discipline. The mother, although subordinate, controlled the day-to-day activities in the home and was often responsible for the family budget.

Once in America, the children broke free of parental control due to economic necessity. Although sons had always been allowed some independence, daughters soon gained their freedom as well, because they were expected to work outside the home like their brothers. Education eventually also changed the family. Early immigrants repeatedly denied their children schooling, sending them to work instead. However, by 1920, education was considered an important stepping stone for Italian Americans.

Spanards. In the traditional Spanish family, the father spent much of his time working and socializing outside the home, while the mother devoted her life to her children. Typically one daughter would choose not to marry and would care for her parents as they aged. In the United States, Spanish American families are usually limited to immediate members, although the obligation to parents remains stronger than for most Americans. An elder may live part of the year with one child, then part of the year with another child. Independent living and retirement homes are also common.

The Basque family was customarily extended. Basques in Spain are prohibited from marrying non-Basques, but in the United States many Basques marry other nationalities. Basques accept all members who marry into their family.

Spanish women hold unique status among southern Europeans. Class distinctions are more important than gender when it comes to educational and professional attainment. The Basque women are historically recognized for their equality. Since ancient times their duties have been as valued as those of men, and jobs are often not gender-specific.

Portuguese. Like the Italians, the Portuguese have close family solidarity and have had some success in maintaining the traditional family structure. Grown sons and daughters often live close to their parents and family members try to care for the sick at home. Family structure is threatened, however, when women must work or generational values change. Men tend to dominate the family and, as a result, some Portuguese American women marry outside the group.

Traditional Health Beliefs and Practices

Traditional Italian health beliefs include concepts common in the American majority culture as well as concerns associated with folk medicine. Fresh air is believed necessary to health, and some older Italian-Americans maintain that the "heavy" air of the United States is considered unhealthy compared to the "light" air of Italy. Well-being is defined as the ability to pursue normal, daily activities. There is the expectation that health declines with age.

Many Italian Americans believe that illness is due to contamination (through an unclean or sick person) or heredity ("blood"). Older immigrants may also think that sickness occurs because of drafts (surgery may be avoided so that organs will not be exposed to air), the suppression of emotions (i.e., anxiety, fear, grief), or

Pastas, such as spaghetti and rotini, are traditionally topped with tomato-based sauces in southern Italy. Filled pastas, such as ravioli, are more common in northern Italy. (Corbis/Michelle Garrett.)

supernatural causes. Some Italians believe that a minor illness can be attributed to the "evil eye" and that serious conditions result from being cursed by a malicious person or God (Ragucci, 1981; Spector, 1991). Although many Italians do not profess a belief that God punishes sin with a curse, there is often a fatalistic approach to terminal illness as being the result of God's will.

Problems in pregnancy are sometimes thought to be due to diet by Italian Americans, especially those of older generations. Unsatisfied cravings are reputed to cause deformities and, if a woman does not eat a food she smells, she may suffer a miscarriage. Little has been reported regarding Spanish and Portuguese health practices.

Traditional Food Habits

Although the foods of the southern European countries are similar, as detailed in the Cultural Food Groups list (Table 6.2), there are notable differences in preparation and presentation. Many Americans think of Italian cooking as consisting of pizza and spaghetti. In reality, these dishes are only a small part of the regional cuisine of southern Italy, the ancestral homeland of most Italian Americans. Spanish food is mistakenly equated with the hot and spicy cuisine of Mexico. Although Mexico was a colony of Spain, foods and food habits of the two countries differ substantially. Portugal and Spain have very similar cuisines, but most of the Portuguese immigrants to the United States are from the Azore Islands and the island of Madeira. Their diet was less varied than that of the mainland Portuguese.

Ingredients and Common Foods

Foreign Influence

The Phoenicians and Greeks, who settled along the Mediterranean coast in ancient times, are believed to have brought the olive tree and chickpeas (garbanzo beans) to the region. In addition, fish stew, known as *bouillabaisse* in France and *zuppa di pesce alla marinara* in Italy, may be of Greek origin. The Muslims brought lemons, oranges, sugar cane, rice, and a variety of sweetmeats and spices. *Marzipan,* a sweetened almond paste used extensively in Italian desserts, and rice flavored with saffron, as in the northern Italian dish *risotto alla Milanese,* are both believed to have Muslim origins. In Spain the Muslim influence is seen in saffron-colored rice and the use of ground nuts to thicken sauces, and in candies and desserts.

Table 6.2
Cultural Food Groups: Southern European

Group	Comments
PROTEIN FOODS	
Milk/milk products	Most adults do not drink milk but do eat cheese. Dairy products are often used in desserts. Many adults suffer from lactose intolerance.
Meat/poultry/fish/eggs/legumes	Dried salt cod is eaten frequently. Small fish, such as sardines, are eaten whole, providing substantial dietary calcium.
CEREALS/GRAINS	Bread, pasta, or grain products usually accompany the meal.
FRUITS/VEGETABLES	Fruit is often eaten as dessert. Fresh fruits and vegetables are preferred.
ADDITIONAL FOODS	
Seasonings	Dishes using similar ingredients in Italy, Spain, and Portugal often differentiated by distinctive use of herbs and spices. Seasoning in Azore Islands and Cape Verde Islands usually very mild.
Nuts/seeds	Nuts commonly used in desserts and added to some entrees and side dishes.
Beverages	
Fats/oils	Olive oil flavors numerous dishes; used for deep-frying in Spain.
Sweeteners	

Table 6.2
Cultural Food Groups: Southern European—continued

Common Foods	Adaptations in the United States
Cheese (cow, sheep, buffalo, goat), milk	It is assumed that second- and third-generation southern Europeans drink more milk into their adulthood than their ancestors did.
Meat: beef, goat, lamb, pork, veal (and most varietal cuts)	More meat and less fish is eaten than in Europe.
Poultry: chicken, duck, goose, pigeon, turkey, woodcock	
Fish: anchovies, bream, cod, haddock, halibut, herring, mullet, salmon, sardines, trout, tuna, turbot, whiting, octopus, squid	
Shellfish: barnacles, clams, conch, crab, lobster, mussels, scallops, shrimp	
Eggs: chicken	
Legumes: broad beans, chickpeas, fava and kidney beans, lentils, white beans	
Cornmeal, rice, wheat (bread, farina, a variety of pastas)	
Fruits: apples, apricots, bananas, cherries, citron, dates, figs, grapefruit, grapes, lemons, medlars, peaches, pears, pineapples, plums (prunes), pomegranates, quinces, oranges, raisins, Seville oranges, tangerines	First- and second-generation southern Europeans generally eat only fresh fruit and vegetables. Fruit and vegetable consumption tends to reflect general American food habits by the third generation.
Vegetables: arugala, artichokes, asparagus, broccoli, cabbage, cardoon, cauliflower, celery, chicory, cucumber, eggplant, endive, escarole, fava beans, fennel, green beans, lettuce, lima beans, kale, kohlrabi, mushrooms, mustard greens, olives, parsnips, peas, peppers (green and red), pimentoes, potatoes, *radicchio,* swiss chard, tomatoes, turnips, zucchini	
Basil, bay leaf, black pepper, capers, cayenne pepper, chocolate, chervil, cinnamon, cloves, coriander, cumin, dill, fennel, garlic, leeks, lemon juice, marjoram, mint, mustard, nutmeg, onion, oregano, parsley (Italian and curley leaf), rosemary, saffron, sage, tarragon, thyme, vinegar	
Almonds, hazelnuts, *pignolis* (pine nuts), walnuts, lupine seeds	
Coffee, chocolate, liqueurs, port, Madeira, sherry, flavored sodas (e.g., *orzata*), tea, wine	
Butter, lard, olive oil, vegetable oil	Use of olive oil has decreased.
Honey, sugar	

It was the food of the New World colonies, however, that shaped much of Italian, Spanish, and Portuguese cuisine. Chocolate, vanilla, tomatoes, pimentos, pineapple, white and sweet potatoes, maize (corn), many varieties of squash, and turkey were brought back from the Americas. The tomato is of particular importance to the character of southern European cooking. Asian ingredients have had a significant impact in the fare of Portugal and, to a lesser degree, the dishes of Spain and Italy. From India and the Far East came coconuts, bananas, mangoes, sweet oranges, and numerous spices, such as pepper, nutmeg, cinnamon, and cloves.

Staples

Italy. Although the cooking style and ingredients vary from region to region in Italy, some general statements can be made about the differences between the foods of northern and southern Italy. Pasta is a common dish throughout Italy, but in the north it is usually made with eggs and in the shape of flat ribbons. In the agriculturally poorer south the pasta is made without eggs, generally in a tubular form, such as macaroni. In the north, pasta (e.g., *ravioli*) is commonly stuffed with cheese or bits of meat, then topped with a cream sauce, while in the south it is usually served unfilled, with a tomato sauce. Other differences are that northern fare uses more butter, dairy products, rice, and meat than the south, which is notable for the use of olive oil, little meat, and more beans and vegetables, such as artichokes, eggplants, bell peppers, and tomatoes. Seasonings common to all of Italy are garlic, parsley, and basil.

Spain. The rugged terrain in Spain is suitable for raising small animals and crops, such as grapes and olives. In fact, Spain is the largest producer of olives in the world.

Entrees usually feature eggs, lamb, pork, poultry, or dried and salted fish, especially cod, called *bacalao.* Sausages, such as the paprika- and garlic-flavored *chorizo,* are common, and seafood is popular in coastal regions. Meats are often combined with vegetables in savory stews. Each region has its own recipe for *paella,* which typically includes saffron-spiced rice topped with chicken, mussels, shrimp, sausage, tomatoes, and peas. *Cocido,* a stew of chickpeas, vegetables (e.g., cabbage, carrots, potatoes), and meats (e.g., beef, chicken, pork, meatballs, sausages), also varies from area to area, but is always served in three courses. The strained broth with added noodles is eaten first, followed by a plate of the boiled vegetables, and concluded with a plate of cooked meats. Crusty bread is served with the meal.

Garlic and tomatoes flavor many Spanish dishes, for example, *gazpacho,* a refreshing pureed vegetable soup that is usually served cold, and *zarzuela* (meaning "operetta"), a fresh seafood stew. Olive oil is also a common ingredient used in almost all cooking, even deep-frying pastries, such as the ridged, cylindrical doughnuts known as *churros.* Sauces accompany many dishes. *Alioli* is made from garlic pulverized with olive oil, salt, and a little lemon juice. It is served with grilled or boiled meats and fish. Another popular sauce, called *romescu,* is sometimes mixed with alioli to taste by each diner at the table. It combines pureed almonds, garlic, paprika, and tomatoes with vinegar and olive oil. Spain's best-known dessert is *flan,* a sweet milk and egg custard topped with caramel.

Wine usually accompanies the meal. *Sangria,* made with red wine and fresh fruit juices, is served chilled in the summer. Spain is probably most famous in the United States for its sherries.

Portugal. Portuguese fare shares some similarities in ingredients with Spanish cuisine but a more generous addition of herbs and spices, including cilantro, mint, and cumin, distinguishes the cooking. Fish dominates the diet of the Portuguese; they are said to prepare *bacalhau* (dried salt cod) in 125 different ways. Sardines are often grilled or cooked in a tomato and vegetable sauce. *Chouriço,* similar to the Spanish pork sausage, chorizo, and *linguiça,* a pork and garlic sausage, are often eaten at breakfast. Other typical dishes are *cacoila,* a stew made from pig hearts and liver, then served with beans or potatoes; *isca de figado,* beef liver seasoned with vinegar, pepper, and garlic, then fried in olive oil or lard; and *assada no espeto,* meat roasted on a spit. A common soup

■ An Italian proverb states that after age 40, a person can "expect a new pain every morning."

■ *Spaghetti* comes from the Italian word for string, *spago.*

■ A *tortilla* in Spain is an egg omelet, not the cornmeal flatbread eaten by the Mexicans. It is believed that the Spanish called the Mexican bread by that name because of its similar shape.

■ Rice was brought to Italy by the Muslims; the Italians eat more rice than any other Europeans. Thomas Jefferson supposedly smuggled rice out of Italy to the United States, where his first attempts to cultivate it were unsuccessful.

is *caldo verde,* or green broth, made from kale or cabbage and potatoes. A unique "dry" soup of bread moistened with oil or vinegar and topped with anything from meat, chicken or shell fish and vegetables is called *açordu.* Fava beans, chickpeas, and lupine seeds (*tremocos*) are added to some dishes. Rice and fried potatoes are so popular they are often served together. Crusty country breads and, in the north, a cornmeal bread called *broa* also accompany the meal. Portuguese sweet breads, *pan doce,* and doughnuts, *malassadas*, are also specialties.

Regional Variations

Italy. Some of the regional specialities in the northern area around Milan are *risotto,* a creamy rice dish cooked in butter and chicken stock, flavored with Parmesan cheese and saffron; *polenta,* cornmeal mush (thought to have been made originally from semolina wheat), often served with cheese or sauce; and *panettone,* a type of fruitcake. The cheeses of the region include *Gorgonzola,* a tangy, blue-veined cheese made from sheep's milk, and *Bel paese,* a soft, mild-flavored cheese. The area is also known for its apperitifs, such as bittersweet *Vermouth.*

Venice, located on the east coast and known for its romantic canals (although the city actually consists of 120 mud islands), has a cuisine centered around seafood. Its best-known dish is *scampi,* made from large shrimp seasoned with oil, garlic, parsley, and lemon juice. Inland is Verona, famous for its delicate white wines, such as *Soave.* Turin, the capital of the western province of Piedmont, is known for its *grissini,* the slender breadsticks popular throughout Italy, and *bagna cauda* (meaning "hot bath"), a dip for raw vegetables consisting of anchovies and garlic blended to a paste with olive oil or butter. A summer favorite is *vitello tonnato,* braised veal served cold with a spicy tuna sauce. Located on the northwest coast of Italy, Genoa is known for its *burrida,* a fish stew containing octopus and squid, and *pesto,* an herb, cheese, and nut paste (usually made with basil), which has become popular in the United States.

Moving southward, the city of Bologna is the center of a rich gastronomic region. Pasta specialties of the area include *lasagne verdi al forno,* spinach-flavored lasagna noodles baked in a *ragū* (meat sauce) and white sauce, flavored with Parmesan cheese; and *tortellini,* egg pasta stuffed with bits of meat, cheese, and eggs, served in soup or a rich cream sauce. It is traditionally served on Christmas Eve. A similar stuffed pasta is *cappelleti,* named for its shape, a "little hat." Other specialities are *mortadella,* a pork sausage (similar to American bologna); *prosciutto,* a raw, smoked ham served thinly sliced, often as an appetizer with melon or fresh figs; and *Parmesan* cheese, a sharply flavored cow's milk cheese with a finely grained texture.

Florence, the capital of Tuscany, has a long history of culinary expertise. In 1533, Catherine de' Medici (the Medici family ruled Florence) married into the royal family of France. She is often credited with introducing Italian fare, at the time the most sophisticated cuisine in Europe, to France. Florence is renowned for its green spinach noodles that

- Pasta is served three ways in Italy: with sauce (*asciutta*), in soup (*en brodo*), or baked (*al forno*).
- Roman soldiers are credited with sowing garlic throughout their conquered lands.
- There are reputedly more than 300 different varieties of olives.
- The name *gazpacho* may have come from the vinegar-and-water drink called *posca,* reportedly offered to Christ on the Cross.

Sample Menu

AN ITALIAN LUNCH

Antipasto: fennel salad, artichoke omelet, olives, prosciutto

Minestre: *pasta e fagiole* or Asciutta: fettucini Alfredo

Osso buco or *Pollo alla cacciatora*

Green salad

Fresh fruit and cheese (e.g., Gorgonzola and Bel paese) or *Gelato*

Wine

are served with butter and grated Parmesan cheese. The term *alla Fiorentina* refers to a dish garnished with or containing finely chopped spinach. Tuscany is also famous for its full-bodied red wine, *Chianti,* and its use of chestnuts, which are featured in a cake eaten at Lent called *castagnaccio alla Fiorentina.*

Rome, the capital of Italy, has its own regional cooking and is probably best known for *fettucine Alfredo,* long, flat egg noodles mixed with butter, cream, and grated cheese. Another dish is *saltimbocca,* meaning "jumps in the mouth"—thin slices of veal rolled with ham and cooked in butter and Marsala wine. *Gnocchi,* which are dumplings, are eaten throughout Italy, but in Rome they are made out of semolina and baked in the oven. Fried artichokes are popular at Easter time, as is roast baby lamb or kid. *Pecorino romano* is the hard sheep's milk cheese of Rome, similar to Parmesan but with a sharper flavor.

The capital of Campania in southern Italy is Naples, considered the culinary capital of the south. Pasta is the staple food, and a favorite way of serving it is simply with olive oil and garlic, or mixed with beans, in the soup *pasta e fagiole.* Pizza is native to Naples and is said to date back to the sixteenth century. Another form of pizza is *calzone,* which is pizza dough folded over a filling of cheese, ham, or salami, then baked or fried. The area's best-known cheeses are *Mozzarella,* an elastic, white cheese originally made from buffalo milk; *Provolone,* a firm smoked cheese; and *Ricotta,* a soft, white, unsalted cheese made from sheep's milk and often used in desserts.

Sicily and other regions of southern Italy use kid and lamb as their principal meat. Along the coast, fresh fish, such as tuna and sardines, are used extensively; *baccalo,* dried salt cod, is often served on fast days. The North African influence shows up in Sicily in the use of *couscous,* called *cuscus* in Italy, which is commonly served with fish stews. Southern Italy's cuisine is probably best known for its desserts. Many examples can be found in Italian American bakeries and espresso bars: *cannoli,* crisp, deep-fried tubular pastry shells filled with sweetened ricotta cheese, shaved bittersweet chocolate, and citron; *cassata,* a cake composed of sponge cake layers with a ricotta filling and a chocolate- or almond-flavored sugar frosting; *gelato,* fruit or nut (e.g., black currant or pistachio) ice cream; and *granita,* intensely flavored ices. *Spumoni* is chocolate and vanilla ice cream with a layer of rum-flavored whipped cream containing nuts and fruits. Another popular sweet is *zeppole,* a deep-fried doughnut covered with powdered sugar. The sweet dessert wine *Marsala* is also a specialty in the region.

■ Olive oil is labeled according to method of processing and the percent of acidity, from *extra virgin* to *virgin* (or "pure") olive oil. In the United States, only the oils derived from the first press of the olives can be called virgin; a blend of refined and virgin olive oils developed to reduce acidity must be labeled "pure."

■ *Linguiça* comes from the Portuguese word meaning "tongue," a reference to the shape of the sausage.

■ In Piedmont, the northern Italian state that borders France, the dish *chicken Marengo* was supposedly created for Napoleon after his defeat of the Austrians. Legend has it his cooks invented the dish from locally available ingredients, including chicken, eggs, crawfish, tomatoes, and garlic. Napoleon liked it so much he ordered it to be served after every battle.

■ Italian folklore has it that basil can only develop its full flavor if the gardener curses daily at it.

Spain. Most Spanish dishes prepared in the United States reflect the cooking of Spain's southern region, with its seafood, abundant fruits and vegetables, and Muslim influence. Central Spain has a more limited diet; roast suckling pig and baby roast lamb are specialties. In the Basque provinces, lamb is the primary meat and charcoal-grilled lamb is a specialty. Seafood is a favorite in some Basque areas, such as *bacalao al pil-pil* (dried salt cod cooked in olive oil and garlic), *bacalao a la vizcaina,* (dried salt cod cooked in a sauce of onions, garlic, pimento, and tomatoes), and *angulas* (tiny eel spawn cooked with olive oil, garlic, and red peppers). Other popular dishes include garlic soup, *babarrun gorida* (red beans with chorizo) and *pipperrada vasca* (eggs with peppers). Simple rice puddings or fruit compotes are typical desserts.

Portugal. Though Portuguese cuisine varies from north to south, from hearty soups and stews to a more refined, lighter style of entrée, the largest regional differences occur between the mainland and the islands. The foods of Madiera, the Azores, and the Cape Verde Islands includes tropical ingredients imported from both Africa and the Americas. In Madeira, which attracts many tourists from throughout Europe, avocados, cherimoya, guava, mango, and papaya are featured in its dishes. Corn is common, as is tuna. Honey cakes and puddings reflect the influence of other European nations. In the Azores and Cape Verde Islands, fare varies significantly from island to island and even city to city. Bananas, corn, cherimoya, passion fruit, pineapples, and yams are prominent. *Acorda d'azedo* is one specialty, a mixture of cornbread, vinegar, onions, garlic, saffron, and a little lard boiled together and eaten for breakfast. Beef is the preferred meat, and seafood such as cockles, limpets, crab, lobster, and octopus is eaten in many areas. Little fat or oil is added to dishes, and spicing is mild, often limited to onion, garlic, salt, and pepper. Tea is the preferred beverage.

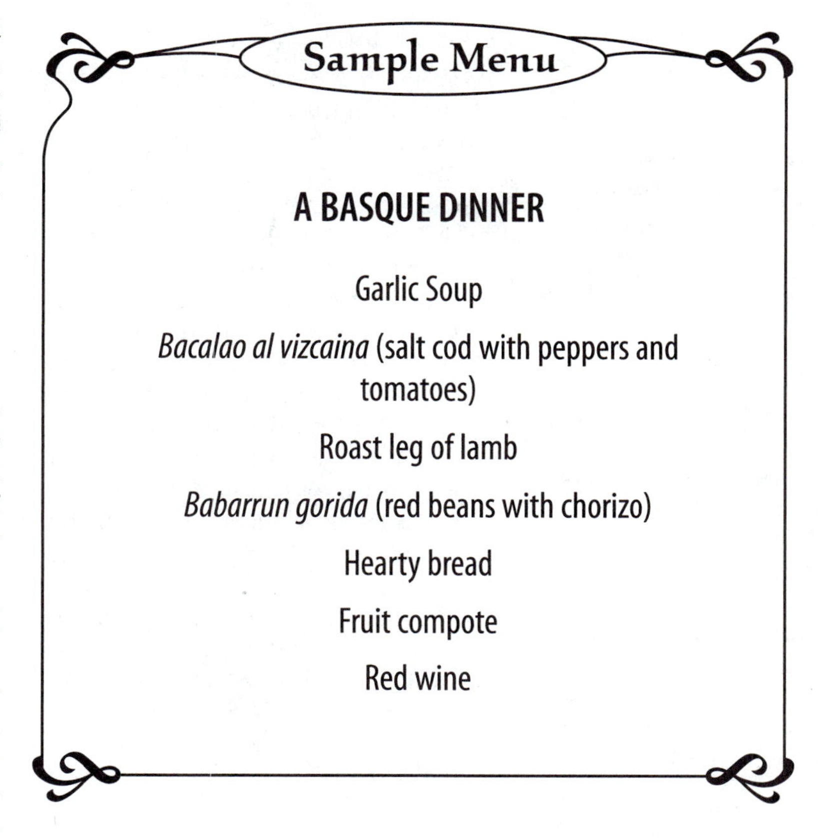

Sample Menu

A BASQUE DINNER

Garlic Soup

Bacalao al vizcaina (salt cod with peppers and tomatoes)

Roast leg of lamb

Babarrun gorida (red beans with chorizo)

Hearty bread

Fruit compote

Red wine

Meal Composition and Cycle

Daily Patterns

Italy. A traditional Italian breakfast tends to be light, including coffee with milk (*caffe latte*), tea, or a chocolate drink, accompanied by bread and jam. Lunch is the main meal of the day and may be followed by a nap. It usually starts with *antipasto,* an appetizer course, such as ham, sausages, pickled vegetables, and olives. Next is *minestra* (wet course), usually soup or *asciutta* (dry course) of pasta, risotto, or gnocchi. The main course is fish, meat, or poultry, such as sausages with lentils, veal shank stew called *osso puco,* or *pollo alla cacciatora* (chicken prepared the way hunters braise venison, with wine, vinegar, garlic, and rosemary). It is served with a starchy or green vegetable, followed by a salad. Dessert often consists of fruit and cheese; pastries and ice creams are served on special occasions. Dinner is served at about 7:30 P.M. and is a lighter version of lunch. Wine usually accompanies lunch and dinner. Coffee or *espresso* is enjoyed after dinner, either at home or in a coffeehouse. Marsala may be drunk after dinner, and is often used in the preparation of desserts. One such sweet, now prepared all over Europe, is *zabaglione,* a wine custard.

Spain. By American standards, the Spanish appear to eat all the time. The traditional pattern, four meals plus several snacks, is spread across the day. A light breakfast (*desayuno*) of coffee or chocolate, bread, or churros is eaten about 8:00 A.M., followed by a midmorning breakfast around 11:00 A.M. of grilled sausages, fried squid, bread with tomato, or an omelet. A light snack, *tapa,* is consumed around 1:00 P.M. as a prelude to a three-course lunch (*comida*) at around 2:00 P.M., consisting of soup or salad, fish or meat, and dessert, which is often followed by fruit and cheese. Many businesses close for several hours in the afternoon to accommodate lunch and a nap. Tea and pastries (*merienda*) are eaten between 5:00 and 6:00 P.M. and more tapas are enjoyed at 8:00 or 9:00 P.M. Finally supper, including three light courses such as soup or omelets and fruit, is consumed between 10:00 P.M. and midnight. Tapas are usually served in bars and cafés and are accompanied by sherry or wine; the variety of tapas is tremendous; it is not unusual for more than 20 kinds to be offered on a menu. The evening meal may be skipped if a substantial number of tapas are eaten at night. The main meal of the day is lunch, which is generally eaten at home and consists of three hefty courses.

Portugal. Portuguese meal patterns are similar to those of Spain, although the evening meal is usually eaten earlier. As in Spain, red wine usually accompanies the meal. Portugal is famous for its sweet, rich wines, Port and Madeira; the latter is often drunk with dessert.

- The city of Bologna is nicknamed *la grossa* (the "fat") and is proud of it.

- Catherine de' Medici's son, Henry III, is said to have introduced the fork to the French court, after he observed it being used in Venice. Although originally from Constantinople, the three-pronged fork was created in Italy for eating pasta.

- The art of making ice cream is believed to have been invented by the Chinese, who brought it to India; from there it spread to the Persians and Arabs. The Muslims brought it to Italy, and it was a Sicilian, Francisco Procopio, who introduced ice cream to Paris in the 1660s. The British discovered it soon after and later brought ice cream to America.

■ Lamprey is a popular food in northern Portugal, where it is often prepared with currylike seasonings.

■ Bread may be served during the meal in Italy with olive oil and a strong, aged, red wine vinegar known as balsamic vinegar. Each person mixes the oil and vinegar as desired, then dips the bread.

■ *Espresso,* which means "made expressly for you," is made from finely ground dark roast coffee through which water is forced by steam pressure. *Cappuccino* is espresso topped with frothy, steamed milk.

■ Zabaglione was supposedly created to increase male vigor by a Franciscan monk who tired of hearing the confessional complaints from women about their tired husbands.

Special Occasions

Italy. Italy celebrates few national holidays, probably because of its divided history. Most *festas* are local and honor a patron saint. Other significant religious holidays are usually observed by families at home, although some cities, such as Venice, have a public pre-Lenten carnival. In some areas of the United States, where southern Italians predominate, St. Joseph, the patron saint of Sicily, is honored during Lent. Breads in the shape of a cross blessed by the parish priest, pasta with sardines, and other meatless dishes are featured.

Among Italian Americans it is traditional to serve seven seafood dishes on Christmas Eve. During the Easter holidays, Italian American bakeries sell an Easter bread that has hard-boiled eggs still in their shells braided into it. Special desserts may accompany the holiday meal, such as *panettone, amaretti* (almond macaroons), and *torrone* (nougats) at Christmastime and *cassata* at Easter. Colored, sugar-coated Jordan almonds, which the Italians call *confetti,* meaning "little candies," are served at weddings.

Spain. The most elaborate of Spanish festivals is Holy Week, the week between Palm Sunday and Easter. It is a time of Catholic processions; confections and liqueurs such as coffee, chocolate, and *anisette* (licorice flavored) are served. Holiday sweets include *tortas de aceite*, cakes made with olive oil, sesame seeds, and anise; *cortados rellenos de cidra,* small rectangular tarts filled with pureed sweetened squash; *torteras*, large round cakes made with cinnamon and squash and decorated with powdered sugar; and *yemas de San Leandro,* a sweet made by pouring egg yolks through tiny holes into boiling syrup. It is often served with marzipan.

Special dishes are also prepared for Christmas and Easter. The Basques eat roasted chestnuts and *pastel de Navidad,* individual walnut and raisin pies, at Christmas; an orange-flavored doughnut, called *causerras,* is featured on Easter. At New Year's, it is customary for the Spanish (and the Portuguese) to eat 12 grapes or raisins at the 12 strokes of midnight. It is believed to bring luck for each month of the coming year.

Portugal. Christmas Eve typically features two meals in Portugal: dinner and a post–midnight Mass buffet in the early hours of Christmas morning. Dinner often includes a casserole of bacalhua and potatoes, as well as meringue cookies known as *suspiros* ("sighs"). The buffet offers mostly finger foods, such as fried cod puffs and sausages.

In America, the Holy Ghost (Spirit) Festival is the most popular and colorful social and religious event in the Portuguese community. Although the origins of the event are obscure, it is believed to date back to Isabel (Elizabeth) of Aragon, wife of Portugal's poet-king, Dom Diniz (1326). One story is that the festival derives its character from the belief that Isabel was particularly devoted to the Holy Ghost and that she wanted to give an example of charity in the annual distribution of food to the poor. In the United States the festival is usually scheduled for a week after Easter and before the end of July and is held at the local church or Hall of the Holy Ghost (also called an IDES Hall). The main event of the festival takes place on the last day, Sunday, with a procession to the church and the crowning of a queen after the service. The donated food, which was originally distributed to needy persons on Sunday afternoon but is now often served at a free community banquet, is blessed by the priest. The most traditional foods at the feast are a Holy Ghost soup of meat, bread, and potatoes, and a sweet bread called *massa sovoda.* The bread is sometimes shaped like little doves, called *pombas.*

Also celebrated in the United States is the Feast of the Most Blessed Sacrament, which was started in New Bedford, Massachusetts by four Madeirans in gratitude for their salvation from a shipwreck en route to the United States in 1915. It attracts over 150,000 visitors on the first weekend in August for music, dancing, and traditional foods such as linguiça, bacalhau, fava beans, assada no espeto, and cacoila. The Festa de Sennor da Pedra is held later in the month. This Azore Island tradition includes a parade and similar traditional foods. Other festivities are associated with the Madeiran cult of Our Lady of the Mount (a shrine on the island of Madeira).

Therapeutic Uses of Food

Many Italians, particularly older immigrants, categorize foods as being "heavy" or "light," "wet" or "dry," and "acid" or "nonacid" (Ragucci, 1981). Heavy foods, such as fried items and red meats, are considered difficult to digest; light foods, including gelatin, custards, and soups, are believed to be easy to digest and appropriate for people who are ill. Wet and dry refer to how foods are prepared (with or without ample broth or fluid) as well as to their inherent qualities. For example, leafy greens such as escarole, spinach, and cabbage are considered wet. A wet meal is served once a week by some Italian Americans to "cleanse out the system." Wet meals, especially soups, are considered necessary when a person is sick, because illness is associated with dryness in the body. Citrus fruits, raw tomatoes, and peaches are thought to be acidic foods that may cause skin ailments and are avoided if such conditions exist (see also Table 7.1, "Selected European Botanical Remedies").

Other Italian beliefs about foods are that liver, red wine, and leafy vegetables are good for the blood and that too many dairy products make the urine "hard" (kidney stones).

Contemporary Food Habits in the United States

Adaptations of Food Habits

It is generally assumed that second- and third-generation Americans of southern European descent have adopted the majority American diet and meal patterns, preserving some traditional dishes for special occasions. These assimilated Americans consume more milk and meat but less fish, fresh produce, and legumes than their ancestors. Olive oil is still used often, although not exclusively; pasta remains popular with Italians.

Nutritional Status

Nutritional Intake

Little research has been conducted on the nutritional intake of southern European Americans. It can probably be assumed that they suffer from dietary deficiencies and excesses similar to those of the majority of Americans. However, descendants of southern Europeans may have a higher incidence of lactose intolerance than other European groups.

In a study of elderly, Portuguese immigrants in Cambridge, Massachusetts, it was found that dinner was the main meal of the day and that the subjects had moderate intake of breads and grains and low intake of fruits, vegetables, and dairy products. Although dairy intake was low, many of the subjects ate sardines, a rich source of calcium. The subjects reported low consumption of sweets and alcohol, although the researcher stated that the Americanized Portuguese diet tends to be high in sugar and fat (Poe, 1986).

A comparison of the nutritional status for rural and urban residents in Spain found that the diet was high in protein and low in vegetable intake in both groups, yet no vitamin or mineral insufficiencies were identified (Varela et al., 1985). Another study of European elders concluded Spanish women had the healthiest diet (Schroll et al., 1997). In general, the Mediterranean diet, which is typified in Spain and southern Italy, has been characterized as health promoting due to a high intake of complex carbohydrates and a low intake of fat with a higher proportion of monounsaturaed fats from olive oil to saturated animal fats (Ferro-Luzzi & Branca, 1995). The greater emphasis on grains, legumes, vegetables, and fruits; lower intakes of meat and dairy foods; and promotion of wine in moderation differentiate the Mediterranean diet from that recommended by U.S. health officials (Willet et al., 1995). However, a study by the Italian

■ It was the Spanish in the 1600s who first added sugar to the bitter chocolate beverage native to Mexico. Its popularity spread quickly through Europe, even though certain clerics tried to ban it due to its association with the "heathen" Aztecs.

■ The word *tapas* means "lid," and the first tapas were pieces of bread used to cover wine glasses to keep out flies. Tapas are differentiated from appetizers in Spain in that they are strictly finger foods, such as olives, almonds, stuffed mushrooms, shrimp, sausage bits, and so on.

■ Although the Spanish are fond of chocolate, it is used mostly as a beverage and is rarely added to pastries or confections.

■ The Holy Ghost Festival is not widely celebrated in Portugal. It was most likely brought to the United States by immigrants from the Azores.

■ In the Portuguese town Amarante, the Festa de São Gonçalo is held the first weekend in June. Dating back to pagan times, it is traditional for unmarried men and women to exchange phallus-shaped cakes as tokens of their affection for each other.

■ Both balsamic vinegar and olive oil are believed to be health-promoting foods in Italy.

■ Some Italians believe that wine mixed with milk in the stomach causes too much acid, so milk is avoided at meals and consumed mostly with snacks.

■ Some diet books advocate a traditional Mediterranean diet to promote weight loss and lessen the risk of heart disease.

Association for Cancer Research has found that cancer rates increased as food habits changed in Italy; pasta consumption has fallen and meat intake has quadrupled since 1950. One study suggests similar changes in the diets of southern Italian Americans have occurred (Ragucci, 1981).

Counseling

Conversational style of southern Europeans is animated, warm, and expressive. Feelings are more important than objective facts in a discussion. Shaking hands with everyone in the room in greeting and leaving is appropriate; some men include pats on the back, and women may quickly embrace or kiss on the cheeks. Eye contact among elders tends to be frequent and quick, whereas younger people may prefer steady eye contact. Touching is very common, especially between members of the same sex.

It has been noted that Italian American clients are open and willing to detail symptoms, although some women may demonstrate high levels of modesty and may resist discussing personal topics (Spector, 1991). Italian Americans may seek medical advice from family and friends before consulting a health professional. They express preference for providers who are warm and empathetic (*sympatico*) and disdain those who are perceived as arrogant and unapproachable (*superbo*).

Recent Italian immigrants or those who are elders may be very concerned about the qualities of their blood or may have many gastrointestinal complaints. There may be confusion regarding hypertension, which is considered "high" or "too much" blood, and anemia or low blood pressure, which is associated with "low" blood (Ragucci, 1981).

Dietary requirements should be carefully detailed for some Italian Americans. Restrictions recommended for clients with diabetes may be ignored if daily social activists (i.e., coffee and pastries with friends) must be modified. Language difficulties may occur among elders or new immigrants.

Information regarding the counseling of Spanish Americans or Portuguese Americans is limited. A high rate of illiteracy has been reported in the Portuguese population (40 percent of surveyed elders; 15 percent of recent immigrants).This should be taken into consideration when preparing educational materials.

When interviewing persons of southern European ancestry, the nutritionist should assess the client's degree of acculturation and traditional health practices, if any. Personal food preferences should be determined.

7

Central Europeans, Russians, and Scandinavians

The European settlers from central Europe, Russia, and Scandinavia were some of the earliest and largest groups to come to the United States. Though many arrived as early as the 1600s and most had come before the beginning of the 20th century, the upheavals of two world wars and the collapse of the Soviet Union have led to continuous immigration from these regions during the last century (see the map in Figure 7.1). The influence of central Europeans, Russians, and Scandinavians on American majority culture, especially in the area of cuisine, are substantial. Bread baking, dairy farming, meat processing, and beer brewing are just a few of the skills these groups brought with them. Their expertise permitted the expansion of food production and distribution that encouraged nationwide acceptance of their ethnic specialties, leading to the creation of a typical American cuisine. This chapter focuses on the traditional and adapted foods and food habits of Germans, Poles, and other central European groups, Russians, Danes, Swedes, and Norwegians.

Central Europeans and Russians

Central Europe stretches from the North and Baltic Seas, south to the Alps, and east to the Baltic States. It includes the nations of Germany, Austria, Hungary, Romania, the Czech Republic, Slovakia, and Poland, as well as Switzerland and Liechtenstein. Most of the countries share common borders; Austria, Hungary, the Czech Republic, Romania, and Slovakia are situated south of Germany and Poland. Switzerland, an isolated nation, is surrounded by Germany, Austria, France, Italy, and Liechtenstein. The climate of central Europe is harsher and colder than that of southern Europe, but much of the land is fertile. Russia and its former territories, including the Commonwealth of Independent States (CIS) and the Baltic States, extend east to the border with China and the Pacific Ocean. Its vast geography includes the Arctic and parts of the Middle East. Except in the southern republics, the harsh winters of Russia affect agricultural capacity.

The large number of immigrants from central Europe made significant contributions to the literature, music, and cuisine of the United States. Many central European foods have become standard fare. What would a baseball game be without hot dogs and beer, or a picnic without potato salad? Croissants, originally from Budapest, are now a common item at fast-food restaurants. This next section explores these and other food customs of central Europe and Russia, and their impact on the American diet.

Cultural Perspective

History of Central Europeans and Russians in the United States

Immigration Patterns

Germans. For almost three centuries, Germans have been one of the most significant

■ "Eastern Europe" is the term sometimes used to define the region that is also called European Russia (the western half of the country; east of the Ural mountains is known as Siberia or Asian Russia). Before the breakup of the USSR, "Eastern Europe" was sometimes used to describe those countries under Soviet control (e.g., Czechoslovakia, Hungary, East Germany, etc.).

■ According to legend, *croissants* were first created by the bakers of Budapest, who expressed their joy regarding their city's victory over the Turks in 1683 with a roll in the shape of the Islamic crescent. Crescent-shaped rolls symbolizing the moon were made as early as the eighth century in Europe, however.

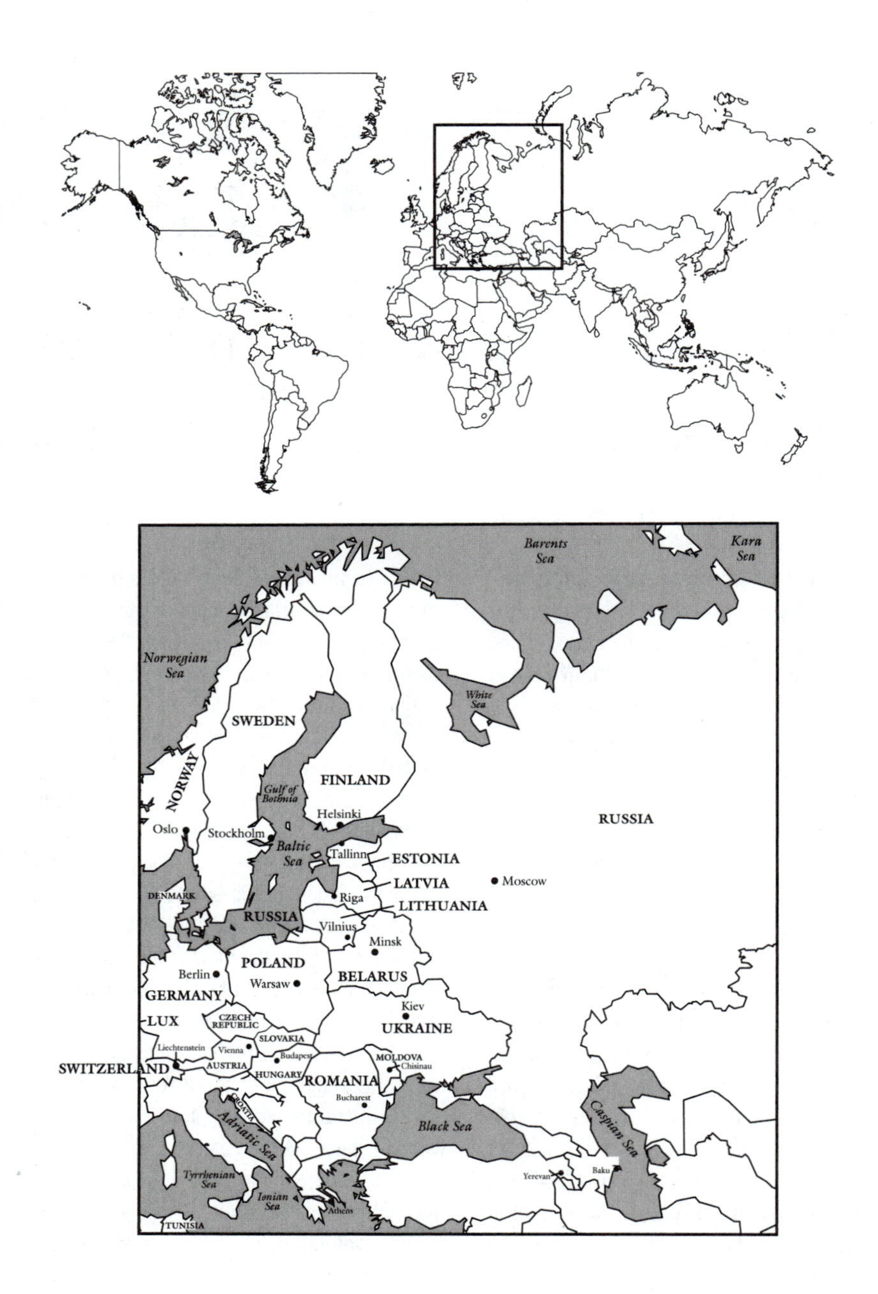

Figure 7.1
Central Europe, Scandinavia, Russia

■ **The word *Dutch* is a corruption of *Deutsch*, meaning "German," and has nothing to do with the Netherlands.**

elements in the U.S. population. In the 1990 U.S. census, it was found that 58 million people identified themselves as being of German descent. Germans are also one of the least visible of any American ethnic group.

The earliest German settlement in the American colonies was Germantown, Pennsylvania, founded in 1681. By 1709, large-scale immigration began, primarily from the Palatinate region of southwestern Germany. Most of the immigrants settled in Pennsylvania. The majority were farmers who steadily pushed westward searching for new lands for their expanding families. Those in Pennsylvania, Ohio, and Indiana became known as the *Pennsylvania Dutch.*

Immigration dropped off between 1775 and 1815, but an economic crisis in Europe once again prompted numerous Germans to come to the United States. It is estimated that approximately 5 million Germans immigrated to America between 1820 and 1900. Like the earlier settlers, most were farmers who arrived with their families, although by the end of the century there were increasing numbers of young, single people who were agricultural

Traditional foods of central Europe and Russia. Some foods typical of the traditional diet of central Europe and Russia include beets, cabbage, ham, herring, *kasha,* potatoes, rye bread, sausages, and sour cream. (Photo by Laurie Macfee.)

laborers and servants. Many of these settled in the Mississippi, Ohio, and Missouri River valleys, the Great Lakes area, or the Midwest. Most Germans avoided the southern United States, but there are sizeable German settlements in Texas and New Orleans.

A third phase of immigration began after the turn of the 20th century, when approximately 1.5 million Germans came to America. Many were unmarried industrial workers seeking higher pay, and others were the descendants of Germans who had settled in ethnically isolated colonies in Russia as early as the 16th century. Discrimination and the revolution of 1917 led to their departure. Most of these third-phase immigrants joined growing numbers of second- and third-generation Germans living in urban areas. Cities with considerable German populations included Cleveland, New York, Toledo, Detroit, Chicago, Milwaukee, and St. Louis. Russian Germans, however, tended to settle in rural areas, especially in Colorado.

During the 1930s, many of the German immigrants were Jewish refugees. After World War II, displaced persons of German descent and East German refugees made up the sizeable German immigrant group who settled in the United States.

Poles. Although the Poles have arrived in the United States continuously since 1608, the largest wave of immigration occurred between 1860 and 1914, mostly for economic reasons. The early phase was dominated by Poles (approximately 500,000) from German-ruled areas of Poland (Pomerania and Poznan) or by Poles who worked in western Germany. German Poles often became part of the German or Czech communities or established farming settlements in the Southwest and Midwest.

The number of Polish immigrants from Germany began to decline after 1890, but the slack was taken up by the arrival of more than 2 million Poles from areas under Russian and Austrian rule. The German Poles left their homeland to become permanent settlers, but the Russian and Austrian Poles came as temporary workers. Although 30 percent returned to Poland, many eventually moved back to the United States permanently. The Austrian and Russian Poles tended to settle in the rapidly developing cities of the middle Atlantic and midwestern states, especially Chicago, Buffalo, and Cleveland.

Polish emigration after World War I was usually not for economic reasons. Most (more than 250,000) left because of political dissatisfaction: government instability and dictatorship in the 1920s and 1930s, the German invasion and occupation from 1939 to 1945, and the pro-Soviet communist government after 1945. Many settled in urban areas in which there were substantial existing Polish populations. More recently, small numbers of younger Poles have taken advantage of the freedom resulting from post-communist rule to come to the United States.

■ Both the Czechs and the Slovaks are descendants of Slavic tribes.

■ Swiss immigrants of the early 19th century were often Mennonites, who established religious communities in Ohio, Pennsylvania, and Virginia.

■ The Gypsies are an insular ethnic group found throughout the world. When they first arrived in Europe in the 1300s from India, they were mistaken for Egyptians. Their name derives from this error. Those from eastern and southeastern Europe are the *Rom,* and some Gypsies prefer the name *Roma.*

Other Central Europeans. Nearly 4 million Austrians, Hungarians, Czechs, Slovaks, and Swiss have come to the United States, primarily during the late 19th and early 20th centuries, for economic and political reasons.

Austrian immigration patterns are not entirely known because Austrians and Hungarians were classified as a single group in U.S. statistics until 1910. It is believed that more than 2 million Austrians came to the United States in the decade following 1900, searching for economic opportunities. Most were unskilled and many were fathers who left families in Austria with the hopes of making their fortune. Many Austrians never found the advancement they were seeking, and approximately 35 percent returned home. A second, smaller wave of immigration occurred in the 1930s, when 29,000 well-educated, urban Austrian Jews fled Hitler's arrival.

The first group of Hungarians to arrive in the United States were several thousand political refugees following the revolution of 1848. Most were men—well educated, wealthy, and often titled. Later Hungarian immigrants who immigrated at the turn of the century were often poor, young, single men who found job opportunities in the expanding industrial workplace. Many worked in the coal mines of eastern Ohio, West Virginia, northern Illinois, and Indiana. Cities that developed large Hungarian populations were primarily located in the Northeast and Midwest. More than 50,000 additional Hungarians entered the United States as refugees after World War II and the 1956 uprising against the communist government. They first settled in the industrial towns populated by earlier Hungarian immigrants, but many, mostly professionals, soon moved to other cities that offered better jobs.

Czech immigrants initially tended to be farmers or skilled agricultural workers who settled in the states of Nebraska, Wisconsin, Texas, Iowa, and Minnesota, often near the Germans. Later Czech immigrants were skilled laborers; they settled in the urban areas of New York, Cleveland, and especially Chicago.

The majority of the Slovak immigrants were young male agricultural workers who arrived before World War II. Those who decided to remain in America later sent for their wives and families. The majority settled in the industrial Northeast and Midwest; they labored in coal mines, steel mills, and oil refineries.

Immigrants from Switzerland came to the United States for economic opportunities. The majority arrived prior to World War I, seeking jobs as artisans or professionals in the urban areas of New York, Philadelphia, Chicago, Cincinnati, St. Louis, San Francisco, and Los Angeles.

Another group without national boundaries found throughout central Europe (as well as in northern and southern Europe) are the Gypsies. Gypsy immigrants to the United States are not counted in U.S. census figures. Estimates are that there are between 200,000 and 1,000,000 Gypsies in America, from a variety of Gypsy groups, and speaking different dialects. Although they originate from numerous European countries, it is believed that the majority living in the United States came from central Europe.

Russians. Russian immigrants originally came to Alaska and the West Coast, rather than to the eastern states. Most of their settlements were forts or outposts, used to protect their fur trade and to shelter missionaries. When Russia sold Alaska to the United States in 1867, half of the settlers returned home and many of the others moved to California. Subsequent immigration was primarily to the East Coast, although some Russians (Molokans, followers of a religion that had rejected the Russian Eastern Orthodox Church) immigrated to the West Coast in the early 20th century.

Russians, mainly impoverished peasants seeking a better life, began to arrive in large numbers during the 1880s. A second wave of immigrants came after the 1917 revolution, when more than 2 million people fled the country; 30,000 settled in the United States. After World War II, only small numbers of

Eastern Orthodox priest blessing the Coca-Cola plant in Moscow, 1996. (AP/Wide World Photos)

Soviet refugees, primarily Jews, were allowed to emigrate. Following the breakup of the Soviet Union, nearly 200,000 Russians settled in the United States between 1990 and 1993.

The settlement patterns of Russians are similar to those of other immigrants from central Europe. For the later wave of immigrants the port of entry was New York City. Many remained in New York, and others settled in nearby industrial areas that offered employment in the mines and factories.

Current Demographics and Socioeconomic Status

Germans. Germans differ little from the national norms demographically, although they are slightly higher in economic achievement and are generally conservative in attitudinal ratings. Only the Pennsylvania Dutch, and the rural-dwelling Germans from Russia who settled in the Midwest and in a few concentrated communities in Texas, retain some aspects of their cultural heritage.

The high degree of German acculturation is attributed to their large numbers, their occupations, and the time of their arrival in the United States. Furthermore, entry of the United States into World War I created a storm of anti-German feeling in America. German-composed music was banned, German-named foods were renamed, and German books were burned. As a result, German Americans rapidly assimilated, abandoning the customs still common in other ethnic groups, such as ethnic associations and use of their oral and written language.

Poles. Polish Americans form one of the largest ethnic groups in the United States today. In 1990, it was estimated that there were more than 9.5 million Americans of Polish descent; many still live in the urban areas of the Northeast and upper Midwest where their ancestors originally settled. Economically the

■ The changing socioeconomic systems of Russia have created marketing opportunities for U.S. food firms. In 1992 the National Peanut Council gave away 50,000 pounds of peanut butter in Moscow to introduce the product.

■ Nearly one in every four Americans is of German descent, the largest ethnic group in the United States.

■ It is estimated that there are 8 million Canadians of German descent and 787,000 of Polish ancestry. In addition, 1 million Canadians list their heritage as Ukranian.

■ 73 percent of the population in Switzerland speaks German Swiss, 20 percent speaks French Swiss, and 5 percent speaks Italian Swiss; in addition, most Swiss speak one or two other languages.

third-generation Polish American has moved modestly upward, but the majority of Polish Americans still live just below or solidly at middle-class level. 45 percent of the males have white-collar jobs, and 40 percent of working Polish women are semiskilled or unskilled laborers. Poles have been active in the formation and leadership of U.S. labor unions. More recent immigrants usually possess higher occupational skills and educational backgrounds than earlier immigrants.

Other Central Europeans. There is continued confusion over Austrian ethnicity dating back to changing national boundaries and names. It is believed that although only 950,000 Americans identify themselves as being of Austrian descent in the 1990 U.S. census, as many as 4 million U.S. citizens may actually be of Austrian ancestry. Though early immigrants settled mostly in the Northeast, the largest populations of Austrian Americans are now found in New York, California, and Florida. At the turn of the century, Austrians were involved in clothing and tailoring, mining, and the food industry, including bakeries, meatpacking operations, and restaurants. Today Austrians are found in a diverse range of occupations.

In the 1990 U.S. census, it was estimated that there were more than 1.5 million Hungarian Americans in the United States. Most settled originally in the Northeast, but younger generations have migrated to California and Texas, while many Hungarian retirees have moved to Florida. Economically the Hungarians differ little from other central European immigrants. Most live in urban areas and work mostly in white-collar occupations. First- and second-generation Hungarian Americans encouraged their children to become engineers, a science that was respected by the Hungarian aristocracy at the turn of the century.

More than 1,296,000 Americans of Czech descent were identified in the 1990 U.S. census. As of 1970, most Czechs lived in cities or rural nonfarm areas; statistics show that 450,000 of them grew up in homes where Czech was spoken. Cities and states with large Czech populations are California, Chicago, Iowa, Minnesota, Nebraska, New York City, Texas, and Wisconsin. Occupationally, only a small number of Czech Americans are still farmers; a majority now hold sales, machinist, or white-collar jobs. Many Czechs have been successful in industry, founding businesses in the production of cigars, beer, and watches.

There are nearly 2 million Americans of Slovak descent, according to the 1990 U.S. census. The first two generations of Slovaks grew up in tightly knit communities anchored by work, church, family, and social activities. The third and fourth generations have sought higher education, work in white-collar jobs, and live in the suburbs; the 1990 U.S. census revealed only 26 percent of Slovaks hold jobs in areas such as manufacturing and construction. As of 1989 the median family income was $40,000, far above the national median. Less than 4 percent of Slovaks live in poverty. Cultural ties are still strong among the later generations of Slovaks.

More than 600,000 citizens declared Swiss ancestry in the 1990 U.S. census. Most Swiss were multilingual and often multicultural when they arrived, assimilating quickly into American culture. The few Swiss who come to the United States today work mostly in the U.S. branches of Swiss companies.

Gypsies retain their reputation for roving after arrival in the United States; their exact numbers are unknown and they are a very mobile population, often living in trailer parks. Many renovate apartments and houses to accommodate large social gatherings, then pass the homes on to other Gypsy Americans when they move. It is estimated that approximately half live in the rural areas of the south and west, and half live in urban regions. The cities with the largest concentrations of Gypsies are Los Angeles, San Francisco, New York, Chicago, Boston, Atlanta, Dallas, Houston, Seattle, and Portland. Traditionally tinkers and traders, Gypsies have been very successful at independent trades, such as housepainting and asphalt paving, and service work such as body–fender repair and dry cleaning. Gypsies have also entered the car dealership profession in large numbers. Women are a strong presence in the mystical arts, including fortune-telling. The Gypsies divide the urban regions of America to minimize competition between Gypsy-owned businesses.

Russians. In 1990 approximately 3 million Russian Americans were living in the United States. They have mostly moved out of the inner-city settlements to the suburbs. Approximately 181,000 Russian Jews immigrated to the United States between 1971 and 1991. Most of these recent émigrés have settled in urban areas, including New York, Chicago, Los Angeles, and San Francisco. Immigrants who have arrived since 1920 have come from relatively high educational and professional backgrounds.

Since World War II the relations between the Russian American community and American society have largely been dependent on the political climate between the United States and Russia. During the 1950s, anti-Soviet and anticommunist sentiments in the United States caused many Russian Americans to assume a low profile that hastened their acculturation.

Worldview

Religion

Germans. The majority of German immigrants were Lutheran; a minority were Jewish or Roman Catholic. Today the Pennsylvania Dutch and the rural Germans from Russia faithfully maintain their religious heritage. Both groups are primarily Protestant, mostly Lutheran or Mennonite. Mennonites are a religious group derived from the Anabaptist movement, which advocated baptism and church membership for adult believers only. They are noted for their simple lifestyle and rejection of oaths, public office, and military service. The Amish, a strict sect of the Mennonites, follow the Bible literally. They till the soil and shun worldly vanities such as electricity and automobiles. Their life centers on *Gelassenheit,* meaning submission to a higher authority through reserved and humble behavior, placing the needs of others before the needs of the individual.

Poles. Most Polish immigrants were devout Catholics; they quickly established parish churches in the United States. The Catholic Church is still a vital part of the Polish American community, although Polish Americans have been found to marry outside the church more than other Catholics.

Other Central Europeans. Austrians are mostly Roman Catholic and have been actively involved in promoting Catholicism in America. In 1829 the *Leopoldine Stiftung* was founded in Austria to collect money throughout Europe to introduce religion to the U.S. frontier, resulting in more than 400 Catholic churches established in the East, the Midwest, and in what was known as "Indian" country. There are also small numbers of Austrian Jews. The majority of Hungarians are Catholic, although in the United States, nearly 25 percent are Protestant.

In Europe, most Czechs were Roman Catholics, but one-half to two-thirds of 19th-century Czech immigrants from rural areas left the church and were considered "free thinkers" who believed in a strong separation of church and state. Subsequent generations now belong to a variety of faiths.

Religion is still an important factor in the lives of Slovaks. Most are Roman Catholics who attend services regularly. First- and second-generation Slovaks usually send their children to parochial schools supported by the ethnic parish.

Traditional spirituality for the Gypsies is derived from Asian Indian religions, such as Hinduism and Zoroastrianism (see chapter 2 regarding Eastern religions and Chapter 14 for more information on Asian Indian faiths). While traditions and customs vary by tribe and to a certain degree by the host culture, Gypsies are thought to be united in their worldview, called *romaniya.* Many believe in God, the devil, ghosts, and predestination. Most of all they adhere to the concept that persons and things are either pure or polluted. Gypsy culture is structured to preserve purity and to avoid contamination through contact with non-Gypsies. Some Gypsy Americans are Christians, often members of fundamentalist congregations, and several churches have specifically combined Gypsy spiritual concepts with Christian practices.

Russians. Except for the Soviet Jews, the primary organization of the Russian American

■ Policy in the former Soviet Union required aging parents to leave if their children wished to emigrate. One result is that Russian immigrants are one of the oldest populations in the United States: 20 percent are older than 65 years.

■ The Amish and Mennonites are referred to as the "Plain People." The more liberal and worldly members of the Lutheran and Reformed Churches are called the "Gay Dutch," "Fancy Dutch," or "Church People."

■ The Amish live by the acronym JOY—Jesus first, Others in between, and Yourself last.

community today is the Russian Orthodox Church. Religion has always played a central role in the Russian community, and the Orthodox church has tried to preserve the culture. However, the largest branch of the Eastern church, officially known as the Orthodox Church in America (formerly the Russian Orthodox Church outside Russia), now includes people from other central European countries and the Russian traditions have been deemphasized.

Family

■ Many Amish families are finding it difficult to maintain traditional values due to growing contact with the majority culture through suburban sprawl, outlet malls, and so forth. Recent arrests of Amish youths for dealing drugs at hoedowns exemplifies the problem.

Germans. The traditional German family was based on an agricultural system that valued large families in which every member worked in the fields to support the household. Even when German immigrants moved to urban areas, family members were expected to help out in the family business. It is assumed that most German families today have adopted the smaller American nuclear configuration. The exception may be among the Pennsylvania Dutch, particularly the Amish, who continue to have large families of 7 to 10 children. It is not unusual for an Amish person to know as many as 75 first cousins or for a grandparent to have 35 grandchildren (Kraybill, 1995).

Poles. Traditionally the Polish American family was patriarchal and the father exerted strong control over the children, especially the daughters. The mother took care of the home, and if the children worked, it was near the home or the father's workplace. Since the 1920s the overwhelming majority of Polish American families have been solely supported by the father's income; wives and children have rarely worked.

Other Central Europeans. Tight nuclear families typify traditional Austrian households. Although the father is in charge of family finances, it is the mother who rules home life. Assimilation in the United States has led to a deterioration of the nuclear family, including an increased divorce rate. Traditional Czech and Hungarian families were male dominated and included many relatives. In the United States, participation in church activities, fraternal societies, and political organizations often served to replace the extended family for both men and women. The role of women has become less circumscribed; children are typically encouraged to pursue higher education and professional careers. Family ties are strong among the Slovaks. Parents are respected; they are frequently visited and cared for in their old age. Weddings are still a major event, although they are not celebrated for several days as they once were.

Gypsies customarily maintain extended families, although in the United States more nuclear families have been established. When traveling was common, multifamily groups (smaller than tribes) would temporarily band together. Affiliation with this group, called a *kumpania,* often continues today. The father is in charge of all public matters, but, at home, women often make most of the family income and manage money matters. Women also retain some power through their ability to communicate with the supernatural world. Usually Gypsies do not date, and arranged marriages are still common.

Russians. Traditionally Russians lived in very large family groups with women legally dependent on their husbands. This structure changed, however, with the education and employment opportunities offered to women during the communist rule of the Soviet Union. Most women worked, and families became smaller. Even when employed full-time, however, women remained responsible for all household chores.

In the United States, Russian family structure has shrunk even further. Russian couples have significantly fewer children than the national average for American families. Education is emphasized, especially if it can be obtained at a Russian-language school. Many first-generation immigrants attempted to maintain ethnic identity by restricting their children to spouses from their immediate group, but marriage to non-Russians is now the norm.

Traditional Health Benefits and Practices

German biomedicine makes extensive use of botanicals (see Table 7.1 "Selected European

Botanical Remedies"), though continued use is not documented in German Americans. A study of elderly German Americans in Texas shows that many believe that illness is caused by infection or stress-related conditions (Spector, 1991). Some Germans believe that sickness is an expected consequence of strenuous labor. Health is maintained by dressing properly, avoiding drafts, fresh air, exercise, hard work, and taking cod liver oil. A few respondents mentioned the importance of religious practices and that suffering from illness is a blessing from God. Numerous home remedies—such as chicken soup for diarrhea, vomiting, or a sore throat; tea for an upset stomach; and milk with honey for a cough—are common.

The Pennsylvania Dutch customarily believed that cold drinks were unhealthy and eating meat three times per day was the cornerstone of a good diet. Many use home remedies and healers to treat illness. Sympathy healing is especially well-developed. This traditional folk practice uses charms, spells, and blessings to cure the symptoms of disease. It is called either *powwowing* (though not related to Native American beliefs and practices) or by its German name, *Brauche* or *Braucherei.* There is a strong religious foundation to the practice, and the healer acts as God's instrument and requests God's direct assistance in treatment. Powwow compendiums still in use today offer everything from household tips to home remedies for warts, burns, toothache, and the common cold. The Amish in particular subscribe to sympathy healing, the laying on of hands to diagnose illness, reflexology (foot massage thought to benefit other areas of the body, such as the head, neck, stomach, and back), as well as the use of herbal home cures, especially teas (Hostetler, 1976; Yoder, 1976).

Elderly Polish Americans in Texas (Spector, 1991) have reported that a shortage of medical supplies in Poland led to the widespread use of faith healers. Although such healing practices are not documented in the United States, many Polish Americans are deeply religious and believe that faith in God and the wearing of religious medals will help to prevent illness. Other health maintenance beliefs include avoidance of sick people, a healthy diet, sleep, keeping warm, exercise, a loving home, and avoidance of gossip. Homemade sauerkraut is believed good for colic, as is tea and soda water. Other remedies noted were chamomile tea for cramps, mustard or oatmeal plasters or dried raspberries and tea with wine for colds, cooked garlic for high blood pressure, and warm drinks (milk, tea, or lemonade) for coughs.

Gypsies have unique health beliefs (Sutherland, 1992). Health is maintained through *marimé,* a system of purity and pollution that may be related to Asian Indian beliefs (see chapter 14). The separation of clean from unclean dictates much of Gypsy life. The body is an example of this dichotomy. The upper body is pure, as are all its secretions, such as saliva. The lower half is impure and shameful. Care is taken to avoid contamination of the upper body by touching the lower body (only the left hand is used for personal care). Menstrual blood is especially impure. Purity is also maintained by avoiding public places, where non-Gypsies (who are considered unclean) frequent and by not touching contaminated surfaces, as well as use of disposable utensils, cups, and towels when in impure locations (e.g., hospitals).

Gypsies divide illnesses into those that are due to contact with non-Gypsies, and those that are Gypsy conditions caused by spirits, ghosts, the devil, or through breaking cultural rules. Home remedies and Gypsy healers (usually an older woman versed in medicinal lore) are considered best for Gypsy illnesses. Non-Gypsy conditions are suitable for treatment by non-Gypsy physicians, though a non-Gypsy folk healer such as a curandero may be consulted as well.

In the Siberian region of Russia, sickness was traditionally attributed to a spiritual crisis due to theft of the soul, intervention of evil spirits, breach of taboos, or injury caused by cursing or the deliberate sending of harmful objects into the body. Shamans, who are healers with magicoreligious powers (see chapter 2 for more information), would attempt to cure a person through realignment of the life forces or retrieval of the soul by using visualization techniques, singing, chanting, prognostication, dream analysis, and séances. These beliefs are often in conflict with the biomedical approach favored by most Russian health practitioners (Balzer, 1987). Recent elderly

■ The Pennsylvania Dutch traditionally considered oysters to be an aphrodisiac and warned young women that eating them would lead to a wanton heart.

■ Symptoms of colic in Amish infants are attributed to a condition known as *livergrown,* which can only be cured with sympathy healing.

■ The Poles say "a doctor's mistakes are covered in earth."

■ Gypsy women are traditionally prohibited from touching food, water, or utensils intended for other family members during their period or following childbirth.

Table 7.1
Selected European Botanical Remedies

Scientific Name	Common Name	Parts	Traditional Use
Achillea millefolium	Yarrow; milfoil; thousand leaf; green arrow; wound wort	Root; leaves; flowers	Cramps; colds; indigestion; menstrual irregularities; anorexia; kidney disorders; gallbladder conditions; wounds
Althea officinalis	Althea root; marshmallow	Root; leaves; flowers	Gastrointestinal infections; stomach pain; irritable bowel syndrome; colitis; cystitis
Angelica archangelica ☹	Angelica; European angelica; garden angelica; wild parsnip	Root; leaves; stems; seeds	Indigestion; colic; flatulence; cramps; anorexia
Arctium lappa	Burdock; cockle buttons	Root; leaves; fruit	Gout; kidney stones; gastrointestinal infections; rheumatism; arthritis
Artemsia absinthium ☠	Absinthe; wormwood; madderwort	Leaves	Indigestion; gallbladder problems; anorexia
Betula pendula	Silver birch; weeping birch; European white birch	Leaves; sap; oil	Kidney disorders (especially stones); bladder stones; urinary tract infections; rheumatism
Calendula officinalis	Pot marigold	Flowers	Stomach ailments; gastrointestinal infections; chronic wounds
Carduus marianus; Silybum marianum	Milk thistle; St. Mary's thistle	Flowers; seeds	Liver toxins; liver ailments; indigestion
Chelidonium majus ☠	Celadine; rock poppy; wartweed; Jacob's ladder; devil's milk; cocksfoot	Whole plant	Gallbladder problems (especially stones) gastrointestinal cramps; liver conditions; warts
Cnicus benedictus ☠	Holy thistle; blessed thistle	Leaves; flowers	Indigestion; stomach pain; arteriosclerosis; obesity
Crataegus oxyacantha; C. laevigata; C. monogym ☹	Hawthorn	Flowers; fruit	Heart aliments; hypertension; memory problems
Digitalis purpurea ☠	Purple foxglove; common foxglove; fairy bells; lion's mouth; dog's finger	Leaves; flowers; seeds	Heart conditions; diuretic
Echinacea angustifolia, E. purpurea ☹	Purple cone flower; narrow leaf cone flower; hedgehog cone flower; Sampson root	Root; flowers	Fatigue; colds; flue; rhinitis; sore throat; inflammation; chronic infections; abscesses; chronic wounds
Foeniculum vulgare ☠	Fennel	Seeds; oil	Stomach pain; colic; gastrointestinal disorders; bloating; flatulence; kidney ailments (especially stones); urinary tract infections; anorexia;
Fragaria vesca	Wood strawberry; wild strawberry	Leaves; fruit	Anemias; kidney conditions; liver ailments; bladder stones; gout; gastrointestinal problems; dysentery; diarrhea; obesity; fever; TB; rheumatism; rashes; nervousness
Glycyrrhiza glabra ☠	Licorice root	Root	Peptic ulcers; bronchitis
Humulus lupulus	Hops	Flowers (female only)	Indigestion; colic; anxiety; tension; irritability; insomnia

Table 7.1
Selected European Botanical Remedies—continued

Scientific Name	Common Name	Parts	Traditional Use
Hypericum perforatum ☹	St. John's wort; hypericum	Flowers	Indigestion; anxiety; depression; menopausal problems; insomnia; kidney conditions; wounds; burns
Hyssopus officinalis ☹	Hyssop	Flowers; oil	Indigestion; colic; flatulence; pleurisy; asthma; chronic coughs; bronchitis
Matricaria recutita; Chamomilla recutita	German chamomile	Flowers	Stomach pain; colic; peptic ulcer; Crohn's syndrome; irritable bowel syndrome; gingivitis; bronchitis; menstrual cramps; vaginitis; allergies; rashes
Plantago afra; P. psyllium; P. ovata ☹	Psyllium; blond psyllium	Seeds	Indigestion; peptic ulcers; colitis; constipation; hemorrhoids
Primula veris	Primrose; cowslip	Root; leaves; flowers	Bronchitis; chronic cough; hyperactivity; insomnia; allergies; asthma
Rhamnus spp. ☹	Buckthorn; alder buckthorn	Bark; fruit	Constipation; hemorrhoids
Rosa canina	Dog rose; wild briar; hiptree	Flower hips	Indigestion; colic; diarrhea; kidney conditions; urinary tract infections; gallstones; edema; lung complaints; colds; flu; rheumatism; sciatica
Rubus idaeus	Raspberry; European raspberry; red raspberry	Leaves; fruit	Diabetes; heart conditions; stomach pain; diarrhea; menstrual irregularities; childbirth; bronchitis; flu; rashes; eye ailments
Solidago virgaurea	Goldenrod	Flowers	Urinary tract infections; gastrointestinal disorders; kidney stones; bladder stones; thrush; vaginal yeast infections
Tanacetum parthenium ☹	Feverfew; ague plant; devil daisy; midsummer daisy	Flowers	Migraine headaches; fever; amenorrhea; arthritis; rheumatism
Uritica dioica ☹	Nettle	Root; leaves; flowers	Kidney problems (including stones); urinary tract infections; edema; diuretic; prostate enlargement; TB; coughs; allergies; nosebleeds; wounds
Vaccinium myrtillus	bilberry	Leaves; fruit	diabetes; stomach pain; kidney disorders; urinary tract infections; diarrhea; hemorrhoids; arthritis; rheumatism
Valeriana officinalis ☹	Valerian	Root	Indigestion; nausea; stress; tension; insomnia
Viburnum opulus	Crampbark; European cranberry bush; whortleberry	Bark	Hypertension; colic; constipation; irritable bowel syndrome; menstrual pain; back pain; arthritis
Vitex agnus-castus ☹	Chaste tree; monk bark; agnus castus	Fruit	Menstrual irregularity; PMS; infertility; milk stimulation

Note: Data on some plants are very limited; adverse effects may occur even if not indicated.
Key: ☠, *all or some parts reported to be harmful or toxic in large quantities or specific preparations;* ☹, *may be contraindicated in some medical conditions/with certain prescription drugs.*

Russian immigrants report that illness is caused by social factors, such as war, political problems, poor medical are, or starvation. An authoritarian health care system with little patient autonomy that emphasized environmental causes for disease accounts for this view. Preventative medicine was uncommon (Brod & Heurtin-Roberts, 1992).

Traditional Food Habits

Ingredients and Common Foods

Staples and Regional Variations

The regional variations in central European and Russian cuisine are minor. Ingredients in traditional dishes were dictated by what could be grown in the cold, often damp climate.

Common ingredients are potatoes, beans, cabbage and members of the cabbage family, beets, eggs, dairy products, pork, beef, fish and seafood from the Baltic Sea, freshwater fish from local lakes and rivers, apples, rye, wheat, and barley (see Table 7.2 for the cultural food groups). Foods were often dried, pickled, or fermented for preservation—for example, cucumber pickles, sour cream, and sauerkraut.

Bread is a staple item and there are more than 100 different varieties. Because the climate makes wheat harder to grow, bread is often made with rye and other grains; thus it is darker in color than bread made from wheat flour. Another common breadlike food in the region is dumplings (called *knedliky* in Czech, *Knödel* in German, and *kletski* in Russian), which can be made with flour or potatoes and with or without yeast. Stuffed dumplings, filled with meat, liver, bacon, potatoes, or fruit, are called *pierogi* (Polish), *pelmeni* (Russian), or *varenyky* (Ukranian). Dumplings are usually boiled in water. Related to the dumpling is stuffed pastry dough, which is baked or fried. It is customarily filled with meat or cabbage. Small individual pastries are called *pirozhki* in Russian, and a large oval pie is known as a *pirog* (also called as *kulebi*). One elaborate version, *kulebiaka,* usually includes a whole salmon with mushroom and rice filling.

Next to bread, meat is probably the most important element of the diet. Pork is the most popular. *Schnitzel* is a meat cutlet, often lightly breaded and then fried. Ham is served fresh or cured; Poland is famous for its smoked ham, and in Germany, *Westphalian* ham is lightly smoked, cured, and cut into paper-thin slices. Beef is also common. In Germany, *Sauerbraten,* a marinated beef roast, is the national dish. Poultry is well liked. Germans often eat roast goose, stuffed with onions, apples, and herbs, on holidays. In Russia, chicken is stewed on special occasions, and breaded chicken cutlets, called *kotlety pozharsky,* are common. A famous Russian dish is *chicken Kiev,* breaded, fried chicken breasts filled with herbed butter.

In the past, meat was often scarce and expensive; thus many traditional recipes stretched it as far as possible. Dishes common throughout the region are ground meat that has been seasoned, mixed with a binder such as bread crumbs, milk, or eggs, formed into patties, then fried. In Germany, ground beef (and sometimes pork or veal) is served raw on toast as *steak Tartar.* Ground meat is also used to stuff vegetables (such as stuffed cabbage) or pastry, or is cooked as meatballs. Cut-up meat is often served in soups, stews, or one-pot dishes. In Germany a slowly simmered one-pot dish of meat, vegetables, potatoes, or dumplings is called *Eintopf.* Hungary is known for its *gulyás,* a paprika-spiced stew known as goulash in the United States. Sweet Hungarian paprika is ground, dried, red chile peppers to which sugar has been added. As chiles are a New World food, it is thought the Hungarians used black pepper to season their food before the discovery of the Americas.

Ground meats are also made into sausages. In Germany there are four basic types of sausage (*Wurst*). *Rohwurst,* similar to American-style liverwurst, is cured and smoked by the butcher and can be eaten as is. Examples include *Teewurst,* a raw, spiced pork sausage that is spreadable like pâté, and *Mettwurst,* a mild, sliceable pork sausage. *Bruhwurst* (the frankfurter or *Wienerwurst* is one type) is smoked and scalded by the butcher; it may be eaten as is or heated by simmering. *Kockwurst,* which is like a cold cut, may be smoked and is fully cooked by the butcher. *Leberwurst* (liverwurst), *Blutwurst* (blood sausage), and *Süize* (head cheese) are examples. *Bratwurst,* similar to sausage links,

■ Legend has it that it was the Mongols who showed the central Europeans how to broil meat, make yogurt and other fermented dairy products, and preserve cabbage in brine.

■ *Pumpernickel* is a very dark rye bread. Its German name derives from its supposed gas-producing qualities: *Pump,* meaning a small hollow sound, and *Nickel,* a dwarf.

■ *Spätzle* are tiny dumplings common in southern Germany. They are made by forcing the dough through a large spoon with small holes into boiling water. They are usually served as a side dish.

■ A *frankfurter* means a sausage from the German city of Frankfurt. There are differing opinions as to how the frankfurter was introduced to America, but it soon became a favorite at baseball games. The cartoonist T. A. Dorgan drew a sausage with a tail, legs, and a head so that it looked like a dachshund, leading to the nickname "hot dog."

is sold raw by the butcher and must be pan-fried or grilled before eating.

The Polish are famous for *kielbasa,* a garlic-flavored pork sausage. In Austria, some sausages are called *wieners.* Two popular sausages with both the Czechs and Slovaks are *jaternice,* made from pork, and *jelita,* a blood sausage, which can be boiled or fried.

Fresh- and saltwater fish and seafood are often eaten smoked or cured. In Germany, herring is commonly pickled and eaten as a snack or at the main meal. Eel is also popular. In Russia, smoked salmon and sturgeon are considered delicacies, as is *caviar,* which is roe from sturgeon. Caviar is classified according to its quality and source. *Beluga,* the choicest caviar, is taken from the largest fish and has the largest eggs; its color varies from black to gray. *Sevruga* and *osetra,* taken from smaller sturgeon, have smaller eggs and are sometimes a lighter color. *Sterlet,* or imperial caviar is from a rare sturgeon with golden roe. The finest caviar is sieved by hand to remove membranes and is lightly salted. Less choice roes are more heavily salted and pressed into bricks.

Dairy products are eaten daily. Cheeses may be served at any meal, from the fresh, sweet varieties to the strongly flavored aged types like Limburger. Fresh milk is drunk; butter is the preferred cooking fat. Buttermilk, sour cream, and fresh cream are also common ingredients in sauces, soups, stews, and baked products. In Austria and Germany, whipped cream is part of the daily diet, served with coffee or pastries.

In much of central Europe sweets are enjoyed daily. They are eaten at coffeehouses in the morning or afternoon, or bought at the local bakery and served as dessert. There are numerous types, such as cheesecakes, coffee cakes, doughnuts, and nut- or fruit-filled individual pastries. Austria is reputed to be the home of apple strudel, made from paper-thin sheets of dough rolled around cinnamon-spiced apple pieces. Germany is known for *Schwarzwälder Kirschtorte* (Black Forest cake), a rich chocolate cake layered with cherries, whipped cream, and *Kirsch* (cherry liqueur). *Dobosch torte,* a multilayered sponge cake with chocolate filling and caramel topping, is a favorite in Hungary.

In central Europe the common hot beverage is coffee. In Russia, strong tea, diluted with hot water from a *samovar,* is drunk instead of coffee. A samovar is a brass urn, which may be very ornate, heated by charcoal inserted in a vertical tube running through the center.

Although southwestern Germany, Austria, and Hungary produce excellent white wine, the most popular alcoholic drink in the region is beer. The Czechs are known for *pilsner* beer, which is bitter tasting but light in color and body. German beers can be sweet, bitter, weak, or strong, and are typically bottom-fermented (meaning the yeast sinks during brewing). *Lager,* a bottom-fermented beer that is aged for about 6 weeks, is the most common type. *Bock* beer is the strongest flavored, traditionally made once a year when the brewing tanks are cleaned. *Märzenbier*, a strong beer midway between a pilsner and a bock beer, is served at Oktoberfest (see section on "Special Occasions") in Munich. In Russia, a sour beer fermented from rye bread called *Kvass,* is popular.

Vodka, which is commonly drunk in Poland and Russia, is a distilled spirit made from potatoes. It is served ice cold and is often flavored with seasonings, such as lemon or black pepper. In Poland, one vodka, *goldwasser,* contains flakes of pure gold.

Meal Composition and Cycle

Daily Patterns

Central Europe. In the past, people of this region would eat five or six large meals a day, if they could afford it. The poor, and usually the people who worked the land, had fewer meals, which were often meatless. Today modern work schedules have affected the meal pattern, resulting in three meals with snacks each day.

In Germany and the countries of central Europe the first meal of the day is breakfast, which consists of bread served with butter and jam. Sometimes it is accompanied by soft-boiled eggs, cheese, and ham. At midmorning, many people have their second breakfast, which may include coffee and pastries, bread and fruit, or a small sandwich. Lunch is the main meal of the day. Traditionally people ate lunch at home, but today they are more likely to go to a cafeteria or restaurant. A proper lunch begins with soup, followed by a fish course, then one or two meat dishes served with

■ With the collapse of state control over the countries of the former Soviet Union, caviar production is now unregulated. Quality control has disappeared, and overharvesting threatens to destroy the sturgeon population.

■ The national dish of Switzerland is cheese fondue (chunks of bread dipped into melted cheese). The Swiss are known for their zesty cheeses, with holes, such as *Emmenthal* (the original Swiss cheese) and *Gruyère.*

■ The dense chocolate cake known as *Sachertorte* was the subject of a 7-year court battle in Vienna over who had the right to claim the original recipe—the Hotel Sacher or Demel's, a famous bakery. The Hotel Sacher won.

■ *Malzbier* is a German beer (1% alcohol) that is considered appropriate for young children and nursing mothers.

■ In Germany, cutting potatoes, pancakes, or dumplings with a knife is an insult to the cook or host because it suggests that these items are tough.

Table 7.2
Cultural Food Groups: Central European and Russian

Group	Comments
PROTEIN FOODS	
Milk/milk products	Dairy items, fresh or fermented, are frequently consumed.
	Whipped cream is popular.
Meat/poultry/fish/eggs/legumes	Meats are often extended by grinding and stewing.
	Russians tend to eat their meat very well done.
CEREALS/GRAINS	Bread or rolls are commonly served at all meals. Dumplings and *kasha* are also common.
	Numerous cakes, cookies, and pastries are popular.
	Rye flour is commonly used.
FRUITS/VEGETABLES	Potatoes are used extensively, as are all the cold-weather vegetables.
	Cabbage is fermented to make sauerkraut.
	Fruits and vegetables are often preserved by canning, drying, or pickling.
	Fruit is often added to meat dishes.
ADDITIONAL FOODS	
Seasonings	Central Europeans tend to season their dishes with sour-tasting flavors, such as sour cream and vinegar.
Nuts/seeds	Poppy seeds are often used in pastries; caraway seeds flavor cabbage and bread.
Beverages	Central Europeans drink coffee; Russians drink tea.
	Many varieties of beer are produced.
	Hungarians and Austrians tend to drink more wine than other central European people.
Fats/oils	
Sweeteners	

Table 7.2
Cultural Food Groups: Central European and Russian—continued

Common Foods	Adaptations in the United States
Milk (cow, sheep) fresh and fermented (buttermilk, sour cream, yogurt), cheese, cream	Milk products are still frequently consumed.
Meat: beef, boar, hare, lamb, pork (bacon, ham, pig's feet, head cheese), sausage, variety meats, veal, venison	Consumption of meat and poultry has increased; use of variety meats has decreased.
Fish: carp, flounder, frog, haddock, halibut, herring, mackerel, perch, pike, salmon, sardines, shad, shark, smelts, sturgeon, trout	Sausages are often eaten.
Shellfish: crab, crawfish, eel, lobster, oysters, scallops, shrimp, turtle	
Poultry and small birds: chicken, cornish hen, duck, goose, grouse, partridge, pheasant, quail, squab, turkey	
Eggs: hens, fish (caviar)	
Legumes: kidney beans, lentils, navy beans, split peas (green and yellow)	
Barley, buckwheat (kasha), corn, millet, oats, potato starch, rice, rye, wheat	More white bread, less rye and pumpernickel bread are eaten.
Fruits: apples, apricots, blackberries, blueberries, sour cherries, sweet cherries, cranberries, currants, dates, gooseberries, grapefruit, grapes, lemons, lingonberries, melons, oranges, peaches, pears, plums, prunes, quinces, raisins, raspberries, rhubarb, strawberries	
Vegetables: asparagus, beets, broccoli, brussels sprouts, cabbage (red and green), carrots, cauliflower, celery, celery root, chard, cucumbers, eggplant, endive, green beans, kohlrabi, leeks, lettuce, mushrooms (domestic and wild), olives, onions, parsnips, peas, green peppers, potatoes, radishes, sorrel, spinach, tomatoes, turnips	
Allspice, anise, basil, bay leaves, borage, capers, caraway, cardamom, chervil, chives, cinnamon, cloves, curry powder, dill, garlic, ginger, horseradish, juniper, lemon, lovage, mace, marjoram, mint, mustard, paprika, parsley, pepper (black and white), poppy seeds, rosemary, rose water, saffron, sage, savory (summer and winter), tarragon, thyme, vanilla, vinegar, woodruff	Saffron is a popular spice in Pennsylvania Dutch fare.
Nuts: almonds (sweet and bitter), chestnuts, filberts, pecans, walnuts	
Seeds: poppy seeds, sunflower seeds	
Beer, hot chocolate, coffee, syrups and juices, fruit brandies, herbal teas, milk, tea, kvass, vodka, wine	
Butter, bacon, chicken fat, flaxseed oil, goose fat, lard, olive oil, salt pork, suet, vegetable oil	
Honey, sugar (white and brown), molasses	

Romanian Fare

Romania is a nation poised between the West and the East. Some describe Romanian food as "pastoral" with Turkish and Hungarian overtones. However, there are also many Italian and central European influences. Beef, veal, mutton, lamb, pork, chickens, geese, and ducks are popular. Freshwater fish such as pike and catfish are harvested from the Danube and other rivers. Cabbage, red and green peppers, leeks, tomatoes, onions, radishes, and lettuce are common vegetables. Temperate fruits, particularly grapes, plums, and berries, are eaten. Other common foods include walnuts, filberts, olives, sour cream, and sheep and goat cheeses. The national bread is *mamaliga,* which is like the Italian *polenta.* It is sliced and spread with butter or topped with cheese, meats, or fruit for dessert. Another specialty is *pastramă* (from the Turkish meaning "to keep") which is lamb, beef, pork, or goose cured (spicing varies, from garlic and black pepper to allspice, nutmeg, and hot red pepper) and then smoked. Ground meats are also popular, made into patties, stuffed into cabbage leaves, or as sausages. One dish meals such as stews and soups are eaten with whole grain bread; one example is *ciorba,* a soup made with vegetables (e.g., peppers, onions, sauerkraut, tomatoes) and meat (usually ground) or fish, then flavored with sour ingredients (e.g., sauerkraut juice, pickle juice, or vinegar). Cake is a traditional dessert, but custards (including one similar to Italian *zabaglioni*) and souffles are also eaten. Romanian beverages include wine (red, white, sweet, dry) and *ţuica,* a plum and wheat brandy. Most Romanians belong to the Eastern Orthodox Church, and adhere to the numerous feasting and fasting days of the church calendar. There are an estimated 365,000 Americans of Romanian descent.

■ Sharing food is essential to Gypsy culture. The harshest punishment that can be imposed on an individual is to be banned from communal meals.

■ The light supper of sausages, cheese, and salads is called *Abendbrot,* in Germany, meaning "evening bread."

■ Buckwheat (groats) is a grain native to central Asia.

■ Hazelnuts are also called filberts because they ripen at the end of summer near St. Philibert's Day (August 22).

vegetables, and perhaps stewed fruit. Dessert is the final course and is usually served with whipped cream. A quicker and lighter lunch may consist of only a stew or a one-pot meal.

A coffee break is taken at midafternoon, if time permits. It typically includes coffee and cake or cookies. The evening meal tends to be light, usually salads and an assortment of pickled or smoked fish, cheese, ham, and sausages eaten with a selection of breads. Guests are usually not invited for dinner but may come for dessert and wine later in the evening.

Gypsies customarily eat two meals each day, first thing in the morning and in the late afternoon. Meals are typically social occasions, featuring the dishes common in their adopted homeland; stews, fried foods, and unleavened breads are especially popular.

Russia. In czarist times the aristocracy ate four complete meals per day; dinner was the largest. The majority of the population never ate as lavishly or as often. One part of the traditional czarist evening meal, *zakuski* (meaning "small bites"), is still part of dinner in Russia today. The meal starts with zakuski, an array of appetizers, which may range from two simple dishes, such as cucumbers in sour cream and pickled herring, to an entire table spread with countless hors d'oeuvres. An assortment of zakuski usually includes a variety of small open-faced sandwiches topped with cold, smoked fish; anchovies or sardines; cold tongue and pickles; and ham, sausages, or salami. Caviar, the most elegant of zakuski, is served with an accompanying plate of chopped hard-boiled eggs and finely minced onions. Other zakuski include marinated or pickled vegetables, hot meat dishes, and eggs served a variety of ways.

The meal that peasants ate after a long day's work is still the basis of a typical dinner in present-day Russia. The staples are bread; soup made from beets (*borscht*), cabbage (*shchi*), or fish (*ukha*); and *kasha* (cooked grain, usually buckwheat). Kvass or beer usually accompanies the meal.

Special Occasions

The majority of central European holidays have a religious significance, although some traditions date back to pre-Christian times. The two major holidays in the region are Christmas and Easter. Many of the American symbols and activities associated with these holidays, such as the Christmas tree and the Easter egg hunt, were brought to the United States by central European immigrants.

Germany. Germany is a land of popular festivals: Nearly all are accompanied by food and drink. Probably the best-known celebration is Munich's Oktoberfest, which lasts for 16 days from late September through early October. Founded in 1810 to commemorate the marriage of Prince Ludwig of Bavaria, it is now an annual festival with polka bands and prodigious sausage-eating and beer-drinking.

Advent and Christmas are the holiest seasons in German-speaking countries. The Christmas tree, a remnant of pagan winter solstice rites, is lit on Christmas Eve when the presents, brought not by Santa Claus but by the Christ Child, are opened. The Christmas tree is not taken down until Epiphany, January 6. A large festive dinner is served on Christmas Day, and it is customary for families to visit one another. Foods served during the Christmas season include carp on Christmas Eve and roast hare or goose accompanied by apples and nuts on Christmas Day. Brightly colored marzipan candies in the shape of fruits and animals are traditional Christmas sweets. Other desserts prepared during the season are spice cakes and cookies (*Pfeffernüsse* and *Lebkuchen*), fruit cakes (*Stollen*), cakes in the shape of a Christmas tree (*Baumkuchen*), and gingerbread houses.

On Easter Sunday the Easter bunny hides colored eggs in the house and garden for the children to find. Ham and pureed peas are typically served for Easter dinner. Candy Easter eggs and rabbits are also part of the festivities.

Sample Menu

GERMAN ABENDBROT

A selection of sausages, ham, and cheeses (Westphalian ham, *Teewrust,* etc.)

Pickled herring and/or Smoked eel

Pumpernickel bread, small rolls

Potato salad, pickled beet salad, green pickles

Apple strudel with whipped cream

Beer or white wine

Poland. Christmas and Easter are the two most important holidays in Poland, a predominantly Catholic country. On Easter the festive table may feature a roast suckling pig, hams, coils of sausages, and roast veal. Always included are painted hard-boiled eggs, grated horseradish, and a Paschal lamb sculptured from butter or white sugar. Before the feasting begins, one of the eggs is shelled, divided, and reverently eaten. The crowning glory of the meal is the *babka,* a rich yeast cake. All the foods are blessed by the priest before being served.

On Christmas Eve, traditionally a fast day, the meal consists of soup, fish, noodle dishes, and pastries. One popular soup, *barszcz Wigilijny,* a cousin to Russian borscht, is made with mushrooms as well as beets. Carp is usually the fish served on Christmas Eve. A rich Christmas cake, *makowiec,* is shaped like a jelly roll and filled with black poppy seeds, honey, raisins, and almonds.

Jelly doughnuts, *paczki,* are eaten on New Year's Eve, while on New Year's Day, *bigos,* "hunter's stew," made with a variety of meats and vegetables (some form of cabbage is required), is washed down with plenty of vodka.

Austria, Hungary, the Czech Republic, and Slovakia. At Christmastime the Czechs eat carp four different ways: breaded and fried, baked with dried prunes, cold in aspic, and in a fish soup. The Christmas Eve meal might also include pearl barley soup with mushrooms, fruits, and decorated cookies. Christmas dinner features giblet soup with noodles, roast goose with dumplings and sauerkraut, braided sweet bread (*vanocka* or *houska*), fruits and nuts, and coffee. *Kolaches,* round yeast buns filled with poppy seeds, dried fruit, or cottage cheese, are served at the Christmas

- *Marzipan,* a paste of ground almonds and sugar, is commonly used in desserts and candies throughout central Europe.
- In the Polish Easter meal the dairy products, meats, and pastries symbolize the fertility and renewal of springtime, while the horseradish is to remember the bitterness and disappointments in life.
- Polish tradition is to invite the poor and the ill to Christmas Eve supper.
- One customary Christmas Eve dish in Poland is *karp po zydowsku,* chilled slices of carp in a sweet-and-sour aspic with raisins and almonds. It is of Jewish origin, dating to when Poland was a haven for Jews in the 14th century.

The custom of the Christmas tree comes from Germany. Traditionally the tree was decorated with cookies (such as *Lebkuchen* and *Pfeffernüsse*) and candy ornaments. (© Corbis)

■ An ancient Slavic saying is "A guest in the house is God in the house."

meal and on most festive occasions. For Easter a baked ham or roasted kid is served with *mazanec* (vanocka dough with raisins and almonds shaped into a round loaf).

The Slovaks break the Advent fast on Christmas Eve by eating *oplatky,* small wafer-like Communion bread spread with honey. The meal may contain wild mushroom soup, cabbage and potato dumplings, stuffed cabbage (*holubjy*), and mashed potato dumplings covered with butter and cheese (*halusky*). A favorite dessert is *babalky,* pieces of bread sliced, scalded and drained, then rolled in ground poppy seeds, sugar, or honey. Mulled wine usually accompanies this meal, as do assorted poppy seed and nut pastries and a variety of fruits. For Easter the Slovaks prepare *paska,* a dessert in the form of a pyramid containing cheese, cream, butter, eggs, sugar, and candied fruits, decorated with a cross. The meal, blessed by the priest on Holy Saturday, includes ham, sausage (*klobása*), roast duck or goose, horseradish, an Easter cheese called *syrek,* and an imitation cheese ball made from eggs (*hrudka*).

In Hungary the most important religious holiday is Easter. Starting before Lent, pancakes are traditionally eaten on Shrove Tuesday; sour eggs and herring salad are served on Ash Wednesday. During Easter week, new spring vegetables are enjoyed, as well as painted Easter eggs. The Good Friday meal may include a wine-flavored soup, stuffed eggs, and baked fish. The biggest and most important meal of the year is the feast of Easter Eve, which consists of a rich chicken soup served with dumplings or noodles, followed by roasted meat (ham, pork, or lamb), then several pickled vegetables, stuffed cabbage rolls, and finally a selection of cakes and pastries served with coffee. The Christmas Eve meal, which is meatless, usually features fish and potatoes. The Christmas Day meal often includes roast turkey, chicken, or goose accompanied by roast potatoes and stuffed cabbage, followed by desserts of brandied fruits or fruit compote and poppy seed and nut cakes.

In addition to Christmas and Easter the Austrians celebrate *Fasching* (a holiday also known in southern Germany). Originating as a pagan ceremony to drive out the evil spirits of winter in which a procession would parade down the main street of a town ringing cow bells, it developed into a multiday carnival associated with Lent.

Russia. Before the 1917 revolution, Russians celebrated a full calendar of religious holidays. Today on fast days the observant do not eat any animal products (see chapter 2 for the fast days in the Eastern Orthodox Church). Of all the holidays, the most important is Easter, which replaced a pre-Christian festival that marked the end of the bleak winter season.

The "Butter Festival" (*maslenitas*) precedes the 40 days of Lent. One food eaten during this period is *blini,* raised buckwheat pancakes. Blini can be served with a variety of toppings such as butter, jam, sour cream, smoked salmon, or caviar. Butter is the traditional topping because it cannot be eaten during Lent.

Traditional foods served on Easter after midnight Mass are *pascha,* similar to the Slovak paska but decorated with the letters "XB" ("Christ is risen"); *kulich,* a cake made from a very rich sweet yeast dough baked in a tall, cylindrical mold; and red or hand-decorated hard-boiled eggs. On Pentecost (Trinity) Sunday (50 days after Easter), kulich left over from Easter is eaten.

Twelve different dishes are served during the Christmas Eve meal. One of the traditional dishes may be *kutia,* wheat grains combined with honey, poppy seeds, and stewed dried fruit. A festive meal is served on Christmas Day. On New Year's Day, children receive gifts, and spicy ginger cakes are eaten. A pretzel-shaped sweet bread, *krendel,* is eaten on wedding anniversaries and name days (saint's day celebrated as birthdays in the Eastern Orthodox faith).

Sample Menu

A RUSSIAN DINNER

Pirozhki (meat turnovers)
or Smoked salmon and caviar appetizers

Shchi (cabbage soup)

Kotlety pozharsky (chicken cutlets)

Kasha (buckwheat groats)

Mixed vegetables in sour cream dressing

Fruit compote

Vodka and beer

Therapeutic Uses of Foods

A study of elderly German Americans found that they sometimes use home remedies to treat minor illnesses. Chicken soup is used for diarrhea, vomiting, or sore throat. Tea is taken for an upset stomach, and milk with honey is commonly used for a cough (Spector, 1991). Traditionally the Pennsylvania Dutch believe that cold drinks are unhealthy and that eating meat three times a day is the cornerstone of a good diet. Herbal teas are consumed for a variety of complaints (Hostetler, 1976; Yoder, 1976). Elderly Polish Americans reportedly believe that sauerkraut is good for colic, as are tea and soda water. Chamomile tea is used for cramps, tea with dried raspberries and wine for colds, cooked garlic for high blood pressure, and warm beverages (milk, tea, or lemonade) for coughs.

■ The term *pascha* comes from the Hebrew word for Passover, *Pesach*.

Vanocka, a Christmas bread popular in the Czech Republic. (Courtesy of Florida State News Bureau.)

■ Ghost vomit *(Fuligo septica)* is a myxomycete or slime mold. The thin, yellow, creeping mass is found on rotting wood and other decaying material, where it eats bacteria.

■ In the 1800s, fried oysters with a glass of Schnapps was a popular Pennsylvania Dutch breakfast for men.

■ "Sauerkraut Yankees" was the derogatory nickname for the Pennsylvania Dutch during the Civil War.

Gypsies have unique health beliefs, many involving food. For example, Gypsies traditionally think that fresh food is the most nourishing and that leftovers are unwholesome. Canned and frozen items may be mistrusted as not being fresh. Many Gypsies believe that non-Gypsies carry disease; and they may insist on using disposable plates and utensils anytime they must eat in a public place. Insufficient intake of "lucky" foods, such as salt, pepper, vinegar, and garlic, can predispose a person to poor health. Home remedies are common, such as tea with crushed strawberries, asafetida (called Devil's Dung), and ghost vomit (*Fuligo septica*) (Hancock, 1991; Sutherland, 1992).

Contemporary Food Habits in the United States

Adaptations of Food Habits

Ingredients and Common Foods

The central European and Russian diet is not significantly different from American fare. Immigrants made few changes in the types of foods they ate after they came to the United States. What did change was the quantity of certain foods. Most central European immigrants were not wealthy in their native lands, and their diets had included meager amounts of meat. After immigrating to the United States, they increased the quantity of meat they ate considerably.

The people of eastern Pennsylvania, where there is a large concentration of German Americans, still eat many traditional German dishes adapted to accommodate available ingredients. Common foods include *scrapple* (also called *ponhaus),* a pork and cornmeal sausage flavored with herbs and cooked in a loaf pan, served for breakfast with syrup; sticky buns, little sweet rolls thought to be descended from German cinnamon rolls known as *Schnecken*; *schnitz un knepp* (apples and dumplings), a one-pot dish made from boiled ham, dried apple slices, and brown sugar, topped with a dumpling dough; *boova shenkel,* beef stew with potato dumplings; *hinkel welschkarn suup,* a rich chicken soup brimming with tender kernels of corn; *apple butter,* a rich fruit spread, very much like a jam; *schmierkaes,* a German cottage cheese; *funnel cake,* a type of doughnut; *shoofly pie,* a molasses pie thought to be descended from a German crumb cake called *Streuselkuchen; fastnachts,* doughnuts, originally prepared and eaten on Shrove Tuesday to

Pennsylvania Dutch funnel cake is an unusual "doughnut" made by pouring the batter through a funnel or from a pitcher into the hot oil in a swirled pattern.

use up the fat that could not be eaten during Lent; and *sweets and sours,* sweet-and-sour relishes, such as coleslaw, crabapple jelly, pepper relish, apple butter, and bread-and-butter pickles, served with lunch and dinner. Saffron crocuses were cultivated in parts of Pennsylvania and the seasoning was used to color many dishes dark yellow, from soups to a traditional wedding cake known as *Schwenkfelder.*

Much of German cooking has been incorporated into American cuisine. Many foods still have German names, although they are so common in the United States that their source is unrecognized (e.g., sauerkraut, pretzels). Other foods contributed by the Germans are hamburgers, frankfurters, *braunschweiger* (liver sausage), *thuringer* (summer sausage), liverwurst, jelly doughnuts, and pumpernickel bread. Beer production, especially in Milwaukee, was dominated by the Germans for more than 100 years. German immigrants created a lager-style beer that is milder, lighter, and less bitter than typical German beer; it can now be described as American-style beer (see also "Regional Americans," chapter 15).

Meal Composition and Cycle

Third- and fourth-generation central European and Russian Americans tend to consume three meals a day (with snacks), and meal composition is similar to that of a typical American meal, although more dairy products and sausages may be eaten. Some traditional foods and ingredients may not be available in areas without large central European or Russian populations.

Many central European and Russian American families serve traditional foods at special occasions. Polish Americans celebrate Pulaski Day on October 11, a national day of remembrance that features a large parade in traditional apparel down the streets of New York City. The Austrians and Czechs typically observe St. Nicholas Day (December 6), when apples and nuts are put in the stockings of well-behaved children and coal is given to the naughty ones. Hungarian Americans observe three unique holidays in the United States with a combination of patriotic and religious activities. The first is March 15, commemorating the revolution of 1848; the second is August 20, St. Stephen's Day; and the attempted revolt known as the Revolution of 1956 is honored on October 23.

Among the Amish, many national holidays, such as the Fourth of July and Halloween, are not observed. A second day of celebration is added to Christmas, Easter, and Pentecost: the first day is reserved for sacred ceremonies; the second day for social and recreational activities. Most Amish celebrate all holidays quietly with family.

Nutritional Status

Nutritional Intake

Very little has been reported on the nutritional intake of central European- or Russian Americans. A Polish study (Kulesza et al., 1984) examined the dietary habits of male Poles and Slovaks. They found that 50 percent of the Slovaks ate pork or beef at least four times a week, compared to 10 percent of the Poles; 62 percent of the Poles and 44 percent of the Slovaks drank milk daily. Both groups ate vegetables and fruit frequently.

In a joint study, nutrient intake and its association with high- and low-density lipoproteins in selected American and Soviet subpopulations was reported by The U.S.-USSR Steering Committee for Problem Area I (1984). They found that Soviet men living in Moscow and Leningrad had lower intakes of protein and polyunsaturated and monounsaturated fatty acids, and higher intakes of total carbohydrate, complex carbohydrates, and sucrose than the U.S. sample. The American men had higher polyunsaturated-to-saturated fatty acid ratios. Soviet men consumed on the average 508 milligrams of cholesterol per day. (The American Heart Association recommends that adults consume fewer than 300 milligrams of cholesterol per day.) Based on the results of both studies, it is reasonable to assume the Polish, Slovak, and Russian Americans now consume a nutritionally adequate diet but that they possibly may be at risk of developing heart disease. A comparison of other nutritional studies in central and eastern Europe suggests that significant gaps in data prevent any clear understanding of current food intake or nutritional adequacy in the region (Charzewska, 1994).

■ Prevalence rates of hepatitis B are considered intermediate in Russia. Outbreaks of hepatitis A and E have been reported.

■ It is thought that nearly 80 percent of immigrants from Russia are from the regions most affected by the Chernobyl nuclear power accident of 1979. Increases in leukemia and thyroid cancer in this population have been noted (Ackerman, 1997).

A study of recent immigrant Russian mothers found strong support for breast-feeding. All but one of 90 participants breast-fed their infants exclusively or partially for an average of 28 to 30 weeks, regardless of whether the babies had been born in Russia or the United States (Knapp & Houghton, 1999). This contrasts with health reports from Russia, where breast-feeding is discouraged and infant nutrition has been compromised by inadequate supplies of formula. High rates of iron deficiency anemia (affecting 25 to 50 percent of children) and endemic goiter have been reported (Ackerman, 1997).

A small study of Gypsies in Boston found high rates of hypertension, Type II diabetes, and vascular disease, affecting between 80 and 100 percent of the population over age 50. Approximately 85 percent were obese. Chronic renal insufficiency was also a problem. Genetics as well as environment may be a factor: European research has found Gypsies at risk for several metabolic conditions, including phenylketonuria, galactokinase deficiency, citrullinaemia, Wilson's disease, and metchromatic leucodystrophy. Life expectancy is estimated to be between 45 and 55 years (Thomas, 1985, 1987).

Counseling

Communication difficulties may occur with recent or older central European or Russian immigrants. Language barriers may require the use of a competent translator.

Good manners and formality are expected in German conversations. Education is respected, and use of titles is important. Honesty and directness are appreciated. Germans are monochronistic and prefer to deal with one topic at a time. Direct eye contact is a sign of attentiveness and trust. A handshake is used in greeting, but there is little other touching between acquaintances.

Poles tend to speak more quietly than Americans and feel uncomfortable with loud behavior. Discussions about politics are avoided. Direct eye contact and a handshake of greeting and when leaving are appropriate.

Germans, as well as other central Europeans, emphasize self-reliance and may avoid health care. Compliance problems develop if medication is seen as a last resort, taken only until symptoms disappear.

Among the Amish, modern health technology is not in conflict with traditional religious precepts and the Amish are willing to seek help, especially for emergency care. The Amish have a practice of "changing doctors" often, however, in an effort to secure the maximum healing knowledge in a situation. A survey of physicians with experience treating the Amish (Hostetler, 1976) noted more problems with digestive disorders, obesity, and chronic bed-wetting than with non-Amish clients. It was the doctors' opinion that the Amish eat a diet higher in fat and salt than the non-Amish. The Amish showed fewer symptoms of heart disease and alcoholism than non-Amish patients. Hereditary diseases may be more common among the Amish than the U.S. population as a whole.

In working with Gypsy American clients, health care providers should be aware that English is usually a second language and that illiteracy is common. However, Gypsies are often very adaptive and may use many forms of communication, depending on the situation. Gypsies believe that the measure of a man's worth is in his girth—weight gain is associated with health, weight loss with illness (Heimlich, 1995b). Gypsies will often not seek treatment until an emergency develops. Numerous family members will come with a client to provide support (Sutherland, 1992).

Russians expect formality between acquaintances, and first names are saved for close friends and relatives. Russians may initially respond to any question that requires an affirmative or negative answer with a no. Direct eye contact is the norm. A quick three kisses on the cheeks or a handshake is used in greeting. Touching becomes more prominent with familiarity. Russians may consider it impolite to sit with the legs splayed or with an ankle resting on the knee.

Counseling and mental health issues were avoided by elderly, Russian immigrants in one study (Brod & Heurtin-Roberts, 1992), perhaps due to the social and economic stigma associated with psychological disorders in Russia. These Russian clients expected extensive help with personal problems and information on social services from their primary provider. An increased perception of pain and somatic symptoms are common. Russian clients are often very assertive in their requests because aggressive behavior was necessary to receive attention in the Russian health care system.

Many Americans of Central European and Russian decent are highly acculturated. An in-depth interview can determine any communication preferences or traditional health practices.

Scandinavians

The Scandinavian countries include Sweden, Norway, Denmark, Finland, and Iceland. With the exception of Denmark and the island of Iceland, they are located north of the Baltic and North seas and share common borders with each other and Russia. Most of the population in Scandinavia is concentrated in the warmer southern regions; the harsher northern areas extend above the Arctic Circle. Norway's weather is more moderate than that of Finland and Sweden because its long western coastline is washed in the temperate North Atlantic Drift. Denmark juts into the North Sea to the north of Germany, and its capital, Copenhagen, is directly opposite from Sweden. The majority of Scandinavian immigrants to the United States arrived in the 1800s. This next section reviews the traditional foods of Scandinavia and the Swedish, Norwegian, and Danish contributions to the American diet.

Cultural Perspective

History of Scandinavians in the United States

Legend is that the Norsemen (ancient Scandinavians), renowned seafarers and explorers, first discovered North America and colonized as far west as Minnesota in the 13th and 14th centuries. The documented presence of Scandinavians in America dates back to the 17th century. Jonas Bronck, a Dane, arrived in 1629 and bought a large tract of land from the Native Americans that later became known as the Bronx in New York City.

Immigration Patterns

The majority of Scandinavians arrived in the United States in the 1800s, led by the Norwegians and the Swedes. During the 19th century, no other country, except Ireland, contributed as large a proportion of its population to the settlement of North America as Norway. During the 1900s an additional 363,000 Norwegians, more than 1,250,000 Swedes, 363,000 Danes, and 300,000 Finns entered the United States.

The peak years of Scandinavian immigration to the United States were between 1820 and 1930. Overpopulation was the single most important reason for emigration. The population of all the Scandinavian countries had increased substantially, resulting in economies that could not absorb the unemployed and

- Most Gypsies will not eat food prepared by non-Gypsies because it is considered impure.

- Gypsies may reject injections for fear that something impure from the outside will contaminate the pure inner body.

- Health care providers working with Russian immigrants report that many Russians believe potatoes cause non-insulin dependent diabetes.

- Few Icelanders have made their home in the United States; the 1990 U.S. census indicates approximately 40,000 Americans claim Icelandic descent.

- The Scandinavians introduced the cast iron stove to the United States. Cooking was done previously in fireplaces and brick ovens.

Danish Americans in Texas, slicing apples, 1915. (From Cynthia Thyrsen Preismeyer, Danevang, TX, copy from Institute of Texan Cultures at San Antonio.)

landless agrarian workers. In Sweden the problem was magnified by a severe famine in the late 1860s. For the Norwegians there was the additional lure of freedom that America offered, and the chance for emancipation from the peasant class. The Scandinavians settled in homogeneous communities.

The Norwegians and Swedes often moved to the homestead states of the Midwest, especially Illinois, Minnesota, Michigan, Iowa, and Wisconsin. One-fifth of all Swedish immigrants settled in Minnesota. Pockets of Finns and Danes also settled in this region, but they were fewer in number. Norwegians and Swedes migrated to the Northwest, working in the lumber and fishing industries. The shipping industry attracted Norwegians to New York City, where they still live in an ethnic enclave in Brooklyn.

Although Swedes and Norwegians are often associated with the rural communities of the Midwest, by 1890, one-third of all Swedes lived in cities and many Norwegians were seeking opportunity in the urban areas. Chicago and Minneapolis still have large Scandinavian populations.

The Danes, in an effort to preserve their ethnicity, developed 24 rural communities between 1886 and 1935 in which, for a set number of years, land could be sold only to Danes. The best known of these communities are Tyler, Minnesota; Danevang, Texas; Askov, Minnesota; Dagmar, Missouri; and Solvang, California. Today most Danes live in cities, primarily on the East or West coasts. The largest concentrations of Danes are in the Los Angeles area and in Chicago.

Following World War I, U.S. immigration from Finland dropped significantly as Finns chose other countries for emigration. As the Finnish population stagnated, ethnic identity became very difficult to maintain. Second- and third-generation Finns are highly acculturated.

Current Demographics and Socioeconomic Status

According to the 1990 U.S. census, there are approximately 1.6 million Danes, 4.7 million Swedes, 3.9 million Norwegians, and 650,000 Finns and their descendants now living in the United States. Most Scandinavians assimilated rapidly into American society, rising from blue-collar to white-collar jobs within a few generations. Many Norwegians and Swedes have continued farming in the Midwest, although many Norwegian Americans are employed in management or specialty professions, and Swedish Americans have made significant contributions in writing and publishing, the arts, education, the ministry, and politics. Danes entered a variety of occupations, but they were most prominent in gardening, raising livestock, and dairy farming. Urban Danes are not associated with specific occupations. Finnish American men have been active in fields such as natural resources management, mining engineering, and geology, while women have been attracted to nursing and home economics.

Worldview

Religion

The majority of Scandinavians who immigrated to the United States were Lutheran, though each nationality had its own branch of the church. As with other immigrant groups, the church helped ease the adjustment of recent arrivals. In this century, many of the Scandinavian and German Lutheran churches joined together to create the Evangelical Lutheran Church in America (ELCA).

Family

The nuclear Scandinavian family was at the center of rural life. Families were typically large and the father was head of the household. Kinship ties were strong: families were expected to pay the way for relatives remaining in Scandinavia to come to the United States, where they would be given a room, board, and help in finding employment. The power of the father diminished and family size decreased as the Scandinavian Americans became more integrated in mainstream society. Among the Finns, it has been noted that when both parents work they may choose to have only one child.

Traditional Health Beliefs and Practices. Information on traditional Scandinavian health beliefs and practices is very limited. Fish was considered necessary for good health. Women were protected following childbirth until they regained their strength. The Finns believe in natural health care, practicing massage, and cupping (bloodletting). Further, the sauna (a traditional steambath) is reputed to have therapeutic qualities. It is used by Finns when ill, and even by midwives during childbirth. It is considered a remedy for colds, respiratory or circulatory problems, and muscular aches and pains (Wargelin, 1995).

Traditional Food Habits

The fare of Scandinavia is simple and hearty, featuring the abundant foods of the sea and making the best use of the limited foods produced on land. Most Scandinavian cooking and food processing reflect preservation methods of previous centuries. Fish was dried, smoked, or pickled, and milk was often fermented or allowed to sour before being consumed. Scandinavians still prepare a large variety of preserved foods and prefer their food salty. The basics of the Scandinavian diet are listed in the cultural food groups list (Table 7.3).

Ingredients and Common Foods

Staples and Regional Variations

The Scandinavians are probably best known for their use of fish and shellfish, such as salmon, sardines, and shrimp. In Norway the fish-processing industry is believed to date back to the 9th century. Today Scandinavian dried salt cod is exported all over the world. Other fish dishes include salmon marinated in dill, called *gravlax;* smoked salmon, known as lox; and the many varieties of pickled herring.

Though Scandinavians drink milk and cook many of their dishes with cream and butter, they also use considerable quantities of fermented dairy products, such as sour cream, cheese, and buttermilk. Popular Danish cheeses are semifirm, mellow, nutty-tasting *Tybo* (usually encased in red wax); firm and bland *Danbo;* semisoft, slightly acidic *Havarti;* rich, soft *Crèma Dania;* and Danish blue cheese.

Common fruits and vegetables are apples, potatoes, cabbage, onions, and beets. Pea soups are a winter specialty throughout Scandinavia, often served with pancakes. Several varieties of berries (particularly lingonberries) and wild mushrooms are local specialties.

Meat, traditionally in limited supply, was stretched by chopping it and combining it with other ingredients. Today the Scandinavians still eat many vegetables, such as onions and cabbage, stuffed with ground pork, veal, or beef. The Swedes are known for their tasty meatballs and the Danes for *fricadeller,* which are patties of ground pork and veal, breadcrumbs, and onion fried in butter.

Bread is also a staple food item and is often prepared from rye flour (white bread is called French bread in Scandinavia). The

■ In Scandinavia they say, "Danes live to eat, Norwegians eat to live, and Swedes eat to drink."

■ The cooking of Finland mixes Swedish and Russian elements, such as *smörgåsbords, piroshkis* (meat turnovers), and *blini* (thin buckwheat pancakes). Vodka is preferred over *aquavit*.

■ A traditional Norwegian dish that is still eaten in some rural areas is *lutefisk,* dried salt cod soaked in a lye solution before boiling.

■ *Veal oscar,* veal topped with a béarnaise sauce, white asparagus, and lobster or crab, is named after Swedish King Oscar II (1872–1907), a renowned gourmet.

Table 7.3
Cultural Food Groups: Scandinavian

Group	Comments
PROTEIN FOODS	
Milk/milk products	Dairy products, often fermented, are used extensively.
Meat/poultry/fish/eggs/legumes	Fish is a major source of protein, often preserved by drying, pickling, fermenting, or smoking.
CEREALS/GRAINS	Wheat is used less than other grains.
	Rye is used frequently in breads.
FRUITS/VEGETABLES	Fruits with cheese are frequently served for dessert.
	Preserved fruits and pickled vegetables are common.
	Tapioca (from cassava) is eaten.
ADDITIONAL FOODS	
Seasonings	Savory herbs and spices preferred. Cardamom is especially associated with Scandinavian sweets.
Nuts/seeds	Marzipan (sweetened almond paste) is used in many sweets.
Beverages	
Fats/oils	Butter is often used.
Sweeteners	

Scandinavian breads may or may not be leavened, are often hard, and vary in size and shape. A thin round bread called *lefser*, made of potatoes cooked on an ungreased griddle, is usually eaten with butter and sugar and folded like a handkerchief.

Desserts, whether they are served after a meal or at a coffee break, are rich but not overly sweet. Most are made with butter and also contain cream or sweetened cheese and the spice cardamom. *Aebleskivers* are spherical Danish pancake puffs, sometimes stuffed with fruit pre-

Table 7.3
Cultural Food Groups: Scandinavian—continued

Common Foods	Adaptations in the United States
Buttermilk, milk, cream (cow, goat, reindeer); cheese, sour cream, yogurt	
Meat: beef, goat, lamb, hare, pork (bacon, ham, sausage), reindeer, veal, venison	More meat and less fish are consumed.
Fish and shellfish: anchovies, bass, carp, cod, crab, crawfish, eel, flounder, grayling, haddock, halibut, herring, lobster, mackerel, mussels, oysters, perch, pike, plaice, roche, salmon (fresh, smoked, pickled), sardines, shrimp, sprat, trout, turbot, whitefish	
Poultry and small birds: chicken, duck, goose, grouse, partridge, pheasant, quail, turkey	
Eggs: chicken, goose, fish	
Legumes: lima beans, split peas (green and yellow)	
Barley, oats, rice, rye, wheat	More wheat used, fewer other grains.
Fruits: apples, apricots, blueberries, cherries, cloudberries, currants, lingonberries, oranges, pears, plums, prunes, raisins, raspberries, rhubarb, strawberries	A greater variety of fruits and vegetables are obtainable in the United States than in Scandinavia but may not be eaten.
Vegetables: asparagus, beets, cabbage (red and green), carrots, cauliflower, celery, celery root, cucumber, green beans, green peppers, nettles, kohlrabi, leeks, mushrooms (many varieties), onions, parsnips, peas, potatoes, radishes, spinach, tomatoes, yellow and white turnips	
Allspice, bay leaf, capers, cardamom, chervil, cinnamon, cloves, curry powder, dill, garlic, ginger, horseradish, lemon juice, lemon and orange peel, mace, marjoram, mustard, mustard seed, nutmeg, paprika, parsley, pepper (black, cayenne, white) rose hips, saffron, salt, tarragon, thyme, vanilla, vinegar	
Almonds, chestnuts, walnuts	
Coffee, hot chocolate, milk, tea, ale, aquavit, beer, vodka, wine, liqueurs	
Butter, lard, margarine, salt pork	
Sugar (white and brown), honey, molasses	

serves. During Advent, they are filled with whole almonds. Another popular dessert is pancakes served with preserved berries or jam. The Scandinavians use almonds, almond paste, or *marzipan* in desserts as often as Americans use chocolate. The Danes are best known for their pastries or, as they call them, *Wienerbrød* (Vienna bread). The pastries were brought to Denmark by Viennese bakers 100 years ago when the Danish bakers went on strike. When the strike was over, the Danes improved the buttery yeast dough by adding jam and other fillings.

Besides milk and other dairy drinks, common beverages are coffee, tea, beer, and

Danish *smørrebrød:* Danish Fontina and Havarti cheeses, ham, salami, and smoked salmon are a few of the toppings typical of the open-face sandwiches known in Denmark as smørrebrød. (Courtesy of Denmark Cheese Association.)

■ When Scandinavians toast, they say *skoal,* which probably derives from the word for skull. Ancient Norsemen used the empty craniums of their enemies as drinking vessels.

■ *Øllebrød,* a Danish rye bread porridge made with nonalcoholic beer, is a favorite breakfast for children.

■ One theory about the origin of the *smørrebrød* is that it dates back to the time before people had dishes, when rounds of bread were used for plates.

■ In Sweden, on the morning of December 13, St. Lucia's Day, the eldest daughter in the home, wearing a long white dress and a crown of lingonberry greens studded with lit candles, serves her parents saffron yeast buns and coffee in bed.

aquavit. Aquavit, which means "water of life," is a liquor made from the distillation of potatoes or grain. It may be flavored with an herb, such as caraway, and is served ice cold in a *Y*-shaped glass. Beer is customarily consumed everyday; occasionally it is drunk with aquavit.

Meal Composition and Cycle

Daily Patterns

The Scandinavians eat three meals a day, plus a coffee break midmorning, late in the afternoon, or after the evening meal. Breakfast is usually a light meal that may consist of bread or oatmeal porridge, eggs, pastries, cheese, bread, fruit, potatoes, or herring. Fruit soups may be served in the winter, topped with thick cream.

Lunch in Denmark is frequently a *smørrebrød,* which means "buttered bread," an open-faced sandwich eaten with a knife and fork. Buttered bread is topped with anything from smoked salmon to sliced boiled potatoes with bacon, small sausages, and tomato slices. Smørrebrød may also be served as a late afternoon or bedtime snack.

A buffet meal in Sweden is the *smörgås-bord* (bread and butter table), a large variety of hot and cold dishes arrayed on a table and traditionally served with aquavit. Ritual dictates the order in which foods are eaten at a smorgasbord. The Swedes start with herring, followed by other fish dishes, such as smoked salmon and fried fins. Next are the meats and salads (pâtés and cold cuts), and the final course before dessert is comprised of hot dishes, such as Swedish meatballs and mushroom omelets.

Dinner is complete, often including an appetizer, soup, éntrèe, vegetables, and dessert. Potatoes are usually served with the evening meal. Coffee is served with the dessert course.

Special Occasions

December is the darkest month of the year in Scandinavia, and Christmas celebrations are a welcome diversion. The Christmas season lasts from Advent (4 weeks before Christmas) until January 13, Saint Canute's Day. However, the climax of the season is on Christmas Eve, when the biggest, richest, and most lavish meal of the year is eaten.

Traditional foods eaten on Christmas Eve are rice porridge sprinkled with sugar and cinnamon; and lutefisk served with a white sauce, melted butter, green peas, boiled potatoes, and mustard. Buried in the rice porridge is one blanched almond; the person who receives it will have good fortune in the coming year. Another common food served for Christmas is pork or ham, often accompanied by sauerkraut or red cabbage. In Denmark a goose is preferred.

Dozens of cookies and cakes are prepared for the Christmas season. The cookies are often flavored with ginger and cloves; the Christmas tree may be hung with gingerbread figures. Deep-fried, brandy-flavored dough, known as *klejner, klener,* or *klenätter,* is also popular. The traditional holiday beverage is *glögg,* a hot alcoholic punch.

Midsummer's Day (June 24) is a popular secular Scandinavian holiday. It features maypoles, bonfires, and feasting. In Sweden, fish accompanied by boiled new potatoes and wild strawberries are eaten; in Norway, *rommegrot,* a cream pudding sprinkled with cinnamon and sugar, is served; and in Finland, new potatoes with dill and smoked salmon are typical festive fare.

Contemporary Food Habits in the United States

Adaptations of Food Habits

The Scandinavians assimilated rapidly into American society, yet their diet did not change significantly. Many of their food habits are similar to the diet of the American majority, such as three large meals per day containing dairy products and animal protein. Because many Scandinavians settled in the Midwest, their consumption of fresh seafood declined, but they eat freshwater fish.

Scandinavian Americans still prepare traditional holiday foods such as rice porridge and lutefisk. Some Scandinavian foods, such as meatballs, Swedish butter cookies, and Danish pastries, are commonly eaten by many Americans.

Nutritional Status

Nutritional Intake and Counseling

Very little has been published on the nutritional status of Americans of Scandinavian descent. It is assumed that since both their traditional diet and the well-accepted typical American diet are high in cholesterol and fat, Scandinavian Americans may be at increased risk of developing heart disease. Scandinavians have a high rate of stomach cancer attributed in part to their high intake of salted fish. These rates decline with acculturation in the United States, however. Finns purportedly have high rates of heart disease, stroke, alcoholism,

■ The tradition of serving the two simple foods (rice and fish) for Christmas dates back at least four centuries when the Scandinavians were Catholic and Christmas was a fast, instead of a feast.

■ Eating lutefish has become a symbol of ethnic identity in some Scandinavian American communities.

■ Midsummer's Day is still observed by many Scandinavian Americans. In some areas of the United States, it has become *Svenskarnas* Day (Swede's Day), celebrating Swedish culture and solidarity.

■ St. Urho's Day (March 16) was invented by Finnish Americans as a spoof on St. Patrick's Day. It commemorates the driving of the grasshoppers from Finland by the saint.

depression, and are often lactose-intolerant (Wargelin, 1995).

Scandinavians tend to be highly analytical. Emotions are controlled; superficiality and personal inquiries are avoided. Swedes and Finns are comfortable with silence during a conversation. As a rule, Danes are a little more informal than other Scandinavians or northern Europeans and may use first names. Danes and Swedes make and maintain direct eye contact, whereas Norwegians and Finns make direct eye contact intermittently. A brief, firm handshake is used in greetings. Other touching is infrequent, reserved for friends and relatives.

Scandinavians are likely to avoid discussion of illness until necessary. Some may consider sickness indicative of either physical or moral weakness. Finnish Americans are likely to pursue natural therapies, including chiropractic therapy and acupuncture (see also Table 7.1, "Selected European Botanical Remedies"). An in-depth interview should be used to establish any traditional health beliefs, as well as the foods clients consume.

8

Africans

African Americans are one of the largest cultural groups in the United States, including nearly 30 million people, more than 12 percent of the total American population. The majority are blacks who came originally from West Africa, although some have immigrated from the Caribbean, Central America, and, more recently, from the famine- and strife-stricken East African nations. A small number of Americans of African heritage are white, primarily immigrants from the nation of South Africa.

Most African Americans are the only U.S. citizens whose ancestors came by force, not choice. Their long history in America has been characterized by persecution and segregation. At the same time, blacks have contributed greatly to the development of American culture. The languages, music, arts, and cuisine of Africa have mingled with European and Native American influences since the beginnings of the nation to create a unique American cultural mix.

African Americans live with this difficult dichotomy. They are in many ways a part of the majority culture, because of their early arrival, their large population, and their role in the development of the country. Much of their native African heritage has been assimilated and their cultural identity results more from their residence in the United States than from their countries of origin. Yet they are often more alienated than other ethnic groups from white American society. This chapter discusses sub-Saharan African cuisines (North African fare is more similar to that of the Middle East. See chapter 13) and their contributions to U.S. foods and food habits. The historical influence of West African, black slave, and southern cuisines on current African American cuisine is examined.

■ The terms *African American, black,* and *black American* are used interchangeably in research literature. *African American* is usually the preferred term because it emphasizes cultural heritage. However, *black* or *black American* is used by many African Americans who feel these terms more accurately reflect their current identity. Some recent immigrants from Africa resent the use of *African American* by persons who have lived in the United States for generations.

Cultural Perspective

Africa is the second largest continent in the world and has a population estimated in the mid-1990s at more than 670 million people. It straddles the equator, and much of its climate is tropical, yet rainfall varies tremendously (see Figure 8.1). Rainforests, grassland savannas, high mountain forests, and temperate zones are found in the far south and along the Mediterranean. In the north the Sahara, the largest desert in the world, stretches from the Atlantic to the Red Sea, separating the Arabic northern African nations (Morocco, Algeria, Tunisia, Libya, and Egypt) from the sub-Saharan western, eastern, and southern regions.

Numerous ethnic groups have evolved in Africa, and it is estimated that between 800 and 1,700 distinct languages are spoken. Cultural identity is strong. The long history of conflict and conquest on the continent has never completely eliminated tribal affinity; most destabilization in individual nations today arises over ethnic issues.

History of Africans in the United States

The arrival of black indentured servants taken forcefully from West Africa preceded the arrival of the Mayflower in America. In 1619, Dutch

Figure 8.1
Sub-Saharan Africa

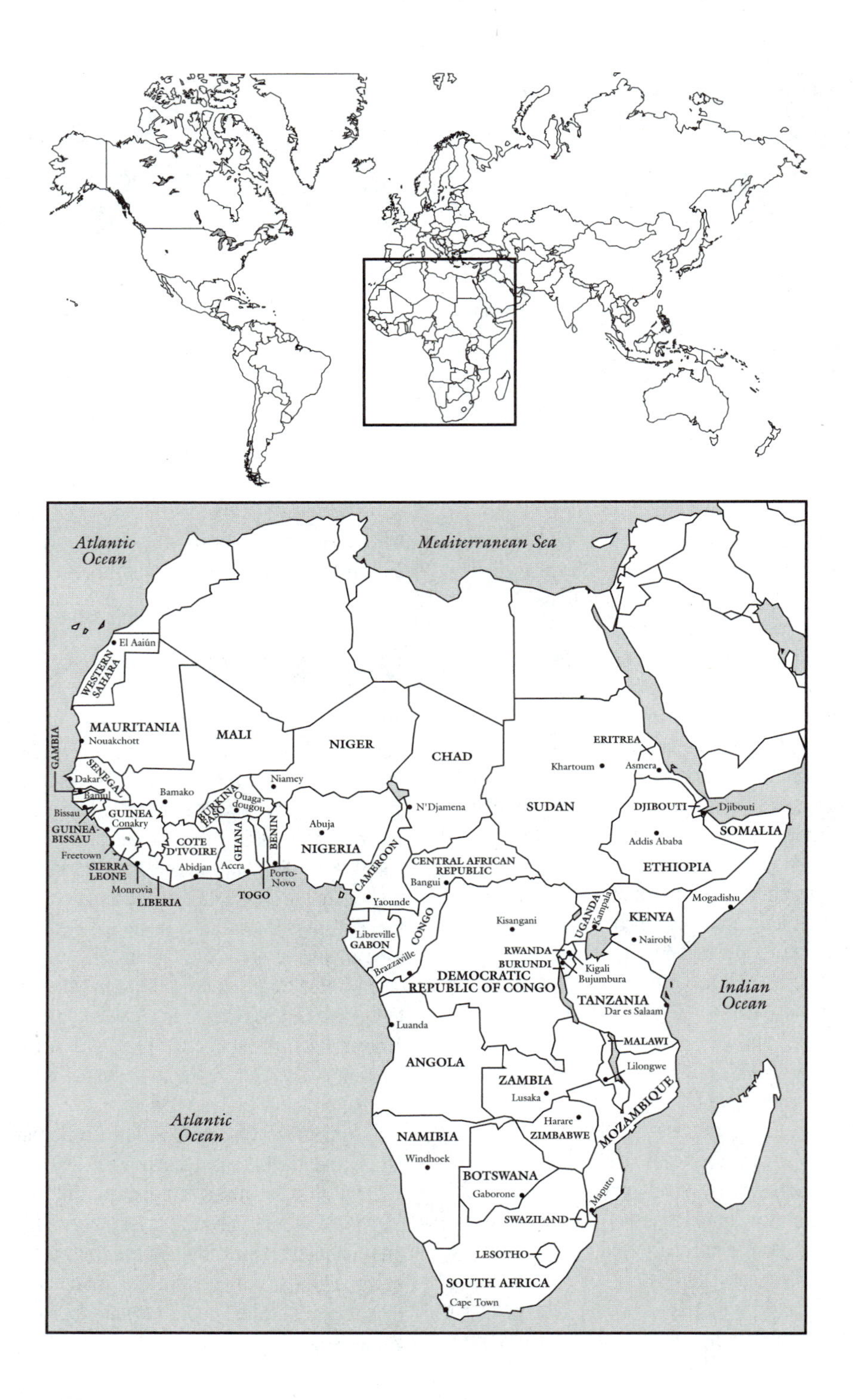

traders sold 20 West Africans to colonists in Jamestown, an early English settlement. More than 425,000 slaves were subsequently imported legally, ancestors of the majority of black Americans residing in the United States today.

Enslavement

The institution of slavery was well established before the first blacks were brought to North America. European slave traders negotiated with African slave suppliers for their human

Traditional African American foods. Some foods typical of the southern black diet include bacon, black-eyed peas, *chayote* squash, corn, greens, ham hocks, hot sauce, okra, peanuts, watermelon, and sweet potatoes. (Photo by Laurie Macfee.)

cargo; which West Africans became slaves and which sold slaves depended on intertribal conditions.

The European traders kept some records of tribal affiliation, but none of place of origin. It is believed that more than one-half the slaves in the United States came from the coastal areas of what are now Angola and Nigeria. Others came from the regions that are today Senegal, Gambia, Sierra Leone, Liberia, Togo, Ghana, Benin, Gabon, and the Democratic Republic of Congo (formerly Zaire). These political identities are relatively recent, however, and the slaves identified with their tribal groups, such as *Ashanti, Bambara, Fulani, Ibo, Malinke,* or *Yoruba,* rather than with a specific country or Africa as a whole.

The tribal villages of West Africa were predominantly horticultural. Individuals viewed their existence in relation to the physical and social needs of the group. The extended family and religion were the foundations of tribal culture. It was especially difficult for individuals to be separated from their tribe because identity was so closely associated with the group. It is perhaps for this reason that enslaved individuals held on tenaciously to their African traditions. African language, ornamentation (i.e., scarification and teeth filing), and other customs were very threatening to slave owners. New slaves, usually in small groups from 8 to 30, were often housed at the perimeter of plantations until they became acclimated. They learned English in 2 or 3 years through contact with Native Americans or white indentured servants. When they became sufficiently acculturated, they would be allowed to work in positions closer to the main plantation house.

This initial period of separation allowed slaves to maintain many cultural values despite exposure to slaves from other tribal groups, indentured servants of different ethnic groups, and the majority culture of the white owners. At the same time, frontier farming was sufficiently difficult that slave owners were quite willing to learn from the slaves' agricultural expertise; therefore intercultural communication was inevitable. Instead of becoming totally acculturated to the ways of the owners, slaves developed a black Creole, native-born culture during the early slave period, combining both white and West African influences.

After the end of slave importation in 1807, the black Creole population swelled. By law, slavery was a lifelong condition and the children of female slaves were slaves as well. Although most

■ Slaves working in kitchens built outside plantation living quarters (due to fire danger) were asked to whistle as they brought food to the main house to prevent them from sampling along the way.

■ The word *Creole* has numerous definitions, mostly describing Africans and Europeans who moved to the U.S. South and Latin America during the colonial period. *Creole Negro* (typically translated to "black Creole") was the term used for Africans who developed their own American culture influenced by the British, Spanish, and French settlers in the region.

slaves worked on farms and on cotton, tobacco, sugar, rice, and hemp plantations, many others worked in mines and on the railroads. A large number of slaves were found in the cities, doing manual labor and service jobs. It was during this period of rapid growth in the African American population that separate racial group identities began to form in the United States.

Emancipation

The movement to free the slaves began with the American Revolution. Many of the northern states banned slavery from the beginning of independence. By the 1830s, there were 300,000 "free persons of color" living in urban areas outside of the Deep South. Tension between states that supported slavery (the Confederacy) and those that opposed it (the Union) was one factor that led to the Civil War (called the War between the States in the South) in 1861.

In 1862, President Lincoln signed the Emancipation Proclamation. Union victory over the Confederacy in 1865 and the subsequent ratification of the Thirteenth Amendment gave all blacks living in the United States their freedom. Some left the South immediately, searching for relatives and a better life. However, most remained in the South because they lacked the skills needed to begin an independent life; for instance, fewer than 2 percent of former slaves were literate.

After emancipation, former slave owners continued to exploit African American labor through tenant farming and sharecropping. Under this system, black farmers were perpetually in debt to white landowners. Slowly, as competition for skilled farm labor increased, work conditions for black Americans in the South improved. Literacy rates and political representation increased. At the same time, racial persecution by white supremacists, such as members of the Ku Klux Klan, became more frequent.

At the turn of the 20th century, depressed conditions in the South and industrial job opportunities in the northern states prompted more than 750,000 African Americans to settle in the Northeast and Midwest. Most were young men, and the majority moved to large metropolitan areas, such as New York, Boston, Detroit, Chicago, and Philadelphia. The influx of southern blacks was resented by both whites and the small numbers of middle-class blacks who had been well accepted in the northern cities. Laws that established racial segregation were enacted for the first time in the early 1900s, resulting in inner-city African American ghettos.

Because of poor economic conditions throughout the country, there was a pause in African American migration north during the Great Depression. The flow increased in the 1940s, and in the following 30 years more than 4 million African Americans left the South to settle in other regions of the country. This migration resulted in more than a change in regional demographics. It meant a change from a slow-paced rural lifestyle to a fast-paced, high-pressured, urban industrial existence.

In the 1960s the movement against the injustices of "separate but equal" laws (which permitted segregation as long as comparable facilities, such as schools, were provided for African Americans) gained momentum under the leadership of blacks such as Martin Luther King, Jr. Violent riots in city ghettos underscored the need for social reform. Civil rights activism resulted in the repeal of many overtly racist practices and passage of compensatory laws and regulations meant to reverse past discrimination, as typified by federal affirmative action requirements.

Current Demographics

Today just more than one-half of African Americans live in the U.S. South, most in suburban areas. Since 1988, more blacks have been moving to the South than to the northern states, reversing the demographic trend northward established in 1900. The remaining African American population is found predominantly in northeastern and midwestern urban areas. As of 1993, approximately 1 percent of Americans of African descent are identified as recent immigrants; nearly 100,000 from the West African nations of Nigeria, Ghana, and Senegal; 70,000 from the more eastern countries of Ethiopia and Kenya; and 44,000 from South Africa. Less than 1 percent of the black population is of Caribbean or Central American descent.

Socioeconomic Status

African Americans continue to suffer from the discriminatory practices that began with their enslavement, yet it is estimated that two-thirds are making steady economic progress. The black middle class is growing and the economic gap between blacks and whites is narrowing. According to 1995 census figures, high school graduation rates are equal among blacks and whites. Nevertheless, one-third suffer severe economic constraints (Bigelow, 1995). The African American unemployment rate is more than double that of whites, and average income is substantially lower than for whites. In 1997 over one-fourth (26.5 percent) of black families lived below the poverty line. While life expectancy rates have increased over all in the United States, they declined for male African Americans in the 1980s. Currently life expectancy is lower for black men than for any other group in the United States, primarily because of the social problems found in many African American communities, particularly access to health care.

Many African Americans believe they are not completely accepted in American society. Blacks isolated in urban ghettos frequently experience alienation. Frustration, hopelessness, and hostility often result. At the same time, discrimination has promoted ethnic identity among African Americans, due in part to a shared history of persecution. Although Americans of African descent are geographically, politically, and socioeconomically diverse, there is a strong feeling of ethnic unity, as demonstrated by the black pride movement of recent years.

Worldview

Religion

Spirituality was integral to African tribal society, and indigenous religious affiliations were maintained by most slaves despite attempts to convert blacks to Christianity. Although the first black church, a Baptist congregation, was founded in the 1770s in South Carolina, it was only after American religious groups became involved in the anti-slavery movement that the black Creole community responded with large numbers of conversions.

Religion is as essential to African American culture today as it was to African society. For many black Americans the church represents a sanctuary from the trials of daily life. It is a place to meet with other African Americans, to share fellowship and hope. More than 75 percent of African Americans belong to a church. The largest denomination is the National Baptist Convention of the U.S.A. Others with large African American followings include the Methodist Episcopal churches and Pentecostal denominations, such as the Church of God in Christ. A small percentage of African Americans are Muslims who are members of either the World Community of Al Islam or the Nation of Islam (see chapter 4 for more information about this religion).

Recent immigrants from Africa adhere to a variety of faiths. Approximately one-quarter of Africans today are Muslim (the fastest growing religion on the continent) and one-quarter are Christian, mostly Protestant. Many Ethiopians follow an Eastern Orthodox faith that is similar to, but separate from, the Egyptian Coptic Church. In addition, it is believed that nearly one-half of the African population participate in traditional tribal religions or combine elements of several faiths.

■ On October 16, 1995, African American men gathered in Washington, D.C., as a demonstration of solidarity and atonement. Called the Million Man March, between 650,000 and 1,000,000 black men and their supporters participated in the event.

■ The Eastern Orthodox church in Ethiopia prohibits the consumption of pork, a practice that may reflect early Judaic influences in the region.

Family

The importance of the extended family to African Americans has been maintained since tribal times. It was kinship that defined the form of African societies. During the early slave period the proportion of men to women was two to one, and the family structure included many unrelated members. As the black Creole population increased, nuclear families were established but were often disrupted by the sale or loan of a parent. The extended family provided for dislocated parents and children.

In 1990, 46 percent of African American families were headed by women. More than one-half of black children are born to single mothers and nearly one-half of those to teenage girls. The family network often includes grandparents, aunts, uncles, sisters, brothers, deacons or preachers, and friends. Such extended kinship still supports and protects individuals, especially children, from the problems of a discriminatory society. The extended family has been found to be equally valued by both wealthy and poor African Americans.

■ Even though interracial marriages have quadrupled in the past 30 years, as of 1992, fewer than 1 percent of African Americans wed members of other races.

■ The percentage of single black women giving birth hit a 40-year low in 1996. Yet due to steep declines in the birthrate of married blacks since 1970, the overall percentage of black children born outside marriage has increased, giving the unwarranted appearance of skyrocketing illegitimacy.

■ The concept "you are what you eat" has led to numerous taboos about local game meat in Africa.

■ *Acokanthera* is an African shrub that traditionally provided the poison for African arrows. It is so deadly that just a minute amount penetrating the skin can cause cardiac arrest within minutes. It is used by some herbalists for snakebite, parasitic infection, and other conditions (Iwu, 1993).

■ There has been a resurgence in voodoo since the 1960s as part of a general interest in alternative health practices. Pharmacies and mail-order companies in the South do a brisk business in voodoo supplies.

Traditional Health Beliefs and Practices

Africans view life as energy rather than matter. A person lives transitionally on earth, interacting with all environmental forces, from those of the gods and nature to those exerted by the living and dead. Life events can be influenced by these forces and an individual, can in turn, influence these forces toward good or evil. Health is maintained through harmony. Disharmony and illness occur when someone (living or dead), the gods, or nature is intentionally malevolent. As described by one African expert:

> Even if it is explained to a patient that he has malaria because a mosquito carrying malaria parasites has stung him he will still want to know why that mosquito stung him and not another person. The only answer which people find satisfactory to that question is that someone has "caused" (or "sent") the mosquito to sting a particular individual, by means of magical manipulations. (Mbiti, 1970, p. 169)

A traditional African healer must first diagnoses the illness, determine the supernatural cause of the illness, then dislodge the evil and take measures to prevent reoccurrence. The healer often uses herbs and other natural prescriptions to treat the symptoms. African healers claim that when they are in a trance, the spirits of their ancestors who are familiar with botanical cures come to them in the dead of the night to impart knowledge. The spirits visit in a variety of forms, such as an alligator or a crippled man. Encountering such a spirit can provide useful information on the curative powers of local plants (Sofowora, 1982). Bleeding, massage, dietary restrictions, chants, and charms may complete the cure (see Table 8.1, "Selected African Botanical Remedies").

The health beliefs and practices of some African Americans reflect traditional African concepts as well as those encountered through early contact with both Native Americans and whites. It is often difficult to determine the origins of a specific practice, and it is also likely that both blacks and whites adhere to the same beliefs. Many Americans of African heritage maintain health by eating three meals each day, including a hot breakfast. Laxatives are used regularly and cod liver oil may be used to prevent colds. A copper or silver bracelet also may be worn for protection; if it is removed, harm will occur. If the skin darkens around the bracelet, illness is impending and precautions, such as more rest and a better diet, should be undertaken (Spector, 1991).

Home remedies are common; prayer is the most important. Tea made from the yellowroot shrub (*Xanthorhiza simplicissima*) is thought to cure stomachache and fever. Sassafras tea or hot lemon-flavored water with honey is considered good for colds. Vicks VapoRub may also be ingested for colds. Garlic placed next to an ill person will help remove evil (Burke & Raia, 1995; Jackson, 1976).

Some African Americans believe that illness is a punishment from God. Stress is considered the cause of hypertension by some blacks, and "worriation" results in diabetes (Jackson, 1981). Others, especially in the rural South, believe that illness is due to evil spirits or witchcraft. A person may be "hexed," "fixed," "mojoed," or "rooted" by someone with supernatural skills. Healers and conjurers are needed to "fix" or "trick" the evil. The resulting illness may be cured by herbal treatments, incantations, or magical transference. For example, by placing a toad on the head of someone with a headache, when the toad later dies, the headache will disappear (Jackson, 1976).

Best known of traditional healers are the practitioners of *voodoo,* also called *hoodoo.* This combination of African and Catholic beliefs is thought to have originated in the Caribbean (see chapter 10); where it is still practiced in the South, it was also likely influenced by European witchcraft (Brandon, 1976). The men and women practitioners can use voodoo magic for good or evil. They cure "unnatural" illnesses (those of supernatural cause) through casting spells, the use of magic powders and *gris-gris,* bags worn around the neck with powders, animal bones or teeth, stones, and/or herbs. Other healers include traditional herbalists or root doctors and spiritual, sympathy, or faith healers who derive their powers from God. A patient may choose to use one or all such healers to treat an illness, and the specialty of one may overlap into another. A root doctor, for example, may apply home remedies or may use charms like a conjurer to remove (or even send) evil. In most cases,

Table 8.1
Selected African Botanical Remedies

Scientific Name	Common Name	Parts	Traditional Use
Abrus precatorius ☠	Mongaluchi; mtipitipi; iwere-jeje; crab's eye; love bean; wild licorice	Leaves, fruit, seeds	Colic; constipation; coughs; venereal disease; cancer; abortifacient
Alchornea cordifolia	Bambami; ububo; eweepa; Christmas bush	Leaves; bark, fruit	Urinary tract problems; stomach ailments; hemorrhoids; constipation; cough; skin infections
Allium satvium ☹	Tafanuwa; kitunguusumu; ayu; garlic	Bulbs	Diabetes; respiratory infections; parasites
Alstonia boonei	Awun; pattern wood; stool wood	Leaves, bark, latex	Hypertension; malaria; fever; rheumatism; lactation stimulant; parasites; snakebite; arrow poison
Anthocleista nobilis, A. parviflora ☠	Awudifoakete; bontodi; cabbage tree	Leaves, bark	Diabetes; liver conditions; stomach pain; colic; hemorrhoids; parasites; purgative; uterine problems
Artemsia afra ☠	Umhlonyane; African wormwood; als	Leaves	Stomach problems; indigestion; constipation; malaria; colds; coughs; fever; parasites
Azadirachta indica; Melia azadirachta ☹	Dogon yaro; oforo-oyinbo; neem; nim; nimba; margosa tree; Indian lilac tree	Root, bark, flowers, seeds	Malaria; liver ailments; gastric ulcers; constipation; urinary tract conditions; fever; skin problems
Bridelia ferruginea	Kirni; ira	Root, bark, leaves	Diabetes; stomach complaints; gastric ulcers; urinary tract disorders; arthritis; children's illnesses; cough
Carica papaya	Gwanda; papayu; ibepe; sayinbo; papaya; pawpaw	Leaves, fruit, seeds	Hypertension, digestive problems; stomach pain; diuretic; malaria; parasites
Catha edulis ☠	Mulung; mlonge; khat; kat; African tea; Arabian tea; Bushman's tea; cafta	Leaves	Nerve problems; fatigue
Catharanthus roseus	Madagascar periwinkle; pervenche de Madagascar	Leaves	Diabetes; indigestion; constipation
Cocculus pendulus	Tiati; tiahat; mboum sehel; liligui;	Root, leaves	Hypertension; diuretic; fever; menstrual irregularities; fertility
Costus afer	Kakizuwa; tete-egun; ginger lily; bush cane	Root, stems	Diabetes; hypertension; stomach pain; sleeping sickness; cough; parasites
Cryptolepis sanguinolenta	Ouidoukoi; delboi; ganga-mau; Ghana quinine; yellow dye root	Root	Hypertension; urinary tract infection; malaria; fever
Glycyrrhiza glabra ☠	Licorice	Root	Appendicitis; TB; cough
Holarrhena floribunda ☠	Fufu; nofu; kedan; ako-ire; are-ibeji; areno; isai	Root, leaves, bark	Malaria; fever; dysentery; fertility; snakebite
Jateorhiza palmata	Agbihu; atutu; calumba	Root, leaves; whole plant	Hypertension; stomach upset; lung infections; male impotence
Jatropha curcas ☠	Bi ni da zugu; chi ni da zugu; kwolkwelaje; botuje; lapalapa; seluju; polopolo; Barbados nut; physic nut; termite plant	Root, leaves, stems, seeds	Dysentery; indigestion; gout; diarrhea; incontinence; purgative; cancer; venereal disease; abortifacient

continued

Table 8.1
Selected African Botanical Remedies—continued

Scientific Name	Common Name	Parts	Traditional Use
Momordica charantia ☹	African cucumber; balsam pear; bitter melon	Root, leaves, fruit	Colic; gastrointestinal disorders; parasites; purgative; abortifacient
Morinda lucida; M. citrifolia	Sangongo; oruwo; owuru; brimstone tree	Leaves, bark	Hypertension; stomach ailments; constipation; dysentery; purgative; abortifacient
Picralima nitida	Erin; akuamma seeds	Leaves, bark, seeds	Hypertension, stomach pain; liver problems; pneumonia; malaria; sleeping sickness; yellow fever
Portulaca oleracea ☹	Adwere; missidi kumbare; baba jibji; halsen saniya; tanguipeta; papasan; amalenyane; purslane; pigweed	Leaves; whole plant	Diabetes; diuretic; emetic; parasites
Rauwolfia spp. ☠	Mwembemwitu; mkufi; quinine plant	Root; leaves; whole plant	Heart palpitations; indigestion; purgative; emetic; rheumatism; convulsions; malaria; smallpox; insomnia; manic conditions; toothache
Rincinus communis ☠	Tomontigi; zurma; kula kula; mbono; nyonyo; lapa-lapa-adete; lara; ilara; castor bean	Root, leaves, oil	Purgative
Sclerocarya birrea	Emaganu; muyombo; msewe; marula; hog plum; cider tree	Fruit; kernels	Diabetes
Solanum nigrum ☠	Goutan kadji; mtura; ndura; odu; ogumo; igba yirin elegun; black nightshade	Leaves; fruit	Stomach conditions; constipation; diuretic
Strychnos icaja ☠	K'ok'ihmo; mwavi; umkuhlu; strichnine	Root; whole plant	Gastrointestinal disorders; hernia; malaria
Tamarindus indicus	Tombi; tsamia; ol massamburai, ukwaju; mkwayo; ajagbon; tamarind tree	Leaves; bark; fruit	Heart disease; constipation; fevers; postchildbirth fatigue; leprosy
Trigonella foenumgraecum ☹	Helba; fenugreek	Leaves, seeds	Diabetes; stomach disorders; indigestion; liver ailments; rheumatism
Uvaria chamae	Boile; atore; mgandasimba; eru; finger-root; bush banana	Root, leaves	Liver conditions; acute stomach pain; fever; cough; dysentery
Vernonia amygdalina ☹	Mponasere; siwakewai; shiwaka; ndumburghai; ewuro; e jije; bitter leaf	Leaves	Hypertension; gastrointestinal disorders; diarrhea; venereal diseases; morning sickness; abortifacient

Note: Data on some plants are very limited; adverse effects may occur even if not indicated.
Key: ☠, all or some parts reported to be harmful or toxic in large quantities or specific preparations; ☹, may be contraindicated in some medical conditions/with certain prescription drugs.

healers of all kinds use a holistic approach and spend a great deal of time on a patient, providing a feeling of spiritual as well as physical well-being.

Traditional Food Habits

What are traditional African American foods—foods of Africans in the 17th and 18th centuries, foods of the slaves, or foods of the black South since emancipation? African American cuisine today includes some elements from each of these diets.

Ingredients and Common Foods

Historical Influences

African American foods offer a unique glimpse into the way that a cuisine develops. Even before West Africans were brought to the United States, their food habits had changed significantly due to the introduction of New World foods such as cassava (*Manihot esculenta,* a tuber that is also called manioc), corn, chiles, peanuts, pumpkins, and tomatoes during the 15th and 16th centuries. The slaves brought a cuisine based on these new foods and native West African foods, such as watermelon, black-eyed peas, okra, sesame, and taro. Adaptations and substitutions were made based on what foods were available. Black cooks added their West African preparation methods to British, French, Spanish, and Native American techniques to produce American southern cuisine, emphasizing fried, boiled, and roasted dishes using pork, pork fat, corn, sweet potatoes, and local green leafy vegetables. The cuisines of other African regions have had little impact at this point on the typical American diet, although recent immigrants may continue to prepare and consume traditional fare.

African Fare

West African. Knowledge of West African food habits before the nineteenth century is incomplete. It is mostly based on the records of North African, European, and American traders, many of whom considered the local cuisine unhealthy. Most West Africans during the slave era lived in preliterate, horticulturally based tribal groups. There was a heavy dependence on locally grown foods, although some items, such as salt and fish (usually salt-cured), could be traded at the daily markets held throughout each region.

Historically, staple foods varied in each locality. Corn, millet, and rice were used in the coastal areas and Sierra Leone. Yams were popular in Nigeria (where they were customarily boiled and pounded into a paste called *fufu*). Cassava (often roasted and ground into a flour known as *gari*) and plantains were the dietary foundation of the more southern regions, including the Congo and Angola. The arid savanna region of West Africa bordering the Sahara Desert was too dry for cultivation, so most tribes were pastoral, herding camels, sheep, goats, and cattle. In the north, these animals were eaten; in other regions, local fish and game were consumed. Chickens were also raised, though in many tribes the eggs were frequently traded, not eaten, and the chicken itself was served mostly as a special dish for guests. Chicken remains a prestigious meat in many regions today.

There were many similarities in cuisine throughout West Africa. Most foods were boiled or fried, and then small chunks were dipped in a sauce and eaten by hand. Palm oil was the predominant fat used in cooking, giving many dishes a red hue. Peanut oil, shea oil (from the nuts of the African shea tree), and occasionally coconut oil were used in some regions. The addition of tomatoes, hot chile peppers, and onions as seasoning was so common that these items were simply referred to as "the ingredients." Most dishes were preferred spicy, thick, and "sticky" (mucilaginous).

Legumes were popular throughout West Africa. Peanuts were especially valued and were eaten raw, boiled, roasted, or ground into meal, flour or paste. Cow peas (*Vigna ungulculata* neither a standard pea nor a bean—black peas are one type of cow pea) were eate substitute for meat, often combined with a s starch such as corn, yams, or rice. *Bamb groundnuts,* similar to peanuts, were also con mon. Nuts and seeds were frequently used to flavor and thicken sauces. Mango seeds (called

■ "Goober," the nickname for peanut, comes from the Angolan word for the legume, *nguba.*

■ Throughout Africa, women cultivate nearly all food that is consumed, while men grow export crops.

■ Taboos against eating eggs still exist in some areas of Africa, including beliefs that they make childbirth difficult and that they excite children.

■ *Kola nuts* are a bitter nut with twice the caffeine content of coffee beans. The original recipe for Coca-Cola, invented in 1886, included extracts from coca leaves and kola nuts.

■ Pods from the flowers of roselle (*Hibiscus sabdariffa*) are used to make *bissap rouge,* a sweet beverage popular in Senegal. They are also used in the popular herbal tea called "Red Zinger" (Harris, 1998).

■ Insects such as termites (often called "white ants") and locusts are consumed in many regions of Africa.

■ Ethiopians claim that coffee originated in the region of Kaffa, where it derived its name. Historically the beans were made into wine.

agobono, og bono, or *apon*), cashews, *egusi* (watermelon seeds, usually dried and ground), *kola nuts,* and sesame seeds were popular.

Many varieties of tropical and subtropical fruits and vegetables were available to West Africans, but only a few were widely eaten. *Akee apples, baobab* (both the pulp and seeds from the fruit of the baobab tree), guava, lemon, papaya (also called *pawpaw*), pineapple, and watermelon were the most common fruits. Many dishes included coconut milk. In addition to starchy staple roots and the flavoring ingredients of onions, chile peppers, and tomatoes, the most popular vegetables were eggplant, okra, pumpkin, and the leaves from plants such as cassava, sweet potato, and taro (also called *callaloo* or *cocoyam*).

West African cuisine today remains very similar to that of the past. Although fish is favored, little meat is consumed. A mostly vegetarian fare has developed based on regional staples such as beans, yams, and cassava. *Gari foto* is a popular Nigerian specialty often eaten for breakfast; it combines *gari* (cassava meal) with scrambled eggs, onions, chiles, and tomatoes. It is sometimes served with beans. Stews featuring root vegetables, okra, or peanuts and flavored with small amounts of fish, chicken, or beef are common. Curries are popular in Nigeria, often served with dozens of condiments and garnishes, such as coconut, raisins, chopped dates, peanuts, dried shrimp, and diced fruits. *Pili-pili,* a sauce of chile peppers, tomatoes, onion, garlic, and horseradish, is usually offered at the table so that each diner may spice dishes to taste.

Deep-fried fish, fried plantain chips, and balls made from steamed rice, black-eyed peas, yams, or peanuts are snack foods available at street stalls in urban areas. The favorite West African sweet is *kanya,* a peanut candy. Bananas are commonly baked and flavored with sugar, honey, or coconut for dessert. Sweetened dough balls prepared from millet or wheat flour are a specialty in many areas; in Ghana they are called *togbei* ("sheep balls") and are brightly colored with food dye before being deep-fried. They are served for special occasions such as birthdays and weddings.

Ethiopian. Mountainous plains and lowland valleys cover much of Ethiopia and the climate is mostly arid. Millet, including a variety unique to Ethiopia called *teff,* sorghum, and plantains are the staple foods produced, and coffee is the leading export crop. Other foods common to the region include barley, wheat, corn, and potatoes, as well as peanuts and other legumes. Some chicken, fish, mutton, and beef is available.

Market scene in Ghana, West Africa. (Photo courtesy of the World Health Organization/P. Almasy.)

Ethiopian cuisine has been little changed by outside influences, and the restricted intake of animal products required in the Ethiopian Eastern Orthodox religion has facilitated the development of vegetarian fare. Examples include *yataklete kilkil,* a garlic- and ginger-flavored casserole, and *yemiser selatta,* a lentil salad. A mixture of ground legumes called *mitin shiro* is added to most vegetarian stews. *Wat,* meaning "stew," is the national dish of Ethiopia. It is thick, spicy, and features whole, hard-boiled eggs with lentils, chickpeas, peanuts, and root vegetables and chicken, fish, or beef. It is served with rice or the traditional Ethiopian flat bread called *injera.* Injera is prepared with a spongy, fermented dough made from teff, then cooked on a griddle in a very large circular loaf. Ethiopian foods are frequently flavored with a hot spice mixture known as *berbere,* which includes allspice, cardamom, cayenne, cinnamon, cloves, coriander, cumin, fenugreek, ginger, nutmeg, and black pepper. *Niter kebbeh,* clarified butter with onions, garlic, ginger and other spices, is added to other dishes. Honey (sometimes consumed with the bee grubs) is popular as a sweetener; it is also fermented to make *tej,* a meadlike beverage. *Tella,* home-brewed millet or corn beer, is commonly consumed, as is coffee, especially espresso (introduced by the Italians).

Sample Menu

A WEST AFRICAN MEAL

Coconut crisps

Chicken in peanut sauce or Okra stew

Rice balls

Adalu (mashed vegetables)

Mangoes

Ginger beer

East African. The climate and topography of Kenya, Tanzania, and Uganda are well suited to farming and ranching. Cassava, corn, millet, sorghum, peanuts, and plantains are the foundation of the diet. Crops grown for export include coffee, tea, cashews, and cloves. Cattle are raised in the northern plateaus of Kenya; they are considered a gift of the gods (especially among the Maasai tribe), and they indicate wealth. The abundant game animals are also often sacred, although specific taboos vary from region to region. Eating fish and seafood is common along the coast.

The cuisines of East Africa are predominantly vegetarian, influenced in part by Arab, Asian Indian, and British fare. In Kenya the national dish is *ugali,* a very thick cornmeal porridge. Mashed beans, lentils, corn, plantains, and potatoes are also popular. Coconut milk, chile peppers, and curry spice blends flavor many dishes. In Uganda, peanuts are a staple food and are used in everything from stews to desserts. Plantains are the core food of Tanzania. They are used in soups (with or without beef), stews, fritters, custards, and even to make wine. Coconut milk is a frequent flavoring, as is curry powder. Throughout the region, dishes made with taro greens or other leafy vegetables and side dishes of local grains and produce, such as eggplant and papaya, round out the cuisine.

South African. South Africa has a very temperate climate favorable to many fruits and vegetables uncommon in the rest of the continent, such as cucumbers, carrots, apricots, tangerines, grapefruit, quinces, and grapes. The cuisine has been strongly influenced by the European settlers of the region, including the Dutch, British, and French. Muslim slaves imported from Malaysia and India have also had a significant impact on South African fare. Mutton, beef, pork, fish, and seafood are popular.

South African meat specialties include sosaties, skewered, curried mutton; *bredie,* a mutton stew that may include onions, chiles, tomatoes, potatoes, or pumpkin; *frikkadels,* braised meat patties; *bobotie,* a meatloaf

■ Somali food is very similar to Ethiopian. For example, a typical meal includes a spicy stew, often with beef or lamb, eaten with *anjeero,* a large flat bread like *injera.* One unique favorite is spaghetti, adopted during Italian occupation.

■ *Sukuma wiki* is Swahili for "stretch the week" and is a stew of leftover meats and vegetables popular in Kenya.

■ Zanzibar, an island just off the coast of Tanzania, was known as the Spice Island in the 1800s. It was a trade center between East and West and supplied most of the cloves used worldwide.

■ The Sudan, which bridges the desert regions of North Africa and the tropical forests of West and East Africa, has a cuisine that reflects both Middle Eastern and African influences. For example, fava beans or a salad of cucumber and yogurt might be served at the same meal as an okra stew and *kisra*, the Sudanese staple bread that is nearly identical to Ethiopian *injera*.

■ The nutritional results of the limited slave diet were seen in the many cases of pellagra, beri-beri, and "sore-mouth" caused by deficiencies in vitamin A and many of the B vitamins. Malnourishment also left African Americans vulnerable to malaria, yellow fever, cholera, and other diseases common in the South.

■ A slave folktale describes the land of Diddy-Wah-Diddy, where roast hogs and chickens run about calling "Eat me!" and there are fritter ponds of oil.

■ Agricultural scientist George Washington Carver conducted numerous culinary experiments on peanuts. He is credited with the discovery of peanut butter and peanut oil, although both were in use in West Africa for centuries before his find. Today Americans eat more than 2 billion pounds of peanuts annually as butter, snacks, and candy. It takes about 548 peanuts to produce one 12-ounce jar of peanut butter.

flavored with curry and topped with a custard mixture when baked; and *biltong,* meat strips dried and preserved over smoke. Grape-stuffed chicken or suckling pig is sometimes served for special occasions. Spicy fruit or vegetable relishes called *chutney* (for more information on Asian Indian foods, see chapter 14); *atjar,* unripe fruit or vegetables preserved in fish or vegetable oil with spices like tumeric and chile powder; and fresh grated fruit or vegetable salads flavored with lemon juice or vinegar and chiles accompany the dishes.

Sweets are very common. Dried fruits, fruit leathers called "planked" fruit, and fruit preserves or jams are popular. Many pastries are available, too, such as tarts made with raisins, sweet potatoes, coconut, or custard. Cookies are a favorite. *Koeksister* are braided crullers that are deep-fried and dipped in a cinnamon syrup, and *soetkoekies* are a spice cookie flavored with the sweet wine Madeira.

The Slave Diet

When West Africans were forcefully taken from their tribes, they were not immediately separated from their accustomed foods. Conditions on the slave ships were appalling, but most slave traders did provide a traditional diet for the tribal members on board. The basic staples of each region, plus dried salt cod (which was familiar to most West Africans), were fed to the slaves in minimal quantities. Chile peppers and the native West African *malagueta* peppercorns were used for seasoning because they were believed to prevent dysentery. It wasn't until the Africans were sold in America that significant changes in their cuisine occurred.

The diet of the field workers was largely dependent on whatever foods the slave owners provided. Salt pork and corn were the most common items. Sometimes rice (instead of corn), salted fish, and molasses were included. Greens, legumes, milk, and sweet potatoes were occasionally added. The foods provided, as well as their amount, were usually contingent on local availability and agricultural surplus. Hunger was common among the slaves. Some slave owners allowed or required their slaves to maintain garden plots or to plant needed vegetables around the periphery of the cotton or tobacco fields. Okra and cow peas from Africa were favored, as well as American cabbage, collard and mustard greens, sweet potatoes, and turnips. Herbs were collected from the surrounding woodlands, and small animals such as opossums, rabbits, raccoons, squirrels, and an occasional wild pig were trapped for supplementary meat. Children would often catch catfish and other freshwater fish.

During the hog-slaughtering season in the fall, variety pork cuts, such as *chitterlings* (intestines, pronounced *chitlins), maw* (stomach lining), tail, and hocks, would sometimes be given to slaves. Some slaves were encouraged to raise hogs and chickens. The eggs and the primary pork cuts were usually sold to raise cash for the purchase of luxury foods. Chickens, a prestigious food in West Africa, continued to be reserved for special occasions.

West African cooking methods were adapted to slave conditions. Boiling and frying remained the most popular ways to prepare not only meats but also vegetables and legumes. Bean stews maintained popularity as main dishes. Corn was substituted for most West African regional staple starches and was prepared in many forms, primarily as cornmeal pudding, cornmeal breads known as *pone* or *spoon bread, grits* (coarsely ground cornmeal), and *hominy* (hulled, dried corn kernels with the bran and germ removed). Pork fat (lard) replaced palm oil in cooking and was used to fry or flavor everything from breads to greens. Hot pepper sauces were used instead of fresh peppers for seasoning. No substitutions were available for many of the nuts and seeds used in West African recipes, although peanuts and sesame seeds remained popular.

Food for the slave field workers had to be portable. One-dish vegetable stews were common, as were fried cakes, such as *hushpuppies* (perhaps named because they were used to quiet whining dogs), and the cornmeal cakes baked in the fire on the back of a hoe, called *hoecakes.* Meals prepared at home after a full day of labor were usually simple.

The West Africans who cooked in the homes of slave owners enjoyed a much more ample and varied diet. They popularized fried chicken and fried fish. They introduced "sticky" vegetable-based stews (thickened with okra or the herb sassafras, which when ground is called *filé* powder), such as the southern specialty

Many traditional southern foods, such as fried chicken, corn bread, spicy stews, bean dishes, and simmered greens, reflect West African influences. (© Bonnie Kamin/PhotoEdit.)

gumbo z'herbes, nearly identical to a recipe from the Congo. Green leafy vegetables (simply called "greens") became a separate dish instead of being added to stews, but they were still cooked for hours and flavored with meat. Ingredients familiar to West Africans were used for pie fillings, such as nuts, beans, and squash.

Foods after Abolition of Slavery

The food traditions of African Americans did not change significantly after emancipation, and they differed little from those of white farmers of similar socioeconomic status. One exception was that pork variety cuts and salt pork remained the primary meats for blacks, while whites switched to beef during the post–Civil War period.

African American Southern Staples

The traditional southern African American cuisine that evolved from West African, slave, and postabolition fare emphasizes texture before flavor; the West African preference for "sticky" foods continues. Pork, pork products, corn, and greens are still the foundation of the diet. The Cultural Food Groups list (Table 8.2) includes other common southern black foods. (For information about the food habits of blacks from the Caribbean, see chapter 10; for more information on foods of the South, see chapter 15 on regional Americans.)

Pork variety cuts of all types are used. Pig's feet (or knuckles) are eaten roasted or pickled; pig's ears are slowly cooked in water seasoned with herbs and vinegar and then served with gravy. Bits of pork skin (with meat or fat attached) are fried to make *cracklings*. Chitterlings also are usually fried, sometimes boiled. Sausages and head cheese (a seasoned loaf of meat from the pig's head) make use of smaller pork pieces. Barbecued pork is also common. A whole pig (or just the ribs) is slowly roasted over the fire. Each family has its own recipe for spicy sauce, and each has its opinion about whether the pork should be basted in the sauce or the sauce should be ladled over the cooked meat.

Other meats, such as poultry, are also popular. Occasionally the small game that was prevalent during the slave period, such as opossum and raccoon, is eaten. More often the meal includes local fish and shellfish, such as catfish, crab, or crawfish. Frog legs and turtle are popular in some areas. Meats, poultry, and fish are often combined in thick stews and soups, such as *gumbos* (still made sticky with okra or filé powder), that are eaten with rice. They may also be coated with cornmeal and deep-fried in lard, as in southern-fried chicken and catfish.

■ In the years following emancipation, chickens were known as "preacher's birds" because they were commonly served to the minister when he came calling on Sunday.

■ The phrase "living high on the hog" comes from the postabolition period, meaning that a family was wealthy enough to eat the primary pork cuts, such as chops and ham.

Sample Menu

A TRADITIONAL BLACK SOUTHERN SUPPER

Ham hocks and black-eyed peas or Southern-fried chicken with gravy

Greens

Macaroni and cheese or Potato salad

Sweet potato pone or Spoon bread

Peach cobbler

Fruit juice or iced tea

■ Ethiopians generally eat one meal in the early evening, snacking throughout the rest of the day.

■ Traditionally, African Americans in the South believed that heavy meals would stay with a person; light meals were appropriate only for infants and invalids.

The vegetables most characteristic of southern African American cuisine are the many varieties of greens. Food was scarce during the Civil War, and most southerners were forced to experiment with indigenous vegetation, in addition to cultivated greens such as chard, collard greens, kale, mustard greens, spinach, and turnip greens. Dockweed, dandelion greens, lamb's quarter, marsh marigold leaves, milkweed, pigweed, pokeweed, and purslane were added as acceptable vegetables. Traditionally the greens are cooked in water flavored with salt pork, fatback, bacon, or ham, plus hot chile peppers (or hot-pepper sauce) and lemon. As the water evaporates, the flavors intensify, resulting in a broth called *pot likker*. Both the greens and the liquid are served; hot sauce is offered for those who prefer a spicier dish.

Other common vegetables include black-eyed peas, okra, peas, and tomatoes. Onions and green peppers are frequently used for flavoring. Corn and corn products are as popular in southern black cuisine today as they were during the slave period. Corn bread and fried hominy are served sliced with butter. Wheat flour biscuits are also served with butter or, in some regions, gravy. Dumplings are sometimes added to stews and greens.

Squash is eaten as a vegetable (sometimes stuffed) and as a dessert pie sweetened with molasses. Sweet potatoes are also used both ways. Other common desserts include bread pie (bread pudding), crumb cake, chocolate or caramel cake, fruit cobblers, puddings, and shortcake, as well as sesame seed cookies and candies.

Meal Composition and Cycle

Daily Patterns

Two meals a day were typical in West Africa, one late in the morning and one in the evening. Snacking was common; in poorer tribes, snacks would replace the morning meal and only dinner would be served. Food was eaten family style or, more formally, the men would be served first, then the boys, then the girls, and last the women. Sometimes men would gather together for a meal without women. Mealtimes often were solemn; people concentrated on the attributes of the food and conversation was minimal.

The West African tradition of frequent snacking continued through the slave period and after emancipation. Meals were often irregular, perhaps due to the variable hours of agricultural labor.

The traditional southern-style meal pattern was adopted as economic conditions for both blacks and whites improved. Breakfast was typically large and leisurely, always including boiled grits and homemade biscuits. In addition, eggs, ham or bacon, and even fried sweet potatoes would be served. Coffee and tea were more common beverages than milk or juice.

Lunch, called dinner, was the main meal of the day. It was eaten at midafternoon and featured a boiled entrée, such as legumes or greens with ham, or another stew-type dish. Additional vegetables or a salad may have been served, as well as potatoes and bread or biscuits. Dessert was mandatory and was usually a baked item, not simply fruit. In some homes a full supper of meat, vegetables, and potatoes was served in the evening. Poorer agricultural families often ate only two of these hearty meals a day.

Today few southern African Americans, or whites, continue this traditional meal pattern in full. The southern-style breakfast might be served just on weekends or holidays, for example. As in the rest of the country, a light lunch has replaced the large dinner on most days, and supper has become the main meal.

Special Occasions

Sunday dinner had become a large family meal during the slave period, and it continued to be the main meal of the week after emancipation. It was a time to eat and share favorite foods with friends and kin, a time to extend hospitality to neighbors.

Many southern African Americans still enjoy a large Sunday dinner. It is usually prepared by the mother of the house, who begins cooking in the early morning. The menu would probably include fried chicken, spareribs, chitterlings, pig's feet (or ears or tail), black-eyed peas or okra, corn, corn bread, greens, potato salad, rice, sweet potato pie, and watermelon. Homemade fruit wines, such as strawberry wine, might also be offered.

Other holiday meals, especially Christmas, feature menus similar to the Sunday meal, but with added dishes and even greater amounts of food. Turkey with cornbread stuffing and baked ham are often the entrees; other vegetable dishes, such as corn pudding, sweet peas, and salads, are typical accompaniments. A profusion of baked goods, including yeast rolls, fruit cakes and cobblers, custard or cream pies, and chocolate, caramel, and coconut cake, round out the meal.

Southern black cuisine is particularly well suited to buffet meals and parties. A pan of gumbo, a pot of beans, or a side of barbecued ribs can be stretched to feed many people on festive occasions. Informal parties to celebrate a birthday, or just the fact that it's Saturday night, are still common.

The African American holiday of *Kwanzaa* has gained popularity in recent years. Created in southern California in 1966, Kwanzaa recognizes the African diaspora and celebrates the unity of all people of African heritage. It begins on December 26 and runs through New Year's Day. Each day a new candle is lit to symbolize one of seven principles: unity, self-determination, collective work and responsibility, cooperative economics, purpose, creativity, and faith. The holiday culminates with a feast featuring dishes from throughout Africa, the Caribbean, the U.S. South, and any other region where Africans were transported.

Role of Food in African American Society

Food was central to African village life. Human fertility, crop fertility, and spiritualism were

■ Ethiopians in rural regions of Africa may fast every Wednesday and Friday (no food from sunrise to noon; no animal products of any type until sunset) according to their Eastern Orthodox faith.

■ "Juneteenth" celebrations, featuring traditional southern fare, are held in many African American communities to commemorate the emancipation of the slaves.

■ In the Dan tribe of Liberia and the Ivory Coast, special *wunkirle* spoons in a symbolically female shape are presented to the most generous woman in the village.

Kwanzaa, the African American holiday celebrated from December 26 through January 1 each year, culminates with a feast featuring dishes from throughout Africa, the Caribbean, the U.S. South, and other regions where Africans now live. (© Merritt Vincent/PhotoEdit.)

Table 8.2
Cultural Food Groups: African American (Southern United States)

Group	Comments
PROTEIN FOODS	
Milk/milk products	Dairy products are uncommon in diet (incidence of lactose intolerance estimated at 60–95 percent of the population). Milk is widely disliked in some studies; well accepted in others. Few cheeses or fermented dairy products are eaten.
Meats/poultry/fish/eggs/legumes	Pork is most popular, especially variety cuts; fish, small game, poultry also common; veal and lamb are infrequently eaten. Bean dishes are popular. Frying, boiling are most common preparation methods; stewed dishes preferred thick and "sticky." Protein intake is high.
CEREALS/GRAINS	Corn is primary grain product; wheat flour is used in many baked goods. Rice is used in stew-type dishes.
FRUITS/VEGETABLES	Green leafy vegetables are most popular, cooked with ham, salt pork, or bacon, lemon and hot sauce; broth is also eaten. Intake of fresh fruits and vegetables is low.
ADDITIONAL FOODS	
Seasonings	Dishes are frequently seasoned with hot-pepper sauces. Onions and green pepper are common flavoring ingredients.
Beverages	
Nuts/seeds	Nuts often used in ways similar to traditional West African dishes, such as nut- or seed-based desserts.
Fats/oils	
Sweeteners	

often interlinked. Many tribes existed at subsistence level yet were generous in sharing available food among tribal members.

In the American South, food has traditionally been a catalyst for social interaction, and southern hospitality is renowned. Food is lovingly prepared for family and friends. The sharing of food is considered an important factor in the cohesiveness of African American society.

Therapeutic Uses of Food

Many African Americans maintain health by eating three hearty meals each day, including a hot breakfast. Numerous other beliefs about food and health are found among poor African Americans living in the rural South. Some of these dietary concepts were brought to other regions of the country during the great migrations.

Table 8.2
Cultural Food Groups: African American (Southern United States)—continued

Common Foods	Regional Adaptations in the United States
Milk (consumed mostly in desserts, such as puddings and ice cream), some buttermilk; cheese	Blacks in urban areas may drink milk more often than rural blacks.
Meat: beef, pork (including *chitterlings,* ham hocks, sausages, variety cuts)	Pork remains primary protein source; prepackaged sausages and lunch meats are popular.
Poultry: chicken, turkey	Small game is rarely consumed.
Fish and shellfish: catfish, crab, crawfish, perch, red snapper, salmon, sardines, shrimp, tuna	Variety cuts are considered to be *soul food* and eaten regardless of socioeconomic status or region.
Small game: frogs, opossum, raccoon, squirrel, turtle	Frying is still popular, but more often at evening meal; boiling and baking are second most common preparation methods.
Eggs: chicken	
Legumes: black-eyed peas, kidney beans, peanuts (and peanut butter), pinto beans, red beans	
Biscuits; corn (corn breads, grits, hominy); pasta; rice	Store-bought breads often replace biscuits (toasted at breakfast, used for sandwiches at lunch).
Fruits: apples, bananas, berries, peaches, watermelon	Fruits are eaten according to availability and preference; intake remains low.
Vegetables: beets, broccoli, cabbage, corn, greens (chard, collard, kale, mustard, pokeweed, turnip, etc.), green peas, okra, potatoes, spinach, squash, sweet potatoes, tomatoes, yams	Green leafy vegetables (*greens*) are popular in all regions; other vegetables are eaten according to availability and preference; intake remains low.
Filé (sassafras powder), garlic, green peppers, hot-pepper sauce, ham hocks, salt pork or bacon (added to vegetables and stews), lemon juice, onions, salt, pepper	
Coffee, fruit drinks, fruit juice, fruit wine, soft drinks, tea	
Peanuts, pecans, sesame seeds, walnuts	
Butter, lard, meat drippings, vegetable shortening	
Honey, molasses, sugar	Cookies (and candy) are preferred snacks.

The conditions known as "high blood" and "low blood" are one example. High blood (often confused with a diagnosis of high blood pressure) is thought to be caused by excess blood that migrates to one part of the body, typically the head, and is caused by eating excessive amounts of "rich" foods or foods that are red in color (beets, carrots, grape juice, red wine, and red meat, especially pork). Low blood, associated with anemia, is believed to be caused by eating too many astringent and acidic foods (vinegar, lemon juice, garlic, and pickled foods) and not enough red meat. Other blood complaints include "thin" blood that cannot nourish the body, causing a person to feel chilly; "bad" blood, due to hereditary, natural, or supernatural contamination;

"unclean" blood when impurities collect over the winter months and the blood carries more heat; and "clots," when the blood thickens and settles in one area, associated with menstruation or stomach and leg cramps (Jackson, 1981).

Some African Americans believe that peppermint candies or yellowroot tea can cure diabetes. Others recommend a mixture of figs and honey to eliminate ringworm and goat's milk with cabbage juice to cure a stomach infection. In some areas, eggs and milk may be withheld from sick children to aid in their recovery.

Pica, the practice of eating nonnutritive substances such as clay, chalk, and laundry starch, is one of the most perplexing of all food habits practiced by African Americans, whites, and other ethnic groups. Studies have determined that pica is most often practiced by black women during pregnancy and the postpartum period, and that rates are unchanged since the 1970s (information on pica among other ethnic groups or age groups in the United States is limited). It is common in the South, where anywhere from 16 to 57 percent of pregnant African American women admit to pica. But pica is also found in other areas of the country where there are large populations of African Americans. In rural regions the substance ingested is usually clay. In urban areas, laundry starch is often the first choice, though instances of women who ate large amounts of milk of magnesia, coffee grounds, plaster, ice, and paraffin have also been reported. Many causes for pica have been postulated—a nutritional need for minerals, hunger or nausea, a desire for special treatment, and cultural tradition are the most common hypotheses. Reasons for pica reported by women include flavor; anxiety-relief; texture and the belief that clay prevents birthmarks or that starch makes the skin of the baby lighter and helps the baby to slip out during delivery (Boyle & Mackey, 1999; Horner et al., 1991; Hunter, 1973).

■ *Geophagy,* or clay eating, is common among the men, women, and children of Africa. It is done to alleviate hunger, to soothe the irritation of intestinal parasites, for spiritual purposes in connection with the swearing of oaths, and for medical reasons. Active trade in clay tablets and disks is found at some markets.

■ *Pica* is the Latin word for magpie, a bird known for its indiscriminate diet. More unusual pica items reportedly consumed include cigarette butts and toilet air-freshener blocks.

■ Aunt Jemima Pancake Mix was named after a tune popular at turn-of-the-century minstrel shows. Uncle Ben's Converted Rice was named for a real-life African American rice farmer in Texas renowned for his quality product.

Contemporary Food Habits in the United States

Most researchers have noted that the food habits of African Americans today usually reflect their current socioeconomic status, geographic location, and work schedule more than their African or southern heritage. Even in the South, many traditional foods and meal patterns have been rejected because of the pressures of a fast-paced society.

Adaptations of Food Habits

Ingredients and Common Foods

Food preferences do not vary greatly between blacks and whites in similar socioeconomic groups living in the same region of the United States. Comparisons do show that African Americans choose items such as fried chicken, barbecued ribs, corn bread, sweet potato pie, collard greens, and fruit-flavored drinks and juices as favorite foods more often than whites do. In addition, cookies or candy are preferred snack items (versus soft drinks for whites). African American households purchase cereal and bakery products, dairy products, sugar, and other sweets less often than do white households, however.

The popularity of *soul food* is notable, considering the similarities between African American and white diets. Traditional southern black cuisine was termed "soul food" in the 1960s. African Americans adopted this cuisine as a symbol of ethnic solidarity, regardless of region or social class. Today soul food serves as an emblem of identity and a recognition of black history for many African Americans.

Meal Composition and Cycle

While the common foods that African Americans eat reflect geographic location and socioeconomic status, meal composition and cycle have changed more in response to work habits than to other lifestyle considerations.

The traditional southern meal pattern of the large breakfast with fried foods, followed by the large dinner with boiled foods and a hearty supper, has given way to the pressures of industrial job schedules. One study (Jerome, 1969) showed that southern breakfast habits were maintained for only 18 months after migration to the North and then were replaced with a meal typically consisting of biscuits or toast, sausage, and coffee.

Recent research indicates that some African Americans living outside the South no longer identify certain foods, such as okra and yams, as African in origin. Many of the items known to be traditional fare are not eaten often, including pig's feet and chitterlings (some of these foods are also associated with the poor, and respondents may be reluctant to admit eating such items). The exception was greens, which were identified most often as a traditional African American food and were also the most popular with respondents: 78 percent eat greens at least once a month (Byars, 1996).

African Americans throughout the country now eat lighter breakfasts and sandwiches at a noontime lunch. Dinner is served after work, and it has become the biggest meal of the day. Snacking throughout the day is still typical among most African Americans. In many households, meal schedules are irregular and family members eat when convenient. It is not unusual for snacks to replace a full meal.

Frying is still one of the most popular methods of preparing food. An increase in consumption of fried dinner items suggests that the customary method of making breakfast foods has been transferred to evening foods (which were traditionally boiled) when time constraints prevent a large morning meal (Jerome, 1969). Boiling and baking are second to frying in popularity (Wheeler & Haider, 1979). African Americans use convenience foods and fast foods as income permits.

Nutritional Status

The nutritional status of African Americans is difficult to characterize because only a limited number of studies have addressed this population. Most research has shown that African American' nutritional intake is associated with socioeconomic status, since poor families of any ethnic group often depend on subisdies such as food stamps, school breakfast and lunch programs, and emergency food programs.

Nutritional Intake

Nutrient deficiencies are prevalent among the large number of African American families at or near poverty levels in the United States. A study looking at the variety of foods consumed daily found that the percentage of African Americans reporting zero servings of dairy, meat, grain, fruit, or vegetable was in almost all cases higher than for whites. Variety increased with higher levels of income and education for both groups (Kant et al., 1991).

The most frequent insufficiencies are of calories, iron, and calcium. There is some question, however, regarding required iron and calcium intake for African Americans, based on evidence that blacks have lower hemoglobin levels and higher serum ferritin levels independent of iron intake, and a substantially lower risk of osteoporosis than whites (Kumanyika & Helitzer, 1985; Lazebnik et al., 1989). Deficiencies in vitamins A, D, E, C, B_6, niacin, folacin, pantothenic acid, magnesium, and zinc have also been reported. The nutrient density of the diet of poor urban African Americans was found to be greater than that of the diet of whites of similar socioeconomic status, suggesting that some nutritional deficiencies may be caused by insufficient food intake, not poor diet (Emmons, 1986). Vitamin and mineral supplementation is lower among African Americans than among whites; in addition, laxative use is common and may be a factor in nutritional status.

Many African Americans' diets are low in fruits, vegetables, and whole-grain products, indicating a low intake of fiber. Even as income increases, fresh produce is ignored in favor of increased expenditure on meat and other protein foods. However, it is noteworthy that many dietary comparisons use food frequency data without defined portion sizes. A study of rural blacks found that when portion size was explained, participants ate on average larger portions of fruits and most vegetables than standard definitions, increasing their daily intake of fruits and vegetables by a significant two-thirds serving (Campbell et al., 1996).

The percentage of calories from animal proteins for blacks is consistently greater than for whites. Sodium intake is high if convenience foods and fast foods are frequently eaten. Poor diet, the large number of teenage pregnancies, and inadequate prenatal care may contribute to a greater incidence of prematurity and low birth weight infants. Low mean daily folate intake is of particular concern, because it is associated with a greater risk of both preterm delivery and low birth weight (Scholl et al., 1996). Overall,

■ It is estimated that 60 to 95 percent of adult Americans of African descent are lactose-intolerant. Some studies show that milk is widely disliked and avoided; others indicate milk is consumed as often by blacks as it is by whites. One trial found that lactose digestion in African American adolescent girls improved on a dairy-rich diet (Pribila et al., 2000).

■ Researchers have suggested that increasing calcium intake among poor African American children may help reduce toxic lead absorption (Mahaffey et al., 1986).

■ A review of the literature on black elders found low nutrient intake compared to white elders; however, study samples were often small. Anthropometric, clinical, and biochemical data were sparse (Fahon & Seaborn, 1998).

African American women are three times more likely than whites to have a preterm infant of very low birthweight (less than 1,500 grams). Morbidity and mortality rates for black mothers and their infants are also disproportionately high, due in part to nutritional deficiencies caused by low socioeconomic status.

African American mothers are thought to breast-feed their infants at a lower rate than white mothers. In a nationwide survey, 33 percent of blacks breast-fed exclusively, compared to 65 percent of whites (Martinez & Krieger, 1985). Another study that included small numbers of African Americans reported that ethnicity was not a significant factor in breast-feeding among well-educated respondents. 63 percent of black infants and 73 percent of white infants had been breast-fed (Serdula et al., 1991). Blacks differ somewhat from whites in what solid foods they feed infants. One-third of low-income mothers offered non-milk liquids or solids to their infants at 7 to 10 days, 77 percent did so by 16 weeks, and 93 percent by 16 weeks (Bronner et al., 1999). Another study showed that high-sodium foods were more frequently offered to African American children than to white children (Schaefer & Kumanyika, 1985).

Obesity is a common problem for Americans of African descent in adolescence and adulthood. It is estimated that 48 percent of black women and 30 percent of black men are overweight (compared to 27 percent of white women and 25 percent of white men). Fat patterning has also been shown to differ between African American and whites. African Americans may have more upper-body and deep-fat depositions than whites (Zillikens & Conway, 1990).

Excess weight gain may be attributed to many factors, including socioeconomic status, poor eating habits, genetic predisposition, and more permissive attitude regarding obesity (Kumanyika, 1987). A study of adolescents has shown that while African American and white boys have similar dieting and purging behaviors, race was an important factor with girls. African American girls were more likely to use laxatives and diuretics; white girls were more likely to use vomiting (Emmons, 1992). Furthermore, standard anthropometric measures may be inappropriate for African Americans. Weight-for-height growth charts as indicators of percentage of body fat and the use of waist-to-hip ratios in defining heart disease risk have been found misleading in some studies (Croft et al., 1995; Hoerr et al., 1992).

Concurrent with obesity is a high rate of non-insulin-dependent diabetes mellitus among African Americans, affecting approximately 20 percent of the population. This is three to four times the incidence in the total U.S. population. A higher rate of death due to the disease has been reported among blacks, and diabetes has also been found to increase the rate of deaths due to ischemic heart disease, especially among African American women (Will & Casper, 1996).

Hypertension is a leading health problem for African Americans. Their incidence of high blood pressure is nearly twice that of the white population. Although statistics on the ratio of normotensives to hypertensives by ethnic group are limited, the data suggest that blacks have the highest proportion of hypertensives of any group. In addition, African Americans with hypertension are 5 times more likely to suffer chronic heart failure and 10 times more likely to develop kidney failure.

Rates of iron deficiency anemia among African Americans are significantly higher than for whites at every age, regardless of sex or income level, ranging from a low of about 5 percent of 45- to 59-year-olds living above poverty level to a high of nearly 30 percent of persons older than 60 living below poverty level. This incidence remains excessive even after adjustments for differences in hemoglobin distributions are made (Kumanyika & Helitzer, 1985). Pica may result in anemia among pregnant women and newborns. Hookworm can also be a cause in the rural South. Other blood disorders resulting in hemolytic anemia that are prevalent in African Americans include sickle-cell disease and glucose-6-phosphate dehydrogenase deficiency.

Mortality rates among Americans of African heritage are significantly higher than the U.S. average. African American men in particular are less likely to reach the age of 65 than men in the developing countries of Bangladesh or El Salvador (McCord & Freeman, 1990; Murray, 1990). Cardiovascular dis-

■ Studies suggest that African American women do not necessarily equate being overweight with being unattractive and that they are less preoccupied with dieting than are white women (Kumanyika, 1993; Stevens et al., 1994).

■ Iron overload due to excessive amounts of iron in the diet may affect up to one-third of sub-Saharan Africans (Gordeuk et al., 1992)

■ A study that reviewed black mortality rates for cardiovascular disease (CVD) in New York City and place of birth found that both men and women born in the South had death rates 30 percent higher than those of blacks born in the Northeast, and four times higher than those of blacks born in the Caribbean. In each age and gender category, Caribbean-born blacks living in New York City had CVD mortality rates lower than those of whites (Fang et al., 1996).

Hypertension in African Americans: Slaves, Racism, Genes and Diet

Black Americans have the highest prevalence of hypertension in the world. Yet black Africans have substantially lower rates. What happened in the migratory process to produce this change?

Nearly one out of every four adults in the United States has hypertension, also known as high blood pressure (HBP). It is defined as having a systolic (heart-pumping) pressure of at least 140 mm Hg (millimeters of mercury) and a diastolic (heart-resisting) pressure of at least 90 mm Hg. At these HBP readings, a person has mild hypertension. Severe hypertension readings are those over 180 mm Hg systolic and 110 mm Hg diastolic. The age-adjusted prevalence in non-Hispanic black men and women is nearly 37 percent, compared to 25 percent for non-Hispanic white men and 21 percent for women. By age 50, nearly 60 percent of African Americans suffer from HBP. They develop it earlier in life, have higher rates of severe hypertension, and have a greater risk of death from renal failure, stroke, and heart attack associated with HBP. Furthermore, a study of African Americans with HBP found hypertension associated with poorer general health and lower physical functioning (American Heart Association, 1999; de Forge et al., 1998; Gilliam, 1996).

The cause of hypertension is unknown in nearly 95 percent of all cases. Researchers are especially puzzled over the prevalence among black Americans. Studies show sub-Saharan African blacks with the lowest prevalence of HBP, though rates increase with urbanization (Steyn et al., 1996; Wilson et al., 1991). Intermediate to high prevalence has been reported for Caribbean blacks (Dominguez, 1999; Grim et al., 1990). One hypothesis for the differences between Old and New World blacks is the increase in negative stressors associated with elevated blood pressure. Numerous studies have demonstrated an association between hypertension and low socioeconomic status, including low levels of education. Research also suggests that chronic discrimination and the struggle for social acceptance can lead to continuous stress and an increased prevalence of HBP (Adams et al., 1999; Dressler, 1996; Krieger, 1997; Wilson et al., 2000).

Another, more controversial, theory has been proposed regarding differing rates of HBP between black Africans and black Americans. Researchers note that most slaves brought to the United States came from regions of Africa that had meager salt resources and that over generations their bodies may have adapted to the deficiency by becoming efficient at salt retention. When these blacks were forcibly brought on overcrowded ships to the Caribbean and the United States, those who were best able to conserve salt may have avoided death due to dehydration or mineral-depleting fever, vomiting, and diarrhea. Because up to 70 percent of slaves died within the first 4 years of captivity, survival from these endemic illnesses could hypothetically have assured a genetic selection for salt retention (Wilson, 1986; Wilson & Grim, 1991). While some scientists believe such evolution is impossible in such a short period of time, and others state that the hypothesis has racist overtones, some scientists find the idea of physical adaptation for low salt intake plausible (Fackleman, 1991; Kurokawa & Okuda, 1998).

What is known with more certainty is that many African Americans today retain a relatively high amount of the salt they ingest and that this salt sensitivity is a factor in HBP, even among adolescents (Wilson et al., 1999). Other minerals may also play key roles: Low intakes of calcium, magnesium, and potassium are associated with hypertension. (In a biological twist of fate, it is believed that salt retention is associated with calcium retention, which may account for why many black women have high bone mineral density and low rates of osteoporosis despite low calcium intake.) Several studies have suggested that lowering salt intake and increasing calcium, magnesium, and potassium intake may reduce hypertension (Dwyer et al., 1998; Ford, 1998; Kawano, 1998; Kawasaki et al., 1998).

Given the data on mineral intake, it is reasonable to assume that diet is as important in hypertension among African Americans as is chronic stress or genetic adaptations. Recent research findings from the DASH (Dietary Approaches to Stop Hypertension) project confirm this concept. A diet low in total fat, saturated fat, and cholesterol while high in fruits, vegetables, and low-fat dairy foods was found effective in reducing blood pressure, especially in African Americans, even with no reduction in salt intake or weight loss. When salt restrictions were added, the diet was still more effective, lowering blood pressure in both people with and without hypertension. Whatever the causes behind the hypertension in African Americans, food habits are proving a significant part of the answer (Adrogue et al., 1996; National Heart, Lung, and Blood Institute, 2000; Svetkey et al., 1999).

■ A study of African American women's preceptions about health care revealed that some believe that it is the intent of the dominant culture to "eliminate" all African Americans (Ndidi Uche Griffin, 1994).

■ Chewing *khat* (*Catha edulis*) is an ancient practice in parts of the Middle East and Africa. The bitter leaves contain the alkaloid cathinone that has amphetamine-like affects. It is a social drug, often chewed at weddings, funerals and other group functions in Moslem Ethiopian life. it has recently spread beyond the Islamic community and is widely used at social occasions with beer chasers. It is estimated that nearly half the urban population in parts of Ethiopia chew khat regularly. Khat may delay gastric emptying, resulting in lack of appetite and constipation (Heyman et al., 1995).

ease, cancer, cirrhosis, diabetes, and infections account for most of the excess deaths. Most research has focused on associations between socioeconomic status and excess mortality, reporting that while differences in excess mortality rates between advantaged and disadvantaged populations are often vast, they do not fully account for racial disparities (Geronimus et al., 1996; Lillie-Blanton & Laveist, 1996). Efforts to link dietary intake differences to these discrepancies have been inconclusive; one study suggests that poor nutritional status may contribute to reduced length of survival for African American women with breast cancer (Coates et al., 1990). It is estimated that more than one-third of African American deaths are due to inadequate health care.

Little has been reported on the nutritional adequacy of the traditional diet of recent African immigrants from Ethiopia, Somali, and the Sudan. Studies in Israel of Ethiopian immigrants found deficiencies in vitamin D, iodine (due in part to food goitrogens), and calcium. Vitamin A deficiency has led to xerophthalmia and blindness in many regions. Among Sudanese immigrants, blindness due to trachoma is also common. High rates of extreme malnutrition, malaria, thyphoid hepatitis B, tuberculosis, syphilis, and parasitic infection have also been reported (Ackerman, 1997; Fogelman et al., 1995).

Dietary changes of Ethiopians in Israel are marked. A survey of teens found that within 18 months of arrival, only 30 percent maintained a traditional diet, 60 percent consumed a mixed diet, and 15 percent ate only Israeli foods (Trostler, 1997). More than half of daily calories came from snacks and fast foods, especially sweets and soft drinks. Most milk products were disliked, with the exception of hot chocolate, a favorite with the youth. Fat intake increased, while fruit and vegetable intake decreased. Though obesity is unusual, glucose intolerance is common, and the prevalence rate of Type II diabetes increased from 0.4 percent to between 5 and 8 percent within a few years. Over 20 percent of men also developed hypertension after immigration (Bursztyn & Raz, 1995).

A study of Ethiopian women in the United States found that most breast-fed their infants on average 4 months. Going back to work and reduction in mother's milk were the primary reasons given for cessation (Meftuh, et al., 1991).

Counseling

Many African Americans have limited access to health care. Cost, including time off from work, is often an issue (20 percent of African Americans have no health insurance). Self-reliance is highly valued and may lead to delay in seeking care or minimization of symptoms. Other African Americans feel patronized by nonblack providers; some choose to suffer at home. Furthermore, an attitude that fate determines wellness may restrict medical visits. When a doctor's care is sought, it is usually for treatment of symptoms rather than for prevention of illness and health maintenance.

African American conversational style is fully engaged and very expressive. Interjections of agreement or disagreement are frequent. Words are often spoken rhythmically and passionately. Response time is very quick. A direct, but respectful, approach is appreciated. Eye contact is made while speaking, but prolonged eye contact is considered rude. African Americans may avert their eyes while listening attentively. A firm handshake and smile are the expected greeting; hugging and kissing may also be included. Touching is common, and reluctance to touch may be interpreted as personal rejection. When counseling African Americans, it is helpful to be direct yet respectful. Attentive listening is more important than eye contact to many blacks, although they may interpret rapid eye aversion as an insult. Some clients may be suspicious or hostile when working with nonblack health care professionals. Such attitudes are rarely directed specifically at the health care worker, instead they are an adaptation to what is perceived as a prejudicial society.

African Americans may not consider themselves active participants in their interaction with providers and may not communicate needs or questions. This is sometimes done to test the competency of the provider, who is expected to diagnose without help from the client (Jackson, 1981). Another study suggested, however, that although some African Americans fear diagnosis of a serious disease, such as cancer, individuals prefer to receive the news from the doctor personally and to be involved in treatment decisions (Blackhall et al., 1995).

Nonnutritive food intake during pregnancy may be missed unless information about pica is solicited during the interview. Most women who eat clay, laundry starch, or other nonfood items will willingly list the items consumed when asked directly about the habit. The nutritional effects of pica are uncertain. Possible problems include excessive weight gain (from laundry starch), aggravated hypertension (from the sodium in clay), iron-deficiency anemia, and hyperkalemia (Kumanyika & Helitzer, 1985). Furthermore, over-the-counter remedies for the gas and constipation that can accompany pica may be harmful during pregnancy (Boyle & Mackay, 1999).

Traditional health practices, such as using diet to cure "high" and "low" blood, may complicate some nutrition counseling. Pregnancy is sometimes considered to be a "high" blood state, and pregnant women will avoid red meats. Patients who confuse hypertension with "high" blood may east astringent foods, which are often high in sodium, to balance the condition. Home remedies for diabetes, such as peppermint candies or yellowroot bush tea, should also be investigated. It is unlikely that a client will mention any use of other healers. It may be useful to directly inquire if a rural African American believes an illness is due to outside forces or witchcraft, and what additional treatment is being sought.

Counseling recommendations should be action- or task-oriented (Randall-David, 1989), and whenever possible fit within existing health care systems used by the client. Although the research is inconclusive, some studies suggest that African Americans may be more sensitive to pain than whites. Practitioners should not discount such possible differences when taking histories or providing care.

Little information is available regarding counseling recent immigrants from Africa. Recommendations for Ethiopians include use of an interpreter from the client's community; a warm, personable communication style; a positive outlook; and disclosure of poor prognosis or terminal illness to the patient's family (preferably not the wife or mother) or friend, who will then inform the patient (Beyene, 1992).

An in-depth interview is especially important with African American clients. Variability in diet related to region, socioeconomic conditions, and degree of ethnic identity should be considered. Information regarding country of origin may be significant: Clients from the Caribbean or Central America are more likely to identify with the foods and food habits of Latinos than with those of blacks in the United States. Religious affiliation may also be important, such as among Americans of African descent who are members of Islam or an Eastern Orthodox faith.

■ Ethiopians receive a personal name at birth and use the father's personal name as a surname. Women do not change their names with marriage; thus, within a family, the husband, the wife, and the children may all have different last names.

■ Traditional healers in Somalia reportedly burn the skin to relieve pain and perform uvulectomy or tooth removal for respiratory infections (Ackerman, 1997). Similar practices in Ethiopia have been noted, in addition to eyelid incision to cure conjunctivitis (Hodes, 1997).

■ Ethiopians may prefer an injection to taking pills.

9

Mexicans and Central Americans

Latinos are one of the largest ethnic groups in the United States, representing nearly 11 percent of the total population. If current growth trends continue, the number of Latinos in America will double within 25 years, and by early in this century they will surpass African Americans as the largest nonwhite minority. Yet Latinos are not a single group, arriving in the United States from the diverse nations of Latin America. Though a majority share Spanish as a common language of origin, others speak English, French, Portuguese, or a native Indian dialect as their mother tongue.

Immigrants from Mexico and the countries of Central America bring a rich cultural history (see Figure 9.1). The Olmec culture, known for its sophisticated sculpture, existed in southeastern Mexico as early as 1200 B.C.E. The great Aztec, Mayan, and Toltec civilizations thrived while Europe was in its dark ages. Their independent mastery of astronomy, architecture, agriculture, and art astonished later explorers. Spanish occupation of Mexico and Central America introduced new ideas and traditions, most notably Roman Catholicism, British, French, and Austrian intrusions also provided minor contributions. The foods of Mexico and Central America reflect the native Indian and European heritage of the region. This chapter examines Mexican cuisine and the food habits of Americans of Mexican descent. An overview of recent immigration from Central America and the traditional fare is also included. The following chapter reviews Latinos from the Caribbean Islands and South America.

Mexicans

Estados Unidos Mexicanos, the United Mexican States, is the northernmost Latin American country. It is more than one-fourth the size of the United States, with 756,065 square miles of territory. The varied geography includes a large central plateau surrounded by mountains except to the north. Coastal plains edge the country along the Gulf of Mexico and the Pacific. The separate Baja peninsula is found in the west and the Yucatán peninsula juts out in the southeast. Snow-capped volcanoes, such as Orizaba, Popocatépetl, Ixtacchiuatl, and El Chichón, and frequent earthquakes also affect the landscape. The climate ranges from arid desert in the northern plains to tropical lowlands in the south.

Almost two-thirds of Mexicans are *mestizos*—that is, of mixed Indian and Spanish ancestry. Thirty percent are native Indians, and about 10 percent are whites of Spanish descent. Spanish is the official language. Only 1.5 percent of Mexicans speak a single Indian language, mostly Nahuatl.

■ Though Mexico seems logically a part of Central America, it is geographically a part of North America.

■ The term *Latino* is used to describe people originally from Mexico, the Caribbean and Central and South America. It suggests culture of Latin heritage, not exclusively of Spanish background. For example, Haitians and Brazilians are both Latinos but speak French and Portuguese, respectively. *Hispanic* is preferred by some, though there is no clear definition of this term. It can mean "born in Latin America," "ancestors born in Latin America," "Spanish surnames" or "Spanish speaking."

Cultural Perspective

History of Mexicans in the United States

Immigration Patterns

Mexican immigration patterns have changed over the years since the Mexican-American War clearly defined the U.S.–Mexican border. Today

Figure 9.1
Mexico and Central America

■ The term *Chicano* came from the Aztec word for Mexicans, *Meshicano.*

■ Mexican Americans account for more than 60 percent of Latinos in the United States.

Mexicans living in America and their descendants can be classified in the following groups: (1) *Chicanos:* those who are born in the United States (from the descendants of the wealthy Mexican landowners who controlled the area from California to Texas in the 18th and early 19th centuries to the children of the most recent arrivals), as well as those who immigrated from Mexico and became U.S. citizens; (2) *Braceros:* those who work here legally but remain Mexican citizens; and (3) undocumented residents: those who enter the country illegally.

Mexicans lived in what is now the American Southwest for hundreds of years before the United States declared its independence in 1776. Although they welcomed American settlers, they soon found themselves outnumbered and their economic and political control of the region weakened. At the end of the Mexican-American War in 1848 the 75,000 Mexicans living in the ceded territories became U.S. citizens.

Between 1900 and 1935, it is estimated that 10 percent of the Mexican population, approximately 1 million persons, emigrated north to the United States. Then, during the Great Depression, tens of thousands of undocumented aliens, plus those admitted legally under the 1917 contract labor laws, were "repatriated" and sent back to Mexico.

After the Depression, the need for cheap labor increased. The Bracero program was created to meet this need. Thousands of Mexicans were offered jobs in agriculture and on the railroads. Following World War II, the continued need for migrant farmworkers encouraged more than 1 million Mexicans to immigrate between 1951 and 1975. In the past 30 years, Mexicans have been the largest single group of legal immigrants to the United States.

Current Demographics

In 1997, it was estimated that there were more than 18 million Chicanos and Braceros in this country, of whom nearly 5 million were born in Mexico. The census also accounted for approximately 3.5 million undocumented Mexicans living in the country, although this figure may be

Traditional Latin American foods. Some foods typical of the traditional Latin American diet include *achiote,* avocado, *bacalao,* black beans, cassava *(yuca)*, chile peppers, cilantro, corn tortillas, *jícama,* papaya, plantains, pinto beans, pork, *tomatillos,* and tomatoes. (Photo by Laurie Macfee.)

low due to the reluctance of undocumented residents to become involved in government procedures. Economic pressures in Mexico have increased the number of Mexican immigrants entering the United States each year. Many are attempting to escape the life-threatening poverty that affects many of the populace.

The majority of Chicanos live in California and Texas. Other states, such as Illinois, Arizona, and New Mexico, also host large Chicano populations. 80 percent of the immigrants from Mexico settle in U.S. cities such as Los Angeles, San Antonio, and Chicago. Other recent immigrants, legal and undocumented, settle in the urban Latino neighborhoods called *barrios.*

A continuing decline in the Mexican economy may encourage even greater emigration in the near future, both legal and undocumented. The U.S.–Mexican border is 1,931 miles long, most of it in unpopulated desert regions. More than 1.5 million Mexicans are apprehended each year as they attempt to enter the United States illegally. Some estimates suggest that twice that number may cross successfully.

Political pressures in the United States regarding taxpayer support of services for undocumented residents have led to immigration limits and social program restrictions. It is unknown how effective these changes will be long-term in preventing illegal entry into the country or how it will affect immigration patterns. What is certain is that the economic disparities between the two nations will continue to draw Mexicans to the United States.

Socioeconomic Status

Chicanos, Braceros, and undocumented aliens occupy three main socioeconomic classes. There are the migrant farmworkers, who maintain a culturally isolated community; the residents of the urban barrios, who also are segregated from much of American society; and a growing number of acculturated middle-class Chicanos.

Undocumented aliens tend to move in with family members who already reside in the United States. Usually this is in a predominantly Latino neighborhood, where they can live inconspicuously among the residents while becoming familiar with the American social and economic systems.

Despite a rapidly growing Chicano middle to upper class, many Americans of Mexican descent have low socioeconomic status. In 1996, nearly one in every four families fell below the poverty level. This is due in part to a

■ The 1993 Canadian Census listed more than 176,000 residents of Latin American heritage.

■ In 1996, several states sued the U.S. government over inaccurate minority population counts; lower-than-actual census numbers mean reduced funding for immigrant programs. The courts ruled that the census was reasonably diligent in its count and therefore was not responsible for the faulty figures.

■ Some of the most recent immigrants from Mexico have been Mexican Indians (mostly Mixtecos from Oaxaca), who are replacing mestizos in some migrant agricultural jobs. They speak neither English nor Spanish.

■ When a girl turns 15 years old, her family hosts a *quincinera,* an elaborate coming-out party with music, feasting and dancing.

disproportionate number of Chicanos employed in unskilled or semiskilled labor. Although only 3 percent of Mexican Americans are currently working in agricultural jobs, more than 50 percent are employed in manufacturing or service occupations. Approximately 16 percent hold professional or managerial positions. Just more than one-half of Chicanos graduate from high school, but this figure rises to 65 percent among second-generation Mexican Americans. As many as two-thirds of Mexican American girls do not finish high school. Nearly 5 percent of adults older than 35 have completed 4 or more years of college; more than 15 percent have less than 5 years total of education.

Worldview

Chicanos, Braceros, and undocumented immigrants often live in culturally homogeneous communities. Their ethnic identity is proudly maintained; they speak Spanish and prefer Mexican music and food. The concept of *la Raza* (meaning "the people") was first promoted in the 1960s as a pride and solidarity movement for all persons of Latin American heritage.

When exposed to the mainstream American society, immigrants from Mexico can be highly adaptive. Many Chicanos are completely assimilated. They may speak no Spanish and consider themselves "white." Others are successfully bicultural. Cross-cultural marriages are becoming more common, especially in the northern regions of the United States.

Religion

It is estimated that between 75 and 97 percent of Americans of Mexican descent are Roman Catholics. Traditional religious ceremonies, such as baptism, communion, confirmation, marriage, and the novenas (9 days of prayer for the deceased) are important family events (see chapter 4). Yet despite their numbers within the laity, they are underrepresented within the institutional church; as of 1992, only 5 percent of U.S. bishops were Latinos.

A strong faith in the will of God influences how immigrants from Mexico perceive their world. Many believe they have no direct control over their own fate. Nearly all Mexican Americans who are not Catholic practice Protestant faiths; evangelical churches are particularly popular in urban areas.

The Chicano Family

The most important social unit in the Chicano community is the family. In contrast to American majority society, the well-being of the family comes before the needs of the individual. The Chicano father is typically the head of the household. He is the primary decision maker and wage earner. In traditional Mexican society the wife is a homemaker. In America, this role is changing. One-half of Mexican American women work outside the home, are responsible for household management, and are likely to be involved in family decisions. Men rarely increase involvement in chores as women increase their hours of employment. Some women find that their new roles are occasionally in conflict with their self-concept as mother and caregiver.

Children are cherished in the Chicano family, and families are usually large. Children are taught to share and to work together; sibling rivalry is minimal. When possible, an extended family is the preferred living arrangement. Grandparents are honored and are often involved in child care. Because of space limitations in the United States, however, many Chicano elders live in separate apartments. During periods of hardship, other relatives such as aunts, uncles, and godparents willingly accept the care of children. Girls were traditionally raised differently from boys and were kept at home to learn household skills; they were carefully chaperoned in public. Although such strict supervision diminished with successive generations born in the United States, family expectations may limit the educational and professional attainments of some young women.

Traditional Health Beliefs and Practices

Traditional health care in Mexico includes elements of Indian supernatural rituals combined with European folk medicine introduced from Spain. Beliefs and practices are closely interrelated with the culture, resulting in a health system with that is widely shared throughout Latin America. Most Mexican Americans are familiar with the conditions specific to the culture, and many use some of the associated treatments.

Health is a gift from God, and illness is almost always due to outside forces (unless one is being punished by God for one's sins). An individual must endure illness as inevitable. Prayer is appropriate for all illness, and beseeching the saints for intervention through the lighting of candles on behalf of a sick person is common. Pilgrimages may be made to religious shrines, especially those devoted to the Virgin Mary or St. Francis.

Health care is family based. The condition of a family member is discussed with mothers, grandmothers, and wives, who are the health experts in each family. Home remedies are nearly always tried first before outside help is sought. Mint, chamomile, and anise teas are especially popular, used to treat nausea, gas, diarrhea, and colic. Over-the-counter remedies, such as Pepto-Bismol and Alka Seltzer, are also used (Kay, 1977; Mikhhail, 1994). Laxatives and enemas are common. Traditional medications are available at herbal pharmacies called *botánicas*.

When an ailment is unresponsive to home cures, another health practitioner may be consulted. Often the services of a healer known as a *curandero* (or *curandera* if the healer is female) are sought. A curandero has healing powers that may be God given at birth, learned, or received through a "calling" (Graham, 1976; Spector, 1991). Although curanderos treat some symptoms, mostly with herbal remedies, they specialize in somatic ailments and are essential to curing illnesses due to supernatural causes. In regions where witchcraft is practiced, a curandero can counteract the hexes or spells of a *brujo* (a person who works on behalf of the Devil). Faith is crucial to the success of a curandero. Prayer is his or her primary treatment; the lighting of candles or the use of wood or metal effigies formed in the shape of the afflicted body part (called *milagros* or *exvoto*) may also be used. Cleansing rituals and massage are sometimes applied in certain conditions.

Illness is believed due to (1) excessive emotion, (2) dislocation of organs, (3) magic, (4) an imbalance in hot or cold, or (5) is considered an Anglo disease, such as pneumonia and appendicitis (Granger, 1976; Maduro, 1983; Spector, 1991). Treatment is based on the cause of the disorder.

Susto is an ailment thought due to excessive emotion, such as smoldering anger or shame. A type of susto known as *espanto,* which occurs when an individual is so frightened by a ghost that the soul leaves the body, is the most typical form of the disorder. Susto is a serious condition resulting most often in general malaise and depression. Mild susto is sometimes treated at home with sugar or sugar water. More serious susto, particularly when the soul is involved, must be cured by a curandero and may require lengthy treatment. If a person feels too much rage and suffers from revenge fantasies, *bilis* can occur. In this condition, excess bile is thought to spill into the blood, causing symptoms such as loss of appetite, vomiting, headaches, nightmares, and inability to urinate. *Envidia* is another ailment, taking the form of various illnesses (some terminal) caused by the emotion of envy among one's friends and neighbors. A person's success may be tempered by the misfortune of envidia.

→ Jealousy

A problem caused by the displacement of organs in infants is *caida de la mollera,* or fallen fontanel. It occurs from a fall, yanking the nipple out of a baby's mouth too quickly, or if a baby is too young when it is held vertically. The fontanel appears depressed, and it is believed that the palate drops, preventing the infant from feeding. The inability to suckle or a change in stools are symptoms. Tight caps can help to prevent the condition, and the application of salt poultices or olive oil (followed by a dip in water accompanied by prayers) may be used to treat it. The baby may be held upside down and shaken gently, the hair pulled, the fontanel sucked, or the palate may be pressed up with a finger or thumb to reposition the fontanel.

Mal de ojo (evil eye) is a condition with supernatural origins. Children are thought to be especially vulnerable to the ailment, which has flulike symptoms, including fever and headache. It is caused when one person, usually inadvertently, casts a strong, admiring look on another person. Irrational behavior and mental disabilities are often attributed to mal de ojo. This condition can result in death, so prompt diagnosis and cure is imperative. A curandero is required for treatment. A cleansing ritual that includes "sweeping" over the ill individual with an egg, then breaking the egg

- Aztec medicine was a highly developed system featuring an elite group of certified practitioners with access to a zoo and an herbarium for research. It was abolished during the Spanish conquest.

- Traditional health care supplies are often available at religious fiestas.

- Curanderos customarily do not charge clients for their services, but they do accept gratuities.

into a saucer, is performed. The egg is "read" to see whether the cure has been effective. It may be read immediately, or left under the bed overnight before examination. Prayers, herb teas, and sweeping with herb bundles may also be part of the treatment. *Mal aire* (bad air or wind) and *mal puesto* (witchcraft) account for certain other disorders, such as swelling, trembling, or paralytic twitching.

Empacho, a digestive ailment characterized by nausea, gas, and weakness, is widely known throughout Latin America. It is sometimes classified as an illness due to eating too many hot or cold foods, or a hot-cold imbalance in the stomach due to emotional upset. The direct cause is believed to be a ball or wad of food adhered to the stomach. Herb teas are administered at home, and if they are ineffective, a curandero is employed. Treatment consists of prayer, pinching the spine, and stomach massage to restore a proper hot-cold balance.

Traditional Food Habits

Mexicans are very proud of their culinary heritage. It is a unique blend of native and European foods prepared with Indian (mostly Aztec) and Spanish cooking techniques. There are even some French and Viennese influences from the Maximilian reign. The resulting cuisine is both spicy and sophisticated.

Ingredients and Common Foods

Many people associate the cooking of Mexico with chile peppers. Although chiles are used frequently, not all Mexican dishes are hot and spicy. Other New World foods such as beans, cocoa (from the Aztec word for "bitter"), corn, and tomatoes add equally important flavors to the cuisine. These indigenous ingredients were the basis of Indian fare throughout Mexico before the arrival of the Spanish.

■ Another New World food is the potato, thought to have been first cultivated high in the Andes Mountains of Peru and Bolivia, where it was too cold to grow corn. Potatoes are a staple in those regions today.

■ The scientific name for the cacao tree is *Theobroma,* meaning "food of the gods."

■ Corn is believed to have been domesticated from extinct wild varieties in southern Mexico somewhere between 8000 and 7000 B.C.E. It spread south into Central America and north into what is now the United States.

Aztec Foods

The Aztec empire had an estimated population of 25 million people at its peak in the 15th century. About one-quarter of the population were an elite class of nobles who were supported by the remaining 75 percent of the Indian slave populace. The capital city of Tenochtitlán was surrounded by lakes on which were built *chinampas,* rich agricultural fields of mud scooped from the lake bottoms. It is believed that these drought-resistant fields produced enough food to feed 180,000 people annually. The monarchy stored surplus crops to protect against famine. The Aztecs were also known for their animal husbandry and game protection laws.

Documents from early Spanish expeditions recorded the enormous variety of foods enjoyed by the Aztec nobility. More than 1,000 dishes are described. Montezuma II reportedly ate up to thirty different items per meal, each kept warm on a pottery brazier. These included hot tortillas of several types; turkey pie; roast turkey, quail, and duck; fish, lobster, frog, newt, and insect dishes garnished with red, green, or yellow chiles; squash blossoms; and sauces of chiles, tomatoes, squash seeds, or green plums. *Chocolatl,* a hot unsweetened chocolate drink made from native cacao beans, was the most popular beverage.

Corn was the staple grain. Legumes, fruits, and vegetables were plentiful; turkeys and dogs were domesticated for meat; and some small game was available. The notable deficiency of the Aztec diet was a consistent source of fat or oil.

Spanish Contributions

The Spanish arrived in Mexico with cinnamon, garlic, onions, rice, sugar cane, wheat, and, most important, hogs, which added a reliable source of domesticated protein and fat to the native diet. These additions combined with indigenous ingredients produce the classic flavors and foods of Mexican cuisine, such as corn tortillas with pork filling; tomato, chile, and onion sauces or *salsas;* rice and beans; and boiled beans fried in lard, known as *frijoles refritos,* or, as they are incorrectly called in English, refried beans. The Spanish also introduced the distillation of alcohol to native Mexican fermented beverages; *tequila* and *mescal* were the result.

Staples

The cuisine of Mexico is very diverse, and many inaccessible regions have retained their native diets. Others have held onto traditional foods and food habits despite Aztec or Spanish domination. The diets of still other areas differ because of the availability of local fruits, vegetables, or meats. The majority of poor Mexicans have little variety in their diet; some subsist almost entirely on corn, beans, and squash. This divergence makes it difficult to typify Mexican foods in general (Table 9.1). Nevertheless, some foods are found, in varying forms, throughout Mexico.

Tortillas are the flat bread of Mexico. Traditionally they are made by hand. Corn kernels are heated in lime solution until the skins break and separate. The treated kernels, called *nixtamal,* are then pulverized on a stone slab *(metate).* The resulting flour, *masa harina,* is combined with water to make the tortilla dough. Small balls of the dough are patted into round, flat circles, about 6 to 8 inches across. The tortillas are cooked on a griddle (often with a little lard) until soft or crisp, depending on the recipe.

Beans are ubiquitous in Mexican meals. They are served in some form at nearly every lunch and dinner and are frequently found at breakfast, too. They are often the filling in stuffed foods and are a common side dish, such as simmered *frijoles de olla* ("out of the pot") and frijoles refritos. They are popularly paired with rice as well.

One-dish meals are typical, almost always served with warm tortillas. Hearty soups or stews called *caldos* are favorite family dinner entrées. Casseroles, known as *sopas-secas,* using stale tortilla pieces, rice, or macaroni, are eaten as main dishes.

Stale tortillas can also be broken up and softened in a sauce to make *chilaquiles,* which are served as a side dish or light entrée. They can also be soaked in milk overnight, then pureed to make a thick dough. This dough is used to prepare *gordos,* which are fat-fried cakes, or *bolitos,* which are added to soup and are similar to dumplings.

Meats are normally prepared over high heat. They are typically grilled, as in *carne asada* (beef strips), or fried, as in *chicharrónes* (fried pork rind). Slow, moist cooking (stewing, braising, etc.) may also be used. These techniques help to tenderize the tough cuts that are generally available, as does marinating, another common preparation method. Nearly all parts of the animal are used, including the variety cuts and organs. Sausage is especially popular, such as spicy pork or beef *chorizo.*

Mexico is famous for its "stuffed" foods, such as *tacos, flautas, enchiladas, tamales, quesadillas,* and *burritos.* These are found throughout the country, with regional variations. Tacos are the Mexican equivalent of sandwiches. Tortillas, either soft or crisply fried, are filled with anything from just salsa to meat, vegetables, and sauce. Flautas ("flutes") are a variation on the taco, with tortillas tightly rolled around the filling, then fried until crispy. They may be served with a red or green chile sauce or guacamole. Enchiladas are tortillas softened in lard or sauce, then filled with meat, poultry, seafood, cheese, or egg mixtures. The tortilla rolls are then baked covered with sauce. Tamales are one of the oldest Mexican foods, dating back at least to the Aztec period. A dough made with either masa harina or leftover *pozole* (hominy) is placed in corn husks (in the north) or young leaves of avocados or bananas (in the south). The leaves are folded and then baked in hot ashes or steamed over boiling water. The tamale may contain plain dough, be filled with a seasoned meat or vegetable mixture, or be sweetened for a dessert. After cooking, the husk or leaf is unfolded, revealing the aromatic tamale. Quesadillas are tortillas filled with a little cheese, leftover meat, sausage, or vegetable, then folded in half and heated or crisply fried. Burritos are popular in northern Mexico. They are similar to tacos, but large, thin, wheat flour tortillas are used instead of corn tortillas. The most common filling is beans with salsa.

Vegetables are usually part of the main dish or served as a substantial garnish. Potatoes, greens, tomatoes, and onions are most common. Chile peppers are used extensively in seasonings, sauces, and even stuffed, as in *chiles rellenos* and the Independence Day dish *chiles en nogada,* garnished with the colors of the Mexican flag—white sauce, green *cilantro*, and red pomegranate seeds.

Sugar cane grows well in Mexico, and sweets of all kinds are popular. Dried fruits and vegetables, candied fruits and vegetables, and sugared fruit or nut pastes are eaten alone and used in more complex desserts. The Spanish make many desserts with eggs, and some of these

■ The snack food Fritos, based on an old Mexican recipe for fried masa harina, was invented in San Antonio, Texas.

■ *Epazote,* a pungent herb with minty overtones, is added to many dishes, especially those with beans, because it is believed to reduce flatulence.

■ *Menudo* is a tripe and hominy soup that is believed to have curative properties, particularly for hangovers. It is a popular weekend breakfast dish.

■ Vanilla, made from the bean of a vine native to Mexico and Central America, flavors many desserts. Legend is that the goddess Xanat fell in love with a mortal Mexican youth. Distraught that she could not marry him, she became a vanilla plant to provide him happiness.

Table 9.1
Cultural Food Groups: Mexican

Group	Comments
PROTEIN FOODS	
Milk/milk products	Few dairy products are used (incidence of lactose intolerance estimated at two-thirds of the population). Dairy products are used more in northern Mexico than in other regions.
Meats/poultry/fish/eggs/legumes	Vegetable protein is the primary source for majority of rural and urban poor.
	Pork, goat, poultry are common meats.
	Beef is preferred in northern areas and seafood in coastal regions.
	Meat is usually tough, prepared by marinating, chopping, grinding (sausages are popular), or sliced thinly. It is cooked by grilling, frying, stewing, or steaming and is usually mixed with vegetables and cereals.
CEREALS/GRAINS	Corn and rice products are used throughout the country; wheat products are more common in the north.
	Principal bread is *tortilla;* European-style breads and rolls are also popular.
FRUITS/VEGETABLES	Vegetables are usually served as part of a dish, not separately.
	Semitropical and tropical fruits are popular in most regions (limited availability in north).
ADDITIONAL FOODS	
Seasonings	Food is often heavily spiced; 92 varieties of chiles are used.
	Regional sauces are typical.
Nuts/seeds	Seeds are often used in flavoring.
Beverages	
Fats/oils	Traditional diet is relatively low in fat.
Sweeteners	Spanish-influenced pastries, candies, and custards/puddings are popular.

recipes have been adopted in Mexico. *Flan,* a sweetened egg custard topped with carmelized sugar, is the most common. *Huevos reales* is another popular dessert, made with egg yolks, sugar, sherry, cinnamon, pine nuts, and raisins.

The most common beverage in Mexico is coffee, which is grown in the south. Soft drinks and fresh fruit blended with water and sugar, called *aguas naturales,* are also favored. Adults drink milk infrequently, except in sweet-

Table 9.1
Cultural Food Groups: Mexican—continued

Common Foods	Adaptations in the United States
Milk (cow, goat), evaporated milk, hot chocolate; *atole;* unaged cheeses	Aged cheese is used in place of fresh cheese; more milk (usually whole) is consumed; ice cream is popular.
Meats: beef, goat, pork (including *chicharrónes* and variety cuts)	Traditional entrées remain popular.
Poultry: chicken, turkey	Fewer variety cuts are used.
Fish and seafood: camarónes (shrimp), *huachinango* (red snapper), other firm-fleshed fish	Protein intake may decline in second-generation Mexican Americans.
Eggs: chicken	Beans are eaten less frequently.
Legumes: black beans, chickpeas (garbanzo beans), kidney beans, pinto beans	
Corn (*masa harina, pozole, tortillas);* wheat (breads, rolls, *pan dulce,* pasta); rice	Wheat tortillas are used more than corn tortillas; convenience breads are used. Increased consumption of baked sweets, such as doughnuts, cake, and cookies, is noted. Increased consumption has occurred of sugared breakfast cereals.
Fruits: avocados, bananas, *carambola, cherimoya,* coconut, *granadilla* (passion fruit), *guanábana,* guava, lemons, limes, *mamey,* mangoes, melon, oranges, papaya, pineapple, plantains, strawberries, sugar cane, *tuna* (cactus fruit), *zapote* (fruit of the *sapodilla* tree)	Fruit remains popular as dessert and snack item; apples and grapes are accepted after familiarization.
Vegetables: cactus (*nopales or nopalitos*), chiles, corn, *jícama,* onions, peas, potatoes, squashes (*chayote,* pumpkin, summer, etc.), squash blossoms, *tomatillos,* tomatoes, *yuca* (cassava)	
Anise, *achiote* (annatto), chiles, *cilantro* (coriander leaves), cinnamon, cocoa, cumin, *epazote,* garlic, mace, onions, vanilla	Use of spices depends on availability.
Piñons (pine nuts), *pepitas* (pumpkin seeds), sesame seeds	
Atole, beer, coffee (*café con lèche*), hot chocolate, soft drinks, *pulque, mescal, tequila,* whiskey, wine	Noted are increased consumption of fruit juices, Kool-Aid, soft drinks, and beverages with caffeine; and decrease in use of hard spirits.
Butter, *manteca* (lard)	Fat intake increases, including use of mayonnaise and salad dressings.
Sugar, *panocha* (raw brown cane sugar)	

ened, flavored beverages such as hot chocolate with cinnamon. The most popular alcoholic beverage in Mexico is beer. The Mexican wine industry is also developing rapidly, in part due to a 1982 ban on the import of foreign wine. Men drink alcoholic beverages at all occasions when they gather socially. In addition to tequila and mescal, whiskey is typically served at these times.

Hand-made *tortillas,* made from *masa harina* (a type of cornmeal) or wheat flour, are the staple bread of Mexico. In this photograph, taken in a tortilla factory, the lime-soaked corn kernels are pulverized on a stone *metate.* On the right a tortilla is patted into a flat circle by hand. (Courtesy of San Antonio Light Collection; copy from the Institute of Texas Cultures at San Antonio.)

■ *Caesar salad,* romaine lettuce with a tangy dressing of lemon, vinegar, garlic, anchovy, romano cheese, and raw egg, was invented in northern Mexico.

■ Mayauel was the ancient goddess of pulque. She was believed to bring both intoxication and death to mortals.

■ A worm from the maguey plant in a bottle of mescal or tequila is a signature of authenticity.

Regional Variations

Mexican Plains. The northern and west-central regions of Mexico consist of mostly arid plains. The Indians who originally inhabited the area were called *Chichimecs* by the Aztecs ("sons of the dog") because of their seminomadic lifestyle. It is believed that their diet consisted of corn, beans, squash, greens, cactus fruit (*tuna*), and young cactus leaves (*nopales*). They also hunted small game and ate domesticated turkeys and poultry. When the Mexican Indians mixed with the Indians of the American Southwest, they adopted piñon nuts (also called pine nuts or pignolis), pumpkin, and plums.

The Spanish introduced longhorn cattle to the northern plains, as well as wheat. Today this is the only region of Mexico where people eat substantial amounts of beef and cheese. A favorite way to prepare beef is to air-dry it in thin slices called *cecinas,* similar to American chipped beef but more assertive in flavor. Cecinas can be used in stews or soups; it can be fried or used as fillings for other foods. The most popular Mexican cheeses are unripened. There is *queso blanco,* which is similar to mozzarella, and creamy *queso fresca,* which is similar to ricotta or pot cheese. Wheat is also used in more products in the north than in the south.

The sap of the maguey cactus (century plant) is credited with being a reliable substitute for fresh water in the arid countryside; it is called *aguamiel* or "honey water." Tequila is probably the best-known beverage of the region, made from the distillation of fermented aguamiel, known as *pulque.* Pulque is the sour, mildly alcoholic beverage that was drunk throughout Mexico before the arrival of the Spaniards. The Spaniards, lacking grain, tried their distillation methods on the pulque, producing *mescal,* a harsh brew. Tequila is the more refined, twice-distilled version of mescal. It is produced in the central-western state of Jalisco around the towns of Tequila and Tepatitlan from a maguey subspecies *Agave tequiliana.*

Tropical Mexico. Fruits and vegetables are featured foods in the southern coastal regions of Mexico. Tomatoes, green tomato-like *tomatillos,* chayote squash, onions, *jícama* (a sweet, crispy root), sweet and starchy plantains, carambola (star fruit), *cherimoya* (custard apple), *guanábana* (soursop), *mamey* (a type of plum), pineapple, *yuca* (a tuber also called *cassava* or *manioc*), and *zapote* (the fruit of the *sapodilla* tree) are just a small sampling of the produce available in this area.

More than 90 varieties of chile peppers are found in the region, varying enormously in degree of hotness. The chemical heat of chiles comes from the alkaloid capsaicin, found mostly in the fleshy ribs and seeds inside the fruit. It is not known why the sting of chile-spiced foods is enjoyed. It may stimulate the appetite and digestion through irritation of the stomach lining, or it may cause the body to release pain-killing endorphins creating a comfortable, gratified feeling.

Avocados, grown in the tropical climate, vary in size from 2 to 8 inches across, in skin color from light green to black, and in flavor from bland to bitter. Their succulent, smooth flesh is added to soups, stews, and salads. They are most popular in *guacamole,* mashed avocado with onions, tomatoes, chile peppers or chili powder, and cilantro, the pungent leaf of the coriander plant. It is used as a side dish, a topping, or a filling for tortillas.

Yucatán. The cuisine of the Yucatán peninsula reflects its unique history. It was isolated from the rest of the country by dense, mountainous jungles until modern times. Many of the residents are descendants of the Mayas, the early Indian dynasty of the region. Some regional favorites date from this time. For example, one popular preparation method is to steam foods, called *píbil.* Traditionally food was cooked this way in an outdoor pit, but today it is prepared more often in a covered pot. Shrimp are a local specialty; the long coastline of the Yucatán along the Gulf of Mexico provides ample seafood. Sauces made with toasted squash seeds or the bright red spice *achiote* are also common to the area.

■ The type of still used to produce mescal and tequila is a design that originated in Southeast Asia. It may have made its way to Jalisco through Spanish trade between the Philippines and Mexico (Perry, 1999).

■ *Chile* (with an *e*) comes from the Nahuatl word *chilli.* Foods like the powders, sauces, and stews made with chiles are conventionally called *chili* (with an *i*).

■ Avocado comes from the Nahuatl word for the fruit, *ahuactl,* meaning "testicle," which they resemble hanging from the tree.

■ *Achiote* is called annatto in the United States. It is made from the seeds of a tropical tree and used sometimes to color Cheddar-style cheeses, ice creams, margarine, and some baked goods.

Tropical fruit from Latin America. (Courtesy of the Florida Division of Tourism.)

Southern Mexico. The foods of southern Mexico are similar to those of the Yucatán in that they are more tropical and more Indian influenced than the foods of other regions. They are renowned for their complex, spicy sauces that use numerous ingredients such as chiles, nuts, raisins, sesame seeds, spices, and even chocolate. These sauces come in many colors, including red, yellow, green, and black, and are called *moles,* probably from the Aztec word for "sauce with chiles," *molli.* Poultry, goat, and pork are the most popular meats, although some specialties of the region are more unusual, such as *chalupines,* a cricket that is found in the corn fields.

Meal Composition and Cycle

Daily Patterns

In familes where income is not limited, the preferred meal pattern is four to five daily meals: *desayuno* (breakfast), *almuerzo* (coffee break), *comida* (lunch), *merienda* (late afternoon snack), and *cena* (dinner). Most meals are eaten at home and served family-style. If there are too many people to sit at the table, each one is served individually from the stove.

Desayuno is a filling breakfast, often including tortillas, eggs, meat, beans left over from the previous night, *bolillos* (wheat rolls), *pan dulce* (sweet bread, pastry, or cake), and fresh fruit. Coffee and hot chocolate are the preferred beverages. Near 11:00 A.M. is almuerzo, which features pan dulce or fruit, served with *café con lèche* (coffee with milk).

Comida is traditionally the largest meal of the day, eaten about 1:00 or 2:00 P.M. A complete comida includes six full courses, from soup through dessert. When possible, an afternoon rest period (siesta) follows this meal. Merienda is a light meal of sweet rolls, cake, or cookies eaten around 6:00 P.M. Coffee, hot chocolate, or a warm drink of thin, sugared cornmeal and milk gruel, called *atole,* accompanies the sweets. A light supper, cena, follows between 8:00 and 10:00 P.M. This meal may be skipped entirely or expanded into a substantial feast on holidays or other formal occasions. Recently many Mexicans have adopted the American habit of eating a light lunch and a heavy supper, eliminating merienda altogether.

Snacking is frequent in urban Mexico; munching occurs from morning to midnight. *Antojitos,* or "little whims," include foods such as *tostadas* (called *chalupas* in northern Mexico), fried tortillas topped with shredded lettuce, cheese, or meat. Nearly every block offers streetside food vendors, providing everything from fresh fruits to grilled meats. In addition, many neighborhoods feature an open-air market that also offers ready-to-eat foods.

Sample Menu

A TRADITIONAL MEXICAN COMIDA

Meatball soup

Chilequiles (tortilla casserole)
or *Chiles rellenos* (stuffed chiles)

Pescado Yucateco(red snapper in achiote sauce)

Simmered beans and rice

Warm tortillas

Fresh papaya, *cherimoya, granadilla,* or *zapote*

Fruit juice or beer

Special Occasions

In Mexico, many foods are associated with holidays. For example, the Día de los Santos Reyes (also called Día de los Reyes Magos, or Three Kings Day) on January 6 is customarily celebrated with *rosca de reyes,* a raisin-studded, ring-shaped loaf of bread. Baked inside the bread is a figurine of the infant Jesus, and the person who receives it is obligated to give a party on February 2.

During Lent, *capirotada* is a popular dessert. Each family has its own recipe for this holiday bread pudding. Another holiday food is bread decorated with a skull and crossbones (*pan de muerto*), eaten on All Soul's Day (November 2) as part of a large feast honoring the deceased. Christmas festivities, called *posadas,* frequently feature *piñatas,* brightly decorated papier-mâché animals and figures that are filled with sweets. Blindfolded children take turns swinging a large stick at the hanging piñata until it breaks and candies fly everywhere. In some regions, thin, fried, anise-spiced cookies called *buñuelos* are eaten at Christmastime. They are drenched in syrup and served in pottery bowls, which when empty are smashed on the street for good luck. On Christmas Eve a salad of fruits, nuts, and beets is served. Turkey, tamales, and *arroz con leche* (rice pudding) are special foods eaten at holidays throughout the year.

Role of Food in Mexican Society

In family-centered Mexican society, food-related activities facilitate interactions between family members and help delineate family roles. Meal planning is usually the wife's responsibility. Depending on economic status, food is prepared by the wife or by servants supervised by her, because Mexican foods can be laborious to prepare. The final dishes are greatly appreciated by all who partake in the meal, and it is considered an insult not to eat everything that is served. In rural areas, food sharing is an important social activity, reflecting the Indian worldview. To reject offered food or drink is a severe breach of social conduct.

Therapeutic Uses of Food

Some Mexican Indians and rural poor practice a hot-cold system of diet and health. It is believed to have derived from the Arab system of humoral medicine brought to Mexico by the Spanish, combined with the native Indian worldview. Although it has parallels with other classification systems, such as the Asian practice of yin-yang (see chapter 11 for details), the Mexican system is only applied to foods and to the prevention and treatment of illness. It does not encompass moral or social beliefs. The Mexican hot-cold theory is based on the concept that the world's resources are limited and must remain in balance. People must stay in harmony with the environment. Hot has the connotation of strength; cold, of weakness. When the theory is applied to foods, items can be classified according to proximity to the sun, method of preparation, or how the food is thought to affect the body. Meals balanced between hot and cold foods are considered to be health promoting. Unbalanced meals may cause illness. Thus a typical comida in a rural village would consist of rice (hot), soup (made with hot and cold ingredients), and beans (cold).

Although the hot-cold classification of foods does vary, items generally considered hot are alcohol, aromatic beverages, beef, chiles, corn husks, oils, onions, pork, radishes, and tamales. Cold foods include citrus fruits, dairy products, most fresh vegetables, goat, and tropical fruits. Some foods, such as beans, corn products, rice products, sugary foods, and wheat products, can be classified as either hot or cold depending on how they are prepared.

Illnesses are also believed to be hot or cold and are usually treated with a diet rich in foods of the opposite classification. Certain conditions, such as menstruation and childbirth, are considered to be hot, and some women refuse hot foods at these times, such as pork, or very cold foods that might create a sudden imbalance, including cucumbers, tomatoes, and watermelon (Kay, 1977; Maduro, 1983). Sour foods are believed to thin the blood and are avoided by menstruating women because they are thought to increase blood flow; acidic foods may also be shunned because they are said to cause menstrual cramps.

Food and herbal remedies are also used in some regions. For example, chamomile is believed to cure colic, menstrual cramps, anxiety, insomnia, and itching eyes. Garlic is chewed for yeast infections in the mouth, toothache pain, and stomach disorders; boiled peanut broth is used to cure diarrhea; boiled cornsilk is taken for kidney pain; honey and water are given to infants for colic; oregano is used for fever, dry cough, asthma, and amenorrhea; and papaya is thought to help cure digestive ailments, diabetes,

- In the city of Oaxaca, artisans carve intricate nativity scenes and representations of the Virgin of Guadelupe into large (up to 2 feet long) radishes in a state-sponsored competition the night before Christmas Eve. The origins of the event are lost in history.

- Some Mexicans avoid cold air and drafts after eating chiles (which are classified as hot) to avoid causing an imbalance in their bodies.

- Mexican women sometimes eat cilantro after giving birth because it is considered a cooling food for postpartum women.

- Some Mexicans believe a sprig of parsley tucked above the ear will cure a headache.

asthma, tuberculosis, and intestinal parasites. (See Table 9.2, "Selected Mexican American Botanical Remedies")

■ *Chili con carne* (beans with beef) is not a Mexican dish, although similar combinations of ingredients are found throughout Mexico. Chili is believed to have originated in Texas, after the Mexican-American War.

■ Hispanics frequent restaurants more than any other ethnic group in the United States. Among Mexican Americans' favorite establishments (in order of popularity) are fast food, pizza, Mexican fast food, Chinese, Coffee shops, and full service Mexican (Elder et al., 1999).

Contemporary Food Habits in the United States

The foods of Mexico have influenced cooking in the American Southwest and California. In Texas, Tex-Mex cuisine features many modified Mexican dishes, such as chili, tamale pie, and nachos. Barbecued chile-spiced meat kebobs called *anacuchos* and *capriotada* with whiskey sauce known as "drunken pudding" are other examples of Tex-Mex creativity. (See chapter 15 on regional Americans for more information.) Chicanos, Braceros, and undocumented immigrants living in the United States may eat these and other American foods or may eat a more traditional diet, depending on length of time in the country, location, income, or other factors.

Adaptations of Food Habits

Many Mexicans in the United States eat a diet similar to that of their homeland. Recent immigrants, those who live near the U.S.–Mexican border, and migrant workers are most likely to continue traditional food habits.

Chicanos and Mexicans who are well established in the United States often become more acculturated. In one hospital, it was found that "Mexican" patients overwhelmingly preferred American foods, even when offered traditional Mexican meals (Smith, 1979). A marketing study (Wallendorf & Reilly, 1983) noted that Mexican immigrants living in the southwestern United States are more likely to eat a diet with a high intake of red meats, white bread, sugared cereals, caffeine-containing beverages, and soft drinks than their socioeconomic counterparts in Mexico or white neighbors. This suggests that, rather than adopting a diet that fell somewhere between the food habits of Mexico and those of the United States, the Mexican immigrants in the survey accepted the stereotypical American consumption patterns of a decade ago.

Ingredients and Common Foods

Some American foods are accepted in even the most traditional Mexican American families. Doughnuts, pie, cake, cookies, ice cream, and Popsicles are popular as desserts and snacks. Sugared cereals are becoming more common at breakfast for children, and American cheeses are often used in place of unaged Mexican cheeses.

A preference for sweet or carbonated beverages usually increases in the United States. Soft drinks, Kool-Aid, and juices are popular with meals and as snacks. Beer and coffee are also consumed with meals or snacks throughout the day. Whole milk is considered a "superfood" for children, but adults, especially men, consider it to be a juvenile drink. Milk is often flavored with chocolate, eggs, and bananas in a drink called *licuado;* mixed with coffee; or mixed with cornmeal to make atole. Some adults reject milk completely, claiming that they are allergic to it (it is estimated that two-thirds of Latinos are lactose-intolerant).

More acculturated Mexican Americans buy many prepared and convenience foods. Baked goods are usually purchased, including tortillas (often wheat tortillas are chosen over corn), breads, pan dulce, and even special desserts like flan. Extra income is usually spent on meats.

Meal Composition and Cycle

Daily Patterns. Studies of low-income Latinos and Mexican migrant farmworkers indicate that, for most meals, traditional foods are preferred. These include eggs, beans or meat, and tortillas or pan dulce for breakfast; a large lunch of beans, tortillas, and meat, or a soup or stew; and a lighter dinner of tortillas, beans or meat, and rice or potatoes. As in Mexico, vegetables tend to be served as part of a soup or stew. Fruit remains a typical snack and dessert, especially familiar varieties such as bananas, oranges, mangoes, guava, pineapple, strawberries, and melon. One study determined that the largest change is in increased consumption, of both basic items and new foods, due to increased income and increased availability (Dewey et al., 1984).

Table 9.2
Selected Mexican American Botanical Remedies

Scientific Name	Common Name	Parts	Traditional Use
Aloe spp. ☹	Sábilla; aloe vera	Leaves, juice	Indigestion; heartburn; gastric ulcers; constipation; rheumatism
Archtostaphylos uvaursi, A. manzanita, A. pungens ☹	Manzana; manzanita; corallino; bearberry; uva ursi	Leaves, berries	Kidney problems, bladder infections; bronchitis
Artemsia absinthium ☠	Arenjo; artemsia; wormwood	Leaves, stems	Indigestion; heartburn; gallbladder problems; parasites; menstrual cramps
Betula spp.	Abedul; alamo blanco; birch	Leaves, bark, resin	Kidney problems; urinary tract disorders
Borago officinalis ☠	Borraja; borage	Leaves, flowers	Diuretic; kidney disorders; bladder problems, fever; sore throat; cough
Brickellia spp.	Prodigiosa; amula; mala mujer; bricklebush	Leaves, flowers	Diabetes; liver problems; digestive upset; diarrhea; *bilis*
Casimiroa edulis	Zapote blanco; sapodilla; cochizápotl; white sapote	Leaves	Sedative; hypertension; *empacho*
Chenopodium ambrosiodes ☠	Epazote; Mexican tea; goosefoot	Root; whole plant	Indigestion; dysentery; parasites; delayed menstruation
Crataegus pubescens; C. mexicanus ☹	Tejocote; manzanilla; manzanita; christé; Mexican hawthorne	Root, leaves, bark, fruit	Kidney inflammation; cough; colds
Croton niveus, C. tiglium, C. reflexifolius	Copalchi; quina blanc; garañona; chichiquáhuitl	Leaves, bark	Diabetes; kidney disorders; fever; purgative
Cynara scolymus ☹	Alcachofa; quahtlahuitzquilitl; artichoke	Leaves, flowers	Diabetes; liver problems
Ephedra torreyana, E. trifurca, E. viridis ☠	Popotillo; itamo real; cañutillo del campo; Mormon tea; Mexican tea; desert tea	Stem	Diuretic; kidney disorders; bladder infections; stomach problems; diarrhea
Erynigium spp.	Yerba del sap; sea holly	Root, leaves, flowers	Kidney problems; emaciation; aphrodisiac
Eysenhardtia polystachya; Caesalpinia bonducella	Palo azul; palo dulce; palo cuate; cualaldulce; kidney wood	Wood	Diuretic; hypertension; kidney disorders; urinary tract disorders; stomach pain; fever; toothache
Glycyzrrhiza glabra ☠	Orozús; yerba dulce; regaliz; palo cuate; coahtli; licorice root	Root	Gastric ulcers; sore throat and coughs; menstrual problems
Heterotheca spp. ☠	Arnica Mexicana; camphor weed	Root; whole plant	Stomach problems; diarrhea; hemorrhoids; toothache; bronchitis; menstrual cramps
Ipomoea purga ☠	Jalapa; brionía; michoacán; chicícamolli; xtabentum; jalap; Mexican morning glory	Root	Purgative
Ipomoea stans ☠	Tumba vaqueros; riñona; morning glory	Root	Kidney disorders; diarrhea; menstrual cramps; epilepsy; hysteria
Montanoa tomentosa ☠	Zoapatle	Leaves	Oral contraceptive; labor induction
Oenothera spp. ☹	Yerba del golpe; evening primrose	Flowers	Kidney disorders; urinary tract infections; menstrual cramps

continued

Table 9.2
Selected Mexican American Botanical Remedies—continued

Scientific Name	Common Name	Parts	Traditional Use
Opuntia spp.	Nopal; tuna; cholla; tlatocnochtli; prickly pear cactus; Indian fig	Root, paddle, fruit	Diuretic; diabetes; kidney stones; urinary tract infections
Phytolacca americana ☠	Fitolaca; pokeroot; pokeberry	Root, leaves, berries	Laxative, cathartic; inflammations
Psidium spp.	Guayaba; guava	Leaves	Diarrhea; dysentery
Rhamnus purshiana, R. californica ☠	Cáscara sagrada; buckthorn	Bark	Laxative; cathartic; fatigue; weakness
Sechium edule	Chayote; choyotl; vegetable pear	Leaves	Diuretic; kidney problems; hypertension
Smilax spp.	Zarsaparilla; cocolmeca; sasparilla; red China root	Root, leaves	Diabetes; excessive menstrual bleeding; skin problems; diarrhea; laxative
Taraxacum officinale ☹	Diente de léon; chicória; dandelion	Root, stem, leaves	Diuretic; kidney disorders; liver problems; skin rash
Tecoma stans; stenolobium stans	Tronadora; trumpet bush; yellow elder	Root, leaves	Diabetes; diuretic; hangover nausea

Note: Data on some plants are very limited; adverse effects may occur even if not indicated.
Key: ☠, all or some parts reported to be harmful or toxic in large quantities or specific preparations; ☹, may be contraindicated in some medical conditions/with certain prescription drugs.

Chicanos more often adopt the American meal pattern of small breakfast, small lunch, and large dinner. Although consumption of tortillas remains relatively stable, breakfast cereals have become popular with all family members and sandwiches are a common lunch item. Meats and cheese become more prevalent at meals, beans are eaten less frequently, and vegetables are served as side dishes such as salads (Romero-Gwynn et al., 1993).

Changes in preparation methods may also occur. Recent immigrants may not know how to use the baking and broiling apparatus on an oven and may continue to fry and grill foods outdoors. Newer immigrants sometimes avoid canned and frozen foods because they do not know how to prepare them.

Special Occasions. The Mexican custom of saving dishes that require extensive preparation, such as tamales and enchiladas, for Sunday and holiday meals is continued in the United States. In one study (Bruhn & Pangborn, 1971), migrant workers were asked what was the main dish served at holidays. No preference was found for Easter; tamales were preferred for Christmas. Turkey with mashed potatoes was the most popular Thanksgiving entrée, indicating that this American holiday was adopted along with its traditional foods.

In addition to religious holidays, two secular celebrations are significant in the Mexican American community. The first is Mexican Independence Day on September 16, commemorating the war of liberation from Spain. Observations emphasize ethnic unity including *mariachis* and traditional clothing. Foods the color of the Mexican flag, such as white rice, green avocado, and red or green chilis are eaten. *Cinco de Mayo* (May 5) is the second holiday, more widely recognized by all ethnic groups throughout the United States even though the meaning of the event (remembrance of a historic victory over France) is often forgotten amidst the parades, piñatas, and Aztec dancing that typify the day.

Nutritional Status

Nutritional Intake

It can be difficult to determine health statistics on Chicanos, Braceros, and undocumented

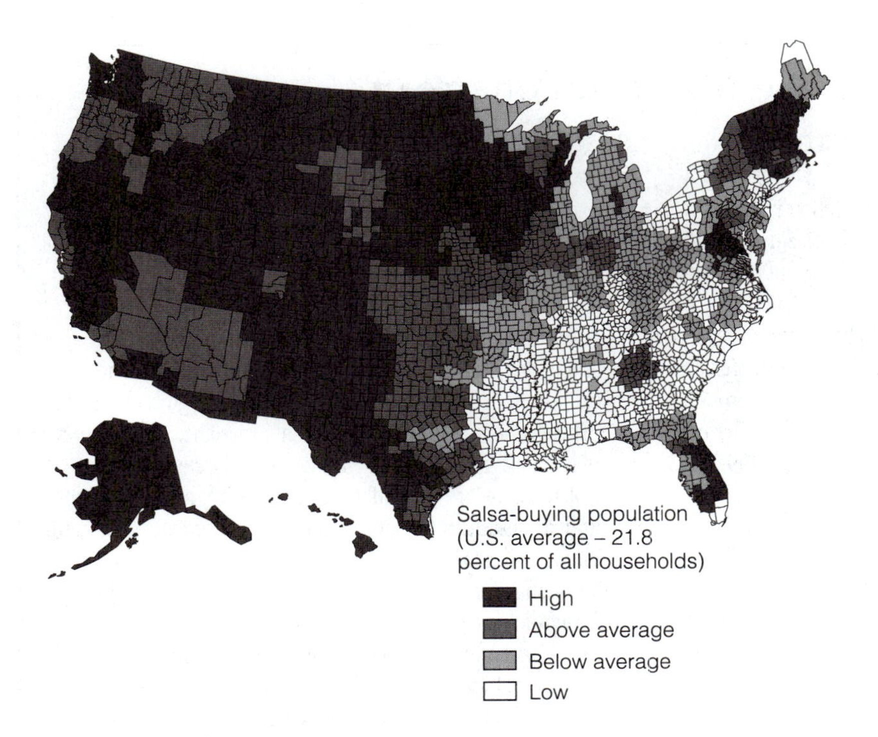

Figure 9.2
Salsa sales in the United States
Source: Atlantic Monthly, May 1997.

aliens because they often grouped with whites or other Latinos in collected data. Information from research on Latinos can be used cautiously, since Mexicans comprise such a large percentage of the total Latino population. Nevertheless, some nutritional problems have been identified in both new and acculturated immigrants from Mexico through studies of Spanish-surnamed patients, especially those residing in the Southwest.

The birth rate among women of Mexican descent who are 15 to 44 years of age is 114 births per 1,000 women, nearly double the figure for white women. Among unmarried teenage girls, the pregnancy rate is twice that of whites; one-half of teenage Latino mothers do not receive prenatal care during the first trimester. Despite these risk factors, infant mortality is lower for Mexican American infants than for white infants (see the box "Breaking the Mold).

Approximately 20 to 50 percent of Spanish-surnamed women breast-feed their newborn infants, and reported increases in the percentage of women who breast-feed parallel those in the total U.S. population. In one study of migrant farmworkers in California, it was found that an infant was less likely to be breast-fed after the parents came to the United States, even if the baby was born in Mexico (Kokinos & Dewey, 1986). Breast-fed babies are usually weaned from the breast to the bottle. Long-term use of the bottle, with both milk and sweetened liquids (i.e., Kool-Aid, fruit juice, tea) is sometimes a problem, resulting in iron-deficiency anemia and extensive tooth decay among toddlers (baby-bottle tooth decay).

In Mexico, the prevalence of malnutrition resulting in stunting of children varies between 10 and 35 percent depending on region (Romero, 1997). In the United States, poverty adversely affects the diet of 30 to 40 percent of the total Latino population.

In general, Mexican Americans have been found to consume more fat, often as fried foods, and fewer fruits and vegetables than Anglos of similar background (Bartholomew et al., 1990; DiSogra et al., 1994). Acculturation often exacerbates these intake differences. Second-generation Mexican Americans eat more meat and cheese and fewer beans; eat vegetables as side dishes (instead of in soups and stews) with added butter, dressings, or mayonnaise; and choose sweetened fruit drink mixes (i.e., Kool-Aid) over fresh, blended fruit drinks (Romero-Gwynn et al., 1993).

Some research indicates total protein intake may also decline with the length of stay in the United States (Chávez et al., 1994; Guendelman

■ Salsa sales in the United States surpassed those of catsup in 1997, an example of the Mexican American influence on the American diet (see Figure 9.2).

■ A study of Tarahumara Indians in Mexico was done to observe the effects of a high-calorie, high-fat, low-fiber diet on a population that traditionally consumes a low-fat, high-fiber diet. After 5 weeks, blood cholesterol levels increased 31 percent and triglyceride levels increased 18 percent; all subjects also gained weight. Researchers hypothesize that adopting the typical American diet would have serious health consequences for the native population over time (McMurry et al., 1991).

Breaking the Mold: The Mexican American Immigrant Experience

One of the assumptions regarding assimilation is that life improves for immigrants the longer they reside in the United States. Yet recent studies of the largest immigrant population contradict this model. Foreign-born Mexican Americans are found to he healthier overall, to eat slightly better diets, and to have lower rates of infant mortality than U.S.-born Mexican Americans with foreign-born parents and for U.S.-born Mexican Americans with U.S.-born parents—despite higher rates of poverty and less access to medical care (Harris, 1999; Landale et al., 1999; Schaffer et al., 1998).

The scientific community expressed little interest about differences in immigrant health status associated with place of birth until researchers in the late 1980s discovered a startling trend. Poor, disadvantaged women who had immigrated from Mexico were giving birth to babies who were as healthy as those of white U.S. women with overall higher levels of income and education. Rates of premature births, low birth weight rates (an indicator of infant health within a population and a predictor of neonatal mortality rates), and newborn death rates among the immigrant women were equal to or less than that for whites. The foreign-born women also demonstrated better birth outcomes than Mexican American women who were born in the United States (Collins & Shay, 1994; Scribner & Dwyer, 1989).

Suddenly, the assimilation model of health was in question. Other research on foreign-born and U.S.-born Mexican Americans have offered similarly perplexing data. Studies of adolescents and pregnant women have reported higher intakes of grain products, fruits, and vegetables and lower intakes of fat in foreign-born subjects than in those born in the United States (Mendoza & Dixon, 1999; Schaffer et al., 1998). Studies on the affect of acculturation on food habits of Mexican Americans find parallel trends toward adopting a higher-fat diet while reducing grain, fruit, and vegetable intake (Romero-Gwynn et al., 1993). Marketing and restaurant analyses support the popularity of American foods among Mexican Americans (Elder et al, 1999; Wallendorf & Reilly, 1983). In another, comprehensive review of statistics on the overall health and health risk behaviors of adolescents, the data were even more revealing. In nearly every ethnic group studied, the longer a subject's family had lived in the United States (as calculated according to whether the subject was foreign born, born in the United States with foreign-born parents, or born in the United States with U.S.-born parents), the poorer the subjects health and the more likely the subject was to engage in risky behaviors, even after controlling for neighborhood, family, education, and income variables. Mexican Americans who were born in the United States with U.S. parents had significantly higher rates of health problems (including obesity, asthma, learning difficulties, missing school due to illness or emotional problems, and psychological distress) compared to those who were foreign born. Health risk behaviors, determined by sexual experience, delinquency (e.g., painting graffiti, shoplifting, burglary, stealing a car), violent behavior (especially use of weapons), and use of controlled substances (cigarettes, chewing tobacco, alcohol, marijuana, and hard drugs) were more than double in U.S.-born Mexican American adolescents with U.S. parents than in foreign-born Mexican American youth (Harris, 1999).

What accounts for such significant differences? Most hypotheses have addressed the disparities in birth outcomes, suggesting that selective migration occurs (only healthy women come to the United States), that deaths during pregnancy may be greater (thus skewing the data), or that infant deaths are underreported in the foreign-born Mexican American community (which may include high numbers of undocumented residents). Other theories emphasize the protective factors of the Mexican culture (Guendelman, 1995). While information on the relative health of those who choose to immigrate and hidden deaths is difficult to obtain, research does suggest that pregnant foreign-born Mexican American women behave in ways significantly different from those born in the United States. Nutrient intake, including protein, folate, vitamin C, iron, and zinc, is better; smoking and alcohol consumption rates are substantially lower; and mental health problems are fewer (Acevedo, 2000; Landale et al., 1999; Schaffer et al., 1998). Other factors considered important to positive pregnancy outcomes, such as adequate weight gain and prenatal care, are less likely in foreign-born Mexican Americans. Researchers suggest these negatives may be compensated for by greater family and spousal support (due to lower rates of single motherhood), fewer numbers of pregnant women under age 20, and less accumulative acculturation stress (Collins & Shay, 1994; Landale et al., 1999; Scribner & Dwyer, 1989). Religious faith may also play a role (Magana & Clark, 1995). Determining the reasons for why place of birth is so significant will help researchers devise a new assimilation model and suggest approaches for improving the diet, pregnancy, and health outcomes for all Mexican Americans.

& Abrams, 1995). Low protein consumption combined with low iron intake sometimes results in low hemoglobin levels among young children and pregnant women. However, excessive rates of low iron intake or low blood iron status among Americans of Mexican descent have not been confirmed. Further, some studies suggest that within the subgroup of Mexican American children, protein consumption is mostly adequate (Dewey et al., 1984b; Murphy et al., 1990).

Deficiencies of calcium and riboflavin are common; often due to low consumption of dairy products. Although the traditional Mexican diet has good sources of vitamins A and C, thiamin, niacin, B_6, folate, phosphorus, zinc, and fiber, low intakes of these nutrients by Mexican Americans have been reported; inadequate income or lack of traditional ingredients may limit consumption of these nutrients (Ballew & Sugarman, 1985; Fanelli-Kuczmarski et al., 1990; Knapp et al., 1985; Zive et al., 1995). These problems may be increased by intestinal infections (the leading cause of death in Mexico), such as amoebic dysentery, especially among new arrivals.

Nutritional inadequacies may contribute to other diseases. In the southwestern United States, the mortality rates for pneumonia and influenza are higher for persons with Spanish surnames than for whites. Deaths of Latinos from tuberculosis in New Mexico and Los Angeles have been nearly double those of whites in those regions. Mexican American children living in poverty are also at risk for lead toxicity (Carter-Pokras et al., 1990).

In contrast to the problem of undernutrition is the prevalence of obesity in persons of Mexican descent, who are two to four times more likely than whites to be overweight (Stern et al., 1983). One study found 42 percent of Mexican American women and 31 percent of Mexican American men to be obese (Fanelli-Kuczmarksi & Woteki, 1990); another reported 55 percent of women and 43 percent of men overweight (Dewey et al., 1984a). Cultural ideal weight is greater for Mexicans than for Anglo-Americans. Many Latino women believe it is normal to gain weight after marriage. Extra weight indicates health and well-being, not only for adults but also for children. Obesity is sometimes perceived as a status symbol.

Several studies have shown that the rate of non-insulin-dependent diabetes mellitus is two to five times higher in persons of Mexican descent than in the white population and increasing rapidly. Complications, including kidney failure and blindness, are also more prevalent. It is considered the number one health problem of Chicanos by many health professionals. This high rate is not explained by the incidence of obesity, age, or education, but may be related to percentage of Indian heritage (Burke et al., 1999; Haffner et al., 1991; Markides & Coreil, 1986).

A higher prevalence of gallbladder disease and elevated triglyceride levels have been found among Mexican Americans; rates of elevated serum cholesterol and cardiovascular disease are lower than among whites. Data on hypertension in Mexican Americans is inconclusive.

Cavities are common among Americans of Mexican descent, as is gingivitis. Studies show that nearly one-third of all immigrants from Mexico never receive any dental care. Migrant workers and their children are especially at risk (Nurko et al., 1998).

Counseling

Access to biomedical health care may be limited for Americans of Mexican descent, especially for Braceros and undocumented aliens. Income may restrict doctor visits, and transportation to clinics may be unavailable. Over 37 percent of Mexican Americans do not have health insurance (USHHS, 1997). Further, Spanish-speaking clients may be uncomfortable with interviews conducted in English.

The communication style of Latinos is nonconfrontational; a warm, dignified relationship is most effective, and crucial in difficult health care situations. Touching a client with a handshake is important. Avoid prolonged eye contact, which is considered rude. Most Latinos are present oriented and polychronic (able to do several things at once). Inflexible appointments are problematic for many Latino clients, who may prefer walk-in clinics (Spector 1991). Latinos may be uninterested in lengthy indirect discussion of a condition and prefer a direct, action-oriented approach.

Attitudes may differ from American biomedical beliefs. For example, it is considered

■ A study on the prevalence of iron overload disorders in Hispanics suggests the rate may be slightly higher than among whites (Felelti et al., 1996).

■ Early researchers were perplexed by the absence of the niacin deficiency disease, pellagra, in Mexicans who consumed a corn-based diet. Pellagra was common in the southern United States, where corn was also a staple. It was found that when the corn kernels were prepared for *masa harina,* the alkaline lime solution used to soften them released the niacin that was bound to a protein. The lime is also believed to add calcium to the diet, although bioavailability is questionable (Looker et al., 1993).

■ Rural Hispanic elders were more likely to report inadequate intake of vegetables and problems meeting basic nutrition needs than were rural white elders in one study (Marshall, 1999). Another study of more urban elders, however, found similar intakes of calories and nutrients, though food choices varied between Hispanics and whites (Pareo-Tubben, 1999).

■ Chagas' disease, caused by *Trypanosoma cruzi,* may be found in some Mexican immigrants who have consumed *chinche de compostela* (an insect with reputed aphrodisiac properties) (Salazar-Schettino, 1983).

inappropriate for men to acknowledge illness. People who go on working despite bad health are respected. Modesty and privacy are highly valued; thus a woman may wish to be treated by a female caregiver and a man by a male caregiver.

Studies suggest that anywhere from 20 to 81 percent of clients use home remedies (Council on Scientific Affairs, 1991; Mikhail, 1994). Traditional healers, such as curanderos, are consulted by anywhere from 4 to 21 percent of the population (Mikhail, 1994; Risser & Mazur, 1995, Skaer et al., 1996). Although these practices are most common in poor, rural regions, most Mexican Americans are knowledgeable about folk conditions: one study in Texas reported that mal de ojo had been diagnosed and treated in 63 percent of Mexican American homes surveyed, susto was known in 62 percent, empacho in 48 percent, and caída de mollera in 34 percent of homes (Trotter, 1991). Those who consult curanderos believe that healers are most effective for folk illnesses.

■ *Queso fresco,* traditionally made from raw milk, is responsible for more food-borne illness than any other cheese in the United States (Bell et al., 1999).

■ Although not widely used, medicinal substances such as *vibora de cascabel* (dried rattlesnake powder can be a source of botulism (Algert & Ellison, 1989).

■ Sixty-one percent of Latino clients and health providers at a California clinic had "heard" that during the postneonatal exam physicians cut or break the hymens of female infants (Cronin et al., 1996).

Most traditional health beliefs and practices among Mexican Americans support the emotional well-being of a client and do not interfere with therapy. It has been suggested by many researchers that folk conditions provide an important release valve in Latino cultures, especially for men who are expected to endure pain. Disorders due to outside causes are not blamed on an individual, and the resulting irrational behavior or lethargy is excused (Granger, 1976; Spector, 1991). Several potentially harmful situations are noteworthy. Digestive complaints, such as empacho (see "Traditional Health Beliefs and Practices" section) are sometimes treated with toxic lead- or mercury-based medication. *Greta, azarcón,* and *asogue* are still available today in Mexico. The condition of caída de la mollera in infants has been associated by some health practitioners with severe diarrhea and dehydration, resulting in the depressed fontanel. Providers should be aware of this possibility when presented with this disorder. In some regions, a tea made from the psychoactive wormwood (the toxic ingredient formerly found in the alcoholic beverage absinthe) is used for diarrhea. Finally, babies may be given home remedies made with honey, a known cause of infant botulism.

Family participation in health care is common, and members should be consulted in both making a diagnosis and in prescribing treatment. They may have specific ideas about the cause of an illness and the best approach for a cure; their confidence and cooperation can help ensure client compliance. One study found that family involvement in serious choices about issues such as life support is more important to Mexican Americans than patient autonomy (Blackhall et al., 1995). Although the husband has authority over most situations regarding the family, serious health care decisions may be the responsibility of the mother or grandmother (Randall, 1991).

A study of Mexican American families living on the Texas–Mexico border (Yetley et al., 1981) found that the husband also exercised control of the food budget and food purchases. The wife did the actual meal planning, shopping, and preparation. Women identified strongly with their food-related tasks within the family structure. Because their self-concept and status in the family and community is related to their abilities as a cook and homemaker, nutrition intervention and advice may be perceived as an accusation of inadequacy.

Children are also an important influence on food habits in some households. Those raised in the United States may be the only English-speaking members of the family and may be responsible for translating in the market. It has been found that these children prefer foods that they have seen advertised on television. The adoption of new foods is influenced by the presence of bilingual children in the family (Dewey et al., 1984b). Researchers studied newly immigrated Latinos in the San Francisco area and found that the importance of the family unit can be used to motivate changes in food habits (Ikeda & Gonzales, 1986). Adults unwilling to make changes that would benefit their own health may make those same changes to improve the well-being of their children.

As with all clients, an in-depth interview is crucial in effective nutrition counseling. Experts in the health care of Latinos recommend that health professionals who work often with Latinos learn Spanish. Familiarity with Spanish medical terminology is the minimum proficiency needed for meaningful communication.

Central Americans

The seven nations of Belize, Guatemala, El Salvador, Honduras, Nicaragua, Costa Rica, and Panama make up Central America, an isthmus connecting North America to South America. The eastern coastal region edges the Caribbean Sea. An 800-mile chain of active volcanoes and mountains, beginning at the Mexican border in the north and continuing with only one break into central Panama in the south, forms the temperate backbone of the region. Central America is similar to the rest of Latin America in history of foreign intervention, and heterogeneous culture.

Cultural Perspective

History of Central Americans in the United States

Immigration Patterns

Central American immigrants to the United States have arrived in two distinct waves. Early records are inexact because separate statistics on Central Americans were not kept until the 1960s by the U.S. Census Bureau. Even today they are usually listed with immigrants from South America in most demographic studies.

Until the early 1980s, immigrants to the United States were of two groups. The first were well-educated professional men who arrived in search of employment opportunities. The second were women, who often outnumbered the men two to one, coming in search of temporary domestic jobs. These Central American immigrants were largely urban residents and settled mostly in New York, Los Angeles, San Francisco, Miami, and Chicago, where they blended into existing Latino communities.

The second major wave began in the late 1970s and early 1980s, with the exodus of refugees from the brutal civil wars in El Salvador, Guatemala, and Nicaragua. It is estimated that millions of residents have been displaced in these countries, about one-third of whom have emigrated. Many moved to Mexico, and a substantial number have continued on to the United States. They are known as the "foot people" because many have literally walked to the United States (Melville, 1985). Less is known about this group, except that they are often younger, poorer, and less educated than the previous immigrants from the region.

Current Demographics and Socioeconomic Status

Over 6 percent of U.S. Latinos (approximately 1.5 million) are from the seven nations of Central America, one-half of whom arrived between 1980 and 1990. Exact figures are unknown, however, because it is believed that many Central Americans may enter the United States illegally at the border with Mexico and are undistinguished from undocumented Mexicans. Most of these recent immigrants have settled in California, Texas, and Florida. The largest known populations are the approximately 500,000 Salvadorans, 270,000 Guatemalans, and 200,000 Nicaraguans. It is believed that there may be an equal number of illegal residents from these countries living in the United States as well. Approximately 115,000 Americans of Honduran descent were identified in the 1990 U.S. census. Immigration from the other Central American nations, Costa Rica, Belize, and Panama, is minimal.

Central American immigrants, even those from the first wave, are slow to naturalize; only 35 percent of Central Americans in the United States have obtained citizenship. Those who are not refugees often return to Central America for visits and maintain active contact with their homeland.

Most socioeconomic data on Central Americans are combined with statistics on South Americans and do not account for recent arrivals, many of whom may be undocumented residents. Information on the first wave of Central American immigrants suggests that they are a middle-class population with income and education levels well above most Mexican Americans and Puerto Ricans (about equal to the Cubans). American-born children of these first immigrants graduate from high school in numbers greater than whites. But

- **Central American identity is sometimes more related to race and class than to country of origin. The majority of immigrants before the 1980s were white professionals; more recent refugees are predominantly *Ladino* (mixed Spanish and Indian heritage) or Indian *campesinos* (peasants).**

- **Guatemalan Americans prefer to be called *Chapines*. The term was originally a derogatory term for residents of Guatemala City but has new meaning in the United States, reflecting ethnic pride.**

A Mayan chocolate container. (© The Bettmann Archive.)

these figures do not reflect the large numbers of recent immigrants employed as migrant farmworkers, gardeners, domestic cleaners, dishwashers, and foodservice workers, as well as those placed in other low-skilled jobs. Undocumented residents often face difficulties in obtaining employment and education opportunities. Disposable family income may be impacted by the money that is sent to support relatives still living in the homeland.

Worldview

The large numbers of recent immigrants from Central America suggest that ethnic identity is preserved by many new residents in the United States. For example, Salvadorans establish highly insular neighborhoods within the larger Latino community, where an immigrant can live and conduct business exclusively with other Salvadorans. Guatemalans are a more diverse population of immigrants, and it is the Mayan communities that are most likely to keep traditional beliefs and practices. In contrast, Nicaraguans are dispersed among other Latinos and are adapting more to the pan-Latino community than retaining their own heritage exclusively.

Religion

Most Central Americans are Roman Catholic. Some Guatemalans observe Catholic practices while adhering to Mayan religious beliefs; participation in native religions declines in the United States because they are usually dependent on sacred locations in Guatemala. Evangelical and fundamentalist denominations, such as the Pentecostal Church, have attracted many Central Americans after they arrive in the United States. Small "storefront" congregations that involve active participation and those churches that offer traditional Central American social activities in addition to worship have been especially successful.

Family

Central Americans highly value family and extended kinship. It has been noted that some apartment buildings in Latino neighborhoods are rented entirely to several families from the same village in Central America. The father is the undisputed head of the household and provider for the family. Children are carefully controlled, especially daughters.

The traditional roles of the men and women become less delineated in the United States, where women are sometimes more easily employed and husbands must take on some domestic responsibilities. Family disintegration has taken place in some refugee camps prior to immigration to the United States, where overcrowding and unemployment led to intergenerational conflict (Miralles, 1989). In other situations, family members were forced to immigrate separately. Some married outside the Central American community for immigration benefits; others found that when their families were reunited, children had become more independent (Mumford, 1995).

Traditional Health Beliefs and Practices

A good diet, fresh air, and regular hours are thought necessary to preserve health by many Central Americans (Kuster & Fong, 1993), although the concept of structured exercise is often unfamiliar (Boyle, 1989). Salvadorans believe that being too thin can cause sickness, and Americans are considered at risk for ill health because they are so thin.

Some Central Americans view health as a balance between the spiritual and social worlds. For most, prayer is important in maintaining balance. Nicaraguans may believe in witchcraft, practiced by *brujos* or *brujas* who can assume the shape of animals and have the power to cure illness. Guatemalans consider illnesses to be caused by outside forces and include disease sent by Satan to punish unbelievers and sickness due to witchcraft.

A balance of hot and cold is also necessary to health, and can be disrupted by strong emotions. In addition to susto and mal de ojo, other folk conditions related to emotional states include *bilis* or *cólera,* which in extreme cases precipitates stroke. Exposure to sudden extremes of hot or cold is avoided (Miralles, 1989).

In the culturally diverse region of Nicaragua's east coast, more than 200 plants with traditional medicinal uses have been identified. These include coca leaves, the source of cocaine. In Nicaragua, rural ethnic groups were found to use traditional healing practices more often than urban residents of mixed heritage (Barrett, 1995).

Guatemalans believe that strength is maintained through the quantity and quality of a person's blood. In urban regions of Guatemala, researchers have noticed the emergence of new categories of food items, such as "strong" or "health-promoting" foods, perhaps due to the influence of modern health promotion concepts (Miralles, 1989).

Over-the-counter remedies, such as analgesics and cough suppressants, are commonly used by Guatemalan Americans, although they are considered weak by Guatemalan standards. Medications (including antibiotics) and herbs, such as chamomile, are sometimes brought to immigrant families by new arrivals from Guatemala.

■ Cilantro is hung in some Central American kitchens as protection from evil.

■ Health care is politicized in some Central American countries. Modern clinics and health promotion were associated with leftist reforms in Nicaragua, for example. In the 1980s, rightist rebels targeted health facilities and urged citizens to return to traditional health practices.

Traditional Food Habits

Ingredients and Common Foods

Central American cuisine offers many of the foods common throughout Latin America. The native Indian dishes remain prominent in the highland areas, Spanish influences are found in the lowland regions, and the cooking of the multicultural eastern coast shares many similarities with Caribbean Islander fare. The northern nations have foods similar to those of southern Mexico; the southern countries have been more greatly influenced by European and African cuisines.

Staples

Early Mayan records indicate that the foundation of their diet was corn and beans, supplemented with squash, tomatoes, chiles, tropical fruit, cocoa, and some game. Indian foods were particularly important in the development of Guatemalan cuisine but gradually become less significant in the south of Central America. Rice, introduced by the Spanish, has become a staple in most regions. (See the cultural food groups listed in Table 9.3.)

Beans are eaten daily. Black beans are especially popular in Guatemala, while red beans are common in other nations. Beans are served simmered, pureed, or fried and are often paired with rice. In Nicaragua, red beans and rice fried

■ The Mayan word for "corn," *wah,* also means "food."

■ In Guatemala, refried black beans (*fríjoles volteados*) are fondly called "Guatemalan caviar."

Table 9.3
Cultural Food Groups: Central Americans

Group	Comments
PROTEIN FOODS	
Milk/milk products	Milk is not widely consumed as a beverage, but evaporated milk and cream are popular in some regions.
Meat/poultry/fish/eggs/legumes	Legumes are important in the cuisine and are often served with rice. All types of meat/poultry are eaten, but pork is popular throughout the region. Eggs are commonly served. Fish and shellfish are consumed in the coastal regions. Sea turtle eggs are popular.
CEREALS/GRAINS	Rice and corn are the predominant grains of the region. Wheat flour breads are common.
FRUITS/VEGETABLES	Tropical fruits are abundant. Some temperate fruits such as grapes and apples are also available. Salads and pickled vegetables are popular.
ADDITIONAL FOODS	
Seasonings	Cilantro (fresh coriander) and *epazote* are important herbs. Sour orange juice gives a tang to some food; coconut milk flavors others. *Achiote* is used to color foods orange.
Nuts/seeds	
Beverages	Hot chocolate and coffee, grown in the region, are favorite hot beverages. *Refrescas*, cold drinks, are made with tropical fruit flavors. *Boj, chicha,* and *venado* are locally made alcoholic beverages.
Fats/oils	Lard is the most commonly used fat.
Sweeteners	Honey and sugar are used as sweeteners.

with onions are called *gallo pinto* ("painted rooster") due to the colors of the dish.

Corn is eaten mostly as tortillas. Enchiladas in Central America are open-faced sandwiches similar to Mexican tostadas. They typically feature meat covered with pickled vegetables such as cabbage, beets, and carrots. Known as *mixtas* in Guatemala, the tortilla is spread first with guacamole, then topped with a sausage and pickled cabbage. In El Salvador a stuffed specialty is called *pupusas.* A tortilla is filled with chicharrónes, cheese, or black beans, then completed with another tortilla; the edges are sealed and the pupusa is then fried. They are traditionally served

Table 9.3
Cultural Food Groups: Central Americans—continued

Common Foods	Adaptations in the United States
Milk (evaporated), cream; cheese (aged and fresh—crumbly farmer's cheese type)	Milk and hard cheeses are often disliked.
Meat: beef, iguana, lizards, pork (all parts, including knuckles, tripe, and skin), venison	
Poultry: chicken, duck, turkey	
Fish and shellfish: clams, conch, flounder, mackerel, mussels, sea snail, shark, shrimp, sole, tarpon, turtle	
Eggs: poultry, turtle	
Legumes: beans—black, chickpeas, fava, kidney, red, white	
Corn (tamales, tortillas), rice, wheat (bread, rolls)	
Fruits: apples, bananas, breadfruit, coconut, grapes as well as raisins, guava, *mameys,* mangoes, *nances,* oranges (sweet and sour types), papaya, passion fruit, *pejibaye,* pineapples, prunes, tangerines, *zapote* or *sapodilla*	Increased intake of potato chips has been reported.
Vegetables: asparagus, avocados, beets, cabbage, carrots, cauliflower, chayote squash, chile peppers, corn, cucumbers, eggplant, green beans, hearts of palm, leeks, lettuce, onions, *pacaya* buds (palm flowers), peas, plantains, potatoes, spinach, sweet peppers, tomatillos, tomatoes, watercress, yams, yuca (cassava)	
Achiote (annatto), chile peppers, cilantro, cinnamon, cloves, cocoa, *epazote,* garlic, onions, mint, nutmeg, thyme, vanilla, Worcestershire sauce	
Palm tree nuts, *pepitoria* (toasted squash seeds)	
Coffee, chocolate, tropical fruit drinks, alcoholic beverages (rum, beer, and fermented or distilled fruit, sugar cane, and grain drinks)	Increased intake of soft drinks has been noted among Guatemalan refugees in Florida.
Butter, lard, vegetable oils, shortening	
Honey, sugar, sugar syrup	Increased intake of candy is noted.

with pickled cabbage. Tamales are also common, often stuffed with poultry or pork. They are called *nactamal* in Nicaragua, where the dough is flavored with sour orange juice and the filling includes meat, potatoes, rice, tomatoes, onions, sweet peppers, and mint. Black tamales are served on special occasions in Guatemala, stuffed with a mixture of chicken, chocolate, spices, prunes, and raisins. *Empanadas,* small turnovers made with a wheat flour dough and filled with a savory meat mixture, are popular.

French bread, introduced from Mexico, is eaten regularly in the form of small rolls in Honduras and Guatemala. In El Salvador,

■ In Guatemala, eggs poached and served with a seasoned broth are used to treat hangovers.

■ Chocolate was so prized in Mayan culture that cocao beans were used as currency.

■ Though iguana tail is eaten throughout Central America, parts of South America and the Caribbean, it is especially popular with Indians in Nicaragua.

■ The Guatemalans say, "Full stomach, happy soul."

French bread is used with native turkey and pickled vegetables to make sandwiches. Coconut bread is a specialty on the Caribbean seacoast. Rice is often cooked with coconut milk or fried or, in Costa Rica, served as pancakes.

Soups and stews are popular throughout Central America. Beef, plantains, and cassava in coconut milk; spicy beef stew; beef in sour orange juice; pork and white bean stew; chicken cooked in fruit wine; *mondongo* (Nicaraguan tripe soup); and *sopa de hombre* ("a man's soup") made with seafood and plantains in coconut milk are a few specialties. In Guatemala the Mayan meat stew called *pepián* is thickened with toasted squash seeds. Meat, poultry, and fish are frequently roasted as well.

Fruits and vegetables are numerous. Although bananas, coconut, plantains, yuca (cassava), tomatoes, sweet peppers, chayote squash (known as *huisquil* in Guatemala), mangoes (considered an aphrodisiac in Guatemala), oranges, and avocado predominate, cabbage, cauliflower, carrots, beets, green beans, lettuce, spinach, breadfruit, passion fruit (*granadilla*), pineapples, mameys, and *nances* (similar to yellow cherries) are also common. *Pejibaye* (peach palm) is a fruit especially popular in Costa Rica. Onions and garlic flavor many dishes. Salads and pickled vegetables are common as appetizers, as side dishes, and on sandwiches.

Coffee, grown throughout the region, is a popular drink, as is hot chocolate. *Refrescas,* cold beverages, are made in tropical fruit flavors, such as mango and pineapple. *Tiste,* a Nicaraguan favorite, is made with roasted corn, cocoa powder, sugar, cold water, and cracked ice. Beer is widely available. Fermented beverages such as *boj* (from sugar cane) and *chicha* (a wine made from fruit or grain, fortified with rum) are consumed. *Venado* is a common distilled drink made from sugar cane. Sweets, such as the praline-like candy called *nogada,* sweetened baked plantains, ices made with fruit syrups, custards, and cakes or fritters flavored with coconut or rum, are eaten as snacks and for dessert.

Regional Variations

Although many foods of Central America are similar, they are often flavored with local ingredients for a unique taste. Coconut milk flavors many dishes in Belize and Honduras; seafood specialties include conch and sea turtle. The foods of El Salvador feature many indigenous flavors including corn, beans, tomatoes, chiles, and turkey. Achiote is common in mild seasoned Guatemalan fare. The juice of sour oranges is mixed with sweet peppers or mint in many Nicaraguan recipes. Costa Ricans prefer foods simmered with herbs and seasonings such as cilantro, thyme, oregano, onion, garlic, and pimento; rice is also frequently consumed. Panamanian fare is more international in flavor; one specialty is *sancocho,* a stew of pork, beef, ham, sausage, tomato, potato, squash, and plantains.

Meal Composition and Cycle

Daily Patterns

As in other Latin American regions, beans and corn are the cornerstones of the daily diet, eaten at every meal by the poor. Rice is also common. *Queso blanco* (a fresh cheese) or meat is added whenever resources permit. Dinner in wealthier areas usually includes soup, meat or poultry (sometimes fish), tortillas or bread, and substantial garnishes such as avocado salad, fried plantains, and pickled vegetables. Appetizers, such as slivers of broiled beef, bites of meat- or cheese-filled pastry, and soft-boiled turtle eggs, are eaten in some urban regions before dinner; dessert may also be served, typically including custards, ice creams, cakes, or fritters.

Special Occasions

Celebrations in Central America are focused on Catholic religious days. Christmas, Easter and Lent, saints' days (including All Saint's Day), and even Sundays may mean a change in fare. Special dishes include the cheese-flavored batter bread called *quesadilla* that is served in El Salvador on Sundays; *sopa de rosquillas,* a soup made with ring-shaped corn dumplings traditionally eaten on the Fridays of Lent in Nicaragua; *gallina rellena Navidena,* a Nicaraguan Christmas dish of chicken stuffed with papaya, chayote squash, capers, raisins, olives, onions, and tomatoes; and plantains served in chocolate sauce during *Semana Santa* (the Holy Week before Easter) in Guatemala.

Chicken in tomato sauce (*guisado*), chicken served with a cornmeal porridge, or stews thickened with masa harina are Indian specialties eaten at ceremonial occasions. In some areas the stews are provided by the village headman to serve the community. In Guatemala, All Saints' Day is celebrated with a unique salad called *fiambre*. As described by one author (Marks, 1985), these enormous salads involve a family social event at which as many as 50 friends and relatives share the creation. They feature vegetables (e.g., green beans, peas, carrots, cauliflower, beets, radishes, cabbage) mixed with chicken, beef, pork, and sausages, then artfully garnished with salami, mortadella, cheese, asparagus, pacaya buds, and hard-boiled eggs. The dressing is either a vinaigrette or a sweet-and-sour sauce.

Sample Menu

A GUATEMALAN DINNER

Guisado (Mayan stew) or
Sopa de pesce (fish soup)

Frijoles volteados (refried black beans)

Guacamole, Pickled cabbage, and Tomato salad

Coconut candy

Hot chocolate or coffee

Therapeutic Uses of Foods

Central Americans consider a good diet essential to health, as is a balance of hot and cold. Illness can occur if one eats or touches something that is too hot or too cold or through strong emotions. Guatemalan Americans commonly believe that diarrhea is caused by hot weather and can be alleviated by consuming cold drinks, such as Kool-Aid or Gatorade (Miralles, 1989). Herbal remedies are popular throughout Central America, including chamomile and coca leaves, the source of cocaine (Barrett, 1995).

Contemporary Food Habits in the United States

Adaptations of Food Habits

There is scant information on current Central American food habits. It is assumed that low rates of assimilation among many Central American immigrants result in preservation of many traditional food habits. Most Central American ingredients are available in the Latino communities where they settle.

Salvadoran refugees report that, in general the quality of their diet has declined since arriving in the United States. They state that in El Salvador, more foods were made at home from fresh ingredients; they believe that in the United States, more processed items and "junk" foods are eaten, and some nutritious foods are too costly to consume (Boyle, 1989).

Health workers in Florida report that Guatemalan refugees believe that if a food is tasty and does not cause stomach discomfort, it must be "good" to eat. High intake of candy, soft drinks, and potato chips has been noted. Milk, which is often not well tolerated, is avoided. WIC (Supplemental Food Program for Women, Infants, and Children) nutritionists found that some food supplements, including milk and cheese, are disliked because of their taste or texture and are sometimes discarded (Miralles, 1989).

■ Anecdotal reports suggest that hamburgers, hot dogs (served in tortillas topped with pickled cabbage) and pasta are popular with many Guatemalan Americans.

■ Lactose intolerance may be prevalent among Central Americans.

Nutritional Status

Nutritional Intake

Limited data on the nutritional status of Central American immigrants have been published. Those who arrive after spending time in refugee camps may suffer high rates of malnutrition resulting in diseases such as beriberi, pellegra, scurvy, and vitamin A deficiency problems, especially in children younger than the age of 5. Infectious disease often follows; tuberculosis and parasites are common (Boyle, 1991).

Endemic infections may cause problems as well. Chagas' heart disease, resulting from infections with *Trypanosoma cruzi* (found in most of Central America) presents symptoms similar to other coronary artery conditions and may be overlooked in diagnosis (Hagar & Rahimtoola, 1991). U.S. outbreaks of cyclosporiasis due to contaminated raspberries imported from Guatemala in 1996 and 1997 suggest another source of infection. Sickle-cell anemia was found to be high (5.7 percent) among mostly Central American adolescents in Los Angeles and appears to be associated with this population independent of African heritage (Hamdallah & Bhatia, 1995).

Low birth-weight infants were not found to be a problem among Central Americans in a Chicago study. Even those at significant personal or environmental risk (i.e., living in low-income, urban neighborhoods) showed no excessive low birth weight (Collins & Shay, 1994). Researchers report that Guatemalans consider breast-feeding healthy for infants but impractical. Breast-feeding often is used as supplementation to formula and solid foods for the first 2 to 3 years of a child's life (Miralles, 1989).

An occupational hazard for many Central Americans employed as U.S. farmworkers is pesticide or herbicide poisoning. Exposure occurs when labor codes are unenforced or through worker mishandling of the dangerous products.

Counseling

Access to biomedical heath care can be especially difficult for Central Americans. Many are economically and linguistically isolated within their communities; others are undocumented residents avoiding detection by authorities.

Most immigrants from Central America, as with other Latinos, are present oriented. They typically view health from day to day, believing that they have no control over the future. The concept of scheduled appointments may be unfamiliar, and there is little interest in arriving on time.

Touching is used to communicate feelings. Men usually embrace close friends, and women are likely to hug all acquaintances. Salvadorans, Guatemalans, and Nicaraguans all prefer a light handshake that is lingered over. Eye contact and smiling are expected. Salvadorans use their hands expressively, but it is considered impolite to point with the fingers or the feet. Guatemalans prefer a soft voice in conversation. Most Central Americans have a different sense of personal space than do Anglos, and prefer to sit and stand closer than is comfortable for many whites. Backing away may be seen as an insult, however (Axtell, 1991).

Illness is a sign of weakness among Guatemalans, and a person may be stigmatized because he or she is unable to fulfill responsibilities. Culturally based descriptions of symptoms were found among Guatemalans in Florida (Miralles, 1989). "Weak heart" referred to palpitations or dizziness; "weak stomach" meant indigestion; and "weak nervous system" was applied to headaches or insomnia. Taking blood samples was very anxiety provoking because of Guatemalan belief in the need for strong, ample blood. Anemia was associated with weak blood, to be cured by "eating iron." Guatemalans believe injections are the most effective treatment for illness, more potent than pills. Treatment is successful if symptoms are alleviated.

A study of Salvadorans found that the most important source of emotional support came from family and friends. Single immigrant men preferred living with other Salvadorans when possible (Boyle, 1989). Posttraumatic stress syndrome is prevalent among refugees from El Salvador and other Central American nations; it is especially acute among those who are here illegally and suffer continuous anxiety regarding the possibility of deportation (Molesky, 1986).

Central Americans often assimilate into other Latino communities. It is believed that a cross-cultural exchange of health beliefs and practices may occur in some areas, and practitioners should be aware of possible Mexican or Caribbean Islander concepts adopted by Central Americans.

The in-depth interview is crucial in counseling Central Americans because so little data about food habits, health practices, and nutritional status are available. In addition, information from family members or community experts may be needed.

10

Caribbean Islanders and South Americans

Latinos from the Caribbean Islands and South America often seem more different from one another than similar. Their homelands vary from the tropics of the islands and northern Brazil to the highland plains of Argentina and the snow-topped mountains of Peru. Their ethnic backgrounds include native Indian, Spanish, Portuguese, French, British, Danish, Dutch, African, Asian Indian, Chinese, Italian, German, and Japanese. And though Roman Catholicism is practiced by a majority, smaller numbers follow Protestant faiths, Judaism, and numerous indigenous Afro-European religions including voodoo, santeria, and candomblé.

One commonality between Caribbean Islanders and South Americans is a variety of regional fares with few national cuisines. Dishes typically combine native ingredients with foods introduced from Europe, Africa, and Asia, with a broad preference for strong, spicy flavors. This chapter reviews Caribbean Islanders and their fare, focusing on Puerto Ricans and Cubans. A summary of South Americans is also presented. Other Latinos are covered in chapter 9.

Caribbean Islanders

More than 1,000 tropical islands in the Caribbean stretch from Florida to Venezuela. They include the Bahamas, the Greater Antilles (Jamaica, Cuba, Hispaniola, and Puerto Rico), and the Lesser Antilles. The largest island is Cuba, and the smallest islands are barely more than exposed rocks. Most were claimed at one time by Spain, Britain, France, the Netherlands, Denmark, or the United States, but now include the independent nations of Antigua/Barbuda, the Bahamas, Barbados, Cuba, Dominica, Dominican Republic, Grenada, Haiti, Jamaica, St. Christopher/Nevis, St. Lucia, St. Vincent/Grenadines, and Trinidad and Tobago, as well as the U.S. territory of Puerto Rico. Many islands, such as the Virgin Islands (U.S.) and Martinique (France), are still under foreign control.

The islands are uniformly scenic. The tropical warmth and torrential rains provide the ideal climate for a lush plant cover that includes numerous indigenous fruits and vegetables. Later immigrants found the region suitable for imported crops. The Caribbean Islands share a history of domination by foreign powers and political turmoil. Native Indians, Europeans, blacks from Africa, and Asians from China and India have intermarried over the centuries to produce an extremely diverse population.

Cultural Perspective

History of Caribbean Islanders in the United States

Immigration Patterns

It is estimated that the total Latino population in the United States is more than 26 million people. Immigrants from Puerto Rico constitute approximately 11 percent of the total;

Figure 10.1
Caribbean Islands and South America

■ In 1995, more than 500 Puerto Ricans renounced their U.S. citizenship, saying it was imposed against their wills and saying that they are *Boricua* (the Taino Indian word for Puerto Rican), not American. The "Boricua First" movement advocates independence for the island.

those from Cuba about 5 percent. In addition, there are small groups of immigrants from other Caribbean nations, most significantly from the Dominican Republic and Haiti.

Puerto Ricans. Puerto Ricans differ from most other people who come to the United States in that they are technically not immigrants. They come to the mainland as U.S. citizens and are free to travel to and from Puerto Rico without restriction. Nearly half of the population of Puerto Rico resides on the mainland of the United States, and the number of Puerto Ricans who live in New York City is more than double the number living in San Juan, the largest city in Puerto Rico.

Small numbers of political exiles from Puerto Rico arrived in America in the 1800s, but most returned home when Puerto Rico became a U.S. possession. Others arrived when unemployment increased in the depressed agricultural economy of the island during the

1920s and 1930s. It was after World War II that the largest number of Puerto Ricans moved to the mainland. Unlike other immigrants, the Puerto Rican population in the United States is in continual flux. Many Puerto Ricans live alternately between the mainland and the island, depending on economic conditions. In many years the number of Puerto Ricans leaving the mainland is greater than the number arriving.

Cubans. Cubans have immigrated to the United States since the early 19th century. In the early years, the majority were those who found economic conditions disadvantageous or who were politically out of favor with the current government.

The majority of Cubans came to the United States after Fidel Castro overthrew the dictatorship of Fulgencio Batista in 1959. In the 3 years following the revolution, more than 150,000 Cubans arrived in America. Most of these were families from the upper socioeconomic group fleeing the restraints of communism. Commercial air travel between Cuba and the United States was suspended after the Cuban missile crisis in 1962. Airlifts of immigrants from 1965 to 1973 increased the total number of Cubans in the United States to nearly 700,000. Due to the political differences between the two countries, most Cubans have not been subject to the usual immigration quotas.

Immigration from Cuba slowed with the end of the airlifts. In 1980, another large group of 110,000 Cubans arrived in Florida in private boats (the Mariel boatlift) seeking asylum. Unlike earlier immigrants, these recent arrivals, called *marielitos,* were mostly poor, unskilled laborers. They were often single and black. Today a trickle of exiles continues to come to the United States. Some arrive by boat and others go through legal immigration channels from a neutral third country.

Current Demographics

Puerto Ricans. Figures from the 1990 U.S. census show that almost 2 million Puerto Ricans have settled on the U.S. mainland, the majority in New York. In the 1930s, Puerto Ricans began moving into East Harlem, which became known both as *El Barrio* and as "Spanish Harlem." More than 900,000 Puerto Ricans now reside in New York City, and because Puerto Ricans are a young population, they now constitute nearly one-quarter of the students in the city's public schools. Boston, Chicago, Philadelphia, Newark, Miami, San Francisco, and Los Angeles also have Puerto Rican communities.

■ Puerto Ricans living in Puerto Rico are not counted in the U.S. census. If all Puerto Ricans were included, they would constitute nearly one-quarter of all Latinos in the United States.

Cubans. More than 1 million Cubans were counted in the 1990 U.S. census, one-third of them noncitizens. Americans of Cuban descent

Cuban cafeteria, Miami, Florida. (Courtesy of the Metro Dade Department of Tourism.)

prefer to live in the Miami area, which is sometimes called "Little Havana." The climate is similar to that of their homeland, and the Cuban population in the region is the largest in the United States. Efforts to resettle exiles in other areas have been only moderately successful. Many Cubans living in Los Angeles, Chicago, New York City, and Jersey City and Newark, New Jersey choose to move to Miami after gaining job skills. Those Cubans who remain in other cities are most likely to be employed in technical or professional occupations. Nearly all Cubans who live in the United States are urban dwellers.

■ Jamaican Americans often refer to themselves as *Americans.*

Other Caribbean Islanders. In the 1990 U.S. census, approximately 500,000 residents from the Dominican Republic, over 400,000 Jamaicans, and 280,000 residents from Haiti were identified. Unofficial estimates suggest that several hundred thousand undocumented immigrants from the three nations have also settled in the United States; most live in the urban areas of New York, New Jersey, Massachusetts, and Florida, with smaller numbers in Illinois. Dominicans often immigrate to Puerto Rico first in search of employment, then come to the mainland. Jamaicans and Haitians come directly to the United States; many Haitians are political refugees seeking asylum.

Socioeconomic Status

Immigrants from the Caribbean vary tremendously in both economic and educational attainment. Among Latinos, Puerto Rican Americans have the highest rate of unemployment (in 1990, it was 31 percent of men and 59 percent of women) and Cuban Americans have the lowest. In 1987 the median annual income of Puerto Ricans in the mainland United States was approximately half that of the U.S. Cuban population, again the lowest and highest among Latinos, respectively. Almost 25 percent of Puerto Ricans live in poverty. Although nearly half of Cuban Americans are employed in professional, technical, managerial, sales, or administrative support positions, one-third of Cuban American families still earned less than $20,000 in 1990.

Even though large numbers of Dominican professionals have immigrated to the United States, most obtain lower-paying jobs in manufacturing and service industries. More than 22 percent of Dominican families live at the poverty level. Though some Jamaicans are also limited by their lack of skills, a majority hold professional, management, technical, or sales jobs. Haitian immigrants have usually come from poor segments of their society and have entered migrant labor and service positions. It is estimated that nearly one-third of Haitians in Miami are unemployed.

Educational rates for first-generation immigrants from the Caribbean are very similar to those of other Latinos. Nearly half of all adults from Puerto Rico and Cuba did not attend high school. Differences in educational attainment appear with second-generation Americans of Caribbean descent, however. Only 27 percent of Puerto Ricans living on the mainland complete high school; 83 percent of U.S.-born Cuban Americans graduate from high school and one-quarter complete some college. Cuban Americans are more likely than any other Latino group to pay for education; nearly one-half have attended private schools. Dominicans in New York have demonstrated their commitment to education through political involvement, running for control of local school boards. Statistics of school attainment of Dominicans and Haitians in the United States are usually combined with those of other Latinos and are not generally available for individual groups. Over two-thirds of Jamaican Americans are thought to be high school graduates.

Socioeconomic differences among Latino groups from the Caribbean are due to many factors. Puerto Ricans living on the mainland are free to travel between the United States and their homeland, and the frequent changes in residence may hamper socioeconomic improvement. Puerto Ricans have also chosen to settle primarily in New York City, where as a new ethnic minority they took over the decayed neighborhoods previously occupied by African Americans. In general, their low education levels and undeveloped job skills translate into unemployment and low-level employment. Furthermore, some Puerto Ricans, Dominicans, Jamaicans and Haitians often face discrimination similar to that experienced by African Americans because many are of African heritage.

In contrast, Cubans in the United States are mostly political refugees who emigrated out of necessity, not choice. There are a disproportionate number of Cubans in the United States from the upper socioeconomic levels, and although many lost all their material goods when they emigrated, they brought upper-class values, including the importance of educational and financial success. Instead of displacing any ethnic minority in Florida, they immigrated in such numbers that they immediately became the dominant ethnic group. This is not to say that there are no well-educated, wealthy Puerto Ricans living on the mainland or no poor Cuban Americans. However, circumstances surrounding their immigration have influenced the general socioeconomic status of both groups.

Worldview

Ethnic identity is strongly maintained in both the Puerto Rican and Cuban communities in the United States. Puerto Ricans on the mainland continue close ties with the island; Cubans believe it is important to retain their heritage because they cannot return. Spanish may be spoken exclusively in the home and used frequently to conduct business within the community.

In contrast, immigrant populations with large numbers of undocumented residents, such as Dominicans, Jamaicans, and Haitians, often try to blend into existing Latino communities or assimilate into mixed ethnic neighborhoods. It has been noted that Dominicans and Jamaicans (who speak English) appear to acculturate quickly.

Religion

The majority of Caribbean Islanders are Roman Catholics, although many are not faithful followers of the church. The role of the Catholic Church has been less important in the Caribbean than in other regions of Latin America; it is estimated that only 63 percent of Cuban Americans born in the United States are Catholic and that 25 percent have no religious affiliation.

A number of other religions are practiced in the islands, including Protestantism and Judaism, as well as some folk religions. The best known of these is *voodoo,* a unique combination of West African tribal rituals with Catholic beliefs and local customs. Saint Patrick is associated with the African snake deity Damballah, for example, and St. Christopher is identified with Bacoso, the god responsible for infectious illness. Certain rites, such as repeating the Hail Mary, making the sign of the cross, and baptism, are practiced in conjunction with ancestor worship, drums, and African dancing. Worship is family based, and there is no central leadership or organization of activities. Typically ceremonies are conducted for annual events such as Christmas and the harvest and for funerals. Voodoo originated in Haiti, although very similar Afro-Catholic cults are found on the other islands. In Cuba and Puerto Rico, they are called *santeria*. Many followers of voodoo or santeria are also members of Christian faiths and do not believe that there is any contradiction in practicing both religions simultaneously. *Rastafari* is another Afro-Caribbean faith, indigenous to Jamaica. Rastas practice a natural, simple lifestyle typified by bare feet, loose clothing, dreadlocks, and sacramental use of marijuana. It is also considered a political movement due to Rastafari opposition to traditional government and support for repatriation of blacks to Africa.

Family

The Puerto Rican family is based on the concept of *compadrazco,* which means coparenting. Grandparents, aunts and uncles, cousins, and godparents are all considered part of the immediate family, responsible for the care of children (Green, 1995).

Men are the heads of households as well as being in charge of community matters. The oldest boys in the family are expected to help with supervision of younger siblings, particularly daughters. Women maintain the home. Men are expected to be aggressive; women are traditionally reserved. As in most Latino cultures, age is respected; even younger children are taught to defer to older children.

Traditional Cuban families are also patriarchal and extend to include relatives. Godparents are significant in childrearing. Children are deferential to elders and well chaperoned in public.

Caribbean Islander families often change in the United States. Economic pressures,

■ **Acculturation has been especially difficult for a group of Cuban teenagers who in 1994 abandoned their families for the United States, where they believed they would be given houses and money to live on their own.**

American values of individualism and equality, and intergenerational stress are often cited as responsible for nontraditional adaptations. Studies have shown that one-third of Dominicans, for example, live in nuclear family groupings in the United States even though only 1 percent did so in the Dominican Republic. Dominican women in the United States also have fewer children than those on the island (Buffington, 1995b). In the United States, Caribbean Islander women assume more authority in the household with their increased involvement in the workforce, and Caribbean Islander children gain greater autonomy.

■ A small study of Latinos found that faith and prayer are especially important to women in healing therapies (Zapata, Shippee-Rice, 1999).

Traditional Health Beliefs and Practices

Many Caribbean Islanders hold health beliefs that are similar to those of other Latin American cultures. For example, prayer, the lighting of candles to saints, and home remedies such as teas, have been found to be the most common health practices of Puerto Ricans living in New York City (Freidenberg et al., 1993). The conditions of empacho, susto, and mal de ojo are also well known. Haitians consider eating well, cleanliness, and regular sleep essential to health. Fat is associated with well-being; thinness is thought to be indicative of poor health due to emotional or psychological conditions.

Folk conditions reported by Puerto Ricans (in addition to those listed previously) include *pasmo,* a type of paralysis due to an imbalance of hot and cold; *ataque,* hysterical frenzy or sudden onset of illness; and *fatique*, acute breathing difficulties. Both pasmo and ataque are cured through folk remedies, while fatique responds to emergency care provided by a physician. Haitians are especially concerned with blood irregularities,

Serious conditions, such as sickness due to supernatural causes, may require the cures associated with the healing practice of *santeria.* (© Francoise de Mulder/Corbis)

which are classified as hot, cold, weak, thin, thick, dirty, and yellow.

The traditional healing practices common in the Caribbean are more closely related to African beliefs than the Arab-Spanish humoral system used in hot-cold applications. Mild conditions are treated through an informal system of older women (mothers, grandmothers, or neighbors) who are knowledgeable about the use of herbs, amulets, and charms. More serious conditions, such as those due to supernatural causes, particularly witchcraft, require the cures associated with voodoo and santeria. Spiritualist healers, known as *espiritos* and *santeros.* intervene with the saints on behalf of a bewitched person (practices related to those of the American South; see chapter 8 on African American "Traditional Health Beliefs and Practices"). Santeros specialize in soul possession and mental disorders. Dreams may play an important role in health care, because they are a connection with the supernatural world. Ancestors provide instructions to an individual regarding health behaviors through dreams.

Traditional Food Habits

Ingredients and Common Foods

Caribbean food habits are remarkably similar for an area influenced by so many other cultures. The indigenous Indians, the Spanish, French, British, Dutch, Danes, Africans, Asian Indians, and Chinese have all had an impact on the cuisine. Although each island has its specialties, the basic diet is the same throughout the region.

Indigenous Foods

Columbus likened the West Indies to paradise on Earth. The islands are naturally laden with fresh fruits and vegetables that originally came from South America, including the staple *cassava* (two varieties of tuber, bitter and sweet, that are also known as *manioc* and *yuca;* tapioca is a starch product of manioc), avocados, bananas and plantains, some varieties of beans, cashew apples (fruit surrounding the cashew nut), cocoa, coconuts, guavas, *malanga* (a mild tuber sometimes called *tanier*), papayas, pineapple, *soursop* (a fruit with a cottonlike consistency), several types of squash (including *chayote,* called *chocho* or *christophene* in the islands), sweet potatoes, and tomatoes. Fish and small birds are also plentiful.

As in other Latin American areas, many varieties of chile peppers grow profusely in the West Indies. The native cuisine makes frequent use of these for flavoring, especially in pepper sauces, such as *coui,* a mixture of cassava juice and hot chile peppers. Other herbs and spices include allspice and annatto.

Because of the abundance of fresh fruits and vegetables year-round, traditionally there was little need for preparation or preservation of foods. Consequently, cooking techniques were underdeveloped in the native populations. Cassava was baked most often in a kind of bread, made from pressed, dried, grated cassava that was fried in a flat loaf. Fish and game were either covered with mud and baked in a pit or grilled over an open fire.

Foreign Influence

The Europeans who settled in the Caribbean were impressed with the abundant supply of native fruits and vegetables. Yet they longed for the accustomed tastes of home. The Spanish brought cattle, goats, hogs, and sheep to the islands, in addition to introducing rice. Plants introduced for trade by the Europeans included breadfruit, coffee, limes, mangoes, oranges, and spices such as nutmeg and mace. The African slaves brought in to work the sugarcane fields cultivated *akee* (a mild, apple-sized fruit), okra, and *taro* (also called *eddo* or *dasheen;* both the roots and the leaves are eaten). The demand for Asian ingredients by later immigrants resulted in the introduction of soybean products, Asian greens, and tamarind to the Caribbean.

Staples

Legumes are eaten throughout the Caribbean; the rice and kidney beans of Puerto Rico are also found in the Dominican Republic and Jamaica. In Cuba, black beans are preferred; in Haiti, black-eyed peas are combined with the rice. The legumes in all countries are prepared similarly, flavored with lard and salt. Onions,

- *Cassava* contains hydrocyanic acid, which is toxic in large amounts. The acid must be leached out and the tuber cooked before it can be eaten safely.

- The American word barbecue is probably derived from *barbacoa,* the Arawak term for grilling meats.

- Nearly all parts of the akee fruit contain hypoglycins, which can cause fatal hypoglycemia. Most akee products are banned in the United States.

- The black beans and rice popular in Cuba are called *Moros y Cristianos* (Moors and Christians), a reference to Spanish history.

- Pickled fish, called *ceviche,* is popular throughout Latin America. Believed to have originated in Peru, it features scallops or cubes of raw fish marinated in salted lemon or lime juice mixed with chiles and onions until "cooked," meaning the fish becomes opaque and flaky.

sweet peppers, and tomatoes are added in some variations.

Examples of other foods common throughout the West Indies are found in the Cultural Food Groups list (Table 10.1). They include Indian foods such as cassava bread, chili sauces, and pepper pot (a meat stew made with the boiled juice of the cassava, called *cassarep*). European-influenced items such as *escabeche* (fried, marinated fish, seafood, or poultry) and *morcillas* (a type of blood sausage), as well as fried corn cakes, are popular in most Caribbean countries. Foods from Africa found throughout the region include *callaloo* (a dish of taro or malanga greens cooked with okra), dried salt cod fritters (called *bacalaitas* in Puerto Rico, these cakes have a different name on nearly every island, from *arcat de marue* to "stamp and go"), *foofoo* (okra and plantain), and *coocoo* (cornmeal-okra bread). Dishes from India and Asia are also common on many islands, although they are better known in the areas where cheap labor was most needed: the French-, British-, and Dutch-dominated islands (few Asians immigrated to Puerto Rico, Cuba, or the Dominican Republic). Curried dishes, called *kerry* in the Dutch-influenced islands and *colombo* on the French-influenced islands, and variations of pilaf are considered Caribbean foods. Chinese cuisine is also popular and Chinese-owned restaurants are omnipresent.

The most popular beverage in the Caribbean is coffee. It is often mixed with milk and is consumed at meals, as a snack, and even as dessert, flavored with orange rind, cinnamon, whipped cream, coconut cream, or rum. Some of the most expensive coffee in the world is produced in the Blue Mountains of Jamaica, where the cool, moderately rainy climate is ideal for coffee cultivation. Most of the rich beans are exported to England and Italy, although small amounts can be found in the United States.

The most important beverage in the Caribbean, at least historically, is the spirit distilled from fermented molasses—rum. This alcoholic drink is believed to have originated on the island of Barbados in the early 1600s as a by-product of sugarcane processing. Molasses is the liquid that remains after the syrup from the sugarcane has been crystallized to make sugar. It is fermented, naturally or with the addition of yeast, then distilled to make a clear, high-proof alcoholic beverage. Rum can be bottled immediately or aged in oak casks from a few months to 25 years. Caramel is added to achieve the desired color. Nearly every island produces its own variety of rum.

The molasses produced in the West Indies was crucial to the development of the region during the 17th and 18th centuries. The Caribbean Islands were one corner of the infamous slave triangle, formed when molasses was shipped to New England for distillation into rum, rum was shipped to Africa and exchanged for slaves, and slaves were shipped to the West Indies to work in the sugarcane fields.

■ Many well-known mixed drinks were developed in the Caribbean with rum. Planter's punch is an adaptation of the recipe from India for *panch* (lime juice, sweetening, spirit, water, and spices) and was probably invented by the British in the 18th century. The *daiquiri*, a popular mix of lime juice, sugar, and rum, was created in the 1890s and named after a Cuban town. The *cuba libre* (rum and coke with a slice of lemon) is popular throughout the Caribbean and Central America.

■ *Pickapeppa* sauce is a Jamaican specialty, made from tomatoes, mangoes, raisins, tamarind, vinegar, and chiles. The sweet and tangy condiment is so popular that it was hoarded when a strike halted production, causing a worldwide shortage (Walsh & McCarthy, 1995).

Regional Variations

Despite the similarities in foods throughout the West Indies, most islands have their own specialties. Haiti is known for its banana-stuffed chicken dish called *poulet rôti à la créole* and barbecued goat with chile peppers (*kabrit boukannen ak bon piman*); the Dominican Republic for *sancocho,* a pork intestine stew; Jamaica for akee and salt cod, curried goat, and cassava bread (bammies); Curaçao for its orange-flavored liqueur of the same name; Dominica for its "mountain chicken," a large, tasty frog; and Barbados for its unusual seafood dishes made with flying fish, green turtles, and sea urchins.

In addition to rice and kidney beans, Puerto Rican fare is notable for its use of *sofrito,* a combination of onions, garlic, cilantro, sweet peppers, and tomatoes that are ground together in a wooden bowl, then lightly fried in lard colored with annatto seeds. Sofrito serves as a base for many dishes, as well as an all-purpose sauce. Some foods are seasoned with *adobo,* a mixture of lemon, garlic, salt, pepper, and other spices. More pork, beef, and goat is eaten in Puerto Rico than in most Caribbean countries. Plantains are stuffed with spicy beef mixtures or *chicharrones* (fried pork rind). Chicken is frequently prepared with rice, *arroz con pollo,* which is usually served with marinated beans or peas known as *abichuelas guisada,* or in a stew called *asopao.* Land crabs and *ostiones,* a type of oyster that grows on the roots of mangrove trees, are popular, but the fish most commonly eaten is dried salt cod, called *bacalao.* It is soaked and drained before use to remove

Specialty Cooking of Jamaica

The nearly half-million Americans of Jamaican ancestry have had significant impact on U.S. pop culture, in part because they speak English. Calypso, reggae, the Rastafari religion (an Afro-Caribbean faith), and dreadlocks are among the many cultural additions. In cuisine, two regional specialties have piqued American interest: *jerk* and *i-tal.*

Jerk is believed to be related to the dried meat called *jerky.* Legend is that the technique was created by escaped African slaves known as Maroons (from the Spanish word for "untamed"—*cimarron*), who spiced and smoked wild pig meat to preserve it. Today, the word *jerk* is used to identify the wet spice mixture used as a barbecue seasoning. It includes allspice, black pepper, cinnamon, ginger, nutmeg, thyme, scallions, and Scotch bonnet chile pepper (considered by some to be the hottest chiles in the world). Some recipes call for garlic, onions, ground coriander, bay leaves, brown sugar, or other seasonings; most moisten the paste with a little oil, lime juice, or soy sauce. Traditionally, the meat is rubbed with the jerk blend and marinated for several hours. It is then grilled in a pit over Jamaican pimento (allspice) wood, covered with banana leaves. The intense heat of the jerk rub is ameliorated by the length of the cooking, typically 1 to 4 hours, depending on the meat. Though pork and chicken are the meats found at every street jerk stand in Jamaica, more recently the cooking technique has been applied to turkey, fish, seafood, and even vegetables. Jerk pork is used to make jerk sausage in some parts of Jamaica. For a complete meal, rice and peas, cassava bread, or cornsticks accompany the meat (Lalbachan, 1994; Walsh & McCarthy, 1995).

I-tal, meaning "vital," is the Rasta way of life. Applied to food it emphasizes simple, unprocessed vegetarian fare. Fruit, vegetables, and grains are permitted, while pork, red meat, salt, and artificial additives are prohibited. Some Rastas will eat chicken or fish (but shun bottom feeders such as shrimp and lobster, scaleless fish such as shark, and any fish more than 12 inches long). In general, milk, coffee, soft drinks, and alcohol are not consumed. Only cookware and utensils made from natural products, such as wood and earthenware, are used in the preparation and consumption of i-tal foods, which are ideally eaten raw or cooked over a fire (microwave ovens are avoided by many Rastas). A woman is not allowed to prepare food for others when she is menstruating. Typical i-tal dishes include rice and peas (made with either pigeon peas or kidney beans) cassava bread, baked yams, vegetable stews, cornmeal porridge, sautéed plantains, and freshly squeezed juices. Thyme, cinnamon, allspice, coconut, and reputedly marijuana are used to flavor foods (Murrell, 1995; Walsh & McCarthy, 1995).

some of the salt and then added to numerous dishes including *serenata,* a mixture of cod and potatoes. Starchy vegetables such as potatoes and plantains are eaten almost daily; leafy green vegetables are uncommon.

Cuba is unique in the West Indies for the preferred use of black beans in its cuisine. In addition to black beans and rice, spicy black bean soup is very popular. *Picadillo* is a type of beef hash that is flavored with traditional Spanish ingredients of olives and raisins, as well as Caribbean tomatoes and chile peppers. It is served with fried plantains or boiled rice, or topped with fried eggs. Other Spanish-influenced beef dishes are *ropa vieja* ("old clothes"), which is spicy beef strips, and *brazo gitano,* a cassava pastry filled with corned beef. Roast pork is also popular. A chicken specialty is *chicharrones de pollo,* small pieces of chicken that are marinated in lime juice and soy sauce, breaded, then fried in lard. As in Puerto Rico, starchy vegetables, such as *foo-foo* (cassava balls), are preferred over leafy green vegetable side dishes.

Meal Composition and Cycle

The most typical aspect of a Caribbean meal is its emphasis on starchy vegetables with some meat, poultry, or fish. Meats are frequently fried. Leafy vegetables are often an ingredient in soups, stews, and stuffed foods; fruits are enjoyed fresh, frequently as snacks or dessert.

Ethnic heritage determines which dishes are served. A poor native Indian may eat mostly cassava, tomatoes, and chiles with a bit of meat or fish at every meal. An Asian Indian may serve typically Asian Indian meals adapted to Caribbean ingredients, such as a curried dish garnished with coconut, fried

Table 10.1
Cultural Food Groups: Caribbean Islands

Group	Comments
PROTEIN FOODS	
Milk/milk products	Few dairy products are used; incidence of lactose intolerance is assumed to be high. Infants are given whole, evaporated, or condensed milk.
Meats/poultry/fish/eggs/legumes	Traditional diet is high in vegetable protein, especially rice and legumes; red and kidney beans are used by Puerto Ricans, black beans by Cubans. Pork and beef are used more in Spanish-influenced countries. Dried salt cod is preferred over fresh fish; some seafood specialties. Eggs are a common protein source, especially among the poor. Entrées are often fried in lard or olive oil.
CEREALS/GRAINS	Breads of other countries are well accepted. Fried breads are popular.
FRUITS/VEGETABLES	Starchy fruits and vegetables are eaten daily; leafy vegetables are consumed infrequently. Great diversity of tropical fruits is available, eaten mostly as snacks or dessert. Lime juice is used to "cook" (a marinating method called *escabeche* or *ceviche*) meats and fish.
ADDITIONAL FOODS	
Seasonings	Chile-based sauces are often used to flavor foods.
Beverages	Teas of all sorts are common and are often thought to have therapeutic value. Rum is especially popular, and is often added for flavoring to foods and beverages.
Fats/oils	
Sweeteners	

plantains, and pineapple. Most menus, however, consist of a multicultural mix, such as European blood sausage and *accra,* West African–style fritters made from the meal of soybeans or black-eyed peas. The list of dishes served at a meal on a French-influenced island may be identical to that on a Spanish-influenced island, yet the two meals will taste different because butter was used for cooking the former and lard the latter.

Daily Patterns

Breakfast usually consists of coffee with milk and bread. If income permits, eggs are the

Table 10.1
Cultural Food Groups: Caribbean Islands—continued

Common Foods	Adaptations in the United States
Cow's milk (fresh, condensed, evaporated), *café con leche, café latte*); aged cheeses	More milk is consumed.
Meats: beef, pork (including intestines, organs, variety cuts), goat	More meat and poultry are eaten as income increases.
Poultry: chicken, turkey	
Fish and shellfish: bacalao (dried salt cod), barracuda, bonito, butterfish, crab, dolphinfish (*dorado*), flying fish, gar, grouper, grunts, land crabs, mackerel, mullets, *ostiones* (tree oysters), porgie, salmon, snapper, tarpon, turtle, tuna	Less fresh fish is consumed. Traditional entrees remain popular.
Eggs: chicken	
Legumes: black beans, black-eyed peas, chickpeas (garbanzo beans), kidney beans, lima beans, peas, red beans, soybeans	
Cassava bread; cornmeal (fried breads, *surrulitos,* puddings); oatmeal; rice (short-grain); wheat (Asian Indian breads, European breads, pasta)	Short-grain rice is still preferred. More wheat breads are eaten.
Fruits: acerola cherries, *akee,* avocados, bananas and plantains, breadfruit, *caimito* (star apple), cashew apple, citron, coconut, cocoplum, custard apple, gooseberries, *granadilla* (passion fruit), grapefruit, guava, *guanábana* (soursop), kumquats, lemons, limes, *mamey,* mangoes, oranges, papayas, pineapple, pomegranates, raisins, *sapodilla,* sugar cane, tamarind	Temperate fruits are substituted for tropical fruits when latter are unavailable. More fresh fruit is eaten. Starchy fruits and vegetables are still frequently consumed.
Vegetables: arracacha, arrowroot, black-eyed peas, broccoli, cabbage, *calabaza* (green pumpkin), *callaloo* (malanga or taro leaves), cassava (*yuca,* manioc), chiles, corn, cucumbers, eggplant, green beans, lettuce, *malangas,* okra, onions, palm hearts, peppers, potatoes, radishes, spinach, squashes (*chayote,* summer, and winter), sweet potatoes, taro (*eddo, dasheen*), tomatoes, yams	Low intake of leafy vegetables are often continued.
Anise, annatto, bay leaf, chiles, chives, *cilantro* (coriander leaves), cinnamon, *coui* (chiles mixed with cassava juice), garlic, mace, nutmeg, onions, parsley, scallions, *sofrito* (fried onions, garlic, sweet peppers, tomatoes, and cilantro, from Puerto Rico), thyme	
Beer, coffee (*café con leche*), teas, soft drinks, milk, rum, Irish moss (seaweed extract), sorrel	Fruit juice and soft drink consumption may increase.
Butter in French-influenced countries; coconut oil; *ghee* (Asian Indian clarified butter); lard in Spanish-influenced countries; olive oil	
Sugar cane products, such as raw and unrefined sugar and molasses	

most common addition, then cereals and fruits. Urban residents are more likely to skip breakfast than are people in rural households.

Rice and beans with meat, if affordable, is the most popular lunch. In rural regions a starchy vegetable, such as potatoes, plantain, breadfruit, taniers, taro, or yams, with dried salt cod is also common. Meats, milk, and salad vegetables are extra items added to the basic foods whenever possible.

Dinner is similar to lunch. Rice and beans provide the core of the meal, with meats, vegetables, and milk added when available. In

■ On Good Friday in Trinidad and Tobago, the white of an egg is added to warm water, and the future is predicted by its shape.

■ Rum cake (also known as black cake) is a fruitcake specialty of the Caribbean, especially in Jamaica, where it is served at weddings, Christmas, and other special occasions.

upper-income households the meal may consist of three or four different dishes. Milk and coffee with milk (*café con leche*) are the preferred beverages. Soft drinks and beer are also popular. Ice cream and pastries are typical desserts for those who can afford them. Snacking is frequent among some groups, especially children. Fruits, sweetened fruit juices poured over shaved ice, and coffee with milk are common.

Special Occasions

The early European dominance in the West Indies resulted in an emphasis on Christian holidays. Christmas is important, especially in the Spanish-influenced islands that are predominantly Catholic. In Puerto Rico, *pasteles* are prepared to celebrate Christmas. Similar to Mexican *tamales,* pasteles are a savory meat mixture surrounded by cornmeal or mashed plantains, wrapped in plantain leaves, and steamed. Carolers traditionally stop at houses late at night to request hot pasteles from the occupants. Other holidays reflect the multicultural history of the islands. *Carnival* is celebrated in some Caribbean countries, such as Trinidad and Tobago, and is similar to Mardi Gras in the United States. The pre-Lenten festivities feature parades of dancing celebrants; many are elaborately costumed as traditional European or African figures. Food booths that line the parade route provide a day-and-night supply of carnival treats. Fried Asian Indian fritters are particularly popular. Examples of nonreligious events include the day-long birthday open house on Curaçao for friends, relatives, and acquaintances. Thanksgiving is observed in Puerto Rico. Turkey, stuffed with a Spanish-style meat filling, is the main course.

Therapeutic Uses of Food

Some Caribbean Islanders adhere to a hot-cold classification system of diet and health similar to that found in Mexico (see previous section on Mexican therapeutic uses of foods). In addition to the categories of hot and cold, Puerto Ricans add cool. Imbalances in hot and cold—for example, sitting in the shade of a tree after being out in the sun—can cause illness even years after the imbalance has occurred (Spector, 1991). Haitians believe that women are warmer than men and that a person cools as he or she ages (Laguerre, 1981).

The hot-cold theory of foods practiced in the Caribbean includes not only the category of cool foods, but also those that are considered "heavy" or "light." A balance of hot-cold elements is attempted at meals, and heavy foods, such as starches, are consumed during the day, while light foods, such as soup, are eaten in the evening. Although the specific classification of items varies from person to person, one guideline for Puerto Ricans (Harwood, 1981) indicates bananas, coconuts, and most vegetables are cold; chiles, garlic, chocolate, coffee, evaporated milk and infant formula, and alcoholic beverages are hot. Cool foods include fruit, chicken, bacalao, whole milk, honey, onions, peas, and wheat. Excessive intake of cool or cold foods can make a cold condition, such as a cough, develop into a chronic illness, such as asthma.

Pregnancy, which is defined as a hot condition by most Puerto Ricans, is a time when a hot-cold balance is practiced carefully and hot

A typical Cuban meal includes savory *picadillo* (right), black beans and rice, bread, and *flan* (a Spanish-style custard) for dessert. (Courtesy of the Florida News Bureau.)

foods are avoided. When infants suffer from hot ailments, including diarrhea or rash, infant formula may be replaced with whole milk, or cooling ingredients such as barley water, mannitol, or magnesium carbonate may be added to the formula. High-calorie tonics (eggnogs and malts are popular types) are taken by some Puerto Ricans to stimulate the appetite and provide strength or energy. These are considered especially appropriate for pale children and for pregnant or postpartum women.

Research with Dominican American mothers suggests that nutritional practices during lactation may sometimes include avoidance of certain protein foods and increased intake of fluids such as malt beer, milk, orange juice, chocolate milk, and noodle soup (Sanjur, 1995). Formula may be withheld from sick infants and tea provided instead.

In Haiti, pregnant women are encouraged to eat red fruits and vegetables to strengthen the blood of their babies (Laguerre, 1981). Haitians also believe that certain illnesses in infants can be caused if a nursing mother's milk is too thick or too thin. Further, if a woman is frightened while breast-feeding, her milk goes to her head, causing her a headache and diarrhea in the baby. *Gaz,* another condition, causes pain in the shoulder, back, legs, or appendix and headaches, stomachaches, or anemia. Foods such as corn or a tea made from garlic, cloves, and mint are home remedies for gaz (Unaeze & Perrin, 1995).

Contemporary Food Habits in the United States

Adaptations of Food Habits

Traditional food habits are easily maintained in the self-sustaining immigrant communities of Spanish Harlem in New York and Little Havana in Miami. Ingredients for Caribbean cuisine are readily available through Puerto Rican and Cuban American markets. Cubans, for example, consider drinking strong coffee a way to maintain their ethnic identity; by contrast, Americans drink "weak" coffee (Pasquali, 1985). Changes do occur, however, as immigrants settle into

■ Rum was believed to cure many types of disease and injury in the 17th and 18th centuries. As late as 1892 a physician prescribed Cuban rum for the 6-year-old future king of Spain when he suffered from a fever. The royal family was so impressed by his quick recuperation that they granted the maker of the rum, the Bacardi family, the right to use the Spanish coat of arms on their label.

■ Marijuana may be smoked, added to foods, or prepared as a tea to treat conditions such as asthma, vision problems, and diabetes in Jamaica.

■ In a study of elderly Cubans living in the United States, it was found that convenience items were sometimes used to prepare traditional foods: aluminum foil in place of banana leaves for *tamales,* and packaged crescent rolls for *pastele* dough (Pasquali, 1985).

■ Fruit-flavored beverages, considered a source of vitamin C, are infrequently consumed by Cubans (Fanelli-Kuczmarski, et al., 1995).

■ In Haiti, protein foods are served first to the father in the household; leftovers go to the wife and children. It is believed this pattern continues in Haitian American homes.

■ The West Indian Carnival has been held annually for over 65 years in New York City. It celebrates the cultures of the Caribbean, featuring an enormous parade, music competitions, and traditional food such as curried goat and Jamaican jerk barbecue.

■ Major reggae music festivals with ample Caribbean food are held throughout the United States on February 6—Bob Marley's birthday.

culturally mixed communities and children grow up as Americans.

One study compared the diet of three groups of women from Puerto Rico: (1) those living in New York (forward migrants), (2) those who had lived on the mainland but later returned to the island (return migrants), and (3) those who never lived on the mainland (nonmigrants). It was found that nonmigrants and return migrants ate more starchy vegetables, sugar, and sweetened foods than did forward migrants. Forward migrants ate a greater variety of foods, including more beef, eggs, bread, fresh fruit, and leafy green vegetables. Puerto Rican women who had lived on the mainland quickly reverted to their traditional food habits when they returned to the island (Immink et al., 1983).

Ingredients and Common Foods

Research on the food habits patterns of Caribbean immigrants in the United States is limited. It is thought that rice, beans, starchy vegetables, sofrito, and bacalao remain the basis of the daily diet of many Puerto Ricans who live on the mainland. Poultry is used when possible, and egg intake decreases. Because of Cuban Americans' greater discretionary income, their diet usually includes additional foods, such as more pork and beef. Recent poorer immigrants from Cuba, including the marielitos, are more restricted in what foods they purchase, however, and may follow a diet that is closer to the subsistence-level Puerto Rican regimen.

A study of Puerto Rican teenagers in New York found that, consistent with traditional food habits, there is little diversity of food items in their diet (Duyff et al., 1975). However, in the study mentioned previously, Puerto Rican women on the mainland were reported to eat a greater variety of foods than those on the island. According to marketing studies, Caribbean Islanders accept some American foods, especially convenience items, and purchase frozen and dehydrated products when they can afford to do so. The proportion of meat in the diet often increases on the mainland, as does the consumption of milk (and other dairy foods) and soft drinks (Chávez et al., 1994; Harwood, 1981; Sanjur, 1995). Intake of leafy vegetables continues to be low. Local fruits often replace the tropical fruits of Puerto Rico.

Recent data regarding Dominicans in the United States indicate that only small changes have occurred in consumption patterns. Protein and fat intake has increased slightly, mostly due to eating more meat, while carbohydrate intake has decreased. Dominican women reported that their diet was more varied and abundant than in their homeland (Sanjur, 1995).

Meal Composition and Cycle

The meals of Puerto Ricans residing on the mainland are similar to those of people on the island, with a few changes. A light breakfast of bread and coffee may be followed by a light lunch of rice and beans or a starchy vegetable, with or without bacalao. Often this traditional midday meal becomes a sandwich and soft drink, however. A late dinner consists of rice, beans, starchy vegetable, meat if available, or soup. Salad is included in some homes. Many researchers have reported an increase in the amount of snacking between meals, mostly on high-calorie foods with little nutritional value (Harwood, 1981; Sanjur, 1995).

Interviews with Haitians living in New York City suggest that some traditional dietary practices are discontinued; for example, the main meal is eaten in the evening instead of at noon. Although some Haitians adhere to hot-cold classifications of food, they may differ from those used in Haiti (Laguerre, 1981). Dominicans have also started eating lighter lunches (Sanjur, 1995).

One study (Haider & Wheeler, 1979) reported that many low-income Latina women living in New York did not plan menus far in advance and that this hampered their ability to add variety to their diets. It was found that food shopping serves as a social occasion for many of the women and is one of the few opportunities they have to get out of the house; thus they may go to the grocery store more often than is really necessary. The investigators reported that nearly half of the women questioned preferred to fry main dishes. Boiling was the second choice, baking third. Broiling food was a distant fourth choice. The researchers note that in many low-income households the oven or broiler element may not work, restricting food preparation methods to frying and boiling.

Nutritional Status

Nutritional Intake

There is even less information on the nutritional status of Caribbean American immigrants to the United States than on Chicanos (see chapter 9). The few studies available suggest several health trends in these immigrants that have important nutritional implications.

Although the birthrate for Puerto Ricans is slightly lower than that for Americans of Mexican descent, it is still higher than for the non-Latino population. In addition, nearly 10 percent of Puerto Rican infants born on the mainland are of low birth weight. This undoubtedly contributes to the high infant mortality rate: among Puerto Ricans in New York, the infant mortality rate is reported to be 70 percent higher than that of the total population; the numbers are even higher in Puerto Rico. Risk factors including poverty, young maternal age, low educational attainment, and inadequate prenatal care are positively correlated with these figures (Collins & Shay, 1994).

Breast-feeding is reportedly uncommon among Puerto Rican, Cuban, and Haitian women in the United States. Those few who start breast-feeding may switch to bottle feeding after 2 to 4 weeks. Whole milk, condensed milk, and evaporated milk are frequently fed to infants, as are juices. Solid food typically is introduced at a young age (DeSantis & Halberstein, 1992; Kumanyika & Helitzer, 1985). Observation of other Caribbean Islanders suggests that breast-feeding may be prevalent, however (Gupta, 1987).

When traditional Caribbean diet is limited because of low income, it inevitably results in low intake of many vitamins and mineral. The emphasis on carbohydrates and vegetable protein, with a low consumption of leafy vegetables and fruit, provides inadequate intake of calories, vitamins A and C, iron, and calcium. A study indicated that the only foods consumed by at least one-half of recent Cuban immigrants were eggs, rice, bread, legumes, lard and oils, sugar, and crackers (Gordon, 1982). Of the same group, 79 to 100 percent reported never eating leafy green vegetables or fresh fruits. Deficiencies of B_1, B_{12}, folate, and sulfur amino acids have been reported (Ackerman, 1997). Anthropometric measurements and physical observation suggested that 20 percent of the children under 15 years of age showed signs of malnutrition, 37 percent of the men and 17 percent of the women had adipose tissue measurements consistent with adult marasmus, and 12 percent of the immigrants suffered from anemia. Evidence from a Florida study suggests many recent immigrants, especially refugees from Haiti, are malnourished (DeSantis & Halberstein, 1992).

■ Tropical sprue and intestinal parasites are endemic in Haiti, accounting for both nutritional deficiencies and anemia.

Many native Puerto Ricans suffer from parasitic diseases, including dysentery, malaria, hookworm, filariasis, and schistosomiasis. As many as one-third of Puerto Rican school children in New York City were found to be infected with parasites (Harwood, 1981), and it is believed that 60 percent of the rural population in Cuba has also contracted parasitic illnesses. These conditions contribute to general poor health and nutritional deficiencies among poorer immigrants.

In contrast to the malnutrition evident in some areas of Puerto Rico and Cuba, high rates of overweight and obesity have been found in the Puerto Rican and Cuban American population. National data suggest that 40 percent of Puerto Rican women and 26 percent of men living on the mainland are overweight. Among Cuban Americans, the figures are 31 percent of women and 28 percent of men (Fanelli-Kuczmarski & Woteki, 1990). While calorie and fat intake is reportedly high in many Caribbean Islander groups, some research suggests high risk for overweight among Puerto Rican populations with low energy consumption (Haider & Wheeler, 1979; Immink et al., 1983), perhaps due to low levels of activity. Cultural norms regarding weight and health may be significant factors in overweight among Caribbean Islanders. Plumpness is a sign of well-being among Puerto Ricans; in Haiti, obesity is associated with health and happiness.

The prevalence of non-insulin-dependent diabetes mellitus is two to three times higher among Puerto Ricans than among whites; Cuban Americans develop the condition at rates similar to whites. Socioeconomic, behavioral, and genetic factors may account for the differences in risk between the two groups (Flegal et al., 1991). However, renal failures among men due to diabetes are reportedly higher for both Hispanic blacks and Hispanic whites than for non-Hispanic blacks and whites (Chiapella & Feldman, 1995).

■ Lactose intolerance is thought to be a problem among many Caribbean Islanders, although estimated incidence has not been reported.

■ Puerto Rican children living in New York City have been found to be at high risk for lead toxicity (Carter-Pokras et al., 1990).

■ Researchers suggest that the Rastafari movement may provide an affirmation of black identity for some followers (Hickling & Griffith, 1994).

Chronic liver disease and cirrhosis are ranked as the sixth most common cause of death among Hispanics. Men are more likely to drink heavily than women. Among Puerto Rican men, 17 percent reported drinking one or more ounces of alcohol each day; 9 percent of Cuban men have similar drinking habits (Bassford, 1995).

The incidence of dental caries among Caribbean immigrants is higher than that of Mexican Americans. Cuban men were found to have a higher rate of caries, 73 percent, than Puerto Rican men, 62 percent (Gordon, 1982). Puerto Ricans have a high prevalence of periodontal disease (Ismail & Szpunar, 1990).

Counseling

Counseling Americans of Caribbean descent is similar to counseling other Latino clients and patients (see the previous chapter). Access to medical care may be limited (one study found 20 percent of Puerto Ricans and 25 percent of Cuban Americans uninsured) (Trevino et al., 1991), health beliefs may differ from those of the provider, and language problems may make interviews difficult. Illness may be viewed as a sign of personal weakness among Haitians, revealed only to family members for as long as possible.

Caribbean Islanders use an expressive conversational style. Respect and politeness are practiced, but it is not considered rude to interrupt a speaker. Shaking hands in greeting and in leaving is customary, and touching is very common, especially between members of the same sex, who may hug freely. Direct eye contact is expected throughout the Caribbean, with some variations: Looking away suggests disrespect or dishonesty among Cubans; Dominicans maintain eye contact depending on the situation, and men are more likely to look directly at one another than are women; Haitians may avert their eyes from authority figures.

Puerto Ricans are open about physical and emotional complaints, although they may hesitate to ask questions because this might be interpreted as disrespectful. Adequate time and consideration of symptoms is necessary for both diagnosis and the client–provider relationship. High degrees of modesty are found in both women and men, who may prefer health providers of the same gender.

In contrast, Haitians may expect a quick physical examinations (with a stethoscope) and a fast , accurate diagnosis. Lack of a prescription to address a complaint may be seen as incompetence.

A few traditional Puerto Rican hot-cold practices may be problematic. Pregnant women may avoid iron supplements, which are classified as hot. Infant formula diluted with mannitol or magnesium carbonate to cool it (See "Therapeutic Uses of Foods" section) may cause diarrhea in infants. A mother may cool a hot medication, such as vitamins or aspirin, by providing a cool beverage with it, possibly fruit juice or milk of magnesia (Harwood, 1981).

Puerto Ricans also may interpret high blood pressure as being "too much" blood or "thick" blood; conversely, that low blood pressure means "weak" blood or anemia. Some Puerto Ricans believe that ulcers lead to cancer.

A study of Long Island Cuban Americans showed that 20 percent acknowledged some degree of belief in santeria. It is believed that many more may be unwilling to admit to these folk practices, or use them only in time of stress (Pasquali, 1994).

Biomedical prescriptions may also present difficulties. It has been noted that many medications available only by prescription in the United State are readily available over the counter or through the black market in Latin American countries. Overuse, inappropriate use, and addiction can occur (Miralles, 1989; Pousada, 1995).

Although an immigrant from the Caribbean may be of African, Asian, European, or Indian descent, the island of origin is as important in determining diet as ethnic identity. A black from Puerto Rico, for example, is more likely to eat Latino foods than just African dishes or soul food. Researchers working with low-income Latinas and black women in New York (Wheeler & Haider, 1979) recommended that significant attention be placed on the client's socioeconomic status. Diet therapy for a poor Puerto Rican immigrant is likely to require a different approach than that for a wealthy Cuban immigrant within the context of the basic Caribbean cuisine. Haitians reportedly have difficulty with dietary compliance (Laguerre, 1981).

Puerto Rican family members will gather to support a hospitalized individual, often bringing favorite foods. Dietary restrictions may require careful explanation.

Researchers have reported that 92 percent of Puerto Rican teenage girls obtain nutrition information from their mothers, 51 percent from home economics classes, 43 percent from health professionals, and 37 percent from friends (Duyff et al., 1975). These data suggest that, as in Chicano households, mothers are the primary source of food and nutrition beliefs and should be respected by health professionals as the experts in their roles.

An in-depth interview should be conducted with each client of Caribbean descent to establish country of origin, ethnic identity, and length of stay in the United States. The client's degree of acculturation, socioeconomic status, use of traditional heath practices, and personal food preferences also should be determined.

South Americans

South America is a vast land that features the rugged ridge of the Andes Mountains stretching from north to south. Highland plains, tropical rainforest, temperate valleys, and desert dunes extend from where the mountain peaks slope toward the coastal edges of the continent. Extremes in the terrain and climate limit agriculture in many areas. The continent contains 12 independent nations: Argentina, Brazil, Bolivia, Chile, Colombia, Ecuador, Guyana, Paraguay, Peru, Suriname, Uruguay, and Venezuela. In addition, France retains control of the territory called French Guyana, and Great Britain claims the Falkland Islands.

Numerous native Indian groups populated the continent prior to settlement by the Europeans. Although the Spanish were the first to arrive, significant numbers of Portuguese, Italians, and Germans also settled in South America. Forced labor from West Africa introduced blacks to the continent, followed by Asian Indian workers after slavery was outlawed. Indian and mixed Indian-European populations live in the tropical highlands, Creoles (descendants of the Europeans) have concentrated in the southern, temperate regions of the continent, and parts of northeastern Brazil are populated primarily by blacks and mulatto black Europeans. In more recent times, Japanese immigration to South America has become notable.

Cultural Perspective

History of South Americans in the United States

Immigration Patterns

The first documented South Americans in the United States were Chileans, who came to California to participate in the gold rush. It is believed that several thousand worked in the mines. Approximately half returned to Chile, and those who remained quickly intermarried and were absorbed into the general population.

Prior to the 1960s, all South American immigrants were counted as "Other Hispanics" in the U.S. census. Specific figures regarding the individual nations before this time are uncertain but are thought to be minimal. Most South American immigration has occurred in the past 20 years, during periods of land reform, economic hardship, or political repression. Jobs and educational opportunities are the primary attractions for the majority of immigrants, though there are significant numbers of political exiles from Argentina and Chile.

Current Demographics and Socioeconomic Status

South Americans account for only 5 percent of Latinos in the United States. As reported in the 1990 U.S. census, there were approximately 350,000 Americans of Colombian ancestry, 200,000 from Ecuador, 162,000 from Peru, 81,000 from Guyana, 65,000 from Brazil, 63,000 from Argentina, 61,000 from Chile, 40,000 from Venezuela, 34,000 from Bolivia; 15,000 from Uruguay, 5,000 from Paraguay; and fewer than 2,000 from Suriname. Immigration since 1990 has been steady. It is believed by some population experts that South Americans are seriously undercounted in the United States. Many enter the country illegally, by overstaying their visas, obtaining falsified documents, or entering via Mexico or Puerto Rico. There may be as many as 350,000 undocumented Brazilians and 300,000 Ecuadorans living in the United States.

- When Chileans came to San Francisco to participate in the gold rush of the mid–19th century, they established a community called *Chilecito,* complete with its own newspaper. In the smaller mining towns, Chileans lived together in "Chilitowns" where they spoke Spanish and cooked traditional meals.

- Canada offered special entry to Chileans seeking political exile in the 1970s and 1980s, so large populations are found in Toronto and Montreal.

- Several thousand Chinese Ecuadorans (approximately 1 percent of the total Ecuadoran American population) live in New York, where they settle mainly in Latino neighborhoods, speaking mostly Spanish and some Chinese.

Most South Americans settle in the Northeast, especially New York and New Jersey. In New York City, Colombians, Ecuadorans, and Peruvians have established ethnic enclaves in Queens, and "Little Brazils" are found in both Queens and Manhattan. Miami and Los Angeles also host large South American populations from most nations. In addition, Brazilians are found in Pennsylvania and Washington, D.C.; Chileans have settled in Texas; Colombians have clustered in Stamford, Connecticut, and the urban areas of Illinois; and Peruvian neighborhoods have developed in Houston, Chicago, and Washington D.C. Second- and third-generation South Americans often leave homogeneous neighborhoods and relocate into mixed communities.

Few data are available on the socioeconomic status of Americans of South American descent. Most immigrants come to the United States in search of employment opportunities. Many coming from Argentina and Chile and smaller numbers from the other nations are well-educated professionals. However, they frequently find that their credentials are not accepted after arrival, forcing them to accept positions in the sales, service, trade, and labor fields, such as restaurant work, construction, child care, or textile and garment industry jobs. It is believed that many second- and third-generation Americans of South American ancestry obtain advanced education and work in professional occupations.

■ A large number of *costeño* immigrants from the coastal regions of Colombia have settled in Chicago. *Costeños* are typically well educated, often holding professional degrees in medicine, engineering, accounting, and architecture.

■ Among the most visible South American immigrants are the Otaveleño Indians of Ecuador. Colorful wool clothing and blankets are exported throughout the Americas, and itinerant Otaveleño salespeople travel to major cities to promote the products. The street vendors wear traditional Indian apparel, often operating illegally, and after a period of time return to Ecuador to attend college or begin a career (Mumford, 1995).

Most South American immigrants are proud of their heritage and differentiate themselves from Americans of Mexican, Caribbean, or Central American background. Brazilians in particular resent being mistaken as Hispanics who speak Spanish (Portuguese is their official language). Some South Americans suffer from discrimination directed toward Mexican Americans or Latinos in general. Others, who are mostly of European heritage, are not recognized as Latinos but may continue the prejudices between some South Americans. For example, there may be lingering hostilities between Bolivians and Chileans.

Worldview

Religion

South Americans are overwhelmingly Roman Catholic, a legacy of the European conquest. In most nations, approximately 90 to 95 percent of the population is a member of the Catholic Church, and its influence is seen in many South American institutions. The constitution of Argentina protects Catholicism and requires that the president of the nation be Catholic. In Ecuador, political leadership is established through sponsorship of local fiestas in honor of the saints. The Catholic faith is taught in Peruvian public schools.

In some regions the practice of Roman Catholicism is often blended with belief systems. In Peru, Incan gods may be included in Catholic rites, for example, and in Venezuela, the *Cult of Maria Lionza* mixes indigenous, Catholic, and African practices. Religious syncretism is greatest in Brazil. *Spiritism,* which was imported from France originally, combines Christian precepts with scientific principals. Popular with the upper middle class of the country, adherents communicate with the dead through spiritual mediums. *Umbanda* is very common in rural areas and among the urban poor, combining several Afro-Brazilian faiths with spiritism and the idea of Christian charity. *Candomblé* is probably the best known of the mixed religions, an Afro-Brazilian faith founded by blacks in the Bahia region that is now practiced nationwide by followers of all ethnicities. African derived beliefs dealing with earthly matters such as health and wealth are combined with Catholic cosmology. Yoruba deities called *orixá* or *orisha* are venerated with rites of worship that include animal sacrifice, feasting, and dancing. Over 20 orixás are recognized in Brazil, and most are correlated with a Catholic entity: *Oxalá,* god of creation, with Jesus Christ; *Exú,* the messenger god, with the Devil; *Ogun,* god of war and ironcraft with St. Anthony or St. George; *Oxoosi,* god of affluence, with St. Sebastian; *Omolu,* god of plagues and illness, with St. Lazarus; and *Oxum,* the fertility goddess (also called the goddess of vanity), with the Virgin Mary. Each orixá is associated with certain personality traits, day of the week, color, plants, animals, foods, and drinks, and each person has an orixá "owner" of his or her head, who influences individual temperament and behavior.

Protestant missionaries have been active in South America during the 20th century and have been especially successful in Ecuador, where in some regions as many as 40 percent practice some Protestant faith. In Brazil, the Baptist, Pentacostal, Seventh-Day Adventist, and Universalist denominations are most popular. Chilean Protestants are usually members of the Pentacostal or Seventh-Day Adventist churches, though those of German ancestry often follow the Lutheran or Baptist faiths. Small numbers of Jews and Buddhists are also found in South America, totally less than 1 percent of the population together.

Most South Americans who immigrate to the United States are Roman Catholic and very involved with their local parish. In some areas, particularly the Bronx in New York City, tension between parishioners of Irish or Italian descent and South American immigrants caused the South Americans to leave the traditional Roman Catholic faith. Colombians, for instance, formed a church based on charismatic Catholicism led by a Colombian priest. It is estimated that up to one-third of Ecuadoran Americans belong to Protestant denominations. It is not known how many South Americans who come to the United States practice blended religions.

Family

Family life is important in all South American societies. In Argentina, Spanish and Italian traditions have shaped family structure. The extended family usually gathers together at least once a week and on holidays as well. Grandparents are involved in most family decisions, and children often stay at home until marriage. In Brazil, extended family members typically live close to one another, and daily visits are common. Relatives mentor children through rites of passage such as confirmation, graduation, the start of a career, and marriage. The father is the head of the household in Chilean homes, but the mother makes all decisions regarding the family. In Colombia, the father holds all authority, and children are taught to obey their parents. Ecuadoran families follow two models: Spanish-influenced families are ruled by the father, who has few responsibilities to the home other than financial support; in Indian-influenced families, the father and mother share more power and household responsibilities. In Peru, extended families typically include godparents, who sponsor baptisms and provide both social and economic assistance. Families are predominantly patriarchal, though more so in the Spanish-speaking upper and middle classes than in poor, rural Indian homes. In contrast to most of South America, the Venezuelan family has changed rapidly in the past decades due to increased national prosperity. Much of the population has relocated to urban centers. Many families have declined in size, and the extended family is less common.

In many areas of South America it is unacceptable for women to work outside the home. Even those with a profession traditionally stay at home after marriage. Among some Indian groups, however, women contribute to the well-being of the family through farm work, and in the urban areas of Venezuela many women have outside jobs but remain responsible for household chores. In Chile, women are usually involved in local social and political issues.

Most South Americans prefer to immigrate as family groups, though financial pressures often demand that a single family member become established in the United States before the rest of the family follows. Individual immigrants commonly move to neighborhoods where relatives, godparents, or friends have settled. They depend on these contacts for housing and support. This system of mutual assistance is maintained after the immediate family arrives, bringing more relatives into the extended family. Colombians and Ecuadorans often broaden their relationships beyond national boundaries to form strong bonds with other Latinos.

Many families suffer from the stresses of American informality and freedom. Men lose some authority over wives and children, and women find it difficult to adjust to working outside the home. Furthermore, many upper- and middle-class women, who had paid help with the housework in South America, must learn to balance a job with responsibility for running a home.

■ When Argentinean families gather on Sundays, they typically enjoy *asado* (an Argentinian barbecue) or pasta.

Traditional Health Beliefs and Practices

Most Brazilians associate faith with health. Catholics may believe in fate and seek intervention from patron saints when ill. Spiritists employ homeopathy, exorcism, past-lives therapy, acupuncture, yoga therapy, and

■ African *feiticeiros* (practitioners of the occult) were traditionally sought by European settlers in Brazil for spells and potions to cure snakebite, repel the evil eye, and enhance sexual experience (Voeks, 1997).

■ It is thought that the candomblé *patuás,* amulet bags hung around the neck to ward off evil (containing plant pieces, sacred items, and devotional writings), were introduced to Bahia in the late 19th century by black Muslims who wore similar pouches that held inscriptions from the Koran (Voeks, 1997).

■ Brazilians often attribute bad health to liver problems or an imbalance between hot and cold, such as drinking a glass of cold water on a hot day or taking a cool shower after eating a hot meal.

■ Many South Americans self-diagnose or seek health advice from their mothers or friends. They then visit a pharmacist where they can purchase many medications, such as antibiotics, by the pill.

chromotherapy to cure sickness (Jefferson, 1995). Followers of candomblé believe that health is maintained by achieving balance between the earthly and spiritual spheres. The *pai-de-santo* or *babalorixá* (high priest) or the *mäe-de-santo* or *ialorixá* (high priestess) may be hired to read the oracle of a personal orixá, for example, so that an individual can improve his or her relationship with the deity. Harmonious relations with one's orixá can maximize *axé* (vital force). Spiritual equilibrium is maintained by observing the preferences and prohibitions of one's orixá, including certain food and beverages, colors, therapeutic herbs, beaded necklaces, and other limitations. The priest or priestess also serves as the local *curendiero,* diagnosing physical and spiritual problems, prescribing healing herbal baths or botanicals (see Table 10.2, "Selected Candomblé/Brazilian Botanical Remedies"), and manipulating occult forces. In Ecuador, either a healer, called a *curandero,* or a witch-doctor, called a *brujo,* treats many illnesses in small villages. In Peru, urban residents typically obtain biomedical health care, but in rural regions home remedies and ritual magic are often preferred (Packel, 1995).

Traditional Food Habits

Ingredients and Common Foods

Staples

The cooking of South America is similar to that of other Latin American regions in that it combines some native ingredients and preparation techniques with the foods of colonial Europeans. The diet is largely corn based and spiced with chile pepper (see Table 10.3). Tomatoes are common and, in tropical areas, cassava (called *yuca*) is a popular tuber. Pumpkins, bananas, and plantains are consumed often. Beef, rice, onions, and olive oil, introduced by the Spanish and the Portuguese, are eaten regularly. However, South American fare also features a number of ingredients used infrequently in the dishes of other Latin American areas. Potatoes, which were first cultivated by the Incas on mountain terraces, are particularly important in the highlands of Peru and Ecuador. Sweet potatoes (the orange-fleshed root vegetable usually called "yams" in the United States) are also native to the region. A white root similar to a mild carrot, known as *apio* or *arracacha* (*Arracacia xanthorrhiza*), is found in Colombia, Peru, and Venezuela; oca (*Oxalis tuberosa*) and yacón (*Polymnia sanchifolia*) are commonly eaten raw and cooked in Bolivia, Brazil, Colombia, Ecuador, and Peru; and the tuber known as *ahipa* (called jicama in the United States and Mexico) is native to the Amazon River basin. Beans, a foundation food in many Latin American regions, are common in most South American countries, yet not eaten at every meal. Other legumes and nuts such as peanuts and cashews are used often in dishes. Local meats, including llama, deer, rabbit, wild pig, and *cuy* (guinea pigs raised for consumption), are favored in some areas. Fish, such as anchovies and tuna, and shellfish, particularly shrimp, crab, spiny lobster, oysters, clams, giant sea urchins (*erizos*), and giant abalone (*locos*), are significant foods in the extensive coastal regions.

Numerous tropical fruits and vegetables are found in the cuisines of South America. (© Wolfgang Kaehler/CORBIS)

A favorite way to prepare meats in South America is grilling. Traditionally sides of beef, whole lambs, hogs, and kids (goats) are hung over smoldering wood to slowly cook for hours in a method called *asado*. Today a grill is used more often. Steaks and marinated kebobs (which often include organ meats) are favorites. Another, even older cooking tradition is to steam foods in a pit oven. In Peru, this method is called a *pachamanca* and typically includes a young pig or goat, with guinea pigs, chickens, tamales, potatoes, and corn tucked around layers of hot stones and aromatic leaves and herbs. In Chile, a *curanto* is closer to an elaborate coastal clam bake, including

Table 10.2
Selected Candomblé/Brazilian Remedies

Scientific Name	Common Name	Parts	Traditional Use
Achyrocline satureoides	Macela; marcela	Whole plant	Hypertension; colic; stomach complaints; liver disorders; diarrhea; dysentery; headache
Ambrosia artemisiaefolia	Artemesia; ragweed	Leaves	Diuretic; stomach pain; liver problems
Annona muricata	Graviola; guanabana; apa oka; soursop	Leaves	Diabetes; rheumatism; fever; parasites
Artemsia absinthium ☠	Losna; wormwood	Leaves	Fever; menopause symptoms
Bauhinia ovata, B. forficata	Unha-da-vaca; pata de vaca; abafé	Leaves	Diabetes; menstrual problems
Bidens pilosa	Carrapicho; pic̃ao; ewe susa	Leaves	Kidney problems; liver ailments
Centropogon cornutus	Bico-de-papagaio; crista-de-peru; ewe akuko	Leaves	Gastric ulcers
Coutoubea spicata	Papai nicolau	Whole plant	Stomach disorders
Hybanthus calceolaria	Purgo-de-campo	Root	Indigestion; ulcers
Mentha pulegium	Poejo; pennyroyal	Whole plant	Stomach problems; nausea
Miconia spp	Canela-de-velho	Leaves	Stomach complaints; parasites
Passiflora alata, P. edula	Maracujá; granadilla; passion fruit	Leaves, fruit	Flatulence; constipation; asthma; bronchitis; hysteria; insomnia
Paullina cupana ☹	Guaraná	Seeds	Fatigue; obesity
Peperomia pellucida	Alfavaquinha-de-cobra; oriri; iriri	Whole plant	High blood pressure; heart palpitations; prostate enlargement
Persia americana	Abacate; avocado	Leaves	Retained urine
Phyllanthus amarus, P. niruri	Quebra-pedra	Whole plant	Diabetes; liver problems; kidney conditions (especially stones); gallstones; urinary tract infections; prostate conditions
Plantago major	Transagem; plantain	Leaves	Stomach disorders; menstrual problems; sore throats
Plectranthus amdoinicus	Tapete-de-Oxalá; Spanish thyme; Cuban oregano; Indian borrage	Leaves	Poor digestion; stomach problems; liver disorders
Ruta graveolens ☠	Arruda; rue	Leaves	Postpartum pain, uterine problems; abortive
Solanum nigrum ☠	Maria preta; erva santa maria; black nightshade; American nightshade	Leaves	Coughs
Solanum paniculatum ☠	Jurubeba; agog ogum	Root; leaves; fruit	Liver conditions; stomach problems; irritable bowel syndrome; anemias; hangovers
Tabebula impetiginosa ☹	Ipe roxo; pau d'arco; taheebo	Bark	Diabetes; Hodgkin's disease; leukemia; venereal diseases; rheumatism; malaria; yeast infections
Veronica spp.	Alumã; ewe auro; speedwell	Leaves	Indigestion; stomach pain; liver problems; kidney complaints
Vitex spp. ☹	Alfazema; paqu d'arco; chasteberry; monk's pepper	Leaves	Diuretic; stomach pain; liver problems; cancer

Note: Data on some plants are very limited; adverse effects may occur even if not indicated.
Key: ☠, all or some parts reported to be harmful or toxic in large quantities or specific preparations; ☹, may be contraindicated in some medical conditions/with certain prescription drugs.

Table 10.3
Cultural Food Groups: South Americans

Group	Comments
PROTEIN FOODS	
Milk/milk products	Milk is not usually consumed as a beverage but used in fruit-based drinks and added to coffee. Many milk-based desserts are enjoyed.
Meat/poultry/fish/eggs/legumes	Beef is a foundation of the diet in parts of Argentina, Brazil, Paraguay, and Uruguay. Some game meats are consumed. Fish and seafood are significant in coastal regions, popular as *ceviche* in Ecuador and Peru. Beans are commonly consumed.
CEREALS/GRAINS	*Cuzcuz,* made from cornmeal, is prepared in parts of Brazil; *arepa,* cornmeal bread, is a staple in Venezuela. Pasta is popular in Argentina, Paraguay, Uruguay. Rice and corn puddings are a favorite.
FRUITS/VEGETABLES	Tropical and temperate fruits are plentiful and popular, added to savory and sweet dishes. Fruit compotes and fruit pastes are enjoyed. Potatoes are a staple in the Andes. Cassava flour and meal are common in many areas; tapioca is used in desserts.
ADDITIONAL FOODS	
Seasonings	Toasted cassava meal, farinha, is sprinkled over foods in Brazil. Spicy hot foods are preferred in many areas; salsas are common.
Nuts/seeds	Coconut and coconut milk are added to numerous dishes. Peanut sauces flavored with chiles are common in the Andes.
Beverages	Coffee is often served concentrated, then diluted with evaporated milk or water. Maté is more popular than coffee or tea in parts of Argentina, Brazil, and Paraguay.
Fats/oils	Dendê oil flavors and colors many dishes in the Bahia region of Brazil.
Sweeteners	

shellfish, suckling pig, sausages, potato patties, peas, and beans layered with seaweed.

Stuffed foods are also common, including pastry turnovers filled with savory meat, fish, or cheese fillings. They are called *empanadas* in Argentina, which specializes in a turnover with a flaky, Spanish-style dough enriched with indigenous ingredients such as mashed potatoes, cassava, or corn. The turnovers are usually baked, but sometimes they are fried. Fillings are as many as there are cooks. Chopped meat, olives, raisins, and onions are popular. In Chile, the turnovers may be filled with abalone, and in Brazil, where they are known as *empadinhas,*

Table 10.3
Cultural Food Groups: South Americans—continued

Common Foods	Adaptations in the United States
Cow's, goat's milk; evaporated milk; fresh and aged cheeses	Available cheeses are sometimes substituted for unavailable traditional cheeses
Meat: beef (including variety cuts), frog, goat, guinea pig (*cuy*), llama, mutton, pork, rabbit *Poultry:* chicken, duck, turkey *Fish and shellfish:* abalone, bass, catfish, cod (including dried salt cod), crab, eel, haddock, lobster, oysters, scallops, shrimp, squid, trout, tuna *Eggs:* chicken, quail, turtle *Legumes:* beans (black, cranberry, kidney), black-eyed peas	Less acceptable meats such as guinea pig may no longer be eaten
Amaranth; corn, rice, quinoa; wheat	
Fruits: abiu, acerola, apples, banana/plantains, cashew apple (*cajú*), *caimito,* cherimoya, guava, grapes, *jabitocaba* lemons, limes, *lulo (naranjillo),* mammea, mango, melon, olives, oranges (sweet and sour), palm fruits, papaya, passion fruit, peaches, pineapple, *pitanga;* quince, raisins, roseapple, *sapote,* soursop, strawberries, sugar cane *Vegetables: ahipa* (jicama), *arracacha* (*apio*), avocado, *calabaza* (pumpkin), cassava (*mandioca; yuca*), green peppers, hearts of palm, kale, okra, *oca,* onions, *roselle,* squash (chayote, winter), sweet potatoes, tomatoes, *yacón,* yams	
Achiote, allspice, chiles (*aji,* malagueta, pimento), cilantro, cinnamon, citrus juices (lemon, lime, and sour orange), garlic, ginger root, oregano, paprika, parsley, scallions, thyme, vinegar	
Brazil nuts, cashews, coconut, peanuts, pumpkin seeds	
Batidas (tropical fruit juices, sometimes made with alcoholic beverages), coffee, *guaraná,* soft drinks, sugarcane juice, tea, *yerba maté* and alcoholic beverages: beer, *cachaça* (sugarcane brandy), *pisco* (grape brandy), *chicha* (distilled corn liquor), wine	
Dendê (palm) oil, olive oil, butter	Vegetable or peanut oil is substituted for dendê oil.
Sugarcane, brown sugar, honey	

a spicy shrimp mixture is traditional. In Bolivia, where they are called *salteñas*, the turnovers are filled with cheese. Tamale-like steamed packets of dough-wrapped fillings are also popular throughout South America. In Peru, *chapanas* are made with cassava dough, while in Ecuador, *bollos* are formed around cooked chicken meat with plantain dough. In Brazil, a freshly grated corn kernel dough is mixed with coconut and cassava (and no filling) to prepare *pamonhas.* A ground cornmeal dough flavored with achiote and tomatoes is preferred for Venezuelan *ballacas.* A favorite in Argentina, Bolivia, Brazil, Chile, and Ecuador are *humitas,* which feature fresh kernel or ground cornmeal dough wrapped around a variety of savory or sweet meat, fish, or vegetable fillings.

Regional Variations

National differences exist, although there are few clearly distinctive divisions in South American fare. Several countries share similar dishes, and only a few nations have well-developed regional cuisines.

Peru and Ecuador. The cooking of Peru and Ecuador is divided into the highland fare of the Andes and the lowland dishes of the tropical coastal regions. The cuisine of the mountain areas is among the most unique in South America, preserving many ingredients and dishes of the Inca Indians. Potatoes are eaten at nearly every meal, and often for snacks. Over 100 varieties are cultivated. *Ocopa,* boiled potatoes topped with a cheese sauce and chile peppers or peanuts, is a typical dish in Peru. In Ecuador, fried potato and cheese patties called *llapingachos* and potato cheese soup served with slices of avocado known as *lorco* are common. Traditionally the tubers are preserved by freezing in the cold night air and then drying in the hot daytime sun. *Papa seca* are boiled first, then dried until the potatoes are rocklike chunks that must be rehydrated before consumption; *chuño* are not cooked before drying and are often ground into a fine potato starch. Corn is also grown in the mountains. Some varieties have kernels the size of small strawberries that when prepared as hominy are known as *mote* and are a popular snack item. Bananas and plantains are cooked as savory chips and made into flour for breads and pastries.

- Peruvians are especially fond of rabbit dishes.

- *Llapingachos* is traditionally served with crunchy, fried corn kernels (sprinkled with salt) called *maiz tostadas.*

- In the rural villages of Ecuador, a popular alcoholic drink is *costa,* traditionally made from the yucca plant by women who chew and spit out the pulp to facilitate fermentation.

- In Chile, the favorite beans are *porotos,* called "cranberry beans" in English.

The foods of Peru and Ecuador are preferred *picante* and feature abundant use of chile peppers in both the highlands and along the coast. *Salsa de ají,* a combination of fresh chopped chile, onion, and salt, is served as a condiment at most meals. Orange- or yellow-hued dishes are favored; along the coast, achiote colors foods, and in the Peruvian highlands an herb known as *palillo* is used.

Charqui, dried strips of llama meat, is a specialty of the Andes. *Anticuchos,* chunks of beef heart maninated in vinegar with chiles and cilantro, then skewered and grilled, are a spicy Peruvian favorite also from the Andes. Along the coast, seafood dominates the diet. The region is famous for its *ceviches,* a method of preparing fresh fish, shrimp, scallops, or crab by marinating small raw chunks in citrus juice (lemon and lime in Peru; sour orange and lime in Ecuador). The acidity of the juice "cooks" the fish and turns it opaque. At many beaches, *cevicherias* offer the dish as a snack or light meal with beer. Chopped onion, tomato, avocado, and cilantro are often added. In Peru, *ceviche* is typically garnished with sliced sweet potato. *Pisco,* a grape brandy that originated in Peru, is a national favorite, often mixed with orange juice to make the refreshing drink called *yugeno.*

Argentina, Chile, Bolivia, Uruguay, and Paraguay. Hearty, ample fare with an emphasis on beef exemplifies the cooking of these southern nations. Argentina is a major beef-producing region, and its people eat more beef per capita than in any other country worldwide. The temperate weather permits the cultivation of numerous fruits and vegetables, notably strawberries, grapes, and Jerusalem artichokes (known as *topinambur* in Chile). The cooking of Argentina, Chile, Paraguay, and Uruguay has been influenced more by their immigrant populations than by the numerous small Indian groups native to the area. The Spanish introduced cattle, and the Italians brought pasta. Smaller numbers of Germans, Hungarians, and other central Europeans have added their foods as well.

The national dish of Argentina is *matambre,* which means "to kill hunger." A special cut of flank steak is seasoned with herbs, then traditionally rolled pinwheel fashion around a filling of spinach, whole hard-boiled eggs, and whole or sliced carrots, and then tied with a string and poached in broth or baked. *Matambre* can be served as a main course, but it is often chilled first and offered as a cold appetizer. Grilled steaks are particularly popular in Argentina and surrounding nations. In Paraguay, steaks are typically served with *sopa Paraguay,* a cornmeal and cheese bread. In Uruguay, beef is eaten nearly as often as in Argentina, although mutton and lamb are also common.

Robust soups and stews are everyday fare. In Bolivia, beef stew is made with carrots, onions, hominy, and *chuño*. The stews of Argentina often pair meat with fruits as well as vegetables, such as *carbonada criolla* (beef cooked with squash, corn, and peaches) or *carbonada en zapallo* (veal stew cooked in a pumkin). In Paraguay, soups reveal European inspiration, such as *bori-bori,* beef with cornmeal and cheese dumplings, and *so' o-yosopy,* beef soup with bell peppers, tomatoes, and vermicelli or rice, topped with Parmesan cheese. Fish soups and stews are popular in Chile, which has an extensive coastline and plentiful seafood. Clam or abalone chowder with beans (*chupe de loco*) and *congrio* (an elongated, firm-fleshed fish that looks a little like an eel) cooked with potatoes, onions, garlic, and white wine are specialties.

National favorites include pasta (e.g., spaghetti, ravioli, and lasagna), which is served in many homes on Sundays in Argentina. It is considered auspicious to eat it on the 29th of every month as well. In Chile, beans are especially popular, and seafood is eaten regularly. Wines from the temperate midlands of the country are considered some of the best on the continent. *Pisco* is consumed in both Bolivia and Chile, where it is mixed with lemon juice, sugar, and egg whites to make a *pisco sour.* In Bolivia, legs from the giant frog found in the Andean Lake Titicaca are a specialty, and *chicha,* a distilled corn liquor, is popular.

Although coffee is consumed throughout the area, another caffeinated beverage is equally popular in some regions. Called *maté,* it is an infusion made from the leaves of a plant (*Ilex paraguariensis*) in the holly family native to Paraguay. Served hot or chilled, *maté,* is consumed nearly every afternoon with small snacks in Paraguay and in parts of Argentina. The dried, powdered leaves are called *yerba* and are traditionally mixed in a gourd with boiling water. A special metal straw is inserted to drink the brew.

Colombia and Venezuela. The fare found in Colombia and Venezuela is colonial Spanish in character, cooked with olive oil, cream, or cheese and flavored with ground cumin, achiote, parsley, cilantro and chopped onions, tomatoes, and garlic. Yet native tastes are still evident. *Guascas,* or *huascas* (*Galinsoga parvilora Lineo*), an herb native to Colombia, provides a flavor similar to boiled peanuts in soups and stews. Hot chile pepper sauces are served on the side of most dishes. Tropical fruits and vegetables, including avocados, bananas and plantains, *naranjillo* (a small, orange fruit related to tomatoes used for its tart juice), pineapple, and coconut milk or cream are other common regional ingredients.

In Colombia, *Bogatá chicken stew* (made with chicken, two types of potatoes, and cream) and *sancocho* (a boiled dinner traditionally made with beef brisket or other roast and ample starchy vegetables like potatoes, sweet potatoes, plantains, or cassava) are typical Spanish-influenced dishes. Examples in Venezuela include *ropa vieja* ("old clothes"); shredded flank steak served in a sauce made with tomatoes, onions, and olive oil; and *pabellón caraqueño,* flank steak served on rice with black beans, topped with fried eggs, and garnished with fried plantain chips. Dishes with more indigenous flavors include *arepa,* the staple cornmeal bread of Venezuela that is formed into 1-inch thick patties and cooked on a griddle (it is sometimes stuffed with meat or cheese before it is fried), *cachapas,* tender cornmeal crepes, and mashed black beans, known as *caviar criollo* or "native caviar." Tropical fruits, such as guavas and pineapple, are often sweetened and dried to make fruit leathers and fruit pastes that are favorite snacks.

Brazil. The cooking of Brazil is very different from that of other South American countries due to Portuguese and African influences. The Portuguese arrived in the 16th century, looking for land to cultivate sugar cane. They contributed dried salt cod and linguiça to the diet, stews known as *cozidos* made with many different meats and vegetables (known as *cocido* in Portugal), and a variety of exceptionally sweet desserts based on sugar and egg yolks, such as caramel custards and corn (*canjica*) or rice (*pirão de arroz*) puddings flavored with coconut. African slaves put to work on the sugar plantations brought foods unknown in nearby countries, such as dendê oil (a type of palm oil) and okra. Spicy dishes were preferred. In West Africa, malagueta peppercorns, a small, hot grain, was used to season foods; in Brazil, Africans adopted a very small, mouth-searing chile pepper indigenous to the area and also called it malagueta. It is typically minced and

■ Restaurants specializing in grilled beef are called *parrillas* in Argentina.

■ Pizza is popular in both Argentina and Uruguay.

■ One of the more unusual specialties of Colombia is *hormiga culona,* a dish made from the big-bottomed ants found in the Santander region. Toasted ants, which taste similar to popcorn, are favorite roadside treats during the June season for insects.

■ Avocado *batidas,* made with the ripe vegetable, milk, sugar, and lime are favorites in Colombia and in Brazil but are disliked in other South American nations (Novas & Silva, 1997).

added to dendê oil, often with dried shrimp and grated ginger root, to make a hot sauce.

Although Indian, Portuguese, and African tastes and textures have influenced cooking throughout Brazil, nowhere are they more prominent than in the state of Bahia. Known as Afro-Brazilian fare, or *cozinha baiana,* it is famous for fritters made from dried shrimp, dried salt cod, yams, black-eyed peas, mashed beans, peanuts, and ripe plantains fried in dendê oil. *Vatapá,* another specialty, is a paste made with smoked dried shrimp, peanuts, cashews, coconut milk, and malagueta chiles. It is used as a filling for black bean fritters called *aracanjá* and sometimes served with rice as an entree.

The national dish of Brazil is *feijoda completa,* which originated in Rio. Black beans cooked with smoked meats and sausages are served with rice, sliced oranges, boiled greens, and a hot sauce mixed with lemon or lime juice. Toasted cassava meal, called *farinha,* is sprinkled over the top like Parmesan cheese. *Farinha* is served with most dishes and is often mixed with butter and other ingredients such as bits of meat, pumpkin, plantains, or coconut milk to create crunchy side dishes called *farofa.* Rice or cornmeal porridge called *pirão* is another type of side dish. Middle Easterners who immigrated to the southeastern areas of Brazil brought the concept of couscous to the country and adapted the dish to native ingredients. *Cuzcuz paulista* is prepared with cornmeal in a *cuzcuzeiro,* which looks like a colander on legs that is inserted over a pot of boiling water to steam. The basket of the *cuzcuzeiro* is first lined with seafood or poultry and vegetables, which flavors the cornmeal as it cooks and looks decorative when the *cuzcuz* "cake" is inverted.

In the far south the cuisine has been influenced by the foods of Argentina. Grilled meats are a favorite in Brazil, especially in the south, home of the frontiersmen known as *gauchos,* who herded cattle on the grassland plains. Sides of beef were traditionally staked at the edges of a bonfire for slow cooking in a method called *churrasco.* The popularity of the outdoor barbecue led to *churrascaria rodizio,* restaurants located in cities throughout the nation that specialize in spit-roasted beef, pork, lamb, and sausages brought to the table on large skewers and carved to taste. Specialties include *picanha* (rump roast) and beef heart. Assorted side dishes such as salads and potatoes, condiments, and desserts round out the meal. Brazilians in the south also drink *maté,* which they call *chimmarão.*

Coffee, rum, and beer are common beverages in Brazil, but several other drinks are also popular. *Guaraná* is a stimulating carbonated soft drink made from the seeds of the native *guaraná* fruit, which contain caffeine. *Cachaça* is an alcoholic beverage distilled from sugar cane that is often compared to brandy. It is used to make *batidas,* a refreshing punch made with cachaça and fruit juice.

■ *Churrrascaria rodizio* restaurants have opened in many U.S., European, Japanese, and Australian cities in recent years—anywhere steak is appreciated.

■ Coffee in Ecuador is preferred boiled to a thick consistency, then mixed with hot water or milk and sugar to taste.

■ In Chile, it is considered good luck if the first word said by a person in the new year is "rabbit" (Burson, 1995).

Meal Composition and Cycle

Daily Pattern

Breakfast is typically light, often bread or a roll with jam and a cup of coffee. A more complete meal features fresh fruit or pastries and occasionally ham or cheese. Lunch is usually the main meal, consumed leisurely with family or friends. Appetizers such a fritters, *humitas,* or *empanadas* may start the meal, followed by a meat or seafood stew or a grilled meat dish. Side dishes may include rice, beans, farofa, fried potatoes and greens such as kale. Salads are popular in some areas, including Brazil. Dessert can be fruit or a very sweet custard or pudding. In Argentina, the time spent relaxing and socializing after lunch is called *la sombremesa;* sometimes it includes a nap. Dinner is traditionally lighter, sometimes just cold cuts, a seafood salad, or a serving of stew, usually eaten around 9 P.M. each evening, often continuing past midnight. An afternoon break is typical in much of South America: coffee is typically consumed in Argentina, Colombia, Ecuador, and Brazil; tea is served in the late afternoon in Chile and Uruguay; and maté is popular in parts of Argentina, Paraguay, Uruguay, and Brazil. Snacks eaten with the beverage are often fruit, *cachapas* or *arepa,* sandwiches, or a pastry. Street vendors offering coffee, fruit juice, and snacks throughout the day are common in urban areas. A single, substantial meal each day may be all that is available to the poor. The meal may consist of soup or a serving of stew with a side dish of potatoes, plantains, cassava, corn, or rice and beans, depending on the region.

Special Occasions

Catholic traditions have influenced many South American holidays. A rich Christmas Eve dinner is traditional in most nations, often with a roast, such as lechón (suckling pig) in Brazil and cuy or lechón in Ecuador. Italian specialties including torrone and panettone are Christmas items in Argentina, where Epiphany is another significant religious holiday. Easter is important in many homes, and Carnival (Carnaval) festivities featuring dancing, parties, and traditional fare are popular in Brazil, Ecuador, Peru and Uruguay. Americans from these countries sometimes celebrate with parties during the three days before Lent. Saint John's Day is a favorite in Brazil, featuring foods made with corn and pumpkin, and is also celebrated by the Otavaleños of Ecuador with all-night feasting and dancing. All Soul's Day on November 2 includes gifts of food and family picnics at the gravesites of deceased kin in Peru.

Also significant for many Americans of South American descent are the Independence Days observed in various nations. Brazilian Americans honor their Independence Day on September 7 with day-long festivities in Boston, New York, and Newark, New Jersey. Americans from Chile sponsor traditional food and craft booths for fairs to celebrate their Independence Day on September 18. Colombian Americans consume tamales, empanadas, arepas, and other specialties on their Independence Day, July 20. In Ecuador, *primer grito* ("first cry" of independence) is held on August 10 and is officially marked as Ecuador Day in New York City. Independence Day in Peru is July 28. The Day of Tradition is popular in the Argentinian American community, with traditional foods, folk music, and equestrian displays by men dressed as gauchos.

Therapeutic Uses of Foods

Herbal teas are a favorite remedy throughout most of South America (see Table 10.2 for several examples). Candomblé orixás are associated with certain foods, and followers honor their deity by eating those items. Examples include white corn, white beans, rice, porridge, yams, and water with Oxalá; rice, black beans, black-eyed peas, and roasted corn with Omolu; black beans with Ogun; farofa made with dendê oil, black beans, honey, steak with onions, and chachaça with Exú; tapioca, pudding, cooked corn, and a ginger-flavored drink called *aluá* with Oxoosi; and pudding, banana, *ximxim* (chicken stew), and champagne with Oxum. A hot-cold system of medicine, most likely introduced by European immigrants, has also been adapted by candomblé healers, treating hot conditions associated with hot orixás with cool prescriptions associated with cool orixás. Classification is inconsistent, however, and cold conditions are rarely treated with hot remedies (Voeks, 1997).

Sample Menu

LUNCH IN BRAZIL

Aracanjé (spicy bean fritters) and *Empadinhas de camarões* (shrimp turnovers)

Cozido à Brasileira (stew with beef, pork, sausage, sweet potatoes, plantains, and other vegetables)

Feijão de côco (puréed black beans in coconut milk)

Farofa (cassava meal browned in butter)

Couve à mineira (kale cooked with onions and garlic)

Compote de frutas (fruit compote) or *Canjica* (corn–coconut pudding)

Fruit juice or beer

■ *Pan de jamón,* a soft, white bread made with swirls of sliced ham, green olives, and raisins, is traditionally served at Christmas and New Year's in Venezuela.

■ The cornmeal cheesebread known as *sopa Paraguay* is enjoyed by Paraguayans at nearly all special occasions, including birthdays and weddings.

■ *Los Quince,* a party held when a girl becomes an adult at age 15 in Argentina, typically includes a fancy dinner and dancing.

■ Therapeutic herbal teas are called "little waters" in Chile (Burson, 1995).

Contemporary Food Habits in the United States

Adaptations of Food Habits

Very little has been reported on the adapted food habits of South Americans living in the United States. Many continue cooking their favorites

■ *Guasacaca,* a creamy avocado sauce similar to Mexican guacamole except for the addition of olive oil, is often used instead of mayonnaise on sandwiches and hamburgers by Venezuelan Americans (Novas & Silva, 1997).

■ Chileans commonly use both their paternal and maternal surnames.

■ The Surui Indians of the Amazon were encouraged by the Brazilian government to switch from subsistence gardening to growing the cash crop of coffee when they first came in contact with modern civilization in 1969. The necessary constant weeding, performed with machetes, stirred up a fungus in the soil (*Paracoccidiodes brasiliensis*) that infects the lungs when inhaled, sometimes causing death (Discover, 1995).

from home, though recipes are often adapted to accommodate U.S. ingredients or to improve acceptability (e.g., cuy is not often prepared). Substitutions for unavailable ingredients. such as feta cheese for fresh farmer's cheese or peanut oil for dênde oil, are common. Sometimes, the fact that certain traditional items are unobtainable makes other dishes that can be prepared traditionally more popular in the United States than these dishes are in their countries of origin. For instance, *llapingachos* is probably eaten more often by Peruvian Americans than by Peruvians (Novas & Silva, 1997). Among Chileans, it is believed that many find it difficult to adapt to typical American schedules with a work day that begins earlier than in Chile (difficult after a late dinner) and has a short lunch period that precludes a leisurely meal. For poorer Chileans, their diets may improve in the United States where food is less expensive (Burson, 1995).

Nutritional Status

Nutritional Intake

There are minimal data on the nutritional status of Americans of South American descent. Parasitic infection, iron-deficiency anemia, protein-calorie malnutrition are common in many rural areas of South America and in some crowded urban neighborhoods as well. Chronic Chagas' disease involving the esophagus and colon is endemic in some regions, and is considered a risk factor in cardiovascular disease in Brazil (Dominguez et al., 1999). Hypertension rates are believed similar to those in the United States; however, rates in Paraguay are exceptionally high, affecting 39 percent of women and 27 percent of men. Obesity rates are increasing throughout the region. Heart attack mortality in young Brazilian men is reportedly three times that in the United States or Canada.

Counseling

Direct eye contact is common throughout South America, except in Colombia, where eye contact may be avoided with authority figures, elders, or in embarrassing situations. Most South Americans are present oriented and may have relaxed attitudes toward appointment and treatment schedules. Generally speaking, immediate interventions are valued more than either preventive or long-term care. In Argentina, a patient is often protected from a negative prognosis. In Chile, preventive health practices are uncommon, and care is often not sought except in emergency situations.

It is not known how many South Americans continue traditional health practices in the United States. It is thought that the homeopathic remedies and over-the-counter antibiotics sent from family members still in South America may be obtained, and faith healers may be sought. South Americans who are living in the United States illegally usually have no medical insurance and avoid contact with any government bureaucracy, seeking biomedical health care only in emergency situations.

11

Chinese, Japanese, and Koreans

Asia, one of the largest continents, is believed to have been named by the Greeks, who divided the world into two parts: Europe and Asia. It includes China (the People's Republic of China), Taiwan (Republic of China), Japan, the Democratic People's Republic of Korea (North Korea), the Republic of Korea (South Korea), the Mongolian People's Republic, and the countries of Southeast Asia, stretching from its cool northern borders with Europe to the tropical islands of the Pacific (see Figure 11.1).

Immigrants from Asia, particularly China and Japan, have been coming to the United States since the 1800s. Many settled on the West Coast, where the majority of their descendants still live. In recent years, large numbers from throughout the region have arrived in the United States; many are refugees from political oppression, while others seek education and employment opportunities. This chapter introduces the peoples and cuisines of northern and central Asia: the Chinese, Japanese, and Koreans. Southeast Asians are discussed in chapter 12. Peoples of the Middle East and Asian Indians, who reside in parts of Asia Minor and South Asia, respectively, are considered in chapters 13 and 14.

Chinese

Chinese civilization is more than 4,000 years old and has made numerous significant contributions in agriculture, the arts, religion and philosophy, and warfare. Silk and embroidered brocade cloth, intricate jade sculpture, Chinese porcelain and lacquerware, book printing, Confucianism, Taoism, and gunpowder are just a few examples. The name China, meaning "middle kingdom," or center of the world, is probably derived from that of a dynasty that ruled in the 3rd century B.C.E.

China's landscape is dominated by the valleys of two great rivers, the Huang (Yellow) River in the north and the Chang Jiang (Yangtze) in the south. The climate is monsoonal, with most of China's rainfall occurring in the spring and summer months.

The northern plain through which the Huang River flows is agriculturally very fertile. The area is cold and sometimes the severe winter results in a growing season of only 4 to 6 months. In the south the Chang Jiang River starts in Tibet, traverses the southern provinces, and eventually empties into the China Sea near the city of Shanghai. South of the mouth of the Yangtze delta is a rugged and mountainous coastline, off which are located the islands of Hong Kong and Taiwan. The southern provinces are warmer, wetter, and have a growing season that is longer (6 to 9 months) than in the north.

The population of China is estimated to exceed 1 billion people, more than four times as large as the population of the United States. The Chinese have a heterogeneous society with numerous ethnic and racial groups. The Chinese language is equally diverse, with many dialects, some of which are incomprehensible to people of other Chinese regions.

- Nearly half of all U.S. citizens naturalized between 1981 and 1993 were born in Asia.

- Preserved ginger has been found in ancient Chinese tombs, where it was probably placed to nourish the dead.

- The most famous European traveler to China during the rule of the Mongols was Marco Polo, who is said to have brought Chinese noodles to Italy. The Italians, however, were undoubtedly making pasta long before the times of Marco Polo.

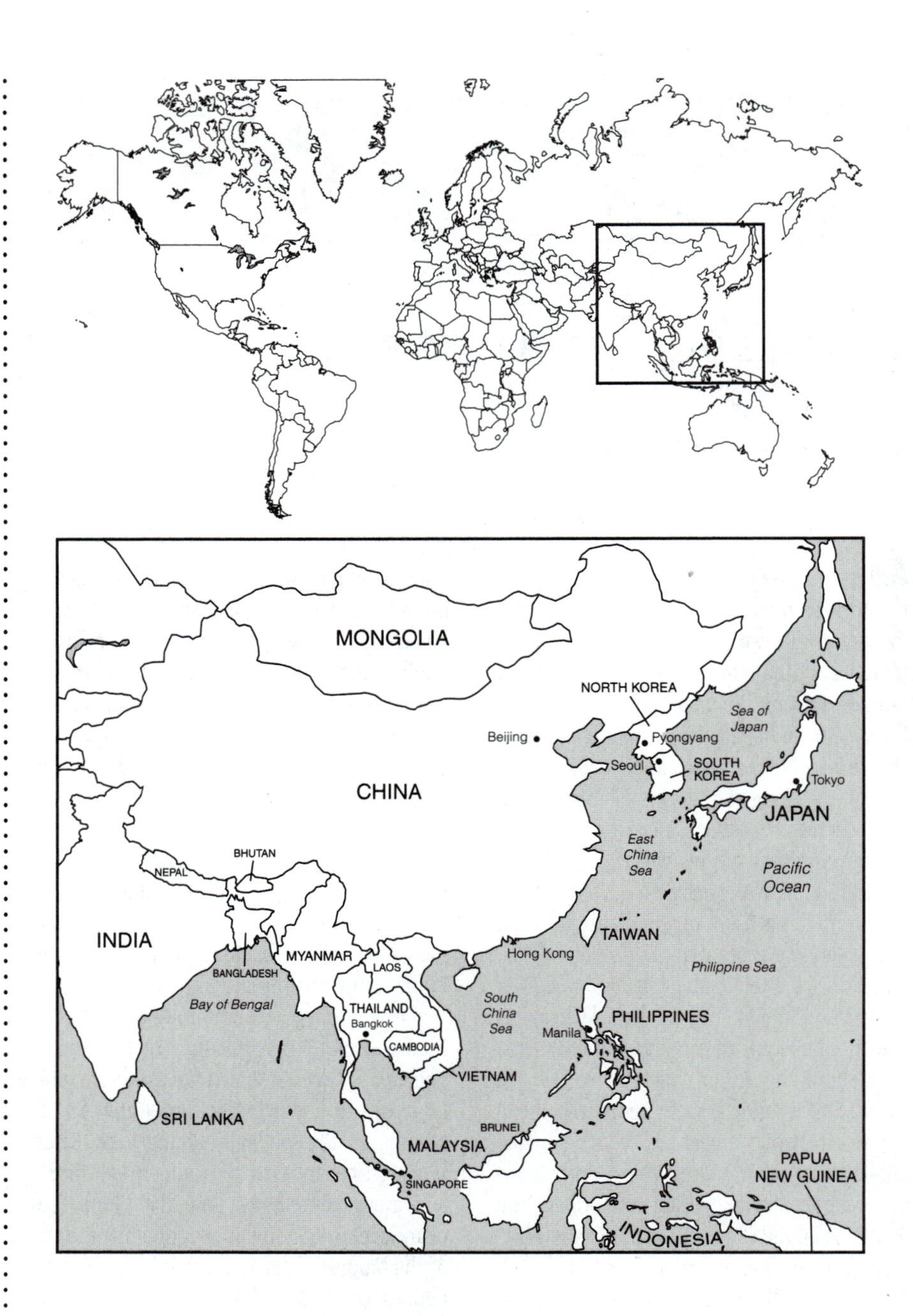

Figure 11.1
China, Japan, and Korea

■ *Chow-chow* is a vegetable relish made with mustard brought to the United States by Chinese railroad workers. Its name may derive from the Chinese word for "mixed," *cha.*

■ It was not until 1970 that the number of Chinese American women equaled the number of men.

Cultural Perspective

History of Chinese in the United States

Immigration Patterns

The first major surge in Chinese immigration to the United States occurred in the early 1850s when the Chinese joined in the gold rush to California; many Chinese still refer to America as the "Land of the Golden Mountain." As mining became less lucrative, the Chinese opened their own businesses, such as laundries and restaurants, but also found employment in other occupations. The Central Pacific Railroad, which joined the Union Pacific as the first cross-country line, was built primarily by 10,000 Chinese workers.

By 1870, there were 63,000 Chinese, mostly male, in the United States, nearly all on the West Coast. It is estimated that another 120,000 Chinese entered the United States during the following decade. Racial discrimination against Asians increased as their numbers

Traditional foods of China. Some foods typical of the traditional Chinese diet include bitter melon, *bok choy,* Chinese eggplant, ginger root, long beans, lotus root, mushrooms, oyster sauce, pork, long-grain rice, shrimp, soy sauce, and water chestnuts. (Photo by Laurie Macfee.)

swelled. By 1880, Chinese immigration slowed to a trickle due to exclusion laws directed against Asians. The Chinese also immigrated to Hawaii, and when the islands were annexed by the United States in 1898, approximately 25,000 Chinese were living there.

Most early Chinese immigrants were from the southeastern Guang-dong province of China, usually referred to as Canton. Most were young men (94 percent) with no intention of staying; they came to make their fortune and then return to China and their families. Many married before coming to the United States, and more than half returned to China. By the 1920s, the Chinese population in the United States had dropped to 1870 levels.

In each city where the Chinese settled, they usually lived within a small geographic area known as Chinatown. Large Chinatowns evolved in San Francisco, New York, Boston, Chicago, Philadelphia, Los Angeles, and Oakland, California. These neighborhoods offered protection against a sometimes hostile social and economic environment; assimilation was not an option. Conditions were often crowded and unusual in the predominance of men, but were tolerated with the expectation of eventual return to China. It was not until 1943 that Chinese could become naturalized U.S. citizens.

Current Demographics and Socioeconomic Status

When the exclusion laws were repealed in 1943, people from many Asian countries once again entered the United States. Chinese immigrants who have arrived since World War II are usually not from Canton. They are urban dwellers from other regions and are generally better educated than earlier immigrants.

Political instability in the People's Republic of China has led to a recent exodus of Chinese. Following the pro-democracy uprising in Tiananmen Square in 1989, U.S. immigration laws were changed with the Chinese Student Protection Act of 1992. More than 41,000 Chinese residents were granted visas and are eligible for citizenship under the provision. Furthermore, the return of Hong Kong to mainland China and the uncertainty of Taiwan's future have led to increased immigration from these islands.

The 1990 U.S. census reports more than 1.6 million Chinese living in the United States,

- In 1852, California governor John Bigler declared the Chinese in his state "nonassimilable."
- Early immigrants and their descendants, as well as many current immigrants from Hong Kong, speak Cantonese, a dialect that is difficult to understand by Chinese immigrants who speak Mandarin.
- Census data report over 922,000 Canadians of Chinese heritage.

with approximately two-thirds being foreign-born; nearly half arrived in the 1990s. Large numbers of Chinese Americans are found today in California, New York, Hawaii, Texas, New Jersey, Massachusetts, Illinois, Washington, and Maryland.

Four types of Chinese American households have been identified (Gould-Martin & Ngin, 1981). First are the sojourners, mostly men born in the early 20th century, who came to the United States with the intent of returning after accumulating some wealth. Political changes in China have stranded this group in urban Chinatowns throughout America. The second type are sojourners who were successful in bringing their wives to the United States. These elderly first-generation couples still live in Chinatown neighborhoods, although their children have often moved from the area.

The third type of household is established by new immigrant families. Often one member will arrive to establish residency, and then family members will follow. These immigrants usually settle with other Asians, often in Chinatowns, where the parents speak Chinese and children learn English in school. Although many families include two wage earners or combine the incomes of extended members, they may remain poor due to low salaries, large families, and support of relatives in the homeland.

Chinese American households of the fourth type include acculturated suburban families. These families include both new immigrants who are white-collar professionals and those Americans of Chinese descent who have lived in the United States for several generations. Both parents may be college educated, often with specialization in engineering and the sciences. Unlike other immigrants of previous generations, these affluent Chinese families live in areas with university or defense research facilities, including the metropolitan areas of the Silicon Valley (California), Houston, Seattle, Pittsburgh, San Diego, and Dallas.

■ Confucius also developed protocols for cooking and eating that were practiced in China until the 19th century.

Chinese Americans value education, and there are disproportionately large numbers with college and graduate degrees. High levels of educational attainment often translate into well-compensated professional employment. Americans of Chinese descent are so well known for their successful transition in the United States that a stereotype has developed, obscuring what is in reality a bipolar population. Although many Chinese Americans are in the upper and middle classes, large numbers, especially recent immigrants, are poorly educated and work in low-paying service jobs.

Worldview

Religion

Most Americans of Chinese descent are not affiliated with a specific church. Religious practices are often eclectic, a combination of ancestor worship, Confucianism, Taoism, and Buddhism. Many early Chinese immigrants were not formally schooled in any religion; instead, beliefs and practices were passed orally from generation to generation. Spirituality is integrated into family and community life. Daily living includes avoiding any actions that might offend the gods, nature, or ancestors.

Early Religion. The ancient faith of China was probably a mixture of ancestor worship and respect for the forces of nature and the heavenly bodies. The supreme power was either *Tien* (Heaven) or *Shang Ti* (the Supreme Ruler or the Ruler Above). One gained favor with the spirits by the correct performance of ceremonies. These beliefs and practices were later incorporated into subsequent Chinese religions.

Ceremonies for the dead are a prominent Chinese religious practice. The dead are supposed to depend on the living for the conditions of their existence after death. In turn, the dead can influence the lives of the living.

Confucianism. Confucius was a sage, one of many who gave order to Chinese society by defining the ways in which people should live and work together. Confucianism incorporated the ceremonies of earlier religions, with the following cornerstones

1. Fatherly love and filial piety in the son (i.e., children are expected to obey their parents and adults are expected to take care of their children)
2. Tolerance in the eldest brother and humility in the younger
3. Proper behavior by the husband and submission by the wife

4. Respect for one's elders and compassion in adults
5. Allegiance to rulers and benevolence by leaders

Inherent in these relationships is the ideal of social reciprocity, which means that one should treat others as one would like to be treated. To enhance harmony in the family and in society as a whole, one must exercise self-restraint. An individual must never lose face—meaning a person's favorable name and position in society—because that would defame the whole family. Many of these values influence Chinese behavior today.

Taoism. The Taoist believes, like the Confucianist, that heaven and humanity function in unison and can achieve harmony, but under Taoism, people are subordinate to nature's way. There is a fundamental duality within the universe of interacting, opposite principles or forces—the *yang* (masculine, positive, bright, steadfast, warm, hard, and dry; sometimes referred to as *Shen*) and the *yin* (feminine, negative, dark, cold, wet, mysterious, and secret; also called *Kwei*) (Figure 11.2). Everything in nature contains both yin and yang, and a balanced unity between them is necessary for harmony. This balance occurs when Tao, the way of nature, is allowed to take its course unimpeded by human willfulness. Taoism advocates a simple life, communion with nature, and the avoidance of extremes.

Buddhism. Buddhism flourished during the T'ang dynasty but then suffered a slow decline. The Mahayana sect dominated in China, blending with traditional Chinese beliefs and resulting in a unique Chinese form of Buddhism. Ten schools of Buddhism flourished in China at one time, but only four were left by the 20th century. The two dominant schools in China are Ch'an (Zen Buddhism in Japan) and Pure Land (see chapter 2 for more information).

Chinese American Spirituality. Both Catholic and Protestant churches were established in the early Chinatown neighborhoods, usually organized by the Chinese dialect spoken in the area. Few first-generation Chinese Americans joined Christian religions, but converts were found in subsequent generations.

Family

Confucian teachings about correct relationships are still important for many Chinese American families, even if they have become Christians. Chinese American families are usually patriarchal. Women are traditionally taught to be unassuming and yielding. They live by the formula of "thrice obeying": Young girls are submissive to their fathers, wives are subordinate to their husbands, and mothers obey their sons. Children are expected to be quiet, acquiescent, and deferential to their elders. Harmony in the family is the ideal, so children are taught not to fight or cry. Showing emotion is discouraged. Chinese parents may be very strict, and children are commanded to honor the family. Many of these ideas conflict with American ideals of equal rights and freedom of speech and lead to intergenerational conflict in the Chinese American home.

Traditional Health Beliefs and Practices

Chinese medicine includes a complex humoral system of professional practice by physicians as well as correlated folk remedies used by laypersons at home. Health beliefs and practices have developed over generations, incorporating Confucian, Taoist, and Buddhist concepts regarding the interdependencies of man and nature and the need for harmony, balance, and moderation of life. The influence of ancient spiritual and magical beliefs is minimal in mainland China, although more often practiced in other regions of Chinese populations, such as Hong Kong and Taiwan.

Professional practice follows texts prepared between approximately 2500 B.C.E. to the 3rd century B.C.E., outlining the dynamic equilibrium of forces necessary for health. These include the five elements, or five evolving phases, of fire, earth, metal, water, and wood, each of which may become unbalanced, much as fire consumes wood, or wood (as a tree) absorbs the earth. These elements correspond with five organs: the heart, spleen, lungs, kidneys, and gallbladder, respectively. Associations with secretions (perspiration, saliva, mucous, spit, and tears); the seasons (summer, late summer, autumn, winter, and spring); colors (red, yellow, white, blue, and green); tastes (bitter, sweet, pungent, salty, and sour); and

Figure 11.2
Yin-yang symbol. This symbol represents the fundamental duality of the universe and the balance between the forces of yang (light) and yin (dark). Each force has a little of the other in it (indicated by the dot of the opposite color).

■ Most Chinese believe in *fengsui,* the way in which a home should be situated and its furnishings arranged to promote optimal flow of energy and personal well-being.

■ Chinese physicians were traditionally paid for their services when the client was healthy. Payment stopped if the client became ill.

directions (south, center, west, north, and east), as well as times of day, odors, sounds, and emotions may also occur (Gould-Martin & Ngin, 1981; Ludman et al., 1989; Sheikh & Sheikh, 1989).

This system was further elaborated by the adoption of Buddhist principles of hot and cold humoral medicine somewhere between the third and sixth centuries, which were congruent with the Taoist system of yin and yang (Anderson, 1987). The concept of harmony was refined to include a balance of these opposites; illness develops when disequilibrium occurs. Organs such as the liver, heart, spleen, kidneys, and lungs are yin, as is the outside and the front of the body. The gallbladder, stomach, intestines, and bladder are yang, as well as the body surface and the back. Outside forces, such as the seasons, are also defined as yin (winter/spring) and yang (summer/fall) and illnesses associated with these times may fall into corresponding categories.

Symptoms of disease usually reflect an imbalance between yin and yang. When there is an excess of yang, acne, rashes, conjunctivitis, hemorrhoids, constipation, diarrhea, coughing, sore throat, ear infections, fever, or hypertension may occur. Anemia, colds, flu, frequent urination, nausea, shortness of breath, weakness, and weight loss suggest that an excess of yin is the problem. Also associated with yin and yang is the condition of the blood. "Weak" blood (yin) may develop during growth, pregnancy, postpartum, and in old age; treatment includes yang therapies, particularly the intake of herbs and certain foods.

The vital force of life is *qi* and is equated with "energy," "air," and "breath." Qi flows along defined meridians in the body, and some conditions are related to the disruption of qi or to excessive qi. Other types of energy that must be balanced for health include *jing,* sexual energy, and *sheng,* spiritual energy (Koo, 1984).

■ Jade charms are worn to keep children safe and to bestow health, fertility, long life, power, and wisdom on adults.

Other lesser forces that may influence health include "wind" (including natural drafts and those resulting from fans, air conditioners, exposure, or a symptom that rhymes with the Chinese word for wind); "poison," which is somewhat related to the western concept of allergies; and "fright," a condition in children that includes listlessness, anorexia, low fever, and crying. Fright is believed to occur when the soul becomes scattered and is mostly limited to Chinese from Taiwan and Hong Kong (Gould-Martin & Ngin, 1981).

A major difference between Chinese medicine and U.S. biomedicine is the idea that the body and mind are unified, governed by the heart. There is no American word to describe the concept. Emotions are often somaticized, meaning that feelings are related to specific conditions. More than 500 symptoms corresponding to emotions have been identified, each characteristic of one or more organs. For example, *tou yun* (or *tou hun*) is vertigo, the most common complaint made by Chinese patients worldwide. Dizziness or a confused state of mind indicates significant imbalance and serious illness. It is a nonspecific condition thought to originate from anger or anxiety manifested in liver, heart, or kidney dysfunction (if the patient is a young man, too much sexual intercourse or masturbation may be believed to be the cause). Liver disorders develop from suppressed hostility. Anger is discouraged in Chinese culture and may accumulate in the liver, causing it to expand and attack other organs. This diagnosis is common for many gastrointestinal complaints. Generalized stomachaches are believed to be due to eating "bitterness" in life, often including an inadequate diet when one is young. Anxiety, nervousness, or the stronger emotion of fear results in heart palpitations (Ots, 1990).

The Chinese maintain health through a properly balanced diet, moderation in activities and sleep, and avoidance of sudden imbalance that might be caused by forces such as wind. Qi must flow freely and blood must be strengthened through nourishment. For instance, if an individual is imprudent and celebrates too much by eating excessive yang foods (see the section "Therapeutic Uses of Foods"), indigestion or a hangover may occur. Eating yin foods, which are often bland, is the remedy (Anderson, 1987).

When home remedies are ineffective, advice from a Chinese physician is traditionally sought. Treatment for nearly all illness involves the restoration of harmony. Therapy may emphasize dietary and lifestyle changes, or attempt to balance the organs so that emotional balance results. Nearly every visit to the doctor results in a botanical remedy; most medicinal herbs are only available through prescription. After asking about symptoms, examining the tongue, and evaluating the pulse, the doctor will determine the proper mixture of plant, animal, and mineral products to cure the disease. The prescription is boiled and the tea is consumed in a single dose. If symptoms continue, another visit to the practitioner is warranted. Patent medicines, especially those imported from China, are recommended for some illnesses (see Table 11.1, "Selected Botanical Remedies in Traditional Chinese Medicine").

Acupuncture is another traditional Chinese treatment. It involves the use of nine types of exceptionally thin metal needles inserted at various points on the body where the qi meridians surface. Meridians are considered yin or yang, and correspond with specific organs. The needles are placed to facilitate a balanced flow of qi, restoring harmony to the afflicted organ, mostly for symptoms of excess yang. Acupuncture may be performed by a Chinese doctor or by a specialist.

Another, less common therapy, is moxibustion. Small bundles of dried wormwood are heated and carefully applied to certain meridians, usually to balance a yin condition. Moxibustion is particularly used during labor and delivery. Massage or therapeutic exercise are other traditional therapies, found more often in China and rarely in the United States.

Chinese doctors practicing in the United States do not have specified training and competency varies. Word-of-mouth recommendations are common within the Chinese community. Chinese doctors may use first aid on injuries or broken bones, prescribe herbs, perform acupuncture or moxibustion, or they may diagnose the condition and provide a recommended course of therapy by a specialist in one of these practices. Asians interested in traditional medicine use Chinese doctors (often concurrently with biomedical therapies) and in recent years these practitioners have attracted a growing multi-ethnic clientele.

Traditional Food Habits

The Chinese eat a wide variety of foods and avoid very few. This may have developed out of necessity, as China has long been plagued with

Botanical remedies are usually combined in formulary mixtures in traditional Chinese medicine.

recurrent famine caused by too much or too little rainfall. Chinese cuisine largely reflects the food habits and preferences of the Han people, the largest ethnic group in China, but not to the exclusion of other ethnic groups' cuisines. For example, Beijing has a large Muslim population, whose restaurants serve lamb, kid (baby goat), horse meat, and donkey, but no pork. Foreigners have also introduced ingredients that have been incorporated into local cuisines. Some foods now common in China, but not indigenous, are tomatoes, potatoes, and chile peppers.

Ingredients and Common Foods

Staples

Traditional Chinese foods are listed in Table 11.2. In China, few dairy products, whether fresh or fermented, are eaten. Rice, the staple food of southern China, is used throughout Asia and is believed to have originated in India (it wasn't introduced to China until the 1st century B.C.E.). There are approximately 2,500 different forms of the grain, but the Chinese prefer a polished, white long-grain variety that is not sticky and that remains firm after cooking. Sticky glutinous rice is used occasionally, mainly in sweet dishes. Although it is usually steamed, rice can also be made into a porridge called *congee,* eaten for breakfast or as a late-night snack, with meat or fish added for flavor. Congee is also fed to people who are ill. Rice flour is used to make rice sticks, which can be boiled or fried in hot oil.

■ It is said of the Cantonese that they will eat everything with four legs except a table and anything with wings except an airplane.

■ The Chinese raise many kinds of insects for consumption, such as scorpions, which are prepared fried or in soups.

■ The Chinese, like most Asians, rinse their rice before cooking it.

■ The custom of throwing rice at newlyweds is believed to come from China, where rice is a symbol of fertility.

Table 11.1
Selected Botanical Remedies in Traditional Chinese Medicine (mostly used in formulary mixtures)

Scientific Name	Common Name	Property	Parts	Traditional Use
Aconitum carmichaeli ☠	Fu zi; aconite	Hot	Root	Metabolic problems; weak heart
Alisma plantago-aquaticae; Alismatis plantago-aquaticae	Ze xie; water plantain	Cold	Bulbs	Diabetes; kidney stones; urinary tract problems; diarrhea; abdominal bloating; pelvic infections
Anemarrhena asphodeloidis	Zhi mu	Cold	Root	Diabetes; hypertension; urinary tract infections; AIDS; anorexia
Angelica sinensis ☹	Dong quai; tang gui; angelica	Warm	Root	Blood toxins; menstrual irregularities; headaches
Arctium lappa	Niu bang zi; budock	Cold	Fruit	Cough with phlegm; rashes
Arisaema amurense, A. consanguineum, A. heterophyllum ☠	Tain nan xing; jack-in-the-pulpit; cobra lily	Warm	Root	Stroke; seizures; dizziness; numb hands/feet; spasms of hands/feet; excessive phlegm; insomnia
Aristolochia spp. ☠	Guang fang ji; snakeroot		Root	Coughs
Asparagus lucidus, A. cochinchinesis	Tian men dong; wild asparagus	Cold	Root	Diabetes; coughs; poor memory
Astragalus membranaceus ☹	Huang qi; milk vetch	Warm	Root	Diabetes; urinary tract infections; obesity; edema
Atractylodes lancea, A. macrocephala, A. ovata	Cang zhu; bai zhu	Warm	Root	Diabetes; hypertension; urinary tract infection; indigestion; anorexia
Bupleurum chinense	Chai hu	Cool	Root	Hypertension; liver ailments; gallstones; obesity; metabolic problems
Chrysanthemum indici	Ye ju hua; wild chrysanthemum	Cool	Flowers	Hypertension
Citrus reticulata ☹	Chen pi; tangerine	Warm	Peel	Hypertension; poor digestion; obesity
Codonopsis pilosula, C. lanceolate	Dang shen; bastard ginseng; poor man's ginseng	Neutral	Root	Fatigue; anorexia; diarrhea; blood toxins; lactation stimulant
Ephedra sinica ☠	Ma huang; ephedra	Warm	Root; stems	Asthma; allergies; bronchitis; night sweats
Gentiana scabra ☹	Long dan cao; Chinese gentian; ryntem root	Cold	Root	Hypertension; urinary tract infections; indigestion; genital pain; arthritis; conjunctivitis; fever; colds
Gingko biloba ☹	Ying xing; gingko nut	Neutral	Seeds	Asthma; coughs; TB; urinary tract infections
Glycyrrhiza uralenis ☠	Gan cao; licorice	Neutral	Root	Spleen ailments; anorexia; fatigue; cough; sore throat
Leonuris heterophylli ☹	Yi mu cao; motherwort	Cold	Whole plant	Spleen ailments; anorexia; fatigue; cough; sore throat
Lycium chinensis; L. barbarum	Gou qi zi: wolfberry; matrimony vine; boxthorn	Neutral	Fruit	Cancer; diabetes; hypertension; kidney ailments; lung problems; blood toxins; fatigue; blurred vision; impotence
Nelumbo nucifera; Nymphea lotus	Lian zi; he ye: lotus	Neutral	Root, leaves, flowers, seeds	Heart palpitations; dysentery; diarrhea; dizziness; insomnia; premature ejaculation
Ophiopogonis japonica	Mai men dong	Cold	Root	Diabetes; thirst; insufficient urine; bronchitis
Panax ginseng	Ren shen; ginseng	Warm	Root	Stress; fatigue; urinary tract infections; lung problems; nausea

Table 11.1
Selected Botanical Remedies in Traditional Chinese Medicine (mostly used in formulary mixtures)—continued

Scientific Name	Common Name	Property	Parts	Traditional Use
Phellosdendron amurense ☠	Huang bai; amur corktree	Cold	Bark	Urinary tract infections; dysentery; liver ailments; hypertension; symptoms of menopause; dermatitis
Pinella ternata, P. tuberifera	Ban xia	Warm	Root	Hypertension; indigestion; stomach pain; nausea; obesity
Plantago asiatica ☹	Che qian zi; che qian cao	Cold	Seeds; whole plant	Hypertension; lung conditions; diarrhea; urinary tract infections; edema; venereal diseases
Polygonum multiflorum, P. chinesis	He shou wu	Warm	Root	Cancer; weakness; anemia; dizziness; constipation; insomnia; hair loss
Prunella vulgaris	Xia ku cao; selfheal; all heal	Cold	Flowers	Hypertension; cancer; headache; vertigo; goiter
Prunus armeniaca ☠	Xing ren; ku xing ren; apricot pit	Warm	Kernel	Cancer; obesity; constipation; asthma
Pueraria lobata	Ge gen; kudzu	Cold	Root	Diabetes; hypertension; fever; thirst; headache
Rehmannia glutinosa	Sheng di; shu di huang; Chinese foxglove	Cold (raw); warm (steamed)	Root	Heart palpitations; hypertension; diabetes; liver conditions; kidney ailments; hemorrhage; irregular menstruation; fever; thirst; vertigo; lumbago; irritability; insomnia; nosebleed
Rheum palmatum, R. tanguticum	Da huang; rhubarb	Cold	Root	Hypertension; liver ailments; gallstones; constipation; diarrhea; amenorrhea; hemorrhage
Salvia miltiorrhiza ☠	Da shen; salvia	Cold	Root	Heart disease; heart palpitations; menstrual irregularities; insomnia
Schizandra chinensis ☹	Wu wei zi	Warm	Fruit	Diabetes; heart palpitations; asthma; urinary tract problems; diarrhea; insomnia; premature ejaculation
Scutellarea baicalensis ☠	Huang qin; Chinese skullcap; golden root	Cold	Root	Hypertension; gallstones; constipation; threatened miscarriage; nausea; headache; coughs; insomnia
Stephania tetrandra	Han fang ji	Cold	Root	Hypertension; obesity; edema; arthritis; rheumatism
Tribulus terrestris; T. lanuginosa	Bai ji li; ci ji li; ji li; puncture vine; goat's head	Warm	Fruit	Hypertension; chest pain; liver conditions; anorexia; headache, dizziness
Uncaria rhynchophylla ☹	Gou teng; gambir	Cool	Stems	Hypertension; liver ailments; fever; convulsions; headaches; dizziness; blood toxins during pregnancy
Xanthium strumarium ☠	Can er zi; cocklebur	Warm	Fruit	Headache; arthritis; rhinitis; lumbago
Zingiberis officianalis ☹	Shen jiang; ginger root	Hot	Root	Nausea; vomiting; headache
Ziziphus jujuba; Z. spinosa	Da zao; suan zao ren; jujube; Chinese date	Neutral	Fruit, seeds	Stress; anxiety; anorexia; poor digestion; anemia; fatigue; insomnia; night sweats; hysteria

Note: Data on some plants are very limited; adverse effects may occur even if not indicated.
Key: ☠, all or some parts reported to be harmful or toxic in large quantities or specific preparations; ☹, may be contraindicated in some medical conditions/with certain prescription drugs.

■ Chinese recipes were recorded as early as the 9th century B.C.E.

■ The Chinese consider the dark meat of poultry tastier and more tender than the white meat.

■ Cabbage sales in China have plummeted as the standard of living has increased. Tomatoes, cucumbers, and celery now outsell the formerly number one vegetable.

■ In some regions, two longans or litchis are placed under the pillows of newlyweds because their names sound like the words "to have children quickly" (Newman, 1998b).

■ In China a marriage is consummated when the new in-laws officially accept a cup of tea from the bride. If the groom dies before the wedding, the bride-to-be is said to have "spilled her tea."

Wheat is also common throughout China, although it is used more often in the north than the south. It is popular as noodles, dumplings, pancakes, and steamed bread. Thin, square wheat-flour wrappers are used to make steamed or fried *egg rolls* with a meat, vegetable, or mixed filling and *wontons* (in which the wrapper is folded over the filling), served either fried or in soup. *Spring rolls,* which are similar to egg rolls, are made with very thin, round wheat-flour wrappers. Buckwheat is grown in the north and commonly made into noodles.

The Chinese eat a variety of animal protein foods. Although they eat all kinds of meat, fish, and poultry, they eat less at any one meal than is customary in the West. In China, soybeans are known as the poor man's cow, as they are made into products that resemble milk and cheese. Soybeans are transformed into an amazing array of food products that are indispensable in Chinese cooking:

Soy sauce, made from cooked soybeans that are first fermented and then processed into sauce. The Cantonese prefer light-colored soy sauce in some dishes over the darker, more opaque kind used in Japanese and some Chinese cooking.

Soy milk, prepared with soaked soybeans that are first pureed, then filtered, and then boiled to produce a white, milklike drink.

Bean curd, or *tofu,* made by boiling soy milk and then adding gypsum, which causes it to curdle. The excess liquid is pressed from the curd, producing a firm, bland, custardlike product.

Black beans, made with cooked fermented soybeans preserved with salt and ginger. Black beans are usually added as a flavoring in dishes.

Hoisin sauce, a thick, brownish-red sweet-and-sour sauce that combines fermented soybeans, flour, sugar, water, spices, and garlic with chiles and is often used in Cantonese cooking.

Oyster sauce, a thick brown sauce prepared from oysters, soybeans, and brine that is also used in Cantonese fare.

Other beans are also popular, made into pastes, flour, or even thin, transparent noodles known in the United States as cellophane noodles or bean threads.

Chinese cuisine makes extensive use of vegetables. Many are those common in other regions of the world, such as asparagus, broccoli, cabbage, cauliflower, eggplant, green beans, mushrooms, onions, peas, potatoes, radish, and squash. Chinese varieties may differ, however. For instance, leafy *bok choy* and wrinkled *napa* cabbage are preferred over European types; long beans, small purple eggplant, and the large white icicle radish are featured in many dishes. One popular squash variety is called a winter melon; it is pale green and mild in flavor. Mushrooms of all types, including the tiny *enoki,* grayish oyster mushrooms, *shiitake,* straw mushrooms, and dried kinds such as cloud (or wood) ears flavor numerous dishes. Lily buds, snow peas (pea pods), bamboo shoots, water chestnuts, bitter melon, and lotus root are other, more distinctively Asian vegetables found in Chinese cuisine.

The Chinese eat fresh fruit infrequently, occasionally for a snack or for dessert, and it is preferred slightly unripe or even salted. Chinese dates (*jujubes*), persimmons, pomegranates, and tangerines are favorites. A few fruits are typically preserved in syrups, such as pungent kumquats, yellow-orange loquats, longans (dragon eyes) and litchis, a tropical fruit with creamy, jellylike flesh.

Traditionally people cooked with lard if they could afford it. In recent years, soy, peanut, or corn oil is more common. Until recently sugar was not used in large quantities; many desserts were made with sweetened bean pastes.

Hot soup or tea is the beverage that usually accompanies a meal. Tea, used in China for more than 2,000 years, was first cultivated in the Chang Jiang valley and later introduced to western Europe in the 17th century. There are three general types of tea: green, black (red), and *oolong* (black dragon). Green tea is the dried, tender leaves of the tea plant. It brews a yellow, slightly astringent drink. Black tea is toasted, fermented leaves that have a black color; it makes a reddish drink. Black tea is commonly drunk in Europe and America. Oolong tea is made from partially fermented leaves.

Chinese alcoholic drinks are often called wines, but they are not usually made from grapes. They are typically distilled alcohols made from grains or from fruit, like plums. A few examples are bamboo leaf-green (95 proof), *fen* (made from rice, 130 proof), *hua diao* (yellow rice wine), *mou tai* (made from sorghum, 110 proof), and red rice wine. Beer is also very popular.

Most Chinese food is cooked, and very little raw food, except fruit, is eaten. Cooked foods may be eaten cold. Common cooking methods maximize the limited fuel available, including stir-frying, steaming, deep-fat frying, simmering, and roasting. In stir-frying, foods are cut into uniform, bite-size pieces and quickly cooked in a *wok* (a hemispherical shell of iron or steel) in which oil has been heated. The wok is placed over a gas burner or in a metal ring placed over an electric burner. Food can also be steamed in the wok. Bamboo containers, perforated on the bottom, are stacked in a wok containing boiling water and fitted with a domed cover. Roasted food is usually bought from a commercial shop, not prepared in the home.

The Chinese usually strive to obtain the freshest ingredients for their meals, and in most American Chinatowns it is common to find markets that sell live animals and fish. However, because of seasonal availability and geographic distances, many Chinese foods are preserved by drying or pickling.

Regional Variations

China is usually divided into five culinary regions characterized by flavor or into two areas (northern and southern) based on climate and the availability of foodstuffs. In recent years, however, regional differences have diminished due to increased global influences, particularly television (Newman, 1998a).

Northern. This area includes the Shandong and Honan regions of Chinese cooking. The Shandong area (Beijing is sometimes included in this area, sometimes considered a third division of northern cooking) is famous for *Peking duck* and *mu shu pork,* both of which are eaten wrapped in Mandarin wheat pancakes topped with hoisin. Honan, south of Beijing, is known for its sweet-and-sour freshwater fish, made from whole carp caught in the Huang River. Much of the north is bordered by

■ Thousand-year-old eggs (also called hundred-year-old, century, and pine flower eggs) are duck or chicken eggs cured for 3 months in a lime, ash, and salt mixture. The whites become black and gelatinous; the yolks turn greenish.

Mongolian Fare

The Mongolians once ruled an empire that stretched from China to Europe. In most recent times, it has been colonized by Russia and China. Today, it is an independent nation reestablishing its cultural identity through shared language, customs, and cuisine.

Historically, Mongolians consumed red foods (meat) and white foods (dairy), and this tradition continues today with the addition of some grain products. Meats, especially mutton, goat, and beef, are favorites (camel meat is eaten when available, though it is banned in some areas). Meat is enjoyed barbecued on a grill or over charcoal in a specially designed pot that sits on the table. It is also added to soups, stuffed into pancakes, and served on sesame seed buns (Newman, 2000b). Dairy foods are numerous, prepared from cow, sheep, goat, or camel's milk. There are three types of butter (liquid, yellow, and white), a type of milk "tofu," sour milk (similar to yogurt), milk leather (made from the milk film skimmed off boiled milk and air-dried), and fresh cheese. Milk is added to tea, called Mongolian tea, sometimes with a little salt or fried millet (Ang, 2000). Cheese is mixed with sugar and flour, then baked, to make a dessert known as milk pie. *Kumys,* a wine distilled from fermented milk (traditionally mare's milk), is a Mongolian specialty and consumed at many occasions.

Millet is the staple grain in Mongolia. It is cooked like a porridge or roasted until it pops like popcorn. Flour is made from millet, buckwheat, or wheat and cooked as fried pancakes or steamed flat breads. Tea is consumed at every meal and with snacks. Three meals a day are typical, consumed with the fingers. Special occasions include Lunar New Years and the *Naadam* festival, a three-day event featuring wrestling, archery, and horse races.

Table 11.2
Cultural Food Groups: Chinese

Group	Comments
PROTEIN FOODS	
Milk/milk products	Dairy products are not routinely used in China. Many Chinese are lactose-intolerant. Traditional alternative sources of calcium are tofu, calcium-fortified soy milk, small bones in fish and poultry, and dishes in which bones have been dissolved.
Meat/poultry/fish/eggs/legumes	Few protein-rich foods are not eaten. Beef and pork are usually cut into bite-size pieces before cooking. Fish is preferred fresh and is often prepared whole and divided into portions at the table. Preservation by salting and drying is common. Shrimp and legumes are made into pastes.
CEREALS/GRAINS	Wheat is the staple in the north, white rice in the south. *Fan* (cereal or grain) is the primary item of the meal; *ts'ai* (vegetables and meat or seafood) makes it tastier. Rice is washed before cooking.
FRUITS/VEGETABLES	Many nonAsian fruits and vegetables such as bananas and chile peppers are popular. Potatoes, however, are not well-accepted. Vegetables are usually cut into bite-size pieces before cooking. Slightly unripe fruit is often served as a dessert. Both fresh fruits and vegetables are preferred; seasonal variation dictates the type of produce used. Many vegetables are pickled or preserved. Fruits are often dried or preserved.
ADDITIONAL FOODS	
Seasonings	Complex, sophisticated seasoning combinations common. Various tastes appreciated, such as the moldy flavor of lily flower buds. Spice and herb preferences distinguish regional cuisines.
Nuts/seeds	Nuts and seeds are popular snacks and may be colored or flavored.
Beverages	In northern China the beverage accompanying the meal may be soup. In the south, it is tea. Alcoholic drinks, usually called wines, are rarely made from grapes. They are either beers or distilled spirits made from starches or fruit.
Fats/oils	Traditionally lard was used if it could be afforded. In recent years, soy, peanut, or corn oil is more commonly used.
Sweeteners	Sugar was not used in large quantities; many desserts were made with bean pastes.

Table 11.2
Cultural Food Groups: Chinese—continued

Common Foods	Adaptations in the United States
Cow's milk, buffalo milk	Most Chinese, even elders, consume dairy products, especially milk and ice cream, some cheese. Some alternative sources of calcium may no longer be used.
Meat: beef and lamb (brains, heart, kidneys, liver, tongue, tripe, oxtails); pork (bacon, ham, roasts, pig's feet, sausage, ears); game meats (e.g. bear, moose) *Poultry:* chicken, duck, quail, rice birds, squab *Fish:* bluegill, carp, catfish, cod, dace, fish tripe, herring, king fish, mandarin fish, minnow, mullet, perch, red snapper, river bass, salmon, sea bass, sea bream, sea perch, shad, sole, sturgeon, tuna *Eggs:* chicken, duck, quail, fresh and preserved *Shellfish and other seafood:* abalone, clams, conch, crab, jellyfish, lobster, mussels, oysters, periwinkles, prawns, sea cucumbers (sea slugs), shark's fin, shrimp, squid, turtle, *wawa* fish (salamander) *Legumes:* broad beans, cowpeas, horse beans, mung beans, red beans, red kidney beans, split peas, soybeans, white beans, bean paste	
Buckwheat, corn, millet, rice, sorghum, wheat	Chinese Americans eat less *fan* and more *ts'ai.* The primary staple remains rice, but more wheat bread is eaten.
Fruits: apples, bananas, custard apples, coconut, dates, dragon eyes (*longan*), figs, grapes, kumquats, lily seed, lime, litchi, mango, muskmelon, oranges, papaya, passion fruit, peaches, persimmons, pineapples, plums (fresh and preserved), pomegranates, pomelos, tangerines, watermelon *Vegetables:* amaranth, asparagus, bamboo shoots, banana squash, bean sprouts, bitter melon, broccoli, burdock root, cassava (tapioca), cauliflower, celery, cabbage (*bok choy* and *napa*), chile peppers, Chinese long beans, Chinese mustard, chrysanthemum greens, cucumbers, eggplant, flat beans, fuzzy melon, garlic, ginger root, green peppers, kohlrabi, leeks, lettuce, lily blossoms, lily root, lotus root and stems, luffa, dried and fresh mushrooms (black, button, cloud ear, wood ear, *enoki, shiitake,* straw, oyster, monkey's head), mustard root, okra, olives, onions (yellow, scallions, shallots), parsnip, peas, potato, pumpkin, seaweed (agar), snow peas, spinach, taro, tea melon, tomatoes, turnips, water chestnuts, watercress, wax beans, winter melon, yams, yam beans	More raw vegetables/salads are eaten.
Anise, bird's nest, chili sauce, Chinese parsley (cilantro), cinnamon, cloves, cumin, curry powder, five-spice powder (anise, star anise, clove, cinnamon or cassia, Szechuan pepper), fennel, fish sauce, garlic, ginger, golden needles (lily flowers), green onions, hot mustard, mace, monosodium glutamate (MSG), mustard seed, nutmeg, oyster sauce, parsley, pastes (*boisin,* sweet flour, brown bean, Szechwan hot beans, sesame seed, shrimp), pepper (black, chile, red, and Szechwan), red dates, sesame seeds (black and white), soy sauce (light and dark), star anise, tangerine skin, tumeric, vinegar	Many Chinese restaurants use MSG, but it is not usually used in the home.
Almonds, apricot kernels, areca nuts, cashews, chestnuts, ginkgo nuts, peanuts, walnuts; sesame seeds, watermelon seeds	
Beer, distilled alcoholic spirits, soup broth, tea	
Bacon fat, butter, lard, corn oil, peanut oil, sesame oil, soybean oil, suet	
Honey, maltose syrup, table sugar (brown and white)	Sugar consumption has increased mainly because of consumption of soft drinks, candy, and American desserts.

Mongolia, whose people eat mainly mutton. Grilling or barbecuing is a common way of preparing meat in this area. One specialty is the Mongolian hot pot, featuring sliced meats and vegetables cooked at the table in a pot of broth simmering over a charcoal brazier. The food is eaten first, and the broth is consumed as a beverage afterward. One hundred people were killed in China during 1995 by exploding Mongolian hot pots in restaurants. A government crackdown is addressing safety issues and the use of opium in hot pot dishes. (See also the box on "Mongolian Fare.")

Northern China has a cool climate, limiting the amount and type of food produced. Traditionally, foods were often preserved, resulting in a preference for salty flavors. In general, its staples are millet, sorghum, and soybeans. Winter vegetables are common, such as cabbage, turnips, and onions. A delicacy from this area is braised bear paw. Hot clear soup is the beverage that usually accompanies a meal.

Southern. Southern China is divided into three culinary areas: Szechwan (Sichuan)-Hunan, Yunnan, and Cantonese (with Fukien and Hakka regional specialties). Szechwan-Hunan, which is an inland region, features fare distinguished by the use of chiles, garlic, and the Szechwan pepper *fagara.* Typical dishes include *hot and sour soup, camphor-* and *tea-smoked duck,* and an oily walnut paste and sugar dessert that may be related to the nut halvah of the Middle East. Yunnan cooking is distinctive in its use of dairy products, such as yogurt, fried milk curd, and cheese. Dishes are often hot and spicy, and some of the best ham and head cheese in China are found in this area.

Cantonese cooking is probably the most familiar to Americans, because the majority of Chinese restaurants in the United States serve Cantonese-style food. It is characterized by stir-fried dishes, seafood (fresh and dried or salted), delicate thickened sauces, and the use of vegetable oil instead of lard. The Cantonese are known for *dim sum* ("small bites," such as steamed or fried dumplings stuffed with meat or seafood) that are served with tea. Tea is the beverage served with meals. The staple foods of the south are rice and soybeans. As in the north, a variety of vegetables from the cabbage family are used, as well as garlic, melon, onions, peas, potatoes, squashes, tomatoes, and a range of rootlike crops, such as taro, water chestnuts, and lotus root. Southern cooking uses mushrooms of many types to enhance the flavor of the foods and takes advantage of an abundance of fruits and nuts. Fish, both fresh and saltwater, is popular. Also important are poultry and eggs. Pork is the preferred meat.

Along the coast, Fukien provincial fare includes numerous seafood dishes and clear broths. Paper-wrapped foods and egg rolls are thought to have originated there. In the city of Shanghai, chefs specialize in new food creations and elaborate garnishes.

A southern regional specialty is Hakka cuisine, sometimes called the soul food of southern China. The Hakkas fled to the south in the 4th century B.C.E., when the Mongols invaded the north. They remained an insular ethnic group, preserving their traditional language, dress, and foods. Their fare is hearty and robust, featuring dishes made with red rice wine and pungent seasonings, cooked for a lengthy time, often in clay pots. Salt-baked chicken, greens simmered with pork fat, and meat-stuffed tofu are examples.

Tibetan. The Chinese-controlled province of Tibet has a unique fare due to the isolation provided by its locale in the Himalayan mountains. The foundation of the diet is *zampa,* a toasted flour produced from barley or buckwheat. It is traditionally mixed with the butter obtained from yak's, cow's, or sheep's milk (called "crispy oil"), sugar, milk or cream, and sometimes tea to make flattened balls consumed with tea or soup. The zampa can also be used to make *momos,* a Tibetan dumpling filled with meat. Yak and mutton are common, but most Tibetans who are Buddhist do not eat pork, poultry, or fish (Newman, 1999b). Dairy products are also prevalent. Butter-tea, made by churning crispy oil, milk, and salt with brewed tea, is consumed throughout the day. Sour milk, milk solids preserved from the crispy oil process, and the milk film skimmed from boiled milk, then dried, are all consumed. Cabbage, radishes, onions, garlic, leeks, and potatoes are available. Wine, made from barley or buckwheat, is served at special occasions.

■ In southern China, people greet each other by asking, "Have you had rice today?"

■ Bird's nest soup is served sometimes at Chinese banquets. It is made from the cleansed nests of swallows from the South China Sea. The nests are made of predigested seaweed and are bland in flavor yet very expensive. They are reputedly an aphrodisiac.

■ Dumplings shaped like animals, such as birds or frogs, are specialties in Xian. A dim sum banquet may feature several dozen different varieties.

■ "Red foods" are a specialty in the Fukien province due to the use of red wine paste (a sediment remaining after the fermentation of rice wine) on pork, poultry, in soups and even dumpling dough (Newman, 1999a).

Meal Composition and Cycle

Daily Patterns

The Chinese customarily eat three meals per day, plus numerous snacks. Breakfast often includes the hot rice porridge congee, which in southern China may be seasoned with small amounts of meat or fish. In northern China, hot steamed bread, deep-fried crullers, dumplings, or noodles are served for breakfast. In urban areas, lunch is a smaller version of dinner, including soup, a rice or wheat dish, vegetables, and fish or meat. Sliced fruit may be offered at the end of the meal.

Although the Chinese are receptive to all types of food, the composition of a meal is governed by specific rules—a balance between yin and yang foods and the proper amounts of *fan* and *ts'ai.* Fan includes all foods made from grains, such as steamed rice, noodles, porridge, pancakes, or dumplings, which are served in a separate bowl to each diner. Ts'ai includes cooked meats and vegetables, which are shared from bowls set in the center of the table. Fan is the primary item in a meal; ts'ai only helps people eat the grain by making the meal more tasty. A meal is not complete unless it contains fan, but it does not have to contain ts'ai. At a banquet the opposite is true. An elaborate meal must contain ts'ai, but the fan is usually an afterthought and may not be eaten.

Etiquette

The traditional eating utensils are chopsticks and a porcelain spoon used for soup. Teacups are always made out of porcelain, as are rice bowls. At a meal, all diners should take equal amounts of the ts'ai dishes, and younger diners wait to eat until their elders have started; it is rude to reject food. It is also considered bad manners to eat rice or noodles with the bowl resting on the table; instead, it should be raised to the mouth. It is rude to pick at your food or to lick your chopsticks. Laying your chopsticks across the top of the rice bowl or dropping them brings bad luck. It is also improper to stick chopsticks straight up in a rice bowl because in some areas this symbolizes an offering to the dead.

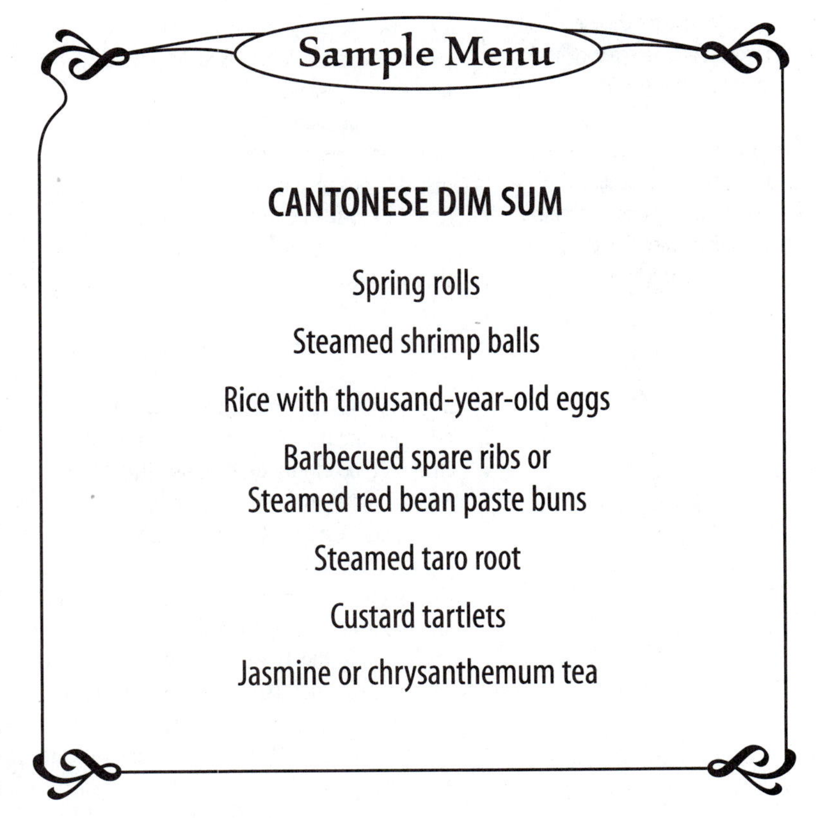

Special Occasions

Traditionally the Chinese week did not include a day of rest. Consequently, there were numerous feasts to break up the continuous work days. Chinese festival days do not fall on the same day each year because their calendar is lunar. Celebrations are traditionally yang occasions, because heat symbolizes activity, noise, and excitement in China (Anderson, 1987). Yang foods, such as meats, fried dishes, and alcoholic beverages, are featured at feasts (see "Therapeutic Uses of Foods").

The most important festival is New Year's, which can fall any time from the end of January to the end of February. Traditionally the New Year was a time to settle old debts and to honor ancestors, parents, and elders. The New Year holiday season begins on the evening of the 23rd day of the last lunar month of the year. At that time the Kitchen God, whose picture hangs in the kitchen and who sees and hears everything in the house, flies upward to make his annual report on the family to the Jade Emperor. To ensure that his report will be good, the family smears his lips with honey or

■ One story about the origin of chopsticks is that they were invented after an ancient Chinese emperor banned the use of cutlery at the table. More likely they were invented as an extension of the fingers. They are made from bamboo, ivory, or plastic. Chopsticks are used in most countries that have been influenced by China, including Japan (where the chopsticks are generally shorter and have rounded rather than squared sides and more pointed tips) and Korea (where the chopsticks are typically made of metal, the same length as the Japanese type, but flatter). Chopsticks are found frequently in Vietnam (the Chinese type), although forks, spoons, and fingers are also commonly used. Other Southeast Asian cultures use chopsticks occasionally, mostly for rice or noodles.

■ According to the Chinese, a child is 1 year old at birth and becomes 2 years old after the New Year.

■ The New Year's dragon dance and firecrackers are thought to inhibit the yin element and promote the yang forces. Red, the color of yang, is used throughout the New Year's season.

■ Eating crab and persimmons together is one food taboo practiced by some elderly Chinese Americans because these foods represent extreme hot and extreme cold and may be "poisonous" if mixed.

■ During the postpartum period a woman is especially susceptible to disease-causing "wind." She can help prevent it from entering her body by eating chicken cooked in wine or pig's feet simmered with ginger and vinegar.

sweet rice before they burn his picture. A new picture of the Kitchen God is placed in the kitchen on New Year's Eve.

Food preparation must be completed on New Year's Eve, as knives cannot be used on the first day of the year because they might "cut luck." Deep-fried dumplings, made from glutinous rice and filled with sweetmeats, and steamed turnip and rice flour puddings are usually included in the New Year's Day meal. During the New Year festivities, only good omens are permitted and unlucky-sounding words are not uttered. Foods that sound like lucky words, such as tangerine (good fortune), fish (surplus), chicken (good fortune), chestnuts (profit), and tofu (*fu* means riches), are eaten.

Friends and relatives visit each other during the first 10 days of the new year, and good wishes, presents, and food are exchanged. Children receive money in small red envelopes. Traditionally the Feast of Lanterns, the 15th day of the 1st month, ends the New Year's season and is marked by the dragon dancing in the streets and exploding firecrackers to scare away evil spirits.

Ch'ing Ming, the chief spring festival, falls 106 days after the winter solstice. Families customarily go to the cemetery and tend the graves of their relatives. Food is symbolically fed to the dead, then later eaten by the family. Sweets and alcoholic beverages are popular offerings. *Duan wu,* the Dragon Boat Festival, is held on the 15th day of the 5th month to commemorate the drowning death of a famous 3rd-century B.C.E. poet. A boat race and special dumplings are traditional. The Moon Festival occurs at the end of September on a full moon (15th day of the 8th lunar month). Because the moon is a yin symbol, this festival was traditionally for women, but today it also symbolizes the togetherness of the family. It is sometimes called the harvest festival or moon's birthday. Large round cakes filled with spices, nuts, fruit or red bean paste, called moon cakes, are typically eaten during this event.

Therapeutic Uses of Food

The Chinese believe that a good diet is critical for physical and emotional harmony and necessary to strengthen the body against disease. One study in Hong Kong reported that three-quarters of respondents listed dietary prescriptions in the prevention of disease (Koo, 1984). Health is generally maintained by eating the proper balance of yin and yang foods (see "Therapeutic Use of Foods" in chapter 1). Hot foods generally include those that are high in calories, cooked in oil, and irritating to the mouth and those that are red, orange, or yellow in color. Examples include most meats, chile peppers, tomatoes, mushrooms, eggplant, persimmons, onions, leeks, garlic, ginger, and alcoholic beverages. Cold foods are often low in calories, raw or boiled/steamed, soothing, and green or white in color. Many vegetables and fruits are considered cold items, as are some legumes. Pork, duck, and honey also are classified as cold in some regions. Staples, such as boiled rice and noodles, are typically placed in a third, neutral category (Anderson, 1987; Ludman & Newman, 1984). Some food preparations can make foods hotter or colder by the infusion or removal of heat. Which foods are considered yin or yang varies from region to region. Chinese Americans who are acculturated may be uncertain about some categorizations and thus identify many foods as neutral.

Typically, hot foods are eaten in the winter, by menstruating, pregnant, and postpartum women (especially during the first month following childbirth), and for fatigue. Cool foods are consumed in the summer, for dry lips, and to relieve irritability. It is believed that as a person grows older, the body cools off and more hot foods should be eaten. Extra care should be taken with children's diets, because they are more susceptible to imbalance.

In addition to yin and yang, some foods are believed to affect the blood or promote wound healing and are labeled *pu,* or *bo,* meaning "strengthening." It is a classification that is separate from the concept of yin and yang but often used in conjunction with it; most strengthening foods are also categorized as hot. The yin condition of "weak" blood (most associated with pregnancy, postpartum, and following surgery) is treated with specific hot items such as protein-rich soups made with chicken, pork liver, eggs, pig's feet, or oxtail. Other health-promoting foods identified by Chinese Americans include royal jelly (made from honey), bee pollen, *lin chih* (edible fungus), rattlesnake meat, dog meat, roasted beetles, barley juice, garlic, *dong gwai* (angelica, a celery-like herb), fruit juice, and milk. Too many of certain yang items, however, can cause the blood to "thin," and these foods are avoided for conditions such as hypertension (Anderson, 1987; Ludman & Newman, 1984; Ludman et al., 1989; Sun, 1996).

Ginseng is the best-known health-promoting Chinese food. It is made from an herb (genus *Panax*) found in Asia and the Americas. The root is boiled until only a sediment remains, then powdered for use in teas and broths. Ginseng reputedly cures cancer, rheumatism, diabetes, sexual dysfunctions, and complaints associated with aging. It is most often used as a restorative tonic. Taro root is also thought to have therapeutic properties, such as improving eyesight, curing vaginal discharge, reducing weakness, and promoting multiple births; it will also bring good luck if eaten on the 4th day of the first lunar month (Newman, 1998b). Other popular remedies include deer antlers, rhinoceros horns, and pulverized sea horses (Spector, 1991).The concept of "like cures like" (sympathy healing) is seen in many food cures for specific illnesses (Koo, 1984). Hong Kong residents identify chicken with walnuts (which resemble brains) as a remedy for headaches; soups made with bones for treating broken bones; and male genital organs from sea otters, deer, or other animals to cure impotence.

Some food taboos have been noted during pregnancy. Soy sauce may be avoided to prevent dark skin, and iron supplements may not be taken because they are thought to harden the baby's bones and make birth difficult. Shellfish may also be shunned. Ginseng is used as a general tonic (Campbell & Chang, 1981).

Contemporary Food Habits in the United States

Adaptations of Food Habits

Generally changes in eating habits correlate with increasing length of stay in the United States, particularly with subsequent generations. Dinner remains the most traditionally Chinese meal, whereas breakfast, lunch, and snacks tend to become more Americanized. Younger persons are also more likely to accept American fare.

Ingredients and Common Foods

Most Americans of Chinese descent regularly consume several Chinese foods, such as rice, soybean products, cooked vegetables, and fruit (Chau et al., 1990). One preliminary study suggests that the majority (88 percent) of foreign-born immigrants prefer Chinese fare at home, although younger respondents (ages 20 to 34) expressed preference for American foods (Sun, 1996). Meat and poultry intake increases, while some traditional protein items like pig's liver and bone marrow soup often remain popular. Traditional fruits and vegetables may be replaced by more commonly available American items, such as potatoes, lettuce, apples, peaches, and watermelon (Grivetti & Paquette, 1978; Sun, 1996). Sugar intake also increases, mainly through consumption of soft drinks, candy, and pastries.

Even though milk is not a familiar item in the typical Chinese diet, several studies suggest that milk is consumed by nearly half of Chinese Americans in the United States (Chau et al., 1990; Schultz et al., 1994; Sun, 1996). Cheese, yogurt, and ice cream have also been found to be well accepted.

One study found that U.S.-born Chinese women have a more varied diet than Chinese American women who were foreign born. They ate more breads, cereals, dairy foods, meats, vegetables and ethnic items, such as Italian and Mexicans foods (Spindler & Schultz, 1996).

Meal Composition

Surveys of Chinese elders living in northern areas of California suggest that an almost traditional diet is eaten (Chau & Lee, 1987; Chau et al., 1990). Their lunches and dinners consisted mainly of Chinese-style foods; their breakfasts varied. Approximately half of the respondents attempted to balance hot and cold foods. Other studies suggest that the use of yin and yang in the diet may diminish over time and that Chinese Americans may practice some aspects of it, but without knowledge as to

■ In 1718, a Jesuit missionary in Quebec discovered an American species of ginseng that is nearly identical to the Chinese variety. Growing demand in China led many Americans, including Daniel Boone, to hunt the root for export.

■ A study of Asian college students found fast foods and sweet/salty snacks very popular. Intake of fats, sweets, dairy products, and fruit increased, while intake of meats and vegetables decreased (Pan et al., 1999).

why certain food combinations are preferred (Chau & Lee, 1990; Ludman et al., 1989). Stir-frying, simmering, and steaming remain the favored cooking methods (Sun, 1996; Zhou & Britten, 1994).

Americans of Chinese descent usually celebrate the major Chinese holidays of New Year's and the Moon Festival with traditional foods. Chinese American Christians sometimes combine the spring festival of Ch'ing Ming with Easter festivities. In addition, some Chinese Americans recognize the founding of the People's Republic of China (mainland China) on October 1 (on the solar calendar) or the establishment of the Republic of China (Taiwan) on October 10 with cultural performances and banquets.

■ Hypertension is considered a *yang* condition and is often treated by the consumption of *yin* foods (Newman, 2000a).

Nutritional Status

Nutritional Intake

The traditional Chinese diet is low in fat and dairy products and high in complex carbohydrates and sodium; research with elderly foreign-born Americans of Chinese descent indicates that this dietary pattern is mostly continued in the United States (Chau et al., 1987; Choi et al., 1990). As length of stay and the number of generations living in the United States increases, the diet becomes more like the majority American diet—higher in fat, protein, sugar, and cholesterol, and lower in complex carbohydrates. One study of women including foreign-born Chinese Americans, U.S.-born Chinese Americans, and white Americans revealed that all three groups consumed more than recommended levels of fat in their diets, suggesting that some changes in food consumption may occur very quickly (Schultz et al., 1994). The U.S.-born cohort also demonstrated high levels of nutrition knowledge and their diet contained a higher concentration of nutrients than either the foreign-born Chinese Americans or the white Americans. Further analysis showed that U.S.-born Chinese women eat a more varied diet than the foreign-born Chinese women. The increased consumption of breads and cereals, dairy foods, ethnic dishes (e.g., Italian, Mexican), meats, and vegetables was associated with an improved intake of riboflavin, iron, folacin, and calcium (Spindler & Schultz, 1996).

Some Americans of Chinese descent continue to avoid fresh dairy products because of lactose intolerance, which may be found in as many as 75 percent of Asians. Low calcium intake has been reported (Kim et al., 1993; Schultz et al., 1994). Alternative calcium sources are soybean curd, soy milk, if fortified with calcium, and soups or condiments made with vinegar in which bones have been partially dissolved. However, as noted previously, many Chinese Americans do consume milk, cheese, and yogurt, as well as leafy green vegetables, and calcium deficiency should not be presumed. Low vitamin A and C intake has been observed in some Americans of Chinese descent, but iron intake is satisfactory, perhaps due in part to the use of iron-containing cooking tools, such as woks (Zhou & Britten, 1994).

It is generally assumed that the Chinese eat a diet high in sodium, which may contribute to hypertension. Hypertension rates among Chinese Americans in California are significantly lower than for whites, but there is limited knowledge about hypertension by members of the Chinese American community (Choi et al., 1990; Stavig et al., 1988).

There is low risk for heart disease in China, and low rates of cardiovascular disorders have been found among Americans of Chinese descent. However, stroke risk is very high in China, and some researchers speculate that similar prevalence rates may be found among Chinese Americans (Choi et al., 1990). Cancer risk, especially for colorectal and breast cancers, has been found to increase with length of stay in the United States, and cancer is now the leading cause of death in Chinese Americans. Dietary changes, including lower intake of protective foods (i.e., soybean products) and higher intake of saturated fats are thought to be factors. Liver cancer among men and cervical cancer among women is significantly higher than among whites (Jenkins & Kagawa-Singer, 1994).

Obesity and overweight are found to be low among Chinese Americans, while thinness and extreme thinness are more common. Research on anthropometric measures indicates that weight for height ratios underestimate obesity in Chinese Americans when skinfold measurements are also taken (Netland & Brownstein, 1985). Calculated energy requirements may differ as well. Predictive equations for basal metabolic rate (BMR) and for resting energy expenditure (REE) are found to overestimate BMR and REE in adult Chinese Americans (Case et al., 1997; Liu et al., 1995).

Infant mortality rates for Chinese Americans are the lowest of any ethnic group in the United States (Singh & Yu, 1995), although researchers caution that the published figures may underestimate deaths due to racial classification errors (Lin-Fu, 1988). Most studies report low birth weight rates among Chinese Americans similar to those of whites. However, some researchers suggest that the definition of low birth weight may be inappropriate for Chinese infants who are typically smaller than the U.S. average (Gardner, 1994). Breast-feeding is common in China (Spector, 1991). Some researchers report that Chinese American babies may be weaned to unfortified gruel at about 5 months of age, resulting in some vitamin deficiencies (Ling et al., 1975).

High rates of turberculosis, parasitic infection, and hepatitis B have been found in many recent Asian American immigrants. Limited information regarding prevalence among Chinese immigrants specifically has been reported. Clonorchiasis, a liver fluke infection of the biliary passage or pancreatic ducts, has been identified in 25 percent of recent immigrants from Hong Kong and a smaller number of those from China (Hann, 1994).

Counseling

Americans of Chinese descent accept personal responsibility for their health; keeping healthy is considered an obligation to family and society. However, Western health care is underutilized. Language barriers, low income, long work hours, inconvenient location, and lack of insurance are some of the reasons believed to limit access. In addition, it has been reported that the concept of preventive checkups is unfamiliar to many Chinese Americans (Mo, 1992). Hospitals are frequently believed to be the place where a person goes to die; hospitalization rates of Chinese Americans are lower than for any other ethnic group in America. Blood tests, which are thought to permanently diminish the blood supply, are of particular concern to some Chinese clients who may avoid all biomedical health care for this reason.

Some Chinese Americans favor biomedical providers of Chinese heritage, although effective treatment is the primary concern for most clients. Preferred communication style is formal and includes unrushed dialogue (focusing on time is considered offensive), detailed explanation of the origins and symptoms of any condition in understandable terms, simple treatment, and a positive outlook. It is considered important for Chinese patients to maintain hope. When possible, terminal illness should be discussed first with family members to determine how and when a client is informed. Medical confidentiality is not widely practiced (Gould-Martin & Ngin, 1981).

The Chinese have a quiet conversational approach. Some speakers may pause during conversation as a sign of thoughtfulness. Interruptions should always be avoided. Many Chinese avoid confrontation and will initially say yes to questions that require a positive or negative response. Asking questions can be interpreted as disrespect, a sign that the person speaking is being unclear. Surprise or discomfort may be expressed by quickly and noisily sucking in air (Axtell, 1991). Glancing around and averted eyes are common; direct eye contact with elders is expected, however.

Elder or less acculturated Americans of Chinese descent may show deference to authority by means of acceptance and submission. In the hospital setting, patients are often silent rather than voicing complaints; providers should not necessarily accept this as compliance but instead should actively seek information about patient satisfaction. Emotional displays are considered immature (Randall-David, 1989), but most Chinese patients are willing to discuss feelings in conjunction with somatic symptoms (see the section "Traditional Health Beliefs and Practices").

■ **In Asia, nurses perform only medical procedures. Family members stay at the hospital to provide feeding, bathing, and general care for the patient.**

The traditional Chinese greeting is a nod or a slight bow from the waist while holding palms together near the chest, often without a smile. Touching between strangers and acquaintances is uncommon. Even handshaking may be inappropriate (wait for the extended hand, especially with women), except for westernized Chinese Americans and people from Hong Kong. Hugging, kissing, and backpatting should be

■ More than 90 percent of obstetricians and gynecologists are women in China; the disinterest of men in the specialty may date back to the time when menstrual blood was considered inherently dirty and polluting.

■ In 1996, the U.S. Food and Drug Administration warned consumers about the legal Chinese herbal "highs" containing *ma huang* (ephedra), a methamphetamine-like stimulant with possibly serious side effects such as heart attack, stroke, seizures, and psychosis.

avoided. Good posture is expected, and slouching or putting one's feet on a desk is considered rude. Personal space is typically less among the Chinese, and some clients may be comfortable sitting or standing close to a practitioner.

Chinese American women are often very modest; traditionally, Chinese women were never touched by male health care providers. Symptoms would be discussed by pointing to an alabaster figurine (Spector, 1991). If a male must do an examination, a formal, polite attitude, explanations of all procedures, and avoidance of tension-relieving jokes or comments will help the client feel more at ease (Gould-Martin & Ngin, 1981). Furthermore, within the family, sons receive more concern and attention over minor symptoms than daughters. Women consequently often do not believe that their complaints warrant care.

Researchers are unsure how many Chinese Americans practice traditional Chinese medicine. It is believed that the majority first self-diagnose and self-prescribe at home before seeking outside care, although the reasons for why certain foods or medications are consumed for an illness (particularly the complementary use of yin and yang) are often lost through acculturation (Chau et al., 1990; Ludman et al., 1989). Biomedicine is completely accepted by many Chinese Americans; one unpublished study found that 88 percent of foreign-born Chinese Americans subjects preferred biomedical care for the treatment of illness (Sun, 1996). Others consider it best in the treatment of acute symptoms. For example, a client with non-insulin-dependent diabetes may consult a biomedical physician regarding symptoms, but when he or she finds that no cure is offered may seek a Chinese doctor to restore balance to the body and treat the actual cause of the disease.

Doctors who practice Chinese medicine are often consulted concurrently with biomedical care in an effort to maximize the chances of a cure. Few conflicts in therapies have been identified, although the active agents in most herbal medicines remain unidentified in Western terms. A prescription from a Chinese medical doctor remains in possession of the client, who may reuse it if symptoms reoccur, or who may share it with family and friends. Some prescriptions are passed along from generation to generation; others are obtained directly from China. Providers should encourage traditional practices if desired by a client, but ask for information regarding herbal medicines consumed. Occasionally, a client may present multiple burn marks from moxibustion, and this treatment should be determined before presumption of abuse.

Translation can be a significant barrier in effective health care. There are numerous Chinese dialects, and few translators are available. If necessary, family members may need to be recruited. A Chinese client expects that the provider will perform few tests and ask a limited number of questions during an examination. Recommendations on diet, relaxation, and sleep are desired as an integral part of treatment. Long-term therapy intended to cure a disease is preferred over short-term surgical solutions or invasive treatment, even at the expense of pain or discomfort from symptoms. Most Chinese clients are resolved to die at home (many actually return to China), and their wishes should be accommodated.

Few compliance problems have been noted. One difficulty that sometimes arises is the issue of lengthy or continuous medication. Many Chinese are accustomed to single-dose Chinese remedies and may discontinue a pre-

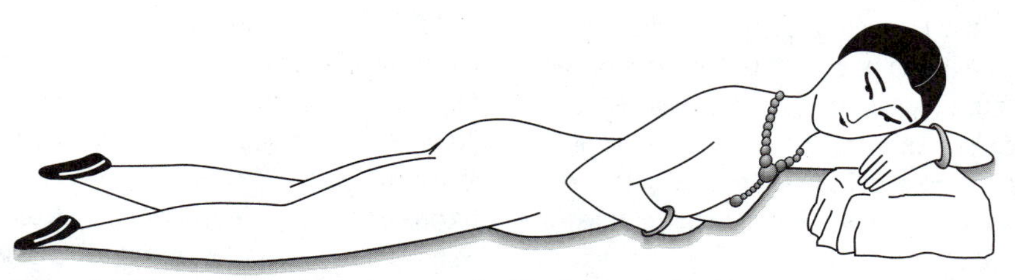

Traditionally, Chinese women were never touched by their male health care providers. Symptoms would be discussed by pointing to an alabaster figurine, like the one shown.

scription if directions are not thoroughly explained. Many researchers have remarked on the difficulty of eliminating high-salt items (i.e., soy sauce) from Chinese fare and recommend reduction as a goal in cases where a low-sodium diet is required (Chew, 1983). Dairy products may be accepted when clients become familiar with the foods (Schultz et al., 1994).

An in-depth interview should tactfully be conducted to identify which traditional practices, if any, are followed by a client. Even when concepts such as yin and yang are considered primitive or rustic by some Chinese Americans, they may still adhere to certain food combinations learned at home without knowing why. Birthplace and acculturation may significantly impact dietary intake of Chinese Americans (Schultz et al., 1994); individual preferences should be determined.

Japanese

The multi-island nation of Japan is off the coast of East Asia and has approximately the same latitude and range of climate as the East Coast of the United States. The capital of Japan is Tokyo, located on the island of Honshu. Today Japan is a prosperous country of 125 million people that has overcome the limitations of a mountainous geography, a rugged coastline, and few mineral resources. Perhaps Japan's greatest natural resource is the sea, which provides one of the richest fishing grounds in the world.

■ Among Japanese Americans, first-generation immigrants born in Japan are called *Issei*, second-generation Japanese Americans born in the United States are known as *Nisei*, and the third and fourth generations are known as *Sansei* and *Yonsei*, respectively.

Cultural Perspective

History of Japanese in the United States

Immigration Patterns

Significant Japanese immigration to the United States occurred after 1890 during the Meiji era. The immigrants were mostly young men with 4 to 6 years of education from the rural southern provinces of Japan. Most came for economic opportunities and many eventually returned to Japan. They settled primarily in Hawaii and on the West Coast of the United States and often worked in agriculture, on the railroads, and in canneries. Like the Chinese

Traditional foods of Japan. Some foods typical of the traditional Japanese diet include *daikon*, dried sardines, Japanese eggplant, Japanese pickles, *nori* (seaweed), red beans, *shiitake* mushrooms, short-grain rice, shrimp, soy sauce, and wheat noodles. (Photo by Laurie Macfee.)

■ *Heike-gani* crabs are believed to be reincarnations of drowned warriors due to the facelike markings on their shells.

■ Economic competition has renewed some anti-Japanese sentiments in the United States. In 1982 a Chinese American man was murdered in Detroit by two non-Asian autoworkers who mistook the victim for being of Japanese descent.

■ Over 68,000 residents of Japanese ancestry were living in Canada in 1993.

before them, Japanese immigrants opened small businesses, such as hotels and restaurants, to provide services for their countrymen. In contrast to the Chinese, many Japanese became farmers, ran plant nurseries, and were employed as gardeners. The Japanese prospered within their ethnic communities.

Most Japanese women came to the United States as "picture brides": Their marriages were arranged by professional matchmakers and they were married by proxy in Japan. They did not usually meet their husbands until they disembarked from the ship in the United States.

Discrimination against the Japanese was commonplace. The Issei (first-generation immigrants) were classified as aliens who were ineligible to become naturalized U.S. citizens, and in 1913, land ownership became illegal in California. Although many bought land in the names of their children, who were Americans by birth, Japanese-owned and leased land was reduced by half in the 1920s. In 1907 the Japanese government had informally agreed to limit the number of emigrants, and in 1924 the Japanese Exclusion Act halted Japanese immigration completely.

World War II heightened the prejudice against the Japanese on the West Coast. After Japan attacked Pearl Harbor, all West Coast Japanese, even if they were U.S. citizens, were "evacuated" to war relocation camps, and many remained there for the duration of the war. Most lost or sold their businesses as a result of internment. Nevertheless, many Nisei (second generation) volunteered for combat duty and fought in Europe.

Current Demographics and Socioeconomic Status

After the war, most Japanese Americans resettled on the West Coast, and the most discriminatory laws were repealed or ruled unconstitutional. The successful postwar recovery of Japan resulted in reduced emigration to the United States, usually far below the quota allotted under current immigration laws.

According to 1990 U.S. census figures, approximately 866,000 Japanese Americans live in the United States; 70 percent of them live in California or Hawaii. Many West Coast

Japantown, San Francisco. (Kevin Fleming/Corbis)

cities have a section of town called Little Tokyo or Japantown, and a small number of older Japanese still live in these homogeneous neighborhoods. Most Japantowns contain Japanese American–owned restaurants, markets, and other small businesses, as well as Chinese churches or Buddhist temples.

More than 95 percent of Japanese Americans live in culturally mixed urban and suburban areas. Americans of Japanese descent are unique for a nonwhite ethnic group in the degree of assimilation and economic mobility they have experienced. Family incomes are often above the national average and only 7 percent live in poverty. Of the third-generation Sansei, many of whom were born in the relocation camps, 88 percent have attended college and most hold professional jobs. Some Japanese Americans have noted that few Asians are found in top private and public sector positions, however, and believe that they suffer from the stereotype of being too unassertive for upper management.

Worldview

Religion

Early Japanese immigrants usually joined a Buddhist temple (Pure Land sect) or a Christian church after arriving in America. The church frequently provided employment and an opportunity to learn English. Today, it is thought that there are more Japanese Americans who belong to Protestant faiths than there are who follow Buddhism.

Shintoism, the indigenous religion of Japan, does not have a formal organization, but its beliefs are a fundamental part of Japanese culture. The Shinto view is that humans are inherently good. Evil is caused by pollution or filthiness, physical as well as spiritual; goodness is associated with purity. Evil can be removed through ritual purification.

Shinto deities, called *kami,* represent any form of existence (human, animal, plant, or geologic) that evokes a sense of awe. Kami are worshipped at their shrines as ritual expression of veneration and thankfulness. Prayers are also said for divine favors and blessings, as well as for avoidance of misfortunes and accidents.

Family

Until World War II the structure of the Japanese American family had its roots in Japan and was similar to that of the Chinese due to the strong influence of Confucianism. In addition, the rigid pattern of conduct that evolved in Japan during the 16th century resulted in the following practices among the Issei and their descendants:

1. *KoKo.* Filial piety defines the relationship between parents and children, between siblings, and between individuals and their community and rulers. (See the Chinese religion section on Confucianism in the first part of this chapter for further explanation.) One outcome is that the Issei expect their children to care for them in their old age.
2. *Gaman.* The Japanese believe it is virtuous to suppress emotions. The practice of self-control is paramount.
3. *Haji.* Individuals should not bring shame on themselves, their families, or their communities. This Japanese cultural concept exerts strong social control.
4. *Enryo.* There is no equivalent word in English, but the Japanese believe it is important to be polite and to show respect, deference, self-effacement, humility, and hesitation. Thus, many older or less acculturated Americans of Japanese descent are neither aggressive nor assertive.

Japanese clan or village affiliation has traditionally been much weaker than in China, and Japanese immigrants arrived in the United States prepared to raise nuclear families similar to those in white America. Typically picture brides were trapped in unhappy marriages but persevered on behalf of their children. Most Issei women worked alongside their husbands to support the family financially.

The internment of Japanese Americans during World War II brought further changes in family structure and accelerated acculturation into mainstream society after the war. In the camps, very low wages were paid and the pay was the same for everyone; thus the father could no longer be the primary wage earner. The camps were run democratically, but positions could be held only by American citizens, so the younger generation held positions of authority. The camps also allowed the Japanese to work in a wider range of jobs than those available to them on the outside. After their internment the Nisei no longer had to follow the few occupations of their parents.

Sansei couples generally form dual-career households. Nearly 50 percent marry outside their ethnic group. The societal problems prevalent in majority American homes, such as spousal abuse, have surfaced among Japanese Americans as well (Easton & Ellington, 1995), and it is not known if the family values that have thrust Americans of Japanese descent into educational and financial success will continue in the fully assimilated fourth generation of Yonsei.

■ Japanese physicians were polled to discover why they began prescribing traditional herbal remedies; a majority said they were influenced by pharmaceutical company salespeople and advertising.

■ An estimated 10,000 men in Japan die annually from *koroshi* (literally "death from overwork"). Stress-reducing therapies, called *iyashi,* include herbs, teas, and a chain of 10-minute massage parlors.

■ A small grill in Japan is called a *hibachi,* meaning "fire bowl."

Traditional Health Beliefs and Practices

Early Japanese health beliefs involved Shinto concepts of purity and pollution. Health was maintained through cleanliness and avoidance of contaminating substances such as blood, skin infections, and corpses (Hashizume & Takano, 1983). Botanical remedies were used, particularly purgatives, in the prevention and treatment of disease.

When Buddhism was introduced in the 6th century, the concept of harmony was applied within the context of Japanese culture to mean a person's relationship with nature, family, and society. Imbalance resulting from poor diet, insufficient sleep, lack of exercise, or conflict with family or society disrupts the proper flow of energy within the body, leading to illness. Chinese practices such as acupuncture, moxibustion, and massage were accepted as ways to restore the energy flow along the meridians of the body (see the section "Traditional Health Beliefs and Practices" in the Chinese part of the chapter). The application of yin and yang in health and diet was limited in Japan.

The more complex herbal medications of China were brought to Japan as *kanpo.* However, the numerous plants, animals, and minerals necessary for kanpo were not widely available on the islands, so its use was confined to the elite, urban aristocracy until recent times. Practitioners of the profession were called *kanpo-i* and underwent rigorous training. Kanpo-i approached each case individually, reviewing symptoms carefully and in detail before determining the best combination of therapies and medications for the specific patient. Diagnosis was an art that recognized symptoms may present differently in every consultation (Lock, 1990).

Western biomedicine was introduced to Japan in the 16th century with the arrival of the Portuguese. It was widely embraced; Japanese *kanpo-i* were required to retrain if they wished to continue working as doctors. The majority of Japanese Americans migrated to the United States during the time when kanpo was rarely practiced, and they were often unfamiliar with its therapies.

Since 1960, Japan has been in the middle of a "kanpo boom" (Lock, 1990). Concerns about the side effects of biomedical therapeutics and a growing interest in holistic and herbal healing has prompted the resurgence. Kanpo-i take a generalized approach, using natural medications with broad affect to stimulate the immune system. Some herbs also have known bacteriostatic action or anti-inflammatory properties. Small doses of the medications are taken for lengthy periods of time to promote gradual improvement. Physicians trained in biomedicine are also prescribing kanpo for many clients (though without the extensive education of kanpo-i); mass production of herbal medications by pharmaceutical companies began in the 1970s.

Traditional Food Habits

Japanese ingredients, as well as cooking and eating utensils, are very similar to those of the Chinese, due to the strong influence China has had on Japan. Yet Japanese food preparation and presentation are unique. The Japanese reverence for harmony within the body and community and with nature has resulted in a cuisine that offers numerous ways to prepare a limited number of foods. Each item is to be seen, tasted, and relished. The Japanese also place an emphasis on the appearance of the meal so that the visual appeal reflects balance among the foods and the environment. For example, a summer meal may be served on glass dishes so that the meal looks cooler, while a September meal may include the autumn colors of reds and golds.

Ingredients and Common Foods

Japan's mountainous terrain and limited arable land have contributed historically to a less than abundant food supply. Even today much of Japan's food supply is imported.

Staples and Regional Variations

The basic foods of the Japanese diet are found in the Cultural Food Groups list (Table 11.3). Several key ingredients were adopted from China, including rice, soybeans, and tea. Rice (*gohon*) is the main staple, and it is eaten with almost every meal. In contrast to the Chinese, the Japanese prefer a short-grain rice that contains more starch and is stickier after cooking. Rice mixed with rice vinegar, called *su,* is used in *sushi,* one of the most popular Japanese specialties in both Japan and abroad. Sushi rice is formed with fish and seafood to make decorative, bite-sized mounds served with soy sauce for dipping. Types include the following:

- *Nigirisushi,* which features rice topped with items such as sliced raw fish or squid (called *sashimi*), cooked octopus, crab, or shrimp; omelet strips or roe of salmon (*ikura*); sea urchin (*uni*) or flying fish (*tobikko*); and may be wrapped in a strip of seaweed
- *Makisushi,* a roll of sushi rice, often including cucumber (*kappamaki*), tuna, mushrooms, or other fillings, then wrapped in a sheet of seaweed and sliced into individual pieces
- *Chirashisushi,* with the topping ingredients literally "scattered" over the top of a large mound of rice, eaten with chopsticks

Rice is also made into noodles, although those made from wheat (known as *udon, somen,* and *ramen*) or buckwheat (*soba*) are more commonly consumed.

Soybean products are an important component of Japanese cuisine. *Tofu* (bean curd), soy sauce (*shoyu*), and fermented bean paste (*miso*) are just a few. *Teriyaki sauce* ("shining broil") is made from soy sauce and *mirin,* a sweet rice wine. Sugar, shoyu, and vinegar are a basic seasoning mixture. Shoyu and mirin can vary in strength, and the amounts used depend on personal taste.

Green tea is served with most meals. Tea was originally used in a devotional ceremony in Zen Buddhism. The ritual was raised to a fine art by Japanese tea masters; as a result, they also set the standards for behavior in Japanese society. Today the tea ceremony (*kaiseki*) and the accompanying food remain a cultural ideal that reflects the search for harmony with nature and within one's self. The meal features six small courses that balance the tastes of sweet, sour, pungent, bitter, and salty. The tea that is used for the ceremony is not the common leaf tea that is usually used for meals, but rather a blend of ground dry tea or a tea powder. Hot water is added to the tea and the mixture is whipped together using a handmade whisk, resulting in a frothy green drink.

Japanese fare does not use many dairy foods. Soybean products and a wide variety of fish and shellfish (fresh, dried, or smoked) are the primary protein sources. Fish and shellfish are often eaten raw. Beef and poultry are also popular but are very expensive. Only small amounts of meat, poultry, or fish are added to the vegetables in traditional Japanese recipes.

Fresh fruits and vegetables are the most desirable and are eaten only when in season. As in China, many Asian and European varieties are consumed (see the section "Staples" in the Chinese part of the chapter). Favorites include the *daikon* radish (a long white radish similar to the Chinese version, but longer, up to 12 inches in length), chrysanthemum leaves, and the winter tangerines called *mikan.* Pickled vegetables are available year round and are eaten extensively. Fresh fruit is the traditional dessert.

The Japanese use a large amount of seaweed and algae in their cooking for seasoning, as a wrapping, or in salads and soups. There are many types: *Nori* is a paper-thin sheet of algae that is rolled around sushi. *Kombu* is an essential ingredient in *dashi,* or soup stock made from dried bonita fish and seaweed. *Misoshiru* is a popular soup made with dashi and miso. *Wakame* and *hijiki* are used primarily in soups and salads. *Aonoriko* is powdered green seaweed used as a seasoning agent.

Japanese dishes are classified by the way the food is prepared: *suimono,* clear soups; *yakimono,* grilled food; *nimono,* foods simmered in seasoned broth; *mushimono,* steamed foods;

■ *Gohan* means "cooked rice" and is also the Japanese word for "meal."

■ In 1994, Japanese consumers took to the streets to protest the importation of rice from California needed to alleviate a crop shortage.

■ Red rice, made by steaming rice with adzuki beans, is often eaten at celebrations because red is considered a joyous and lucky color.

■ *Kaiseki* meals have recently become trendy in both Japan and the United States, costing up to $100 per person at restaurants featuring the ceremonial menu.

■ *Sansai ryōri* is a style of cooking with fresh wild herbs and vegetables such as goosefoot, mugwort, nettles, ferns, and braken. It is considered the essence of spring.

■ *Kombu* sounds like the word for "happiness," and it is often presented as a hostess gift by guests.

Table 11.3
Cultural Food Groups: Japanese

Groups	Comments
PROTEIN FOODS	
Milk/milk products	Japanese cooking does not utilize significant amounts of dairy products. Many Japanese are lactose-intolerant. Soybean products, seaweed, and small bony fish are alternative calcium sources.
Meat/poultry/fish/eggs/legumes	Soybean products and a wide variety of fish and shellfish (fresh, frozen, dried, smoked) are the primary protein sources in the Japanese diet. Fish and shellfish are often eaten raw. Chicken is used more often than beef; price is probably the limiting factor in meat consumption.
CEREALS/GRAINS	Short-grain rice is the primary staple of the diet and is eaten with every meal. Wheat is often eaten in the form of noodles, such as *ramen, somen,* and *udon.*
FRUITS/VEGETABLES	Fresh fruits and vegetables are the most desirable; usually eaten only in season. Many fruits and vegetables are preserved, dried, or pickled.
ADDITIONAL FOODS	
Seasonings	Sugar, *shoyu,* and vinegar are a basic seasoning mixture. *Shoyu* and *mirin* can vary in strength; amounts used will vary according to taste.
Nuts/seeds	
Beverages	Green tea is the preferred beverage with meals; coffee or black tea is drunk with western-style foods. *Sake* or beer is often served with dinner.
Fats/oils	The traditional Japanese diet is low in fat and cholesterol.
Sweeteners	

agimono, fried foods (e.g., *tempura,* adapted from a dish introduced in the 16th century by the Catholic Portuguese for religious fast days, consisting of shrimp and selected sliced vegetables lightly battered and deep-fried); *aemono,* mixed foods in a thick dressing; *sunomono,* vinegared salad; and *chameshi,* rice cooked with other ingredients. Foods that are cooked at the

Table 11.3
Cultural Food Groups: Japanese—continued

Common Foods	Adaptations in the United States
Milk, butter, ice cream	First-generation Japanese Americans drink little milk and eat few dairy products. Subsequent generations eat more dairy foods.
Meat: beef, deer, lamb, pork, rabbit, veal *Poultry:* capon, chicken, duck, goose, partridge, pheasant, quail, thrush, turkey *Fish:* blowfish, bonita, bream, carp, cod, cuttlefish, eel, flounder, herring, mackerel, porgy, octopus, red snapper, salmon, sardines, shark, sillago, snipefish, squid, swordfish, trout, tuna, turbot, yellowtail, whale *Shellfish:* abalone, *ayu,* clams, crab, earshell, lobster, mussels, oysters, sea urchin roe (*uni*), scallops, shrimp, snails *Legumes: adzuki,* black beans, lima beans, red beans, soybeans	Dried fish and fish cakes are available in the United States, but some varieties of fresh fish are not. Japanese Americans eat more poultry and meat than fish.
Wheat, rice, buckwheat, millet	Rice is still an important staple in the diet and usually eaten at dinner.
Fruits: apples, apricots, bananas, cherries, dates, figs, grapes, grapefruits (*yuzu*), kumquats, lemons, limes, loquats, melons, oranges, peaches, pears, pear apples, persimmons, plums (fresh and pickled), pineapples, strawberries, *mikan* (tangerine) *Vegetables* artichokes, asparagus, bamboo shoots, beans, bean sprouts, broccoli, brussels sprouts, beets, burdock root (*gobo*), cabbage (several varieties), carrots, chickweed, chrysanthemum, eggplant (long, slender variety), ferns, ginger and pickled ginger (*beni shoga*), green onions, green peppers, gourd (*kanpyo,* dried gourd shavings), leeks, lettuce, lotus root, mushrooms (*shiitake, matsutake, nameko*), okra, onions, peas, potatoes, pumpkins, radishes, rhubarb, seaweed, snow peas, shiso, sorrel, spinach, squash, sweet potatoes, taro, tomatoes, turnips, watercress, yams	Fewer fruits and vegetables are eaten; freshness is less critical.
Alum, anise, bean paste (*miso*), caraway, chives, *dashi,* fish paste, garlic, ginger, mint, *mirin*, MSG, mustard, red pepper, *sake,* seaweed, sesame seeds, *shiso, shoyu,* sugar, thyme, vinegar (rice), *wasabi* (green, horseradish-like condiment)	
Chestnut, gingko nuts, peanuts, walnuts; poppy (black and white), sesame seeds	
Carbonated beverages, beer, coffee, gin, tea (black and green), *sake,* scotch	Japanese Americans drink less tea and more milk, coffee, and carbonated beverages.
Butter, cottonseed oil, olive oil, peanut oil, sesame seed oil, vegetable oil	Japanese Americans consume more fats and oils because of increased use of western foods and cooking methods.
Honey, sugar	Increased use of sugar, sweet desserts is noted.

table and one-pot dishes are called *nabe mono. Sukiyaki* is a simmered beef dish usually prepared at the table. The name means "broiled on the blade of the plow" and probably dates back to ancient times. The current version is mislabeled, however, because it is a *nimono-,* not a *yakimono-,* style food. Seafood, fish, fruits, and vegetables that are pickled in a mixture of miso,

■ *Teppanyaki* is a Japanese term for grilling. The style familiar to U.S. diners (typified by the Benihana restaurant chain) was invented to take advantage of the tourist trade. Beef, chicken, shrimp and vegetables are cooked on a hot grill in the center of a large table, then served with *ponzu*, a soy sauce and citrus juice mixture.

■ When a family moves to a new home, they give *soba* noodles to the neighbors on either side and across the street as a gesture of friendship.

■ A Tokyo department store has recently built an exact replica of the famous Viennese bakery, Demels (Siemering, 1999).

soy sauce, vinegar, and the residue from sake production are known as *tsukemono,* and they are served nearly every meal. Japanese foods are usually cut into small pieces if the item is not naturally easy to eat with chopsticks, and dishes are frequently modified for children, as it is believed that adult recipes are too spicy for them.

Cooking style varies from region to region in Japan. Kyoto is known for its vegetarian specialties, and Osaka and Tokyo are known for their seafood. Nagasaki's cooking has been greatly influenced by the Chinese.

Meal Composition and Cycle

Daily Pattern

Traditionally the Japanese eat three meals a day, plus a snack called *oyatsu.* Breakfast usually starts with a salty sour plum (*umeboshi*), followed by rice garnished with nori, soup, and pickled vegetables. Some families may have an egg with their rice.

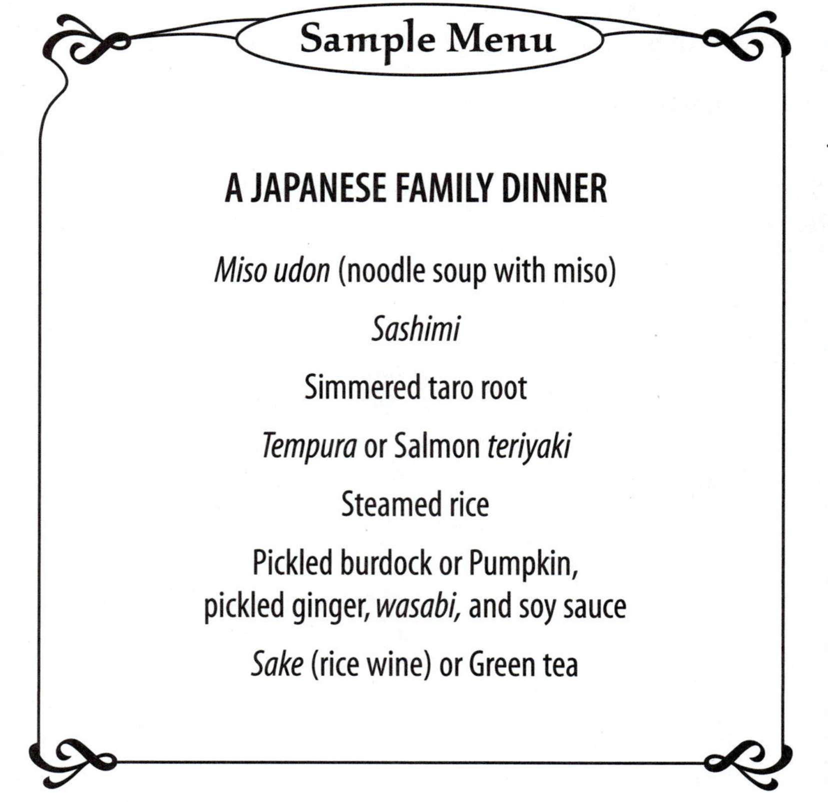

Lunch is simple and often consists only of rice topped with leftovers from the previous night. Sometimes hot tea or dashi is added to the rice mixture. A bowl of noodles cooked or served with meats, poultry, or fish, and vegetables is a popular alternative to leftovers. Dinner typically includes rice, soup, or sukemono and three dishes: a raw or vinegared fish, a simmered dish, and a grilled or fried dish. Pink pickled ginger and the very pungent green horseradish-like condiment called *wasabi* garnish many meals, and soy sauce is also usually available.

The Japanese tend not to serve meals by courses. Instead, all the dishes are presented at the same time in individual portions, each food in its own bowl or plate. The soup, however, is sometimes served last or near the end of the meal. Traditionally desserts were not common in Japan; meals usually ended with fruit. In addition, the Japanese often eat a boxed meal, called *bento.* A pleasing assortment of at least 10 items is packaged attractively for consumption at school, picnics, or even between acts at the theater. Some restaurants specialize in *bento* meals.

Snacks include several kinds of sweets, rice crackers, or fruit. Traditional Japanese confections include *mochi gashi* (rice cakes with sweet bean paste), *manju* (dumplings), and *yokan* (sweet bean jelly). Green tea is served after all meals except when Western-style food is eaten; then coffee or black tea is served. Beer or *sake* (rice wine, usually served warm) may be served with dinner.

Etiquette

The Japanese, like the Chinese, eat with chopsticks and follow many of the same customs regarding their use (see the section "Etiquette" in the Chinese part of the chapter); the rice bowl is not held as closely to the mouth, however. Soups are consumed directly from the bowl; the only dish eaten with a spoon is an unsweetened egg custard known as *chawanmushi.* Traditionally the Japanese eat their meals at low tables where the men sit cross-legged and women sit with their legs tucked to the side. Shoes are removed first. Dishes on the left are picked up with the right hand and dishes to the right are lifted with the left hand. It is impolite to serve sake to oneself.

Special Occasions

In Japan there are numerous festivals associated with the harvesting of specific crops or with local Shinto shrines or Buddhist temples. The most important and largest celebration in Japan is New Year's. The Japanese share many holiday traditions with the Chinese. Homes are cleaned thoroughly and all debts are settled before the New Year; food is also prepared ahead so that no knives or cooking will interfere with the 7-day event.

The Japanese celebrate New Year's on January 1. The New Year's foods consist of 10 to 20 meticulously prepared dishes served in a special set of nesting boxes. Each dish symbolizes a specific value, such as happiness, prosperity, wealth, long life, wisdom, and diligence. For example, fish eggs represent fertility, mashed sweet potatoes and chestnuts protect against bad spirits, and black beans represent a willingness to keep healthy through hard work and sweat.

An important New Year's food is *mochi,* a rice cake made by pounding hot, steamed rice into a sticky dough. Traditionally a Buddhist *o sonae mochi* is set up in many homes. A large rice cake represents the foundation of the older generation and is placed on the bottom, and a smaller rice cake symbolizing the younger generation is placed atop it, followed by a tangerine indicating generations to come. The o sonae mochi is as meaningful to the Japanese as a Christmas tree is to Americans, preserving good fortune and happiness for future generations. Another special food is *ozoni,* a soup cooked with mochi, vegetables, fish cakes, and chicken or eggs. A special rice wine called *otoso* is consumed to preserve health in the coming year.

Japanese Buddhist temples usually hold an Obon festival in the 2nd or 3rd week of July to appreciate the living, honor the dead, and comfort the bereaved. Food and dancing are a traditional part of the holiday.

Certain birthdays are thought to be either hazardous or auspicious in Japanese culture. When a man turns 42 or a woman becomes 33, special festivities are held to prevent misfortune. Age 61 marks the beginning of second childhood and a person dons a red cap for this honor. At age 77 a person puts on a long red overcoat, and at the most propitious birthday of all, age 88, the celebrant may begin wearing both the hat and the coat.

■ At many celebrations the Japanese shout "Banzai! Banzai! Banzai!" The cheer is translated as "10,000 years!" and dates back to the 3rd century B.C.E.

■ Rice was often in short supply in Japan, so eating the round rice cakes, *mochi,* at New Year's represented wealth and prosperity.

■ A crane symbolizes 1,000 years. On special birthdays, 1,000 red *origami* (folded paper sculpture) cranes are displayed as a wish for longevity.

Obon festival, San Jose, California. Dancing and traditional foods are featured at this annual Buddhist celebration.

■ Sea urchin roe, called *uni,* is thought to enhance male sexual potency in Japan, where wholesalers pay up to $100 per pound for it. California imposed strict urchin harvesting laws to prevent extinction along the coast (more than 200 million pounds were gathered for export between 1985 and 1990). Maine suffers from similar overharvesting.

Therapeutic Use of Food

Although the use of yin and yang is not as prevalent among the Japanese as it is among the Chinese, there are many beliefs about the harmful or beneficial effects of specific foods and food combinations. Traditionally certain food pairs, such as eel and pickled plums, watermelon and crab, or cherries and milk, are thought to cause illness. Pickled plums and hot tea, which are customarily eaten for breakfast, are believed to prevent constipation. Both pickled plums and rice porridge are thought to be easily digested and well tolerated during recovery from sickness.

Contemporary Food Habits in the United States

Adaptations of Food Habits

In general, second-generation Japanese Americans eat a typical American diet, but they may eat more rice and use more soy sauce than non-Asians. Traditional foods are prepared for special occasions. Third- and fourth-generation Japanese Americans appear to be totally acculturated to American foodways. Even in Japan a westernized diet is increasingly followed. Bread and butter are becoming staples, and consumption of meat, milk, and eggs is increasing. One survey reported that between 1960 and 1985, fat consumption in Japan went from 11 percent of calories in the diet to more than 25 percent of calories (Lands et al., 1990).

Nutritional Status

Nutritional Intake

The traditional Japanese diet is low in fat and cholesterol. Most cooking fats are polyunsaturated and butter is rarely used. Japanese Americans, however, consume a more typically American diet and this change may contribute to increased incidence of several diseases. According to classic epidemiological studies, mainland Japanese Americans have a higher risk of developing colon cancer and heart disease than the Japanese in Hawaii, and those in Hawaii have a higher risk than the Japanese in Japan. It has been postulated that the increase is caused by diet, since it is correlated to a higher intake of cholesterol and animal fat and a lower intake of dietary fiber (Marmot & Syme, 1976; Wenkam & Wolff, 1970). Although the prevalence of cardiovascular disease among Japanese Americans is only one-half that for white Americans, it is expected that death rates will climb steadily as further acculturation occurs (Gardner, 1994). Other cancers, such as those of the breast and rectum, have also increased in Japanese Americans as their stay in the United States lengthens.

Changes in diet have also been implicated in the high rates of noninsulin-dependent diabetes found among Japanese American men. Among Nisei men in one study, the rates for diabetes were twice that for similarly aged white men living the same region of the United States and four times that for similarly aged men in Japan (Tsunehara et al., 1990). Increased abdominal fat deposits are found associated with later development of diabetes in Japanese American men (Bergstrom et al., 1990). A genetic predisposition for diabetes combined with increased fat consumption, especially animal fats, may account for the disproportionately high rates.

Japanese American elders may have a low intake of calcium because of limited consumption of dairy products (Kim et al., 1993). The Japanese have a high incidence of lactose intolerance. Although seaweed, tofu, and small bony fish contain calcium, they may not be eaten in adequate amounts to provide sufficient intake. Prevalence of osteoporosis may be higher than among whites (see the box). Calorie consumption and meat intake have also been found to decline with age in Japanese Americans.

Traditional diets tend to be high in salt from soy sauce, dashi, miso, monosodium glutamate (MSG), dried preserved fish, and pickled vegetables. Rates of conditions sometimes linked to high-sodium diets, such as hypertension, stroke, and stomach cancer, are extremely high in Japan, but have been dropping as the Japanese adopt western fare. A study of Japanese American men in Hawaii reported that while consumption of all vegetables was found to be protective against stomach cancer, no association was

Asians, Dairy Foods, and Osteoporosis

Osteoporosis, which means "porous bone," affects 10 million women and 2 million men in the United States. Another 16 million have low bone mineral density (BMD), which may put them at risk of developing the disease. Osteoporosis is characterized by reduced height, a stooped spinal deformity best known as "dowager's hump," and over 1.5 million fractures annually. The bones most often broken include those of the spine, the hip, and the wrist. The causes of osteoporosis are mostly unknown. Contributing factors include ethnicity, family history, low calcium intake, insufficient weight-bearing exercise, smoking, high alcohol consumption, and low levels of estrogen in women and testosterone in men (NIH, 2000).

White women have long been considered at highest risk for osteoporosis. In particular, thin white women have been thought most vulnerable because higher body mass is related to better BMD. Over half of white women in the Unites States are estimated to have low BMD, compared to just 38 percent of black women, and body mass may account for a majority of that difference in women (though not in men) (Cundy et al., 1995; Nelson et al., 1991). Data from the National Health and Nutrition Survey III reported that Mexican American women also had a lower prevalence of low BMD than white women (Looker et al., 1995). However, recent findings in a large study of postmenopausal U.S. women cast doubt on the assumption that osteoporosis is a white woman's disease, which raises new questions about the role of low BMD in development of the condition (Reuters, 1999; Siris et al., 1998).

Preliminary data from the National Osteoporosis Risk Assessment Initiative indicate that the highest percentage of the disease occurs in Native Americans (9.5), followed by Asians (8.2), whites (5.2), Hispanics (4.3), and blacks (4.0). Surprisingly, rates of low BMD do not correlate with development of osteoporosis. The prevalence of low BMD indicated above in whites (51 percent) and blacks (38 percent) was less than the low BMD rates found in every other ethnic group: 65 percent of Asians, 59 percent of Native Americans, and 55 percent of Hispanics. Scientists are struggling to understand the relationships between BMD, osteoporosis, and fracture rates. For example, though U.S. Asian women have high rates of low BMD and osteoporosis, they have fewer hip fractures than U.S. white women, with Chinese Americans having the lowest incidence compared to whites, followed by Korean Americans, then Japanese Americans (Lauerdale et al., 1997; Ross et al., 1991). Worldwide, hip fracture rates are rising dramatically in Japan and Hong Kong, and it is estimated that by 2050 half of the projected 6 million plus broken hips will occur annually among Asians (Cooper et al., 1992; Lau et al., 1996; Orimo et al., 2000).

Dietary recommendations regarding osteoporosis have traditionally emphasized a high intake of calcium-rich foods, considered especially important during adolescence when peak bone mass is achieved and in elders who may be less efficient at metabolizing the mineral. Dairy foods are considered one of the best sources because they contain high amounts of calcium as well as vitamin D (which enhances absorption of the mineral). When Asians were not considered at risk for osteoporosis, researchers speculated that while many Asians do not eat dairy foods, they obtained adequate calcium from eating small fish with bones (e.g., sardines), mineral-rich fish sauces, organ meats, and ample dark green leafy vegetables. Some scientists also suggest that soybean intake may be protective, slowing bone mineral loss after menopause (Harrison et al., 1998; Messina & Messina, 2000). Yet, if prevalence for low BMD and osteoporosis among Asians is higher than previously calculated, and if fractures are increasing, do nondairy foods provide adequate calcium intake? Or, if the intake of traditional calcium-rich foods declines with acculturation in the United States (and westernization worldwide), are dairy foods needed to provide the calcium no longer being consumed?

Recently, some physicians and activist groups have suggested that federal nutrition policy and the Food Guide Pyramid is racist because it recommends two to three daily servings of dairy products. They claim this recommendation can be harmful to most U.S. blacks, Asians, Hispanics, and Native Americans because each of these groups has a high percentage of lactose-intolerant individuals. Optional dairy products are advocated (Bertron & Barnard, 1999). It is known that lactase deficiency alone does not affect the BMD of women (Slemenda et al., 1991), and some studies have found that lactose maldigestion can improve with consumption of dairy foods (Pribila et al., 2000; Suarez et al., 1998). More research on the bioavailability of calcium in different foods and the role of diet in the development of BMD and osteoporosis is needed to determine whether dairy food recommendations are sensible for all Americans or simply ethnocentric.

■ A small percentage of Asians lack the ability to metabolize alcohol well. This inherited condition causes immediate skin flushing (reddening) and may even result in heart palpitations when alcohol is consumed. This reaction may contribute to the number of abstainers in the population.

■ Korea is called *Choson* by Koreans, meaning "Land of Morning Calm."

found between sodium intake and the disease, perhaps because the subjects consumed relatively low amounts of high-salt foods such as vegetable pickles and dried fish (Chyou et al., 1990).

A comparison of Asian (Chinese, Japanese, and Korean) alcohol consumption habits found that Americans of Japanese heritage had the most permissive attitude toward drinking, particularly among women. Japanese American men had high rates of heavy drinking (nearly 30 percent) and the fewest abstainers (16 percent). Japanese American women showed similar trends, with the highest rates of heavy drinking (almost 12 percent) and the lowest number of abstainers (27 percent). Although the number of women engaged in chronic heavy drinking is lower than for white American women, the rates for men were comparable. Friends who drink and social occasions where drinking was expected were significant influences on consumption among men (Chi et al., 1989). This study confirms previous work that alcohol consumption may be more frequent than previously assumed among Japanese Americans, although behavior problems related to drinking have not been widely reported (Zane & Kim, 1994).

Infant mortality rates among Japanese Americans are reportedly lower than for whites (Singh & Yu, 1995). Racial classification errors, however, have been found responsible for significant underestimation of infant mortality rates among Americans of Japanese descent (Lin-Fu, 1988). Low rates of premature births and low birth weight infants have been reported. More than 85 percent of Japanese American women seek prenatal care during pregnancy (Easton & Ellington, 1995).

Counseling

Japanese American values such as placing the family before the individual, preserving harmony with society, and respecting and caring for elders may have a positive impact on health (Easton & Ellington, 1995). Illness may be regarded as both a symptom of an unbalanced life, as well as an impediment to fulfilling personal obligations.

Formality and politeness are essential conversational elements in Japan. Addressing Japanese elders or Japanese American Issei, and some Nisei, by their first names is very insulting. Sansei and Yonsei are often more informal (Hashizume & Takano, 1983). Emotional displays are avoided, especially of anger. The Japanese are nonconfrontational and may be reluctant to say no even when the answer to a question is negative. Waving a hand in front of the face with the palm outward indicates "I'm unsure" or "I don't know." Conversational style is often indirect, and frequent pauses are common. Direct eye contact is disrespectful; glancing around or downcast eyes are expected. Smiling can indicate pleasure but is also used to hide displeasure.

The Japanese are extremely high-context, and the slightest gesture may have meaning. (See chapter 3 on intercultural communication for more details.) Broad hand or body gestures may be misconstrued. Touching between strangers or acquaintances is infrequent, though most Japanese Americans are comfortable with a light handshake. The traditional greeting is a bow from the waist with palms against thighs. Slouching and putting one's feet on a desk are signs of disinterest.

Japanese Americans believe that the health care provider is a knowledgeable authority figure who will meet their needs without their assistance. Most Americans of Japanese descent expect to be directed in their health care, yet are insulted if they are ordered to do anything that they feel requires only an explanation. Criticism of a client's health habits can lead to embarrassment and loss of effective communication. Concrete, structured approaches based on information gathered through an unhurried, in-depth interview that determines degree of acculturation and personal preferences are most effective.

Koreans

The mountainous peninsula that forms Korea is suspended geographically, and culturally, between China and Japan. Korea has historically been caught in the middle of both Chinese and Japanese expansionism, yet has maintained a homogeneous population with an independent, distinctive character. Little land is arable and the climate fluctuates between cold, snowy winters and hot, monsoonal summers

Korean street vendors shredding cabbage for the preparation of the spicy vegetable pickle *kimchi*. (Corbis/Michael Freeman.)

that limit agriculture significantly. The peninsula is currently divided into two nations. The Democratic People's Republic of Korea (North Korea), with the capital city of Pyongyang, has a communist government. The Republic of Korea (South Korea) is a democracy supported by the United States. Seoul is the country's capital. Both nations desire the reunification of Korea through political and military domination of the other.

Cultural Perspective

History of Koreans in the United States

Immigration Patterns

Early Korean immigration to the United States was severely restricted by the isolationist policies of Korea. A small number of Koreans arrived before 1900, most of whom were Protestants seeking to escape discrimination and further their education in America.

Between 1903 and 1905, Christian missionaries recruited more than 7,000 Korean men, women, and children to work in Hawaiian sugarcane fields. In 1905, when Korea was under Japanese rule, overseas emigration of Koreans was barred, and in 1907 the United States entered a "Gentlemen's Agreement" with Japan limiting both Japanese and Korean immigration. During the next 17 years, only picture brides and oppressed political activists were permitted entry. In 1924 the Japanese Exclusion Act was applied to Koreans as well, preventing all immigration. Most early Korean immigrants worked as field hands or in domestic service; many first- and second-generation Koreans living in urban areas were barred from professional jobs and established small businesses such as vegetable stands and restaurants.

Between 1959 and 1971, nearly all Korean immigrants were the wives and children of U.S. soldiers who fought in the Korean War. Following relaxation of U.S. immigration laws in 1965, the numbers of Korean immigrants increased, including many college-educated, middle-class professionals and their families.

Current Demographics

The Korean immigrant population has increased dramatically in recent years. As of 1993, approximately 837,000 Koreans were

- The author Jack London contributed to anti-Asian sentiment in the United States when he wrote a newspaper article in 1904 titled "The Yellow Peril" in which he called Koreans "the perfect type of inefficiency—of utter worthlessness" (Nash, 1995).

- The status of Koreans in the United States during World War II was unique. Many remained technically Japanese citizens and as such were declared enemy aliens (although none was interred). Yet many Koreans had been involved for years in anti-Japanese activities. "I am Korean" buttons were worn to distinguish themselves from Japanese residents.

- There are approximately 65,000 Canadians of Korean ancestry.

living in the United States; more than half have arrived since 1980. More than 80 percent of Koreans in America were born in Korea, and most do not seek U.S. citizenship. Large numbers of immigrants to the United States relocated first from North Korea to South Korea, seeking greater freedom. Koreans coming to America hope to find economic opportunity and to avoid any North Korean–South Korean conflict that may arise in the future. Large numbers of Koreans have settled in California, particularly Los Angeles. Other states with significant populations include New York, Illinois, New Jersey, Texas, Washington, Virginia, Pennsylvania, Maryland, and Hawaii.

Socioeconomic Status

■ The economic success of South Korea has lured thousands of second-generation Korean Americans to immigrate there, where they have taken positions in marketing, public relations, and the entertainment industry. Difficulties in adjustment have led them to form expatriate communities in many areas.

Many Koreans must accept temporary or permanent nonprofessional employment due to language difficulties or licensing restrictions. Families working together toward the success of a small business and the purchase of a home are common. Korean American descendants of the 1903–1905 immigrants to Hawaii are securely middle class. Contributing factors include high achievement in education and professionalism; quick mastery of English (faster than Japanese and Chinese immigrants of the same period); and a greater willingness to give up Korean traditions, perhaps due to their experience with nonconformity in Korea as Christians in a Confucian and Buddhist society (Kim, 1980). However, some Koreans do not find the economic prosperity they seek in the United States and so return to Korea.

A notable conflict persists in the immigrant Korean community. Korean Americans may not accept recent immigrants who demand immediate social standing based on prior status in Korea. Recent professional immigrants are sometimes contemptuous of Korean war brides, who are considered uneducated and lower class. The large numbers of new arrivals may serve to coalesce the Korean community because they share similar experiences in acculturation and are a significant percentage of the growing Korean American population.

Worldview

Religion

In South Korea, Buddhism and Confucianism are the majority religions. Approximately 28 percent of South Koreans are Christian, and smaller numbers adhere to shamanism (belief in natural and ancestral spirits) and the national Korean religion *Chundo Kyo,* formerly known as Tonghak (a mixture of Confucian, Taoist, and Buddhist concepts). In North Korea, all religious beliefs other than the national ideology of Marxism and self-reliance are suppressed.

The first Korean immigrants at the turn of the century were Christians attracted to the United States as a religious homeland. Many suffered discrimination in the largely Confucian/Buddhist society of Korea. Although specific figures are unknown, it is believed that many recent immigrants to the United States are also Christian. A study of elder Korean immigrant women in Washington, D.C. indicated that 85 percent were Christian (Baptist, Methodist, Presbyterian, and Catholic), 10 percent practiced ancestor worship (veneration of ancestors often coexists with other religious practices), and 5 percent were Buddhist (Pang, 1991). Other research suggests that larger numbers of Buddhists have immigrated.

Family

Hundreds of years of Confucianism in Korea have significantly influenced family structure regardless of current religious affiliation. Family is highly valued, and loyalty to one's immediate and extended family is more important than individual wants or needs. Generational ties are more important than those of marriage, and parents are especially close to children.

In Korea a male is always the head of the family; if a father is unable to fulfill that role, the eldest son (even if still a child) takes on that responsibility. Birth sequence orders life's events within the family, and older male children are traditionally awarded privileges, such as advanced education, that younger children or daughters are denied.

Elders are esteemed and cared for in Korea. The two major birthday celebrations in

Korean culture occur at age 1, and when an individual reaches age 60, meaning that the person has survived a full five repeats of the 12-year cycle of life and attained old age. The opinions of elders are respected, and after age 60 a person is allowed to relax and enjoy life.

Many changes occur in the Korean family after immigration to the United States. For example, the marriage bond often becomes more important than obligations to one's parents. Few of the elderly maintain traditional arrangements of living with an eldest son's family. Some live with unmarried children or with married daughters, or live alone. Many elders feel that they are now a burden to their children and that old age is a negative experience instead of a privilege (Pang, 1991). Male dominance has declined with increased participation of women in the workplace, yet most married women assume total responsibility for home and their job and often are not paid for their work if employed in a family-owned business. Divorce rates are reportedly high, and intergenerational conflict increases with length of time in the United States.

Traditional Health Beliefs and Practices

Traditional Korean concepts related health to happiness, to the ability to live life fully, to function without impairment, and to not be a burden on one's children (Pang, 1991). A good appetite is a significant indicator of health.

The Korean system of health and illness is closely related to Chinese precepts (see the section "Health Beliefs and Practices" in the Chinese part of the chapter). The proper balance of *um* (yin) and yang must occur to maintain health, influenced by the relationships of the five evolutive elements (fire, water, wood, metal, and earth) and *ki* (vital energy). Too much or too little of these forces result in illness. For example, cold, damp, heat, or wind can enter a body through the pores, which can interfere with ki and weaken yang. Symptoms of the imbalance include indigestion, arthritis, and asthma. Other disruptive causes such as physical exhaustion, eating too much or too little food, and spiritual intervention (by ancestors or supernatural deities) may also result in disease (Chin, 1992).

Both digestion and circulation are prominent in the maintenance of health, because energy is absorbed into the body through the stomach when food is mixed with air or force, and the blood distributes this vital energy. Good quality food is restorative, but too much food can block ki, resulting in cold hands and feet, cold sweats, or even fainting (Pang, 1991). A few Koreans attribute diabetes to eating too much rich food such as meat, sugar, or honey, and getting too little exercise. Blood conditions that interfere with the distribution of vital energy include a "lack" of blood; a "drying" or "hardening" of the blood (typical in old age, causing indigestion, neuralgia, and body aches); and "bad" blood, caused by a sudden fright, which can result in chronic pain.

Korean-specific folk illnesses often include somatic complaints that are an expression of psychological distress (Pang, 1994). Excessive emotion such as joy, sadness, depression, worry, anger, fright, and fear are believed to result in certain physical conditions. *Hwabyung,* attributed most often to anger, is associated with poor appetite, indigestion, stomach pain, chest pain, shortness of breath, weight gain, and high blood pressure among other symptoms (Pang, 1991). *Han,* which causes a painful lump in the throat, occurs when a person suffers disappointments and regrets, such as guilt over the neglect of one's children, parents, or spouse. *Shinggyong-shaeyak* resulting from stress (especially from oversensitivity and lack of happy interactions with family and friends) can cause insomnia, weight loss, and nervous collapse. Traditional cures include use of a shaman or spiritual mediator (*mansin* or *mudang*) to determine whether the cause of an illness is due to disharmony with one's ancestors or natural and supernatural forces (Chin, 1992). Sacred therapeutic rituals to rectify such spiritual disruptions are conducted with the patient, the family, and sometimes the community.

Hanyak is the traditional approach to natural cures in Korea. It is typically practiced by a *hanui.* When a client visits a hanui, he or she obtains a medical history, observes how the patient looks, listens to the quality of the voice, and takes the patient's pulse. More than 24 pulse conditions are defined, from floating to sunken, and smooth, vacant, or accelerated

■ A hanui must determine a client's character to make an accurate diagnosis through questions, such as "Do you like meats or vegetables?" "What season is your favorite?" "Do you worry much?" or "Are you stubborn?" (Pang, 1989).

(Pang, 1989). Hanyak medications are classified according to their plant, animal, or mineral source, and mixed in ways to balance um (yin), yang, and ki. Other physical therapies to restore harmony and vital energy, such as acupuncture, moxibustion, cupping, and sweating (see chapter 2), may also be applied.

In the United States, the hanui may use some biomedical procedures in conjunction with traditional practices. It has been reported that some hanui take blood pressures and body temperatures. Some even offer the convenience of taking hanyak prescriptions in pill form so that bitter-tasting broths or teas are avoided (Pang, 1989).

Many professional Korean health practices have been popularized and may be used as home remedies. In the United States, where access to traditional healers may be limited in some regions, the mother or grandmother in the home often takes responsibility for administering these cures. Some Koreans believe that a person's fate is determined by the forces of um and yang at the moment of birth. Christian Koreans may believe strongly in faith healing and in fate as determined by God.

Traditional Food Habits

Ingredients and Common Foods

Korean cuisine is neither Chinese nor Japanese, although it has been influenced by both styles of cooking. It is a distinctly hearty Asian fare that is highly seasoned and instantaneously recognizable as Korean by its flavors and colors. Sweet, sour, bitter, hot, and salty tastes are combined in all meals, and foods are often seasoned before and after cooking. Five colors—white, red, black, green, and yellow—are also important considerations in the preparation and presentation of dishes.

■ The preparation of *kimchi* every autumn is a special family event. In the past a family's wealth was demonstrated by the ingredients in their kimchi, with rare vegetables and fruits used by the most affluent.

■ One of the most popular black market commodities in Korea is the diced pork and gelatin product Spam.

■ The Koreans say if you eat duck, your feet will become webbed.

Staples

Korean cooking is based on grains flavored with spicy vegetable and meat, poultry, or fish side dishes. Korean staples are listed in Table 11.4.

Rice is the foundation of the Korean diet. Rice cooking is an important skill in Korea; the rice must be neither underdone nor overcooked and mushy. Short-grain varieties are usually preferred, both a regular and glutinous (sticky) type, the latter most often in sweets. Long-grain rice, called Vietnamese rice in Korea, is available but not common. Millet and barley are used, most often as extenders for rice. Noodles are also an important staple and are made from wheat, buckwheat, or mung beans. The buckwheat variety is often used in cold dishes. Fritters, dumplings, and pancakes flavored with scallions, chile pepper, and sometimes fish or meat placed directly in the batter before cooking are other popular grain dishes.

Vegetables are served at every meal. Chinese cabbage (both *bok choy* and *napa*), European cabbage, and a long, white radish (similar but not identical to the Japanese *daikon* radish) are eaten most often. Eggplant, cucumbers, perilla (a mint-like green also called shiso), chrysanthemum greens, bean sprouts, sweet potatoes, and winter melon are also very popular. Vegetables are added to soups and braised dishes and are often served individually as hot or cold side dishes. Pickled, fermented vegetables are included at every meal, usually in the form of *kimchi.* There are many types of kimchi, but most are made with shredded Chinese cabbage and white radish, heavily seasoned with garlic, onions, and chile peppers, then fermented. Cucumber, eggplant, turnip, and even fruits or fish are sometimes added. Some recipes are mild, but most are very hot. Seaweed is eaten as a vegetable, including kelp and laver (called *kim*). Kim is brushed with sesame oil, salted, and toasted to make a condiment. Fruits are eaten mostly fresh. Crisp, juicy Asian pears (known as apple pears in the United States) are very popular; apples, cherries, jujubes (red dates), plums, melons, grapes, tangerines, and persimmons are also common.

Fish and shellfish are eaten throughout Korea. Fresh fish dishes are preferred near the coast or in river regions, but dried or salted fish is more common in the inland areas. *Saewujeot,* a Korean fermented fish sauce, is made from tiny shrimplike crustaceans. It flavors many dishes. Beef and beef variety cuts are especially popular in Korea. Cubes, thin slices, or small ribs of marinated beef are barbecued or grilled at the table over a small charcoal brazier or gas grill. *Bulgogi,* grilled strips of beef flavored with garlic, onions, soy sauce, and sesame oil, is best known. Another Korean specialty is the fire pot (*sinsullo*), similar to the Mongolian hot pot, featuring beef or liver, cooked egg strips, sliced vegetables (e.g., mushrooms, carrots, bamboo shoots, onions), and nuts that are cooked in a seasoned broth heated over charcoal. After the morsels of food have been eaten, the broth is served as a soup. Chicken and poultry are not especially popular in Korea. Soybean products, however, are common, including soy sauce and soy paste. Bean curd, made from soybeans (*tobu*) or mung beans (*cheong-po*), is a favorite. Mung beans, adzuki beans, and other legumes are steamed and added to many savory and sweet dishes. Nuts, such as pine nuts, chestnuts, and peanuts, and toasted, crushed sesame seeds are frequent additions as ingredients or garnish.

Sample Menu

A KOREAN DINNER

Miyak gook (seaweed soup)

Steamed rice

Bulgogi (Korean barbecue beef)

Seasoned *tobu* (bean curd), chrysanthemum leaf salad, and seasoned eggplant

Kimchi

Barley water or Ginseng tea

Seasonings are the soul of Korean cooking. Garlic, ginger root, black pepper, chile peppers, scallions, and toasted sesame in the form of oil or crushed seeds flavor nearly all dishes. Prepared condiments, such as soy sauce, fish sauce (*saewujeot*), hot mustard, and a fermented chile paste called *kochujang*, are also frequently added to foods. Marinades and dipping sauces are common.

Soup or a thin barley water is used as a beverage. Herbal teas are very popular; ginseng tea flavored with cinnamon is a favorite. Ginger, cinnamon, or citron can also be used separately to make a spice tea. A common drink is rice tea, made by pouring warm water over toasted, ground rice or by simmering water in the pot in which rice was cooked. On special occasions, wine might be served. Wines are made from rice and other grains; some include various flower blossoms or ginseng as flavorings. Beer is also well liked. Milk and other dairy products are generally not consumed or used in cooking.

Meal Composition and Cycle

Daily Patterns

Three small meals, with frequent snacking throughout the day, are typical in Korea. Breakfast was traditionally the main meal in Korea, but today is more likely to be something light, caught on the run. Soup is almost always served at breakfast, along with rice (usually as gruel). Eggs, meat or fish, or vegetables may top the meal. Kimchi and dipping sauces are the usual accompaniments.

Lunch is typically noodles served with broth of beef, chicken, or fish and garnished with shellfish, meat, or vegetables. Supper is more similar to breakfast, but with steamed rice. In many modern homes, it has become the main meal of the day. Snacks are widely available from street vendors, including grilled and steamed tidbits of all types. Sweets, such as rice cookies and cakes, or dried fruit (especially persimmons) are also popular snacks.

Rice is considered the main dish of each meal. Everything else is served as an accompaniment to the rice and is called *panch'an*. For dinner, at least one meat or fish dish is included, if affordable, and two or three vegetables are usually served. Kimchi is always offered. Soup is very popular and is served at most meals. Individual bowls of rice and soup are served to each diner, and panch'an dishes are served on trays in the center of the table for communal eating.

■ It is believed that a Portuguese Catholic priest introduced chile peppers into Korea in the 16th century. They were quickly adopted and have become ubiquitous in the cuisine.

■ *Kochujang*, a fermented jamlike chile paste, is prepared by each family on March 3, then traditionally stored for use throughout the year in black pottery crocks.

■ Korean rice wine, *mukhuli*, is customarily served with scallion-flavored pancakes as a snack.

Table 11.4
Cultural Food Groups: Korean

Group	Comments
PROTEIN FOODS	
Milk/milk products	Milk and milk products are generally not consumed or used in cooking.
Meat/poultry/fish/eggs/legumes	Beef and beef variety cuts are especially popular. Barbecuing is a popular method for cooking meat. Fish and shellfish, either fresh, dried, or salted, are eaten throughout Korea. Soybean products are added to many dishes.
CEREALS/GRAINS	Rice is the most important component of the Korean diet. Noodles made from wheat, mung bean, or buckwheat flours are an important staple.
FRUITS/VEGETABLES	A wide variety of fruits are consumed. Vegetables are often pickled and are eaten at every meal.
ADDITIONAL FOODS	
Seasonings	Sweet, sour, bitter, hot, and salty tastes are combined in all meals. Dishes are often seasoned during and after cooking.
Nuts/seeds	
Beverages	Herbal teas are popular, as well as rice tea. They are commonly served after the meal. Soup or barley water is used as a beverage with the meal.
Fats/oils	Animal fat is rarely used.
Sweeteners	Sweets are made for snacks and special occasions.

Wine may be served before the meal with appetizers such as batter-fried vegetables, seasoned tobu, pickled seafood, meatballs, or steamed dumplings. Dessert is seldom eaten, although fresh fruit sometimes concludes the meal. Hot barley water or rice tea is served with the meal.

Drinking is a social ritual, practiced mostly by men. Distilled beverages like *soju,* a sweet potato vodka, are consumed with snacks like spicy squid or chile peppers stuffed with beef.

Etiquette

Chopsticks and soup spoons are the only eating utensils used in Korea. In the past, seating was around a low rectangular table on the floor, and young women were assigned the awkward corner seats (today most tables are set with chairs). Traditionally elders are served first and children are served last at the meal. It is considered polite to fill the soy sauce dish of the people one is sitting beside. Food is always

Table 11.4
Cultural Food Groups: Korean—continued

Common Foods	Adaptations in the United States
	Milk and cheese product consumption increases.
Meat: beef, variety meats (heart, kidney, liver), oxtail, pork *Fish and shellfish:* abalone, clams, codfish, crab, cuttlefish, jellyfish, lobster, mackerel, mullet, octopus, oysters, perch, scallops, sea cucumber, shad, shrimp, squid, whiting *Poultry and small birds:* chicken, pheasant *Eggs:* hen *Legumes: adzuki,* lima beans, mung beans, red beans, soybeans	Beef, pork, and poultry consumption rises. Fish is eaten frequently. Many still consume tofu several times a week. Younger Korean Americans may consume more pork products.
Barley, buckwheat, millet, rice (glutinous and long-grain), wheat	Rice consumption declines, but it is still eaten every day.
Fruits: apples, Asian pears, cherries, dates (*jujubes,* red date), grapes, melons, oranges, pears, persimmons, plums, pumpkin, tangerines *Vegetables:* bamboo shoots, bean sprouts, beets, cabbage (Chinese, European), celery, chives, chrysanthemum leaves, cucumber, eggplant, fern, green beans, green onion, green pepper, leaf lettuce, leeks, lotus root, mushrooms, onion, peas, perilla (*shiso*), potato, seaweed (*kim*), spinach, sweet potato, turnips, water chestnut, watercress, white radish	Increased intake of fruits and vegetables is noted. The majority eat *kimchi* daily.
Chile peppers (*kochujang*—fermented chile paste), chinese parsley (cilantro), cinnamon, garlic, ginger root, green onions, MSG, hot mustard, red pepper sauce, pine nuts, rice wine, *saewujeot* (fermented fish sauce), sesame seed oil, sesame seeds, soy sauce, sugar, vinegar, sea salt	
Chestnuts, gingko nuts, hazelnuts, peanuts, pine nuts, pistachios, sesame seeds, walnuts	
Barley water, beer, coffee, fruit drinks, green tea, honey water, jasmine tea, magnolia flower drink, rice tea, rice water, rice wine, *soju* (sweet potato vodka), spiced teas (ginseng, cinnamon, ginger), wines made from other ingredients	Hot barley water is still the preferred beverage and is served after the meal.
Sesame oil, vegetable oils	
Honey, sugar	

passed with the right hand, and a communal beverage may be passed for all to share.

Special Occasions

Korean cooking was historically divided into everyday fare and cuisine for royalty. The traditions of palace cooking and food presentation, including the use of numerous ingredients in elaborate dishes, are seen today in meals for special occasions. At a meal celebrating a birthday or holiday, or one shared with guests, more dishes are served and both wine and dessert are offered. For special occasions, Koreans offer a thick drink of persimmons or dates, nuts, and spices, or a beverage made with molasses and magnolia that is served with small, edible flowers floating on top.

Both Koreans and Korean Americans celebrate several holidays throughout the year. The first is New Year's, called *Sol,* a 3-day event at which traditional dress is worn and the elders

■ At Korean weddings, sweetened dates rolled in sesame seeds are tossed at the bride to ensure health, prosperity, and numerous children.

Korean Americans celebrate *Sol,* New Year's, on the first full moon. (© Gary Conner/Photo Edit.)

in the family are honored. Festivities include feasts, games, and flying kites. On the first full moon, in a tradition reflecting ancient religious rites, torches are lighted and firecrackers are set off to frighten evil spirits away. Shampoo Day (*Yadu Nal*) is on June 15, when families bathe in streams to ward off fevers. Thanksgiving (*Chusok*) is a fall harvest festival; *duk,* steamed rice cakes filled with chestnuts, dates, red beans, or other items, are associated with the holiday.

A special ceremony is observed on a child's first birthday. He or she is dressed in traditional clothing and placed among stacks of rice cakes, cookies, and fruit. Family and friends offer the child objects symbolizing various professions, such as a pen for writing or a coin for finance, and the first one accepted is thought to predict his or her future career (Nash, 1995).

Therapeutic Uses of Food

Many Koreans follow the um and yang food classification system. Little has been reported about specific foods, although Koreans are believed to adhere to categorizations similar to other Asians: um (cold) foods include mung beans, winter melon, cucumber, and most other vegetables and fruits; meats (e.g., beef, mutton, goat, dog), chile peppers, garlic, and ginger are considered yang (hot) foods.

Preparation of healthy, tasty food is an important way that Korean women show affection for family and friends (Pang, 1991; Marks & Kim, 1993). Good appetite is considered a sign of good health. Foods that are believed to be health promoting include bean paste soup, beef turnip soup, rice with grains and beans, broiled seaweed, kimchi, and ginseng tea. Ginseng products are often used to promote health and stamina and to alleviate tiredness. One very popular tonic, called *boyak,* combines ginseng and deer horn or bear gallbladder (Pang, 1989). More than one-half of Korean American respondents in one survey reported using ginseng, with older women more likely to use such products than men or younger women (Yom et al., 1995). Other home remedies commonly used include ginger tea, *yoojacha* (hot citrus beverage), bean sprout soup, and lemon with honey in hot water. Restorative herbal medicines, vitamin supplements, meat soups, bone marrow soup, and *samgyetang* (game hen soup) also were mentioned by subjects as useful when feeling weak.

A study of pregnant Korean American women (Ludman et al., 1992) suggests that certain foods, such as seaweed soup, beef, and rice, are thought to build strength during this "difficult" time. Food taboos during pregnancy often involve the concept of like causes like. For instance, eating blemished fruit may result in a baby with skin problems. Other such items

■ Dog meat soup is eaten by men to enhance their strength and physical prowess. The South Korean government banned dog meat soup during the 1988 Olympics in Seoul in deference to outside cultural pressure.

■ *Boyak* is often given as a gift, particularly to one's parents, to promote long life.

mentioned are chicken, duck, rabbit, goat, crab, sparrow, pork, twin chestnuts, and spicy foods. Although most food avoidance was attributed to personal preference and availability, many of the women acknowledged familiarity with the traditional beliefs.

Contemporary Food Habits in the United States

Adaptations of Food Habits

Ingredients and Common Foods

A survey of Korean Americans in the San Francisco area showed that many traditional Korean food habits continue after immigration to the United States. Nearly all respondents ate rice at least one meal each day, and two-thirds ate kimchi daily. Beef and beef variety cuts were consumed regularly, and fish was eaten at least once a week. Sesame oil/seeds and vegetable oils were used more often than butter or mayonnaise. Soy sauce, soybean paste, kochujang, and garlic were the most popular condiments. More than 40 percent of Korean Americans surveyed consumed tobu several times a week. Pork and pork products were not commonly eaten, although younger respondents reported more frequent use of these foods than older subjects. Length of stay in the United States had little impact on diet (Yom et al., 1995).

A national study of mostly first-generation Korean Americans found similar trends. Regular consumption continued of some traditional dishes and ingredients, such as rice, kimchi, garlic, scallions, Korean soup, sesame oil, Korean stew, soybean paste, and kochujang. Frequently consumed American foods included oranges, low-fat milk, bagels, tomatoes, and bread. Foods common to both cultures that were regularly eaten included onions, coffee, apples, eggs, beef, carrots, lettuce, fish, and tea. The researchers found that most respondents had access to Korean items but that as participation in American social life increased, acceptance of American foods increased. Persons who had someone who was familiar with Korean fare to cook for them were more likely to continue eating traditional dishes than were those who cooked for themselves (Lee et al., 1999b).

Although milk and dairy products are not consumed in Korea, these foods are often well accepted in the United States. One study indicated that more than one-half of the subjects reported drinking milk or eating yogurt or cheese one or more times each week (Yom et al., 1995). Another study of pregnant Korean American women reported a similar finding that 56 percent of respondents drank milk daily.

■ Korean women traditionally consumed seaweed soup, *miyuk kook,* three times a day for 7 weeks after the birth of a child to restore their strength.

Meal Composition and Cycle

Little has been reported regarding Korean American meal composition and cycle. It is assumed that because Korean meal and snacking patterns are similar to those in the United States, three meals a day remain common. One study found American foods were most popular at breakfast and lunch, but traditional Korean dishes were favored for dinner (Lee et al., 1999a). Survey data suggest that hot barley water is still the preferred beverage of Korean Americans and that it is served after meals (Yom et al., 1995).

Korean Americans observe traditional Korean holidays (see "Special Occasions," previously in this section) as well as other events, such as Buddha's birthday on April 8, Korean Memorial Day (June 6), South Korean Constitution Day (July 17), and Korean National Foundation Day (October 3). Fathers are honored on June 15. In addition, Americans of Korean descent who are Christian celebrate the major religious holidays.

Nutritional Status

Nutritional Intake

Few health studies focusing on Korean Americans have been published; research on Asian Americans often combine heterogeneous Asian and Pacific Islander populations. One comparison of the nutritional status of elderly Asian Americans indicated that Koreans had the poorest diet (Kim et al., 1993). Nearly one-half consumed inadequate amounts of energy, vitamin A, and riboflavin. Almost one-fourth consumed less than two-thirds the recommended daily amount (RDA) of protein, phosphorus, and vitamin C. 20 percent of the men and 25

■ Korean Americans have high rates of glucose 6-phosphate dehydrogenase deficiency disease.

percent of the women had inadequate protein intake. Iron and niacin were the only two nutrients of those studied that were consumed in adequate quantity. Korean Americans reportedly have low calcium consumption as well (Kim et al., 1984, 1993).

Age-adjusted mortality rates for Korean Americans are much lower than for the general population, although infant mortality rates are somewhat higher than that for whites (Gardner, 1994). The leading cause of death among Koreans is stomach cancer; higher incidence of stomach, liver, and esophageal cancer have been found among Korean Americans as compared to whites (Jenkins & Kagawa-Singer, 1994). Factors associated with high risk of these cancers in the Korean diet include low intake of vitamins A (including beta-carotene) and C, and high intake of sodium and chile peppers. Aflatoxins are also sometimes found in Korean soy sauce (Sawyers & Eaton, 1992). In addition, the rate of hepatitis B (responsible for most liver cancer) is very high in Korean Americans.

Information on obesity in Korean Americans is limited. Preliminary data suggest that length of stay in the United States and affluence is positively correlated with being overweight in men; women, however, do not demonstrate this trend. Korean American women show only minor weight gain with length of stay and are more likely to be thin if wealthy. In the study on Asian American elders detailed previously (Kim et al., 1993), 38 percent of the Korean women were found to be thin or very thin. The researchers suggest that this population may be at risk for protein-energy malnutrition. It has also been reported that Korean Americans do not exercise regularly (Crews, 1994).

In general, Asians and Pacific Islanders in the United States develop non-insulin-dependent diabetes mellitus at rates significantly higher than those of the white population. One study in Hawaii found that Korean Americans had a prevalence rate twice that of whites.

Hypertension rates among Korean Americans average 16 percent. Although this is lower than the figure for all adult Americans (30 percent), it should be noted that one study indicated that the rate among Korean American men was higher than 22 percent, while the rate for women was only 3 percent. Family history is the major risk factor; no correlation between diet (including alcohol, spicy food, or sodium intake) has been noted (Tamir & Cachola, 1994). Although rates of hypertension are relatively low, a California study suggests that Korean Americans are less likely than the general population to be aware of their high blood pressure, to receive treatment, or to have their hypertension in control (Stavig et al., 1988). Incidence of death from heart disease also has been found to be very low.

A study of alcoholic consumption patterns in Asians suggests that 25 percent of Korean American men can be characterized as heavy drinkers, while 44.5 percent abstain completely; less than 1 percent of Korean American women are heavy drinkers, and 75 percent are abstainers. This prevalence rate of alcohol abuse among men is equivalent to that of the general U.S. population (Chi et al., 1989). Having friends who drink was a major factor in predicting drinking behavior, confirmed in another study indicating that, for 77 percent of Korean Americans, drinking occurred at social occasions (Zane & Kim, 1994).

Counseling

Korean American clients are similar to many Asians in that language barriers may interfere with counseling. Proficient translators and cultural interpreters are often crucial to effective communication. Cultural attitudes may also interfere with health care. A client may be ashamed of needing help and fear being a burden to other family members during illness. The stresses of acculturation may be especially severe in the Korean American community due to the large number of recent immigrants. Elders Korean Americans in one study were shown to have difficulties with adjustment when they had low levels of educational attainment and when they lived alone (Kiefer et al., 1985).

Koreans use a quiet, nonassertive approach to conversation. Expression of emotions is avoided, and loud talking or laughing is considered impolite (although a Korean may laugh excessively when embarrassed). An indirect approach to topics is appreciated. A Korean may be hesitant to say no or to disagree with a statement. Instead, tipping the head back and sucking air through the teeth are often used to signal dissension. Direct eye contact is expected, and it is used to demonstrate attentiveness and sincerity. Touching is uncommon, except for a light, introductory handshake between men. Hugging, kissing, and backpatting should be avoided, as should crossing one's legs or putting feet on the furniture. It is considered respectful to rise whenever an elder enters the room and to touch the palm of the left hand to the elbow of the right arm when shaking hands or passing an item to an elder (Axtell, 1991; Morrison et al., 1994).

The family is responsible for all its members and is usually involved in health care. In one study it was found that the majority of Korean Americans believed that a patient should not be told of a terminal illness and that the family should make any life support decisions, although younger, better educated, and wealthier Korean Americans were more open to patient autonomy (Blackhall et al., 1995). It also is helpful to determine who is head of the household (often the father, eldest son, or other male member).

Some Koreans find biomedical health systems very unsatisfactory. They become frustrated when a physician relies on lab tests to determine illness, especially if they are told that there is no reason for a specific complaint, or conversely, that some problem exists even though they have no symptoms. They also may disagree with the cause of a complaint; some somatic symptoms may result from emotional distress, yet mental illness is highly stigmatized. Furthermore, they may expect physical treatments (i.e., acupuncture or cupping) from the provider and are likely to want a permanent cure (Pang, 1991). Inconvenient hours and the need for an appointment may also discourage the use of biomedical health care.

In the United States, clients often administer home remedies and consult shamans or hanui before (or while) they seek advice from biomedical health care providers. A study of Korean Americans in Los Angeles revealed that more acculturated and better educated individuals were more likely to use traditional health practices than those of lower socioeconomic status, perhaps due to greater affordability of multiple providers (Sawyers & Eaton, 1992). There is some concern regarding the overmedication of some clients who are using herbal remedies in addition to prescriptive medications (Chin, 1992). For example, ginseng may act as an antihypertensive and multiply the effect of other antihypertensive drugs (Sawyers & Eaton, 1992). Korean Americans must sometimes choose between treatments when traditional and biomedical health practitioners work at cross-purposes; professionals of both systems have been known to advise clients against using the services of the other (Pang, 1989).

Religious affiliation varies in Korean American clients and may have a significant impact on health and nutritional care. An in-depth interview should be used to ascertain religious beliefs as well as any traditional health practices used and personal dietary preferences.

■ **As with other Asians who use moxibustion or cupping, it is important for health care providers to determine the cause of burns and bruises on clients before assuming abuse has occurred.**

12

Southeast Asians and Pacific Islanders

Southeast Asians and Pacific Islanders live in similar tropical regions, and it is believed that they may share some common ancestors, yet their cultures have diverged markedly over the centuries. The countries of Southeast Asia have developed under hundreds of years of Chinese domination. Spanish expansionism in the Philippines and French occupation in Vietnam were also significant. In contrast, Asian and European contact in the Pacific Islands was limited until the eighteenth century, and subsequent foreign influence almost completely overwhelmed the traditional indigenous societies.

Southeast Asians and Pacific Islanders are some of the newest, youngest, and fastest growing populations in the United States. More than 2 million mainland Southeast Asians and Filipinos have arrived since 1975; the number of Pacific Islanders has increased by 50 percent in the last decade. This section discusses the cultures and cuisines of the Southeast Asians who have immigrated in substantial numbers to the United States—Filipinos, Vietnamese, Cambodians, and Laotians—as well as those Pacific Islander groups with significant American populations—native Hawaiians, Samoans, Guamanians, and Tongans.

Southeast Asians

Cultural Perspective

History of Southeast Asians in the United States

Immigration Patterns

Most Filipinos chose to come to the United States for educational and economic opportunities. In contrast, the majority of mainland Southeast Asians who have immigrated to the United States have arrived since the 1970s as refugees from the political conflicts of the region.

The Philippines. Substantial immigration from the Philippines to the United States started in 1898 after the country became a U.S. territory. Approximately 113,000 young male Filipinos traveled to the Hawaiian Islands between 1909 and 1930 to work in the sugarcane fields. Many of these immigrants later moved to the U.S. mainland. These early immigrants were considered U.S. nationals and carried U.S. passports, but most were not allowed to become citizens or own land. Most were uneducated peasant laborers from the island of Illocos. Because

■ The only Filipinos allowed to apply for U.S. citizenship prior to 1946 were those who had enlisted in the U.S. Navy, Naval Auxiliary, or Marine Corps during World War II; had served at least 3 years; and had an honorable discharge.

■ During the Marcos regime, inexpensive air fares were used to lure Filipinos in the United States (and their American dollars) to return to the Philippines for visits.

they were not allowed to bring their wives or families, social and political clubs replaced the family as the primary social structure.

In 1924 the immigration of Filipinos slowed as a result of Asian exclusion laws. After World War II, it became legal for Filipinos to become U.S. citizens, and the number of immigrants increased. Significant numbers of Filipinos arrived in the United States after 1965, when the U.S. immigration laws were changed, and by 1980, more than 350,000 had emigrated to mostly urban areas. Two-thirds of the Filipinos who immigrated after World War II qualified for entrance as professional or technical workers. Yet discrimination against Asians often forced Filipinos into low-paying jobs. "Little Manilas" formed in many California cities, and similar homogeneous neighborhoods were found in Chicago, New York, and Washington, D.C. where some Filipinos opened small service businesses to meet the needs of their community.

■ Approximately 9,000 Thais have immigrated to the United States since the 1970s; most have settled in Los Angeles, New York, and Texas. They are known for their rapid acculturation.

■ During the 1920s and 1930s, Filipinos were often barred from restaurants, swimming pools (and other recreational facilities), and movie theaters because of their dark skin and heavy accents.

■ There are over 234,000 Canadians of Filipino heritage.

Vietnam. Vietnamese immigration to the United States is characterized by three distinct waves. The first occurred when South Vietnam fell to the North in 1975 and 60,000 Vietnamese left the country with the assistance of the United States. Another 70,000 managed to flee on their own. Most of these refugees had been employed by the United States or were members of the upper classes. The latter group immigrated in intact family groups and were able to bring their property with them.

From 1975 to 1977, another wave of Vietnamese left for political or economic reasons, often escaping by sea. Many of these "boat people" left their families and what little money they had to find freedom. The third phase of immigration started in 1978, when increasing numbers of ethnic Chinese living in Vietnam fled the country. This wave of immigration was accelerated by the Chinese invasion of northern Vietnam in 1979. This second group of boat people left with no financial resources, and many lost family members escaping in unseaworthy vessels that were easy prey for pirates. After being rescued, the boat people often lived for several months, even years, in refugee camps in various Southeast Asian countries before immigrating to the United States. In addition, the United States and Vietnam developed the Orderly Departure Program (ODP) in 1979 to bring imprisoned former South Vietnamese solders and the approximately 8,000 Amerasians, children of U.S. fathers and Vietnamese mothers, to the United States.

Cambodia and Laos. Cambodian and Laotian immigration to the United States did not begin until the United States granted asylum to the residents of the refugee camps along the border with Thailand in 1976 to 1979. International concern about the large numbers of refugees and the conditions in the camps prompted accelerated admittance of Southeast Asian immigrants worldwide.

Current Demographics and Socioeconomic Status

Filipino. The Filipino American population quadrupled from 1890 to 1990. Nearly 1.5 million Filipinos now live in the United States. Almost half reside in California, with significant populations of more than 30,000 Filipinos found also in Hawaii, Florida, Illinois, New York, New Jersey, Texas, and Washington.

In general, Filipino Americans earn less than white Americans, yet as a group, they have a poverty rate (6 percent) that is less than half that of the general U.S. population. Many work as migrant field hands or cannery workers. Although more recent immigrants from the Philippines are better educated than those who arrived earlier, their education or professional experience is often not recognized in the United States, forcing them to accept blue-collar jobs when they first arrive. Over time, many professionals do obtain licensing or accreditation and work as physicians, nurses, or lawyers. Others are employed in the sales and service industries, and this group is economically better off than previous immigrants, many of whom are now poor, elderly single men. A schism has developed in the Filipino American community between pre-1965 immigrants who believe the newer immigrants are ungrateful for the advances achieved by the older generation and the more recent immigrants who find the older generation naïve and uneducated (Melendy, 1995).

Vietnamese. More than 550,000 Vietnamese have entered the United States since 1975, and the total Vietnamese population, including those born in America, was more than 614,000 in the 1990 U.S. census; the growth rate between 1980 and 1990 was more than 150 percent. Many were sponsored by American agencies or organizations, which provided food, clothing, and shelter in cities throughout the United States until the Vietnamese could become self-supporting. Many Vietnamese Americans have since relocated in the western and Gulf states, probably because the climate is similar to that of Vietnam.

Vietnamese Americans live primarily in urban areas. Many are young and live in large households with grandparents and other relatives. The first wave of Vietnamese immigrants was well educated, could speak English, and had held white-collar jobs in Vietnam. Many of them had to accept blue-collar jobs initially and had difficulties supporting their extended families. The boat people were even less prepared for life in the United States, because they often had no English language skills, were illiterate, had less job training, and did not have support from an extended family.

Language acquisition has been important in the Vietnamese community, and in 1990, nearly three-fourths of Vietnamese in the United States could speak English well (Bankston, 1995d). Many immigrants have made the decision to assimilate as quickly as possible, even changing their Asian names, such as "Nguyen," to something more Anglicized, such as "Newman." Education is highly valued, and a child's academic achievement is considered a reflection on the whole family. High school dropout rates are very low among Vietnamese Americans, and almost half attend college, compared to 40 percent of white Americans.

Vietnamese have high rates of employment, but often in low-paying jobs. Nearly 25 percent lived in poverty in 1990. They are prominent in the fishing and shrimping industry in the Gulf states, and those who obtain college degrees often prefer technical professions, such as engineering.

Cambodian. The 1990 U.S. census identified approximately 150,000 Americans of Cambodian descent, but it is thought that this figure may underestimate actual numbers due to a low response rate by Cambodians unfamiliar with American society in the census count. Nearly half of Cambodian Americans reside in California, with the greatest populations found in Long Beach and Stockton. Another large Cambodian community has developed in Lowell, Massachusetts. Texas, Pennsylvania, Virginia, New York, Minnesota, and Illinois also have significant numbers of immigrants.

Adjustment to American life has been difficult for Cambodians, most of whom came from rural regions where they worked as farmers. They have a high unemployment rate (although the probability of employment increases with length of stay), and those who work have found employment mostly in service jobs and manual labor. In 1990, 42 percent of Cambodian families lived below poverty level.

Low levels of education have limited economic success for many Cambodian Americans. Approximately one-half of adult men and women have less than 6 years of formal education, and only 5 to 10 percent have graduated from high school. Low dropout rates among younger Americans of Cambodian descent may contribute to better economic outcomes in the future.

Laotian. About 150,000 nontribal Laotians live in the United States, according to 1990 U.S. census figures. Most Laotians have settled in the California cities of Fresno, San Diego, Sacramento, and Stockton. Large populations are also found in Amarillo and Denton, Texas; Minneapolis, Minnesota; and Seattle, Washington. Another 150,000 Hmong and approximately 15,000 Mien have been identified. Refugee resettlement services dispersed the Hmong, irrespective of family clan associations, in 53 American cities. Secondary migration of Hmong Americans has reunited clan groups, mostly in the suburban and rural communities of California's Central Valley—Fresno, Merced, Sacramento, Stockton, Chico, Modesto, and Visalia.

Poor fluency in English and low education levels have hindered economic achievement for Laotian Americans. In 1990, two-thirds reported that they did not speak English well. Only 40 percent of adults have completed high school; current Laotian students are generally believed to do well in school, but higher than average high school dropout rates and lower than

■ Vietnamese immigrants are often stereotyped as either successful overachievers or poverty-stricken refugees.

■ In 1993, the Canadian census reported approximately 172,000 residents of Southeast Asian heritage.

■ Fresno, California, with 30,000 Hmong immigrants, is the largest Hmong community in the world.

■ There are five major tribes among the Hmong—the White Hmong, the Black Hmong, the Blue Hmong, the Red Hmong, and the Flowery Hmong—named according to legend by the colors of clothing they were forced to wear by the ancient Chinese for identification purposes. Most Hmong refugees in the United States come from the White and Blue tribes.

■ The Vietnamese say, "Chew when you eat, think when you speak."

■ In Thailand, miniature Thai temples are built and posted on pedestals next to a home to shelter the family ancestors. Such spirit houses are believed to prevent the ghosts from moving in with the family and causing trouble.

■ Hmong ancestors are fed at every festive occasion with a little pork and rice placed in the center of the feast table.

average college attendance rates have been noted. Many Laotians hold jobs as machine operators, fabricators, and laborers, and the unemployment rate is high. In 1990, nearly one-third of Laotian Americans lived in poverty.

Almost two in every three Americans of Hmong descent lived below the poverty line in 1990; nearly all Hmong were employed in blue-collar jobs or manual labor. Some enterprising Hmong women have been successful in marketing the traditional Hmong needlecraft known as *paj ntaub.* The colorful wall hangings, pillows, and bedcoverings have become popular among collectors of native arts. Low levels of educational attainment have been reported for adult Hmong, but Hmong Americans born or raised in the United States show surprisingly high rates of college attendance, nearly equal to whites and surpassing African Americans.

Worldview

Religion

Many Southeast Asians hold beliefs dating back to the ancient religions prevalent before the introduction of Buddhism, Catholicism, and Islam. Most believe in spirits and ghosts, especially of ancestors, who have the ability to act as guardians against misfortune or cause harm and suffering. Ideas about spirit intervention have often been incorporated into eastern and western religious practices or survive as significant superstitions.

Filipino. The majority of all Americans of Filipino descent are Roman Catholics, although it is estimated that as many as 5 percent are Muslim. Religion significantly affects the worldview of Filipinos, especially elders. Many hold that those who lead a good life on earth will be rewarded with life after death. Human misfortunes come from violating the will of God. One should accept one's fate, because supernatural forces control the world. Time and providence will ultimately solve all problems.

Vietnamese. Nearly 70 percent of Vietnamese Americans are Buddhists, and 30 percent are Roman Catholic. Small numbers of Protestants also are found. Those who are Buddhists follow the Mahayana sect, influenced particularly by the Chinese school of Ch'an Buddhism (Zen Buddhism in Japan) called *Tien* in Vietnam. Buddhists believe that a person's present life reflects his or her past lives and also predetermines his or her own and his or her descendants' future lives. They consider themselves as part of a greater force in the universe (see chapter 4 for more information about Buddhism).

Cambodian. The predominant religion in Cambodia is Theravada Buddhism, which places greater emphasis on a person's efforts to reach spiritual perfection than does the Mahayana sect, which employs the help of deities. "Making merit" through good deeds, the participation in religious ritual, and the support of monks and temples is critical to one's progress through reincarnation. Although some Cambodian Americans have converted to Christian faiths, most still practice Buddhism, often in temples established in apartments or homes.

Laotian. Almost all Laotians are also Theravada Buddhists. "Making merit" for Laotians includes the expectation that every man will devote some time in his life to living as a practicing monk, either before marriage or in his old age. In the United States, men find it difficult to fulfill their obligation to the faith. Women may become nuns for periods in their lives as well, especially if widowed. Most Laotians in the United States worship at Buddhist temples with Cambodians or Thais who share their religious practices.

About half the Hmong population in the United States is now Christian as a result of French and American missionary work in Laos and conversions after arrival. Baptists, Presbyterians, Mormons, Jehovah's Witnesses, and members of the Church of Christ have actively recruited the Hmong. The other half of Hmong Americans practice animism, shamanism, and ancestor worship. They believe that the world is divided into two spheres—that which is visible, containing humans, nature, and material objects, and that which is invisible, containing spirits. The shaman acts as intermediary between the two; some spirits, such as those of ancestors, are available to those who aren't shamans. Women are generally responsible for making contact with these more accessible spirits. A few Americans of Hmong heritage belong to a modern Hmong religion called *Chao Fa* ("Lord of the Sky"), started by a prophet in the 1960s who encouraged the Hmong to break with both Laotian and Western ways.

It has been suggested that many Hmong became Protestants because they believed it was necessary to gain entry to the United States or in deference to their church sponsors. Some try to combine ancestor worship with their new faiths so as not to offend the spirits (Adler, 1995).

Family

Southeast Asians share a high esteem for the family, respect for elders, and interdependence among family members. Behavior that would bring shame to the family's honor is avoided, as is direct expression of conflict. Social acceptance and smooth interpersonal relationships are emphasized.

Filipino. The Filipino family is highly structured. At the center is the extended family, containing all paternal and maternal relatives. Kinship is extended to friends, neighbors, and fellow workers through the system of *compadrazgo.* Lifelong relationships are initiated through shared Roman Catholic rituals, particularly the selection of godparents and baptism of new babies. Community obligations created through this system include shared food, labor, and financial resources.

The first Filipino immigrants were single men hoping to make their fortune and return to the Philippines. They developed surrogate family systems of fellow workers who lived together, with the eldest man serving as patriarch. More recent immigrants have come as whole families and have been able to maintain much of their social organization. For example, as many as 200 sponsors may be appointed in a Filipino American baptism, although individual commitment may be less than if it were held in the Philippines (Melendy, 1995).

Filipino children are adored by the family and are typically indulged until the age of 6. At that time, socialization through negative feedback (e.g., shame) begins. Children are taught to be obedient and respectful, to contain their emotions and avoid all conflict, and to be quiet and shy. Politeness is emphasized (Anderson, 1983). A person must avoid shaming him- or herself, as well as the family; discord is minimized by the use of euphemisms and by sending go-betweens in sensitive situations.

Vietnamese. The extended Vietnamese family has been modified in the United States to adapt to American norms. A nuclear family is more typical, though somewhat larger than the

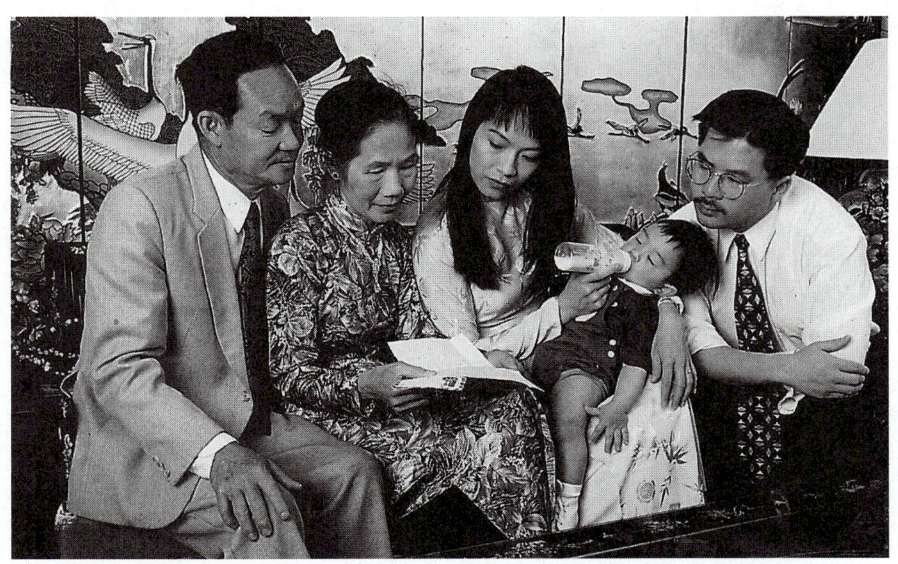

The traditional Vietnamese extended family home is less common in the United States, but relatives often live near each other. (© Michael Newman/PhotoEdit.)

average family in the United States. Close relatives are encouraged to move into homes next door or in the same neighborhood.

Family values are in transition. The father is traditionally the undisputed head of the household, but patriarchy is diminishing as women attain higher levels of education and professional achievement. American-style dating has become common. Although most Vietnamese marry within their ethnic group, women are more likely than men to wed non-Vietnamese mates. Divorce rarely occurs.

Intergenerational conflict is reportedly high. Children are often the first to learn English and acculturate more easily than their parents, causing value conflicts and loss of respect for elders. The role of the elderly is also changing. Old age is valued in Southeast Asia, but in the United States, older relatives are often physically isolated from their peers and even younger family members and may be linguistically isolated within the larger community.

■ A Cambodian proverb states, "If you don't take your wife's advice, you'll have no rice the next year."

■ When Cambodians wed, a Buddhist monk cuts a lock of hair from both the bride and groom to mix together symbolically in a bowl.

■ Hmong marriages are formalized at a 2-day feast featuring a roasted pig.

■ By custom, Hmong men marry the widows of their deceased brothers and support their nieces and nephews.

Cambodian. Large, extended families are common in Cambodia; children are considered treasures. Cambodians are notable in that their traditional kinship system was bilateral, emphasizing both the paternal and maternal lines. The family was primarily a matriarchy until the 1930s, when French influence strengthened the authority of the father. Today men are responsible for providing for their families, while women make all decisions regarding the family budget.

It has been difficult for Cambodian Americans to retain their traditional family structure in the United States. A large percentage of Cambodian homes are headed by single women, due most likely to the large numbers of men killed in recent years during conflicts in Cambodia. Furthermore, Cambodian American women are permitted formal education and their financial contributions are needed to help support the whole family. Differences between immigrants who have lived most of their lives in Cambodia and the children of these immigrants raised in the United States (with no memory of Cambodia) can be enormous in terms of language, acculturation, and values. It has been noted that many Cambodian teenagers suffer from identity problems.

Laotian. Most families are agriculturally based in Laos. Extended members live and work together in the fields to support the whole family. In Laos men represent the family in village affairs, and women run the home. Great significance is given to the site of the family's house; it is believed that as long as the site exists, the family will exist.

Extended families are still important to Laotian Americans due to dependency on relatives for social and economic support. Although nuclear families have become the norm, extended members tend to live nearby each other. Women have attained nearly equal status with men in the United States, and it is not unusual for Laotian men to share responsibility for household chores. Laotians have a notably low divorce rate.

Hmong American families are among the youngest and largest in the United States. According to 1990 figures, the average age of a Hmong American is almost 13 years, compared to 30 years for other Asians combined and 34 for U.S. citizens in general (Bankston, 1995b). More than half of all Hmong are under the age of 18. Typical family size is about 6 members; the Asian average is 3.7 and the white American average is just 3.1. These large families are usually nuclear, reflecting numbers of children rather than relatives. Extended family members are often located nearby, and families frequently congregate with other families from the same traditional clan from Laos.

Men remain the heads of households in America. Women are generally held in high regard in their roles as mothers. Children are the heart of the home, and much of family life revolves around them. As in some other Southeast Asian groups, a generation gap has developed between recently arrived immigrants and their westernized children. Teenage runaways have become increasingly common among the Hmong.

Some Hmong customs have come in direct conflict with U.S. laws. Polygyny, marriage to more than one wife, is unusual but not illegal in Laos. Clan leaders might wed several women to establish political connections, and having several wives was indicative of wealth. The practice that has received the most press in the United States, however, is the kidnapping of young women and enforced marriage. Traditionally girls married between the ages of 14 and 18 after a bridal price (paid by the groom) was agreed to by both families. If no agreement was reached, the couple could elope, and a mediator would help to settle the differences. If the bride was unwilling, the groom could kidnap the girl and the marriage would be recognized after some payment to the bride's family was arranged. Men who have attempted this in the United States have been charged with abduction and sexual assault, often by the young women involved. Many Americans of Hmong descent believe that it is best to wait until a woman is in her late teens or early 20s to wed; arranged marriages are still common.

Conversion to Christianity has been found to split many Hmong families when some members change faiths and others retain their traditional beliefs. Conflict over marriage traditions is especially prevalent.

Traditional Health Beliefs and Practices

Southeast Asian health concepts typically combine facets of multiple belief systems. Indigenous ideas about the origins of illness center on the supernatural world, particularly the intervention of malevolent spirits or the ghosts of angry ancestors. Chinese medical practices involving yin and yang, or the five evolutive elements (see chapter 11) are considerations in some areas of mainland Southeast Asia, while the Mexican hot-cold theory is more prevalent in the Philippines (see chapter 9). Religious precepts regarding rewards for "making merit" or performing good deeds and punishment for violating God's will are also involved in health maintenance. In the most general terms, keeping healthy requires personal harmony with the supernatural world, nature, society, and family fulfilled through one's obligations to one's ancestors, one's religion, and one's kin and community. Illness is usually defined by its cause, not its symptoms.

In the Philippines, health is maintained through the balance (*timbering*) of natural or supernatural elements (Anderson, 1983; Hart, 1981; Montepio, 1986–87). A person is thought to be predisposed to certain illnesses, and the timing of external events contributes to the development of disease. Unbalanced conditions, such as working too much, overeating, excessive drinking, inadequate diet or sleep, unhygienic conditions, infections, accidents, emotional stress (especially fright or anxiety), or loss of self-esteem may increase a person's vulnerability, as do factors such as the season and the weather.

Supernatural illnesses in the Philippines are most often due to the unhappy ghosts of one's ancestors, although witchcraft, or the powers of animal spirits, may also be involved. *Usog* or *tuyaw* occurs when a person transmits illness through the power of the "evil eye" or the use of hands, fingers, words, or even physical proximity (Montepio, 1986–87). It is believed that undesirable traits or conditions can also be transferred magically through contact with a person or object. A pregnant woman will avoid looking at a person with a deformity, for instance, to prevent a similar occurrence in her fetus. Conversely, a pregnant woman who looks at persons or objects with desirable qualities, such as beauty or intelligence, will assure similar characteristics in her child (Orque, 1983a). Many Filipino Americans are familiar with supernatural causes of disease, but do not believe that it can occur in the United States, because ghosts and spirits cannot cross the ocean, nor could they survive in the noisy cities where many Filipinos now live.

Most Filipinos adhere to the concept of *bahala na,* meaning that life is controlled by the will of God and by supernatural forces. If a person behaves properly, shows consideration of others and sensitivity in relationships, fulfills debts and obligations, shows gratitude, and avoids shame, he or she is rewarded with health in this life and eternal life after death (Anderson, 1983). Many Filipinos believe that illness is a punishment for transgressions against God. Religious medals are worn for protection from evil.

Spanish control of the islands administered through Mexico led to the adoption of some aspects of humoral medicine in the Philippines, including the Mexican theory of hot and cold, as well as the condition known as wind or air (*mal aire* in Mexico; see chapter 9, Mexican "Traditional Health Beliefs and Practices" section). Any imbalance in hot and cold, whether occurring through exposure to the elements or

■ **Cleanliness is extremely important to Filipinos, who often bathe twice per day to maintain a proper hot-cold balance in the body.**

■ **Some rural Filipinos will only build their houses on a cold site. One way they determine this is to bury a green coconut (in which the milk cannot be heard when shaken) at the potential location; the next morning it is retrieved, and if the milk is audible, the site is too hot for a home.**

■ The Vietnamese frequently consult astrologists to ascertain their future.

■ Among the ethnic Chinese Vietnamese immigrants, many are traditional Chinese medical practitioners.

■ Laotians often wear copper or silver bracelets or colored strings around their wrists, necks, and ankles to keep their souls from leaving. White string is used by families; red or black strings are tied on by shamans during ritual ceremonies.

through eating too many foods classified as hot or cold, is believed to cause illness. For example, a nursing mother who becomes overheated by too much sun or from exposure to a hot kitchen may find that her milk has become rancid, producing colic or diarrhea in the baby (Hart, 1981). Specifics on the application of hot and cold classifications and treatment vary tremendously from person to person.

Wind is of special concern. It may cause disease directly through drafts or be absorbed through the pores or any wounds, where if it is too cold or too hot it affects the blood, causing increased or decreased circulation that results in a general malaise and increased susceptibility to illness. A postpartum woman avoids bathing for 9 to 40 days after birth of the baby to prevent wind from entering her vagina; a newborn's umbilicus is bound to keep wind from entering that opening; and coconut oil may be rubbed into the skin to block the pores. Whooping cough and mental illnesses are two of the more serious conditions that can be caused by wind.

Also similar to Mexican health beliefs is the idea that strong emotion may cause certain symptoms of illness. Some Filipinos feel that excessive anger or envy are hot conditions, and great fright or joy are cold conditions. Somaticized complaints are common (Orque, 1983a, 1983b).

Three types of traditional healers are common in the Philippines: the midwife, the masseur, and the specialist who cures supernaturally caused illness. In urban regions, where belief in ghosts and spirits is not as prevalent, faith healers are gaining in popularity. Faith healers do not diagnose illness but cure it through prayer, anointing with oil, and the laying on of hands, which transmits a sacred healing energy to the patient (Montepio, 1986–87).

For the Vietnamese, health is related to personal destiny. How one behaved in past lives and the number of good deeds performed by one's ancestors determine one's experiences in this life. Current behavior, such as pleasing good spirits and avoiding evil spirits, can also impact health (Orque, 1983a, 1983b). Similar to the Filipinos, pregnant Vietnamese women may avoid ugly objects or leaving their homes at the times (noon and 5 P.M.) malevolent spirits are active. The use of divination, through fortune-telling, astrology, or physiognomy (the shape of the body, especially the head, as it correlates to the mind), is popular for predicting how a person might expect his or her life to proceed, and interventions that might be needed to prevent certain negative experiences.

Traditionally, the Vietnamese believe that the human body is sustained by three separate souls: one that encompasses the life force, one that represents intelligence, and one that embodies emotions. In addition, nine vital spirits provide assistance to the souls. Soul loss can be an important reason for illness and can be life-threatening. Typically, strong feelings, especially fright, can cause the soul to leave the body (Stephenson, 1995).

The Chinese medical system is commonly used by ethnic Chinese Vietnamese and by some other Vietnamese as well. Maintaining a balance of yin and yang, especially through diet and the treatment of disease, is a primary consideration in health. Like the Filipinos, "wind" (or "air") is sometimes seen as a cause of illness.

Cambodians, Laotians, and Hmong are also concerned with spiritual intervention in health. Laotians identify 32 spirits that oversee the 32 organs of the body (Bankston, 1995c). The Hmong recognize the world of the invisible, where the spirit of every animal, tree, and rock resides, amidst the souls of the living, ancestor spirits, caretaker spirits, and evil spirits. Ancestor spirits require special consideration, because if they become angry they may leave their progeny or fail to protect them from evil. The Laotians have elaborate rituals called *baci,* mostly performed at all special occasions by older men who have been monks, that bind the spirits to their possessor. Among the Hmong, the loss of one's soul, usually due to strong emotional distress, is the single most important cause of illness. It generally results in malaise and weight loss, leading to more serious disease. Related to soul loss is the condition called *ceeb,* or fright illness. It typically occurs to children (although it can happened to adults as well) if they are in an accident, chased by a dog, startled by a noise, or plunged into cold water. The souls becomes disconnected; the blood cools down and slows, resulting in a chilling effect that begins in the extremities and can progress to the vital organs (Capps, 1994).

Unique to Southeast Asians in the United States is the unexplained condition known as sudden unexpected nocturnal death syndrome (SUNDS), when a seemingly healthy person dies in his or her sleep. It is especially prevalent among Laotians, particularly Hmong, although it is thought to occur in all immigrants from the mainland. It was the leading cause of death among Hmong American men ages 25 to 40 years old in 1981 and 1982 (Adler, 1995). Although biomedical hypotheses have been proposed to account for the fatal syndrome, such as heart irregularities or sleep apnea, none have been proven. Some researchers believe that death is caused within the cultural context of the nightmare experience. Specifically, the nightmare spirit, *dab tsog,* enters the room at night and the victim "wakes" to the sensation of the spirit sitting on his or her chest; he or she is unable to move and is terrified. Although many immigrants report having experienced nightmares in Southeast Asia, the attack by the spirit does not usually result in death. It is believed that cultural disruptions have intensified the episodes. A Hmong man in the United States is often unable to perform his obligations to ancestor spirits and may also have difficulties in fulfilling his role as breadwinner and head of household. Guilt and depression create increased vulnerability to fatal nightmare experiences. Posttraumatic stress disorder, exposure to chemical warfare agents, or blood electrolyte imbalances may be other risk factors (Stephenson, 1995).

Traditional healers are typically specialized practitioners among mainland Southeast Asians. They may provide services for broken bones, skin infections, or objects stuck in the throat. Monks may lead religious rituals or shamans may intervene with the spirit world. Among most Southeast Asians, minor illnesses may be treated by anyone with healing experience or a spiritual calling (known as *neng* by the Hmong), typically a grandmother or mother in the home. The family takes responsibility for the illness of an individual and will usually exhaust all remedies available within the house before seeking outside help.

Botanical remedies are very popular with many Filipino and mainland Southeast Asians living in the United States. Cambodians, Laotians, and Hmong sometimes maintain herb gardens for easy access to therapeutic ingredients. Some immigrants frequent Chinese herbalists or will buy imported products from Asia (Gilman et al., 1992). Herbs and other substances are prepared as teas, broths, steam inhalants, or balms (Hoang & Erickson, 1982). Physical therapies may include massage; *cupping* (a heated cup or a cup with a small amount of burning paper is placed over a certain spot on the body until the fire goes out, leaving a round red spot on the skin); moxibustion (burning a small bundle of herbs on the skin or using a lit cigarette); flushing (a Filipino practice that encourages perspiration, purging, flatus, or menstruation); and *coining* (rubbing a coin or spoon dipped in tiger balm or eucalyptus ointment across the skin with pressure), scratching, or pinching affected areas. In most cases, the therapy is used to release any bad wind or excess heat and to restore balance to the body.

Religious rituals may also be used to intervene on behalf of an ill person. The Hmong use shamans in soul-calling ceremonies to return a lost soul to its host body, and the Mien appeal to ancestor spirits to protect family members and assist in healing. These rituals sometimes include animal offerings. A butchered animal, typically a chicken, pig, or occasionally a cow, is purchased from a packinghouse prior to the ceremony, then is cooked and consumed after the rite as part of a feast. In Vietnam, small shrines are sometime constructed to appease ancestor spirits or (in the case of pregnant women) the souls of premature infants who have died and still wander the earth. Offerings may also be made to the Goddess Quang Am for good health. Among Catholic Vietnamese and Filipinos, appeals are made to the Virgin Mary; group prayer has assumed significance for many Protestant Southeast Asians.

■ Mien rituals involving ancestor spirits require a genealogical record of the family going back 10 generations.

Traditional Food Habits

The cuisines of Southeast Asia have many ingredients in common, but food preparation methods and meal patterns reflect the foreign cultures that have influenced each nation. For example, the Vietnamese often serve cream-filled French pastries for dessert, whereas

Traditional foods of Southeast Asia and the Pacific Islands. Some foods typical of the diets of Southeast Asians and Pacific Islanders include coconut, dried anchovies, dried mango, French bread, lemon grass, lime, *nuoc mam*, pineapple, pork, rice, rice paper, rice sticks, taro root, and water chestnuts. (Photo by Laurie Macfee.)

Filipinos frequently have Spanish-style custard flan. As in China and Japan, the staple foods are rice (primarily long-grain), soybean products, and tea. A meal is not considered complete unless rice is included. Instead of soy sauce, however, Southeast Asians often season their food with strongly flavored fermented fish sauces and fish pastes.

Ingredients and Common Foods

Staples and Regional Variations

Filipino. Filipino fare has blended Malaysian, Polynesian, Spanish, and Chinese influences into a distinctive cuisine. There are three principles in Filipino cooking: First, never cook any food by itself; second, fry with garlic in olive oil or lard; and third, foods should have a sour-cool-salty taste. For example, *adobo,* a characteristic Filipino stew, combines chicken, pork, and sometimes fish or shellfish, first fried with garlic in lard and then seasoned with soy sauce and vinegar. Filipinos used a clay pot in the shape of a wok to fry, but they tend to leave the food in longer than the Chinese and allow it to absorb more fat. The common foods of the Philippines are listed in Table 12.1.

■ Filipinos say young persons should never sing in front of the stove or they will marry an old maid or widower.

■ In Filipino culture, sticky, glutinous rice cakes symbolize the cohesiveness of the family.

■ *Puchero* is thought to be an adaptation of the Spanish stew known as *cocido.* It is traditionally served for Sunday supper.

Rice is the foundation of the diet, and the long-grain variety accompanies the meal. It is typically steamed or fried (the preferred method of serving leftover rice). Short-grain, glutinous rice is often used for sweet desserts like *puto,* a sweet, fluffy cake made from rice, sugar, and sometimes coconut milk. A favorite bread, *pan de sol,* is made from rice flour. Noodles are also used extensively. *Pancit* is a popular dish made with rice, wheat, or mung bean noodles mixed with cooked chicken, ham, shrimp, or pork in a soy- and garlic-flavored sauce.

The amount of meat, poultry, or fish a family eats depends on economic status. Pork, chicken, and fish are popular, added as available to mixed dishes such as *sinigang,* a soup of fish or meat cooked in water with sour fruits, tomatoes, and vegetables; *chicken relleño,* a whole chicken stuffed with boiled eggs, pork, sausage, and spices; *puchero,* a beef, chicken, sweet potato, tomato, and garbanzo bean stew with an eggplant sauce; *lumpia,* which are the Filipino version of egg rolls, stuffed with pork, chile peppers, and vegetables like hearts of palm; and *paella,* a Spanish recipe for saffron-flavored rice typically topped with chicken, sausage, pork, seafood, tomatoes, and peas. Fil-

Two traditional Filipino dishes—*lumpia* (similar to an egg roll) and *pancit* (noodles cooked with meat or shrimp in a soy- and garlic-flavored sauce).

ipinos use all parts of the animal in their cooking. For example, *lechon* is a whole roasted pig served on special occasions. The pig variety cuts might show up in various soups or mixed stews, such as *dinu-guan,* consisting of diced pork, chicken, or entrails cooked in pig's blood and seasoned with vinegar and hot green chile peppers or sausage such as garlicky *longaniza.* The skin is commonly fried to make *sitsaron* (similar to the Mexican chicharrónes or American cracklings), which are eaten as snacks or pulverized to top noodle dishes.

The Filipinos drink *caraboa* (water buffalo) milk and use it to make one of the few native cheeses in Asia. Because of U.S. influence in the Philippines, many western dairy products are available, but cow's milk is used infrequently. Evaporated milk is a common ingredient in *leche flan,* a custard, and in *halo-halo,* a liquid dessert that is served in a glass but is eaten with a spoon and consists of shaved ice, coconut milk, mung beans, boiled palm seeds (*kaong*), corn kernels, pineapple jelly, and other ingredients. Halo-halo can be bought premixed with just the shaved ice needed for completion.

A common seasoning, used instead of salt and found throughout Southeast Asia, is fermented fish paste or sauce. In the Philippines the powerful paste is called *bagoong* and tastes somewhat like anchovies, although it can be made from a variety of fish. A similar paste made of shrimp is known as *bagoong-alamang. Patis* is transparent amber fish sauce. To obtain the sour-cool taste that is popular, palm vinegar is commonly used, or a paste made from either the cucumber-like vegetable called *kamis* or the pulp of the tamarind pod. Bagoong, patis, lime (*calamansi*) wedges, and vinegar flavored with chiles are frequently placed on the table so that each diner may add saltiness or sourness to taste.

A principal food in many Pacific Islands is the coconut, and it is widely used in Filipino cooking. In addition, *copra* (dried coconut kernels used for oil extraction) is an important export crop. It takes approximately 1 year for a coconut to mature, but if picked at 6 months, the soft, jellylike coconut meat can be eaten with a spoon and is a popular delicacy. The coconut plant provides several food products, including beverages, cooking liquids, and even a vegetable. The sweet, clear liquid found in young coconuts is the juice or water. It is consumed fresh, but is not used in cooking. Coconut cream, which is used for cooking along with coconut milk, is the first liquid extracted from grated mature coconut meat. After the cream is removed, coconut milk is made by adding water to the meat and then squeezing the mixture. Coconut milk is used primarily in special dishes. Coconut palm blossom sap can be fermented to produce a strong

■ It is difficult to identify a Filipino dish that is without foreign influence. Some Filipinos nominate *kari-kari,* a stew of beef, oxtails, string beans, tomatoes, onions, and ground peanuts flavored with *bagoong* and calamansi limes.

■ The name *coconut* comes from the Spanish, who thought the "nut" had the face of a clown (*coco*).

■ In the Philippines, it is considered good luck if one cleanly splits open a coconut without jagged edges.

Table 12.1
Cultural Food Groups: Filipino

Group	Comments
PROTEIN FOODS	
Milk/milk products	Filipinos make one of the few native cheeses in Asia, from *carabao* (water buffalo) milk. U.S. influence has resulted in the availability of many western dairy products. Many Filipinos may be lactose-intolerant. In desserts, coconut milk is frequently used in place of cow's milk.
Meat/poultry/fish/eggs/legumes	The amount of protein food consumed by a family usually depends on economic status.
CEREALS/GRAINS	Rice is the main staple and is usually eaten at every meal.
FRUITS/VEGETABLES	Vegetables are often consumed in mixed stews, stir-fries and soups. Braised vegetables may be consumed as entree or side dish. Pickled fruits and vegetables are very popular.
ADDITIONAL FOODS	
Seasonings	Food is spicy, but the variety of spices used is limited. Regional cooking is differentiated in part by seasoning preferences.
Nuts/seeds	
Fats/oils	Traditional diet is considerably higher in fat than are other Asian cuisines.
Beverages	
Sweeteners	

alcoholic drink called *tuba* in the Philippines. Hearts of palm, sometimes called palmetto cabbage, is the firm, greenish inner core of the tree; it is used as a vegetable. Bananas, durian (a large, strong-smelling, sweet fruit with a creamy texture), jackfruit, papaya, and pineapples are also popular.

Regional cooking styles are divided into four regions: Luzon (the largest group of islands, also home to the nation's capital,

Table 12.1
Cultural Food Groups: Filipino—continued

Common Foods	Adaptations in the United States
Evaporated and fresh milk (goat or carabao), white cheese	Consumption of milk and other dairy products has increased.
Meat: beef, goat, pork, monkey, variety meats (liver, kidney, stomach, tripe), rabbits *Poultry and small birds:* chicken, duck, pigeon, sparrow *Fish and shellfish:* anchovies, bonita, carp, catfish, crab, crawfish, cuttlefish, *dilis,* mackerel, milkfish, mussels, prawns, rock oyster, salt cod, salmon, sardines, sea bass, sea urchins, shrimp, sole, squid, swordfish, tilapia, tuna *Eggs:* chicken, fish *Legumes:* black beans, black-eyed peas, chickpeas, lentils, lima beans, mung beans, red beans, soybeans, white kidney beans, winged beans	Consumption of fish has decreased; intake of meat, poultry, and eggs has increased.
Corn, oatmeal, rice (long- and short-grain, flour, noodles), wheat flour (bread and noodles)	Rice is not usually eaten at breakfast but is eaten at least once per day.
Fruits: apples, avocados, banana blossoms, bananas (100 varieties), breadfruit, *calamansi* (lime), citrus fruit, coconut, durian, grapes, guava, jackfruit, Java plum, litchi, mangoes, melons, papaya, pears, persimmons (*chicos*), pineapples, plums, pomegranates, pomelo, rambutan, rhubarb, star fruit, strawberries, sugar cane, tamarind, watermelon	
Vegetables: amaranth, bamboo shoots, bean sprouts, beets, bitter melon, burdock root, cabbage, carrots, cashew nut leaves, cassava, cauliflower, celery, Chinese celery, eggplant, endive, garlic, green beans, green papaya, green peppers, hearts of palm, hyacinth bean, *kamis,* leaf fern, leeks, lettuce, long green beans, mushrooms, nettles, okra, onions, parsley, pigeon peas, potatoes, pumpkins, purslane, radish, safflower, snow peas, spinach, sponge gourd, squash blossoms, winter and summer squashes, sugar palm shoot, swamp cabbage, sweet potatoes, taro leaves and roots, tomatoes, turnips, water chestnuts, watercress, yams	More green vegetables are consumed. More raw vegetables/salads are eaten.
Atchuete (annatto), *bagoong, bagoong-alamang,* chile pepper, garlic, lemon grass, *patis,* seaweed, soy sauce, turmeric, vinegar	
Betel nuts, cashews, *kaong* (palm seeds), peanuts, pili nuts	
Coconut oil, lard, vegetable oil	
Soy milk, cocoa, coconut juice, coffee with milk, tea	Chocolate milk is substituted for soy milk. Soft drinks are popular.
Brown and white sugar, coconut, honey	

Manila), Bicolandia, the Viscayan Islands, and Mindanao. Luzon is made up of various ethnic groups, and the cuisine has been strongly influenced by the Spanish. Ocean fish, such as prawns, milkfish (*bangus*), and halibut, as well as the ample use of anchovy sauce and shrimp paste are preferred in the northern areas. Foods are typically boiled or steamed. *Saluyot* ("okra leaves"—not related to okra), a spinachlike green with a slippery texture

Sample Menu

A FILIPINO DINNER

Empanadas (turnovers) or *Lumpia* (egg rolls)

Pork liver soup

Adobo (chicken stew) or *Pancit*

Steamed rice

Halo-halo (coconut and shaved ice dessert) or *Leche flan* (custard)

Beer, coffee, or tea

■ In some regions, raw pork is heavily salted then stored in jars for many months until it ripens. Called *itog,* small amounts are added to other dishes to enhance their flavor.

■ In Vietnam, *com,* meaning "cooked rice," is the same word used for "food."

when cooked, is especially popular in the north. Rice is grown in the central region of Luzon, known for its freshwater fish and richly sauced dishes flavored with onions and garlic. One delicacy is *rellenong manok,* a deboned chicken stuffed with sausage, vegetables, and ground pork mixed with raisins and spices, topped with a tangy red sauce. Stir-frying is the most common cooking technique. Beef cattle are grazed in the more southern areas; coconut products and tropical fruits predominate. Sweetened rice dishes are a specialty, such as *bibingka,* a glutinous rice cake with coconut.

Bicolandia is an ethnically homogeneous region that came in contact with both Malaysian and Polynesian cooking. Foods are spicy hot with chile peppers, balanced by copious use of coconut milk and cream. Taro leaves cooked in coconut milk with ginger and chiles is one example of the unique blend of foods found in this area. Viscayan Islands fare also reflects its heritage: Abundant use of seafood (including a distinctive fermented shrimp paste called *guinamos*) and seaweed, as well as many dessert specialties such as candies and pastries developed due to the sugarcane plantations in the area. The Mindanao region was heavily influenced by the Indonesians and Malaysians. The many ethnic groups living there are predominantly Muslim, so little pork is consumed (see chapter 4 for more information on Muslim food habits). Sauces made from peanuts and chiles are popular, as are curries and other spicy dishes, such as *piarun* (fish spiced with chiles) and *tiola sapi* (boiled beef curry).

Vietnamese, Cambodian, and Laotian. Ingredients are similar in all the mainland Southeast Asian countries, but recipes and meal patterns vary. The French introduced and popularized such items as strong coffee, pastries, asparagus, French bread, and meat pâtés in many of the regions they occupied. The Chinese influence, seen mainly in Vietnam, resulted in the extensive use of chopsticks; stir-frying in the wok; serving long-grain rice separately at the meal, rather than mixed with other ingredients; and the more prevalent use of the yin and yang theory as applied to foods and prevention of disease. In Cambodia, Asian Indian and Malaysian influences are seen by the inclusion of curries in the fare. The common foods of mainland Southeast Asians are listed in Table 12.2.

Rice, both long- and short-grain, is the staple of the diet. Rice products, such as noodles, paper, and flour, are used extensively. In Vietnam, rice paper is used as egg roll or wonton wrappers. In the dish *cha gio,* the moistened paper is wrapped around a variety of meats, fish, vegetables, and herbs and then deep-fried. Often the rice paper is filled with meat, fresh herbs, and vegetables at the table. Dried rice noodles (sticks) are called *pho,* which is also the name of the popular noodle-based soup. In Laos the sticky, glutinous short-grain rice is more prevalent than long-grain types (traditionally formed into small balls to use as scoops for other foods), and the very thin Chinese-style rice noodles are common. Wheat is used to make French bread in Vietnam and noodles throughout the rest of the region. Fried noodles topped with meats and vegetables are a favorite.

Snack vendor in Vietnam. (Courtesy of World Health Organization/P. Almasy.)

Fish and shellfish are the predominant protein food on the mainland. Even land-locked Laos depends on freshwater varieties. Poultry is available in many areas, and pork or goat is eaten in wealthier areas. Beef is used occasionally. Religious prohibitions often influence which meats are consumed.

Like other Asians, the people of mainland Southeast Asia do not use appreciable amounts of dairy products. However, soy milk is a common beverage. Soy products, particularly a chewier version of tofu (soybean curd) called *tempeh,* are common.

The Vietnamese serve many uncooked vegetables, often in the form of salads and pickles. A special-occasion dish is shredded chicken and cabbage salad, *goi go.* Many fresh herbs and spices, including basil, fresh coriander, chile peppers, garlic, ginger, lemon grass, and mint, give their food its distinctive flavor and add color to many dishes. Grilled lemon grass beef, *bo nuong xa,* is usually served at summer picnics and is always found at parties and celebrations. Due to the influence of the French, the Vietnamese also eat asparagus, green beans (*haricots*), and potatoes. Other mainland Southeast Asians also consume large amounts of vegetables and herbs, but usually only cucumbers are eaten raw. Fresh leafy vegetables may top rice or noodle dishes as in Vietnam, but hot broth is poured over the top.

Tropical fruits are available, although in some areas, bananas and plantains are the only fruit widely consumed. Banana leaves are used to wrap rice, vegetables, and meats for steaming

■ Laotians, including Hmong, often eat *lap,* raw ground pork highly seasoned with chiles, ginger, and onions.

■ Vietnamese Buddhists eat soybean products on the 1st, 15th, and last day of the lunar month, when meat is prohibited.

■ *Furr,* a soup containing pork, noodles, garlic, and hemp (marijuana) leaves, is a Laotian specialty.

Table 12.2
Cultural Food Groups: Mainland Southeast Asian

Group	Comments
PROTEIN FOODS	
Milk/milk products	Most Southeast Asians do not drink milk and may be lactose-intolerant. Sweetened condensed milk is used in coffee; whipping cream is used in pastries.
Meat/poultry/fish/eggs/legumes	The traditional Southeast Asian diet is low in protein. Fish, pork, and poultry are common; most parts of the animal are used (brains, heart, lungs, spleen).
CEREAL/GRAINS	Rice is the staple grain and is usually eaten with every meal. French bread is commonly eaten.
FRUITS/VEGETABLES	Hearty garnishes of fresh vegetables are commonly added to dishes. The Vietnamese eat a considerable amount of fruit and vegetables, fresh and cooked. Fruit is often eaten for dessert or as a snack.
ADDITIONAL FOODS	
Seasonings	Fish sauce as well as soy sauce is often used. Fresh herbs are very popular garnishes in Vietnamese dishes; typical Cambodian fare is delicately seasoned; Thai dishes are frequently very hot and spicy, with several types of curry and chile pepper especially popular.
Nuts/seeds	
Beverages	Beverages are usually drunk after the meal or with snacks or desserts.
Fats/oils	
Sweeteners	Sweets are luxury foods.

Table 12.2
Cultural Food Groups: Mainland Southeast Asian—continued

Common Foods	Adaptations in the United States
Sweetened condensed milk, whipping cream	It is expected that younger Southeast Asians will increase their use of dairy products. Ice cream is popular; milk and cheese are often disliked.
Meat: beef, lamb, pork *Poultry and small birds:* chicken, duck, quail, pigeon, sparrow *Eggs:* chicken, duck, fish *Fish and shellfish:* almost all varieties of fresh- and saltwater seafood, fresh and dried *Legumes:* chickpeas, lentils, mung beans (black and red), soybeans and soybean products (*tempeh, tofu,* soy milk), *winged beans*	Meat, lamb, and eggs are eaten more; fish, shellfish, and duck are eaten less because of price.
Rice (long- and short-grain, sticks, noodles), wheat (French bread, cakes, pastries)	Intake of baked goods increases.
Fruits: apples, bananas, cantaloupe, coconut, custard apple, dates, durian, figs, grapefruit, guava, jackfruit, jujube, lemon, lime, litchi, longans, mandarin orange, mango, orange, papaya, peach, pear, persimmon, pineapple, plum, pomegranates, pomelo, raisins, rambutan, roselle, *sapodilla,* star fruit, soursop, strawberries, tamarind, watermelon	Use of fruits and vegetables is dependent on availability and price. It is expected that use of fruits and vegetables will decline.
Vegetables: amaranth, arrowroot, artichokes, asparagus, bamboo shoots, banana leaves, beans (yard-long and string), bitter melon, breadfruit, broccoli (Chinese and domestic), cabbage (domestic, Chinese, savoy, napa), calabash, carrot, cassava (tapioca), cauliflower, celery (domestic and Chinese), chayote squash, Chinese chard, Chinese radish (*daikon*), chrysanthemum, corn, cucumber, eggplant (domestic and Thai), leeks, lotus root, luffa, matrimony vine, mushrooms (many varieties), mustard (Chinese greens), okra (domestic, lady finger), peas, peppers, potato, pumpkin (flowers, leaves), spinach, squash, sweet potatoes (tubers, leaves), taro (root, stalk, leaf, shoots), tomatoes, turnips, water lily greens, water chestnuts, wax gourd, yams	Fresh vegetables and herbs are sometimes grown in backyard gardens.
Allspice, alum, black pepper, borax, cayenne pepper, chile pepper, chives, cinnamon, coconut milk, fresh coriander, curry powder, fennel, galanga root, garlic, ginger, lemon grass, lemon juice, lily flowers, lotus seed, mint, MSG, *nuoc mam,* paprika, saffron, star anise, tamarind juice, vinegar	
Almonds, cashews, chestnuts, macadamia nuts, peanuts, pili nuts, walnuts; locust seeds, pumpkin seeds, sesame seeds, watermelon seeds	Peanut butter is often disliked.
Coffee, tea, sweetened soybean milk, a wide variety of fruit and bean drinks, hot water, hot soup, beer	Carbonated drinks have increased in use.
Bacon, butter, lard, margarine, peanut oil, vegetable oil	The Vietnamese have increased their use of butter and margarine.
Cane sugar, candy	The use of sweetened products has risen in the United States.

■ In Southeast Asia, unripe mangoes are appreciated for their crisp texture and sour flavor; they are sliced and dipped into a mixture of sugar, salt, and pulverized chiles or julienned as a main ingredient in spicy salads with shrimp or cashews.

■ "Without fish sauce or salt, life is nothing," according to a Vietnamese saying.

■ Among the more unusual Filipino specialties, especially popular as snacks is *balut*. These partially developed duck eggs are soft-boiled and sold warm by street vendors. Salt and a little vinegar are added to the embryonic birds before they are popped whole into the mouth.

■ *Turo-turo* are fast-food stands in the Philippines, offering an array of hot items.

in both Cambodia and Laos. Pineapple, papaya, limes, and mangoes are common, as are soursop, star fruit, custard apples, durian, jackfruit, and tamarind (a pod with very tart pulp). In Cambodia and Laos, coconut flavors many dishes. Oranges, lemons, melon, and sugar cane are also popular.

In Vietnam, foods are customarily seasoned with a sauce made from fermented fish, called *nuoc mam*. Also used in cooking, nuoc mam is a salty condiment that is often used instead of soy sauce. It can be transformed into a hot sauce, *nuoc cham*, with the addition of chiles, vinegar, sugar, garlic, and citrus fruit juice. In Cambodia the fermented fish sauce is called *tuk-trey;* a stronger fish paste is also used, known as *prahoc*. The Laotian version of fish sauce is *nam pa; pa dek* is the fermented fish paste.

Tea is the preferred beverage throughout mainland Southeast Asia. In Vietnam, it is served before and after meals, but not during the meal. Tea is often blended with flowers such as rose petals, jasmine blossoms, chrysanthemums, and lotus blossoms (which are especially popular). Coffee is popular in French-influenced areas, usually served with large amounts of sweetened condensed milk added to it. Broth is traditionally consumed at meals, and in poorer, rural regions, such as where the Hmong live, it is the only beverage besides water that is commonly available. In wealthier areas, men may drink beer, and women and children consume soft drinks during meals. Soybean drinks and fruit drinks are common; rice wine is served at special occasions.

Meal Composition and Cycle

Daily Pattern

Filipino. The traditional meal pattern in the Philippines is three meals a day. Breakfast is garlic fried rice with eggs or broiled fish, sausage or meat, plus coffee or hot chocolate; bread may be substituted for rice. Especially popular are sweet, cheesy rolls called *ensaymada*. Lunch and dinner are similar in size and composition. Both are often large meals, including soup, a fish or meat dish (if affordable), rice, vegetable, and fresh fruit or dessert, served in courses (Spanish-style) or all at once (Filipino-style).

In addition to meals, two snacks, called *meriendas*, are consumed in the midmorning and late afternoon. Meriendas may be small or may consist of substantial amounts of food, such as fritters, pastries, fruits, ensaymadas, lumpia, or almost anything else except rice, which is served only at meals.

Vietnamese and Other Mainland Southeast Asians. Mainland Southeast Asians eat two or three meals a day with optional snacks; the number of meals and amount of food are often dependent on income. In Vietnam a traditional breakfast may consist of soup with rice noodles, meat, bean sprouts, and fresh coriander leaves; boiled egg with meat and pickled vegetables on French bread; and either rice and leftover meat or boiled sweet potatoes with chopped roasted peanuts. Lunch and dinner typically include rice, fish or meat, a vegetable dish, and a broth with vegetables or meat. French bread with meat or shrimp pâté may be substituted for a lunch or dinner meal. Snacks may include clear soup, fruit, or pastries. Southeast Asians do not usually associate particular foods with breakfast, lunch, or dinner; thus a variety of foods may be eaten at any meal.

In Southeast Asia, food is typically stir-fried, simmered, steamed, or boiled. Soups are common throughout Southeast Asia and often feature local ingredients. For example, in Vietnam, crab and asparagus soup is popular. Soup may accompany every meal; it is often eaten in restaurants that specialize in *pho bo ha noi* (from Hanoi), a delicate broth to which rice noodles, sliced beef, bean sprouts, herbs, and other seasonings are added immediately before serving. *Mein go* is a chicken noodle soup served in a similar manner.

Etiquette

Chopsticks are used in Vietnam and are the appropriate utensils for some foods, such as noodles, in other mainland Southeast Asian countries. Forks are common everywhere; however, in Vietnam the fingers of the right hand are used for some foods. Both hands customarily rest on the table while dining. Conversation is usually limited during a meal. It is considered polite to leave a little bit of food on the plate to indicate that hunger has been satisfied.

Special Occasions

Filipino. In the predominantly Catholic Philippines, religious festivals and saints' days are numerous (see chapter 4). On all special occasions, it is customary to serve plenty of food buffet-style, with a roasted pig (*lechon*) as the centerpiece. The Filipinos claim to have the longest Christmas season in the world, from December 16 to January 6. The midnight mass celebrated on Christmas Eve is usually followed by the traditional *"media noche,"* a midnight supper of fiesta foods such as roast ham, sweet potatoes, banana flower salad, and *niaga,* a dish made of boiled meat, onions, and vegetables whose name means "good life." Other specialties eaten during the Christmas season are *puto bumbong,* a rice flour delicacy cooked in a whistling bamboo kettle, and *bibingka,* a sweet roll cooked in a clay pan over hot coals.

A midnight mass is also held on New Year's Eve, but many Filipinos attend parties to celebrate the holiday instead. Again, a midnight supper consisting of fiesta foods is traditional. There is also a superstition that eating seven grapes in succession as the clock strikes midnight will bring good luck in the coming year. For birthdays, pancit is eaten to ensure a "long life."

There are numerous Filipino practices and customs associated with Easter, beginning with observances on Ash Wednesday. Late on Easter Eve, young children are awakened to partake of special meat dishes, such as adobo and dinu-guan, in the belief that if they do not do so, they will become deaf.

Sample Menu

A VIETNAMESE LUNCH

Pho (beef soup with fresh vegetable garnish)

Pork with lemon grass

Stir-fried bean curd with vegetables

Steamed rice

Fruit juice or iced coffee with evaporated milk

Vietnamese and Other Mainland Southeast Asians. Of all Vietnamese holidays, Tet, the New Year's celebration, is the most important. It is observed at the end of the lunar year (end of January or beginning of February), just after the rice harvest. In Vietnam the first Tet ritual is an observance at the family grave sites. Offerings of cake, chicken, tea, rice, and alcohol, as well as money, are made at the graves, and then the family picnics on the offerings.

The second ritual, held on the 23rd day of the 12th lunar month, is to celebrate the departure of the Spirit of the Hearth, Ong Tao. He is represented by three stones on which the cooking pots are placed and is honored by a small altar. Like the Chinese Kitchen God, Ong Tao returns to the celestial realm each year and reports on the family's behavior. After the family makes an offering to symbolize his departure, they share a feast including glutinous rice cakes and a very sweet soybean soup. One week later the family celebrates Ong Tao's return to their hearth. The following day is the first day of Tet. Guests (especially those with favorable names, such as Tho, meaning longevity) are entertained with tea, rice alcohol, red-dyed watermelon seeds, candied fruits, and vegetables. Special dishes prepared for the week-long celebration include *banh chung,* glutinous rice cakes filled with meat and beans and boiled in banana leaves, squid soup, stir-fried young seasonal vegetables, pork with lotus root, and sometimes a special shark fin soup.

Many Vietnamese, including those in America, celebrate the Buddhist holiday called Trung Nguyen, or Wandering Souls Day. It occurs in the middle of the seventh lunar month and is celebrated with a large banquet prepared in honor of the lost souls of ancestors.

- Hospitality is very important to the Filipinos, and food gifts to express love or appreciation are common.

- Long, uncut noodles represent a long and prosperous life.

- According to legend, *banh chung* were created by the youngest son of an ancient Vietnamese king. In a contest with his siblings to prepare the best new food, God appeared in a dream and instructed him to make a simple dish of glutinous rice, meat, and mung beans formed into a round (to symbolize the universe) and a square (to represent the Earth). The king was so pleased, he awarded his throne to his inventive son.

■ A feast is held by the Hmong following the birth of a child. Included are two chickens representing the parents, a boiled egg signifying the child, and a small lit candle symbolic of the ancestor spirits whose blessing and protection are sought.

■ Filipinos believe it is inadvisable to eat oranges at the same time as drinking milk because the combination will curdle in the stomach.

■ Some Filipinos believe rice porridge is theraputic for illness (Storz, 1998).

■ Licorice root, which contains *glycyrrhizin,* can cause fluid retention and increase blood pressure if consumed in large amounts.

Traditionally the Vietnamese did not commemorate birthdays but rather honored their ancestors on the anniversary of their death with a special celebration and meal. In the United States, it is now more common to celebrate birthdays.

The largest holiday of the year in Cambodia is also New Year's, *Chaul Chnam,* which begins on April 13 and lasts for 3 days. Prayers and special foods like fried coconut and fried bananas rolled in coconut are offered to the New Year Angel, who descends with either blessings or ill will. The Water Festival, held in November after the seasonal rains have ended, features colorful floats in local rivers.

Most Laotian holidays are religious in origin and are celebrated at local temples. *Pha Vet,* which occurs in the fourth lunar month, commemorates the life of Buddha. *Boon Bang Fay,* held in the sixth lunar month, also honors the Buddha with a fireworks display. Among the Hmong and other Laotians the New Year's celebration is a major event. It begins with the first crow of a rooster on the first day of the new moon in the 12th lunar month, usually in December. The highlight of the festivities is the world renewal ritual, which involves an elder who chants while holding a live chicken. He circles a tree three times clockwise to remove the accumulated evil of the previous year, then circles the tree three more times counterclockwise to invoke good fortune in the upcoming year. It is believed that the bad luck collects in the blood of the chicken, which is traditionally taken to a remote location and slaughtered. New Year's was the one time each year when Hmong from different clans celebrated together, and it is customarily considered a good time to meet future wives and husbands.

Therapeutic Uses of Food

Filipino

When the Spanish came to the Philippines, they introduced the Mexican hot-cold theory of health and diet (see chapter 9, Mexican "Therapeutic Uses of Foods" section). Foods are classified as being hot or cold based on their innate qualities or their effect on the body, not on their spiciness or temperature. Although the classification of certain foods varies regionally, avocados, alcoholic beverages, coconuts, nuts, legumes, spices, chile peppers, and fatty meats are generally considered hot items; tropical fruits, vegetables, milk and dairy foods, eggs, fish, and lean or inexpensive meats are thought to be cold (Claudio, 1994; Orque, 1983a). A balance is attempted at meals between hot and cold elements. It has been suggested that the reason Filipino dishes contain so many ingredients is to ensure this balance.

Some illnesses are characterized as hot or cold and are treated with foods of the opposite category. Diarrhea and fevers are hot; colds and chills are cold. Other food beliefs are based on imitative qualities (like causes like); for instance, pregnant women may avoid dark foods to prevent their babies' skin from being too dark. Sometimes the meaning behind a therapeutic food use is more obscure; horseradish leaves and broth seasoned with ginger are believed to promote milk production in nursing mothers, and fish heads and onions are considered brain food by some Filipinos. Honey, as well as certain herbs such as thyme, marjoram, and chamomile, is used to treat diabetes. Licorice root is considered a general tonic, especially beneficial during times of stress. Some elderly Filipinos have adopted the Asian Indian practice of chewing betel nuts, which is believed to prevent tooth decay, although it leaves permanent stains.

Vietnamese and Other Mainland Southeast Asians

Many Vietnamese follow the Chinese yin-yang theory of health and diet (see chapter 11, Chinese "Therapeutic Uses of Foods" section). As in the hot-cold system, classification is based on intrinsic characteristics rather than temperature or spiciness. Examples of yang (hot) foods are unripe fruit, coffee, and alcoholic beverages. Yin (cold) items include noodles, bananas, oranges, gelatin, and ice cream (Orque, 1983b). Some foods, such as rice, eggs, chicken broth, teas, and sweets, are classified as "neutral" (Muecke, 1983b). Not only must a balance be maintained within a meal, but extremes are also avoided during certain conditions, such as pregnancy. As with Filipinos and other Asians, illnesses are defined as yin and yang, and are sometimes caused by eating too many yin or too many yang foods. Equilibrium

Thai Cuisine

There are 7 times more Vietnamese Americans and 16 times more Filipino Americans than there are Thai Americans in the United States. Yet Thai cooking is more familiar to the general population than either Vietnamese or Filipino fare. Thai restaurants have introduced the distinctive cuisine in many parts of the United States, and dozens of cookbooks published in recent years have further popularized the food.

The country of Thailand is located on the southern end of the archipelago that is Southeast Asia. The hot, monsoonal climate is ideal for rice cultivation. Long-grain rice is the foundation of the diet, and short-grain glutinous rice is used for snacks and desserts. Noodles made of rice, wheat, or mung beans are also common. Tropical fruits and vegetables, such as bananas and plantains, cassava, coconuts, durian, mangoes, pineapple, rambutan, rose apples, sapodilla, taro, and yams, and temperate produce, including carrots, cucumbers, melons, and strawberries, are prominent in the cuisine. Seafood from the lengthy coast, especially shrimp, is popular. Dried herringlike fish (which are sometimes smoked as well) are often flaked into rice for added flavor. Beef, chicken, and pork are common. Duck is a favorite.

Thai food differs from that of its Southeast Asian neighbors because of its flavors. It is one of the hottest cuisines in the world, with lavish use of chile peppers. Several varieties of basil, fresh coriander leaves (cilantro) and root, garlic, ginger root, lemon grass, and tamarind are typical seasonings. In addition, curried dishes are eaten daily. There are three types of curry sauces: yellow, which are smooth, mild, Indian-like sauces that include spices such as cardamon and turmeric; red, which are chunkier, hotter, and typically include ample fresh red chiles and coconut milk; and green, which are prepared with fresh green chiles whose heat is excruciating for all but the most experienced palates. The curry sauces are used with beef, pork, chicken, duck, and seafood. Fermented fish products, such as *nam pla* (similar to the Vietnamese sauce called *nuoc mam*) and *kapi* (a paste made from fish or shrimp), are added to most dishes. *Nam prik,* a sauce that combines *nam pla* or *kapi* with other ingredients like garlic, chile peppers, shallots, lime juice, tamarind, palm sugar, and peanuts, complements dishes like *yam* (fresh vegetables rolled up into a leafy package and dipped into the *nam prik*), salads (*nam prik* is the dressing), noodle dishes, dumplings, fried or grilled foods, and highly spiced raw pork called *nam.*

Noodle dishes are usually eaten for breakfast and lunch. *Phad Thai,* stir-fried noodles cooked with bits of meat, seafood, and vegetables bound with eggs, then topped with peanuts and *nam prik,* is an example. Sweets like coconut custards and fruit jellies are preferred snacks. Thai cooking began as a court cuisine, and this heritage is most obvious in the evening meal. Dinner often includes appetizers, such as deep-fried chicken wings stuffed with ground pork and shrimp, or pastries shaped like delicate flowers. The main meal traditionally features steamed rice; soup; a curried dish, a fried dish; and a salad of raw vegetables and grilled poultry, beef, or seafood. *Mee krob,* a volcanic-looking mound of stir-fried noodles and meats or seafood cooked with sugar until carmelized and garnished with egg lace (beaten egg drizzled into hot oil), bean sprouts, scallion tassels, and fringed chile peppers, is a favorite addition. Various *nam prik* accompany the dishes. All dishes are served at the same time, traditionally on low tables surrounded by pillows; fingers and spoons are the usual implements, and forks are available for pushing food into the spoon. The meal usually concludes with elaborately carved fruits.

is restored by eating foods of the opposite type; during the postpartum period, which is yin, cold drinks are avoided. Yin and yang concepts are less prevalent among Cambodians and Laotians, although some hot and cold beliefs exist regarding specific foods and certain conditions.

The Chinese medical system details other influential elements in health, including the five flavors of sour, bitter, sweet, pungent, and salty; these tastes are harmonized in many Vietnamese dishes.

Vietnamese believe that ingestion of specific organ meats will benefit the like internal organs. For example, consumption of liver will produce a stronger liver. Some Vietnamese believe that eating gelatinous tiger bones (produced by prolonged cooking) will make them strong. Concurrently, some foods may be injurious because they resemble certain disorders. Pregnant women may refuse to eat ginger because the multilobed root is thought to cause polydachtyly (too many digits) in babies.

■ Some young Hmong women avoid eating gizzards because it is believed they cause the placenta to toughen and make birth difficult.

The therapeutic value of some foods is unrelated to yin or yang or how they look. Some Vietnamese eat chile peppers to get rid of worms, or noodles with roasted rice paper and shrimp sauce for curing the flu (Orque, 1983b). Vietnamese women may also consume large amounts of salty foods during pregnancy (Nguyen, 1985). Hmong women eat a diet of rice, chicken broth, black pepper, and herbs for a month after giving birth (Ikeda, 1992), and some clans have specific taboos against eating certain foods, such as heart (Rairdan & Higgs, 1992).

Vietnamese restaurants, especially those featuring the noodle soups called *pho,* have become popular in many communities where Vietnamese immigrants have settled. (© David Weintraub/Stock, Boston.)

Contemporary Food Habits in the United States

Adaptations of Food Habits

Filipino

Little current information on the food habits of Filipino Americans has been reported. Most are able to obtain traditional foodstuffs without much difficulty, although some of the familiar tropical fresh fruits and vegetables are not available. It is believed that most Filipinos still eat rice every day, but not with every meal, and their diets tend to contain a greater variety of foods, especially more milk, green vegetables, meat, and sweets than they did in the Philippines (Lewis & Glaspy, 1975). Meriendas tend not to be eaten as often.

Filipinos born in the United States frequently consume a typically American diet. Breakfast consists of cereal, toast, eggs or meat, juice, and coffee; sandwiches, salads, and sodas are common at lunch; and dinner is usually a meat or fish dish served with rice or potatoes, followed by dessert. Traditional Filipino items may appear at some meals, such as eating longaniza sausage at breakfast or eating halo-halo (sometimes topped with vanilla ice cream) for dessert (Dirige, 1995).

Vietnamese and Other Mainland Southeast Asians

A study (Tong, 1986) conducted in Washington, D.C., found that 30 percent of the Vietnamese households surveyed had changed their eating habits since coming to the United States. Although most continued to eat rice at least once a day, they ate more bread or instant noodles at lunch and more cereal at breakfast. Respondents also reported consuming more meat and poultry and less fish and shellfish than in Vietnam, mainly because of cost. Pork and pork products were still preferred to beef. They also reported consuming fewer bananas and more oranges, fruit juices, and soft drinks. These findings are similar to those reported in a survey conducted in the 1970s among Vietnamese living in northern Florida (Crane & Green, 1980). More than 90 percent of Vietnamese American adolescents in another study were found to prefer their native diet, although a majority listed items such as steak and ice cream as being among their favorite foods. Soft drinks and milk were well liked; cheese and peanut butter were strongly disliked. Only a small percentage of the teens snacked regularly (Story & Harris, 1988).

A survey of Cambodian and Hmong families indicated some similar trends. While traditional items were preferred by the adults, both American and native foods were acceptable to the children. Most-liked items among the adults included steak, oranges, candy, and soft drinks, all of which are prestige foods in Cambodia and Laos.

Least-liked items included cheese, chocolate milk, and milk (Story & Harris, 1989). A detailed study of poor Hmong immigrants in California revealed that the majority of adults (52 percent) consumed two meals each day of rice, greens, and meats. Pork was the preferred meat, although chicken, turkey, fish, and eggs were also eaten. The adults were mostly unfamiliar with baked products, such as bread or cookies, and most strongly disliked both milk and cheese. It was found that many Hmong grow their own vegetables and seasonings (especially varieties difficult to obtain from grocery stores) in backyard gardens. Most children ate three meals, often including a free lunch at school. Snacking was uncommon, although more prevalent among children (Ikeda et al.,1991).

Food purchasing and preparation as well as meal patterns are also changing. Southeast Asian American women report that men frequently help with shopping or cooking. Vietnamese, Cambodian, and Laotian adolescents often are involved in food purchases, and surveys indicate as many as 60 percent of girls and 35 percent of boys have total responsibility for fixing dinner each evening. Southeast Asian women living in the United States are more likely to have a job or to attend adult education classes than in their homeland, relinquishing some household responsibilities to other family members. Further, many families report a significant decline in eating meals together (Ikeda et al., 1991; Story & Harris, 1988, 1989).

Nutritional Status

Nutritional Intake

Filipino. The traditional Filipino diet is higher in total fat, saturated fat, and cholesterol than most Asian diets. Urban Filipinos living in the United States tend to have even higher intakes of these dietary components. Filipino Americans have serum cholesterol levels equal to those of white Americans and have high rates of hypertension. It is hypothesized that Filipinos may have a genetic inability to process large amounts of sodium in their diet (Tamir & Cachola, 1994). Alcohol consumption among Filipinos is also positively associated with increased blood pressure (Stavig et al., 1988).

The impact of high blood cholesterol levels and hypertension on the health of Filipinos is unclear; it is thought that there may be significant errors in identifying Filipinos as an ethnic group in some health statistics. Mortality rates from cardiovascular diseases are conflicting: A 1980 California study reports exceptionally low rates of death from heart disease, a survey in Hawaii at the same time revealed intermediate rates of death, while another California survey in 1989 found relatively high mortality rates (Gardner, 1994; Tamir & Cachola, 1994). Cancer risk is low among Filipinos, although incidence reportedly shifts from that in the Philippines to that in the United States with length of stay (Gardner, 1994). Filipino men have been found at highest risk for cancer of the liver and biliary system (Anderson, 1983), although incidence of these malignancies actually decrease in the United States. While relative risk is low, it is noteworthy that survival rates among Filipino Americans who do develop cancer are lower than for white Americans (Jenkins & Kagawa-Singer, 1994).

The infant mortality rate for Filipino Americans is reportedly 16 percent lower than that for white Americans (Singh & Yu, 1995). Filipino infants have been found less likely to die of perinatal conditions (i.e., problems related to the pregnancy, labor, and delivery) and have exceptionally low numbers of deaths during the postneonatal period.

Filipinos suffer from more non-insulin-dependent diabetes mellitus (rates two to three times the U.S. average), hyperuricemia (resulting in gouty arthritis), and glucose-6-phosphate dehydrogenase deficiency (causing anemia unrelated to iron intake) than white Americans. It should also be noted that alpha thalassemia (hemoglobin H disease) is also prevalent among Filipinos and results in a hypochromic microcytic anemia, especially during an infection or when oxidant drugs are taken.

Although Filipinos consume more milk and cheese than other Asians, many are lactose intolerant (Anderson, 1983) and calcium intake may therefore be limited. Dried fish, fish sauce, and fish paste may provide calcium, but amounts vary depending on the source and quality of the product. One study found some Filipino Americans have poor intakes of calcium and vitamin A (Lewis & Glaspy, 1975). Betel nuts may be chewed by older men, resulting in stained teeth.

■ **Certain Filipino dishes are very high in purines, a concern for patients with gout; *dinu-guan,* for example, often includes pork liver, kidney, heart, and small intestine.**

■ High rates of B_{12} deficiency and iron deficiency anemia have also been reported in Thai vegetarians (Pongstapom & Bunyaratave, 1999).

■ An analysis of broth made with acidified bones reported that 1 tablespoon provided nearly as much calcium as one-half of a cup of milk (Rosanhoff & Calloway, 1982).

■ Many Hmong associate being fat with good health.

■ Some Southeast Asian women avoid taking supplements during pregnancy because they fear the baby will grow too big for delivery.

Vietnamese and Other Mainland Southeast Asians. Food intake data suggest that the calcium intake of mainland Southeast Asians is low, although this observation needs to be verified, since fish sauces and other traditional foods may contain sufficient calcium. Vietnamese have been reported to be lactose intolerant. Riboflavin, magnesium, and zinc consumption was found to be less than 80 percent of the RDA in adults among the Hmong (Ikeda et al., 1991). Deficiencies during pregnancy include riboflavin; vitamins B_6, D, and E; folacin; calcium; phosphorus; potassium; and magnesium (Ikeda et al., 1991; Newman et al., 1991). Iron intake may also be marginal (Ikeda et al., 1991), particularly among children (Sargent et al., 1996). Anemia rates among refugees have been found to vary from 6 to 37 percent (Ackerman, 1997; Brown et al., 1986; Chow & Krumholtz, 1989), although genetic traits such as a high prevalence of hemoglobin E trait-thalassemia syndromes may also cause anemia unrelated to iron intake.

Certain conditions common among recent immigrants from Southeast Asia may compromise their nutritional status. These include tuberculosis, usually the inactive form (approximately 4 percent of refugees are denied entry to the United States due to active cases); between 50 and 60 percent of immigrants in the United States test tuberculin positive (Hann, 1994); intestinal parasites, which can contribute to anemias, fatigue, and weight loss; malaria; liver and renal disease, whose contributing factor is the presence of hepatitis B surface antigens (chronic HBV rates have been estimated at between 5 and 15 percent) (Hann, 1994; Ziegler et al., 1989); and dental problems, caused by chronic malnutrition and, in the United States, excessive consumption of sweets. There is concern that the persistence of continued parasitic infection suggests ongoing transmission; asymptomatic infection is common and poor compliance with treatment may account for the excess cases (Molina et al., 1988). Malnutrition contributes to the prevalence of short stature in recent immigrants (Brown et al., 1986).

Recent immigrants have been reported to have a high incidence of low birth weight infants, as a result of poor maternal weight gain during pregnancy. Traditionally, Southeast Asians gained less weight during pregnancy to have an easier birth. In one study, it was found that the birth weight of babies of Southeast Asians with longer residence are close to the U.S. average; these groups are believed to have benefited from medical and nutritional care (Davis et al., 1982; Li et al., 1990). Infant mortality rates for most Southeast Asians are reportedly low, however, regardless of low birth weight and other potentially adverse risk factors such as nutritional deficiencies, inadequate utilization of prenatal care, and low socioeconomic status. Hmong infant mortality rates are the exception; death rates are slightly elevated in comparison to other Asian groups (Gann et al., 1989; Newman et al., 1991; Rumbaut & Weeks, 1989).

Nearly all babies are breast-fed in mainland Southeast Asia, for periods of about 1 year. Studies on infant feeding practices of Americans of Southeast Asian descent reveal a dramatic decline; breast-feeding is reduced to only 9 to 26 percent of postpartum women. Work, schooling, physical discomfort, embarrassment, and the ready availability of formula through hospitals and the Supplemental Food Program for Women, Infants, and Children (WIC) were cited as reasons for not breast-feeding. Some women mentioned that they believed formula was nutritionally superior to breast milk (Ghaemi-Ahmadi, 1992; Romero-Gwynne, 1989; Serdula et al., 1991). Southeast Asian women generally introduce solid foods later (at about 8 months) than do whites (5 months) or African Americans (about 4 months).

Southeast Asians typically calculate age on a lunar calendar, often starting with being 1-year-old at birth. Reported age may differ as much as 2 years from Western chronological age, which can distort the use of standardized growth curves. Some Vietnamese parents may claim, however, that their children are younger than they are to enroll them in lower school grades; this allows the children to catch up in their schooling.

Counseling

Filipino. Americans of Filipino descent may accept illness as fate, tolerating symptoms until the severity forces them to seek care. Relatives, neighbors, and traditional healers may be consulted first before biomedical service is obtained.

The communication style of many Filipino Americans is formal and polite. Handshaking is the common greeting, although an eyebrow "flash" (quick lifting of the eyebrows) may be used between acquaintances (Axtell, 1991). Health care practitioners are often considered to be authority figures, so responses to questions may be deferential; Filipinos will avoid voicing disagreement. Loud voices are considered rude, and laughter may mask embarrassment (Axtell, 1991). The health provider should not address Filipino elders by their first names, as this is disrespectful. Direct eye contact should not be made because this is an expression of sexual interest or anger. In addition, because Filipinos often avoid situations in which self-esteem may be lost, health care providers should be sensitive in discussing certain subjects, such as socioeconomic background. Modesty may make other topics uncomfortable to discuss as well, such as sexuality (handled best by a provider of the same gender as the client) and shameful conditions such as tuberculosis or mental illness. Filipinos may expect quick results from their health providers and will switch to other healers if they feel progress is too slow (Montepio, 1986–87).

Traditional healers are used by some Filipino Americans, although the numbers are unknown. A study in Los Angeles reported that most respondents, independent of education level, still adhered to many traditional beliefs about the cause of illness, including unbalanced conditions such as eating too much or the wrong combination of foods, working too hard, or that it is a punishment for one's sins against God. Immigrants from rural regions were more familiar with traditional medical practices than immigrants from urban areas, who were more likely to rely on over-the-counter therapeutics (Montepio, 1986–87).

Due to the strong family orientation, relatives play a significant role in a Filipino client's treatment and recovery. For most effective treatment, the provider should discuss diet modifications with family members as well as with the patient. The in-depth interview should be used to determine the patient's degree of acculturation, use of traditional medical practices, and personal food habits.

Vietnamese and Other Mainland Southeast Asians. Trust is a significant issue in Southeast Asian health care. Experiences with medical personnel in refugee camps have left many immigrants suspicious of biomedicine in general. Many Americans of Southeast Asian descent believe that Western practitioners do not understand their medical needs and are disrespectful of their traditional practices; many are fearful of invasive laboratory tests, especially the taking of blood, because this may upset the body's balance. Surgery may be avoided and even autopsies denied due to fears about the relationship between the body and soul. Privacy issues may also be of concern. For example, the need to completely undress or the use of hospital gowns, breast and pelvic exams, and discussions about family planning should be postponed until a client–provider relationship has been established, preferably between a provider of the same gender as the client (Hoang & Erickson, 1982; Muecke, 1983a). Acceptance of invasive procedures, such as having blood taken and pelvic exams, was reportedly increased in one study of pregnant Hmong women by the use of a videotape on prenatal care narrated in the Hmong language (Spring et al., 1995).

Language barriers may be significant. Southeast Asian clients desire a full description of their disorders and therapies, and translators fluent in a client's dialect and culture may be essential to communication. Differences in medical concepts and technologies require careful explanation of procedures to help clients overcome fears and to enhance compliance (Muecke, 1983).

The family is responsible for the health of its members, and a client may exhaust home care before seeking outside help. Furthermore, some Southeast Asian Americans may deny discomfort and pain until it becomes intolerable. As a result, a client may be very ill before deciding to go to a clinic or hospital. Traditionally, prevention of disease occurs primarily through harmonious living; most Southeast Asians have little experience with medical checkups or treating a condition when no symptoms are present.

A very polite, unhurried, and reserved conversational style is appreciated by most Southeast Asians. Excited, informal, or frank speech

- Many Filipino languages do not recognize gender, and there may be some confusion with pronouns.

- Some Filipinos believe that fat is a protection against becoming too "cold" and losing vital energy; thus, being overweight is preferred to being too thin.

- Some Filipino elders prefer their food soft and warm, and will reject beverages with ice (Storz, 1998).

- Southeast Asians philosophically regard quality of life to be more important than length of life, believing that personal illness or suffering will diminish in the next reincarnation.

- A Hmong legend describes a bird who brings death by drinking the blood of its victim. The Hmong and other Southeast Asians are usually upset by blood-testing procedures.

may be considered rude (Rairdan & Higgs, 1992). The Vietnamese place a high value on social harmony; both Confucian and Buddhist belief systems encourage modesty. The clinician should be aware that, in general, Southeast Asian clients will be aggreeable to avoid disharmony or to please the questioner. When angry or embarrassed, Southeast Asian Americans may laugh to hide their emotion. Proper posture and appearance are important. In addition, certain nonverbal forms of communication should be carefully observed. The head is considered sacred, and it is extremely offensive to pat or even touch the head of an adult or child without permission. The feet are the lowest part of the body, thus it is impolite to point with the foot or show the bottoms of one's shoes. It is also rude to snap one's fingers or signal by using an upturned index finger since this is how dogs are called. Respect is shown by giving a small bow of the head when greeting elders and by using both hands to present any item to client.

Numerous studies have noted that refugees from mainland Southeast Asia are at special risk for mental health problems, due to the horrors of war, difficulties in escape, lengthy camp confinement, and the extreme cultural differences between their homeland and United States. Posttraumatic stress disorder is common. One study of the Hmong suggests, however, that levels of depression, anxiety, hostility, and other symptoms of adjustment problems may gradually resolve with length of stay (Westermeyer et al., 1984).

Adherence to traditional health beliefs varies, often according to whether new religious faiths have been adopted; Christian churches strongly discourage ancestor and spirit worship (Capps, 1994). Some studies suggest that the majority of immigrants continues certain practices, such as the use of botanical home remedies and coining (see the section "Traditional Health Beliefs and Practices") for many years after arrival (Gold, 1992; Hoang & Erickson, 1982), although the costs and inconvenience of some traditional cures, including difficulties in obtaining animals and herbs, the disintegration of clan ties, and the scarcity of shamans, present barriers in some communities (Adler, 1995). Southeast Asians frequently develop a medical pluralism, accepting and rejecting theories and therapeutics coherent with their acculturation experiences. It is not unusual for Southeast Asians to consult multiple biomedical and traditional practitioners for relief of symptoms.

The Vietnamese usually characterize U.S. biomedicine as yang (hot) and traditional Vietnamese medicine as yin (cold). Biomedicine is seen as fast-acting, temporary, and likely to have side effects. It is useful for emergency situations. Vietnamese medicine is considered slow-acting, gentle, without side effects, a permanent cure, or useful in prevention. Some immigrants believe it is important to counteract the yang impact of biomedical therapies with yin herbal remedies (Stephenson, 1995). It has been reported that some Vietnamese Americans routinely reduce the dose or duration of prescription medications, because they are thought to be too strong, appropriate for larger people, or because of side effects. Medication is frequently ceased when symptoms are alleviated (Gold, 1992; Stephenson, 1995).

Hospital nurses report that numerous conflicts have developed around traditional Vietnamese birthing practices (Stephenson, 1995). Because the postpartum period is defined as a yin (cold), women avoid exposure to cold and "wind" (see the section, "Traditional Health Beliefs and Practices"). Clients may refuse to get out of bed soon after birth and refuse to take a shower or wash their hair. Yin foods, such as cold beverages and vegetables, may be refused even if prepared in a culturally sensitive manner.

Treating a Southeast Asian American client requires communicating with their relatives as well. Health care decisions may be the responsibility of family elders or the family as a whole. Developing a relationship with relatives and enlisting their cooperation can significantly increase compliance.

Medical histories are often difficult to obtain because patients in Southeast Asia are often not informed of the name of their condition or of what medicines are prescribed (Rairdan & Higgs, 1992). An in-depth interview is critical to determine the patient's country of origin (a patient may be offended if grouped with all other Southeast Asians), length of time

■ Traditional therapies such as coining or moxibustion may leave marks on the skin; abuse should not necessarily be presumed.

■ Little is known about the active agents in Southeast Asian herbal remedies. High levels of lead have been found in *paylooh*, an orange powder ingested for rashes or fever.

■ Vietnamese frequently express concerns about hospital food. Jell-o (which may be described as "fruit gravy") is especially abhorred (Stephenson, 1995).

■ Hmong may prefer unseasoned foods when hospitalized and may desire water that is boiled before drinking it (Rairdan & Higgs, 1992).

■ Southeast Asians may not distinguish between fruit juices and fruit-flavored soft drinks during diet recalls.

in the United States, and any immediate health problems. Degree of acculturation and personal food preferences should also be noted.

Pacific Islanders

Pacific Islanders are the peoples inhabiting some of the 10,000 islands of Oceania. Polynesia, Micronesia, and Melanesia are the three areas that make up the Pacific region. Polynesia includes the major islands and island groups of Hawaii, American Samoa, Western Samoa, Tonga, Easter Island, and Tahiti and the Society Islands. The 2,000 small islands of Micronesia include Guam, Kiribati, Nauru, the Marshall and Northern Mariana islands, Palau, and the Federated States of Micronesia. Although the boundaries of Melanesia are not exact, it includes the nations of Fiji, Papua New Guinea, Vanuatu, the Solomon Islands, and the French dependency of New Caledonia.

Although the area is geographically similar, consisting of mostly small, tropical coral or volcanic islands, the Pacific Islanders are a racially and culturally diverse population. European, American, and Japanese influences have been extensive. Today there are greater numbers of some Pacific Islander groups living in the United States than in their native homelands.

Cultural Perspective

History of Pacific Islanders in the United States

Immigration Patterns

Since the first settlement of the island in Oceania, the migration of population groups has been very fluid between Polynesia, Micronesia, and Melanesia. As conditions on one island grew too crowded, colonization of surrounding islands occurred. That trend continues, with greater economic opportunity the primary motivation for Pacific Islander immigration to the United States and other countries.

Hawaiians. When Cook first arrived in Hawaii, approximately 300,000 native Hawaiians were living on the islands. European diseases introduced by the explorers and missionaries decimated the population, and by 1910, only a little more than 38,500 persons of

■ Hansen's disease (leprosy) was one of the illnesses introduced to Hawaii by the Europeans. In the mistaken belief that the disease was transmitted sexually, the afflicted were ostracized and exiled on the island of Molokai during the 19th century.

A rendering of the *Resolution,* the ship used by James Cook to explore Polynesia. Many Pacific Islander cultural traditions disappeared after European discovery of the islands in the eighteenth century. (© Corbis-Bettmann.)

Hawaiian ancestry remained. A high rate of intermarriage has resulted in a population with few Hawaiians of full native heritage. A large number of Hawaiians have migrated to the mainland United States, but the census figures do not reflect such interstate relocations.

Samoans. In 1951 the Pago Pago naval base that employed many Samoans moved to Hawaii, and many Samoans followed. Increasing population pressures and a deteriorating economy encouraged further immigration. Once in Hawaii, Samoans sometimes move to the mainland United States in search of broader job availability and for wider educational opportunities for their children. A chain of immigration between Hawaii and the West Coast was created, with Samoans established in the mainland helping extended family members to settle nearby.

Guamanians. Following attainment of citizenship in 1950, many Guamanians enlisted in the U.S. armed forces seeking better employment and higher wages. Many moved to Hawaii and the West Coast of the U.S. mainland. By the early 1970s, approximately 12,000 Guamanians had immigrated, mostly Chomorros. Major populations have settled in San Diego, Los Angeles, San Francisco, the Seattle area, and Hawaii.

Tongans. The immigration of Tongans to the United States did not begin until population pressures in the 1960s decreased economic opportunities in Tonga. Under the strict hereditary social structure, only the eldest son in a family may inherit land, leaving younger men with little economic mobility. Unlike most other Pacific Islanders, Tongans usually immigrate directly to the U.S. mainland, rather than settling first in Hawaii.

Current Demographics and Socioeconomic Status

■ Nearly 7,500 Fijians live in the United States, primarily in California. Most are of Asian Indian descent, were originally indentured laborers in Fiji, and follow the Islamic or Hindu faith.

Hawaiians. More than 256,000 U.S. citizens were self-identified as Hawaiians or part Hawaiians in the 1990 U.S. census. A little over half (135,231) live in Hawaii: 20 percent of the total state population is part Hawaiian, and about 1 percent is full Hawaiian. The remaining Hawaiians live primarily on the West Coast, mostly in California. Scant socioeconomic data are available, but reports indicate that Hawaiians occupy the lowest economic strata in the state of Hawaii, along with other Pacific Islander immigrants, living mostly in rural and semirural regions. Hawaiian heritage is less of a handicap on the mainland, where Hawaiians have a cultural advantage over other Pacific Islanders (through extensive exposure to American society) and often enter the middle class.

Samoans. Probably more Samoans are living outside Samoa than in American and Western Samoa combined. The 1990 U.S. census figures showed between 55,000 and 57,000 Samoans living in the United States, three-quarters of whom were born in America. The largest groups are found in Honolulu, Los Angeles, the San Francisco Bay area, and Salt Lake City. Smaller numbers have settled in Laie, Hawaii; Oakland, California; and Independence, Missouri. Religion is a factor for some immigrants: Samoan Mormons tend to migrate to Mormon centers such as Salt Lake City when they move to the United States. Although most Western Samoans emigrate to New Zealand, approximately 15 to 20 percent of Samoan immigrants come from the independent nation.

Samoans in the United States are often employed in unskilled or semiskilled labor such as assembly line jobs, construction, janitorial or maintenance jobs, or security guards. Low pay (per capita income is half that of the national average), large immediate families, and responsibilities for extended family members in Samoa often translate into a poor standard of living. In 1989, one-quarter of Samoan American families were living below the poverty line. Unemployment is thought to be high, perhaps 25 to 30 percent of men on the mainland and as high as 50 percent of men in Hawaii (Howard, 1980; Fitzpatrick-Nietschmann, 1983). Twenty-five percent of young men and 33 percent of young women do not complete high school. Samoans often find social adjustment difficult because their titled rank from Samoa is unrecognized in the United States.

Guamanians. The 1990 U.S. census statistics indicate 47,750 Guamanians residing in the

United States. Many are Chomorros, although self-identification figures are unclear on the exact percentage. The Chomorros often become part of Pacific Islander communities (with Samoan and Tongan immigrants) and are believed to suffer the same discrimination and economic hardships as other Pacific Islanders.

Tongans. Nearly 17,000 Tongans live in the United States according to U.S. census figures. Many are Mormons and are aided in their immigration by their church, and most move to communities of other Pacific Islanders on the West Coast and in Salt Lake City. A small community of Tongans has also located in Hawaii.

Worldview

Religion

Pacific Islanders follow a wide variety of religions, mostly according to which missionary groups were active in their homeland. Hawaiians practice mostly Protestantism, Buddhism, or Shintoism. Samoans are largely Methodist, Catholic, Mormon, and Anglican. Chomorros are primarily Catholic, and Tongans in the United States are mostly Mormon. Religion is often prominent in the lives of Pacific Islanders, and ministers usually are held in high esteem. In Samoa, nightly readings from the Bible are common in most homes, and prayers are offered at every meal (Cox, 1995).

Family

Although most native religions were abandoned, many concepts central to Pacific Islander culture remained within the structure of families. For example, on many islands, social rank and power were established by birth order within the extended kinship system or clan, and even within families, younger siblings deferred to their older brothers and sisters. Elders were respected. The senior male in the group, whether in the village or in a family, managed all group matters. In Hawaii, women and men were segregated under the kapu system, each with specific roles and responsibilities. Extended families were the foundation of society, and children were raised usually by grandparents, aunts and uncles, and even remote kin rather than by just the parents. Household composition was flexible, and all members were obligated to support the extended family, resulting in substantial redistribution of resources. Generosity and sharing were highly valued. Any social transgressions committed by the individual were the responsibility of the whole family. In Samoa, if the violation was severe, the family could be disinherited from their land and stripped of any social title.

Some of these same practices are continued by Pacific Islanders today. Unlike many immigrant groups, they maintain extended families in the United States. Responsibility for child rearing is shared among family members (children may move freely between homes), and household chores are assigned according to age and gender. The oldest man (or occasionally the oldest woman) in the home assumes control, collecting everyone's paycheck (or weekly contribution) and providing for the household needs. The good of the whole family is considered before the benefit to the individual, and most Pacific Islanders are guided by their desire to avoid bringing shame on their family.

The stresses of acculturation in the United States usually occur due to moving from a society in which there is little anonymity (with individual behavior reflecting on the whole family or village) to a society in which Pacific Islanders are often marginalized or even invisible if misidentified as Asians or Filipinos. Most Pacific Islanders maintain close contact with their homeland and fulfill their obligation to family by sending financial support. Trips back to the islands for political and social events are common.

Traditional Health Beliefs and Practices

Religion and medicine were closely linked in Pacific Islander culture, and the loss of many traditional health beliefs and practices occurred with the adoption of nonnative faiths. Although folk healers specializing in herbs, massage, or religious and/or spiritual intervention work in Pacific Islander communities both in Hawaii and on the U.S. mainland, their current use is not well documented (Cox, 1995; Snyder,

■ In Samoa a serious offense by an individual can be ameliorated by an *ifoga* (literally "a lowering"). The extended family positions themselves in front of the victim's home and remains there until invited in and forgiven. They formally apologize through the presentation of gifts and cash.

■ *Waiki* is among the more invasive of traditional Hawaiian therapies. The macerated pulp of the bottle gourd is administered as an enema through the use of a narrow-necked bottle gourd. The procedure was fatal in some cases, but those who survived reportedly lived a long, healthy life. The concept of enemas was probably introduced by Europeans in the 19th century (Whistler, 1992).

■ Hysterical behavior, especially by women, was called "ghost sickness" by the Tongans and attributed to possession by the spirit of a deceased female ancestor. Another traditional illness occurs in infants when the fontanel does not close *(mavae ua)*.

■ Leaves of the *ti* plant reputedly ward off evil spirits in Hawaii, where today ti leaves are waved like pom-poms at football games for good luck.

■ Traditionally, Samoan men obtained full-body tattoos, in part to help them empathize with the prolonged pain of childbirth suffered by Samoan women.

■ Sickness is considered a disruption of the Samoan social order. It is often destructive, especially if it results in death, but can also be creative: pregnancy is believed to be a sickness (Kinloch, 1985).

1984). Traditional practitioners are identified with a broad range of Eastern and Western religious affiliations, and offer a wide spectrum of services. Their clientele are also extremely diverse, often crossing ethnic or religious lines to seek effective care (Snyder, 1981).

Hawaiians mostly practice home remedies. For example, drinking seawater followed by freshwater is believed to be a general tonic. Numerous herbs are used as cures, such as the pepper plant (*Piper methysticum*), which is given in a diluted form to sedate teething infants or blended to produce the intoxicant *awa* or *kava*. It is estimated that more than 58 herbal remedies were available traditionally for respiratory problems (Winters & Swartz, 1995) (see Table 12.3, "Selected Polynesian Botanical Remedies").

Samoans believe that health is maintained through a good diet, cleanliness, and harmony in interpersonal relationships. An individual is at high risk of illness if he or she does not fulfill family obligations or support village life. The concept of balance is essential: disruptions in interpersonal relationships, working too hard, sleeping too little, or eating the wrong foods can cause dislocation of the *to'ala* (the center of one's being located just beneath the navel) to another part of the body, where it induces pain, poor appetite, or other symptoms (Howard, 1986; Kinloch, 1985). Treatment typically requires the restoration of balance. A family may get together to openly air disputes so that harmony can be reestablished or massage by an elder may be used to gently coax the to'ala back into position. A traditional Samoan healer may be consulted to cure certain folk illnesses, particularly those due to supernatural causes, such as spirit possession by malevolent ghosts or the actions of ancestor spirits angered by a person's conduct. One such condition is *musu*, a mental illness in young men and women typified by extreme withdrawal. Healers may be herbalists, masseuses, bone setters, midwives, *taulasea* (a general practitioner familiar with "Samoan sicknesses"), power healers (who provide spiritual interventions), or diviners (specializing in the determination of why an illness has occurred). Massage, herbal remedies, and communication with the supernatural elements are the usual therapies (Cox, 1995; Janes, 1990).

Traditional Food Habits

The cooking of the Pacific Islands developed without benefit of metal pots, pans, and utensils, and many foods were eaten raw. The indigenous cuisine was probably based on breadfruit, taro, cassava, yams, and perhaps pigs and poultry. Fruits were also widely available, although those often associated with the Pacific region, such as coconuts and bananas, were not introduced from Indonesia until approximately 1000 C.E. Other items, including sugar cane and pineapple, were brought by European plantation owners. Pacific Islander fare, especially in Hawaii, has been influenced by European missionaries and traders, as well as by the significant numbers of Japanese, Chinese, Korean, Filipino, and Asian Indian agricultural workers who arrived in the 18th and 19th centuries. Each group added its own cooking to that of the Pacific Islanders, but these foods did not meld into a single cuisine. Instead, each recipe retains its traditional character, and dishes from several different cultures may be served at each meal (see also chapter 15, "Regional Americans"). Foreign flavors are less noteworthy outside Hawaii and are sometimes limited to only a few influences.

Ingredients and Common Foods

Staples

Starchy vegetables are the mainstay of the traditional Pacific Islander diet (Table 12.4). These include the root vegetable *taro,* which is a little denser and more glutinous than the white potato; breadfruit, with a fluffy, breadlike interior; cassava; and yams. These foods were often cooked, and then pounded into a paste. In Hawaii, taro root paste eaten fresh, or partially fermented, is called *poi* (a word that originally referred to the pounding method). When food was scarce, the Hawaiians survived on the purplish-colored poi, sometimes with a little seaweed or fish added to it. Although taro root is also a staple in Samoa, it is usually boiled but not pounded. Arrowroot is used to thicken puddings and other dishes. The Europeans introduced wheat, and bread is eaten in some areas; for instance, Portuguese sweet bread is known as Hawaiian bread in Hawaii. Asian set-

Table 12.3
Selected Polynesian Botanical Remedies

Scientific Name	Common Name	Parts	Traditional Use
Alphitonia zizyphoides	Toi	Bark	Stomach pain; gastrointestinal disorders; lung conditions; amennorhea
Bidens pilosa	Ko'oko'olau; fisi'uli; kofe Tonga; Spanish needles; beggar's tick	Leaves; flower buds	Blood toxins; stomach problems; fatigue; cuts, boils; eye ailments
Bishcofia jaranica	'O'a; toog; koka; tonga-tonga; Bishopwood	Bark; leaves	Gastrointestinal disorders; eye problems; burns
Cananga odorata ☠	Moso'oi; mohokoi; ylang-ylang; perfume tree	Bark	Indigestion; colic; constipation; puerperal infection
Catharanthus roseus ☹	Fetû; mato 'o le lâ; Madagascar periwinkle	Unspecified	Hypertension
Heliotropium anomalum ☠	Hinahina; heliotrope; turnsole	Leaves	Diabetes; menstrual problems; asthma; malaise; fractures, sprains
Hibiscus tiliaceus	Fau; hau; beach hibiscus; cottonwood	Bark; sap; flowers	Ease childbirth; milk stimulation; constipation; appendicitis; circumcision wounds; boils; fractures; eye problems
Ipomoea indica ☠	Fue 'ae puaka; koali; blue morning glory	Root; bark; stems	Gastrointestinal ailments; constipation; purgative
Jasminum simplicifolium ☠	Tutu'uli; wild jasmine	Bark; leaves	Liver ailments
Jatropha curcas ☠	Fiki; physic nut; purging nut	Seed/nut	Constipation; purgative; *mavae ua;* ghost sickness
Lagenaria siceraria	Hue; ipu 'awa'awa; bottle gourd; calabash	Fruit	Emetic
Morinda citrifolia	Noni; nonu; Indian mulberry	Leaves; flowers; fruit	Diabetes; hypertension; anorexia; stomach pain; liver problems; urinary tract infections; sore throat; TB; edema; parasites; elephatiasis; toothache; eye ailments; ghost sickness
Musa nana	Mai'a; dwarf banana	Root; flower sap	Fatigue, weakness; thrush
Neisosperma oppositifolium	Fao	Bark	Diabetes; hypertension; cancer
Oxalis corniculata ☠	'I'i; kihikihi; 'ihi; wood sorrel	Leaves	Asthma; TB; mouth infections; sore throats; *tapitopito; mavae ua*
Pandanus tectorius	Fala; fã; hala; screwpine; pandanus	Roots; male flowers	Malaise; TB; urinary tract infections; constipation; dysentery; *to 'ala* ailments
Peperomia spp.	'Ala 'ala wai nui; kapa; peperomia	Leaves; whole plant	Uterine ailments; tumors; asthma; fatigue
Piper methysticum ☹	'Ava; 'awa; kava; kava; kava-kava; intoxicating long pepper	Root	Urinary tract infections; venereal diseases; headache; backache; rheumatism; bronchitis; asthma; sleep induction; euphoric
Pipturus albidus	Mamaki	Sap; fruit	Malaise; constipation; dysentery; colitis
Plantago major	Lau kahi; filo; common plantain; great plantain; broad-leaf plantain	Leaves; seeds	Diabetes; urinary tract infections; constipation; skin ailments (including sea urchin spine injury)

continued

Table 12.3
Selected Polynesian Botanical Remedies—continued

Scientific Name	Common Name	Parts	Traditional Use
Psidium guajava ☹	Ku'ava; kuawa; guava	Leaves; leaf buds	Digestive tract disorders; dysentery; rashes; wounds; skin infections
Rorippa sarmentosa	A 'atasi; 'akataha; pa'ihi; Polynesian cress	Leaves	Urinary tract infections; gastrointestinal disorders; liver conditions; hemorrhoids; diarrhea; asthma; bronchitis; rheumatism; eye ailments; elephatiasis
Sida spp.	'Ilimi; mautofu	Leaves; flowers	Ease childbirth; milk stimulation; constipation; diarrhea; stomach problems; gall bladder conditions; sore throat; thrush; cough; skin ailments
Solanum nigrum ☠	Magalo; polokai; pōpolo; nightshade, black nightshade	Leaves	Coughs; sore throats; colds; wounds
Spondias dulcis	Vi; vi'ava'ava; ambarella; otaheite apple; Polynesian plum	Bark; leaves; fruit	Stomach ailments; diarrhea; mouth/throat infections; liver conditions
Syzygium corynorcarpum; Eugneia corynorcarpa	Seasea; hehea	Bark; leaves	Urinary tract infections; sore throat; parasites; TB; scrotal edema; fever; tetanus
Syzgium malaccense; Caryophyllus malaccensis; Eugenia malaccensis	Nonu fi'afi'a; fekika kai; 'ohi 'a 'ai; wax apple; Malay apple; mountian apple	Bark; leaves	Gastrointestinal ailments; stomach pain; diarrhea; sore throat; coughs; thrush; headache; venereal diseases
Tournefortia argentea	Lau mafiafia; touhuni; tree heliotrope	Leaves	Food poisoning
Vigna marina	Fue sina; lautolu tahi; nanea; beach pea	Leaves	Urinary tract infections; ease childbirth; vomiting blood; puerperal infection; ghost sickness
Vitex trifolia ☹	Namulega; lala tahi	Bark; leaves	Fever; puerperal infection; headache; sunstroke; cancer; lung problems; TB; cough; skin ailments; eye problems
Wollastonia biflora	Ateate; ate	Stems; leaves	Urinary tract infections; liver ailments; diarrhea; ease childbirth; venereal diseases; wounds
Zingiberis zerumbet ☹	Avapui; angoango; 'awapui; wild ginger	Root; leaves	Stomach pain; gastrointestinal disorders; gastric ulcers; diarrhea; sinus infections; asthma; sore throat; cough; puerperal infection; elephantiasis; toothache

Note: Data on some plants are very limited; adverse effects may occur even if not indicated.
Key: ☠, all or some parts reported to be harmful or toxic in large quantities or specific preparations; ☹, may be contraindicated in some medical conditions/with certain prescription drugs.

■ Fish hooks are symbolic of good luck in Hawaii.

tlers popularized both short- and long-grain rice, as well as noodles.

More than forty varieties of seaweed are consumed. Cooked greens, including the leaves of the taro root, yam, *ti* plant, and sweet potatoes, are very popular.

Fish and seafood are abundant in the Pacific Islands and in some regions were eaten at every meal. Mullet is one of the most popular fish, but many others, including *mahimahi* (dophin fish, not related to the mammal), salmon, shark, tuna, and sardines, are also consumed. A tremendous variety of shellfish is available, such as clams, crabs, lobster, scallops, shrimp, crawfish, and sea urchins, as well as many local species. Eel, octopus, and sea cucumbers are also eaten. Although some fish and seafood was stewed or roasted, some was also eaten uncooked, marinated in lemon or lime juice, which turns the fish opaque much the same as cooking it. A popular Hawaiian specialty is *lomi-lomi,* made with marinated chunks of salmon, tomatoes, and onions, served with or without poi as an appetizer. A similar Samoan dish, called *Oka,* is also made with chunks of raw fish marinated in a mixture of lemon juice and coconut cream.

Pork is the most commonly eaten meat, especially for ceremonial occasions. Traditionally it was cooked in a pit called an *imu* in Hawaii, a *hima'a* in Samoa, and an *umu* in Tonga. A fire was built over the stones lining the pit, and when the coals were hot, layers of banana leaves or palm fronds were added. The pig and other foods, such as breadfruit and yams, were placed on the leaves, then covered with more leaves and sealed with dirt. In some cases, water was poured over the rocks just before the pit was closed, steaming the foods instead of baking them. The pit was left sealed for hours until the food was completely cooked.

Chicken is widely available, as are eggs. Limited grazing land kept beef from becoming a frequently eaten item. Milk and other dairy products are also uncommon. Soybean products are used by Asian residents, and winged beans are a popular legume on some islands.

Fruits and nuts are important ingredients in Pacific Islander cuisine. Bananas, candlenuts *(kukui nuts),* citrus fruits, coconuts, pineapples, guavas, litchis, jackfruit, mangoes, melons, papayas, passion fruit, and *vi* (ambarella) are a few of the widely available varieties. Fruits are eaten fresh or added to dishes such as Samoan papaya and coconut cream soup *(supo 'esi)* and deep-fried dumplings filled with pineapple or bananas *(pani keki).* Coconuts provide juice for drinking, sap for fermentation, and milk or cream used in numerous stewed dishes (coconut milk can also be made into foods resembling cheese and buttermilk). Immature coconuts are considered a delicacy throughout the Pacific. *Haupia,* a traditional gelatin-like Hawaiian dessert, is made from coconut milk sweetened with sugar.

Traditional Pacific Islander fare was not highly seasoned. The flavors of lime or lemon juice, coconut milk or cream, and salt predominate, with occasional use of ginger, garlic, tamarind, and scallions or onions. Coconut oil and lard are the preferred fats, providing a distinctive taste to many dishes. Foreign spices, such as Asian Indian curry blends, and sauces like soy sauce have been incorporated into some dishes.

- Hawaii is a leader in aquaculture. As early as 1778, Captain Cook reported 360 fish farms on the island of Kauai, producing an estimated 2 million pounds of fish annually.

- *Lomi-lomi* was adapted from the technique used to prepare the dried salted fish introduced to Hawaii by New England whalers. *Lomi* means "massage," and the slices of dried fish were rubbed under water to remove some of the salt and to tenderize it.

- The enzyme papain, extracted from papayas, is used in some meat tenderizers. In Hawaii, papaya juice is applied to jellyfish stings to reduce the pain.

- Macadamia nuts, for which Hawaii is famous, were introduced from Australia fewer than 100 years ago.

Pork was traditionally the most commonly eaten meat, particularly for ceremonial occasions. The pig and other foods were cooked in a stone-lined pit over coals. (© Porterfield/Chickering/Photo Researchers, Inc.)

Table 12.4
Cultural Food Groups: Pacific Islanders

Group	Comments
PROTEIN FOODS	
Milk/milk products	Milk and other dairy products are uncommon. Many Pacific Islanders are lactose-intolerant.
Meat/poultry/fish/eggs/legumes	Pork is the most commonly eaten meat. Soybean products are used by Asian residents. Winged beans are a popular legume on some islands.
CEREAL/GRAINS	Europeans introduced wheat bread, and Asians brought rice and noodles.
FRUITS/VEGETABLES	Starchy vegetables are the mainstay of the diet. They were often cooked and pounded into a paste. More than 40 varieties of seaweed are eaten. Cooked greens are popular. Fruits are an important ingredient. Immature coconuts are considered a delicacy. Arrowroot is used to thicken puddings and other dishes.
ADDITIONAL FOODS	
Seasonings	Food is not highly seasoned but often flavored with lime or lemon juice and coconut milk or cream.
Nuts/seeds	Nuts are a core ingredient.
Beverages	Coconuts provide juice for drinking and sap for fermentation.
Fats/oils	Coconut oil and lard are the preferred fats.
Sweeteners	

■ When drinking *kava,* it is considered polite to drip a few drops onto the ground and say *"manuia"* as a blessing before quaffing.

Meal Composition and Cycle

Daily Patterns

Traditional meals included poi or boiled taro root, breadfruit, or green bananas; fish or pork; and greens or seaweed. In Samoa, Guam, and Tonga, most dishes are cooked in coconut milk or cream. Although the evening supper was generally the largest meal, little distinction was made between the foods served at the two or three daily meals. When food was pit-cooked, amounts suitable for 2 or 3 days at a time would be prepared. Fresh fruit was eaten as snacks. Beverages made from coconut juice or sap were common. The Asians introduced various teas, including those made from lemon grass and orange leaves. In Samoa a drink made from ground cocoa beans mixed with water, called *koko samoa,* is traditional. *Kava,* a bland-tasting but mildly intoxicating beverage made from the pepper plant, remains a popular drink in many regions.

Etiquette

Most Pacific Islanders consider hospitality an honor, and outsiders are usually exempt from

Table 12.4
Cultural Food Groups: Pacific Islanders—continued

Common Foods	Adaptations in the United States
	Increased intake of milk has occurred.
Meat: beef, pork *Poultry and small birds:* chicken, duck, squab, turkey *Eggs:* chicken, duck *Fish and shellfish: ahi,* clams, crabs, crawfish, eel, lobster, *mahimahi,* mullet, octopus, salmon, sardines, scallops, sea cucumber, sea urchin, shark, shrimp, swordfish, tuna, turtle, whale *Legumes:* beans (long, navy, soy, sword, winged), cowpeas, lentils, pigeon peas	Dietary changes often occur before immigration. Many are dependent on imported foods such as canned meats and fish.
Rice, wheat	Increased intake of bread and rice is noted.
Fruits: acerola cherry, apples, apricot, avocado, banana, breadfruit, citrus fruits, coconut, guava, jackfruit, kumquat, litchis, loquat, mango, melons, papaya, passion fruit, peach, pear, pineapple, plum, prune, soursop, strawberry, tamarind *Vegetables:* arrowroot, bitter melon, burdock root, cabbage, carrot, cassava, cauliflower, *daikon,* eggplant, ferns, green beans, green pepper, horseradish, jute, kohlrabi, leeks, lettuce, lotus root, mustard greens, green onions, parsley, peas, seaweed, spinach, squashes, sweet potato, *taro, ti plant,* tomato, water chestnuts, yams	The traditional starchy vegetables have decreased in use and may only be consumed at special occasions.
Curry powder, garlic, ginger, mint, paprika, pepper, salt, scallions or green onions, seaweed, soy sauce, tamarind	
Candlenuts (*kukui*), litchi, macadamia nuts, peanuts	
Cocoa, coconut drinks, coffee, fruit juice, *kava* (alcoholic beverage made from pepper plant), tea	Increased consumption of sweetened fruit beverages and soft drinks has occurred.
Butter, coconut oil or cream, lard, vegetable oil and shortening, sesame oil	Use of vegetable oils and mayonnaise has increased.
Sugar	

traditional manners. In Samoa, for example, it is considered rude to eat in front of someone without sharing. When eating a meal, respect should be shown for the food, because it represents the host's generosity; this includes not talking during the meal. Most hosts will not eat until a guest is satisfied. As in many societies, it is impolite to refuse food, although a guest is not obligated to eat every morsel served.

Special Occasions

Throughout the Pacific Islands, special events were commemorated with feasting, often including pit-roasted foods. In Hawaii, weddings, childbirth, completion of a canoe or house, or a prolific harvest or abundant catch was celebrated with a *luau,* featuring a whole pig, poultry, fish, and vegetables cooked in the imu. In Samoa the feast is sometimes preceded by a kava ceremony, in which the beverage is distributed ritualistically to guests who are expected to drain the cup in one gulp. Traditionally kava was offered as a gesture of hospitality and for occasions such as the ordination of a new chief. In Tonga, where umu-cooked food accompanies celebrations like the

■ Kava is made from the chewed or ground root of the native pepper plant, mixed with water in a stone bowl. It looks like dishwater and reputedly tastes a little like dirt or licorice.

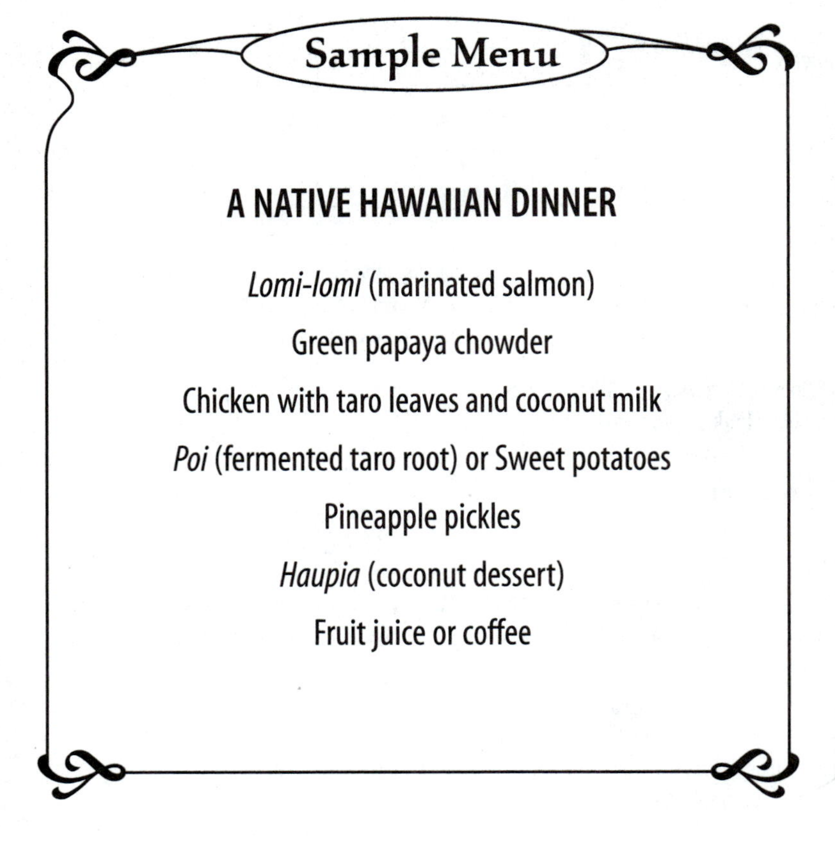

■ Utah is the first state in the United States to prosecute drivers under the influence of kava. A special blood screening is necessary to detect the substance.

commemoration of a royal birthday, special coconut frond stretchers are woven for transporting the massive amounts of food prepared for the occasion.

Holidays celebrated in the Pacific Islands are usually those associated with religious affiliation. Hawaiians also honor Prince Kuhio on March 26 and the Kamehameha Dynasty on June 11. Samoans feast nearly every Sunday, and almost all denominations celebrate "White Sunday" on the third Sunday of October, venerating children. After a special service featuring religious recitations by children, a festive meal is served and children are waited on by the adults in their extended family (Cox, 1995).

Role of Food

■ Certain fish in Hawaii were reserved for royalty, and women rarely consumed fish, although they were permitted to eat shellfish.

Food holds particular importance within most Pacific Island cultures. Sharing food is a way of demonstrating generosity and support for family and village. It is also a way of expressing prosperity or social standing. Many events are celebrated with feasting, and food is eaten to excess as part of the ceremony in some regions. Traditionally gender roles were defined by food interactions. Boys and girls were often raised similarly until the ages of 8 or 9, at which time they were separated for training in food procurement (farming and fishing) or food preparation (cooking and food storage). Throughout the Pacific Islands, gifts of food are given often. Because the gifts are given without expectation of reciprocity, it is a serious affront to reject any item presented.

Therapeutic Uses of Foods

Hawaiians practice many home remedies. For example, drinking seawater followed by freshwater is believed to be a general tonic. Numerous botanicals are used as cures, and it is estimated that traditionally over 58 different herbal remedies were available for respiratory ailments (Winters & Swartz, 1995). Most notable is the pepper plant (*Piper methusticum*—related to the betel vine), which is used to make kava. It is sometimes used medicinally as an analgesic or narcotic. In diluted form it is used as a sedative and given to teething infants.

Samoans believe a good diet is essential to good health. Eating the wrong foods can result in dislocation of the *to'ala* (the center of one's being located just below the navel) to another part of the body, where it induces pain, poor appetite, or other symptoms. Treatment requires the restoration of balance (Howard, 1986; Kinloch, 1985).

Contemporary Food Habits in the United States

Adaptations of Food Habits

Ingredients and Common Foods

The dietary changes made by many Pacific Islanders begin before immigration to the mainland United States. Most Pacific Islanders in their homelands are highly reliant on imported foods, particularly processed items like canned meats and fish, cooking oil, mayonnaise, cookies, breakfast cereals, and soft drinks. One study on Samoa indicated that 80 percent of all foods found in one market were from the United States, New Zealand, or Australia (Shovic, 1994). Consumption of native foods, such as taro and coconut, increases on Sundays (Hanna et al., 1986). In Guam a high intake of meats, fried foods, foods cooked in coconut milk, white rice, and sweetened beverages and a low intake of produce have been noted (Pobocik & Shovic, 1996). Traditional starches, such as taro root and cassava, are often reserved for special-occasion feasts (Fitzpatrick-Nietschmann, 1983).

Little has been reported on Pacific Islander food habits in the United States. A study comparing the intakes of Western Samoans (living in a poorer, less westernized culture) to those of American Samoans (living in a wealthier, more westernized culture) found that the Western Samoans ate a diet higher in total fat due to a reliance on coconut cream compared to a diet higher in protein, cholesterol and salt due to higher consumption of processed foods by American Samoans (Galanis et al., 1999). Limited studies on Samoans who have moved to Hawaii show diets with greater variety of foods; traditional foods only contribute minimally to daily intake and items such as rice, bread, sugar, beef, canned fish, milk, soft drinks, and sweetened fruit beverages make up most of the diet (Hanna et al., 1986). It is assumed from available health statistics that a diet substantially higher in fat and simple carbohydrates and lower in fruits, vegetables, and fiber has been adopted by many immigrants.

Meal Composition and Cycle

Three meals each day are common for most Pacific Islanders. Breakfast is most often cereal with coffee. More traditional meals may be eaten for lunch and dinner; a few Hawaiians still eat poi once or twice a day. Fruit appears more often as part of the meal rather than as a snack. It is believed that Sunday feasting among Samoans is still prevalent.

Nutritional Status

Nutritional Intake

Nutritional deficiencies are rare in most Pacific Islander diets, when a high calorie intake generally guarantees nutritional sufficiency. However, research on Somaoans in Hawaii indicated that fewer calories were consumed than in Samoa and that riboflavin, calcium, and iron intake was low (Hanna et al., 1986).

Mortality rates among Pacific Islander Americans are believed to be high, although figures are limited because as a group, Pacific Islanders are often included with other Asians. Hawaiian data indicate that life expectancy for Hawaiians is lower than that for whites and Asians living in the state. Infant mortality statistics are also limited. One 10-year study in Hawaii indicated that neonatal and postnatal deaths were almost 50 percent higher among Hawaiians and part-Hawaiians than among whites, although Samoan rates were slightly lower than those for whites (Gardner, 1994).

Pacific Islander immigrants do not exhibit excessive risk for contagious ailments or parasitic infection. Chronic disease rates associated with westernization are of primary concern. Obesity rates among Pacific Islanders are some of the highest in the world, regardless of where they live. Native Hawaiian and Samoan adults show average body mass indices that exceed those used to define obesity in the general American population (Crews, 1994). A study of Somoans living in Western Somoa and Hawaii showed large weight gains with increasing modernization. In rural Samoa, the prevalence for being overweight was 33 percent for men and 46 percent for women; in Hawaii the prevalence was 75 percent for men and 80 percent for women. It was noted that high birth weights are reported in American Samoa, as are rapid weight gains in infants throughout the Samoan islands (McGarvey, 1991).

Obesity may be caused directly by overeating (within the context of family and church activities) combined with inadequate physical exercise, and indirectly because overweight was an aesthetic preference in the traditional cultures (Fitzpatrick-Nietschmann, 1983). Some researchers hypothesize that the rate of obesity in Pacific Islanders is partially due to a hereditary predisposition for adipose tissue growth that gave the ancient Pacific Islanders an advantage when migrating long distances in ships (McGarvey, 1991). Other scientists believe that it is the change in the types of carbohydrates consumed that leads to severe overweight. A small study of native Hawaiians who ate as much of traditional

■ Many Pacific Islanders are lactose-intolerant; it is estimated that 50 percent of Samoan adults do not tolerate milk, although one study suggests that environmental factors, not genetics, are the primary determinants on whether a Samoan can or cannot drink milk (Seakins et al., 1987).

■ Samoan Americans have excelled at sports where girth is an advantage, such as football and sumo wrestling.

■ In Guam, an extremely high prevalence of both amyothrophic lateral sclerosis (called *letigo*) and Parkinsonism dementia (known as *bodig*) has been reported. Both are late-onset, cause paralysis, and account for nearly 20 percent of all adult deaths that occur over the age of 25.

■ Samoan women are traditionally treated to a rich coconut drink called *vaisalo* after childbirth.

starchy vegetables and greens as they desired supplemented with small amounts of fish and poultry reduced total calorie intake, lost weight, and lowered their blood pressure and total serum cholesterol. Those who were diabetic reduced or eliminated their insulin requirements (Shintani et al., 1991). A survey of Samoans living in Hawaii indicates attitudes toward obesity are changing; 93 percent of respondents identified being overweight with unhealthiness (Fitzpatrick-Nietshmann, 1983).

Risk for non-insulin-dependent diabetes mellitus, associated with being overweight, is also high for Pacific Islanders. Prevalence estimates are the highest for native Hawaiians (two to six times the 5 to 7 percent prevalence among whites) and from 9 to 16 percent for Samoans. In addition to obesity, researchers suggest that longer life expectancies translate into increased risk of diabetes; there may be between 33,000 and 66,000 Americans of Pacific Islander descent with non-insulin-dependent diabetes (Crews, 1994). Health statistics from California and Hawaii suggest mortality rates from diabetes are also higher for Pacific Islanders than for the general population. It is postulated that Samoans may be especially susceptible to the kidney damage associated with hypertension because end-stage renal failure is a common cause of death in American Samoan diabetes patients (Crews, 1994).

Heart disease is the leading cause of death among all Asian and Pacific Islanders in California; there were 82 deaths per 100,000 in 1989 (compared to rates of 277 for whites, 238 for African Americans, and 63 for Hispanics). However, a more detailed look at the figures reveals differences between population groups. The rate for Samoans was highest, at 100; the rates for Hawaiians and Guamanians were substantially lower, 78 and 56, respectively (Tamir & Cachola, 1994). Excessive mortality rates from heart disease for Hawaiians, more than twice that for the total population, have been reported in Hawaii (Curb et al., 1991).

Hypertension rates are high among Samoans and near the general population average for Hawaiians in California and Hawaii. High total cholesterol levels (and low HDL [high-density lipoprotein] levels) are also noted among Hawaiians and Samoans. The few studies on diet suggest that nearly all fat in the traditional Pacific Islander diet is saturated (36 percent of calories from coconut fats alone) and that saturated fats are also predominant in acculturated diets (Hanna et al., 1986).

Native Hawaiians show a higher incidence rate for cancers of the breast, cervix, uterus, esophagus, larynx, lung, pancreas, stomach, and multiple myeloma than whites. Survival rates are also generally lower (Jenkins & Kagawa-Singer, 1994). Hawaiians also experience higher rates of respiratory ailments, including asthma and bronchitis.

Research on infant feeding practices in Hawaii suggest that the percentage of Hawaiians in Hawaii breast-feeding their babies is more than 50 percent. Another approximately 30 percent mix breast-feeding with a bottle. Native Hawaiian women were more likely to introduce solid foods to their infants before 4 months of age than whites, Filipinos, or Japanese women in Hawaii; sweetened beverages and baby food desserts were common items included in the infants' diets. Tooth decay due to excessive consumption of sweetened beverages in a bottle is three times the U.S. average among children 5 years old and younger in Hawaii (Goldberg et al., 1995).

Counseling

Concepts regarding the role of the individual in health care may be an issue for Pacific Islander American clients. Biomedicine presumes that better health depends on behavior changes made by a client. Pacific Islanders view the role of the individual as interdependent on the group and may not take responsibility for personal health (Fitzpatrick-Nietschmann, 1983). Samoan Americans generally seek care for symptomatic relief and are typically uninterested in long-term approaches to disease prevention or management (Janes, 1990). Low compliance rates in the treatment of hypertension and diabetes mellitus in Pacific Islanders have been noted. A study on native Hawaiian health needs found that cultural differences were a primary reason for underutilization of services (Mayeno & Hirota, 1994).

Language difficulties may occur as well. According to the 1990 U.S. census, more than

40 percent of Samoan Americans reported that they do not speak English well (Pouesi, 1994) and thus linguistic isolation may occur.

Samoans expect exceptional politeness in interactions; showing irritation, anger, or other hostile emotion is considered rude and a sign of weakness (Cox, 1995). Judgmental or accusatory attitudes regarding lifestyle (especially weight gain) may cause Pacific Islanders to avoid further counseling. When entering a room, Samoans walk around to greet each person with a smile and handshake. Eye contact is expected. Both Samoans and Tongans are concerned that all participants in a conversation be at an equal level; for example, everyone should be sitting on mats or in chairs. It is offensive to stand while speaking to someone who is sitting. When seated, legs should be crossed or kept close to the body; extended legs or pointing one's feet at a person is considered poor manners.

Little has been reported on the role of healers by Pacific Islander Americans. An in-depth interview should be used to determine if a client has any traditional health beliefs or practices regarding a specific condition and if religious affiliation is a factor. Due to the paucity of research, coworkers or a client's family members may provide significant information regarding specific Pacific Islander groups.

13

Greeks and Middle Easterners

Greece and the countries of the Middle East are considered the cradle of Western civilization and democracy. Geographically the Middle East is in close proximity to Africa, Europe, and Asia. It has traditionally been a cultural crossroads; ideas and products of the region, including food, have influenced all Western societies. Many immigrants from the Middle East have come to the United States in search of economic opportunity and political stability. Most immigrants in the United States retain a strong ethnic identity, exhibited in their religious faith and in their maintenance of many traditional food habits. This chapter examines the cuisine of the Middle East, its role in the culture, and the changes that have occurred in the United States.

Cultural Perspective

The countries of the Middle East include Bahrain, Egypt, Iran, Iraq, Israel, Jordan, Kuwait, Lebanon, Oman, Saudi Arabia, Syria, Turkey, the United Arab Emirates, and Yemen. In addition, there are the Palestinians (an Arab ethnic group) and the Kurds (an Indo-European ethnic group), each without a homeland. Although Greece is considered a European country, its cuisine resembles that of the Middle East; the geography and climate are also similar. (See Figure 13.1 for a map of the region.) Greece dominated or greatly influenced its Middle Eastern neighbors in ancient times, and in turn it was conquered and ruled by the Turkish Ottoman empire for 4 centuries in the modern era. Several states and countries previously under domination of the former Soviet Union and the nations of North Africa also share certain commonalties with the Middle East, especially fare. Some notable immigrant populations in the United States are the Croatians and Serbians from the war-torn region of Croatia, Serbia, and Bosnia-Herzegovina (former states in the united Yugoslavia), Slovenians, and Armenians from the newly independent Republic of Armenia and from Turkey.

Populations within the nations of the Middle East are diverse in religious affiliation. Judaism, Christianity (particularly Eastern Orthodox), and Islam all have substantial numbers of followers in the region; religious faith often directs daily life.

Geographically most of the area is considered sandy and arid, but distinct units of arable land exist along the seacoasts and in some plains and valleys, such as the Fertile Crescent (a plain in Iraq fed by the Euphrates and Tigris Rivers) and the Nile River valley of Egypt.

History of Greeks and Middle Easterners in the United States

Immigration Patterns

Statistics on immigration from the Middle East are inexact. Until 1900, all immigrants from the area were called "Egyptians." Later arrivals were typically termed "Syrians" or "Turks from Asia." More specific nationalities have been recognized since the 1930s, yet it has only been in recent years that "Palestinian" has been defined as an immigration category.

■ An *Arab* is commonly defined as a person who speaks Arabic; the term does not refer to a particular religious belief. Iranians speak Farsi and call themselves Persians.

■ The Socialist Federated Republic of Yugoslavia, a communist satellite of the former Soviet Union, consisted of six states: Serbia, Croatia, Slovenia, Bosnia-Herzegovina, Montenegro, and Macedonia. In 1992, Serbia and Montenegro became the Federal Republic of Yugoslavia, and the remaining four states became independent republics.

Figure 13.1
Greece and the Middle East

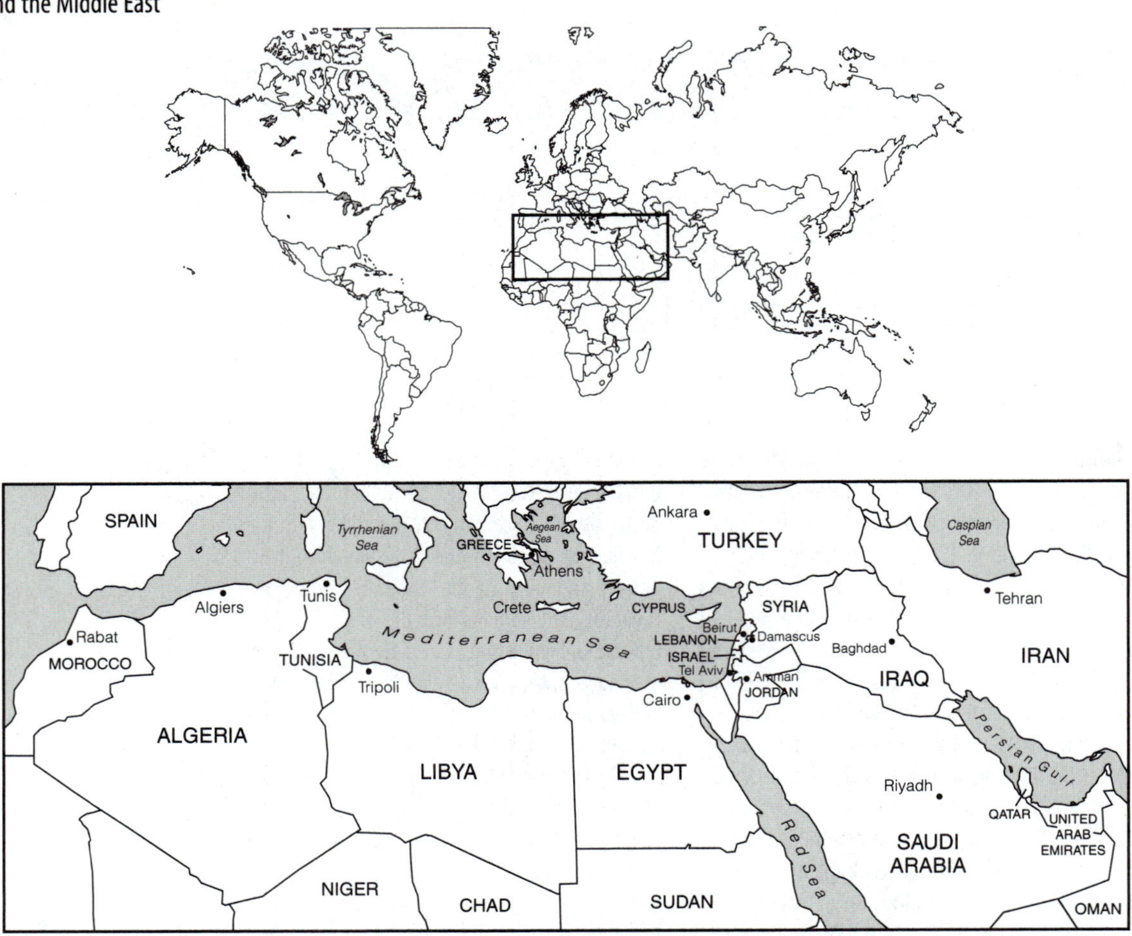

■ Around the turn of the 20th century, an elite group of Arab artists, writers, and poets settled in New York City. They called their literary circle the Pen League; the best-known member was Kahlil Gibran, author of *The Prophet.*

■ So many Dalmatian farmers settled in the Pajaro Valley of northern California that Jack London called it "New Dalmatia."

Immigration from Greece and the Middle East has occurred primarily in two waves. The first lasted from the late 1800s to the 1920s, when the restrictive Immigration Act of 1924 was imposed; the second wave started after World War II and has not yet ended. The early Greek and Arab immigrants were mostly young men from rural agricultural areas who came to America primarily for economic opportunities. The Arabs were mainly Christians from the area that is today Lebanon and Syria, although small numbers of Turkish Armenians and Muslims also arrived during this period. Most settled in New England and the Midwest and were employed in the textile, shoe, and wire factories. Many came to make their fortune and go back to their homeland; approximately 30 percent of early Greek immigrants returned to Greece.

During the 1850s and 1860s, numerous Croatian immigrants from the area of Dalmatia arrived in the United States. Most migrated to the southern and western regions, where they had a substantial impact on the oyster fisheries of Mississippi and the fig, apple, grape, and plum horticulture of northern California. The many Serbs who immigrated to the United States in this period were unskilled laborers who obtained industrial jobs in the Northeast. Slovenian immigrants in the 19th century were mostly farmers seeking economic opportunity. They settled initially in the rural Midwest, forming self-sustaining ethnic communities with Slovenian churches, schools, businesses, and social organizations.

Immigration from Greece and the Middle East after World War II reflected political unrest in the area. In Greece a bitter civil war

Traditional foods of Greece and the Middle East. Some foods typically included in the diets of people in Greece and the Middle East include almonds, chickpeas (garbanzo beans), *couscous*, cracked wheat, dates, eggplant, feta cheese, figs, *filo* dough, garlic, lamb, lemon, olives, *pita* bread, and yogurt. (Photo by Laurie Macfee.)

from 1946 to 1949 and a military coup in 1967 resulted in numerous refugees who sought asylum in the United States. While most settled in New York, Detroit, Chicago, and other cities of the Midwest, a large community developed in Tarpon Springs, Florida, supported by diving for sponges. Some Greeks were attracted to mining and railroad work in the West.

A high percentage of immigrants from Jordan, Egypt, Lebanon, Iraq, and Syria were Palestinian refugees who initially fled Israel when it was declared a state. After first settling in refugee camps, many later emigrated to the United States when Israel won the 1967 war against neighboring Arab countries. Political unrest and Islamic revolution in Iran led to a large exodus of Persians beginning in 1976. Many were members of the wealthy Iranian elite; others were Jewish and Christian minorities. In addition, many Arabs who were sent to the United States for education and training remained for economic reasons.

Turkish, Croatian, and Serbian immigration also increased following World War II, including professionals, such as engineers and physicians, seeking better job opportunities. More than 60,000 Armenian refugees have arrived since 1980, settling primarily in Los Angeles, with small numbers joining the older Armenian American communities in Boston, New York, Chicago, Detroit, and the agricultural region of Fresno, California.

■ The 1993 Canadian census lists over 244,000 residents of Middle Eastern ancestry.

Current Demographics and Socioeconomic Status

More than 1 million persons claimed Greek heritage in the 1990 U.S. census. New York City has the largest concentration of Greek Americans, followed by California, Illinois, Massachusetts, and Florida. In recent years, Greek Americans have moved increasingly south and west in the United States.

Demographic figures for Middle Easterners are more problematic. It is believed that many deny their Arab ancestry due to discrimination, and others refuse cooperation with government-sponsored surveys due to negative experiences in

■ Some Arab Americans face significant discrimination due to ongoing tensions between the United States and countries in the Middle East. The media, especially movies, have vilified Middle Easterners and inflamed anti-Arab sentiments.

■ *Persia* was the conventional European name for Iran used until the early 20th century. Today, concern about prejudice or anxieties about the Iranian State Intelligence Agency prompt some Iranian immigrants to use the designation *Persian American*.

their homelands. The census reports nearly 400,000 self-declared Americans from Lebanon; 308,000 from Armenia; 236,000 from Iran; 130,000 from Syria; 127,000 listed as Arab; 83,000 from Turkey; 80,000 from Israel; 79,000 from Egypt; 48,000 Palestinians; 23,000 Iraqi; 20,000 from Jordan; 7,600 miscellaneous Middle Easterners; 4,500 from Saudi Arabia; and 4,000 from Yemen. Unofficial estimates of these populations are often two to three times larger. Most Middle Easterners are located in the Northeast and Midwest, although significant populations are found in California, particularly in Los Angeles. Smaller communities are developing in Texas, Georgia, Alabama, and other areas of the South.

More than 500,000 Croatian Americans and 100,000 Americans of Serbian descent live in the United States. Most originally settled in the mining regions of Pennsylvania, Ohio, and other Midwestern states; mining also attracted sizable populations to Colorado, Nevada, Arizona, and California. Nearly 125,000 Slovenian Americans were identified in the 1990 U.S. census. Although many Slovenian immigrants also became involved in mining, others moved to industrial urban areas in search of jobs. Forty percent live in Ohio; Cleveland has the largest Slovenian community in the nation. Other states with notable Slovenian populations include Pennsylvania, Minnesota, Wisconsin, California, Colorado, Michigan, Florida, and New York.

Early Greek and Arab immigrants preferred to live in urban areas. The Arabs often became peddlers and later opened family businesses, usually dry goods or grocery stores. Many recent Arab immigrants are well-educated professionals. The Greeks sought economic independence by opening small businesses, especially in candy production and restaurants. A tradition of entrepreneurship among Middle Easterners continues; Iranians are frequently involved in real estate development and Armenians often open their own businesses. Rates of self-employment among Greeks and Middle Easterners are very high, between 12 and 36 percent of all workers,

A large Greek community developed in Tarpon Springs, Florida, based on diving for sponges. (© Van Bucher/Photo Researchers, Inc.)

compared to 7 percent for the national average.

Recent immigrants are often well educated, and there is an admitted "brain drain" in the Middle East that has occurred since the mid-1970s. Even families that started with unskilled employment or small businesses have made substantial economic progress through schooling. Second- and third-generation Americans of Greek and Middle Eastern heritage attend college and graduate in professional fields such as engineering, medicine, pharmacy, dentistry, and law at rates far higher than the national average. Greek Americans and Middle Eastern Americans are among the most prosperous of all ethnic groups in the United States.

Worldview

Religion

Religion is often the defining factor in Greek and Middle Eastern life. Though affiliation varies, strong devotion is common. Many congregations remain very insular in the United States, serving the needs of a specific ethnic group, and there is little interest in proselytizing to outsiders.

The ethnicity of Greek immigrants was affirmed mostly by religion; it was said that a person was not Greek by birth but through the active affiliation with the Greek Eastern Orthodox Church. The first Greek Orthodox Church in America was founded in New Orleans in 1864.

Most Greek Americans today still belong to the church, which continues to be the center of Greek community life. The word *orthodox* comes from the Greek for "correct" (*orthos*) and "worship" (*doxa*). A fundamental belief of the Greek Orthodox faith is that an individual attains complete identification with God through participation in the numerous religious services and activities sponsored by the church. Although the Greek Orthodox Church is conservative and traditionally resistant to change, some accommodations have been made in the United States; for instance, the service typically is conducted both in Greek and English, and modern organ music accompanies the liturgy (see chapter 4 for more information about Greek Orthodoxy).

Early Arab immigrants were primarily Christians belonging to the Eastern Orthodox or the Latin rite Maronite, Melkite, or Chaldean churches. Although Christian Arabs are still a majority in America, more recent immigrants follow Islam, and the number of Arab Muslims in the United States is growing rapidly. Most belong to the Sunni sect. Many Arab Muslims in the United States have made several adaptations to accommodate their religious practices to American society. Most significantly, the Friday Sabbath prayer has been moved to Sunday, and many Muslims cannot fulfill their obligation to five daily prayers due to work or school limitations (see chapter 4 for more information about Islam).

Though the vast majority of Iranians in the United States are members of the Muslim Shi'ite sect, small numbers of Iranian religious minorities have come to the United States seeking freedom from persecution, including Jews, Catholics, Anglicans, Protestants, and members of the Baha'i faith (a Muslim offshoot that has renounced its ties to Islam and preaches gender equality, world brotherhood, and pacifism). Most Armenians are members of the Armenian Apostolic Church (an Eastern Orthodox faith noted for allowing its members to make decisions on issues such as birth control and homosexuality without religious influence), although some Americans of Armenian descent are Protestants or members of the Armenian Rite of the Roman Catholic Church. A majority of Croatians and Slovenians are devout Roman Catholics, and most Serbs belong to the Serbian Orthodox Autonomous Church.

■ The tall white hats worn by professional chefs are thought to have originated with the tall black hats worn by Eastern Orthodox priests.

■ Of the 90,000 Jews living in Iran in 1987, 55,000 have since left, including 35,000 who have immigrated to the United States.

■ In 301, Armenia became the first nation to adopt Christianity as its state religion.

Family

Traditionally both Greek and Arab cultures center around a strong patriarchal family whose honor must be maintained. The family demands conformity and subordination of individual will and interest, but in return the members of the family are protected and can identify with the family's status. The Greek term for this pride and obligation to family is *philotomo*, meaning "love of honor." Families often live with extended

■ Greeks traditionally considered educational or career achievement among women to be unfeminine or even un-Greek.

■ In some Arab American homes, girls are not allowed to play certain sports or ride bicycles.

■ Muslims often have serious concerns about raising their children in America. Permissive attitudes about dating and drinking and pressure to practice self-reliance and independent thought prompt some immigrants to send their children, especially their daughters, to the Middle East for schooling.

■ Growing drug and gang problems among Armenian youth have stressed relations between recent immigrants and immigrants who have lived successfully in the United States for many years without significant social problems.

members in a single home or, for well-to-do Middle Easterners, in a family compound. Elders traditionally live with the families of their sons.

Children are valued in both Greek and Arab families, and sacrifice for the good of the children is common. In Islam the teachings of the Quran state that men and women are considered equal, but different. Although men are expected to provide economic security for children, it is the role of women to socialize the children, including the preservation of religious and cultural values. Because of the strict patriarchy, women are expected to provide the love and comfort in a home as well. The relationship between mothers and daughters is very intimate. There is also a strong bond between mothers and sons, especially the first-born son (Luna, 1994). Generally, formal schooling is not considered as important for girls as for boys among Arabs. Education for women is encouraged by some Middle Easterners, however, such as Iranians and Armenians.

Marriage contracts in the Middle East are often arranged between children to establish political and business alliances. The wealth and status of a family may be dependent on the expansion of kinship. Among Arabs there is a preference for marriage between cousins, except among Egyptians and some Arab Christians. Only after an engagement announcement is made are young men and women allowed to "date," and then only when chaperoned. A family's honor is related to the modesty and chastity of the women in the home: A woman is chaste before her wedding and faithful after. Actual or alleged violations of moral codes by a young woman are considered evidence that her mother has failed in her responsibilities; inappropriate sexual conduct also brings shame on her male relatives.

Although Islamic law allows a man to take four wives, most only wed once. Interethnic marriages are strongly discouraged by both Muslims and Christians, especially for women. It is preferable, however, to marry someone of a different ethnic group and the same religious affiliation than to marry outside the religion.

One of the first effects of immigration to the United States is the breakdown of the extended family and reduction of the father's authority, although filial respect is retained. Family ties are still very strong among Greeks and Arabs, and it is still common for elderly parents to live with the family. Some Arab women in the United States have gradually rejected certain Middle Eastern customs, such as veils for Muslim women. Dating is becoming more prevalent as segregation of the sexes cannot be maintained in the American workplace. Interethnic and interreligious marriages are on the rise, particularly among Greek Americans. Yet, for many Middle Eastern women, their traditional roles in the home tend to isolate them from contact with much of American culture; among Iranian immigrants, women feel that their limited authority has decreased since arriving in the United States, that they spend more time on household chores, and that they are less involved in family decisions (Gillis, 1995). More acculturated Arab women also believe that they face serious setbacks in acceptance of their Americanized conduct every time a large influx of new and traditionally minded Middle Eastern women settle in their area. Intergenerational conflict is less than that seen in other ethnic immigrant groups, although second- and third-generation Greeks and Middle Easterners in America are likely to think that their parents are old-fashioned.

Croatian and Serbian families are also traditionally patriarchal. Extended families are the norm, often including friends as well as relatives. Among Croatians, communal living may involve the taking in of boarders. Both Croatians and Serbians have become well acculturated in the United States. The tradition of older generations caring for children has allowed many Croatian and Serbian women to take advantage of educational and career opportunities, and the authority of the father has lessened. Slovenian Americans also are usually acculturated. The extended family structure typical in Slovenia is rarely found among Slovenians in the United States, who prefer nuclear families. American women of Slovenian descent are active in the home, church, and Slovenian schools and are increasingly involved in politics.

Traditional Health Beliefs and Practices

Cleanliness, diet, and keeping warm and dry are all Middle Eastern factors important to maintaining health. Some Middle Easterners believe the illness is due to "wind" or "air" in certain situations. Lebanese Muslims believe that following childbirth, a woman is especially susceptible to wind; showers and baths are avoided for 10 days to prevent wind from entering the veins and causing sickness. An infant is vulnerable to wind through the umbilicus, so the baby's stomach is wrapped at all times with a band called a *zunaad* (Luna, 1994).

Traditional humoral medicine is important in the health practices of Iranians. Though traditional humoral theories identify four bodily humors, in practice Iranians are only concerned with hot and cold. Each person is born with physiological temperament dependent on the ratio of hot to cold, called *tabi'i.* Women, for instance, are considered colder than men, and younger persons are hotter than older people. Tabi'i can be influenced by diet, climate, geographical location, and certain conditions, such as childbirth (Good, 1980).

Greeks and some Arabs also believe that the "evil eye" of one who envies a person can cause accidents or illness. Greeks use ritual prayer and wear blue amulets with an eye in the center or garlic as a precaution against a jealous gaze; Arabs may place blue beads on infants to protect them (Jurgens, 1995; Luna, 1994). Related to the evil caused by envy are other sicknesses due to strong emotions. In Iran, for example, *narahatiye qalb* (heart distress) is a folk condition typified by pounding or fluttering of the heart due to anxiety, sadness, or anger (Helman, 1990). Iranians often put their health into the hands of God. *Tagdir,* meaning God's will, is thought to determine all aspects of life and, ultimately, death. Throughout the Middle East, illness is sometimes seen as a punishment from God.

A long-standing tradition of home health care exists among Greeks and Middle Easterners. Folk remedies such as drinking herbal teas, dropping a hot coal in water to dispel headaches, or rubbing *ko'hl* on the umbilical cord to help it dry are common. In Iran, foxglove blossoms are used for digestive problems or nervous conditions, and mint tea with coriander seeds is used to promote sleep (Lipson, 1992) (see Table 13.1, "Selected Middle Eastern Botanical Remedies"). Biomedical practice is well established in the Middle East, and for the most part its therapies are considered strong and effective (Lipson & Meleis, 1983).

For more complex conditions a traditional healer may be consulted, such as a bone-setter for a serious fracture. Cupping (see chapter 2 for more information) is done in a manner similar to the Asian method, except that the skin is cut with a razor first to allow "bad blood" to be extracted. It may be used to cure chronic leg pain, paralysis, headaches and obesity. Another therapy that is sometimes applied is called *wasm,* or cauterization. A heated iron rod is used to place symbolic burn marks on the patient, for example below the anus to treat diarrhea and under the ear lobe to cure a toothache. The burns are then treated with special herbal poultices (Ghazanfar, 1994).

Traditional Food Habits

Certain common foods in Greece and the Middle East, such as wheat, olives, and dates, are indigenous to the region. Yet most countries in the area claim one dish or another prepared by all ethnic groups to be their own invention. It is doubtful if the source of any Middle Eastern dish will ever be known, as the political and economic history of the region has resulted in very similar foods and food habits.

Ingredients and Common Foods

Staples

The common ingredients used in Greek and Middle Eastern cooking are listed in Table 13.2. Numerous food items, such as wheat, rice, legumes, and lamb, are eaten in most countries.

Wheat, thought to have been cultivated first in this region, is usually eaten as leavened bread loaves in Greece and leavened flat bread in the Middle Eastern countries. One example is the thin round Arabic bread with a hollow center called *pita* (or *pida* in Turkey); another is *lavosh,* which is Armenian cracker bread. Bread traditionally accompanies every Greek and Middle Eastern meal.

Besides bread, wheat doughs are also used to make pies, prepared in a variety of sizes and shapes. Savory pies may contain meat, cheese, eggs, or vegetables. Desserts are usually filled

■ The Turks begin their meals with the prayer "May what you eat bring you well-being."

■ The Sufis, members of an ascetic and mystical Islamic sect, define health as an existential state of abstinence, patience, and self-examination, resulting in harmony with the universe.

■ The Greeks may respond with *"ptou, ptou,"* to a compliment to keep the evil eye from causing harm.

■ *Ko'hl* is a dark powder made mainly from the chemical element antimony. It is used primarily as a cosmetic in the Middle East.

■ During Communion in the Armenian Apostolic Church, followers are offered small pieces of round bread dipped in wine. As they leave the service, they receive lavash in a bag that has been blessed by the priest.

Table 13.1
Selected Middle Eastern Botanical Remedies

Scientific Name	Common Name	Parts	Traditional Use
Acacia niloctica	Motse; babul; qarat; tulh Egyptian mimosa; Egyptian thorn; sant tree	Leaves; seed pods; resin	Diabetes; colds; diarrhea; edema; typhoid; eye problems
Acacia senegal	Kher; kumta; temmer; gum arabic	Bark; resin	Colds; sore throat; whooping cough; dysentery; diarrhea; urinary tract infections; digestive upset; menstrual disorders; typhoid; leprosy; sore nipples; wounds
Alhagi maurorum	Aqul; al heej; igol; kag; camelthorn; Persian manna; Caspian manna	Whole plant	Liver conditions; migraine; rheumatism; aphrodisiac
Allium satvia ☹	Thōm; garlic	Bulb	Diabetes; stomach pain; colic; diarrhea; flatulence; kidney disorders; TB; eye problems; wounds; dandruff
Aloe dhufarensis ☹	Subr; tuf	Leaves; juice	Diabetes; constipation; purgative; cough; arthritis; headache; rashes; eye conditions
Azadirachta indica; Melia azadirachta ☹	Meraimarah; nib; shereesh; neem	Leaves	Fever; stomach pain; colic
Boswellia sacra	Bakhor; beyo; lubān; maqar; mohor; frankincense; olibanum	Bark; leaves; fruit; gum	Lethargy; stomach ailments; urinary tract infection; kidney stones; diuretic; purgative; parasites; rheumatism; venereal diseases; toothache; eye disorders; emotional distress (especially during pregnancy); morning sickness; forgetfulness/amnesia
Capparis spinosa	Fakouha; lasafa; lussef; shafallah; caper; caperbush	Bark; flower buds	Diabetes; diuretic; flatulence; anemia; arthritis; gout; coughs; earache
Catha edulis ☠	Khat; qat; gad; tschat	Leaves	Depression; lethargy; hunger pains; aphrodisiac; narcotic properties
Coccinia grandis, C. cordifolia, C. indica	Bakhra'a; ivy gourd	Leaves	Diabetes; sore throat; earache; skin infections
Digitalis purpurea ☠	Gol gov zabon; purple foxglove	Flowers	Digestive problems; nervous conditions
Echium angustifolium ☠	Lisān al-thaur	Leaves	Liver ailments; kidney conditions (especially stones); herpes
Foeniculum vulgare	Samār; sheeh; fennel	Fruit; oil	Colic; flatulence; constipation; kidney disease
Glycyrrhiza glabra ☠	Irk al hiel; irk al hiou; ud al sūs; licorice root	Root; leaves	Indigestion; purgative; cough; rheumatism; pain relief
Haloxylon salcornicum	Gatha; remth	Stems	Hypoglycemia
Ipomoea pes-caprae	Derēg; beach morning glory; railroad vine	Seeds	Purgative
Lavandula dhofarensis ☠	Ekulun; heryem; hilbēn; lavendar	Leaves	Kidney ailments; stomach pain; parasites

Table 13.1
Selected Middle Eastern Botanical Remedies—continued

Scientific Name	Common Name	Parts	Traditional Use
Mangifera indica	Amba; mango	Leaves; sap	Diabetes; liver conditions; flu
Myristica fragrans ☠	Bisbasa; djus hendi; gewz; joze qu'wa; qoust; nutmeg	Seeds; oil	Indigestion; stomach cramps; malaise
Papaver somniferum ☠	Abu al noum; afyūm; hishas; opium poppy	Flowers; seed pods	Insomnia; stress; cough; narcotic
Perganum harmala ☠	Harmal; khiyys; Syrian rue; African rue		Kidney stones; intestinal distress; rheumatism; parasites; narcotic
Plantago ovata ☹	Bidr qūtnū; hab zaragh; qurayta; rebla; psyllium	Seeds	Constipation; diarrhea; venereal diseases
Portulaca oleraceae ☹	Al khalēga; arbir; farfena; humda; purslane; pigweed	Leaves; whole plant	Constipation; liver ailments; kidney problems
Punica granatum ☠	Anar; ruman; pomegranate	Seeds; rind	Weak stomach; diarrhea; liver conditions; parasites; skin eruptions; eye problems
Rhazya sticta ☠	Adfīr; harmal; neutral henna	Whole plant	Diabetes; stomach pain; colic; constipation; chest pain; fever; lactation stimulant; parasites
Ricinus communis ☠	Arash; kharwa; zait; castor bean	Root; leaves	Constipation; purgative; rheumatism; halitosis; toothache; ulcers
Solanum melongena	Batangen; eggplant; mad apple	Fruit	Hypercholesterolemia; nausea
Solanum nigrum ☠	Ana al deeb; anamnam; mejaje; black nightshade	Whole plant	Stomach pain; kidney ailments; urinary tract infections; venereal diseases
Tamarindus indica	Sebbar; tamr hindī; umar; tamarind	Seed pods	Nausea; constipation; hemorrhoids
Tamarix aphylla, T. articulata	Athal; tarfa; terfel; salt cedar; athel tamarisk	Leaves	Childbirth; headache; fever; diarrhea
Teurcrium mascatense, T. polium ☠	Ay'ad; ja'ada; misk al jinn; neftah	Leaves; stems	Diabetes; stomach pain; colic; liver ailments; childbirth; malaria
Withania somnifera	Balbul; bābu; ebab; gengeneh; haml; sumāl far; ubab; winter cherry	Leaves	Infertility; edema; wounds; bites; sedative
Zingiberis officinalis ☹	Za'ater, zingībīl; ginger	Root	Anorexia; flatulence; bronchitis; coughs; cataracts
Ziziphus spina-cristi	Nabaq; sidr; Libyan lotus; christ-thorn	Leaves; fruit; seeds	Chest pains; respiratory conditions; rheumatism; mouth/gum problems; headache

Note: Data on some plants are very limited; adverse effects may occur even if not indicated.
Key: ☠, all or some parts reported to be harmful or toxic in large quantities or specific preparations; ☹, may be contraindicated in some medical conditions/with certain prescription drugs.

■ *Pastilla* is a Moroccan dish with Middle Eastern origins featuring filo dough filled with cinnamon-scented chicken.

■ At ancient Greek weddings the bride and groom were showered with wheat kernels.

■ *Couscous,* fine granules of semolina wheat flour (like a very small pasta), is steamed in a *couscousier* (a pan with boiling water and a colander insert that holds the granules) and served with stews throughout North Africa.

■ Béchamel, a white sauce used in French cuisine, is thought to have originated in Greece.

with nuts and covered in a thick sugar and honey syrup. Pie pastry, flaky pastry, paper-thin pastry called *filo* (or *phyllo*), and bread dough can all be used. An example of a meat- or cheese-filled pastry that can be served hot or cold is *sanbusak.* Traditionally half-moon shaped and 3 inches long, it is popular in Syria, Lebanon, and Egypt. A similar pie is called *boereg* by Armenians. *Fatayeh* is another specialty served as a snack, featuring bread dough topped with cheese, meat, or spinach and baked like a pizza. *Spanakopita* is a Greek pie made with filo dough filled with spinach and cheese, and *tiropetas* are flaky turnovers stuffed with cheese. Throughout the Middle East and Greece the dessert made with filo dough, *baklava,* is served at every special occasion. No café or bakery would be without it. Baklava contains numerous alternating sheets of filo dough and a walnut, almond, or pistachio filling, which is soaked in a sweet syrup flavored with honey, brandy, rose water, or orange blossom water. It is frequently cut into diamond shapes.

Raw kernels of cracked whole wheat are also used in a number of Middle Eastern dishes. When the kernels are first steamed and then dried and crushed to different degrees of fineness, the cracked wheat is called *burghul* or *bulgur.* Unripened wheat kernels that are dried but not cracked are known in Arabic markets as *fireek.* All varieties of wheat kernels are typically cooked as side dishes or made into *tabouli,* a popular salad containing onions, parsley, mint, and a variety of fresh vegetables. *Kibbeh,* one of the national dishes of Lebanon and Syria, is a mixture of fine cracked wheat, grated onion, and ground lamb pounded into a paste. This mixture can be eaten raw or grilled, and with a great deal of dexterity it can be made into a hollow shell, then filled with a meat mixture and deep-fried. A few pasta dishes are found in Greek cuisine, including pasta with baked lamb or goat and tomatoes called *yiouvetsi* and macaroni baked with cheese, ground meat, tomato sauce, and bechamel sauce called *pastitsio.*

In addition to wheat, rice is also a staple item in Greek and Middle Eastern cuisine. The long-grain variety is used to make *pilaf,* a dish that commonly accompanies meat. The rice is first sautéed in butter or oil in which chopped onions have been browned. It is then steamed in chicken or beef broth. Occasionally saffron or turmeric is added to give the dish a deep yellow color. *Polo* is the Iranian version of pilaf, but a final step in its preparation produces a

Muslim women purchasing sweets for Ramadan from an Iraqi street vendor. (Corbis & AFP.)

rice with a crunchy brown crust that is known as the *tah dig.* A more fragrant variety of rice (*basmati*) is used in Iran for *khoresh*, rice topped with stewed meat, poultry, or legumes and to make the national dish, *chelo kebab,* which is thin pieces of marinated, charcoal-broiled lamb served over the rice.

A large variety of legumes are another important ingredient in Middle Eastern cooking. Cooked pureed chickpeas (garbanzo beans) are the base for the salad called *hummus,* which is often used as an appetizer or as a dip. Ground chickpeas or fava beans are sometimes formed into small balls and then fried and served as a main course (*ta'amia*) or in a pita bread with raw vegetables (*falafal*). A common breakfast food is *foul,* slowly simmered fava or black beans topped with chopped tomato, garlic, lemon juice, olive oil, and cilantro (fresh coriander leaves). The national dish of Egypt is *ful medames,* cooked fava beans seasoned with oil, lemon and garlic, sprinkled with parsley, and served with hard-boiled eggs. Lentils and peanuts are also eaten.

Fresh fruits and vegetables are preferred, but if they are not available, fruits are served dried or as jams and compotes; fruit syrups also flavor many savory dishes. Vegetables are preserved as pickles. Fruits, many first cultivated in the region, are eaten for dessert or as snacks. Figs, dates, grapes, and pomegranates are favorites. Lemons are often used for flavoring. Although many vegetables are used, eggplant is the most popular. A common cooking method for vegetables (the Greek term is *yiachni;* the Arabic word is *yakhini*) is to combine them with tomatoes or tomato paste and sautéed onions together with a small amount of water, then cook until the vegetables are soft and very little liquid remains. Eggplant and other vegetables are sometimes stuffed with rice or a meat mixture. *Moussaka* is a Greek specialty made with minced lamb, eggplant, onions, and tomato sauce baked in a dish lined with eggplant slices. The Turks prefer *imam bayildi* (meaning "the priest fainted"), eggplant filled with tomatoes, onions, and garlic, stewed in olive oil and served cold. A delicacy favored throughout the region is stuffed grape leaves, *dolmas* or *dolmades.* Vegetables are frequently enjoyed raw, mixed together in a salad.

The olive tree contributes in many ways to Greek and Middle Eastern cooking. Olives prepared in the Middle Eastern way have a much stronger flavor than European or American olives; they often accompany the meal or are served as an appetizer. There are two basic types of olives in Greece and the Middle East. Black olives are ripe when picked; they are small, round, wrinkled, and taste slightly bitter and salty. Green olives are picked unripe, having an oblong shape, firm flesh, smooth skin, and tart flavor. Greek *kalamata* olives are a favorite; they are large and oblong, with soft flesh and smooth purplish skin.

The olive is also a source of oil, which is frequently used in food preparation, although both Greeks and Arabs also use clarified butter (*samana*) and most vegetable oils as well as rendered lamb fat and margarine. Olive oil is generally used in dishes that are to be eaten cold. For most deep-frying, corn or nut oil is used, but olive oil is preferred for deep-frying fish.

Fresh milk is not widely consumed in Greece or the Middle East, although it is used in dessert puddings. Dairy products are usually

■ The Syrians blend bulgar with yogurt, dry the mixture in the sun, and then grind it into a powder to prepare *kish'ka*. It is combined with a little water and onions to make a filling for pita bread or thinned further to make soup.

■ Date trees were an ancient symbol of fertility.

■ In Tunisia, vegetables such as tomatoes, peppers, onions and garlic are grilled, then served as a salad dressed with olive oil and lemon juice in the dish called *mechouia*.

■ The term *yogurt* is Turkish. In Syria and Lebanon the fermented milk product is called *laban;* in Egypt, *laban zabadi;* and in Iran, *mast*.

Sample Menu

A PERSIAN LUNCH

Adas polo (rice layered with lentils and lamb or chicken)

or

Khoresh-e fesenjan ba jujeh (chicken stew with pomegranate) over rice

Cucumber, tomato, and onion salad

Olives and pistachios

Feta cheese and *lavash*

Fresh fruit or pastries such as *baklava*

Tea

Table 13.2
Cultural Food Groups: Greek and Middle Eastern

Group	Comments
PROTEIN FOODS	
Milk/milk products	Most dairy products are consumed in fermented form (yogurt, cheese). Whole milk is used in desserts, especially puddings. High incidence of lactose intolerance is reported.
Meat/poultry/fish/eggs/legumes	Lamb is the most popular meat. Pork is eaten only by Christians; not by Muslims or Jews. Jews do not eat shellfish. In Egypt, fish is generally not eaten with dairy products. Legumes are commonly consumed.
CEREALS/GRAINS	Some form of wheat or rice usually accompanies the meal in Greece and the countries of the Middle East.
FRUITS/VEGETABLES	Fruits are eaten for dessert or as snacks. Fresh fruit and vegetables are preferred, but if they are not available, fruits are served as jams and compotes and vegetables as pickles. Eggplant is very popular. Vegetables are sometimes stuffed with rice or a meat mixture.
ADDITIONAL FOODS	
Seasonings	Numerous spices and herbs are used. Lemons are often used for flavoring.
Nuts/seeds	Ground nuts are often used to thicken soups and stews.
Beverages	Coffee and tea are often flavored with cardamom or mint, respectively. Aperitifs are often anise flavored.
Fats/oils	Olive oil is generally used in dishes that are to be eaten cold. For most deep-frying, corn or nut oil is used; olive oil is preferred for deep-frying fish. Sheep's tail fat is a delicacy.
Sweeteners	Coffee is heavily sweetened. Dessert syrups are flavored with honey or rose water.

Table 13.2
Cultural Food Groups: Greek and Middle Eastern—continued

Common Foods	Adaptations in the United States
Cheese (goat's, sheep's, and cow's), milk (goat's, sheep's, camel's, and cow's), yogurt	More cow's milk and less sheep's, camel's, and goat's milk are drunk. Ice cream is popular. *Feta* is the most common Middle Eastern cheese available in the United States.
Meat: beef, kid, lamb, pork, rabbit, veal, some varietal cuts *Poultry:* chicken, duck, pigeon, turkey *Fish and shellfish:* anchovies, bass, clams, cod, crab, crawfish, flounder, frog legs, halibut, lobster, mackerel, mullet, mussels, oysters, redfish, salmon, sardines, shrimp *Eggs:* poultry, fish *Legumes:* black beans, chickpeas (garbonzo beans), fava (broad) beans, lentils, navy beans, red beans; peanuts	Lamb is still very popular. More beef and fewer legumes are eaten.
Bread (wheat, barley, corn, millet), barley, corn, farina, oatmeal, pasta, rice (long-grain and *basmati*), wheat (*bulgur, couscous*)	Bread and grains are eaten at most meals. *Pita* bread is commonly available.
Fruits: apples, apricots, avocado, bergamots, cherries, currants, dates, figs, grapes, lemons, limes, melons (most varieties), oranges, peaches, pears, plums, pomegranates, raisins, strawberries, tangerines *Vegetables:* artichokes, asparagus, beets, broad beans, broccoli, brussels sprouts, cabbage, carrots, cauliflower, celeriac, celery, corn, cucumbers, eggplant, grape leaves, green beans, green peppers, greens, leeks, lettuce, mushrooms, okra, olives, onions, parsley (flat-leaf), peas, pimientos, potatoes, spinach, squashes, tomatoes, turnips, zucchini	Fewer fruits and vegetables are consumed. Olives are still popular.
Ajowan, allspice, anise, basil, bay leaf, caraway seed, cardamom, chervil, chives, chocolate, cinnamon, cloves, coriander, cumin, dill, fennel, fenugreek seeds, garlic, ginger, gum arabic and mastic, lavender, linden blossoms, mace, *mahleb,* marjoram, mint, mustard, nasturtium flowers, nutmeg, orange blossoms or water, oregano, paprika, parsley, pepper (red and black), rose petals and water, rosemary, saffron, sage, savory, sorrel, sumac, tarragon, thyme, turmeric, *verjuice,* vinegar	
Almonds, cashews, hazelnuts, peanuts, pine nuts, pistachios, walnuts; poppy, pumpkin, sunflower, sesame seeds	
Coffee, date palm juice, fruit juices, tea and herbal infusions, yogurt drinks, wine, brandy, aperitifs	
Butter (often clarified), olive oil, sesame oil, various nut and vegetable oils, rendered lamb fat	Olive oil is still popular.
Honey, sugar	

■ One Greek seafood specialty is *taramosalata,* a salty dip made from fish roe.

■ In Greece, thin slices of lamb are folded into pita bread with tomatoes, cucumber, and yogurt to make the sandwichlike treat *gyros.*

■ The Arabs were the first to mix gum arabic with sugar to produce chewing gum. The Greeks prefer to chew on the licorice-flavored resin *mastic* (source of the verb *masticate*).

■ A Serbian specialty, popular throughout the former Yugoslavia, is *sljivovica,* a very high-proof brandy distilled from plums.

fermented into yogurt or processed into cheese. Yogurt is often eaten as a side dish and is served plain (unsweetened) or mixed with cucumbers or other vegetables. It is even diluted to make a refreshing drink. Cheese is usually made from goat's, sheep's, or camel's milk. The most widely used Middle Eastern cheese is the Greek *feta,* which is a salty, white, moist cheese that crumbles easily. *Myzithra,* a soft pot cheese, is a by-product of the feta process. *Kaseri* is a firm, white, aged cheese, mild in flavor and similar to Italian provolone. Armenian string cheese is a mild cheese sold in small logs; long "strings" can be peeled off it, and it forms mozzarella-like strands when melted. *Lebneh* or *labni* is a fresh cheese made by draining the whey from salted yogurt overnight.

Almost all meats and seafoods are eaten in Greece and the Middle East, with the exception of pork in Muslim countries and pork and shellfish among observant Jews in Israel. Lamb is the most widely used meat. Grilling, frying, grinding, and stewing are the common ways of preparing meat in the region. A popular dish is *shish kabob,* marinated pieces of meat threaded onto skewers with pieces of raw tomatoes, onions, and sweet peppers and then grilled over a fire. *Souvlaki* or *shawarma* is very thin slices of lamb (or chicken) layered onto a rotisserie with slices of fat (resulting in one roast), grilled, then carved and served. Meatballs, called *kofta,* are eaten as snacks or served with stewed vegetables. A whole roasted lamb or sheep is a festive dish prepared for parties, festivals, and family gatherings.

Numerous spices and herbs are used in Middle Eastern seasoning as a result of a once thriving spice trade with India, Africa, and Asia. Common spices and herbs are dill, garlic, mint, cardamom, cinnamon, oregano, parsley, and pepper. *Sumac,* ground red berries from a nontoxic variety of the plant, is sprinkled over salads to give a slightly astringent flavor. Other typical Middle Eastern spices include *mahleb,* made from the ground pits of a cherrylike fruit, and *ajowan,* small, black carom seeds with a thymelike flavor. The juice of unripe lemons, *verjuice,* is used to provide a sour taste to dishes.

Ground nuts are often used to thicken soups and stews. Sesame seeds are crushed to make a thick sauce, *tahini,* which is used as an ingredient in Arabic cooking and in a sweet dessert paste known as *halvah.*

Although observant Muslims do not drink alcohol, several nations (e.g., Greece, Iraq, Israel, Turkey) produce wines and spirits. Best known in the United States are the Greek specialties *retsina* (white wine with a resinous flavor) and *metaxia* (orange-flavored brandy). Distilled aperitifs are particularly popular. Many are anise flavored, such as *arak* and *Ouzo* from Greece and *raki* from Turkey.

Regional Variations

There are two schools of thought about the number of regional cooking areas in the Middle East. One identifies three culinary areas: Greek/Turkish, Iranian, and Arabic. The other makes five divisions: Greek/Turkish, Arabic, North African, Iranian, and Israeli. Certainly every region has some unique recipes and cooking methods, but the similarity in fare throughout the region is striking. Israel probably has the most different foods and food habits because its cuisine blends the indigenous Middle Eastern cooking with that of the many Jewish immigrant groups (Russians, Germans, Austrians, and eastern Europeans) who have settled in the area since the turn of the century. Furthermore, many of Israel's citizens adhere to the kosher laws of the Jewish religion (see chapter 4 for more information on Jewish dietary practices). More generally, Arabic foods feature more grains, legumes, and vegetables than the Greek, Turkish, Armenian, or Iranian diet. Conversely, meats, cheeses, and oils are more prominent in the cuisines of Greece, Turkey, Armenia, and Iran.

The cooking of Croatia, Slovenia, and Serbia has strong Middle Eastern elements due to more than 400 years of Turkish influence. Grilled meats, especially lamb, are popular. Kabobs and ground lamb meat patties are skewered and cooked over grapevine branches or charcoal. Eggplants, tomatoes, and onions are frequently consumed. *Musaka,* a minced meat and vegetable dish, is similar to the Greek version. Large, flat wheat breads are made in the countryside, and *potica,* the national bread of Slovenia, is flavored with nuts, raisins, or tarragon. Wheat kernels are cooked with sugar, dried fruits, and ground nuts to make the Serbian dish called *koljivo.* Pastries similar to Arab sanbusak are called *burek* in Slovenia.

Albanian Fare

Albania was not of interest to most Americans until the civil war in the Serbian province of Kosova focused attention on the plight of the Kosovar Albanians. Albanians, living in a country bordered by Greece, Macedonia, Serbia, and Montenegro, have often been involved in regional discord and shifting national boundaries (Jurgens, 1995). Years of foreign rule have left their mark on Albanian cuisine: *pastitsio* and feta cheese from Greece, versions of *imam biyaldi* and halvah from Turkey, omelets and tomato sauces from Italy, *boereg* from Armenia, and *borscht* from Russia. Other dishes including *dolma, kofta,* shish kebabs, *moussaka,* and *baklava,* are popular throughout the region.

In the poorest, rural regions of Albania, farmers and shepards are often limited to a diet of cornmeal bread, cheese, and yogurt, with added lamb or mutton when affordable. In wealthier areas, three meals a day are typical with a midafternoon snack of thick Turkish-style coffee or tea consumed with pastries, nuts, or fresh fruit, called *sille.* A complete lunch or dinner begins with *meze* (appetizers) such as salads, pickles, fish and seafood, omelets, spit-roasted lamb or entrails, and baked variety meats. Examples include *liptao,* a feta cheese salad garnished with bell pepper, deli meats, sardines, and hard-boiled egg; and souplike *tarator,* yogurt flavored with garlic and olive oil and mixed with vegetables, such as cucumber. These are usually consumed with a glass of the distilled Turkish specialty, raki, or a beverage made from fermented cabbage called *orme.* The meal follows with soups, meat, or cheese-stuffed vegetables or casseroles; pilaf-like dishes or pies filled with vegetables, cheese, and/or ground meats called *byrek;* and an assortment of vegetable side dishes and pickles. Desserts may include pastries but are typically a fruit compote. Few legumes are consumed, but nuts (especially walnuts) are added to numerous sweet and savory dishes.

One of the most distinctive characteristics of Albanian fare is the differentiation made between vegetables and fruit. Only vegetables are pickled and served as side dishes, and fruits are only eaten fresh, as desserts, or as preserves. There are no crossover items, such as a fruit pickle or a vegetable-sweetened filling for a pie. Vegetables and fruits are also prepared separately and not mixed together in dishes (Hysa & Hysa, 1998). In addition, though regional specialties were once common, years of communist rule during the first half of the century encouraged conformity, and many culinary differences have diminished (Dosti, 1999).

Filo dough, known as *pita* in the region, is used in numerous recipes, especially desserts; *baklava* is a favorite. Goat and sheep cheeses and fermented dairy products are common. Turkish coffee, and a variation known as Serbian coffee, is consumed at the end of nearly every meal and accompanies snacks. Croatian, Slovenian, and Serbian fare also features many foods found in central Europe, such as paprika-flavored stews, dumplings, and roast pigs and hams, as well as stuffed cabbage (*sarma*) and sauerkraut dishes (see chapter 7 for more information about central European cuisine).

Meal Composition and Cycle

Daily Patterns

People in Greece and other Middle Eastern countries eat three meals a day. The main meal is at midday, and in the hotter climates a short nap follows. Dinner is lighter and is served in the cooler evening hours. In addition, a wide variety of bite-size snacks, *mezze,* are available in cafés and served at home throughout the day for friends and relatives, either as appetizers or as a meal in themselves.

In Greece the traditional breakfast may consist of a slice or two of bread with cheese, olives, or jam, accompanied by coffee or tea. The main meal, eaten at around 2:00 P.M., might include a meat stew, meatballs, or vegetables stuffed with chopped meat or baked beans, a salad of raw seasonal vegetables, yogurt or cheese, and fruit as dessert. Roasted or baked whole meats are served on weekends, accompanied by cooked vegetables, salad, and dessert.

In other Middle Eastern countries, coffee or tea is often served first for breakfast, followed by a light meal that might include cheeses, eggs, olives, jam and bread, and plain yogurt. Lunch is the main meal of the day, eaten between noon and 2:00 P.M. It is customarily a meat and vegetable stew served with rice and bread. Supper is light, like breakfast, but it may also include sandwiches.

■ In Turkey, tripe soup with vinegar and garlic may be eaten after dinner, served with alcoholic beverages.

■ In the Middle East, coffee is consumed sweetened at happy occasions; it is drunk black and bitter at funerals.

■ Early Turkish immigrants to the United States were mostly young men without families; their social lives were centered around coffeehouses.

Preparing Middle Eastern coffee. The coffee of Greece and the Middle East is made in a long-handled *briki*. It is preferred strong, thick, and sweet and is often flavored with cardamom. (© R. & S. Michaud)

■ Iranians say, *"Namek geer,"* meaning "I have eaten your food, now I am your slave."

Throughout Greece and the Middle East, coffee and tea are the common beverages consumed after the meal. Frequently coffee is flavored with cardamom and tea is flavored with mint. Coffee is also served in cafés. Traditionally the drink is made in a long-handled metal *briki*, producing a strong, very thick, and often sweet brew that is served in small cylindrical cups.

Sample Menu

A GREEK MEZZE MENU

Olives

Cheese such as *kaseri* or *myzithra*

Taramosalata (fish roe dip) *or* *Hummus* (chickpea dip) with bread

Spanakopita (spinach and cheese pastry)

Cucumber in yogurt

Dolmas (stuffed grape leaves)

Lamb shish kebobs

Ouzo or wine

Etiquette

Throughout the Middle East, hospitality is a duty and a family's status is measured by how guests are treated. Guests, even uninvited ones, are made to feel welcome and are automatically offered food and drink. Even if they refuse at first, they must eventually accept, as refusal is considered an insult. Invited guests bring a gift, often candy or other sweets, and the host must open it immediately and serve them some. One is obligated to feed a guest, but which pieces of certain foods the guest is given and in what order the food is served express the recipient's status. Status is based on sex, age, family, and social rank. For example, a dignitary or head of the family is served the best portion first. In some areas, it is customary for women to eat separately from men.

Guests are traditionally entertained in a separate room before the meal, at which time scented water is provided so they may wash their hands. The dining table might be a large round metal tray, resting on a low stool or platform, and the diners sit around it on cushions.

Assorted Egyptian pastries made with filo dough, couscous, and nuts. (Corbis/Dave Bartruff.)

In Iran, food is traditionally served on a rug. The meal is set out in several bowls placed on the table or rug and then shared by the diners. After the meal the guests leave the table, wash their hands, and then have coffee or tea.

Several rules of etiquette apply to eating. One should always wash one's hands before eating, eat with three fingers (usually on the right hand), and lick the fingers after eating. It is considered polite to continue eating until everyone else is finished, because if one person stops eating the others feel compelled to stop too. One should leave a little food on one's plate to indicate satisfaction with the abundance of the meal. Only pleasant and joyful things are discussed while dining, and one must compliment the host and hostess.

Special Occasions

In Greece and the Middle Eastern countries, food plays an important role in the celebration of special occasions (often religious) and in the observance of certain events like weddings and births.

In the Eastern Orthodox Church there are numerous feast and fast days (see chapter 4). The most important religious holiday for the Greeks is Easter. Immediately after midnight mass on Holy Saturday the family shares the first post-Lenten meal. It traditionally begins with red-dyed Easter eggs and continues with *mayeritsa,* a soup made of the lamb's internal organs, sometimes flavored with *avgolemono,* a tart egg and lemon sauce used throughout the Middle East. The Easter Sunday meal ususally consists of roast lamb, rice pilaf, accompanying vegetables, cheese, yogurt, and a special Easter bread called *lambropsomo* that is decorated with whole, dyed eggs. Dessert usually includes sweet pastries made with filo dough and *koulourakia,* a traditional Greek sweet bread cookie, sometimes shaped into a hairpin twist or wreath or coiled in the shape of a snake (which the pagan Greeks worshipped for its healing powers).

In addition to religious holidays, Greek Americans typically celebrate Greek Independence Day on March 25. It is commemorated with parades in traditional dress, folk dancing, songs, and poetry readings.

There are also feasts and fasts connected with Islamic religious observances. Traditional festive foods vary from country to country. *Iftar* is the meal that breaks the fast during Ramadan, the month in which Muslims fast from sunrise to sunset; it is common to dine with relatives and neighbors. The meal usually starts with a beverage, preferably water, followed by an odd number of dates and coffee or tea. A large meal, served after prayers, includes dishes that are moist and hearty. Regular items eaten during Ramadan include soups, fruit juices, fresh or dried fruit, and *kataif,* which can be a sweet, stuffed pancake or a shredded-wheat dessert. The dawn meal is usually light, and salty foods are avoided because water is not allowed during the fast. The holiday Eid al-Fitr follows the end of Ramadan and is

■ Unlike most Middle Easterners, Turks consider it rude to eat with your fingers; a fork and knife are preferred.

■ At Greek baptisms the infant is first rubbed in olive oil, then immersed three times in the baptismal font to represent the 3 days Jesus was entombed.

■ *Fasolatha* is a white bean soup with tomato and garlic popular during Lent in Greece.

■ For New Year's the Greeks prepare a sweet spicy bread called *vasilopitta* that has a coin baked into it. The person who gets the piece with the money has good luck in the upcoming year. The Serbs have the same tradition for Christmas Day.

■ Some non-Christian Arab Americans celebrate the birth of Jesus on Christmas; Jesus is considered a prophet in Islam.

■ "Give the guest food even though you yourself are starving" is an Arabic saying.

■ On the 10th day of the first lunar month of the Islamic calendar, Turks celebrate the martyrdom of Mohammed's grandson and the day Noah was able to leave the ark. They prepare *asure*, or Noah's pudding, made from the ingredients remaining after the flood waters receded: fresh and dried fruits, nuts, and legumes.

■ In Greece and the Middle East, sugared almonds (Jordan almonds) are served at weddings to ensure sweetness in married life.

■ Eggs cooked in garlic and chicken soup are frequently consumed by Lebanese women after childbirth.

■ When a woman gives birth to a girl in Iran, coldness may be neutralized with a diet high in hot foods to ensure a male child in the next pregnancy.

described as a cross between the feasting of Thanksgiving and the festivity of Christmas. Typically family, friends, and neighbors gather to celebrate; in areas with large Muslim populations, Eid al-Fitr may be held at the local fairgrounds with games, rides, and many food vendors. The other major holiday observed by Arab Muslims is Eid al-Adha, the Feast of Sacrifice, which is held in conjunction with the annual pilgrimage (Hadj) to Mecca (see chapter 4).

In Iran the most significant holiday of the year is Muharram, which commemorates the martyrdom of the grandson of Mohammed in the 7th century. It is a time of communal mourning and penitence for Shi'ites. Another celebration marking the spring equinox is Nau Roz, which features a meal, called *haft-sinn,* that must include seven foods that start with the letter s. The number 7 probably relates to the seven days of the week or the seven planets of the ancient solar system. On the 13th day of Nau Roz it is customary to have a picnic.

A special occasion in Turkey is National Sovereignty and Children's Day on April 23. It commemorates the establishment of the Grand National Assembly in 1923 and specifically honors all children. The following day has become Turkish American Day in the United States, featuring parades in traditional dress and other festivities.

In addition to traditional Orthodox holidays, Armenians commemorate the martyr St. Vartan on February 10 and on September 23 celebrate their declaration of independence from the former Soviet Union in 1991. In Croatia, Christmas Eve features a meal of cod, and a stuffed cabbage and sauerkraut dish is customary on Christmas. The Easter meal is typically lamb or ham and *pogaca,* an Easter bread that is similar to the Greek lambropsomo with painted eggs on top. Slovenians celebrate St. Nick's Feast. Gifts are distributed to children by St. Nick, dressed as a bishop, who admonishes the youngsters to be good. The grape harvest and winemaking are traditionally commemorated with numerous vintage festivals and St. Martin's Feast. Among Serbians the most auspicious day of the year is Krsna Slava, Patron Saint's Day. This holiday dates back to the worship of protective spirits in pagan times; today each family honors its self-chosen patron saint with a sumptuous feast and dancing that may last for 2 to 3 days. The family customarily announces the annual open house with a small advertisement in the local newspaper. *Krsni kolac* is a ritual bread prepared for the occasion, decorated with the religious Serbian emblem Samo, Sloga, Srvina, Spasava ("Only Unity Will Save the Serbs") as well as grapes, wheat, birds, flowers, barrels of wine, or other representations made in dough.

Therapeutic Uses of Foods

The amount of food eaten is of special concern in the Middle Eastern diet. Ample meals are needed to prevent illness, and poor appetite is regarded as a disease in itself or as a generalized complaint signifying that one's life is not as it should be. Food deprivation is believed to cause illness (Meleis, 1981; Packard & McWilliams, 1993; Spector, 1991). Some Middle Easterners also believe that illness can be caused by hot-cold shifts in food or by eating incompatible foods together (Lipson & Meleis, 1983; Meleis, 1981). Classifications can vary, but examples of hot foods include lamb, eggs, onions, garlic, carrots, bell peppers, apples, dates, quinces, chickpeas, wheat, almonds, walnuts, pistachios, and honey. Cold foods can include beef, cucumbers, tomatoes, eggplant, grape leaves, grapes, lemons, sour cherries, apricots, rice, yogurt, coffee, and beer. The temperature (not spiciness) can cause a shift in the body from hot to cold and vice versa, and it is believed the digestive system must have time to adjust to one extreme before a food of the opposite temperature can be introduced. Shifts are thought to cause headaches, colds, and the flu. In Iranian tradition, eating too many hot or cold foods can also cause illness. Excessive hot foods result in sweating, itching, and rashes. Excessive cold foods can cause dizziness, vomiting, and weakness (Batmanglij, 2000; Lipson, 1992; Lipson & Hafizi, 1998). Furthermore, fresh foods are considered best and canned or frozen foods must be avoided to preserve health. Certain combinations of foods should not be eaten; for example, Egyptians do not consume fish at the same time as dairy products. Other Middle Easterners avoid eating melon with fish and sour milk with melon.

Middle Eastern market. (Courtesy of World-Health Organization/P. Merchez.)

Contemporary Food Habits in the United States

Adaptations of Food Habits

There is scant information on the adaptation of Greek and Arab diets in the United States. It is assumed that, as in other immigrant groups, increasing length of stay is correlated with some Americanization of the diet, with traditional dishes prepared and eaten only for the main meal or for special occasions.

Ingredients and Common Foods

It has been reported that Greek Americans still use olive oil extensively, although they use less of it than their immigrant relatives (Valassi, 1962). Salads still accompany the meal, and fruit is often served for dessert. Vegetables are prepared in the traditional manner. Lamb is still very popular; for special occasions, roasted leg of lamb is substituted for the whole animal. Consumption of beef and pork has increased, whereas consumption of legumes has decreased. Cereal and grain consumption among Greek Americans remains high, and bread, rice, or cereal products are usually included in every meal. Greek Americans consume more milk than their immigrant parents, and ice cream is very popular.

Meal Composition and Cycle

Greek Americans maintain traditional meal patterns, but the main meal of the day is now dinner (Valassi, 1962). They prefer an American-type breakfast and lunch, but dinner is more traditional. However, they have adapted Greek recipes to make them less time-consuming to prepare and to include fewer fats and spices.

Reportedly many Arab Americans still eat their main meal at midday (Lipson & Meleis, 1983). Members of the extended family may dine together daily, with the women who stay at home cooking for employed female relatives (Packard & McWilliams, 1993).

Nutritional Status

Nutritional Intake

Very little has been reported on the nutritional composition of the Greek American or Arab American diet. However, one aspect of their diet has recently received considerable media attention because of its possible link to

■ The demand for properly slaughtered *(halal)* meat among Muslims in the United States has led to increased numbers of Arab halal markets.

■ Greek weddings in the United States offer a blend of Greek and American foods; for example, the wedding cake is served along with baklava.

■ Some researchers note that Muslim women who, for religious reasons, completely cover their bodies with clothing and veils when outdoors may be at risk for vitamin D deficiency (Packard & McWilliams, 1993).

■ Recent Kurdish and Iraqi refugees to the U.S. have had high rates of malnutrition, parasitic infection, hepatitis B, and TB. The prevalence of glucose-6-phosphate dehydrogenase deficiency is also significant in this population (Ackerman, 1997; Hampl et al., 1997).

■ Many Middle Easterners believe that planning ahead, such as the use of birth control, interferes directly with the will of God and can bring retribution on the family.

■ Female health care providers may face added difficulties in gaining the trust and respect of Greek and Middle Eastern clients due to cultural norms regarding gender.

prevention of cardiovascular disease. Traditional Greek and Middle Eastern diets contain more monounsaturated fatty acids (primarily from olive oil) than American diets (Fordyce et al., 1983); a diet high in monounsaturated fatty acids has been correlated with lower readings of resting blood pressure (William et al., 1987). However, it has been noted that olive oil consumption has decreased in Greece since the 1960s and that alcohol consumption has increased (Kromhout et al., 1989). The role of alcohol consumption in a diet that is high in complex carbohydrates and relatively low in saturated fats, like that of Greece and many Middle Eastern countries, is not yet fully understood (Rimm & Ellison, 1995).

The effects of Ramadan fasting have been explored among Muslims. Increases in uric acid blood levels have been noted when weight loss occurs in nonobese men, which may be related to high rates of kidney stones and angina pectoris reported in some epidemiological surveys conducted during the month-long fast (Gumaa et al., 1978; Nomani et al., 1990).

Cross-cultural research on breast-feeding reported that 82 percent of Iranian American mothers in the study exclusively breast-fed their infants. This high rate was attributed to a strong social network of support for the practice (Ghaemi-Ahmadi, 1992).

Counseling

Considerable discomfort and irritation has been noted between health care practitioners and their Middle Eastern American clients, much of it due to cultural differences (Lipson & Meleis, 1983). Most difficulties evolve from misunderstandings in the provider–patient relationship.

Interactions are highly contextual throughout Greece and the Middle East (Hall, 1979). Meaning that body language and general atmosphere are as important to communication as words, if not more so (see chapter 3, "Intercultural Communication"). People in the Middle East spend time getting to know one another before any "business" is discussed. Offering coffee or tea at the beginning of any interaction helps to establish a warm and hospitable atmosphere with Greek or Middle Eastern clients (Hall, 1979; Lipson & Meleis, 1983).

Direct eye contact is expected and necessary to interpret meaning, so Middle Easterners usually sit or stand quite close when conversing. Greeks may smile when angry. Nodding one's head up and down or back and forth can be very confusing. Traditionally moving the chin up and down meant no and back and forth meant yes or "I don't understand." But many Greeks and Middle Easterners use the American protocol, so it is difficult to know whether the gesture is affirmative or negative.

Touching between members of the same sex is frequent, including handshaking, patting, shoulder slapping, hugging, and kissing. Contact between members of the opposite sex is prohibited in some Middle Eastern Muslim cultures, and extended eye contact between men and women can be considered a sexual overture (staring between members of the same sex is acceptable). The left hand is not used for any social purposes. In general, it is best to wait for Middle Easterners to extend their hand in greeting before making any unwanted contact: Follow their lead. Proper posture is a sign of respect, and crossing one's legs, pointing with the foot, or showing the sole of the shoe is considered impolite. In Turkey, one should stand when an elder enters a room.

Due to the significance of *how* the message is communicated to Americans of Greek or Middle Eastern background, providers may find their clients are more receptive to verbal than to written information. A few minutes for general questions about the well-being of other family members or personal interests of the client should be allowed at the beginning of the interaction (Packard & McWilliams, 1994). Family members, especially an elder male relative, may insist on participating in all conversations, even those that customarily take place in the office between only the practitioner and the patient. Because these family members may make the final decisions regarding care, their presence should be valued and their opinions fully solicited.

Greek and Middle Eastern Americans value biomedicine and have considerable respect for authority figures. They may be hesitant to ask questions when confused, however, or they may provide answers that are designed to please the provider. The health provider may have to assess and give advice about a medical or dietary problem without the client explaining his or her needs. If the provider does not repeat the offer to help, the client may believe that the provider is indifferent. Privacy is strongly protected, and a client may resist disclosing information about themselves and their families to strangers until a trusting relationship is established.

A study of Middle Eastern immigrants from several countries representing five ethnic groups found that, in general, immigrants who perceived themselves as more traditionally ethnic experienced more physical complaints and had lower morale. Immigrants who were more acculturated reported better health (Meleis et al., 1992). Complaints by clients of Greek or Middle Eastern heritage are frequently generalized or nonspecific, sometimes indicating anxiety or depression in a patient who does not distinguish culturally between physical and mental health.

Greek and Middle Eastern Americans may expect the health care provider to make decisions for them and be responsible for the consequences. They may also demand the services of the top expert or the department head because the expectation is that they will receive the best care from the most senior, most powerful person in the system (Lipson & Meleis, 1983). Clients may believe that the more intrusive the medical procedure, the more effective the treatment (Meleis, 1981). A poor prognosis should be discussed with the family first, and revealed in stages. It is considered sacrilegious to presume death, because only God can make that final decision and hope must always be maintained.

In one study (May, 1992), parents of Middle Eastern descent indicated that they did not choose health care providers based on ethnicity. Access, referrals, and effectiveness were factors in their choice. Consistent care from a single provider is most successful and may eliminate many communication difficulties. Practitioners who work with Greek and Middle Eastern clients should recognize that a high-context relationship is often intensive and time-consuming (Lipson & Meleis, 1983). The in-depth interview should be used to determine country of origin, degree of acculturation, and religious faith. Information on traditional health care beliefs that are still practiced should be elicited.

- **Hospitals are often feared as places where a person dies, and there may be serious concerns that if death occurs, the body will not be treated according to religious customs.**

- **There are little data about the therapeutic ingredients in Greek and Middle Eastern home remedies. Ko'hl, used on the umbilical cords of newborns, is high in lead content and may present a danger.**

- **Traditional healers in Saudi Arabia sometimes recommend drinking sheep bile to treat diabetes, a practice that can result in acute toxicity.**

14

Asian Indians

India is a culturally complex country with a population of more than 967 million people, nearly four times that of the United States. The sophisticated civilization began approximately 4,000 years ago and is the source of some of the most influential religions, art, architecture, and foods in the world. The South Asian subcontinent contains the fertile Indus and Ganges river basins, as well as parts of the Himalayan mountain range; it varies in climate from extensive desert regions to jungle forests to the world's largest mountain glaciers. (See the map in Figure 14.1.)

The people of India are as diverse as its geography and climate. People from virtually every racial and religious group have migrated to or invaded India at some time in history, and each brought its own language and customs. As the different races and religions intermingled, other cultures were created. One result is that there are currently 15 separate languages recognized by the Indian government. Nearly 300 languages are actually spoken in India, and there are approximately 700 dialects.

It is estimated that more than 815,000 Asian Indians live in the United States today. They represent numerous racial and religious groups, yet many share high education levels and socioeconomic status. This chapter examines traditional Indian foods and food habits, as well as the dietary changes that occur when Asian Indians live in America.

Cultural Perspective

History of Asian Indians in the United States

Given the complexity of Indian culture, it is not surprising that the Asian Indian immigrants to the United States differ from other immigrant groups in several ways. Most significantly, the majority of Indians arriving in America are not escaping political or economic pressures in their homeland. Since 1965, when the national quota system was temporarily dropped from U.S. immigration laws, the majority of Asian Indian immigrants have been from the upper socioeconomic classes. They were somewhat acculturated at the time of arrival, often fluent in English, and acquainted with many American customs.

■ The term *South Asia* is used for the nations of India, Pakistan, Bangladesh, Sri Lanka, Nepal, and Bhutan.

■ Confusion over *Indian* (does it mean a Native American or an Asian Indian?) has made it difficult to find consensus on a designation for Asian Indians living in the United States. *Indo-American* has emerged as the term preferred by many Asian Indians.

Immigration Patterns

The first immigrants to the United States from India were members of the Sikh religion, who arrived on the West Coast in the early 20th century. Many were employed by the railroads and others established large farms. They faced overt discrimination and persecution. Newspapers warned of the "Hindoo invasion"; anti-Asian Indian feelings brought about the expulsion of Asian Indians from Washington logging communities and in 1907 sparked violent riots in California. Although such extreme bigotry lessened

■ In the early 1900s, it was not unusual for single Sikh men living in the United States to marry Mexican American women and start families.

Figure 14.1
India

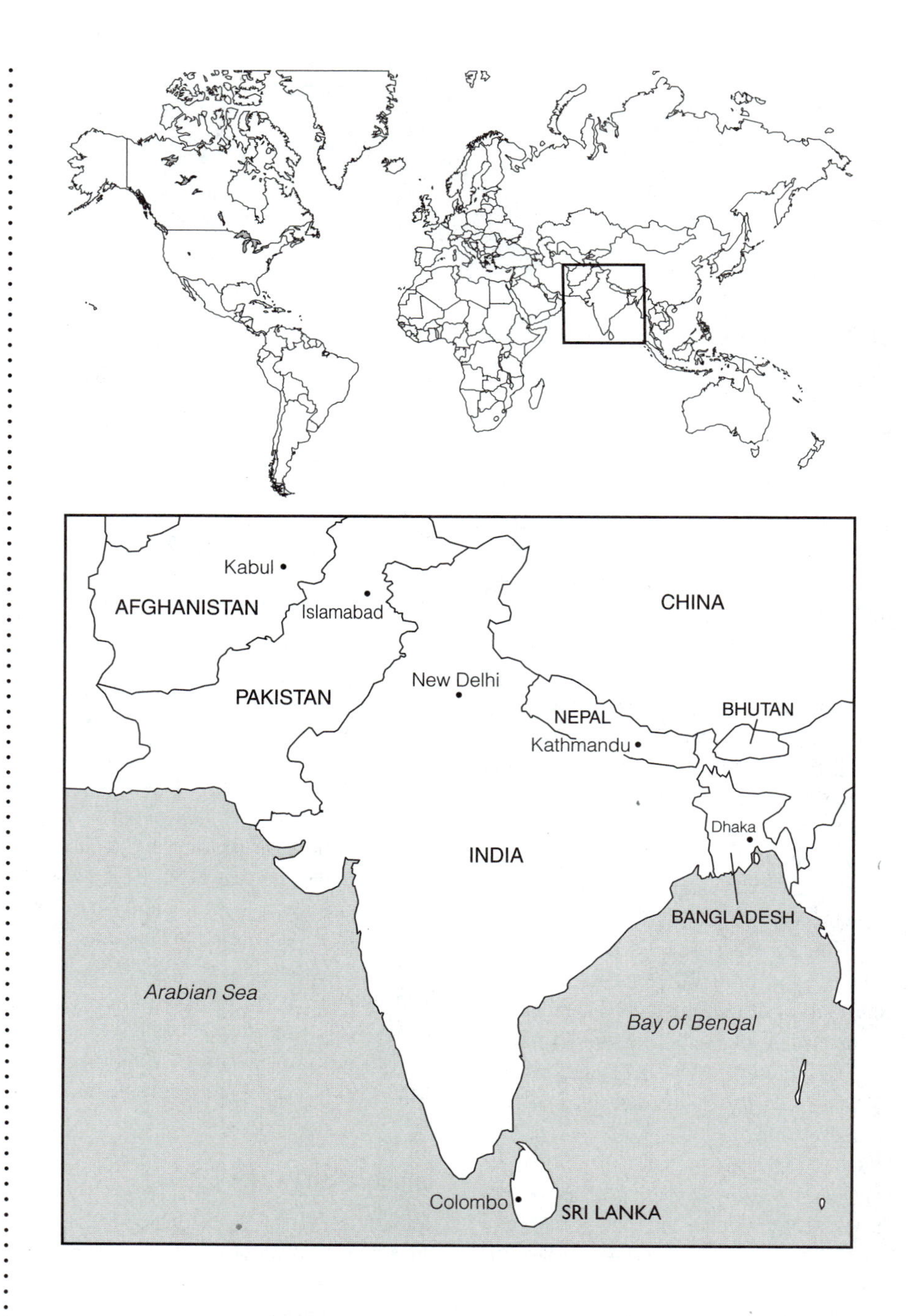

■ Small numbers of Asian Indians coming to the United States in recent years are exiles from regions where Indian immigration in the past has been substantial, such as East Africa and Fiji.

in time, the Asian Indian immigrant population remained small until after World War II.

Relaxed immigration laws encouraged Asian Indians, especially well educated urban professionals, to come to America in the 1960s and 1970s. Economic and social adjustment was a priority for this group, although many Asian Indian traditions continue within the privacy of the home. These immigrants formed a self-reliant community and discouraged comparison or identification with other ethnic groups.

Recent changes in Asian Indian immigration patterns may change this typically high degree of acculturation. Asian Indians coming to America today are more likely to be from rural towns and villages. They are less educated and have experienced less exposure to Western society than previous Indian immigrants. There are also fewer professionals

Traditional foods of India. Some foods typical of the traditional Indian diet include *amchoor* (mango powder), *basmati* rice, broccoli, coconut, cucumber, eggplant, *ghee,* herbs and spices (black pepper, cardamom, chiles, fresh coriander, cloves, coriander seeds, cumin, garlic, ginger root, mint, mustard seeds, nutmeg, tamarind, and turmeric), lentils, peas, plantains, and yogurt. (Photo by Laurie Macfee.)

among them. Most recently arrived Indians are self-employed in jobs that serve their Asian Indian immigrant community, such as restauranteurs, importers, and travel agents.

As a result of the change in the Asian Indian immigrant profile, newcomers now identify more with their immediate ethnic subgroup community than with the wider, pan-Indian community. Their adjustment to American culture is more difficult. The religion, language, and social class of each cultural subgroup retain greater importance for recent immigrants than for previous immigrants from India.

Current Demographics and Socioeconomic Status

Today Asian Indians are one of the fastest-growing immigrant groups in the United States. Their population more than doubled between 1980 and 1990, from 400,000 to more than 815,000. They have settled throughout the United States, but especially in the metropolitan areas of California, New York, and New Jersey, as well as in Illinois, Maryland, Massachusetts, Michigan, Ohio, Pennsylvania, and Texas. There are also several Asian Indian settlements in the agricultural regions of California.

The Indo-American community is very diverse, coming from numerous states and regions with distinctive languages and customs. The most prominent subgroups are the Gujaratis, Bengalis, Marathis, and Tamils.

The relative affluence of Americans of Asian Indian heritage is due to the large percentage of highly educated immigrants. Many were employed in professional or white-collar occupations in India, such as college professors, engineers, physicians, and scientists; most continue in their professions in the United States. Newer immigrants have found success in small business and franchise ownership involving many members of their extended family.

Many Asian Indians come to the United States to complete their college or postgraduate education. They are often unmarried or have left their spouses and children in India. It is not uncommon for the families to join the student in America after he or she has become financially established.

Worldview

The Caste System

The traditional Indian caste system, which influences the social structure of many Asian

- The 1990 U.S. census reported more than 63,000 Americans of Pakistani descent. Most are well educated professionals earning middle- or upper-class incomes.

- Almost 27 percent of all hotels and motels in the United States are owned or operated by Indo-Americans.

- Over 860,000 Canadians are of Asian Indian and Pakistani descent.

Indian groups, is the Hindu method of ordering an individual's role in society. A more encompassing term is *jati,* which is the organization of all aspects of Hindu life, including actions, places, things, and symbols, not just people.

Caste categories are hereditary. There are four main castes associated with certain professions (although members are not necessarily employed in these jobs); the *brahmans* (priests), *kshatriyas* (soldiers), *vaisyas* (merchants or farmers), and the *sudras* (serfs). These castes are divided into more than 1,000 subcastes, usually according to occupation. Existing outside the caste system are individuals who are considered so impure that they are called "untouchables." Although the laws discriminating against untouchability were repealed in 1949, this group of the desperately poor continues to occupy the lowest stratum in Indian society.

■ The untouchables were considered impure because of their contact with cattle carcasses and their consumption of scavenged beef.

■ A reverence for all life, called *ahimsa,* is fundamental to Asian Indian ideology. It is reflected in the religions native to India, as well as in the vegetarian diet that many Indians follow.

■ Nearly all Pakistani Americans are Muslim; small numbers of Hindus, Christians, and Zoroastrians are also part of the community.

The caste system has permeated Indian society despite the fact that it is an exclusively Hindu classification system. Americans of Asian Indian descent often continue to identify proudly with their caste. Most come from the upper castes of brahmans and kshatriyas. As with all cultural practices, it is important to remember that even within a group there is great diversity of individual beliefs and customs.

Religion

The influence of religion on Indian culture is ubiquitous. Every aspect of life and death is affected not only by individual religious affiliation, but also by the Hindu ideology that pervades Indian society.

Hinduism. Nearly 85 percent of Indians are Hindus. Hinduism is an ancient faith, believed to have developed in India between 2000 and 1500 B.C.E. from the Aryan hymns and prayers known as the *Vedas* mixed with elements from traditional Dravidian religion (see chapter 4 for more information about Hinduism and other major Indian religions).

■ Although Sikhs are only 2 percent of the population in India, it is estimated that they make up nearly one-third of Asian Indians living in California.

■ The *kirpan* has become a civil rights issue in some U.S. public schools, pitting religious freedom against provisions restricting "weapons" on campus.

The Hindu Society of India established community organizations to serve the religious needs of early Asian Indian immigrants to the United States. Many Hindu temples now exist in regions where Asian Indians have settled, with services and religious ceremonies conducted by Brahman priests (who are often employed part-time in other occupations). However, temple attendance may be limited to significant religious events. Small shrines are often created in Asian Indian apartments or houses so that prayer and meditation may take place at home.

Islam. Today the Islamic religion is second only to Hinduism in number of followers; one in every nine Asian Indians is a Muslim. Islam was brought to India by traders from Persia, and it expanded with the Muslim invasions of the northern regions beginning about 1000. The Islamic Moghul empire dominated the country for nearly 800 years. The influence of Islam is seen today mostly in northern India.

Buddhism. Buddhism developed as a protestant revolt against Hinduism. Its founder, known as Gautama Buddha, lived in India during the 5th century B.C.E. Although it is a popular religion in other parts of Asia, Buddhism is followed by fewer than 1 percent of Asian Indians today.

Jainism. This branch of Hinduism developed at about the time Buddhism emerged. The Jains believed that all living things have souls. Some wear masks to prevent breathing in insects and sweep a path in front of them to prevent stepping on any creatures. Orthodox Jains are strict vegetarians. Approximately 2 percent of Indians are Jains; in the United States, Jains have established their own temples for worship.

Sikhism. The Sikh religion differs from Hinduism in its belief in a single God. It is best known for its military fraternity, although most Sikhs in India are farmers. Male Sikhs wear a turban and follow the "five *K*s": uncut hair *(kes),* a comb worn in the hair *(kanga),* short pants *(kaccha),* a steel bracelet worn on the right wrist *(kada),* and use of a special saber *(kirpan).* Each has a spiritual meaning; for example, the short pants symbolize self-restraint; the bracelet, obedience; and the comb, purity of mind.

Syrian Christians. One form of Christianity emerged when the Syrians, who migrated to the Malabar coast of southwest India in 345, intermarried with native people. Syrian Christians do not observe Hindu dietary laws, but they do participate in the caste system. Their agricultural community is operated with farm laborers who can be described as serfs.

Goan Christians. Another Christian community developed at the former Portuguese colony of Goa, farther north on the southwest coast. Approximately half of the citizens are Catholic, and the city is dedicated to St. Catherine.

Zoroastrianism. More than 1,200 years ago the Parsis fled from religious persecution in Persia to northern India. The religion they brought is known as Zoroastrianism, an ancient faith that venerates Ahur Mazda, the wise god of fire. The Parsis have adapted many of their practices to blend into Indian society but have maintained their faith through private schooling of their children. They are considered the most westernized of all Indians; significant Parsi communities have developed in New York and Los Angeles.

Judaism. Four small Jewish communities were established in India when Jews fled persecution in Greece, Palestine (under Roman domination), Iraq, and Germany. The largest populations are found in Bombay and Calcutta.

Animism. The oldest religions in India are those practiced by the small tribal populations that live in isolated regions of the Himalayas. They worship spirits associated with natural phenomena, a religious practice known as *animism.* In the past they have practiced such varied social customs as polyandry (having more than one husband) and head hunting.

Family

The husband is the head of the household in the traditional Indian family. The wife usually does not work outside the home and is expected to perform all duties related to housekeeping and child care. She obtains help in these responsibilities from the extended family and, in some homes, from servants. If the wife does hold a job, she can depend on the help of relatives. Children are expected to show respect for their elders; parents may choose what career a child should pursue. Dating is uncommon in India, and many marriages are arranged by families based on similarities in caste, education, religion, and upbringing between prospective husbands and wives.

Strains in the traditional family structure often occur in the United States. Asian Indian women are more likely to work in America than in their homeland, yet they lack the support system of an extended family. Elders may also find themselves cut off from their traditional role of advisors and may not have opportunities for involvement in certain religious activities that would fill their lives in India. Some Asian Indians, especially women raised in India, find it difficult to adjust to these changes. More men than women choose to become U.S. citizens.

Children who grow up in the United States usually insist on making their own career choices. Dating has become more acceptable, but most parents strongly discourage relationships with persons of other ethnic or religious backgrounds. Although most parents do not choose their child's spouse, many Asian Indian children still defer to their parents' opinion; young male students in the United States sometimes ask their families in India to find a suitable mate for them.

The family is seen as the way to preserve Indian values and beliefs while living in America. Even after establishing permanent residence, a majority of Asian Indians regard their national identity as more Indian than American (Sodowsky & Carey, 1988). There is also strong interest in sponsoring the immigration of relatives to the United States. Most Asian Indians have found successful adjustment in the United States through educational and economic achievement in American public life, while maintaining an emphasis on Asian Indian culture within the privacy of their home life.

- It is estimated that Christians make up 3 percent of the population in India.
- The sacred fires of Zoroastrianism are tended in temples protected from the sun and from the eyes of unbelievers.
- In India, marriage is considered the beginning of a relationship from which love develops over time.
- Indo-Americans who are financially secure often assume responsibility for less prosperous relatives.

Traditional Health Beliefs and Practices

Traditional medicine in India has a long and distinguished history. Several systems have

■ Deepak Chopra, an Indian-born physician, has popularized Ayurvedic medicine in the United States through his best-selling books and videos.

■ As early as the 7th century, Ayurvedic physicians were performing surgery to remove kidney stones.

■ Meditation is the quiet consideration of religious teachings in an attempt to personally understand faith and to achieve spiritual enlightenment. It is prominent in Hinduism and Buddhism but also practiced by some Christians and Muslims. Transcendental meditation (TM) is a version associated with yoga that has no affiliation with any particular religion.

■ Homeopathy is well accepted throughout India.

■ Astrology is very popular among Asian Indians.

developed over several thousand years, the most important of which is Ayurvedic medicine, which established the humoral concepts of the body that were later adopted in Greece and eventually evolved into biomedicine as it is practiced today.

Ayurvedic medicine developed into its current form between 500 B.C.E. and 500; it is based on Sanskrit texts and the writings of practitioners. *Ayur* means "longevity" and *veda* means "science or knowledge." The purpose of the Ayurvedic system is to ensure a long and active life so that the wisdom of elders may be passed down to future generations. Ayurvedic physicians, called *vaidyas,* are trained at government-supported schools that grant degrees based on an established curriculum. Their diagnosis focuses on *who* the person is that has the illness, their tastes, their work habits, their character, and their life history. Evaluation of the pulse, the face, the eyes, and the nails provides further data. A person's constitution, including temperament and preferences in food, is believed to be determined at birth.

Ayurvedic therapy uses diet, herbal remedies, and meditation to reestablish equilibrium between the sick person and the universe, including the social, natural, and spiritual worlds. Diet is considered most significant. Foods are classified as hot or cold depending on their effect on the body and must be balanced for each condition (see "Therapeutic Uses of Food" section later in this chapter). In addition, more than 700 plants and animal substances are listed in the Ayurvedic texts for prescriptive use. Because the mind, body, and soul are all considered to be interconnected parts of the whole system, meditation is used to address imbalance in the spirit of a person.

Ayurvedic medicine has declined somewhat in popularity in India as westernized medicine has become more established and is frequently perceived as a paraprofessional practice. Folk beliefs about health and illness are found in some regions. For example, Yunani or Unani-Tibb medicine is common in the northern areas of India. It is an Arabic system that has been modified by Indian practitioners known as *hakim.* It is a humoral system that identified four humors—yellow bile, black bile, phlegm, and blood—and four qualities—heat, cold, moisture, and dryness. Health is sustained through balance of these humors and qualities. Illness is treated by complementary remedies; for example, disease due to too much cold is cured with a hot therapy. Siddha medicine, another humoral system, developed within Tamil culture and is found mostly in southern India. Older practices, such as the use of shamans, bonesetters, and snakebite healers are found in some rural regions.

Home remedies such as herbal infusions and poultices are prevalent in India, often derived from Ayurvedic prescriptions but administered by home diagnosis (see Table 14.1, "Selected Ayurvedic/Indian Botanical Remedies"). Medications available only through prescription in the United States can be purchased over the counter in India; widespread use of antibiotics and mixing of therapeutics has been reported (Bentley, 1988; Ramakrishna & Weiss, 1992; Reeler, 1990).

Traditional Food Habits

It is difficult to generalize about Indian cuisine because of the diverse geography and heterogeneous population of the country. Foods vary north to south, east to west, region to region, and among religious and caste groups.

Ingredients and Common Foods

Staples

Few foods are eaten throughout all of India. Rice is the most commonly consumed. The average Indian eats 8 ounces of it each day. This amount, however, varies considerably by region. Fruits and vegetables predominate in the frequently vegetarian cooking, but the types and ways in which they are prepared also depend on local customs. Dairy foods are also important in most areas. Fermented milk products such as yogurt are found throughout most of the country, as is the cooking fat *ghee,* which is pure, clarified butter (this butter, known as *usli ghee,* is too expensive for daily use in many homes, so vegetable shortening, also called ghee, is often used instead). Seasonings are distinctive. *Masalas* are mixtures of spices and herbs that can be either fresh and "wet" or dried and powdered. Coriander, cumin, fenugreek, turmeric, black and cayenne pepper, cloves,

Table 14.1
Selected Ayurvedic/Indian Botanical Remedies

Scientific Name	Common Name	Parts	Traditional Use
Aconitum felconeri ☠	Mīdhavis; vatsnābh; aconite; monk's hood	Root, leaves, seeds	Indigestion; gout; fever; cough (may be used externally for diabetes)
Adhatoda vasika ☹	Adhosa; vāsāka; malabar nut	Root, bark, leaves, flowers	Diabetes; urinary tract problems; thirst; emaciation; diarrhea; dysentery; nausea; asthma; cough; epilepsy; hysteria; mental disorders
Allium sativum ☹	Laśhan; rasonam; garlic	Bulb, juice	Hypercholesterolemia; hypertension; heart disease; emaciation; colic; hemorrhoids; rheumatism; asthma; cough
Aloe spp. ☹	Gawarpaltra; kumārī; aloe vera; Indian alces	Root, leaves, juice	Digestive disorders; colic; dysentery; hemorrhoids; liver problems; obesity; high or low blood sugar; TB; purgative
Anacyclus pyrethrum ☠	Akarkarā; pellitory; Spanish chamomile	Root	Diabetes; bowel disorders; nerve conditions; rheumatism; sore throat
Azadirachta india; Melia azadirachta ☹	Nimb; nimba; neem; margosa tree; Indian lilac	Whole plant	Diabetes; liver conditions; obesity; nausea; parasites; blood toxins; cough; fever; rheumatism
Balsamodendron mukul; Commiphora wightii ☹	Gugal; gugala; Indian bedellium; guggul	Resin	Hypercholesterolemia; indigestion; obesity; urinary tract infections; hemorrhoids; gout; arthritis
Berberis vulgaris ☹	Kiryat-charayatah; chirāyatā; barberry	Leaves; whole plant	Diabetes; liver problems; urinary tract disorders; gastric/duodenal ulcers
Caltropis procera; C. gigantea ☠	Aka; arka; swallowort	Root, leaves, flowers, juice	Heart problems; stomach complaints; enlarged liver; urinary tract disorders; dysentery; hemorrhoids; purgative
Cissampelos pareira	Patha; harjori; laghu pātā; abuta; velvet leaf	Root, bark, leaves	Urinary tract problems; kidney conditions; bowel disorders; kidney inflammation; indigestion; diarrhea; edema; colic
Coptis teeta; C. trifolia ☠	Mamira; mishamitita; gold-thread; mishmee bitter	Root	Indigestion; flatulence; anorexia; bowel problems; gall bladder infections; fever; skin rashes; ulcers
Croton unctatus; C. coccineuma; C. phillippinensis	Kamala; rechanaka	Fruit	Purgative; parasites
Emblica officianalis	Āmla; āmalakī; Indian gooseberry	Fruit	Diabetes; urinary tract problems; gout; liver diseases; bowel disorders; constipation; hemorrhoids; anemia; osteoporosis
Glycyzrrhiza glabra ☠	Mithi-lakdi; mulathi; yashtīmadhu; licorice root	Root	Indigestion; constipation; emetic; urinary tract problems; stomach pain; colds; coughs; bronchitis; sore throat

Table 14.1
Selected Ayurvedic/Indian Botanical Remedies—continued

Scientific Name	Common Name	Parts	Traditional Use
Gymnema sylvetre ☹	Gudmār; sarpadarushtrika	Root, leaves	Diabetes; high blood sugar; blood toxins; swollen glands; diuretic
Orchis mascula; O. latiflora	Salabmishri; sālam-miśhrī; early purple orchid	Root	Diabetes; emaciation; weakness; diarrhea; dysentery; impotence
Phyllanthus niruri; P. urinaria	Bhūy-ā malakī; niruri; bhūāmalakī	Root, leaves, whole plant	Diabetes; urinary tract disorders; liver problems; dyspepsia; edema; menstrual conditions; gonorrhea
Rauwolfia serpentina ☠	Nakuli; chota-chand; sarpagandha; Indian snake root; quinine plant	Root	Hypertension; bowel disorders, dysentery; insomnia; mental problems; hypochondria
Rheum spp.	Archu; revand-chini; amlavetasa; rhubarb	Root	Purgative; stomach disorders; liver problems; constipation; diarrhea
Ricinus communis ☠	Renda; eranda; vātāri; castor bean	Root, leaves, fruit, seeds, oil	Colic; stomach problems; liver disorders; dysentery; constipation; purgative; rheumatism; sciatica; lumbago
Solanum nigrum ☠	Makoy; kākamāchī; black nightshade	Leaves, fruit	Heart disease; liver problems; diuretic; hemorrhoids; fever; cough; edema
Solanum xanthocarpum; S. virginiamum ☠	Choti katheri; kāntkārī; Thai eggplant	Root, stems, flowers, fruit	Diuretic; kidney disorders; constipation; asthma; cough; TB; fever; gonorrhea
Terminalia arjuna ☹	Arjun; arjuna	Bark	Heart conditions; liver problems; digestive disorders; edema; bone fractures; diarrhea; venereal diseases
Tribulis spp.	Chota-gokhru; gokshura; goat's head; puncture vine	Fruit	Diabetes; urinary tract infections; kidney diseases; gout, hemorrhoids; edema; lumbago; sciatica; impotence; infertility
Valeriana spp.	Bala-tagra; sugandh-bala; tagara; Indian valerian	Root	Colic; flatulence; gastrointestinal fermentation; colon toxins; heart palpitations; migraine; convulsions; insomnia; skin disorders
Withania somnifera	Aśhganda; aśhwagandhā; Indian ginseng; winter cherry	Root	AIDS; cancer; anemia; emaciation; memory loss; rheumatism; MS; fatigue; insomnia; skin problems
Zingiberis officinalis ☹	Adrak/sunth; ginger root	Root	Diabetes; colic; indigestion; flatulence; nausea; incontinence; diarrhea; constipation; liver problems; bowel pain; heart disease; arthritis; rheumatism; fever; colds; cough; laryngitis; headache

Note: Data on some plants are very limited; adverse effects may occur even if not indicated.
Key: ☠, all or some parts reported to be harmful or toxic in large quantities or specific preparations; ☹, may be contraindicated in some medical conditions/with certain prescription drugs.

cardamom, cinnamon, and chile peppers are a common blend that is called curry in western countries. Other typical spices and herbs include *ajwain* (carom or loveage seeds), *asafetida* (a pungent powdered resin), coconut, fresh coriander, garlic, mint, saffron, and *tamarind* (the sour pulp of a bean pod). Beyond these generalities, the staples of the Indian diet are best classified by region.

The greatest division in diet is seen between northern and southern India. Northern cuisine is characterized by the use of wheat, tea, a large number of eggs, garlic, dried or pickled fruits and vegetables, and use of dry masalas that are aromatic rather than hot. These foods are typical of a cooler climate, where wheat grows better than rice and where fruits, vegetables, herbs, and spices are available only seasonally. Boiling, stewing, and frying are the most common forms of cooking.

In the South, steaming is the preferred method of food preparation. Rice, coffee, fresh pickles (some of which are known as *chutney*), *pachadi* (seasoned yogurt side dishes that are called *raytas* in northern India), "wet" masalas that are spicy hot, and fresh fruits, vegetables, herbs, and spices are fundamental to the cuisine. Again, these foods reflect the regional agricultural conditions.

Many Asian Indians are vegetarians; they use some milk products but avoid eggs (see the box). Pork is eaten in some communities in the West, lamb and beef are eaten in many areas of the North, and fish and poultry are eaten in several coastal regions. The cultural food groups list is found in Table 14.2.

Regional Variations

Northern India. In the North, rice is commonly served as a pilaf. Saffron rice with seasoned chicken, lamb, or beef, called *biryani,* is also popular, as are meatballs *(kofta)* and skewered pieces of broiled or grilled meats *(kababs)*. Northern specialties include *korma,* a curried lamb dish with a nut and yogurt-thickened sauce, and masala chicken.

Bread, which is called *roti* in India, is eaten daily. Examples include whole wheat flat breads, such as *chapatis,* which are cooked on a griddle without oil, and *puris,* which are fried. *Paratha,* another fried roti, is used as a wrapping for spiced vegetable fillings; a spicy lamb, potato, or vegetable stuffing is used in the deep-fried turnovers called *samosas.* Milk desserts are favored, such as carrot pudding *(gajar halva)* and rice pudding with cardamom, called *kheer.*

In northwestern India a special cylindrical clay oven heated with charcoal and called a *tandoor* is used. *Tandoori* cooking is identified particularly with lamb and chicken dishes, although the leavened bread known as *naan* is also typically baked in a tandoor.

Coastal India. The coastal regions offer a number of seafood specialties, including curried shrimp, crab, and clam dishes. Fish is prepared in a variety of ways, including fried, steamed, boiled, curried, and stuffed with herbs. Bombay, located on the west coast, boasts a dried salted fish, which is thin, bony, and strongly flavored, known as "Bombay duck".

Southern India. The menus of the South feature numerous steamed and fried rice dishes. Rice is even served puffed, as in a snack called *bhelpuri.* Other grains such as semolina wheat are also popular cooked as a cereal known as *uppama,* which may include vegetables. *Dals,* particularly chickpeas and lentils, accompany nearly every meal in the form of a spiced purée known as *sambar* or as a thin, crisply fried roti called *pappadams.* Fermented lentil flour is used at breakfast for *idli,* which are steamed cakes, and for *dosas,* which are spicy, fried pancakes. Fresh milk curds are also served for breakfast.

Highly spiced vegetable curries, such as *aviyal,* include such southern ingredients as bananas, bitter melons, coconut, green mango, and jackfruit seeds, in addition to potatoes, cauliflower, and eggplant. Refreshing yogurt-based pachadi and spicy pickled fruits or vegetables, such as chutney, accompany the main course. Deep-fried salty foods are favored snacks, as are deep-fried sweets, such as the syrup-soaked, orange-colored pretzels called *jalebis.*

Religious Variations

In addition to region, religious affiliation may greatly influence food habits in India. Religious

- *Ghee* is sometimes poured over sacred images in the worship of certain Hindu deities.

- The word *curry* is believed to be the English adaptation of a southern Indian term for "sauce," *kari.* Curry powder is not a single spice but a complex blend of seasonings that varies according to the cook and the dish. The closest Indian equivalent is *garam masala.*

- Pakistani fare combines elements of Indian and Arab cooking. Halal meats (see chapter 4) are consumed, often prepared in a *tandoor.* Flat breads and pilafs are common. Popular Asian Indian spices, such as cumin, turmeric, and chiles, are combined with more typically Arab flavors of cinnamon, cloves, and cardamom.

- *Dal* is the Hindu word for grains, legumes, or seeds (called *pulses* in India) that have been hulled and split in half—for example, split peas.

- The English words *pepper, sugar,* and *orange* are all derived from Asian Indian terms for those foods.

Table 14.2
Cultural Food Groups: Asian Indian

Group	Comments
PROTEIN FOODS	
Milk/milk products	In general, milk is considered a beverage for children. Fermented dairy products popular.
Meats/poultry/fish/eggs/legumes	Sophisticated vegetarian cuisine; legumes primary protein source. 18 varieties of beans, peas, and lentils commonly used. Hulled, split legumes, grains, and seeds, such as lentils, are known as *dals.* Legumes are typically prepared whole or pureed, or used as flour to prepare baked, steamed, or fried breads and pastries.
CEREALS/GRAINS	More than 1,000 varieties of rice are cultivated. Wheat used mostly in the North; rice in the South. Most breads *(roti)* are unleavened.
FRUITS/VEGETABLES	More than 100 types of fruit and 200 types of vegetables are commonly used. Fruits and vegetables may be used in fresh or preserved pickles, called *rayta* (northern), *pachadi* (southern), or chutney.
ADDITIONAL FOODS	
Seasonings	Aromatic (northern) and hot (southern) combinations of fresh or dried spices and herbs accentuate or complement food flavors.
Nuts/seeds	Nuts and seeds of all types are popular and are used to thicken *korma* sauces.
Beverages	Tea is common in the North, coffee in the South. Coffeehouses are favored meeting places.
Fats/oils	
Sweeteners	

Table 14.2
Cultural Food Groups: Asian Indian—continued

Common Foods	Adaptations in the United States
Fresh cow's, buffalo's, ass's milk; evaporated milk; fermented milk products (yogurt, *lassi*); fresh curds very popular; milk-based desserts, such as *kheer, khir, kulfi, barfi*	More cheese is consumed, ice cream is popular.
Meats: beef, goat, mutton, pork *Poultry:* chicken, duck *Fish and seafood:* Bombay duck, carp, clams, crab, herring, mackerel, mullet, pomfret, sardines, shrimp, turtle *Eggs:* chicken *Legumes:* beans (kidney, mung, etc.), chickpeas, lentils (many varieties and colors), peas (black-eyed, green)	Consumption of legumes decreases; meat intake increases. Meat may be added to traditional vegetarian dishes. Fast foods are popular.
Buckwheat, rice (steamed, boiled, fried, puffed), millet, wheat	Use of American-style breads occurs in place of *roti;* breakfast cereals are popular.
Fruit: apples, avocados, bananas (several types), coconut, dates, figs, grapes, guava, jackfruit, limes, litchis, loquats, mangoes, melon, *nongus,* oranges, papaya, peaches, pears, persimmons *(chicos),* pineapple, pomegranate, pomelos, raisins, starfruit, strawberries, sugar cane, tangerines, watermelon *Vegetables:* agathi flowers, amaranth, artichokes, bamboo shoots, banana rhizome, beets (leaves and root), bitter melon, brussels sprouts, cabbage, carrots, cauliflower, collard greens *(haak),* corn, cucumbers, eggplant, lettuce, lotus root, manioc (tapioca), mushrooms, mustard greens, okra, onions, parsnips, plantain flowers, potatoes, radishes (four types, leaves and roots), rhubarb, sago palm, spinach, squash, sweet potatoes (leaves and roots), tomatoes, turnips, yams, water chestnuts, water lilies	Decreased variety of fruits and vegetables is available; decreased vegetable intake results. More fruit juice is consumed. Salad is well accepted. Use of canned and frozen produce increases.
Ajwain, amchoor, asafetida, bay leaf, cardamom (two types), chiles, cinnamon, cloves, coconut, fresh coriander, coriander seeds, cumin, dill, fennel, fenugreek, garlic, *kewra,* lemon, limes, mace, mint, mustard, nutmeg, pepper (black and red), poppy seeds, rose water, saffron, tamarind, turmeric	Spice use depends on availability.
Almonds, betel nuts and leaves, cashews, peanuts, pistachios, sunflower seeds, walnuts	*Paan* tray may be limited to betel nuts and spices.
Coffee, tea, water flavored with fruit syrups, spices, or herbs; alcoholic beverages such as fermented fruit syrups, rice wines, beer	Increased consumption of soft drinks and coffee is noted. Alcoholic beverages are widely accepted (women may abstain).
Coconut oil, *ghee* (clarified butter), mustard oil, peanut oil, sesame seed oil	Purchased *ghee* is often made from vegetable oil instead of butter.
Sugar cane, *jaggery* (unrefined palm sugar), molasses	Candy and sweets are enjoyed but not overconsumed.

Vegetarianism in India

The ancient Indian diet featured a variety of meats, such as cows, bulls, buffalos, horses, rams, goats, and pigs, in addition to wild game including deer, alligator, and tortoise. The vegetarian ethic entered India slowly, probably beginning with the bulls and barren cows used for sacrifice by the Aryans. Later, prohibitions were extended to the milk cow, the draft bull, as well as the village pig (a useful scavenger) and the village cock. Over time, a more general concern for animal life developed, though meat eating (especially in the upper classes) was difficult for many Indians to forego. Buddhist and, later, Jain doctrines reinforced the concept of *ahimsa,* and vegetarianism became more widely practiced. It is often suggested that only in India, with its enormous variety of available fruits, vegetables, and grains, could such a broad acceptance of a vegetarian diet prevail (Achaya, 1994).

Yet the definition of vegetarianism in India is elusive. It is usually considered a symbol of piety in the Brahmin castes and may be a necessity among the poor. Abstinence from meat and poultry is most common; however, nearly all Indian vegetarians consume milk products, and some eat eggs. Fish is problematic because it is an inexpensive food where available. Except in the state of Gujerat (where the influence of Jainism has been especially strong), a large percentage of people living in coastal regions eat fish, sometimes justifying it as "fruit of the sea." Other Indians practice vegetarianism only on days of religious observance or as they age and become more devout. Some sources suggest that vegetarianism is most prevalent in southern India due to the Muslim influence found in the north. Indian census data show this is not the case. The highest percentages of vegetarians are found in the northwestern and central regions (see map). Overall, it is believed that 30 percent of Indians are strict vegetarians, abstaining from all meat, poultry, fish, and eggs but consuming milk, yogurt, and other dairy products.

Asian Indian vegetarians sometimes become meat eaters when living in America. In one study, one-third of those who were vegetarians in India became nonvegetarians in the United States (Karim et al., 1986).

Three types of nonvegetarian Indo-Americans have been identified: (1) "eggetarians," or those who eat eggs only;

Percentages of vegetarians in different states. *Source:* K.T. Achaya, *Indian Food: A Historical Companion*, ©1994, Oxford University.

(2) semivegetarians, who eat meat, poultry, and fish at restaurants and on social occasions but not in their own homes; and (3) full nonvegetarians, who eat meat, poultry, and fish at any time (Gupta, 1976). Asian Indian vegetarians may pass through all these phases in becoming nonvegetarians. Eating eggs begins soon after arrival in the United States, progressing to chicken, and finally to beef. It is estimated that nonvegetarian Asian Indians take between 2 months and 1 year to accept beef. One survey on acculturation patterns revealed that nearly 75 percent of Indo-Americans believe it is acceptable for nonvegetarians to eat beef; 44 percent indicated that it is acceptable for vegetarians to eat beef (Sodowsky & Carey, 1988). It may take vegetarian Indians 7 years to complete the transition to beef consumption, if it occurs.

As yet, the reasons vegetarians become nonvegetarians have not been stated conclusively. It has been suggested that vegetarianism may lose its social and cultural significance in the United States. Data regarding the influence of factors such as gender, income, region of origin in India, and length of stay in American have been contradictory.

■ Honey is prohibited by Jains because bees might be killed when it is gathered from the hive.

groups have varying dietary practices, yet their cooking is Indian in flavor. Muslims, for example, avoid all pork and pork products but are not vegetarians. Orthodox Jains may only eat "innocent" foods that avoid injury to any life and are therefore strict vegetarians. In addition, there are 22 prohibited foods (e.g., fruit with small seeds or tender new greens) and 32 other items that may have the potential for life to exist (including root vegetables, because insects might be killed when the tubers are harvested). They also refuse to eat any blood-colored foods such as tomatoes and watermelon. Sikh cuisine is noted for its use of wheat, corn, and sugar and the complete abstinence from alcohol and beef (pork is permitted).

The Syrian Christians are renowned for their beef (tenderized by mincing or marinating), duck, and wild boar dishes. Goan Christians are unique in Indian cooking for their use of pork. They make Western-style sausages and have such specialties as a vinegar-basted hog's head stuffed with vegetables and herbs. Most Jews in India keep kosher. The Parsis blend Indian and Persian elements in their cuisine, exemplified by dishes such as *dhansak,* an entrée combining lamb, tripe, lentils, and vegetables. Eggs are especially popular such as *ekuri,* spicy scrambled eggs.

Meal Composition and Cycle

Daily Patterns

Meal patterns in India, though not consistent across regions and classes, vary less than the foods served. Two full meals with substantial snacks are typical. A rich coffee or tea made with milk and sugar is enjoyed by early risers. Breakfast is usually eaten between 9:00 and 11:00 A.M., consisting of rice or roti, a pickled fruit or vegetable, and a sambar, or lentil stew, which may be left over from the previous evening. At 4:00 or 5:00 P.M., similar foods or snack items are eaten with coffee or tea. The main meal of the day follows between 7:00 and 9:00 P.M..

Texture, color, and balance of seasoning are all important factors in an Indian meal. A menu customarily includes at least one rice dish; a curried vegetable, legume, or meat dish; a vegetable legume side dish; a baked or fried roti; a fruit or vegetable pickle; and a yogurt rayta or pachadi. Sometimes a dessert is served, usually fruit. Courses are not presented sequentially in Indian meals; they are placed on the table all at once. Savory dishes are eaten at the same time as sweets. Diners may combine tastes and textures according to personal preference. The meal concludes with the passing of the paan tray. *Paan* is a combination of betel *(areca)* nuts and spices, such as anise seed, cardamom, and fennel, wrapped in large, heart-shaped betel leaves secured with a clove. It is chewed to freshen the breath and to aid digestion.

Snacking is very popular in India. In cities and small towns, snacks are sold in numerous small shops and by street vendors. In villages,

- A French Christian colony on the east coast of India at Pondicherry introduced baguettes, croissants, pâté and French-style desserts into the regional fare.

- Traditionally six tastes (sweet, sour, salty, bitter, pungent, and astringent) and five textures (foods that need to be chewed, those that need no chewing, those that are licked, those that are sucked, and those that are drunk) were balanced in an Indian meal.

- It is illegal to bring betel leaves into the United States. There is a black market for *paan* obtained in Canada and other countries.

Traditional Indian *thali* (individual silver serving tray), featuring a selection of *roti* (flat breads), fruit and vegetable pickles, like chutneys, and yogurt-based *pachadi* or *rayta.*

they are prepared at home. A clear distinction is made between meals and snacks. Many Indian languages have specific words to define each form of eating. In southern India the word *tiffin* is used to distinguish a snack from a meal. The coffee or tea drunk before breakfast or in the late afternoon is considered tiffin. A meal is not a meal unless the traditional staple prepared in the traditional manner, such as boiled rice in southern India (or roti in northern India) is served. This means that no matter how substantial the snack—and some include more food than a meal—it is still called tiffin. Spicy snacks served with chutney often consist of batter-fried vegetables, pancakes with or without a filling, or fried seasoned dough made from wheat or lentils. Snacks sweetened with sugar cane, molasses, or jaggery are usually milk-based, such as the saffron-spiced *khir,* the Indian ice cream called *kulfi,* and the candy *barfi,* although nuts, coconut, sesame seeds, or lentils are also used. A snack may also include a cooling beverage, such as the sweetened, diluted yogurt drink called *lassi* or the fruit juice known as *shurbut.*

Etiquette

The most important aspect of Indian manners is that food is eaten with the hands. Food is served in small individual bowls from serving trays called *thalis.* The thalis may be silver or brass, with matching bowls. Originally the thalis were simply banana leaves and the bowls were earthenware; these are still used today in some rural areas or when disposable trays and bowls are desired.

■ As in other cultures in which the hands are used instead of utensils, it is an unforgivable breach of manners to eat with the left hand; one eats with the right hand and takes care of all other personal needs with the left.

■ Traditionally the male head of the household was responsible for seeing that any guests, pregnant women, or elderly persons were well fed before he could sit down to eat.

■ Sweets represent prosperity because they often include costly ingredients and may be decorated with silver or gold leaf.

Rules of etiquette are important for food that is eaten by hand and prepared with little or no means of refrigeration. Food spoilage and contamination are problems, especially in the warmer regions of the country. Cleanliness is imperative, and many Indian manners clearly support this need. Ancient rules of etiquette promoted healthy food preparation. They included the following: Everyone must clean their hands, feet, and mouth before and after eating, both morning and evening; servants who worked with food had to shave their beards and heads, cut their nails, and bathe before entering the kitchen; food that was cooked the previous night, that turned sour, or that was cooked twice must be discarded; and all grains, vegetables, and meat had to be washed thoroughly before cooking.

During the Buddhist period, other rules were added. Dating from 300 B.C.E., they recommended that one should not talk with food in one's mouth; stuff one's mouth full of food; shake one's hands while eating, as one would scatter rice all over; put one's tongue out while eating; make hissing sounds while eating; lick one's finger or bowl; or accept a drinking cup with hands that are soiled with food. Finally, everyone should start eating at the same time.

Other etiquette rules state that food should be discarded if it comes in contact with the hem of one's gown, if it has been touched by animals, or if it is contaminated by insect or rodent droppings. Indians customarily never serve themselves (especially after their hand is soiled from eating), never offer food from their plates to another person, and rarely save leftovers. Such manners are essential to maintaining sanitary conditions in rural homes. These rules of etiquette may be practiced in even the most westernized families, however, and they remain important symbols of the Indian relationship with food.

Special Occasions

Another aspect of Indian culture that affects daily diet is the concept of feasting and fasting. As with other Indian food habits, feasting and fasting activities are complex and vary greatly from person to person and group to group. No occasion passes in India without some special food observance: regional holidays, community celebrations, and personal events such as births, weddings, funerals, and illness. A devout Hindu may feast or fast nearly every day of the year (see chapter 4).

Feasting. Feasts serve as a method of food distribution throughout the community. They are generally observed by presenting large amounts of everyday foods and sweets of all kinds to the appropriate holy figure; all members of the community then eat the food. Feasts may be the only time that the poor get enough to eat.

Some foods are associated with certain concepts. Rice and bananas both symbolize fertility, for example. Betel leaves represent auspiciousness; ghee, purity; salt, hospitality and pleasantness; mango, hospitality and auspiciousness; and betel nuts and coconuts, hospitality, sacredness, and auspiciousness.

Most festivals are Hindu in origin, and although many are observed nationwide, each is celebrated differently according to the region. *Holi* is a spectacular holiday in the North, featuring reenactments of Krishna's life, fireworks, and colored powders and waters. Celebrants snack at bazaar booths. *Dussehra* is a 10-day holiday observed in both the North and the South. A special dish is prepared each day, and every day that dish is added to those prepared the previous days, culminating in an enormous feast on the last evening, after a torchlight parade of ornamented elephants. *Divali,* the "festival of lights," is the New Year's holiday celebrated everywhere with gifts of sweets.

Non-Hindu harvest festivals also feature feasts. They are dedicated to wheat in the North and rice in the South. At the 3-day rice festival in Pongal, dishes made from the newly harvested rice are ceremonially "fed" to the local cows. The 10-day festival of Onam in Kerali culminates with a feast served by the local women, including 30 to 40 dishes ranging from fiery curries to foods sweetened with a combination of molasses, milk, and sugar.

Asian Indian Muslims may dine with friends on Eid al-Fitr at the end of Ramadan and Eid al-Azha (see chapter 13). Christians celebrate Christmas and Easter in India.

Fasting. Fasting is also associated with special occasions in India. It accompanies both religious and personal events. An orthodox Hindu may fast more days a week than not. However, the term "fast" includes many different food restrictions in India, from avoidance of a single food item to complete abstinence from all food. A person might adopt a completely vegetarian diet for the day or eat foods believed to be spiritually "purer," such as those cooked in milk (see the section "Purity and Pollution" in this chapter). Individuals rarely suffer from hunger because of fasting in India. In fact, more food may be consumed on fast days than on a nonfast day.

Muslims in India also fast, notably during the month of Ramadan. No food or drink is consumed between sunrise and sunset (see chapter 4 for more information).

Sample Menu

AN INDIAN VEGETARIAN DINNER

Samosas (deep-fried potato turnovers)

Aviyal (spicy curried vegetables with coconut milk)

Rice

Sambar (lentil purée)

Yogurt *pachadi* and fruit chutney

Pappadams (spicy fried flat bread)

Bananas, *nongus* (palmyra palm fruit), starfruit

Tea

Role of Food in Indian Society

The importance of food in Indian culture goes far beyond mere sustenance. Sanskrit texts describe how

> From earth sprang herbs, from herbs food, from food seed, from seed man. Man thus consists of the essence of food. . . . From food are all creatures produced, by food do they grow. . . . The self consists of food, of breath, of mind, of understanding, of bliss. (Achaya, 1994, p. 61)

In a society that traditionally experienced frequent famines and chronic malnutrition, food is venerated. Complex traditions have developed around when, how, and why foods are prepared, served, and eaten.

Purity and Pollution

Many Hindu dietary customs are meant to lead to purity of mind and spirit. Pollution is the opposite of purity, and polluted foods should be avoided or ameliorated. To be pure is to be free of pollution.

- Another holiday, *Janmashtami,* commemorates the birth of Krishna. As a boy, Krishna and his friends would steal butter or curds hung high in earthen containers. This story is recreated during the celebration as young boys attempt to break elevated clay pots full of curds.

- The decorative red or yellow dot that Hindu Asian Indian women apply to their foreheads represents joy or prosperity. It is omitted during fast days.

- The Sanskrit scriptures state, *"Annam Brahma"*—"Food is God."

- *Pakka* (meaning "cooked") foods are those that are fried or fat-basted during preparation, preferably in ghee. Pakka foods are relatively unrestricted due to their high degree of purity.

- *Kaccha* (meaning "undercooked") foods are those that are boiled in water, baked, or roasted. Kaccha foods are more susceptible to pollution than pakka foods and must therefore be treated carefully during serving and consumption.

- A Hindu woman is considered impure during her menses and is prohibited from cooking or touching any food that is to be consumed by others.

- The kitchen is considered sacred in many Hindu homes. By custom, no one may enter with shoes on, all who come in must bathe and put clean clothes on first, and children with questionable hygiene are prohibited, to preserve purity.

- Raw foods are considered difficult to digest.

The Hindu classification system of *jati* is used to evaluate the relative spiritual purity of all foods. Purity is determined by the ingredients, how they are prepared, who prepares them, and how they are served. Some foods, such as milk, are inherently pure. Raw foods that are naturally protected by a husk or a peel are less susceptible to pollution. Fried foods (*pakka* foods) are more pure than baked, boiled, or steamed foods (*kaccha* foods), especially if they are fried in ghee. Food that is served in brass dishes is also less vulnerable to pollution than is food in clay dishes. Some foods, such as alcohol and meat, are *jhuta,* meaning that they are innately polluted. *Jhuta* foods are those that are inherently polluted and impure. All leftovers, unless completely untouched by the consumer or by other foods that have been eaten, are jhuta. Those foods that are innately jhuta vary according to religious sect. The term *jhuta* is also used for garbage and offal.

Traditionally only foods cooked and served by a member of an equal or superior caste could be consumed by any Hindu. Only members of the same caste will eat together. Untouched leftovers can be given to a member of a lower caste, such as servants, but polluted leftovers are eaten only by scavengers of the lowest caste. Asian Indians adhere to these commensal rules to various degrees, however. An orthodox Hindu would attempt to follow them at all times. A more modern Hindu might adhere to them only during holy services and holidays. Most westernized Indians eat in restaurants and use convenience products in cooking, ignoring how and by whom the food was prepared.

The consumption of jhuta foods also varies among Hindus. Historically laborers and warriors were allowed to eat meat to help keep up their strength. Some Brahman subcastes permit consumption of impure foods that happen to be plentiful in their region, such as fish (referred to as "fruit of the sea") in the coastal areas and lamb in the North. Other sects are so rigid that even inadvertent intake of polluted food results in spiritual disaster. Members of the International Society for Krishna Consciousness (see chapter 4) believe that if they accidentally eat a prohibited animal food, they will lose human form in their next life and assume the form of an animal that is the prey of the animal they ate.

Asian Indian Women and Food

The role of women in food preparation is extremely important throughout Indian culture. Feeding the family is an Indian woman's primary household duty. She is responsible for overseeing the procurement, storage, preparation, and serving of all meals. Because arranged marriages are common, training in kitchen management is considered essential for a Hindu woman in obtaining good marriage offers.

It is generally believed that a woman cannot be completely substituted for in the kitchen, for she imparts a special sweetness to food. If the wife is unable to perform food-related duties, a daughter or daughter-in-law may substitute. If servants help in meal preparation, it is still important for the woman of the house to serve the food directly from the *chula* (stove) to the table. This often requires many trips. Customarily the wife serves the men and guests first, then the children.

Therapeutic Uses of Food

Ayurvedic medicine is based on the premise that each human is a microcosm of the universe. As such, the body experiences the three inevitable laws of nature (also called universal tendencies) of creation *(sattwa),* maintenance *(rajus),* and dissolution *(tamas)* (Tirtha, 1998). The fundamental elements of fire, water, and wind also have their counterparts in the humors of the body—bile *(pitta),* phlegm *(kapha),* and wind *(vata).* Pitta regulates metabolic activities and resulting heat. Kapha provides structure and support through bone and flesh, and vata represents movement, of muscle and semen. Health is maintained through a careful balance of humors and substances in the body according to each person's internal constitution and external experiences. When pitta is in balance, digestion is comfortable and a person is content; balanced kapha produces physical and emotional stability, strength, and stamina; vitality and creativity are the results of balanced vata (Larson-Presswalla, 1994). Good digestion is critical because food is transformed into the body humors and substances when it is cooked by the digestive *agnis* ("fires"), producing food juices and wastes. Food that is indigestible is harmful because it is believed to accumulate in the intestines and decompose, sending toxins into the bloodstream; excessive waste or too little waste is an imbalance that causes illness (Ramakrishna & Weiss, 1992; Sheikh & Sheikh, 1989).

Foods are classified according to which humors they enhance or inhibit. For example, pomegranate increases vata and reduces pitta and kapha. Molasses does the opposite: it increases pitta and kapha and reduces vata. Some foods are also grouped according to their universal tendencies. Mungbeans, for instance are considered *sattawic,* chile peppers are *rajasic,* and nutmeg is *tamasic.* Furthermore, the hot-cold classification system is used for foods, depending on how they affect the body. The specific identification of an item as hot or cold varies regionally; for example, lentils and peas are considered hot in western India, but cold in northern India. Generally wheat, spices and seasonings (except mustard and sesame seeds), chicken, and oils are classified as hot; rice, leafy vegetables, fruits (except mango, papaya, and jackfruit), dairy products, honey, sugar, pickles, and condiments are considered cold (Achaya, 1994). The hot or cold nature of a food can be altered through the method of preparation. The use of hot spices or roasting may make a cold food hot; conversely, soaking a hot food in water or blending it with yogurt can change it into a cold food (Ramakrishna & Weiss, 1992).

Although a balance of foods according to humoral affect universal tendencies, and hot-cold is essential to health, the exact proportions of each change with age, gender, physical condition, and the weather. Traditionally six seasons are recognized, each with certain dietary recommendations. During winter, when digestion is thought to be strongest, roasted or sour and salty dishes are preferred as well as sweets; in summer and during the monsoons, when digestion is thought to be weak, salty, sour, and fatty foods are avoided (Achaya, 1994).

The way foods are eaten is as important as which foods are consumed. To maximize digestion of foods, a person should eat in a quiet atmosphere, sip warm water throughout the meal, and sit for a short while after dining (Larson-Presswalla, 1994).

Pregnancy is considered to be a normal and healthy condition; however, certain food taboos are sometimes followed. Women especially avoid extremes in foods that are too hot or too cold. Lime juice with honey is a general tonic, believed to prevent excessive bleeding at birth, while cow's milk (particularly with almonds and saffron) and rice porridge are thought to ensure proper development of the fetus. Fenugreek seeds in buttermilk are given for nausea and butter or ghee is believed to make the body supple and ease delivery of the baby.

Numerous dietary remedies are listed for minor illness. Barley water is consumed for a fever; vomiting is treated with milk. Coconut water, buttermilk, anise seed oil, and pomegranate flowers are all considered helpful for diarrhea. A powder called *ashtachooran* (a mixture of asafetida, salt, ginger, pepper, cumin, and ajwain) is added to honey for indigestion. Ginger tea or garlic soup is used to treat colds. Gooseberries and hibiscus flower tea are considered general tonics (Achaya, 1994; Khanna, 1986).

It is widely believed that a highly spiced diet is necessary in the tropical Indian climate to stimulate the liver. One food habit that may result in ill health is the common practice of disguising otherwise undrinkable water with flavorful herbs and spices. Cholera, dysentery, and typhoid are endemic in many regions.

Contemporary Food Habits in the United States

Adaptations of Food Habits

Americans of Asian Indian descent have usually been exposed to American or European lifestyles in India and may be familiar with a westernized diet before immigration to the United States. Yet even the most acculturated Indo-Americans continue some traditional food habits. Most accept American foods when eating out but prefer Asian Indian foods when at home (Sodowsky & Carey, 1988).

Ingredients and Common Foods

Very little research has been conducted on the food habits of Asian Indians in the United States. One study of Asian Indians in Pennsylvania suggested that acculturation takes place

- Many foods are considered incompatible in Ayurvedic medicine, such as honey with ghee, rice with vinegar and honeydew melon with yogurt.

- Cow's milk is thought to increase intelligence; buffalo's milk is believed to strengthen the body.

- In parts of India, garlic is believed to excite the passions and may be avoided.

Samosas, spicy deep-fried turnovers.

■ Asian Indians who are practicing Muslims rarely begin eating pork in the United States. They may drive long distances to purchase holal or Kosher meats to fulfill traditional Muslim dietary laws.

■ Pakistani Americans generally eat at least one traditional meal each day.

in two phases (Gupta, 1976). Typically the first lasts for 2 to 3 years, often while the immigrant is a student. Interaction with mainstream American society may be limited during this period. The recent immigrant prefers to associate with members of the same caste, regional, or linguistic group; experience with American foods often includes only fast foods. Male Asian Indian students are often unable to cook and may rely heavily on purchased meals. Many Asian Indian immigrants will eat hamburgers because of their availability and low cost.

Sometime during the next 10 years, Asian Indians who stay in the United States longer than 4 years enter the second phase of acculturation. They are usually employed by American businesses and are raising families. They keep their social interactions with Americans separate from those with other Indo-Americans. They might serve meat and alcohol to American guests, for example, and vegetarian dishes to Indian guests.

A study of Indians living in Cincinnati found several changes in the types of foods they ate (Karim et al., 1986). Foods that subjects used frequently in India but that were in only "low moderate" use in the United States were ghee, yogurt, dal, roti, rice dishes, and tea. Foods that were in "low moderate" use in India but in frequent use in the United States were fruit juice, canned or frozen vegetables, American bread, dry cereals, cheese and cheese dishes, and soft drinks. Coffee consumption also increased.

Variables that affect acculturation include gender (men tend to change their food habits more readily than women because women are the traditional food preparers in Indian society), age (children raised in the United States prefer American foods), marital status (single unmarried men are the most acculturated, married men with families in India next, and married men with families in the United States are least acculturated), caste (depending on whether caste members used meat or alcohol in India), and region (Asian Indians from rural areas are often stricter vegetarians than those from the cities).

A majority of Americans of Asian Indian descent make an effort to obtain traditional food products. Many markets in the United States specialize in Indian canned and packaged food products, including spices, and many stores provide mail orders. Fresh foods are more difficult to find. Some fruits and vegetables can be bought at Asian specialty markets, and Indian bakeries featuring sweets and tiffin items have opened in some areas. In Cincinnati, 100 percent of Indians interviewed reported that their traditional foods were available (Karim et al., 1986).

Meal Composition and Cycle

Asian Indian eating patterns may become more irregular in the United States, possibly because of the pressures of a faster paced lifestyle. Breakfast is the meal most commonly omitted; snacking occurs between one and three times per day (Karim et al.,1986; Yagalla et al.,1996). Many Americans of Asian Indian descent eat American foods for breakfast and lunch. Traditional Indian evening meals are preferred, if native foods and spices are available. Yet dinners at home may also be influenced by U.S. food habits in that more meat may be eaten and American breads served in place of roti.

Nutritional Status

Nutritional Intake

Little specific information is available about the Asian Indian community, perhaps because most Asian Indian immigrants have come from the higher socioeconomic levels. They more often suffer from the effects of nutritional overconsumption than inadequacy. One researcher has commented that young Indo-Americans, especially students, may rely heavily on fast-food and restaurant meals and are thus susceptible to the same nutritional problems common to most American youth (Eckstein, 1983).

Researchers have noted that Asian Indian immigrants have a significantly higher risk of cardiovascular disease and noninsulin-dependent diabetes mellitus than does the native population in India or the white population in their new homeland. Heart disease rates are estimated to be one and one-half to four times greater among Asian Indian immigrants than among whites; diabetes rates also are calculated to be at least four times higher (Hughes, 1990; McKeigue et al., 1991; Raheja, 1987). This increased risk is found regardless of religious background or adherence to a vegetarian or nonvegetarian diet. Scientists hypothesize that both conditions may be related to hereditary differences in lipid metabolism (Thomas et al., 1986) and increased levels of glucose intolerance (McKeigue et al., 1991; Yagalla et al., 1996) that result in elevated blood triglyceride and insulin levels as well as higher waist-to-hip ratios suggestive of increased abdominal fat (although obesity rates among Asian Indians are very low). Changes in diet are believed to be responsible, including increased fat intake for both vegetarians and nonvegetarians (17 to 22 percent of dietary calories in India compared to 31 to 34 percent in the United States) and a change in the kinds and amounts of carbohydrates consumed (Blesch et al., 1999; Mani et al., 1990; Raheja, 1987; Yagalla et al., 1996).

The arrival of Indians from poorer, rural regions may result in nutritional deficiency problems (it is estimated that 17 percent of the Indian population suffers from moderate to severe malnutrition). These immigrants are also most likely to practice traditional food habits the longest after coming to the United States. It has been found that Asian Indians who live in Great Britain and follow vegetarian diets have marginal intakes of protein, iron, and vitamin B_{12}. Although iron consumption from foods may also be low in the United States, content estimates may not account for the substantial amounts of iron obtained through the use of traditional iron cookware (Kollipara & Brittin, 1996). Research suggests that although zinc deficiency may not be a problem among vegetarian Asian Indian women living in India, bioavailability may be impaired when they come to the United States if these women add significant amounts of dairy products to their diet (Ellis et al., 1987). Nutritional adequacy may be compromised if traditional food habits are maintained in spite of inadequate food sources; if native spices and foods are not available or are too costly, the diet may be poor.

■ One study reported that Indo-Americans are more likely than Asian Indians in India (but less likely than white Americans) to develop breast or colon cancer. (Blesch et al., 1999).

■ Asian Indians may have difficulty digesting fresh milk; it is also estimated 60 percent of adults in Pakistan are lactose-intolerant (Ahmad & Flatz, 1984).

■ Many Indo-Americans in one survey retained traditional beliefs regarding the harmful effects of colostrum for infants and the positive effects of providing sugar water or other non-breast-milk liquids during the first 24-hours after birth (Kannan et al., 1999).

Counseling

The majority of Indo-Americans are very familiar with biomedicine; however, traditional expectations regarding social interactions may interfere with health care (Ramakrishna & Weiss, 1992). A client may not disagree with or correct a provider in situations in which the client believes that the provider is superior in status; compliance may be assumed because the client does not indicate otherwise. If the client does not feel that treatment is satisfactory, he or she will typically change providers rather than confront the problem. Conversely, if a client believes that she or he is an equal with the provider, the client may desire a social rather than professional relationship with the

■ When cost is an issue in obtaining health care, the male head of the household may receive priority over women, girls, or elders in a traditional Asian Indian home.

■ One Ayurvedic remedy for diabetes uses ginger and rock candy.

■ Americans of Asian Indian descent may feel cheated if exhaustive testing procedures or invasive therapies are not provided for their condition (Ramakrishna & Weiss, 1992).

■ Mental illness is highly stigmatized in India. Complaints of headache, leg tingling, or burning on the soles of the feet may be related to psychological distress in recent immigrants.

provider. The client may also feel entitled to special privileges, such as immediate access to the provider and longer appointments.

A leisurely, personal style of communication is preferred by most Asian Indians. Small talk is significant, and rushing is considered rude. A direct no is considered impolite, so an evasive negative response is the norm. Feelings and faith may be more important than objective facts. Some Indian women are less vocal than men.

Several nonverbal communication customs among Asian Indians are significant. Most Indian Hindu, Sikh, and Christian men and women avoid contact with the opposite sex in public, and contact between men and women in public is completely prohibited by Muslims. The traditional greeting is to nod with the head while holding the hands palm to palm beneath the chin and saying *namaste* (nah-mas-tay), but westernized Indians will use a handshake (it is best to wait and follow the lead). Some Asian Indians indicate agreement by shaking their head back and forth the way Americans indicate the negative. Informal smiles are used between equals only—superiors do not generally smile at subordinates or vice versa. The left hand is never used for any social purposes, including handshakes, giving an item to another person, or pointing. The head is considered the seat of the soul; patting or touching another person's head should always be avoided. Conversely, the feet and shoe soles are considered the dirtiest parts of the body, and it is impolite to point with the foot or show the bottom of one's shoe (Axtell, 1991; Morrison et al., 1994). All that said, it should be noted that most Indo-Americans are highly acculturated and can converse in a style comfortable for Americans.

Asian Indian women are often less vocal than men and may be very uncomfortable with a male health care provider. Men do not touch women in either informal or formal situations; physical exams of women are unusual. Female gynecologists and obstetricians may be preferred.

Family members may accompany a patient, especially a woman, as a chaperone during an appointment. The chaperone expects to participate in any discussion of the client and to oversee the exam (privacy among spouses and siblings is limited).Family members may also assume responsibility for all but the most technical care of a patient during hospitalization; it is their obligation to feed, bathe, and support a relative.

Americans of Asian Indian descent may provide more information regarding their condition than what the practitioner may feel is warranted. Ayurvedic medicine focuses on a person's role within the cosmos, so details about what a client is eating, sleep patterns, and changes in the weather may be important to a client. Most clients expect lifestyle advice, particularly regarding diet, as part of their therapy (Larson-Presswalla, 1994; Ramakrishna & Weiss, 1992).

Some clients may expect a complete history and physical exam to be undertaken; others may assume that the practitioner should diagnose simply by taking a pulse, and without prying into what may be perceived as personal issues. The practitioner should proceed cautiously with questioning until the client's expectations have been determined. Indo-Americans have strong opinions about the therapeutic value of medications. Injections are believed most potent; in India, a patient would expect to receive between one and four shots each visit. Capsules are thought to be more effective than tablets, and colorful tablets more potent than white tablets. Furthermore, the medication becomes more effective through the skill of the provider, thus it is imperative that the provider personally hand the prescription to the client.

Little information has been reported regarding the use of ayurvedic medicine or folk remedies by Asian Indians living in the United States. Although most herbal preparations are compatible with biomedical therapies, lead screening tests among children in California revealed high levels of lead in some ayurvedic medications, including *surma* (a black powder used to improve eyesight) and an undetermined oral remedy of small, gray balls (used to improve slow development).

An in-depth interview should be used to establish the client's religious affiliation and degree of adherence, length of residency in the United States, and degree of acculturation or assimilation, as well as vegetarian or nonvegetarian preferences. Clients should also be asked about ayurvedic practices, particularly those regarding diet and home remedies.

15

Regional Americans

In Boston they eat beans. In Philadelphia they eat cheesesteak. In Kansas City they eat barbecue, and in Seattle they drink café lattes. A person from Montana is no more likely to eat grits (ground hominy) than a person from Mississippi is likely to eat Rocky Mountain oysters (deep-fried beef testicles). Just as certain fare is associated with ethnicity and religious affiliation, local food preferences are elemental to American regional identity.

While most Americans expect uniformity in some food products, such as the breakfast cereal they buy in the grocery or the burger they purchase at a fast-food franchise, American cuisine is anything but homogeneous. Consumption data illustrate many differences. People in the Northeast spend the most annually on poultry, bakery products, butter, wine, and tea. Those in the South spend more on pork, sugar, and whiskey. Mutton and beer are above average sellers in the Midwest; eggs, cream, frozen meals, condiments, and coffee are purchased more often in the West. Conversely, westerners spend the least on pork chops and sausages; wine is unpopular in the Midwest; southerners buy fewer frozen juices than average; and bacon is purchased less frequently in the Northeast.

These figures reflect regional tastes and specialties. Tea in the Northeast dates back to colonial times. Pork has long been a foundation of southern fare. Lager-style beer was created in the Midwest, and the ethnic diversity of the West is suggested by the variety of mustards, salsas, soy sauces, fish sauces, and chutneys consumed. This chapter will examine the traditional fare of the Northeast, South, Midwest, and West, noting significant culinary variations and trends in American regional food habits.

American Regional Food Habits

What Is Regional Fare?

Regional fare has traditionally been home-style food prepared with local ingredients, dependent on the seasons and on storage conditions. Most families made do with what they could grow, gather, or barter.

Available indigenous and introduced foods are the most significant of several factors that influence the development of a particular regional cuisine. The spicy cooking of the Southwest with its emphasis on corn, beans, and chiles could not have been created in the upper Great Lakes area, which produces wheat, fish, and dairy foods. Before the advent of food preservation and refrigerated shipping, local items were not only freshest and tastiest, but also most economical. Strong associations with place and food suggest the importance placed on the superior quality of local items. Even today, with a global assortment available, there is a certain cachet to Maine lobsters, Vermont maple syrup, Georgia peaches, Florida oranges, Idaho potatoes, and Washington apples (see Figure 15.1).

Ethnic and religious practices also affect the development of regional fare, particularly specialty foods. Jewish bagels in New York,

■ Clinton, Montana, hosts the Testicle Festival each September: 2 days of feasting on *Rocky Mountain oysters* (marinated, sliced, breaded, and deep-fried) on toasted bread and enjoying live country music. Over 4,500 pounds of beef testicles are consumed annually at the event. Vinita, Oklahoma, sponsors a Calf Fry Festival on the first Saturday after Labor Day, featuring *Prairie oysters* and fries.

■ During colonial times, only 15 percent of the American population lived in cities. The remaining 85 percent were self-sufficient on rural farms.

■ A study of places and foods found that Texas, California, Vermont, Oregon, and Louisiana were the states with the most food associations and that place names enhanced the perception that an item was authentic, distinctive, and high-quality (de Wit, 1992).

Figure 15.1
Foods associated with places

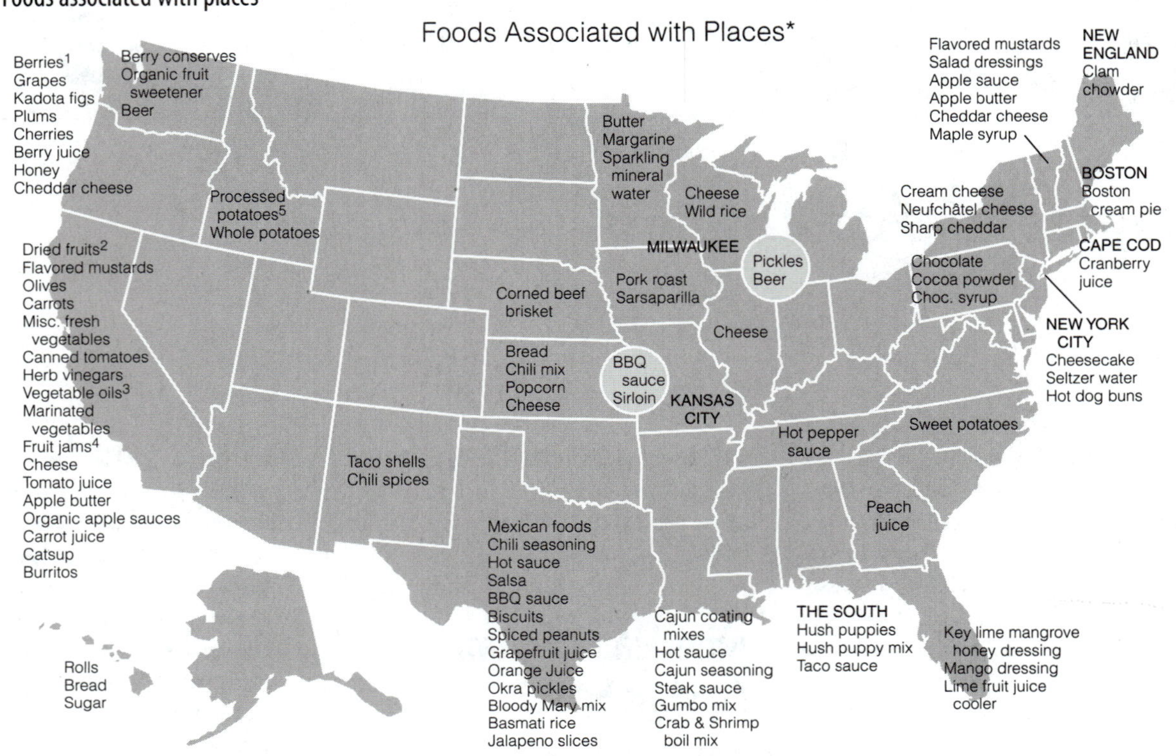

[1]Gooseberries, boysenberries, blackberries, strawberries, blueberries, red raspberries. [2]Apples, peaches, figs, prunes, raisins. [3]Walnut, olive, pecan. [4]Black raspberry, strawberry, apricot, kiwi. [5]Hash browns, tater tots, golden fries, potatoes O'Brien, crinkle cuts, homestyle potatoes, cottage fries, shoestrings, potatoes au gratin, scalloped potatoes, instamash potatoes.

*Listed in order from most to least frequent occurrence.

Source: de Wit, C. W. Food-place associations on American product labels. *Geographical Review, 83,* 323–330. Copyright 1992. Reprinted with permission by American Geographical Society.

German doughnuts called *Fastnachts* in Pennsylvania, Cajun French–style sausages in Louisiana, West African–influenced hoppin' John in South Carolina, southern Italian–flavored pizza in Chicago, Cornish pasties in Michigan, and Mexican-inspired chili con carne in Texas are just a few examples. Most regional cuisine is a blend of several ethnic elements, such as the British–Native American dishes of New England, the African-French-Spanish-British–Native American mélange that is southern fare, and the northern Italian–Mexican-Asian mix found in California cuisine. Religious food habits are a factor in areas where large numbers of a specific faith have congregated. For instance, the majority of Mormons live in Utah, and Mormons do not drink alcohol, tea, coffee, or other stimulating beverages. Alcohol purchases in Utah are strictly controlled through limited availability at state-run outlets. Sweets are allowed, however, and are well integrated into family activities. The people of Utah eat twice as much candy per capita as the U.S. average (Lee, 1992).

A third factor in regional foods is historical heritage. Certain dishes are linked to local tradition. A good example is the Kentucky stew called *burgoo.* Legend has it that the mixture of poultry and red meat with vegetables dates back to the Civil War, when a chef for the Confederate cavalry at Lexington created the stew from native blackbirds, game, and greens. There is no single recipe for the dish, but today it typically includes chicken, pork, beef, or lamb; and cabbage, potatoes, tomatoes, lima beans, corn, and okra. It is seasoned with cayenne. Burgoo is traditionally served at picnics, political rallies, and sporting events, including Derby Day. Current trends can be an influence as well. Some dishes sweep through one region but never gain national acceptance, such as caviar pie (layered hard-boiled eggs, scallions, caviar, and sour cream) in the Southeast (Anderson, 1997), or *loco moco* (a bowl with rice topped with a ground beef patty, then an egg over easy, and gravy) in Hawaii (Kelly, 1983). Other trends start out regionally and then catch on countrywide, such as the salsas of the Southwest.

Economics contribute to the popularity of certain foods. In one study, it was found that some of the best markets for beer in the country are in the poorest areas. Wine is popular in upper-income regions, which are often the worst beer markets despite the growing popularity of local brewpubs (Edmondson, 1998). Households with incomes below $20,000 a year spend about $1,400 per capita on food, buying more white bread, bacon, pork chops, luncheon meats, and eggs per person than the U.S. average. On the other end of the socioeconomic spectrum, households with incomes more than $70,000 annually spend over $2,500 per capita on food, with above average purchases in almost every category, especially expensive cuts of beef, fresh fish and seafood, dairy products, biscuits/rolls, and snack foods (Mogelonsky, 1996). Upscale consumers are more likely to be exposed to unique culinary ideas when dining out in trendy full-service restaurants and to be willing to pay for new or unusual food items.

Regional Divisions

There are many ways to divide the United States. Sometimes regions are delineated by terrain, as in the Great Plains, or marked by major rivers or mountain ranges, as in the Mississippi River Valley or the Appalachians. Sometimes areas are defined by similarities in climate, as in the Sun Belt; by economic affiliation, as in the Steel Belt and the Silicon Valley; or by the characteristics of the population, as in Indian Lands and the Bible Belt. Historical divisions, such as the Mason-Dixon line, or political divisions, including state boundaries, can also be used. The cultural geographer Wilber Zelinsky suggests that traditional regions contain elements of all these variables, characterized as a synergistic relationship between a people and the land that develops over time and is specific to the locality. He calls it *blood and soil* (Zelinsky, 1973). Such regional identity is dynamic, more of a process than a delineation, subject to changes in population, economy, and ecology. Ideally the people who live in an area, through their voluntary affiliation and their mutual agreement on group boundaries, define each region (Brown & Mussell, 1984).

Most often, however, geographic considerations are used to set arbitrary regional divisions independent of cultural relationships. The U.S. Census Bureau and the U.S. Department of Commerce list four regions with nine subdivisions for data collection purposes: Northeast, Midwest, South, and West (see Table 15.1). Although these categories group states with distinctively different cuisines together, such as Florida and Texas, or Alaska and Hawaii, most demographic and food consumption data are presented in this four-region format (see Figure 15.2). It is useful for detecting broad trends, as long as results are not overgeneralized to smaller populations who may observe alternate regional boundaries.

The Northeast

Regional Profile

The states of the U.S. Northeast include those of New England (Connecticut, Maine, Massachusetts, New Hampshire, Rhode Island, and Vermont) and those of the Mid-Atlantic region (New Jersey, New York, and Pennsylvania). The New England area features a rugged, irregular Atlantic coastline with many protected bays. Rolling hills and valleys that gradually become densely forested mountains extend west. The region is noted for its spectacular autumn weather and colorful fall foliage, followed by harsh winters. The Mid-Atlantic states are further south and more temperate in climate. Sandy beaches and estuaries line the long coast. The ridges, river valleys, and fertile plateaus of the Adirondack, Appalachian, Blue, Catskill, and other mountain ranges criss-cross much of the three states. Freshwater lakes dot the region and provide the northern boundary along Lake Ontario and Lake Erie.

Despite differences in climate and geography, the entire Northeast shares a common early history of sophisticated Native American

■ **Fiddlehead fern fronds are traditionally a harbinger of spring in the northern areas of New England. They are usually sautéed in butter.**

Table 15.1
U.S. Regional Divisions

Northeast	
New England	Connecticut, Maine, Massachusetts, New Hampshire, Rhode Island, Vermont
Mid-Atlantic	New Jersey, New York, Pennsylvania
Midwest	
East North Central	Illinois, Indiana, Michigan, Ohio, Wisconsin
West North Central	Iowa, Kansas, Minnesota, Missouri, Nebraska, North Dakota, South Dakota
South	
South Atlantic	Delaware, District of Columbia, Florida, Georgia, Maryland, North Carolina, South Carolina, Virginia, West Virginia
East South Central	Alabama, Kentucky, Mississippi, Tennessee
West South Central	Arkansas, Louisiana, Oklahoma, Texas
West	
Mountain	Arizona, Colorado, Idaho, Montana, Nevada, New Mexico, Utah, Wyoming
Pacific	Alaska, California, Hawaii, Oregon, Washington

Source: U.S. Census Bureau/U.S. Department of Commerce.

societies supplanted by European settlements. The colonial immigrants from England, Germany, the Netherlands, and France were followed by newcomers from Ireland, Italy, Portugal, Poland, and other central and eastern European countries, particularly in the Mid-Atlantic states. African Americans from the South and more recent immigrants from the Caribbean, Central America, Africa, and Asia have added to the diversity of some parts of the Northeast (see each chapter on ethnic food habits for more immigration details).

Although only 19 percent of the total U.S. population is found in the Northeast, over two-thirds of Puerto Ricans living on the mainland reside there, and over one-third of all Asian Indians in the United States have settled in the region. Nonwhite populations total 37 percent in New York and 30 percent in New Jersey, with disproportionately large populations of blacks and Latinos. Yet the rest of the Northeast is still predominantly white (see Figure 15.3), including large percentages of the total U.S. population of Italians (51 percent), Ukrainians (51 percent), Portuguese (49 percent), French Canadians (45 percent), Russians (44 percent), Lithuanians (43 percent), Slovaks (40 percent), and Poles (37 percent). Maine, New Hampshire, and Vermont have the highest-percentage of white populations in the nation, with only 2 to 3 percent nonwhite. Nearly half of all American Jews also live in the Northeast, primarily in New York, New Jersey, and Massachusetts.

While most of the original inhabitants and settlers of the Northeast were farmers, nearly 8 in every 10 persons now living in the Northeast reside in urban areas. The population of the region is also among the oldest in the nation, and the Northeast has the smallest percentage of children under age 18. A disproportionate number of persons over the age of 65 make their homes in the region, more than 11 percent above the national average, with particularly high numbers in Pennsylvania and Rhode Island.

Socioeconomically, families in both New England and the Mid-Atlantic have incomes above the 1997 U.S. median. The percentage of persons falling below the poverty level ($16,400/year for a family of four) in the Northeast was slightly below the U.S. average, with especially low rates in New Hampshire and New Jersey. New York is an exception, with a rate nearly 25 percent higher than average. More than one in every four adults over the age of 25 have a college degree in the region, 10 percent above the national average; Massachusetts and Connecticut have some of the most

Figure 15.2
Regional divisions in the United States

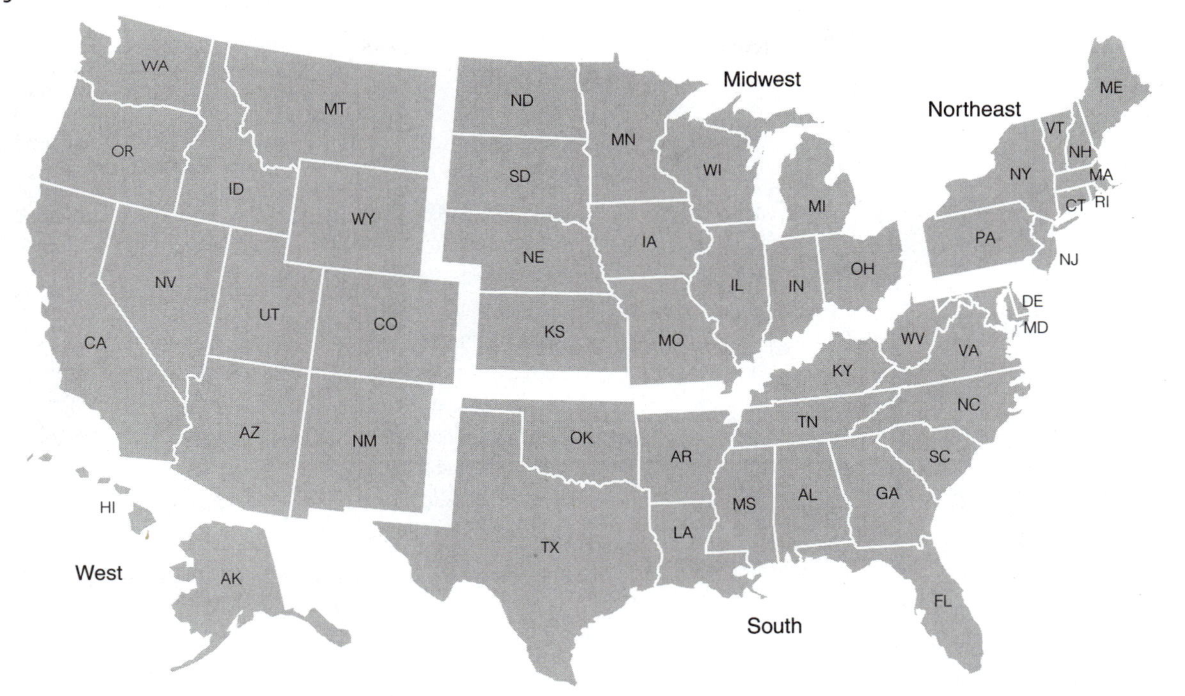

educated populations in the country, with 33 percent and 30 percent having college degrees, respectively.

Traditional Fare

The cooking of the Northeast features the abundance of the Atlantic, the plenty of native and introduced produce, and the freshwater wealth of the many rivers and lakes (see Table 15.2). In New England, seafood, such as clams, lobster, scallops and fish, especially cod, has been prominent. Indigenous game, including wild turkey and venison, supplemented the poultry, pork, and beef brought by early immigrants. The foundation of the diet was traditionally

■ Pickles, relishes, jams, and jellies were commonly prepared to preserve the fruits and vegetables of summer for use in winter.

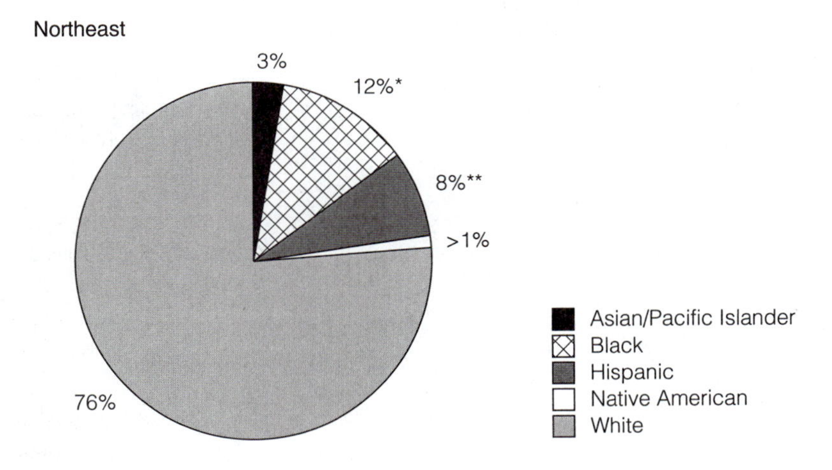

Figure 15.3
Demographics in the Northeast

Table 15.2
Northeastern Specialties

Group	Foods	Preparations
PROTEIN FOODS		
Milk/milk products	Cream; cheddar and cream cheese	Cream soups, sauces, puddings; ice cream
Meats/poultry/fish/eggs/legumes	Native game, particularly venison and turkey	New England boiled dinner; scrapple; red-flannel hash
	Preserved meats, such as corned beef, pastrami, salt pork, ham, bacon, sausages	
	Seafood prevalent, especially clams, lobster, oysters, scallops	Fish stews, soups; clam chowder; clam bakes; oyster or lobster loafs
	Salt and freshwater fish, such as salt cod and shad	Codcakes; shad bakes; *gefilte* fish
	Numerous beans (e.g., cranberry, kidney, lima)	Baked beans; succotash
CEREALS/GRAINS	Corn, wheat, rye	Cornmeal porridges, puddings, and breads
		Dumplings
		Baked goods, such as savory and sweet pies, cakes, doughnuts, waffles, pretzels, bagels
FRUITS/VEGETABLES	Apples, blueberries, cranberries, grapes	Applesauce, apple butter; fruit puddings, pies
	Cabbage, fiddlehead fern fronds, potatoes	Coleslaw; sauerkraut; ferns on toast; potatoes mashed, fried, creamed, baked, scalloped, hashed, as croquettes, salad
ADDITIONAL FOODS		
Seasonings	Salt, pepper, onions, saffron	
Nuts/seeds	Black walnuts, butternuts	
Beverages	Apple cider, hard cider, applejack; ale, beer; rum, whiskey; wine	New York State white, red, sparkling wines; sherry, port
Fats/oils	Lard, butter	
Sweeteners	Maple sugar; molasses	Maple sugar candies, maple syrup pie; chocolates

corn, and many dishes of the region reflect its importance. Beans have also made a substantial contribution. Root vegetables (potatoes, onions, beets, turnips, rutabagas, carrots, etc.) quickly added to the vegetable selection. Wild berries, including blueberries, cranberries, gooseberries, and cloud berries (which look like bleached blackberries), grapes, and beach plums were the main fruits consumed until the apple orchards planted by immigrants became productive. Maple sugar sweetened many foods in New England. Even when molasses and cane sugar became widely available, maple syrup was preferred for many dishes.

The warmer weather and fertile lands of the Mid-Atlantic states have provided more native foods than New England. The coastal waters offer clams, oysters, mussels, scallops, and crabs, while the estuaries shelter ducks, geese, and turtles. Passenger pigeons once darkened the skies with their massive numbers, and bison roamed the area around Lake Erie. Freshwater fish such as catfish, eels, pickerel, salmon, shad, smelt, trout, and whitefish were plentiful; at one time, shad was the most numerous of all freshwater fish in the United States. Both the flesh and roe were commonly eaten.

Introduced foods also flourished in the region. Cabbage, potatoes, yams, carrots, peas, apples, pears, cherries, peaches, and strawberries were easily grown. Tomatoes thrived in the hot summers. Although New Jersey is now one of the most industrialized states in the nation, it is still known as the Garden State due to the success of these early agricultural efforts. Wheat, which was difficult to grow in New England, did well in the Mid-Atlantic. At one time, New York provided all the wheat consumed in the Northeast and much of the South.

New England

The cuisine of New England has been shaped predominantly by Native American preparation techniques combined with British home-style cooking. Roasting, boiling, and stewing are preferred. Dishes are often made with cream, and strong seasonings are avoided. People of the region take pride in simple fare.

The immigrants of the early 17th century were mostly tradespeople, inexperienced in farming and husbandry. History abounds with tales of how the first settlers were dependent on the skills and generosity of local Indians in preventing starvation (see chapter 5 on Native Americans). Corn dishes were especially significant. Cornmeal porridge cooked into a mush-like consistency was a Native American food called *samp* by the early colonists. It was often prepared with another cornerstone of the Indian diet, beans. New Englanders used cornmeal to make an adaptation of the traditional English dish known as *hasty pudding*. The settlers would pour cornmeal porridge into a loaf pan to firm up overnight, then slice it and serve it topped with cream. This new dish, often flavored with maple syrup or molasses, was named *Indian pudding*.

Steamed, baked, and boiled puddings were eaten daily in New England homes. They were known as *grunts* (steamed dough and berries), *slumps* (baked puddings), and *flummeries* (a British molded oatmeal or custard pudding). As in England, a pudding could be savory or sweet and was generally served at the beginning of the meal. Breads were also a mainstay. Many were dense, baked without any leavening (reliable leavenings such as baking powder and baking soda were not available until the mid-1900s). Home-grown yeast from potatoes, hops, or the dregs of beer barrels was used in some recipes. Corn bread cooked in a skillet over the fire was most common. Rye, which grows well in cooler climates, was often combined with cornmeal to make a popular bread called *ryaninjun* (from *"Rye 'n' Injun"*). Stewed pumpkin was sometimes added for a moister loaf. *Boston brown bread* is a traditional recipe of the region, a steamed loaf made with rye flour and flavored with molasses.

Pies made with suet pastry were served at most meals. Savory kinds included an American version of the British steak and kidney pie, chicken pot pie (later topped with biscuits instead of pie crust), a ground pork and onion pie seasoned with allspice called *tourtière* introduced by French Canadians in the region (served traditionally at Christmas or New Year's), clam pie, lobster and oyster pie, and a salt cod "pie" covered with mashed potatoes. Sweet pies were also popular, especially apple pies, made with fresh apple slices, dried rings, or even applesauce. Berry pies were common, and blueberry pie remains a specialty in Maine. Mincemeat, a traditional English treat combining savory and sweet ingredients, was featured at many meals because the filling of meat, dried fruits, nuts, and rum or other alcoholic preservative aged well, becoming tastier over time.

Pork, cod, or beef flavored most main dishes. Long winters required that most meats and fish be preserved, and few recipes called for

■ Fudge was originally a maple sugar candy popular in New England in the 1800s. When cocoa became widely available, the chocolate version was developed.

■ In the old American West, *grunt* was not a pudding but a term for pork, as in "cluck and grunt" (eggs and bacon) for breakfast.

■ At what point pudding was served in the meal became a political issue during the late 1700s when conservative Federalists continued the tradition of eating it first, while more progressive Democrats (known as Republicans at the time) would eat it following the meat course (Lee, 1992).

■ *Anadama bread* is a modern-day corn bread made with mostly wheat flour, a little cornmeal, and sweetened with molasses. It is popular throughout New England. Stories about the name usually involve a woman named Anna who was either too lazy to make bread, so her husband damned her, or an outstanding baker who was damned by her husband when she died.

fresh cuts. Salt pork, bacon, smoked pork, dried salt cod, corned beef, and dried beef were common, usually braised or stewed with vegetables. The *New England boiled dinner* is typical. This one-pot meal is still popular throughout the region and usually includes corned beef brisket simmered for hours with potatoes, onions, carrots, turnips and, traditionally, beets. Cabbage is added toward the end of the cooking time. Seasoning is mild, often just a little black pepper. Leftovers are often chopped up the next day and heated in a skillet with a little cream and sometimes some bacon and more onions to make another New England specialty known as *red-flannel hash,* so-called because the cooked beets would bleed into the other ingredients during frying. A dried salt cod and potato version of the New England boiled dinner is prepared in Massachusetts, called *Cape Cod turkey. Plymouth succotash* is another example. Although this Native American dish (see chapter 5 for more information) is often associated with southern fare today, it was popular in the Northeast during colonial times. It combined corned beef, turkey or chicken, beans, corn, potatoes, and turnips. Other variations featured just vegetables, or combinations of meats, poultry, and fish.

Other popular dishes included dried beef rehydrated in boiling water and served with cream sauce over bread or potatoes (precursor to what is called *chipped beef* today), codfish cakes covered with egg-enriched cream sauce, and fried salt pork topped with cream gravy. In addition to meats and fish, beans were eaten regularly. Best known are baked beans flavored with molasses or maple syrup and salt pork, a recipe adapted from the Native American dish.

No discussion of New England fare is complete without mentioning the importance of shellfish in coastal regions. Lobster is a specialty, especially in Maine. The Indians of the area consumed the meat, used the discarded shells for fertilizer, and formed the claws into pipes. The meat was mostly added to mixed fish dishes by the British settlers. Lobster became more prestigious and more expensive in the following centuries due in part to overharvesting that resulted in scarcity and in part to the difficulty of transporting the fresh crustacean to upscale population centers inland. Clams, oysters, and scallops are also popular in New England. The clambake, in which clams, corn, and other items such as onions, potatoes, or lobster are steamed in a pit on the beach, began as a Native American tradition. Today a wood fire with large rocks stacked on top is built on the sand, and when the fire is embers and the stones are hot, the food is layered with wet seaweed, then covered with a tarpaulin until everything is cooked. Clams fill pies and are featured in the cream-based soup known as *clam chowder.* Those called *steamers* are just that—steamed and served with the broth and melted butter (a bucket of steamers is often the first course of a lobster dinner). Oysters were typically prepared with cream and breadcrumbs in a dish called "scalloped" oysters, or served in oyster stew. Today they are commonly broiled with bacon or breaded and deep-fried. Bay, sea, and Digby scallops are prepared similarly to oysters.

New England desserts are mostly fruit based. In addition to the puddings and pies already discussed, *pandowdies* (baked fruit layered with bread), shortbreads (fruit preserves, biscuits, and cream), and *roly-polys* (fruit rolled up in biscuit dough, then baked) were other favorites. Pound cakes and fruitcakes were enjoyed but were difficult to make before commercial leavening and reliable ovens were available.

Tea and apple cider were consumed daily in colonial times. Hard cider, an alcoholic beverage caused by the fermentation of sugars in apple cider, was also favored. Many New Englanders would start their day with a pint of beer or ale made from barley, corn, pumpkins, persimmons or spruce bark. Rum, as well as whiskey made from rye, was available. Wine from dandelions or gooseberries was a specialty, and an American version of the English drink called a *syllabub,* containing apple cider, sherry or wine, and whipped cream, was served at special occasions.

Mid-Atlantic

Many of the influences on New England fare are seen in the foods of the Mid-Atlantic states as well. Native American fare was combined with immigrant preferences to produce a new regional cuisine. Unlike New England, where most of the colonists were from England, many

■ Boston is still known as "Bean Town" due to its long association with baked beans.

■ Development of the railways led to demand for perishables, such as shellfish, in the rapidly growing cities. The first lobster was shipped to Chicago in 1842—it made it as far as Cleveland alive on ice, at which point it was boiled before further travel (Root & de Rochemont, 1974).

■ One New England dessert popular throughout the nation is chocolate chip cookies, which were created by Ruth Wakefield in 1930 at the Toll House Inn (the building dates back to 1709) in Whitman, Massachusetts.

Lobster is a specialty of Maine, though it is also trapped in other New England coastal states. (Jeff Greenberg/Visuals Unlimited, Inc.)

settlers in the Mid-Atlantic states came from the Netherlands and Germany. They provided a distinctively different flavor to foods, including a greater use of pork (especially sausages) and dairy products, more baked goods, and stronger seasonings. Later immigrants from southern and eastern Europe contributed many specialties. Further, the warmer climate and fertile farmlands offered a greater variety of ingredients to the cooks of New York, New Jersey, and Pennsylvania.

The Dutch in the mid-1600s brought wheat to the New York area, which at the time was known as New Netherland. They also grew barley, buckwheat, and rye. Although these were preferred grains, the Dutch used what they called *turkey wheat* (corn) to make a boiled milk and cornmeal porridge known as *Suppawn* that was eaten daily at breakfast. This same porridge was topped with meats and vegetables for lunch, then baked to make the hearty dish called *Hutspot,* an American adaptation of a stew common in the Netherlands. Dairy cattle provided ample milk, butter, and cheese. The Dutch were among the first settlers wealthy enough to import sugar, brandy, chocolate, and numerous spices, including pepper, cloves, cinnamon, and saffron. Many Dutch specialties of the region have made their way into American cooking, including pickled cabbage; *Kool sla* (from the Dutch word for "cabbage"), now known as coleslaw; and headcheese, a ball-shaped sausage made from the head and feet of the hog. Doughnuts, crullers, pancakes, and waffles were also introduced by the Dutch.

During the same period, German immigrants arrived in the United States. Many were religious outcasts (mostly Mennonites, with smaller numbers of Amish, Schwenkfelders, and other sects) who made their home in the tolerant colony of Pennsylvania (see chapter 7 on Europeans for more information). They became known as the Pennsylvania Dutch, a

■ **The word *cookie* is derived from the Dutch word for a small cake, *Koeckje.***

corruption of the German word *Deutsch,* which means "German." Although some German religious communities remained isolated (even to this day), many German immigrants gradually became integrated into the broader populations of Pennsylvania and surrounding states. Likewise, many German foods of the region have become an indistinguishable part of American cuisine.

Pork was the foundation of the German diet, from ham, pork chops, pork *schnitzel* (pounded into thin slices), bacon, salt pork, pickled pig's knuckles, *souse* (jellied pig's feet loaf), maw (stomach stuffed with meat and vegetables) and a German version of headcheese. Every part of the hog was used, and leftovers would be stretched with lima beans to make a Pennsylvania version of baked beans or with dried green beans and potatoes (*Bohne mit Schinken un'Grumberra*). The best-known leftovers dish is *scrapple,* still popular throughout the state. It is a combination of ground pork or sausage, cornmeal porridge, and spices that is formed into a loaf, sliced into thick slabs when firm, and fried in butter. It is typically served with fried eggs, applesauce, and maple syrup. Additionally, smoked and fresh sausages were consumed daily.

Chicken stews and soups, made substantial with home-made noodles or dumplings, were also popular with the Pennsylvania Dutch. Beef was used in the braised roast known as *Sauerbraten* and in the smoked, cured dried beef called *Bündnerfleisch.*

Asparagus, green peas, sugar peas (called Mennonite pod peas), and rhubarb are a few of the vegetables favored by the Pennsylvania Dutch. Potatoes are eaten mashed, fried, creamed, baked, scalloped, hashed, as croquettes, as dumplings, in stews and soups, and as potato salad. Cabbage is also ubiquitous, mostly as sauerkraut and slaw. Apples are particularly popular—fresh, as applesauce, in pastries, as cider, and in preserves like the thick, sweet spread known as *apple butter.* Many fruits and vegetables are pickled or preserved. Examples include spiced pears, pickled watermelon rind, sweet pickles, and corn relish. Dark rye breads, corn breads, yeast rolls, potato rolls, cinnamon rolls, *streuselkuchen* (coffeecakes with a sugar-crumb topping), doughnuts (called *Fastnachts*), and buckwheat pancakes are just a few of the baked goods found in the region. The Pennsylvania Dutch also make numerous desserts, especially pies (see chapter 7 on central Europeans for more information).

The hearty fare of the Dutch and Germans combined with many traditional items also found in New England, such as puddings, savory pies, and seafood soups and stews, to produce Mid-Atlantic cuisine. Later immigrants to the Mid-Atlantic region introduced foods that have become associated with certain cities and states (see the following "State Specialties" section). Notably, southern Italians in New York and New Jersey brought pizza and spaghetti with tomato–meat sauce; eastern European Jews introduced pastrami, smoked salmon, chopped liver, and other deli specialties.

■ "Better a burst stomach than wasted food" is a Pennsylvania Dutch saying.

■ Small half-moon pies were called *preaching pies* by the Pennsylvania Dutch because they were used to appease restless children during church services.

■ Angel food cake, also known as *angel cake,* is thought to be Pennsylvania Dutch in origin. It is leavened with egg whites.

■ Connecticut is nicknamed the "Nutmeg State" not because it produces the spice or uses it in any specialty but because early Yankee entrepreneurs from the region earned a reputation for selling fake wooden versions of whole nutmeg stored with the real spice so that they smelled authentic.

■ Some American troops of the Revolutionary War reputedly survived on barrels of shad packed in Connecticut.

State Specialties

Connecticut

The earliest European settlers in Connecticut were the Dutch who came from New Amsterdam (what is now New York City) and the English from the Massachusetts Bay Colony. The Dutch planted orchards and founded a dairy industry in the Hartford area, and the English established small farms. Although the state has become increasingly urbanized, some small agricultural concerns remain as a link to the past.

Fishing has provided the foods most associated with Connecticut. Shad, a flavorful relative of herring, was enjoyed by the Native Americans of the region but disdained by the earliest Europeans in the area due to its multitude of tiny, difficult-to-remove bones. Like the salmon, this ocean-dwelling fish spawns annually in freshwater rivers. Every spring the Juneberry plant, also known as the shadbush, blossoms, forecasting the arrival of the shad as they struggle up the Connecticut River. By the mid-18th century, Connecticut residents had changed their opinion of shad, and by the beginning of the 20th century, it had been fished to near extinction. River preservation programs have permitted a comeback, and the shad season is celebrated each June in the towns of the river valley. Traditionally the fish was most valued for its roe, fried quickly in butter. The Native Americans would plank the fish and slowly cook it at the edge of hot coals, a method still practiced today at shad bakes where the fillets are attached to a board with bacon, then smoked.

Lobster rolls are another Connecticut specialty. Lobster meat is mixed with mayonnaise, celery, onions, and lemon juice, then stuffed into toasted, fluffy, white bread buns. A similar dish, sautéed oysters packed into a loaf of bread, was known to the Puritans of Connecticut as *boxed oysters*.

Portuguese immigrants from the Azores arrived in the ports of Connecticut in the early 19th century to find work on whaling ships. They introduced their spicy sausages, such as *linguiça* and *chouriço,* to the region. Also popular are Portuguese white cornmeal breads traditional in the Azores, as well as sweet breads and other baked goods (see chapter 6 on southern Europeans for more information).

Maine

While lobster is undoubtedly the food most associated with Maine, it was fishing and whaling that supported the earliest immigrants from England, France, and later Germany. In successive waves of popularity, cod, mackerel, herring, and sardines were harvested until the schools were fished out or tastes changed. Today commercial fishermen catch mostly haddock and hake.

Lobster was so plentiful along the Maine shoreline in the 17th, 18th, and early 19th centuries that it would pile up 2 feet deep after a strong storm. The Native Americans found it worthy of consumption, but Europeans used it mostly to feed pigs and as fertilizer. Commercial trapping began in the late 1800s with the advent of shipping by train and the development of the canning industry. The lobster supply diminished rapidly, and today trapping regulations are strict. Maine is also known for its mussels, providing nearly two-thirds of the shellfish consumed in the nation, and its sea urchin roe (*uni*), served at local restaurants and sushi bars or exported to Japan (Lee, 1992).

Other Maine specialties include wild or low-bush blueberries, used mostly in baked goods (Maine grows nearly 100 percent of this variety in the United States); potatoes (until the 1950s, Maine was the primary producer of the tuber nationally); and fiddlehead fern fronds. The ferns are eaten fresh on toast or in fiddlehead pie, similar to a quiche (Jones, 1981). They are also sold canned and frozen.

The Maine bean pot is a well-known version of baked beans. It is based on the Indian method of placing the ingredients in a pot that is then buried in a pit over embers, a so-called "bean hole." During the winter, large vats of beans would be placed outside in the snow until frozen, then hung from the rafters of a shed so that just enough for a meal could be sliced off and reheated. Traditionalists insist that a Maine bean pot can only be made with native varieties such as yellow-eye, cranberry, or kidney beans, with one part salt pork for every three parts beans.

Massachusetts

Cod is to Massachusetts what lobster is to Maine. Cape Cod, the peninsula that curls out into the Atlantic, was so named in 1602 for an abundance of the fish. Most were salt-cured for keeping, and many were exported to Europe. Dried salt cod was exchanged for temperate fruit in the Mediterranean and for molasses in the Caribbean (which was used to make rum back in the colonies). Those in charge of the cod trade became wealthy and were nicknamed the "codfish aristocracy." Numerous recipes for the fish were concocted by the Puritan immigrants, including boiled and baked dishes (see Cape Cod turkey, described previously), soups, stews, hash, pies, and, most notably, *codfish cakes.* These cakes, which are also called *codfish balls,* are a sign of regional affiliation for some residents of the Boston area. Another specialty that frequently included dried salt cod in earlier times is chowder, a cream-based soup that may also be made with corn, potatoes, salt pork, and/or seafood. One version, Boston clam chowder, is known nationally. It is typically garnished with crackers, the slightly sweet, small, dry biscuits invented by Massachusetts sea captains for use on long journeys aboard ship. Those used in chowders were called common crackers, Boston crackers, or oyster crackers.

Baked beans are another dish well loved in Massachusetts, especially in Boston. Traditionally the Puritans prepared a large pot of beans on Saturday morning, simmered them over the fire all day, then ate them with Boston brown

■ Lobster is now so costly that many residents of Maine resent the fact that it has become the state icon, because a majority cannot afford to eat it, even lobster trappers (Lewis, 1989).

■ A large wooden replica of a fish, called the Sacred Cod, hangs in the Massachusetts State House.

■ The name *chowder* may come from the French term for an iron pot, *chaudière* or the Welsh word for a fish soup, *jowter.*

■ *Boston cream pie,* a favorite in New England, is not a pie. It is a custard-filled white cake covered with chocolate icing. It is probably derived from a popular colonial dessert called pudding cake that included cake, custard, and usually fruit or jam.

bread for dinner to start the Sabbath (observed from sundown to sundown). Leftovers were kept warm on the hearth for Sunday breakfast with codfish cakes and for Sunday lunch.

Europeans first discovered Jerusalem artichokes in Massachusetts, and the state is home to the hybrid American grape named after the town of Concord. The best-known produce of the region, however, is the cranberry, called fenberry and bounce berry by the English settlers. It was the Dutch who introduced the term *Kranbeere,* meaning crane berry, because the flower resembles the head of a crane. Cranberries grow exceptionally well in the sandy peat bogs of Cape Cod, where they were first cultivated in the early 1800s. Commercial production began in the middle of the century. Most cranberry farms today are still small operations under 10 acres.

■ Indian names for cranberries included *sassamanesh* and *ibimi* (meaning bitter or sour). They were eaten with maple syrup or dried and added to pemmican. Cranberries were also used therapeutically and as a dye.

■ Over 80 percent of the large bush blueberries sold fresh in the United States are grown in New Jersey.

■ Before the advent of inspection and pasteurization of milk, it was necessary to boil it before consumption.

■ Applejack is sometimes called *New Jersey lightning.*

■ One section of New Jersey that remains rural is the Pine Barrens, consisting of sandy soil, scrubby pines, and cedar swamplands. Many residents there depend on hunting and fishing for their daily fare. Venison and clams are specialties (Gillespie, 1984).

■ Delicately flavored Blue Point oysters are native to Long Island.

New Hampshire

The rugged landscape and bitter winters of New Hampshire discouraged extensive farming by early European immigrants to the region. Orchards and dairying were successful, however, and the region became known for its apples and high-quality butter. Apples are still exceptionally popular in the area, made into applesauce, fried apple rings, baked apples, apple butter, cider, and apple pie (sometimes drenched in maple syrup). French Canadian settlers brought pea soup, salmon pie, and a penchant for pork dishes.

New Hampshire was the state where two crops vital to New England were introduced. The first was potatoes, planted in 1719 by Scotch Irish immigrants in Londonderry, New Hampshire. Although potatoes are a New World food (see chapter 5), they only arrived in the United States after widespread acceptance in Ireland. The other transplant was turnips, brought by English settlers in the 1720s.

New Jersey

Over 150 crops are grown commercially in New Jersey, and it has a reputation as being the garden basket of New York City and other urban areas of the Northeast. Asparagus, peas, potatoes, tomatoes, blueberries, strawberries, and peaches are just a few of the successful fruits and vegetables in the region.

The state is famous for discoveries in agricultural research. Scientific work on hybridization has yielded new, improved varieties of peaches, tomatoes, and sweet potatoes. Food-processing techniques developed in New Jersey include condensed, canned soups (the Campbell Soup Company); inspected, bottled milk (the Borden Company); and the first application of pasteurization to milk at a small farm outside Princeton.

New Jersey is also the state where hard apple cider was first distilled to produce the apple brandy known as *applejack.* Another beverage, black tea, was introduced to the nation in the 1880s at Hoboken (the Thomas J. Lipton Company). Green tea had been the brew consumed in the colonies, but it fell out of favor following the Revolutionary War. The heartier flavor and convenience of tea bags reversed the trend.

New York

European settlers in New York discovered an abundance of native foods, such as deer, rabbits, quail, ducks, geese, and turkeys. Fish and seafood were plentiful: Oysters were eaten at every meal by the residents of New York City in the 18th century; littleneck and cherrystone clams were added to soups, stews, and pies; and shad was fished out of the bay, or plucked from the rivers during spawning.

The immigrants also found fertile fields and a climate conducive to many crops. Apples have been especially successful, and grapes grown in the four regions of the Finger Lakes, Lake Erie, the Hudson Valley, and Long Island support a thriving wine industry. New York is second only to California in production, including white (Chardonnay, White Riesling, and Seyval-Villard varietals), red (small amounts of Cabernet Sauvignon and Merlot), and sparkling wines as well as some fortified wines, such as sherry and port. Specialty vegetables like borrage, broccoli rabe, raddichio, and baby squashes can be found on the farms of the Hudson Valley.

The cooking of New York has been forged from immigrant fare. The Dutch brought their rib-sticking porridges, stews, and baked goods. The English and Germans also settled in the region with foods similar to those in New England and Pennsylvania. In particular, the Shakers, an English religious sect, contributed savory herb seasonings; new varieties of beans, squash, and other produce; and an interest in the health-promoting qualities of whole wheat bread. French Canadians followed, bringing *cassoulet* (white beans baked with pork and goose) and onion soup. Italians introduced mozzarella, ricotta, tomato–meat sauces, pizza, *calzone, cannoli, gelato,* and espresso. Many central Europeans arrived during the late 1800s, including Poles who settled in the Buffalo area and Russian Jews who came to New York City. *Pierogi* (stuffed dumplings), *kielbasa* sausage, *kulich* (an Easter specialty cake made with sweet yeast dough) *matzah, gefilte* fish, *blintzes* (thin pancakes served with sour cream or stuffed with sweetened cheese or fruit), and *latkes* (potato pancakes) are a few foods brought by these immigrants. Bagels, introduced by the Poles, were paired with an 1872 New York invention, cream cheese; the chewy, doughnut-shaped roll was popularized throughout the nation during the 1990s. Chinese immigrants at the end of the 19th century brought numerous Asian specialties; Middle Easterners and Caribbean Islanders have added further spice to the cuisine (see chapters on each ethnic group).

New York is noteworthy for the influence of restaurant fare. Taverns, boarding houses, oyster houses, and coffeehouses served the needs of those eating out in the late 18th and early 19th centuries. The first European-style bakery was opened in 1825, and delicatessens serving the Jewish community were established in the 1880s. Full-service continental-style restaurants became popular in the mid-1800s. By the turn of the century, New York City had become the gastronomic center of the nation. Many dishes created for elite diners are now American specialties, such as *Waldorf salad* (originally a mixture of apples and celery in mayonnaise served at the Waldorf-Astoria), *vichyssoise* (chilled leek and potato soup from the Ritz-Carlton), and the dessert *baked Alaska* (from Delmonico's).

Pennsylvania

Nowhere in the nation has German cooking been more significant than in Pennsylvania. Pennsylvania Dutch fare (described previously) was well accepted in the urban areas of the state, such as Pittsburgh, Allentown, Bethlehem, and Reading. Even Philadelphia, which was founded by English Quakers, favored German foods. Scrapple has become so associated with the city that it is often called Philadelphia scrapple, despite its rural origins. It is eaten for breakfast, often drizzled with catsup, and is used to make deep-fried croquettes or to stuff vegetables like green peppers and cabbage for dinner. *Lebanon bologna* is a Pennsylvania Dutch smoked beef sausage that has become a state specialty. It is traditionally sliced, battered or dipped in bread crumbs, fried, and served with sauerkraut and mashed potatoes. Other German foods of the region include sticky buns (a type of cinnamon roll); Christmas spice cookies (called *Pfeffernüsse* and *Lebkuchen*); and pretzels, which are sometimes eaten with vanilla ice cream. Although the origins are lost to history, *Philadelphia pepper pot,* a soup made with tripe, onions, potatoes, and black peppercorns, is most likely a Pennsylvania Dutch recipe and is sometimes served with dumplings. Cheesesteak (grilled strips of beef topped with American cheese and grilled onions in a toasted Italian roll), the quintessential Philadelphia sandwich, was supposedly invented during the 1930s when a hot dog pushcart vendor was accidentally sent beef instead of his standard order of frankfurters.

This is not to say that English foods have been ignored in Pennsylvania. The Quakers prepared many puddings and pies and were famous for their peach desserts. Seafood specialties included creamed oysters and delicately seasoned cream of clam soup. Poultry dishes, such as chicken with green peppers, wild birds baked with oysters, and stuffed turkey, were popular. The residents of colonial Philadelphia were criticized by some Puritans and even by a few members of their own sect for their unseemly interest in fine dining. Numerous private eating clubs opened in the city to accommodate gentlemen interested in socializing over a gourmet meal. French-influenced food was fashionable, including turtle soup (a cream-based soup sometimes flavored with Madeira), chicken or rabbit *fricassee, oeufs à la neige* (the meringue and

■ The duck that has become known nationally as Long Island duckling was first brought to the region by a Chinese sea merchant in 1873.

■ *Buffalo wings,* deep-fried chicken wings drenched in spicy (often using Tabasco sauce) seasoned butter and served with celery and blue cheese dressing, evolved in Buffalo-area bars during the 1960s. No one has established the exact origins of the appetizer.

■ *Vichyssoise* was created by chef Louis Diat around 1917 from a recipe popular during his childhood in the Vichy region of France. American chefs sought to change the name during World War II due to collaboration of the Vichy government with the Nazis, but the new name, *crème Gauloise,* was never accepted because the soup was already too well known by its original title (Jones, 1981).

■ It is said that pepper pot soup was invented out of the only ingredients available by a Pennsylvania Dutch chef working for General George Washington and the American troops at Valley Forge.

■ *Parisian mushrooms* (the white or brown button mushrooms prevalent in modern U.S. markets) were introduced to America by the English in the late 1800s. They were first cultivated in 1926 in Downing, Pennsylvania. Over 75 percent of mushrooms consumed in the nation today are produced in southeastern Pennsylvania.

■ The name *Johnnycake* may have come from corruption of *journey* cake or from *Shawnee* cake. During World War II, the Rhode Island legislature put an end to arguments over spelling and ingredients (yellow or white cornmeal) when it mandated that all fried corn bread in the state be spelled *Jonnycake* and contain only finely ground whitecap flint corn.

■ Both cattle and turkeys were herded on foot the 200 miles between the Green Mountains of Vermont and the meat markets of Boston in the fall.

■ The oldest continually operating cheese factory in the United States was founded in 1822 in Healdville, Vermont. Manufacturers have expanded into the European-style cheese (e.g., Brie, Camembert) market in recent years; the largest producer of feta cheese in the nation is found in Vermont.

custard dessert also known as *snow eggs* or *floating islands*), and the best cognac and fine wines of Bordeaux and Burgundy.

Pennsylvanians are fond of sweets. Philadelphia was one of the first cities in the nation to enjoy ice cream, perhaps as early as 1782. An ice cream parlor with frozen treats, cakes, syrups, and cordials was opened in 1800. The following years produced the first wholesale distributor of ice cream and the first ice cream soda. The city gained a reputation for a high-quality product made with cream, ripe fruit, and sugar, without any eggs, gelatin, arrowroot, or other thickeners. Hot chocolate and butterscotch sauces were popular toppings, and ice cream molded into flowers, fruits, animals, or holiday icons is still a specialty. Another confectionery contribution was affordable chocolate. Commercial production of chocolate for beverages and bonbons began in Pennsylvania during the late 1700s, although it was so costly it was considered a luxury item. Milton Hershey of Derry Church was the first manufacturer of chocolate for the mass market beginning in 1905, when he reduced his expenses by making uniform bars instead of fancy novelties. Two years later he introduced Hershey's Kisses.

William Penn, founder of Pennsylvania, established the first brewhouse in Philadelphia in 1685. It served what was known at the time as American beer, a heavy, top-fermented beverage similar to English ale. Two hundred years later, a German immigrant to Philadelphia founded the first brewery that made a lager-style, bottom-fermented beverage. The new, lighter beer was known as *lager,* or *pilsner* (see chapter 7 for more information). It was an immediate success with local Germans, prompting the opening of beer halls and beer gardens that became the social centers of the immigrant community (Mariani, 1983). Lager soon became synonymous with beer in the United States.

Rhode Island

Rhode Island is the smallest state in the nation and has always had limited arable land. Although it was self-sufficient in the 1700s, over the years it has been best known for its dairies, poultry farms (the meaty Rhode Island Red was developed in Adamsville during the 1850s), breweries, and distilleries. Famous for tolerance since its inception, the state has attracted a variety of immigrants, including English, Irish, French Canadians, Portuguese, Italians, and central Europeans. Many maintained their own food preferences, and a cuisine distinctive to Rhode Island has yet to emerge.

A few Rhode Island specialties date from colonial times. These include *Jonnycake* (often called *Johnnycake*), a fried cornmeal cake made with whitecap flint corn; and tomato-based red clam chowder, a soup that is inaccurately attributed to Manhattan. Clams and oysters were once abundant along the short Rhode Island coastline. Oysters were especially popular among the 19th-century elite, who served them in pies (raw oysters in cream sauce topped with biscuit dough); as patties; creamed, curried; and, for New Year's Eve pickled, with eggnog to wash them down.

Vermont

The earliest European settlers in Vermont were hardy individualists. The terrain was mountainous, the climate cold, and the territory was on the far frontier of the American colonies, claimed by surrounding Native Americans, the French, New York, and New Hampshire. In addition, Vermont was landlocked. It became much more dependent on meats and poultry than nearby states with easy access to seafood. Beef, pork, mutton (a by-product of the wool industry—today sheep are mostly raised for lamb meat), and turkey (domesticated in the early 1800s) are the foundation of Vermont fare. The New England boiled dinner made with corned beef may have originated in the state, and the accompanying fried apple rings or *Vermont fried pie* (fried applesauce turnovers flavored with cinnamon) is definitely a local touch. Ham was popular, made into turnovers, shortcakes (served on biscuits with an egg sauce), and pies.

Dairy foods were also significant in the development of the local cuisine. Cream enriches many dishes and cheese embellished other foods. Cheddar and milder Colby cheeses were most popular, and it is thought that apple pie topped with cheese originated in Vermont.

No food is as associated with Vermont as is maple syrup. The sweet sap of the sugar maple tree had long been used by Indians of the

Maple sap drips into a bucket positioned on a maple tree. Maple syrup is a popular sweetener in New England. (CORBIS/© Jonathan Blair.)

Northeast to cook beans and meats and to flavor cooked fruits and corn (see chapter 5 for more information). The syrup was an everyday sweetener in colonial kitchens throughout the region until cane sugar became more affordable. Maple syrup production peaked in the 1880s, and the sweet has since become a costly item. Vermont specialties include *sugar-on-snow* (syrup poured over fresh snow), maple sugar pie (a double- or single-crust, meringue-covered pie with a filling of cream, eggs, and syrup), and maple sugar candies.

The Midwest

Regional Profile

The Midwest is known as the Great Plains region of the United States. The earliest American settlers and European immigrants in the area found a vast, flat terrain covered by tall prairie grasses. Oak-wooded hills and low mountain ridges ringed the territory. The rich soil irrigated by the extensive Mississippi and Missouri river systems proved ideal for wheat, corn, and numerous fruits. The region is still renowned for its agricultural productivity, which is why it is nicknamed "America's breadbasket."

The Midwest encompasses twelve states with 23 percent of the total land area and just over 23 percent of the total U.S. population. It is divided into the East North Central region (Illinois, Indiana, Michigan, Ohio, and Wisconsin) and the West North Central region (Iowa, Kansas, Minnesota, Missouri, Nebraska, North Dakota, and South Dakota).

The states of the East North Central (ENC) area are bounded by the Great Lakes, which

■ Farming in the United States expanded from the land grant days of the 1860s until the end of World War I, when overseas demand dropped and overproduction reduced prices. By the time the Great Depression began in the 1930s, agriculture was already in a slump. Since 1935, U.S. farmers have increased efficiency and gone through a period of significant consolidation. The number of farms has gone from 6.8 million in 1935 to 1.9 million in 1994 (the lowest number since 1850); the average size of each farm has more than tripled.

■ The Homestead Act and subsequent legislation provided 160 to 480 acres of land for each head of household willing to settle in the new territories. Additional property was available for from $6 to $25 per acre.

temper the climate, ensuring milder weather than that experienced by other areas of the Midwest. Although the French were the first Europeans to explore the region, it was Americans from the Northeast states, as well as Virginia and Delaware, who were the first pioneers. Later immigrants from Germany, Switzerland, Scandinavia, central Europe, and the Cornwall area of England were attracted by the fishing, dairy, mining, lumber, and meat-packing industries. The West North Central (WNC) states are geographically near the center of North America, exposed to long winters, short summers, and extreme temperatures. Most Americans who settled the territory were homesteaders, interested in the inexpensive land and farming opportunities. They came from New England and the Mid-Atlantic states, followed by new immigrants from Germany, Scandinavia, and central Europe, particularly Poland.

As suggested by the history of immigration to the area, the Midwest has the largest percentage of whites in the nation (see Figure 15.4). Over half of all U.S. citizens of Czech and Norwegian ancestry live in the Midwest, as well as large numbers of people of Finnish (47 percent), Croatian (43 percent), Swedish (40 percent), German (39 percent), and Polish (37 percent) heritage. There are below average numbers of blacks throughout the Midwest; the exceptions are in Ohio, Michigan, and Illinois, which have slightly above average African American populations. Native Americans, Latinos, and Asian/Pacific Islanders are also underrepresented, although there is a large population of Latinos in Illinois, approaching the national average, and above average numbers of Native Americans in the Dakotas. Over all, Illinois has the most diverse population in the Midwest; nearly one in every three residents is nonwhite.

One immigrant group that has made the Midwest home is Laotians, including Hmong, who have arrived since the 1970s (see chapter 12 on Southeast Asians for immigration history and food habits). Over 40 percent of all Laotians counted in the 1990 U.S. census lived in the region, with large populations in Minnesota and Wisconsin. Other Southeast Asians have not tended to settle in the area, however. Additional recent immigrant populations of note include Asian Indians in Illinois and Michigan, Koreans in Illinois, and Russians in Illinois and Ohio.

Although almost one in every four Americans lives in the Midwest, the population is not evenly distributed throughout the region: 16 percent of all Americans live in the East North Central states, and only 7 percent live in the West North Central states. The population centers in ENC states are predominantly urban, whereas more than a third of persons living in WNC states are located in rural areas, and the populations of North and South Dakota are almost equally split between urban and rural residences. The 1997 median income level in the ENC states is slightly above the national

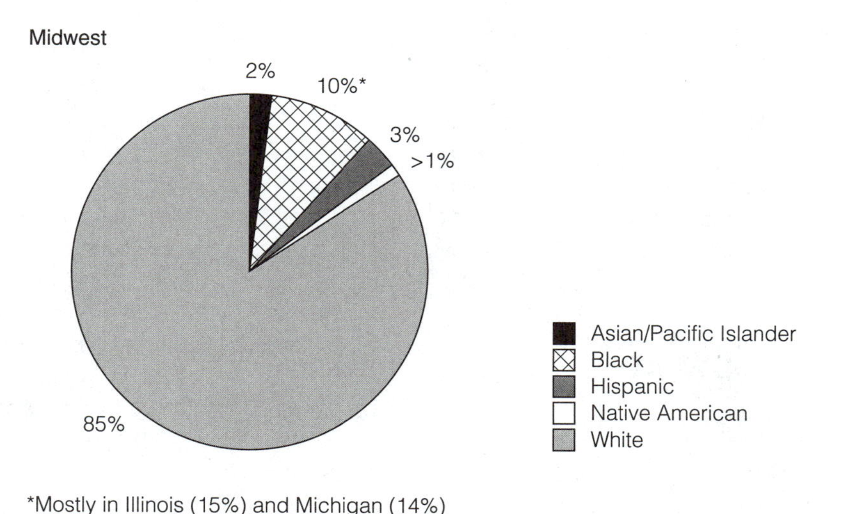

Figure 15.4
Demographics in the Midwest

average, while that of the WNC states is slightly below average. Poverty rates are 27 percent lower in the Midwest than in the country as a whole, especially in Wisconsin, Indiana, and Nebraska, which have averages up to 40 percent below the national figures for the period 1995–1997. More people earn their high school diplomas in the Midwest than among the total U.S. population, but slightly fewer receive college degrees than the national average.

Traditional Fare

Midwestern fare is usually described as no-frills homestead and farm food, exemplifying what is typically called American cuisine (see Table 15.3). Prime meat or poultry is prepared simply, served with vegetables, potatoes, and fresh bread. Hearty breakfasts start the day, robust soups and stews replenish body and spirit, and homespun desserts round out the meal.

The Midwest is a region of food traditions maintained over generations. Beef and pork are preferred meats, although better cuts are affordable today and variety cuts may be consumed less often than in settlement days. Canning (and more recently freezing) to preserve summer's bounty is still a common activity. Bread is sometimes baked at home, and pies make use of seasonal fruits. Midwestern hospitality, which began with festive sorghum pulls, group apple peelings, and canning parties, is continued through buffets, box socials, "pitch-in" or potluck suppers, strawberry socials, corn roasts, and fish boils popular throughout the region.

The earliest American settlement in the ENC region was in Marietta, Ohio in 1788. The people who came to the area from the original colonies were mostly farmers, who survived in their new homes on hogs, corn, beans, squash and pumpkins, cabbage, and potatoes. Corn was eaten at every meal, as porridge or as baked or fried breads. Sun-dried or smoked meat strips called *jerky* were adopted from the Indians of the region, used first for game such as venison, then later for beef. Other wild meats, such as duck, quail, rabbit, squirrel, woodchuck, and raccoon, were widely available. Meat pot pie with biscuit topping was a favorite preparation. Native fruits included persimmons, blueberries, bush cranberries, gooseberries, ground cherries, grapes, and many types of nuts. Later settlers brought wheat and oats, as well as apples, cherries, peaches, and berries. Fishing provided salmon, smelt, trout, and other freshwater fish; dairying, particularly cheesemaking, offered further food variety.

Each group of pioneers in the region brought favorite dishes. Baked beans and succotash were preferred by settlers from New England. The people from New York and Pennsylvania favored sausages, sauerkraut, pickles, and relishes. Those from southern states cooked fried chicken with cream gravy. European immigrants introduced the culinary specialties of their native lands, such as Cornish *pasties* in Michigan and Swiss cheese *fondue* in Indiana (see "State Specialties" sections). In areas where immigrants congregated in numbers, regional ethnic fare developed, including the Michigan Dutch (actually from the Netherlands, not Germans like the Pennsylvania Dutch) and the Ohio Germans.

In the WNC states the settlers of the mid–19th century came to farm the fertile land of Iowa, Kansas, Minnesota, and Missouri. Harsh winters and a scarcity of provisions limited variety in many early pioneer homes. Homemakers of the period describe burying melons in sand, which with luck would stay fresh until Christmas. Other cooks would prepare up to a hundred fresh fruit pies at a time, covering the extras with snow for use throughout the winter months. Parched corn, herbs, bark, or root brews replaced coffee (Luchetti, 1993). Prestige foods were often unavailable, so ample, even excessive, amounts of common foods became symbolic of hospitality in the midwestern frontier. The more western areas of the WNC states, which are drier and less suitable for crops, provided limited agricultural opportunities in wheat and cattle. The region attracted trappers, traders, and prospectors. Wild game, such as bear, buffalo, elk, deer, and small mammals, as well as turkeys, prairie chickens (grouse), quail, doves, and frogs, was hunted; the meat or oil was often sold in settlement towns.

As in the ENC region, pioneers from New England and the Mid-Atlantic states contributed dishes eaten frequently in their eastern homes: Baked beans and apple pies became as common as sausages and sauerkraut. One northeastern specialty that became surprisingly popular throughout the Midwest is oysters. By the mid-1800s, the live shellfish were shipped regularly to

- The "First Law of Midwestern Dining" is "Save room for dessert" (Mandel, 1996).

- Although buffet dining was known in medieval times, the American buffet meal may have originated with the traditional threshers' dinner served to feed migrant farmworkers in the Midwest.

- The highlight of a box social is when lunches or suppers in a basket (the "box") are auctioned off anonymously. The buyer then eats the box's contents with its preparer. In the 19th century, it was a way for single women and men to mingle and meet prospective partners.

- One of the most significant contributions to midwestern food was the founding of lager-style breweries in Cincinnati, St. Louis, Chicago, and Milwaukee by German immigrants to the region.

- *Coon cake* called for a dough made from flour, water, and raccoon fat fried in raccoon oil.

Table 15.3
Midwestern Specialties

Group	Foods	Preparations
PROTEIN FOODS		
Milk/milk products	Milk, buttermilk, butter, cream, cheeses	Cream gravy, fondue, *rømmegrøt, skyr*
Meats/poultry/fish/eggs/legumes	Native game, including buffalo, venison, beaver, raccoon, opossum, turkey, prairie chickens (grouse), pheasant	Jerky, *booyaw, Hasenpfeffer*
	Pork in all forms, especially salt pork, hams (country ham and Westphalian ham), and sausages (bratwurst, weinerwurst, kielbasa); beef	Ham with gravy, pork chops, barbecued pork, hot dogs, *Bubbat* Beef pot pie, stew, barbecued brisket, bierocks, pasties, Cincinnati chili
	Oysters shipped from the East Coast	
	Freshwater fish, especially smelt, sturgeon, trout, and whitefish	Fish boils, fried trout or smelt
	Dried beans	Baked beans with salt pork or bacon
CEREALS/GRAINS	Corn, wheat, rye, oats, wild rice	Corn breads, porridges; oatmeal; *bannocks;* rye breads, pumpernickel; biscuits, dumplings (including stuffed, such as pierogi and verenikas) Baked goods, especially fresh fruit pies (apple, cherry, persimmon, rhubarb), iced cakes, strawberry shortcakes, strudel, *kolaches,* butter cookies, pancakes, *aebelskivers,* Danish pastries
FRUITS/VEGETABLES	Apples, berries (blueberries, elderberries, strawberries), cherries, grapes, peaches, persimmons, rhubarb	Applesauce; apple butter, fritters, bread, salad; fried apples, candied apples; fruit jams and jellies
	Cabbage, onions, peas, potatoes, rutabagas, turnips, wild mushrooms	Sauerkraut, sauerkraut balls; coleslaw; potatoes boiled, fried, baked, as dumplings, salad; onion pie
ADDITIONAL FOODS		
Seasonings	Salt, pepper; parsley, dill; cinnamon, ginger, nutmeg, saffron; molasses	Most foods are preferred mildly spiced.
Nuts/seeds	Almonds, black walnuts, hickory nuts, pecans; poppy seeds	Nut pies, almond paste; nut candies; poppy seed cakes and pastries
Beverages	Apple juice, beer, wine, apple brandy	Lager-style (American) beer
Fats/oils	Butter, lard	
Sweeteners	Sugar, honey, molasses, sorghum	

the region packed in barrels filled with wet straw. One English visitor commented that the rich "consumed oysters and Champagne and the poor [ate] oysters and lager bier" (Mandel, 1996). One recipe from 1859 for small birds, such as magpies, suggested stuffing a breaded oyster into each bird before roasting over a hot fire (Luchetti, 1993). The new European immigrants who arrived were often scattered throughout the region due to the land grant process. Over time, foods such as Czech *kolaches,* Polish *stuffed cabbage* and *pierogi,* Hungarian *goulash,* and German Russian *bierocks* or *runsa* (also known as *pirozhki*) became familiar in midwestern cuisine.

State Specialties

Illinois

The Illinois frontier was known for corn production and hog husbandry. In the beginning of the 1800s the Chicago stockyards were built in city neighborhoods as a fattening-up rest stop for cattle being driven from Texas to the markets of the East Coast for slaughter. By the middle of the century, most beef was being butchered in Chicago for shipment to the rest of the nation. In the early 1930s a fourth major product, soybeans, was added to Illinois agriculture. Progress in soybean processing increased interest in the crop, which was used primarily for oil (to make edible products like margarine, mayonnaise, and salad dressings and for industrial use) and meal (the main source of protein in animal feed).

Meat, poultry, and cheese processing are the primary contributions of Illinois to midwestern fare, and are influential factors in American food habits. Culinary history is replete with stories about Phillip Armour, who made millions in pork sales when he founded the Chicago meatpacking industry; Gustavus Swift, who made his fortune in hams and sausages; Oscar Mayer, a German immigrant who got his start in the hot dog business as a butcher in Chicago; Louis Rich, a Russian immigrant who became involved in poultry processing and founded a turkey luncheon meat empire; and James Lewis Kraft, a grocery clerk who came up with the idea that home-delivered, uniform pieces of cheese would be more popular than freshly cut wedges from a large wheel. He later introduced processed and prepackaged cheeses, including *Velveeta.*

One, more unusual, Illinois food is horseradish. German immigrants brought the eastern European root to the Mississippi River basin area in the southern end of the state in

Sausages, introduced by German and Polish immigrants, have become so much a part of the midwestern diet that they are no longer considered ethnic fare. (Corbis/Sandy Felsenthal.)

■ Chemists working at Ford Motor Company in 1937 trying to create a synthetic wool discovered that soybeans could be made into analog protein products, such as bacon, by spinning vegetable protein fibers and flavoring them to taste like meat.

■ Deep-dish "Chicago-style" pizza is baked in a skillet. It is an American adaptation of the pizza brought to the region by Italian immigrants from Naples.

■ The International Horseradish Festival is held in Collinsville, Illinois, each June. The highlight of the weekend is the horseradish-eating contest, when timed participants eat as many hot dogs heaped with the freshly grated root as possible.

■ Popcorn dates back over 5,000 years in the Americas. It was prepared by Native Americans in New Mexico, the Caribbean, and, according to myth, at the first Thanksgiving in Plymouth.

■ A frontier recipe for *pork cake* dating from 1866 may have been the precursor to the Indiana specialty. It claimed that using lard and salted pork saves the expense of butter, milk, and eggs (Luchetti, 1993).

■ American blue cheese was developed at Iowa State University and first produced in the 1920s at Maytag Dairy Farms in Newton, Iowa.

■ In 1859 a group of German Lutherans, known as True Inspirationists, settled in seven Iowa villages to form what is known as the Amana Colonies. They lived communally, with everyone eating three enormous meals and two coffee breaks each day in a large dining hall. The weekly menu was set and included *Mehlspeisen* (literally "flour desserts," such as simple puddings) on Tuesdays and boiled beef every Wednesday. The Colonies now serve German specialties to visiting tourists in several large restaurants.

the late 19th century. The pungent, gnarly root thrives in the well-drained soil of the region. Over 10 million pounds are grown each year, nearly 85 percent of the global supply.

Indiana

The earliest Indiana pioneers came from Kentucky, North Carolina, and Virginia. They subsisted on acorn cakes and game meats until corn and wheat crops were established and pigs and sheep were plentiful. Agriculture was the primary industry in the state during the 18th and 19th centuries. Today Indiana is best known as the world's top popcorn producer.

Many dishes unique to Indiana feature foods from the state's early history. Whole roasted pigs, barbecued pork, pork roasts, chops, and hams are popular, as is *whole-hog sausage* made from lean cuts. Sausage patties and ham are common for breakfast, typically served with pancakes or biscuits, cream gravy, and fried apples. Batter-fried chicken is served with fried biscuits made with a yeast dough that puffs up into spheres when dropped into hot oil, then slathered with butter while still warm. Hungarians who settled in the South Bend area brought *veal paprika* and poppy seed cake. The Swiss founded the town of Vevay, where they were known for their cheese *fondue.* Vegetable pies, such as green tomato pie and onion pie, are specialties. Small, sweet, indigenous persimmons are used in steamed or baked puddings. Hickory nut cookies and sorghum fudge are other Indiana sweets. *Pork cake,* a moist dessert made with sausage or salt pork, molasses, brown sugar, flour, dried fruits, and spices, is a Christmas tradition in some Indiana homes, and *sweet cream pie,* filled with a heavily sweetened custard, is popular all year.

Iowa

Iowa is corn and pork country. Almost half of the agricultural land in the state is planted in corn, which supports the largest commercial hog production in the nation, with just over 14 million head equaling 25 percent of all U.S. pork. While some corn is exported and some is used for food products, including syrups, sugars, cornstarch, and distilled spirits, most of the grain goes for feed. It takes almost 4 pounds of corn to produce 1 pound of pork meat. Iowa is also the leading producer of soybeans.

Iowa's harsh winters and limited supplies meant that the earliest pioneers survived mostly on fried pork, boiled pork, potatoes, onions, and coffee. Milk was difficult to obtain, and one frontier woman from Norway reported in her diary that she had prepared an all-milk meal as a special treat one evening, featuring *tymelk* (clabbered, or curdled, milk sprinkled with bread crumbs and sugar) and *rømmegrøt* (a milk and flour baked pudding topped with sweetened sour cream). Those more familiar with native prairie foods, such as corn, berries, and plums, made good use of the ingredients. Once established, many settlers were self-sufficient. The Amish who settled around Kalona grew all of their own foods and butchered all of their own meat, traditions still practiced today. Corn bread with tomato juice gravy, stews or hashes with potatoes and peas, fried meats and eggs, and fresh fruit pies are common dishes in the community. The Dutch settlers of the town of Pella serve pea soup, Dutch-style bologna sausage, and special almond paste–filled cookies shaped into strips or S shapes. The Czechs are associated with wild mushroom dishes and *kolaches* filled with sweetened cheese, poppy seeds, or apples. The Swedes gained a reputation for "coffee," a light repast featuring fresh-brewed coffee served with butter cookies. The Germans still prepare Westphalian-style hams (brine-soaked, crusted in black pepper, then slowly smoked) and *Bratwurst.* Today the foods most associated with Iowa are the hearty farm dishes like pot roasts, hams, roasted chickens, mashed potatoes, fresh breads, and seasonal pies traditionally served at harvest meals.

Kansas

Culinary history in Kansas is dominated by the battle between indigenous buffalo herds and introduced cattle. At their peak, it is believed that over 30 million buffalo roamed the Great Plains, providing sustenance, clothing, shelter, and fuel to the Native Americans of the region. American settlers and European immigrants also ate what they called *hump-backed beef,* at least until other meats became widely available. However, the huge herds of buffalo interfered with expanding settlements, the railroads, and Texas cattlemen, who drove their longhorns through the prairie states to the slaughterhouses of the North and East. Abilene, Kansas, became the intersection between the cattle trails and the railroads. From the time the first stockyards were built in the city in the 1860s to the time the cattle drives were halted in the 1880s, over 4 million head came from Texas and were shipped from Abilene to the rest of the country. The buffalo were systematically eradicated, which provided unimpeded access to the grazing lands of the plains. Today Kansas is famous for its corn-fed beef, and cattle are the most important agricultural commodity in the state. Steaks, beef stews, barbecued beef, and hamburgers are all Kansas favorites.

Wheat is a significant crop in Kansas. The first settlers in the region were largely unsuccessful with the grain because their variety required more water than the Kansas climate provided. German Russian Mennonites, who had first migrated to southern Russia, came to the state beginning in 1874. They brought a variety of wheat they had obtained from farmers in neighboring Turkey. They called it Turkey Red and planted small fields of it near their new Kansas homes. It survived the dry conditions and provided hard wheat that was excellent for bread. Hybrids of the original grain, a red winter wheat, are grown throughout the state today.

The German Russian Mennonites also brought some German-style foods familiar in Pennsylvania Dutch areas, such as pancakes, sausages, and buttermilk pie. They introduced *bierocks* and *verenikas* served with cream gravy. Beef rolls stuffed with bacon, onions, and pickles are similar to German *Rouladen,* and sausage-filled buns called *Bubbat* are other Kansas dishes brought by the German Russians. The settlers who came from the Northeast cooked pies daily, when possible, using whatever was available for a filling. In Franklin County, French immigrants improvised with turkey stews and quail in garlic sauce (Lee, 1992).

Michigan

French explorers and traders were the first Europeans in the Michigan territory. They fished the Great Lakes for bass, trout, perch, smelt, and whitefish. The Upper Peninsula offered ample game, particularly deer. A specialty of that region with French roots is *booyaw* or *boolyaw* (perhaps from the French term *bouillon*), a game stew featuring venison or whatever else was available (including rabbit, woodchuck, squirrel, even duck), salt pork, carrots, potatoes, and onions. The French also introduced sour European cooking cherries to the state. Today Michigan is the leading producer of these tart fruits used in pies and preserves.

Pioneers from the Northeast settled in Michigan during the early 1800s. They mostly operated small farms, growing beans, corn, wheat, and garden vegetables. They raised dairy cows and kept a few hogs for meat. Michigan baked beans, which kept many families going during lean years, were topped with a layer of salt pork and baked until firm. The resulting loaf was sliced and served with catsup (Lee, 1992). Over time, farmers diversified, and Michigan became known for its agricultural variety, including celery, cucumbers (many used for pickles), berries, grapes, peaches, and pears. Apples are a significant crop, and local recipes include apple salad, apple meat loaf, apple bread, apple fritters, candied apples, and caramel apples.

In the 1840s the first U.S. mining boom began with the discovery of iron and copper in Michigan's Upper Peninsula. Miners from Cornwall who were attracted by the rush brought their unique lunch specialty called *pasties* to the region (see chapter 6 on Europeans). The complete meal in a turnover often featured venison in the Michigan versions, and turnips were a common vegetable filling. Apples were the most popular fruit used for the dessert end of the pastry. When immigrants from Finland arrived to work the mines in the following years, they adopted the dish, which was similar to *piiraat* and *kukko,* Finnish pastries filled with meat or fish, rice, and vegetables. Local Italians also included pasties in their daily meals. The origins of the dish are sometimes claimed by Finns in the region, although those of Cornish descent point out that the Finn pasties are nontraditional because the filling is mixed instead of layered and edges are crimped instead of using a decorative pastry rope; the Finns also drink buttermilk with their pasties (Lockwood & Lockwood, 1991). Other immigrant contributions to Michigan cuisine include Dutch *apple kock* (a yeast-risen dough topped with apple slices, sugar, and cinnamon), Czech *kolaches*

■ The baked German Russian yeast dough turnover (typically filled with beef, cabbage, and onions) common throughout the Great Plains was derived from the Russian *pirozhki.* It is called a *bierock* in Kansas and the midwestern regions east and south of that state and is known as a *runsa* in Nebraska and the northern Midwest areas.

■ The origins of meat loaf, the quintessential midwestern beef dish, are unknown. It may have come with German immigrants, who sometimes added rye bread as an extender or sauerkraut for moisture (Lee, 1992). It was first documented in an American cookbook in 1902. (Anderson, 1997).

■ Pasties were sometimes warmed in the mines with a "Cornish stove," a shovel held over a candle flame (Lee, 1992). Today's pasty shops, many with untraditional fillings (e.g., pizza ingredients), are common throughout the Upper Peninsula (U.P.) and in cities where "U.P.ers" (*Yoopers*) have settled, such as Detroit.

■ Grape-Nuts does not include grapes or nuts. C. W. Post invented the name because he thought slow-baking the little breadcrumbs that made up the cereal transformed the wheat starches into the sugar dextrose, which he knew as grape sugar. The resulting hard nuggets reminded him of nuts (Wyman, 1993).

■ The Danish community of Askov, Minnesota, has an annual festival commemorating the rutabaga. Although the tuber is commonly called a *Swede* or *Swedish turnip*, it is thought that Danes introduced the rutabaga to the region.

■ The Minnesota Massacre occurred in 1862 when U.S. troops pushed the Ojibwa Indians out of Minnesota, in part to provide access to wild rice stands.

■ Smoked buffalo tongue, a dish favored by some European immigrants, contributed to the precipitous decline of the U.S. buffalo population in the mid-1800s. In 1848 alone, 25,000 buffalo tongues were shipped to St. Louis.

■ One grizzly bear would provide 10 gallons of pure, rendered fat worth $75 in late 19th-century St. Louis (Luchetti, 1993).

and *strudel,* Ukrainian *borscht,* and German *Hollerstriebli* (elderberry flower fritters dipped in powdered sugar).

Michigan is probably best known for its role in the development of the U.S. cereal industry. The city of Battle Creek was home to two health sanitariums during the late nineteenth century. The first was founded by Seventh-Day Adventist leader Ellen Harmon White, who advocated vegetarianism. Her medical director was Dr. John Kellogg, inventor of corn flakes (see chapter 4 on food and religion). C. W. Post, a dissatisfied Kellogg patient, started his own health institute in Battle Creek. He created a coffee substitute, *Postum* (a blend of wheat berries, bran, and molasses) and a cereal based on his own recipe for digestive problems, called *Grape-Nuts.*

Minnesota

Although the cooler, northern climate of Minnesota limited agricultural opportunities, white settlers in the region found game and fish plentiful: deer, elk, moose, bear, goose, duck, trout, bass, walleye, and sturgeon are just a few examples. One pioneer commented on the bountiful, flavorful buffalo roasts and beaver tail jerky (Jones, 1981).

Some Europeans, particularly Germans and Scandinavians, were attracted to the weather, low mountains, and numerous lakes in the state. The Germans brought hogs and dairy cattle and introduced their dark rye breads, including pumpernickel. Sausages, hams, sauerkraut, and specialties such as *Hasenpfeffer* (stewed rabbit), *Spätzle* (tiny dumplings), *Maultaschen* (a sort of German ravioli filled with ground ham, eggs, onions, and sometimes spinach), and *Kuchen* (thin, sweet rectangles of dough covered with fruit, cottage cheese, or streusel) were other German dishes. Preserved fish were a mainstay for the Scandinavians. Pickled fish, smoked fish, and salt-cured fish were popular, particularly the Norwegian dish known as *lutefisk* (see chapter 7 on Scandinavians), served with butter and potatoes. Ham, bacon, Swedish meatballs, and Danish *fricadellar* (fried, breaded ground beef and veal patties) were consumed. Dark breads and the thin Norwegian potato pancake called *lefser* are still common, as are butter cookies (especially at Christmas) and Danish *aebleskivers,* traditionally served with chokecherry or blueberry syrup or jam. The Scandinavian concept of the smörgåsbord was introduced to the nation in Minnesota (see chapter 7 on Scandinavians for more details). Scottish settlers brought shortbread, and immigrants from Croatia, Slovenia, and Serbia added the grilled meats, eggplant, and cabbage dishes familiar in their native lands. Several ethnic communities in Minnesota maintain their culinary heritage at holidays and festivals, including the German Catholic city New Ulm and the Danish town Askov.

Minnesota is famous for its high-quality hard spring wheat and flour. Minneapolis, located on the Mississippi River, became a milling center in the mid-19th century. The replacement of millstones with steel rollers in 1847 increased yield, and the city became the world's largest milling center by the end of the century. It remains a major flour-producing region today. The state is also known for sugar processing, and the Red River Valley is nicknamed "America's Sugarbowl" due to the area's success in growing sugar beets. Minnesota is a major dairying region, one of the top butter and cheese producers in the nation. Wild rice, called a "grain" but actually the triangular-shaped seed of an aquatic grass found in the shallow rivers and lakes of the state, is another Minnesota specialty (see chapter 5 on Native Americans).

Missouri

The early white pioneers who came by covered wagon to the Missouri frontier were mostly from the South. French immigrants from Canada and German settlers from North Carolina, Pennsylvania, Ohio, Louisiana, and other states followed. The southerners brought fried chicken, buttermilk biscuits, cream gravy, and long-simmered greens. The French introduced *crêpes* and *brioche* to the region. They also made hard cider from apples, wine from native grapes, and brandy from peaches. German contributions included sauerkraut, pickled herring, and potato pancakes. The German settlers also established brewing and sausage making in Missouri. *Weinerwurst* (literally sausage from Vienna—probably the first hot dogs in America, also known as *weiners*), *Leberwurst* (liverwurst, a soft, spreadable sausage), and *Blutwurst* (blood sausage) were specialties. Smaller numbers of immigrants from Italy, Greece, Poland, Russia, and other central European countries came to Missouri in the late 1800s, forming ethnic communities throughout the state.

The hilly, wooded terrain found in much of Missouri is unsuited for grain crops or cattle, but excellent for pigs and nut trees. Pork, a meat prized by southerners and Germans alike, was used in numerous dishes. Missouri is well known for its *country hams,* which are cured with salt, then smoked, and hung to age in the cool winter months. The resulting meat is red, salty, and dry in texture. It is traditionally served with biscuits and *red-eyed gravy* made from ham drippings, coffee, and flour. Eastern black walnuts are native to the region. The nuts are strongly flavored with a slightly bitter aftertaste. They are the primary ingredient in black walnut pie. Pecans, too, are indigenous to Missouri. They are popular in pies, candies, cookies, and cakes. Over 7,000 beekeepers also take advantage of the woodlands in the state to provide another specialty of the region—honey.

Some of the most isolated hill country in Missouri is found in the Ozark Mountains. Contrary to immigration trends in urban areas, the people who came to the Ozarks gradually arrived from other states in small groups and were scattered throughout the region. They were known as *backwoodsmen,* and they existed on hunting, fishing, gathering, and cultivation of corn, beans, squash, and various tubers (Matson, 1994). Hogs were let loose to forage until butchering time in December or January. The people of the Ozarks were known for their stews made from opossum, raccoon, or squirrel. Sorghum was used to sweeten foods, ginger root was brewed for beer, and sassafras was steeped for tea. Today the Ozarks are best known as a vacation and retirement destination.

Unlike most of Missouri, Kansas City is renowned for its cattle and wheat, during the 19th century, it was known as "the town built on beef and bread." Yet its reputation was built not on agriculture but on the numerous grain elevators, flour mills, stockyards, and meat-packing plants built to process the wheat and beef shipped to the city. It is famous for its barbecued meats, particularly beef brisket, ribs, and ham, slowly rotisseried over hickory smoke until tender, with a spicy tomato-based sauce served on the side.

Nebraska

The Lewis and Clark expedition listed game, such as buffalo, deer, elk, turkey, grouse, ducks, and numerous fish, as Nebraska food resources in 1804. They also noted wild cherries, grapes, berries, and plums. Other explorers thought that Nebraska was more like a desert, unsuitable for agriculture. Although living off the land was possible for the nomadic Indian tribes of the region and for small groups of traders and trappers, the thousands of homesteaders who arrived with the opening of the railroad in the mid-1800s found pioneer life difficult. Corn was the most successful crop, and cornmeal replaced wheat flour in nearly all recipes. Salt pork and salted beef were staples; milk, eggs, and butter were limited until each settler could purchase livestock and poultry. Dried produce, such as apples, peaches, raisins, pumpkins, and squash, were more common than fresh fruits and vegetables. Irrigation increased agricultural variety in Nebraska. Today it is best known for livestock, including cattle, hogs, sheep, and poultry. Corn is the primary crop, but other grains, such as wheat, oats, barley, and sorghum, are also grown in significant quantities. Potatoes and sugar beets are planted in some areas, and in the driest regions, hay is produced for feed.

As in other midwestern states, immigrants to Nebraska brought their favorite recipes, adapting them to available ingredients. In Nebraska the Swiss introduced plum tarts and a specialty called *Thuna,* breadsticks topped with creamed greens thickened with flour. Czech settlers brought *jaternice* (pork sausage), *jelita* (blood sausage), and *houska* (a sweet, braided bread), as well as the popular kolaches. Swedish yeasted waffles and Hungarian chicken paprika are other examples. Meat stew with dumplings, buttermilk spoon bread, and oatmeal cookies are dishes introduced by homesteaders who came to Nebraska from the U.S. Northeast and South.

North Dakota

White settlement in North Dakota did not really begin until the 1870s, when railroads increased access and inexpensive land attracted farmers. Many immigrants came from Europe, especially Norwegians, Germans, and German Russians. They brought the dishes of their homelands, and today Norwegian food such as

■ Anti-German sentiment during World War II led some people to rename sauerkraut "Liberty Cabbage" (Matson, 1994).

■ In 1910, St. Louis considered itself the U.S. spaghetti capital, due to the 10 pasta factories located in the city.

■ Hostilities broke out between Missouri and Iowa when three hollow trees filled with honey hives were cut down on land claimed by both territories. Missouri won the Honey War and was able to extend its northern boundary into Iowa (Lee, 1992).

■ Arthur Bryant's is the most famous of the Kansas City barbecue restaurants. It was started by African American Charlie Bryant in the 1930s. He called it "a grease house" (Matson, 1994). Food writer Calvin Trillin (1974) has called it "the single best restaurant in the world." It was one of the first integrated businesses in the city.

lefse, *spekejøtt* (smoked, dried lamb), *rullepølse* (cold, spicy rolled beef), *søtsuppe* (fruit soup) and baked goods, including the large pyramid of almond paste and meringue rings called *kransekake.* Other immigrants, such as the Scots in the northern Red River Valley, enjoyed oatmeal porridge served with fresh berries and oatmeal cakes called *bannocks.* A large population of settlers from Iceland smoked mutton, made *skyr* (a sweet, cultured milk product similar to yogurt), fried *kleinur* (doughnuts), and baked *vinarterta* (a multilayered cardamom-flavored cake with fruit fillings) for dessert. The German Russians brought *runsa,* the name for bierocks in this region. The Scotch Irish introduced *colcannon* (mashed potatoes, onions, and cabbage), and the French Canadians came with croissants and cassoulet (see chapter 6 on northern Europeans).

North Dakota crops are grown mostly in the eastern half of the state and include spring and winter wheat, particularly durum wheat, which is used for semolina and durum flour. Barley, sunflowers, potatoes, oats, and beans are other agricultural products. The western half of the state is mostly used for cattle ranching.

■ Cream of Wheat cereal was created in 1893 by Diamond Milling Co. in Grand Forks, North Dakota, to make use of the wheat hearts, or middlings, produced in processing flour.

■ Pierogi is the Polish name for a stuffed, boiled dumpling that is traditionally filled with potatoes, cabbage, onion, and/or meat (it is called *pelmeri* by some Russians; *varenyky* by Ukranians; and *verenikas* by German Russian Mennonites). Pierogi producers in Ohio today offer such innovative fillings as taco and Cajun beef.

■ Cincinnati was nicknamed "Porkopolis" during the 19th century.

■ Some researchers speculate that the unique sweet-hot spicing of Cincinnati chili is a direct result of Tom Kiradjieff's culinary heritage, creating a flavor reminiscent of such eastern Mediterranean dishes as *pastitsio* or *moussaka* (Lloyd, 1981).

Ohio

Ohio was one of the first midwestern frontier regions settled by white pioneers, in part due to its proximity to the original colonial states and in part because land grants were offered to those who had served in the Revolutionary War (bounty lands) or whose homes had been burned by the British (the Firelands). The settlers found a wooded terrain that provided rich soil when cleared for farming. Game and fish were so plentiful that a compilation of recipes from 1880 called *Buckeye Cookery and Practical Housekeeping* included information on how to prepare raccoon, squirrel, opossum, skunks, and sturgeon, as well as how to smoke venison and roast bear meat.

The pioneers from New England brought their baked beans, chicken pot pies, and baked goods to Ohio. They used breadcrumbs to stuff meats, a practice still popular in the region (Lee, 1992). Central Europeans who immigrated contributed their specialties: The Czechs brought fish served with a black sauce made from prunes, raisins, and almonds, and the Slovenians brought *potica,* their national bread, flavored with nuts, seeds, or tarragon. The Poles, one of the largest ethnic groups to settle in the state, introduced favorites such as *kielbasa, pierogi,* and strudels. The most significant influence on Ohio cooking came from the numerous Germans who came to the state, beginning in the early 19th century. They popularized pork, sausage, ham, potato, and cabbage dishes, such as the unusual Ohio specialty called *sauerkraut balls,* which are deep-fried sauerkraut and ham fritters served with mustard sauce. The Germans opened beer gardens, restaurants, and bakeries throughout the region. Not only did they develop the brewing industry in Cincinnati; they also contributed to the first wine industry in the United States along the shores of Lake Erie. Many Ohio cities have retained German communities where German foods are still prevalent.

The state quickly became an agricultural center, producing corn, wheat, and more recently, soybeans. Hogs, cattle, dairy cows, and poultry are significant commodities in some regions. During the 1830s, Cincinnati produced more pork products than any other city worldwide, processing hogs from Indiana and Kentucky in addition to those from within the state. During the same period the state was also known for cheese, particularly the strong-smelling, soft, creamy cheese with a crust called *Liederkranz.* It was invented by the Monroe Cheese Company of Van Wert in 1926 as a substitute for German *Schlosskäse,* but its distinctive flavor has attracted its own devotees.

Ohio is probably best known nationally for a more recent creation, *Cincinnati chili.* In the 1920s, restaurants in the city would prepare chili with beans from the leftover meats of the day. It had a slightly unsavory reputation because one never knew what had gone into it. Tom Kiradjieff, a Macedonian immigrant, created a pure, all-beef version and started serving it over spaghetti in 1923. He boiled the beef first, then simmered it for hours in a tomato sauce flavored with a balanced blend of sweet spices (e.g., cinnamon, allspice, cloves, and nutmeg) and hot spices (garlic, cumin, black pepper, and chiles). It became so popular that chili parlors are commonly found throughout Ohio (and parts of nearby Kentucky) that serve the mild chili *one-way* (just the meaty stew), *two-way* (over spaghetti), *three-way* (spaghetti, chili, topped with grated cheese), *four-way* (spaghetti, chili, topped with cheese and diced onions), or *five-way* (spaghetti, chili, cheese, onions, and finished with kidney beans).

South Dakota

Unlike the farmers who flocked to North Dakota in search of inexpensive land, the first American settlers in South Dakota were miners and fortune hunters, who arrived after the 1870s gold strike in the Black Hills. Although there was abundant game, including deer, mountain goats, and geese, plentiful freshwater fish, and a variety of wild fruit, such as grapes, chokecherries, bush cranberries, juneberries, and buffalo berries, the miners were not hunters or gatherers. Food was scarce, and most survived on salt pork or bacon and beans. A full meal of pork roast, sourdough bread, baked potatoes, and possibly pie was a luxury.

Some immigrants noted the grazing potential of the surrounding land. Cattle ranching quickly became the primary agricultural activity in the state, and it remains so today. The first cattle were longhorns from Texas; they were soon joined by Scottish cattle, including Aberdeen, Angus, and Herefords. Immigrants from Scotland were soon exporting beef to their homeland. Hogs and sheep have also become important commodities in the state. Corn, wheat, rye, and sunflowers are grown in some sections.

Significant numbers of Swedes have settled in South Dakota, introducing the rutabaga to the region, as well as the smörgåsbord, almond cakes, and spice cookies. German Russians and Czechs brought their traditional foods. Other South Dakota specialties include lamb dishes and pheasant roasted with salt pork or prepared with onions in a casserole (Lee, 1992).

Wisconsin

Wisconsin fare assimilates native foods, agricultural abundance, and immigrant preferences. Although the state had been home to small numbers of trappers and traders in the 18th century, the first significant white settlement did not occur until the early 19th century, when farmers from New York, Pennsylvania, Ohio, and Indiana arrived. They established wheat as the primary crop in the region. Farming opportunities, mining jobs, and the lumber industry attracted large numbers of immigrants from Germany, Scandinavia, Great Britain (especially Cornwall, Ireland, and Wales), Poland, Russia, and Switzerland in the mid-1800s. Cities and towns were founded along Lake Michigan (including Milwaukee), many with ethnic neighborhoods. Each group contributed specialties to the cuisine, such as Cornish pasties and saffron cakes, Scandinavian fish boils, and Polish poppy seed pastries.

The Germans established the state as a major pork producer during the 1800s. Hams and sausages have become ubiquitous in the region, so much so that they have lost most of their ethnic associations, and cities such as Sheboygan call themselves the "Bratwurst Capital of the World." Today meat processing is still big business; Patrick Cudahy, Jones Dairy Farms, Hillshire Farms, and Oscar Mayer all operate in the state. Another famous Wisconsin food contributed by the Germans is beer. The first breweries were located in the southwestern section of the state and produced the ales and stouts favored by English settlers. By the middle of the 19th century, ten breweries producing German-style lagers had been founded in Milwaukee, including plants owned by Frederich Miller, Frederich Pabst, and Joseph Schlitz. At the beginning of the 1900s there were over 300 breweries statewide, but Prohibition and consolidation in recent years have reduced that number to 8.

Despite renown in sausages and beer, Wisconsin is reputed nationally for its dairy industry. It is the leading U.S. producer of milk, sweetened condensed milk, butter, and cheese. Dairying was sparked by the arrival of Swiss farmers to the state following the 1844 crop failure in Switzerland. They brought their expertise in breeding livestock and making cheese. *Colby,* a hard cheese similar to Cheddar, is an original Wisconsin cheese that was created in 1885. Another variety developed in Wisconsin is *brick,* a semisoft cheese with holes and a flavor described as sweet, nutty, and spicy. It is mild when fresh but becomes stronger with age. Italian cheeses, including ricotta, mozzarella, provolone, Romano, and Parmesan, are specialties of northern Wisconsin, while blue cheese is made in the caves near Milwaukee.

Wheat has given way to vegetables throughout much of Wisconsin. Cabbage, carrots, corn,

■ The ring-necked pheasant was introduced to the United States from China when a shipment was sent to Oregon. It has naturalized throughout much of the nation and is the state bird of South Dakota.

■ One of the most popular cookbooks in U.S. history, and one of the first to include ethnic and religious specialties, was written by Lizzie Black Kander, president of the Milwaukee Jewish Mission. She taught cooking classes to poor immigrants at The Settlement, the mission's social house. Her compilation of recipes, called *The Settlement Cook Book,* was first published in 1901. The last revision was in 1976; it is now out of print.

■ Fish boils were begun by Scandinavians as an efficient way to feed the workers in lumber camps. A typical recipe includes 100 pounds of fish (usually whitefish), an equal amount of potatoes, 40 pounds of salt, onions—and a pot big enough to boil everything together. Today fish boils at the edge of Lake Michigan are annual tourist events in Wisconsin.

■ Wisconsin produces over 2 billion pounds of cheese each year, 40 percent of all cheese consumed nationally.

cucumbers, green beans, lima beans, and peas are significant crops in the state. One-third of the cabbage grown for sauerkraut in the United States comes from Wisconsin.

The South

Regional Profile

■ One song in the South goes, "South is my blood and South my bone/So haply formed on pork and pone . . . You say:'South? Where can that be?'/Chewing my sugar cane, I repeat:/'Why in all we like to cook and eat.'" (Walter, 1971, p. 19).

Most southerners say the South is more an attitude than a location. This perhaps explains why there are so many definitions of the region, such as those states below the historic Mason–Dixon line or those south of the culinary "grits" line (the divide between where grits are eaten and where they aren't). While no one questions that Alabama, Arkansas, Florida, Georgia, Louisiana, Kentucky, Maryland, Mississippi, the Carolinas, Tennessee, Virginia, and West Virginia are clearly part of the South, the borderline states of Delaware, Missouri, Oklahoma, and Texas can be argued for inclusion either way. Using the U.S. government definition, Missouri is considered part of the Midwest, while Delaware, Oklahoma, and Texas are part of the South.

The lands of the South are varied. They include the fertile coastal plains along the Atlantic and Gulf coasts, the rolling hills leading up to the mountains (called the *Piedmont* in most states), the rugged Appalachian and Ozark mountain territories, the lowlands of the Mississippi Delta, and the high desert plains of the western reaches. The climate also ranges from the warm, moderate Atlantic states to the hot, humid Gulf Coast states to the hot, dry weather in parts of Texas and Oklahoma.

The development of the South was in many ways independent from that of the northern United States. During colonial times, southern states were predominantly agricultural, growing tobacco, wheat, corn, rice, and indigo (a blue dye). The plantation system that emerged in the coastal regions was characterized by commercial farms owned by aristocratic English or French immigrants and worked by African slaves. Each plantation was a self-sufficient, independent operation providing cash crops and food products for use by each household. It was a comfortable, leisurely lifestyle for the upper classes, enlivened by occasional visits to the cultural centers of Atlanta, Charleston, or New Orleans. For the lower classes, which included the slaves and the poor farmers of the inland hill and mountain regions, it was a hand-to-mouth existence.

During the time the northern areas of the country became more urbanized and industrialized, the South remained agricultural, adding cotton as a major crop. Waves of immigrants from central, eastern, and southern Europe who arrived in the United States looking for jobs settled mostly in the North, where manufacturing opportunities existed. The South was largely isolated from changes occurring in the North. Differences of opinion regarding the role of the federal government in state issues, particularly slavery, led to the Civil War in the mid-1800s. After losing the war, the South regrouped in the late 19th century. The traditions and practices that give the South its character became more important than ever. The South continues to preserve its identity, in part, through its cuisine.

Over one-third (35 percent) of Americans make their home in the South, the highest percentage of the U.S. population in any region. It is divided into the South Atlantic states of Delaware, Florida, Georgia, Maryland, North Carolina, South Carolina, Virginia, West Virginia, and the District of Columbia; the East South Central states of Alabama, Kentucky, Mississippi, and Tennessee; and the West South Central states of Arkansas, Louisiana, Oklahoma, and Texas. Over all, the South has below average numbers of Asians and Pacific Islanders, Latinos, and Native Americans, but above average numbers of African Americans: 52 percent of all U.S. blacks live in the South (see Table 15.4). However, the very size of the southern population means that significant numbers of most ethnic groups reside in the region. For example, 30 percent of all Native Americans and Latinos live in the South; both Florida and Texas host above average populations of Latinos. While only 15 percent of U.S. Asians are found in the South, larger numbers of some groups, such as Vietnamese, Pakistanis, and Asian Indians, reside there. In addition to African Americans, groups with disproportionately large representation in the South include persons of British ancestry (39 percent of those in the United States), Scotch Irish ancestry (47 percent), Cuban ancestry (69 percent), and Cajun ancestry (91 percent) (see Figure 15.5).

Although the population of the South is now mostly urban, it ranges from the lowest national percentage of urban dwellers (in West

Table 15.4
Southern Specialties

Group	Foods	Preparations
PROTEIN FOODS		
Milk/milk products	Buttermilk, milk	Cream gravy
Meat/poultry/fish/eggs/legumes	Native game, including buffalo, venison, raccoon, opossum, badger, squirrel, turkey, ducks, alligator, diamondback terrapin	*Brunswick stew,* squirrel stew, possum 'n' taters, turtle soup
	Pork in all forms, especially country-cured and Smithfield hams; beef, mutton, kid	Ham on beaten biscuits, sliced ham and red-eye gravy; barbecued pork; *souse* (head cheese); chitterlings; Texas-style barbecued beef, *chili con carne;* son-of-a-bitch stew; *cabrito*
	Chicken	Fried chicken with cream gravy, chicken and dumplings
	Crab (blue, stone), crawfish, conch, oysters, shrimp, spiny lobster; ocean fish, such as mullet, pompano, shad; freshwater fish, particularly catfish	Crab, shrimp, or crawfish boils; crab cakes; she-crab soup; conch chowder; oyster stew; shrimp *pilau;* shrimp Creole; jambalaya; gumbo; *étouffée;* fish muddle; fried catfish
	Chicken eggs	Scrambled eggs and brains, scrambled eggs and ramps
	Dried beans; peanuts	Baked beans, butter bean custard; peanut soup, peanut brittle
CEREALS/GRAINS	Corn, rice, wheat, buckwheat	Hominy, grits, corn pone, hush puppies, corn bread, spoon bread; rice pilaus; beaten biscuits; buttermilk or sour milk biscuits; buckwheat pancakes
FRUITS/VEGETABLES	Apples, huckleberries, key limes, oranges, mayhaw, peaches, watermelon	Preserves and pickles; fried pies; key lime pie; ambrosia; peach pie
	Wild greens (cochan, creases, dandelion, dock, lamb's quarters, poke, sorrel, and ramp), domesticated greens (e.g., mustard, turnip), black-eyed peas, cabbage, okra, ramps, sweet potatoes	Greens simmered with fat back or salt pork, consumed with pot likker; poke salad (sallet); fried ramps; hoppin' John; coleslaw, fried okra, okra stews; sweet potato pie
ADDITIONAL FOODS		
Seasonings	Chile peppers (especially bird's eye); *filé;* celery, garlic, onions, green peppers; bourbon, sherry, whiskey	Pepper sherry, chili powder, hot sauce; High Holy mayonnaise; barbecue sauce
Nuts/seeds	Black walnuts, hickory nuts, pecans; sesame (benne) seeds	Nut cakes, brittles, glazed pecans, pecan pie, pralines; sesame seed candies and cookies
Beverages	Buttermilk; bourbon, corn whiskey, Sherry, Tennessee whiskey	Whiskey and bourbon are added to barbecue sauces, baked goods, candies
Fats/oils	Lard	
Sweeteners	Sorghum syrup	Used over pancakes, grits, corn bread, in coffee

Virginia, where only 36 percent of residents live in cities), to the highest percentage of urban dwellers in the country (100 percent in the District of Columbia, all in the city of Washington, D.C.). Other states with high urban populations include Florida, Maryland, and Texas. States with low rates of urban residency (where approximately half of homes are rural) are Mississippi, North Carolina, Kentucky, and South Carolina.

Figure 15.5
Demographics in the South

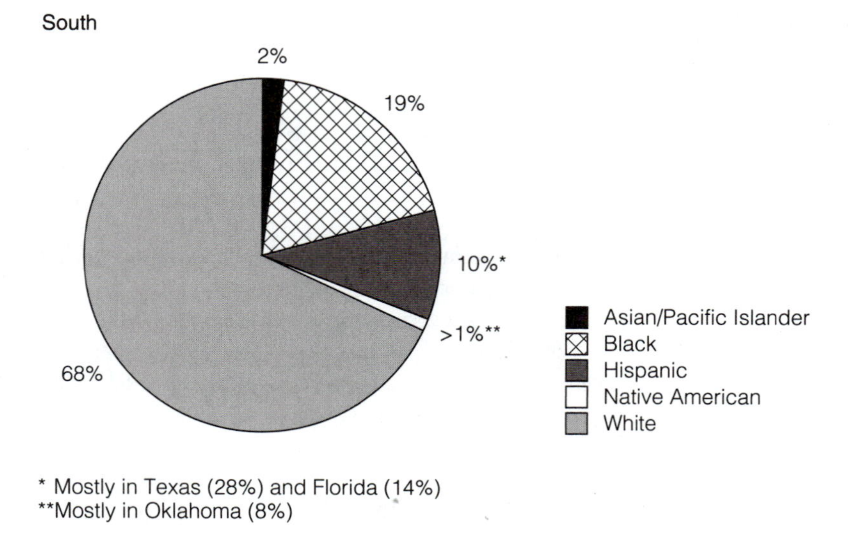

■ "We eat everything but the squeal" is a saying in the South about pigs.

Median income levels in the South fall below those of the nation as a whole. Some states show particularly low median incomes, such as West Virginia (27 percent below average) and Arkansas (26 percent below average). Exceptions in the region with above average median incomes are Maryland, Virginia, and Delaware. As might be expected from these income figures, high rates of poverty exist in the South as well, 10 percent above the national average in 1995 to 1997. Especially high numbers of persons live below the poverty line in the District of Columbia (22 percent), Mississippi (20 percent), and Louisiana (19 percent).

Some of the lowest rates of educational attainment and some of the highest rates are found in the South. The lowest high school graduation rates in the nation are in Kentucky and Louisiana, with Arkansas, Alabama, South Carolina, and West Virginia not far behind. Washington, D.C. has the highest percentage of college completion in the country—over one in every three residents (34 percent) has a bachelor's degree. Maryland and Virginia also have higher than average numbers of persons with college degrees. States with particularly low percentages of college completion include Alabama and West Virginia.

The population of the South is notable for its high numbers of Christians. Over half of all residents (57 percent) belong to a Christian faith, the highest percentage in the nation. Numbers are even higher in the East South Central States and the West South Central states: over 70 percent of persons in Alabama, Mississippi, and Louisiana are Christian.

Traditional Fare

The foods most associated with the South reflect both the bounty of the plantation and the scarcity of the slave diet. Corn dishes, rice, all parts of the hog, sweet potatoes, and greens began as the foundation of southern fare and remain characteristic components today. (See chapter 8 on African Americans for more information on slave foods and their influence on southern cooking.)

The southern lifestyle has fostered a culture of graciousness and cordiality. The isolation of the plantations meant socialization was limited in frequency but lengthy in duration. Hours of travel to nearby homes typically resulted in overnight visits or extended stays. Parties, balls, picnics, barbecues, and seafood feasts were all occasions for get-togethers. For the slaves, Sunday meals with extended kin were the primary way to maintain family connections. In the hills and mountains of the South the difficulties of subsistence farming necessitated friendly relationships between neighbors. Poor families often survived through regular sharing of food. As a result of these conditions, the South has become synonymous with hospitality.

The first European explorers in the South Atlantic states were the Spanish, who arrived in Florida in 1513 and founded St. Augustine in

Fried chicken is the quintessential dish of the South. It is often served with cream gravy and biscuits. (Corbis)

1565. They were soon followed by the English, who started in Virginia and spread north and south along the Atlantic coastline during the 17th and 18th centuries into Delaware, Maryland, the Carolinas, Georgia, and eventually into Florida. The Native American population at the time numbered in the hundreds of thousands, including the members of the Powhatan, Cherokee, Chickasaw, Choctaw, Creek, and Seminole tribes (see chapter 5 on Native Americans).

The white settlers discovered a region with plentiful fruits, nuts, game, fish, and seafood. Native strawberries, blackberries, blueberries, huckleberries, ground cherries, persimmons, Scuppernong grapes, beechnuts, hickory nuts, and pecans covered the land. Bream, catfish, perch, pike, and trout filled the rivers, while oysters, clams, and crab were abundant along the coast. In Florida, pompano, red snapper, shrimp, spiny lobster, and conch were widely available. Diamondback terrapin, sea turtles, and alligators were found in many waterways, and bear, deer, opossum, rabbits, raccoons, squirrels, turkey, grouse, ducks, and quail were prevalent in woodland areas. Indians of the region grew corn, beans, pumpkin, squash, sweet potatoes, and sunflowers.

Most of the first white settlers in the region were farmers who established plantations. They brought wheat, hogs, cattle, poultry, cabbage, potatoes, and fruit trees, including apples and peaches. Africans were imported as laborers. They introduced southern staples, such as peanuts, okra, watermelon, and sesame seeds, and taught the farmers of the lowland coastal areas how to successfully grow and harvest rice (Hess, 1992).

These traditional foods of the Native Americans, the European settlers, and the African slaves combined to create the cooking of the Atlantic South. Hot breads were the cornerstone of every meal, primarily corn breads (see Table 15.5) or biscuits. Country hams, ribs, fatback, cracklings, and chitterlings were produced from hogs. A Native American game dish adopted by the southern settlers, *Brunswick stew,* became a mainstay throughout the region. There are many variations, but most contain chicken, ham or salt pork, corn, beans, potatoes, onions, tomatoes, and lots of black pepper. The quintessential dish of the South is chicken dredged in cornmeal or flour and fried in lard, traditionally served with cream gravy. Long-grained rice was cooked as *pilau* with meat, poultry, greens, peas, beans, or vegetables. (*Hoppin' John,* rice with black-eyed peas, is the most famous dish of this kind.) Oyster roasts (similar to a New England clam bake) and *muddle* (a stew of miscellaneous fish with potatoes and onions) capitalized on coastal resources. Fresh milk, buttermilk, coffee, and tea were the usual beverages.

■ South settlers liked opossum because of its fatty meat. It was typically baked with sweet potatoes, colloquially called *possum'n' taters.*

■ Many sources note that the Native Americans of the South cultivated melons. Melons are not native to the New World, although it is possible they were brought to Florida by the original Spanish explorers, in which case the Indians may have been growing them for perhaps 100 years by the time white settlers arrived from the north.

■ *Pilau* was probably introduced to the South by French Huguenots who settled in South Carolina during the 17th century.

Table 15.5
Southern Corn Breads

Corn bread	Made with white cornmeal, eggs, and water. No sugar is added. Baked in a pan, sliced into squares, served with butter, honey, or sorghum syrup.
Cracklin' bread	Usually yellow cornmeal bread with added pork cracklings for flavor, traditionally cooked in a frying pan on the stove
Spoon bread	Yellow cornmeal bread made with eggs and milk. Baked slowly in a pan until golden crust forms on top and center remains custardlike.
Corn pone	Yellow cornmeal and water (lard added if available) mixed into a stiff dough, formed into sticks (sometimes called *corn sticks*) or patties (sometimes called *hoecakes*) and cooked in a skillet
Hush puppies	Yellow cornmeal and water dough, with added egg and buttermilk if available, formed into balls and deep-fried

■ Corn whiskey became the most popular alcoholic beverage of the Appalachians and in most areas of the South because apples for applejack were in short supply and barley for beer did not grow well in the region.

■ *Poke salad* is a cooked salad, which is an English tradition. Poke greens are parboiled, then fried in bacon or fatback grease until tender. They are seasoned with salt, pepper, and hot sauce or vinegar.

■ *Leather britches* were immature beans in their pods that were strung on string to dry. They would be soaked overnight before cooking, then boiled with a piece of bacon or fatback for flavoring.

Desserts have always had a place on the southern table. Tea breads and cakes (such as *Sally Lunn* cake, popular throughout the region), fruitcakes, and pies (peach, pecan, and sweet potato are most associated with the South) were common. Puddings and custards were an everyday treat. Many were made with leftover cornmeal, rice, or bread; chocolate was a favorite, but costly, so it was used only at special occasions. Candies, such as divinity with nuts and nut brittles, were specialties.

During the early 1800s, Scotch Irish immigrants searching for religious freedom began making their homes in the Blue Ridge, Cumberland and Great Smoky mountains of the Appalachians. They also spread west to the Kentucky and Tennessee frontier. English and some Welsh settlers moved from the coastal Atlantic South states inland to the Piedmont. Germans from Pennsylvania traveled south along the Shenandoah Valley into Virginia and North Carolina. "Hogs 'n' hominy" (pork and corn) kept the pioneers going until they established small farms. Frontier meals were robust. For example, the noon meal might consist of ham, bacon or sausage, chicken or grouse, game meat, dumplings or biscuits, corn bread or grits, gravy, sweet potatoes, and boiled greens served with coffee, milk, or corn whiskey.

Traditionally every bit of the pig was consumed on Appalachian farms, including the snout, or "rooter" (which was roasted), the tail (which was added to stews), and the brains (which were usually boiled, mashed, and scrambled with eggs). Most families kept a dairy cow and a breeding cow for a few calves each year. Fresh beef was preferred. When a cow was slaughtered, it would be shared with neighbors, who would later return the favor. Game of all kinds was popular, but especially squirrel, rabbit, raccoon, opossum, turtle, and frogs. Wild greens were well loved by adults in the Appalachians, but not so popular with children (Page & Wigginton, 1984). Poke, cochan (coneflower), creases (similar to watercress), dandelion, lamb's quarters, dock, sorrel, and ramp (a type of wild onion) are just a few of the varieties. Domesticated greens like mustard and turnip greens were also common. Other than greens, beans, hominy, sweet potatoes, potatoes, okra, and beets were the most frequently consumed vegetables. Corn bread (sometimes with cracklings added), biscuits, dumplings, and/or grits were served at every meal. Watermelon was a favorite, eaten fresh or preserved as pickles or jam. Honey was the most common sweetener, even consumed alone as a dessert. Sorghum syrup was also used. Many of these foods are still favored in the Appalachians today, and some are used as home remedies (see Table 15.6, "Selected Appalachian Botanical Remedies").

The early fare of the East South Central and West South Central states was similar to that of the Atlantic states, but with more significant French overtones. Immigrants from France settled in the Gulf Coast region during the 17th century, and at the end of the 18th

Table 15.6
Selected Appalachian Botanical Remedies Often Prepared Mixed with Whiskey

Scientific Name	Common Name	Parts	Traditional Use
Acer negundo	Box elder; ash-leaved maple	Bark, leaves	Kidney problems
Acorus calamus ☠	Cally root; calamus root; sweet flag; sweet myrtle; sweet rush; rat root	Root	Urinary tract disorders; constipation; stomach pain; colic; colds; asthma
Apocynum cannabinum ☠	Indian hemp; dogbane; north-and-south hemp	Root	Liver problems; stomach ailments; constipation; migraine; colds; arthritis; rheumatism; abortion
Aralia racemosa	Spikenard; goutwort; Indian root	Root	Liver disorders; kidney ailments
Arctium lappa	Burdock	Root; leaves	Kidney complaints; stomach problems; blood toxins; asthma
Aristolochia serpentaria ☠	Snakeroot; snakeweed; serpentary; heartleaf	Root; leaves	Heart conditions; fever; parasites
Chenopodium ambrosioides ☠	Wormseed; wormifuge; vermifuge; Mexican tea	Leaves; seeds	Parasites
Chimaphila maculata	Striped wintergreen; spotted wintergreen; ratsbane; rat's vein	Root; leaves	Kidney disorders; liver ailments; cold; fever; whooping cough (tea may be used as a temporary substitute for breast milk)
Epigaea repens	Mayflower; trailing arbutus; gravel plant; Easter flower	Leaves	Kidney ailments; blood toxins; colds
Eupatorium perfoliatum ☠	Thoughtwort; feverwort; Indian sage; boneset	Root; leaves; whole plant	Kidney problems; rheumatism; cold; flu; measles
Gaultheria procumbens ☹	Wintergreen; mountain tea; checkerberry; tea berry, box berry, deerberry; spiceberry	Leaves	Diuretic; urinary tract problems; kidney disorders; blood ailments; muscle aches; halitosis
Glechoma hederacea; Nepeta glechoma	Gill-over-the-ground; ground ivy; alehoof; haymaids; creeping charlie	Whole plant	Skin rashes; coughs; colds (tea may be used as a temporary substitute for breast milk)
Hydrastis canadensis ☹	Yellowroot; goldenseal; yellow Indian paint; ground raspberry; eyeroot; eyebalm	Root	Kidney complaints; stomach pain; urinary tract infections; liver complaints; weak blood; lung problems; TB; lumbago; bursitis; arthritis; rheumatism; mouth infections; sore throats; eye injuries
Ilex opaca ☠	American holly; hulver bush	Root, bark	Heart conditions; cough
Morus rubra	Red mulberry; mulberry	Bark	Liver problems
Nepeta cateria	Catnip; cat mint	Leaves	Colic; diarrhea; stomach problems; whooping cough; colds; toothache; skin rash (tea may be used as a temporary substitute for breast milk)
Panax quinquefolium	Sang; American gingseng; fivefinger; tartar root; redberry	Root	Hypertension; heart problems; stomach conditions; colic; liver ailments; aging; TB; rheumatism
Phytolacca americana ☠	Poke; poke salat; pokeweed; pokeberry; jalap; cancerroot; inkberry	Root; leaves	Kidney disorders; liver ailments; stomach problems; lung conditions; weak blood; arthritis; bursitis; rheumatism; croup; skin cancer; skin rash; toothache; muscle pain

continued

Table 15.6
Selected Appalachian Botanical Remedies Often Prepared Mixed with Whiskey—continued

Scientific Name	Common Name	Parts	Traditional Use
Plantago lanceolata	Ribwort; he plantain; narrowleaf plantain; English plantain	Leaves	Kidney ailments; urinary tract infections
Podophyllum peltatum ☠	Mandrake; mayapple; devil's apple; hog apple; umbrella plant	Root	Liver conditions; stomach problems; constipation; lung ailments; rheumatism; arthritis; bursitis
Prunus persica ☠	Peach	Bark, leaves	Diabetes; nausea; parasites; allergies; inflammation; cough; fever; induce labor
Prunus serotina ☠	Black cherry; wild cherry	Bark	Hypertension; stomach ailments; nausea; anorexia; liver disorders; lung conditions; cough; croup; blood toxins; rheumatism
Rubus spp.	Blackberry; dewberry; raspberry	Root, berries	Diarrhea; stomach pain; hemorrhage; anorexia; induce vomiting; induce labor
Rumex crispus	Dock; sour dock; yellow dock; dock sorrel; bloodwort	Root, leaves	Hypotension; liver disorders; stomach complaints; lung problems; arthritis; bursitis; rheumatism
Salix spp. ☹	Pussywillow; swamp willow; possum bush	Bark, leaves	Kidney conditions; diarrhea; fever; colds; bedwetting
Sambucus canadensis ☠	Elderberry; common elder; American elder; sweet elder	Bark, leaves, berries	Liver ailments; stomach complaints; lung problems; rheumatism; bursitis; arthritis; bedwetting
Sanguinaria canadensis ☠	Bloodroot; red root; puccoon	Root	Kidney conditions; liver problems; stomach pain; gastric ulcers; weak blood; arthritis; rheumatism; bursitis; lung ailments; cough
Sassafras albidum	Sassifras; sassafac; sassafrax	Root	Hypertension; blood toxins; weak blood; stomach ailments; measles
Taraxacum officinale ☠	Dandelion	Root, leaves	Heart conditions; liver problems; kidney ailments; stomach complaints; lung problems; rheumatism; bursitis; arthritis

Note: Data on some plants are very limited; adverse effects may occur even if not indicated.
Key: ☠, all or some parts reported to be harmful or toxic in large quantities or specific preparations; ☹, may be contraindicated in some medical conditions/with certain prescription drugs.

century, French Acadians from Canada relocated to Louisiana (see chapter 6 on Europeans for more information). They were joined by white American and English settlers arriving from the north. Plantation life in the region was similar to that of the Atlantic South, except that it was more dependent on cotton than on tobacco.

Pork and corn remained key to the cooking of the East and West South Central region. The corn bread in this area is made from white cornmeal without the addition of sugar. The French added their recipes for soups, stews, fricassees, and baked goods to the southern mix, as well as their appreciation for good eating. The resulting cuisine is found in some form throughout the Gulf Coast, from Mobile, Alabama to Beaumont, Texas. The French factor accounts for such adopted and adapted specialties as *bouillabaisse* (a French fish stew);

fish cooked *en papillote* (in paper packets with a velouté sauce); and *sauce mahonaise* (homemade mayonnaise), particularly *High Holy mayonnaise* (a fanciful, Anglicized term for *aïoli*) made with fresh garlic and served with shrimp or cucumbers. Creole cuisine, a blending of French, Spanish, African, English, and Native American cooking, is unique to New Orleans. It is a complex fare with many refined dishes; celery, tomatoes, bell peppers, onions, and garlic are the hallmark flavorings. Cajun fare, created by the French Acadians, is mostly limited to the bayou country of Louisiana, though its gumbos, jambalayas, and étouffées have become popular throughout the region (see chapter 6 on northern Europeans for more information on Creole and Cajun foods). While Oklahoma and Texas are both southern in attitude and enjoy many specialties of the South, such as grits, greens, Gulf Coast seafood, and Brunswick stews, their dishes are also influenced by Native American, central European, and Latino cooking. Beef is the dominant meat; barbecue is prevalent; and hot, spicy seasoning emboldens their dishes.

State Specialties

Alabama

The first Europeans in Alabama came by sea. In the early 1700s, France claimed the Louisiana Territory (including Alabama) and established a fort in Mobile. Young French women sent to marry the soldiers reputedly brought sophisticated culinary skills that they applied to native fish and game. Slaves imported for work on plantations and in city homes added their cooking preferences to the cuisine. Frontiersmen, most of whom were second- and third-generation Americans from Tennessee and the Carolinas, settled the inland hill country during the late 18th and early 19th centuries. The pioneers cleared substantial amounts of forested hillside in hopes of planting profitable crops, but large-scale agriculture in the area was mostly unsuccessful, and the new settlers eked out livings on small, family farms.

French influence was limited to the coast, where seafood was a primary ingredient in the cuisine. Shrimp were especially prevalent, prepared fried, boiled in seasoned water, with rémoulade sauce, and stuffed into *mirleton* (chayote squash), avocados, and other vegetables. Plump, local oysters, called Bon Secour oysters, were plentiful and popular throughout the Gulf Coast. In the 1800s, pushcart vendors sold them fresh on the streets of Mobile, and oyster bars served a dozen at a time: the first nine were eaten with lemon juice, the last three with horseradish.

The foods most associated with Alabama are traditional, inland dishes, however, such as fried chicken, chicken and dumplings, ham balls (fried fritters), Brunswick stew (made with a whole hog's head), and game dishes, especially those made with squirrels or frogs. Biscuits and corn breads were eaten daily. Sweet potato biscuits, crackling bread, and *hoecake bread* (cornmeal and water cooked in a frying pan) were everyday favorites; *beaten biscuits,* which required the dough to be hit repeatedly with a rolling pin to produce pockets of air for leavening, were prepared for special occasions. Hominy, greens, okra, green beans, butter beans, peas, rutabagas, and turnips were typical side dishes. Seasonal pies were popular, especially dewberry (the first ripe fruit of the summer season) and peach. Fried pies, a southern specialty, are thought to have originated in Alabama. Small circles of pie crust are filled with fruit (typically peaches or peach preserves in Alabama), then folded into a half-moon shape, crimped, deep-fried, and sprinkled with powdered sugar. Butter bean custard is an unusual dessert specialty of the region, made with mashed butter beans cooked as a sweet pudding flavored with cinnamon, cloves, and nutmeg. Buttermilk and coffee were preferred before the advent of soft drinks and iced tea—now these later beverages are common. Corn whiskey and wine from native grapes (including Scuppernongs and Muscadines) were consumed often. For parties and other social events, eggnog, punch, and *posset* (a thick eggnog served hot) were popular.

Arkansas

Arkansas is at the crossroads of the South, the Southwest, and the Midwest. The diverse terrain includes the fertile alluvial plains of the Mississippi River in the southeast of the state, the dry pasturelands of the southwest, the orchards and wheatfields of the northwest, and

■ Oyster gumbo in Alabama includes cabbage, tomatoes, and a touch of molasses; no okra or filé is used to thicken the stew.

■ Beef did not become widely available in Alabama until the 1950s. The only beef dishes that were standards in traditional fare made use of inexpensive variety cuts: oxtail soup and calf's head stew (Solomon & Solomon, 1979).

■ Peaches were brought to Arkansas by the Cherokee who had relocated in the state after being forced from their homes in the Atlantic South.

the rocky hills and mountains of the Ozarks in the northeast. The French established a trading post on the Mississippi River in 1686. Trappers made the post their base of operations and earned their living selling bear and buffalo products, such as meat, tallow, oil, and furs, to the residents of New Orleans. A few hardy farmers from Tennessee and Kentucky began settling in the northern parts of the state, and cotton plantations were founded in the southern region by pioneers from Mississippi. Substantial, widespread white settlement did not occur until the 1830s, however, when the Native Americans living in the area were forced into Oklahoma.

Settlers were mostly of English or Scotch Irish heritage, and they brought the foods they prepared in their home states, such as cured hams, sausages, baking soda biscuits, and molasses pies from the north, and fried chicken, buttermilk biscuits, sweet potatoes, and peach cobblers from the south. Barbecued beef and pinto beans are found in the areas of the state adjacent to Texas (Lee, 1992), and in the Ozarks the fare is similar to that found in the Missouri section of the mountains (see previous section), with pork, game meats (especially baked opossum and raccoon), corn, beans, and greens the foundation of the diet. Some Arkansas specialties are pork chops with cream gravy (sometimes made with bits of sausage in it), pan-fried chicken that is then baked with a Creole sauce, and strawberry shortcakes made with crisp biscuits drenched in strawberry juice and topped with fresh berries and whipped cream.

In 1904, rice was introduced in the swampy region between the Arkansas and Mississippi rivers. German Americans from Illinois and Iowa moved into the area to establish paddies, founding cities with names like Stuttgart and Wiener. Today Arkansas leads the nation in rice production. Ducks, which make their winter homes in the dormant rice fields, attract numerous hunters to the state each year. They are a specialty in the region, roasted over a fire, baked with bacon and basted with wine or port, and prepared as gumbo. Catfish have long been an Arkansas favorite, dredged in cornmeal and fried, or in catfish stew. Catfish is traditionally served with *hush puppies* (deep-fried cornmeal biscuits) and coleslaw. The state is the second largest producer of domesticated catfish in the nation.

■ Catfish farming was started in Arkansas to control the quality of the fish (which is affected by what it eats); the Association of Catfish Farmers of America headquarters is in Little Rock.

■ Recent immigrants to Delaware include numerous Central Americans, particularly Guatemalans, who are attracted to jobs in the poultry industry.

■ The three original Native American groups in the region numbered over 350,000 when the Spanish arrived but were completely wiped out by the mid-1800s due primarily to disease. The Seminole, many of whom live on five Florida reservations, were originally Creek Indians who entered the state in the early 1700s.

Delaware

The Dutch and the Swedes both established trading posts in Delaware in the early 1600s. Settlers in the region enjoyed the plentiful seafood, game, and wild fruit available. They planted gardens with vegetables and fruit trees and were particularly successful with corn. They were followed by larger numbers of English and German immigrants, whose influences on Delaware foods can be seen in some of the favorite dishes in the region. Heavily smoked and long-aged hams, sour milk biscuits, corn pudding, and cauliflower with custard sauce are a few specialties. Fish and shellfish were popular, including a local version of the fish stew called muddle, as well as steamed oysters, clams, and crabs.

Today poultry and eggs are the primary agricultural products in Delaware. The first "broilers" in the nation were marketed by Cecil Steele in Ocean View in 1923. Due to a mix-up, she received 10 times more chicks than she ordered for her egg business, and she decided to sell them when they reached 2 to 2½ pounds at about 16 weeks. This was far younger than most chickens were sold and produced a tender bird that could be roasted or broiled instead of stewed. It was the beginning of a national industry, supported by the 1938 founding of the first large poultry processing plant in Frankfort, which made slaughtered, plucked, and cleaned chickens available. Today broiled chicken is the state dish of Delaware.

Florida

The Spanish explorers who arrived in Florida during the 1500s found a region of low rolling hills, massive marshes and swamps, numerous lakes, and dense woodlands. They established several small settlements, including the city of St. Augustine, but did not maintain a presence after the region was ceded to England. While under British rule, ambitious English immigrants came to the state in search of land for tobacco and cotton crops. One entrepreneur recruited workers from the Mediterranean, especially Greece, Italy, and the island of Minorca, to support his indigo plantation in New Smyrna. These immigrants brought eggplant, lemons, and olives to the area, all of which survived in small gardens.

Although tobacco and cotton were commercial failures, the settlers discovered that the groves of orange trees that had taken root after introduction by the Spanish provided a profitable living. Farmers from Alabama, Georgia, and the Carolinas had soon established the Florida citrus industry. Grapefruit were hybridized from pummelos that had been brought from the Caribbean, and other citrus fruits, such as tangerines, tangelos, and Persian (also known as Tahiti) limes, were introduced. Key limes, small, thin-skinned yellow limes with juicy, green flesh, were discovered in the Florida Keys. It is not known where the limes came from, but it is assumed that they drifted to the islands from the Caribbean. They are grown mostly in home gardens and are renowned for their tangy flavor. Today, 70 percent of the U.S. citrus crop is grown in Florida (nearly all the oranges are processed into juice). Florida is also known for other subtropical crops, such as avocados, guavas, kumquats, mangoes, papaya, and pineapples, as well as early ripening crops such as tomatoes and strawberries. Sugar cane is grown in the south of the state, and sabal palmetto palms grow like weeds, providing the delicacy known as hearts of palm.

The Florida waterways and coastline offered a profusion of fish and game to early settlers. Red snapper, pompano (a very large, meaty fish), mullet, and tarpon were a few of the fish commonly available; seafood included shrimp (several varieties), spiny lobster (similar to those of New England, but without claws), conch (a large mollusk), and stone crabs (only the very large claw is eaten—the claw is removed when the crab is caught, and then the crab is thrown back in the water to grow a new one). Green sea turtle and freshwater turtles, frogs, and alligators were additional game meats.

Many of Florida's specialties have developed out of this unique larder. Red snapper fillets are baked with orange juice. Pompano is stuffed with shrimp, seasoned with Sherry, and baked, or prepared *en papillote* (with a nod to the French influence of the Gulf Coast states). Spiny lobster tails are stuffed with fish and grilled, while stone crab claws are traditionally boiled and served with garlic butter or mustard sauce. Rock shrimp, a hard-shelled, white shrimp that tastes like a cross between lobster and shrimp, has become a trendy restaurant item throughout the country. Green turtle steaks are grilled or baked, and the meat is also used in stew and soups. Conch fritters and conch chowder (made with onions and tomatoes, seasoned with Worchestershire sauce, oregano, and bay leaves) are popular. Key lime pie and *ambrosia,* a dessert made from sliced oranges and grated coconut, have become favorites throughout the South.

Other Florida dishes originated with the early and later immigrants to the state. In addition to oranges, the Spanish explorers also brought black beans and chickpeas (garbanzo beans), which were common in Cuba. Few dishes have survived in Florida fare from the Spanish period, but two soups in the region are thought to have Spanish roots: *garbanzo* (dried chickpeas, vegetables, and spicy Spanish chorizo sausage) and *vezada* (beans, potatoes, bell pepper, *morcilla* blood sausage, bacon, and collard greens).

At the beginning of the 19th century, Greek immigrants arrived in Tarpon Springs for sponge fishing jobs (see chapter 13 on Greeks and Middle Easterners). They introduced traditional dishes like *moussaka* (stuffed eggplant), *spanakopita* (spinach- or cheese-filled phyllo dough pastries), and *gyros* (pita bread sandwiches). In recent years, Cuban immigrants to the Miami area have added significant contributions to the cuisine (see chapter 9 on Latinos for more information). *Arroz con pollo* is made with chicken and rice, flavored with the Cuban combination of tomatoes, olives, capers, raisins, and chile peppers. Black beans, traditionally prepared with rice and salt pork or ham, is common. So-called *Cuban sandwiches,* with roast pork, ham, sausage, cheese, and dill pickle filling mounded on Cuban bread, are fast-food favorites. *Flan,* a baked custard with caramel topping (sometimes flavored with orange), has become a popular dessert.

Georgia

Georgia was founded in 1733 by James Oglethorpe, who established the settlements of Savannah and Augusta as refuges for Europe's poor (particularly those in debtors' prison) and persecuted (because of political or religious beliefs). In the following years, thousands of English, Scotch, Irish, northern Italians, Swiss, German Lutherans, French Huguenots, and Sephardic Jews (from Spain and Portugal) arrived to take advantage of freedom and

■ Before the advent of refrigerated box cars in the early 1900s, oranges were such a luxury throughout the rest of the nation that it was traditional to put an orange in children's stockings at Christmas as a special treat.

■ Key lime pie, which has a lime custard filling and is traditionally covered with meringue, can also be made as a chiffon pie (folding the meringue into the custard to lighten it, then topping the pie with whipped cream). Key limes must be used for an authentic pie, but imposters with the juice of other limes often borrow the name.

■ Palmetto palms are so prevalent in Florida that they are known as "swamp cabbage."

■ *Conchas* is the Spanish word for conch mollusk, a term that has become the nickname of residents of the Florida Keys.

■ The barrier islands off the coast of Georgia and South Carolina are the last area where the Gullah dialect is spoken, a patois of English and several West African languages. The food of the islands is typically southern, with pork, greens, and sweet potatoes and specialties like broiled chitterlings (instead of boiled and fried), smoked mullet, and corn bread with sorghum syrup added to it before baking (Carlson, 1997).

■ *Coca-Cola* was invented by an Atlanta pharmacist in 1886 as a headache remedy. "Dope" is the slang term for cola drinks in the South.

■ Vidalia onions are thought to be exceptionally sweet due to the mild Georgia weather and the low-sulfur soil around Vidalia, in Toombs County, Georgia. Over $50 million worth are grown there annually.

■ In Kentucky, over 1.5 million squirrels are killed each year for their meat. The brains, known as *Kentucky caviar,* are a specialty in some areas. They are usually mixed in with scrambled eggs or added to burgoo. Health officials are concerned the brains may be a source of Creutzfeldt-Jacob disease, the human version of mad cow disease.

generous land grants. Oglethorpe expected every family to be hard-working and self-supporting, that none would be richer or poorer than another, and that slavery would be unnecessary. The settlers soon looked to the north and saw that cotton and tobacco were much more profitable than the meager food crops each small family produced, and by 1750 the ideals of Oglethorpe had been discarded in favor of the plantation system worked by slaves. Cotton, tobacco, indigo, and rice became primary crops.

The Georgia settlers quickly adopted a typically southern lifestyle. Plantation hospitality was famous in the region. A description of a company meal from the early 1800s listed turtle soup, trout, ham with sweet potatoes, turkey with a cornmeal and walnut stuffing, rice, asparagus, green beans, followed by orange sherbet to cleanse the diners' palates before continuing with cold venison, cheese, corn fritters with syrup, and sweet potato pie (Lee, 1992). Traditional southern fare was served at family meals as well, such as Georgia squirrel stew (a variation on Brunswick stew), ham, hoecakes, okra with tomatoes, and biscuits served with preserves. Mayhaw jelly, made from the cranberry-like fruit of the native mayhaw tree, is a particular favorite in Georgia and other states along the Gulf Coast.

Fish and seafood were plentiful. Seafood *pilau* (see South Carolina, following) was popular in homes along the Atlantic coast, as were oyster suppers: The informal feast featured oysters cooked over a fire in the moonlight, served with melted butter. A French nuance can be seen in the popularity of dishes such as crab soufflé. German dishes common in the region include sauerkraut and pepper pot soup; the Scots brought scones and *haggis* (hog's stomach stuffed with oatmeal—see chapter 6 on northern Europeans). One unusual dish found in Georgia is *Country Captain Chicken,* invented by a sea captain from Savannah who used Indian spices to liven up his routine fare aboard ship. It is a curried dish that includes tomatoes and green peppers and is served over rice. Many southerners enjoy a bit of heat in their foods: In Georgia, bird's eye peppers, very small (¼ to ½ inch) round, oval, or triangular chiles that are scorching hot, are sometimes crushed and placed at the bottom of a bowl before adding soup or stew.

Peaches thrived in Georgia, and at one time, Georgia was known as the "Peach State." Peach pie is the consummate Georgia dessert, although recipes vary. Some are custard pies topped with sliced peaches, others are two-crust pies, and some are individual deep-fried pies. Today peaches remain an agricultural product associated with Georgia, as do pecans and watermelon. But in recent years, they have been surpassed in economic importance by poultry, peanuts, and cattle.

Kentucky

Kentucky began as the western frontier of Virginia. Explorers had to cross the Appalachians to get into the territory and in the early eighteenth century, it was the domain of hunters and trappers, who profited from the abundant deer, elk, bear, and buffalo in the region. In 1775, Daniel Boone blazed a trail through the mountains and on the western side discovered a large, grassy plain that is now known as "Bluegrass Country." He founded Boonesborough and encouraged emigrants to make their homes in the area. By 1790, there were over 70,000 settlers in Kentucky, most of English, Scotch Irish, or German heritage from surrounding states. They planted tobacco and hemp, clearing the woodlands surrounding the central Bluegrass region.

The settlers found that wheat did not grow well in Kentucky soil, but corn was successful. Corn was used for nearly all baked goods, with the exception of biscuits and pies prepared on special occasions. Game meats were the foundation of the Kentucky diet in the early years. Bear meat was particularly popular because it could be smoked like pork and was fatty enough to provide bacon. *Burgoo,* a stew traditionally made with wild birds and game meats like squirrel, is the signature dish of Kentucky (current versions typically include chicken; pork, beef, or lamb; cabbage, potatoes, tomatoes, lima beans, corn, and okra; and cayenne—some variations add filé powder, curry powder, or bourbon). When supplies of game diminished, pork became the primary meat. Ham was common, and variety cuts were used to flavor vegetables, such as green beans simmered with bacon throughout the day to make a smoky, mushy stew. Root vegetables provided bulk in the winter, while fresh greens were preferred in the spring and summer. Fried chicken was a special dish prepared on Sundays.

The best known of all Kentucky food products is bourbon. The territory was especially suited for the distillation of corn whiskey due to its iron-free, limestone spring water, and some of the Scotch Irish settlers had skills in making the brew. Many farmers found corn whiskey more profitable than cornmeal. It is believed that in 1789 the Reverend Elijah Craig, of Bourbon County, Kentucky, was the first person to produce corn whiskey aged in oak barrels, producing the characteristic flavor of bourbon. In 1860 a further refinement occurred when it was accidentally discovered that charred oak barrels added not only a touch of color, but also a favorable smoky taste. The favorite bourbon drink of Kentucky is the *mint julep* (bourbon sweetened with a touch of sugar or syrup and a hint of fresh mint), traditionally served in a silver cup. Bourbon also flavors stews, hams, pound cakes, fruitcakes, and *bourbon balls* (a candy made with chocolate, crushed vanilla wafers, pecans, corn syrup, and bourbon).

Louisiana

Although Spanish explorers first claimed the Louisiana region in 1541, it was the French who first colonized the state, establishing several fortified settlements along the Gulf Coast, including Nouvelle-Orléans (New Orleans) in 1718. African slaves were brought in to work the cotton, sugarcane, and rice plantations, and thousands of French Acadians from Canada and French Creoles from Haiti seeking refuge from political persecution arrived.

It is sometimes said that Louisiana is divided by a "crawfish curtain," between the south, where the crustacean is revered, and the north, where it is scorned (Gutierrez, 1984). There is indeed a culinary difference in the foods preferred in the coastal areas and those favored in the hilly, upper reaches of the state. For example, the northern region (which was settled mostly by Americans of English, Scotch Irish, and German ancestry) has had little French influence on its cooking and the staples of traditional southern fare prevail. Notably, cornmeal is used most in breads and biscuits. In the Mississippi Delta region, where New Orleans is located, and in the southwestern portions of the state, which contain the bayou country, wheat flour has always been favored, even when it had to be imported from Europe.

Fish and seafood (including the ubiquitous crawfish) are more important than pork in the southern regions of Louisiana. The famous fish and seafood stews of the area, *bouillabaisse, gumbo,* and *jambalaya,* are examples of dishes made from coastal plenty. Shrimp is the main seafood industry in Louisiana, marketed throughout the nation fresh and frozen. It is commonly served boiled with lemon butter or

■ Corn whiskey was often distilled illegally in the secrecy of night to avoid federal tax agents and thus became known as *moonshine.* Other names for the potent brew include "corn likker" and "white lightning."

■ The term *Creole* is often used to describe Europeans born outside Europe and is applied, for example, to the descendants of the original French and Spanish immigrants to New Orleans.

■ Shrimp in Louisiana are sometimes called *Cajun popcorn.*

■ It is said that when a Cajun baby is 9 days old, the mother sticks the child's finger down a crawfish hole to be pinched. This is what makes a Cajun a Cajun (Gutierrez, 1984).

Crawfish is a specialty in southern Louisiana that has become popular in other areas of the South. (Louisiana Tourism.)

with *sauce piquante* (tomatoes, green peppers, onions, bay, vinegar, and hot sauce) over rice, a dish often called *shrimp Creole.* Shrimp is also added to seafood stews and to stuffings for vegetables. Oysters are commonly served raw, on the half shell and by the dozen in the many oyster bars of New Orleans. They are traditionally slurped with a squeeze of lemon juice and a dash of hot sauce or a sauce mixed to taste by each diner with catsup, vinegar, and horseradish. Oysters, too, are added to soups and stews. Crawfish, which look like miniature lobsters, are found in the swamps, riverbeds, and rice paddies of the state. They have become the ethnic emblem of Cajuns and the regional symbol of southern Louisiana. Over 100 million pounds are produced annually. Some are harvested from the wild, but most are cultivated in the approximately 300 crawfish farms operating in the state. Crawfish are typically served at a "boil," where they are cooked in water seasoned with cayenne, salt, and herbs. Potatoes or corn are often added for side dishes. The crawfish are placed in a gigantic mound in the center of the table, and each person takes and peels as many as desired. Only the meat in the tail and the claws is edible, along with the fat found in the head, which is extracted with a finger or sucked out appreciatively. Crawfish are also prepared fried, stuffed, as fritters, in soups and stews, in pies, and as *étouffée* (meaning "smothered") in a spicy tomato sauce.

Game meats continue to be popular in some areas. Wild boar, frog's legs, and alligator are often found. Other regional specialties include rice dishes, such as the fried cakes called *calas,* red beans and rice, and dirty rice (cooked with gizzards). Rice is also the foundation of dishes like gumbo and jambalaya. Baked goods and sweets are specialties, including French *petits fours, crêpes, beignets* (deep-fried squares similar to dougnuts), and pralines (pecan candies). *Café au lait,* a favorite beverage in New Orleans, is a dark roasted coffee (sometimes flavored with chicory root) prepared with equal amounts of hot milk. *Café brûlot* is a sweetened dessert coffee flavored with brandy and curaçao (orange liqueur).

Restaurant fare in New Orleans is renowned. Among the dishes created by local chefs that have obtained national recognition are *oysters Rockefeller* (baked on a bed of salt with a rich spinach sauce), *oysters Bienville* (baked with a béchamel sauce and green pepper, onions, pimento, and cheese), *bananas foster* (sliced bananas cooked in butter, brown sugar, rum, and banana liqueur served over vanilla ice cream—it started out as a breakfast specialty), and *Ramos gin fizz* (a shaken or blended cocktail with cream, gin, lemon juice, orange flower water, and egg whites). Street food is equally tasty in the city. Fried oysters, sliced tomatoes, and onions with tarter sauce on a french bread roll are especially popular. They are called *peacemakers,* from the 19th century when men would bring one home as a surprise for dinner after a fight with their wives. *Po' boy* (for *poor boy*) is another name for the sandwich, although a po' boy may also refer to a sandwich with deli meats, sausages, and cheeses with or without gravy or tomatoes. A *muffeletta* sandwich is yet another version, usually including a chopped olive salad with the meats and cheeses.

■ Absinthe, a liqueur made from wormwood, was exceptionally popular in New Orleans during the 19th century. It was believed to be an aphrodisiac. It was later banned when it was discovered to be toxic and habit forming.

■ Sandwiches with deli meats and cheese on a french bread roll are found throughout the country. In addition to being called *po' boys* and *muffeletta sandwiches* in New Orleans, they are also known as *bombers* (upstate New York), *Cuban sandwiches* (made with roast pork in Miami), *grinders* (New England), heros (New York City), *hoagies* (Philadelphia), *Italian meat sandwiches* (Chicago), and *submarine sandwiches* (from a World War II naval base in Connecticut).

■ Green tomato slices dipped in cornmeal and pan-fried are popular throughout the South. They are served for breakfast in Maryland.

Maryland

In 1634, Cecilius Calvert, Lord Baltimore, was given a land grant on Chesapeake Bay to establish a refuge for 200 fellow English Catholics, many of whom were wealthy "gentlemen adventurers." They were soon outnumbered, however, by Puritan immigrants. In the early 1700s, the English settlers expanded into the inland hill country from the east, established baronial estates, and grew tobacco in the fertile lands previously cleared by the Indians. Their lavish lifestyle was supported by slave labor. Germans from Pennsylvania entered the state from the west, grew wheat in the Piedmont, and produced baked goods (particularly rye and pumpernickel breads) that were sold to markets in the east.

Wild game and pork supported the earliest immigrants. The English, who were partial to beef and mutton, soon imported cattle and sheep as supplemental meats. Herb-roasted veal; lamb stewed with exotic spices, including curry; and sweetbreads were specialties. Hams were produced in every manner: wet-cured in brine or dry-cured with a sea salt or sugar crust, then smoked. One recipe that is still popular, particularly in the southern sections of the state, is stuffed ham. It calls for inserting greens (e.g., cabbage, kale, and/or watercress) flavored with onions, mustard seeds, and cayenne into deep slits of the ham. The ham is served cold and is often the centerpiece of the Easter meal.

Maryland is best known for its seafood, however. The state is indented by the largest estuary in the nation, Chesapeake Bay, which teems with oysters, clams, scallops, and crabs. Many settlers near the bay made their living as watermen, harvesting seasonal seafood, oystering in the winter, and crabbing in summer. It was the abundance of Maryland Chincoteague and Tangier oysters that started the oyster craze in the cities of the Northeast and Midwest in the 1800s. They were eaten three times a day in the state: raw, fried, baked, fricasseed, in seafood stews, in chowder, in oyster stuffing for turkeys, and over steaks. Crabs were equally versatile. A regional specialty is blue crabs, a swimming crab so named because the underside of the large claws is blue. They are traditionally steamed over water flavored with vinegar, and seasoned with salt, pepper, ground ginger, celery and mustard seeds, and paprika. Because they are small, half a dozen or more are served to each diner, with plenty of beer to wash them down. The meat is used to make one of Maryland's most esteemed dishes, *crab cakes*. The crab is mixed with a little mayonnaise, cracker crumbs, and a spicy seasoning of cayenne, dry mustard, and hot sauce, then formed into small patties and fried. They are served with lemon wedges and tarter sauce. Crab soup (with beef stock and bacon) and deviled crabs (baked in the shell and topped with bread crumbs) are other common preparations. Another noteworthy shellfish of the region is softshell crab, which is a blue crab that has shed its hard shell during a molt. The new papery shell is completely edible, but it begins to harden after only a day. Blue crabs are often kept in tanks until they shed their hard shells to time harvesting of the soft-shell crabs. The whole crab is served deep-fried. Clams became an important commodity when the New England supply diminished; many are shipped to the Northeast. Fried clams, deviled clams, and clam fritters are common dishes. Currently over 400 million pounds of seafood are harvested from the Chesapeake Bay each year (Lee, 1992).

Diamondback terrapin is another Maryland specialty found around the Chesapeake Bay. It is the turtle preferred for turtle soup. It is also the main ingredient in terrapin stew, which includes the eggs and liver, all in a gravy made from the turtle stock flavored with butter and Sherry. Beaten biscuits and a glass of sherry are the traditional accompaniments. Atlantic fish, such as butterfish, cod, flounder, herring, mackerel, and stripped bass, are available. Poultry has always been a favorite in Maryland, even more so since broiler production became an important industry in the 1930s (see Delaware section). The early settlers prepared chicken pot pies and chicken–seafood chowders. Fried chicken with cream gravy is still served regularly.

Mississippi

The state of Mississippi is dominated by the Mississippi River, which demarcates its western boundary. Low-lying bottomlands of the Mississippi Delta form the western half of the state, with low, rolling hills found in the eastern half. The French were the first whites to settle in the region. They built forts at Biloxi in 1699, and in Natchez and Vicksburg soon after. The territory was ceded to the British in the mid-1800s and became a haven for Tories escaping the impending conflict of the Revolutionary War in the Northeast. The northern half of the state was settled mostly by pioneers from Georgia.

Both the French and English influenced the cities located along the southern portion of the Mississippi River. Natchez in particular became a wealthy port city supported by the cotton plantations flourishing throughout the state. Cotton was shipped to Europe in exchange for French and English manufactured goods. The big southern breakfast and dinner were the norm for plantation owners and their city associates. Coffee and mint juleps were available for early risers. Late morning repasts included eggs, grits, biscuits, corn breads or muffins, waffles and several meats, such as ham, sausage, or even fried chicken. A large dinner with soups, stews, and dishes similar to those at breakfast was consumed in the early afternoon, supper was a lighter version of dinner. One reason the food was so ample was that the farmers of the Midwest discovered that they could make a profit shipping flour, pork products, and perishable produce to Mississippi cities by flatboat.

Local Mississippi foods include catfish, shrimp, oysters, and blue crabs. Catfish up to 100 pounds can still be caught in the rivers and

- Each female oyster releases about 5 million eggs annually, but only 15 reach maturity. Maryland produces one-third of all oysters consumed in the nation each year.

- *She-crab soup* was a Maryland specialty before crab harvesting was regulated to prohibit keeping egg-bearing females. It included crab meat and roe in a delicate cream base. The eggs gave a slightly sour flavor to the dish. Today it is made from male meat and unfertilized roe.

- In 1893 a taste test was held to determine which turtle soup was superior: the creamy Philadelphia version with turtle meat or the clear turtle consommé with a hint of Sherry preferred in Maryland. The impartial jury selected the Maryland soup as the winner (Jones, 1981).

- Food was not always plentiful on the plantations. During the Civil War, supplies were so short that Mississippi women prepared pie crusts from potatoes and made breads from corncob ashes (Lee, 1992).

■ Oysters were so prevalent in the early days of Mississippi that the streets of Biloxi were paved with oyster shells.

■ Catfish consumption has more than doubled nationally since 1986. It is ranked the fifth most popular fish by Americans.

■ Early white settlements by English immigrants on Roanoke Island in the late 1500s were unsuccessful. Numerous Scuppernong grapevines brought from the mainland are a legacy of the attempts, still used by residents for jam, jelly, and pies.

■ An unusual Appalachian Mountain treat called a *sonker* is preserved in the annual festival named in its honor held at Mount Airy each year. It is a deep-dish fruit or sweet potato pie made with breadcrumbs or biscuit dough or pie crust—there is no agreement on how to prepare a sonker (Carlson, 1997).

lakes of the state, but most are now farmed in ponds. Although the first catfish farms were started in Arkansas, Mississippi is the leading producer in the nation, with over 250 million pounds harvested annually. Traditionally catfish is deep-fried in a cornmeal crust and served with hush puppies and coleslaw. Newer recipes include fried strips served with barbecue sauce or mustard, and catfish pâté (Carlson, 1997).

Shrimp and crab are typically boiled, served with an assortment of seasonings, such as vinegar, lemon juice, bird's eye chiles, and cloves. Oyster bakes are a popular outdoor feast. Pecans are native to the area and a Mississippi favorite, added to sausage stuffing for poultry, added to breads, sugar glazed, orange glazed, and baked in the syrupy pecan pie.

North Carolina

A rugged coast and a line of offshore sandbars known as the Outer Banks protect the state of North Carolina. The state is divided into three distinctive regions: the Tidewater coastal plain, the Piedmont hill country, and the mountainous region that includes the Blue Ridge and Great Smoky ridges of the Appalachians.

In 1663, eight British lords were granted proprietorship of the region known as Carolina (including both North Carolina and South Carolina). Settlement in the northern portion was slow but steady, attracting English, Scotch Irish, and Highland Scots, many of whom came as indentured servants. By the late 1700s, settlers from Virginia and South Carolina were streaming into the region, as well as immigrants from throughout Great Britain, Germany, and Switzerland.

At the time North Carolina was officially recognized as a state in 1789, most residents were small family farmers. Game was plentiful and corn grew easily, supporting residents until garden plots began producing other fruits and vegetables. In the Appalachians and parts of the Piedmont, bacon and cabbage, chicken pot pie, ham with cream gravy, baked beans, biscuits, hush puppies, potatoes, sweet potatoes, and poke salad were favorites. Fried squirrel and grits was a common meal. Scrambled eggs with brains were common at breakfast, and badger, considered by some a dish of last resort, became known as *bombo* in North Carolina. Barbecued pork with spicy hot sauce on the side was prepared for special occasions. *Ramp,* a particularly assertive wild onion, was added to soups, stews, potatoes, or eggs or breaded and fried as a side dish. Thick, caramel-colored sorghum syrup was poured over corn breads or used to sweeten coffee. Wealthier immigrants founded tobacco plantations and could afford roast beef, steaks, beef stew, and breaded pork chops in addition to the omnipresent ham and fried chicken. Along the coast, some residents made their living from the sea, catching mullet, shad, and rockfish and harvesting oysters and shrimp. Coastal specialties include fish muddle and corn dumplings.

One notable ethnic group in North Carolina are the German Moravians, persecuted German Protestants who had immigrated to Pennsylvania originally but moved south in the early 1700s when they discovered that much of that land was already claimed. The Moravians established an insular German community near the Winston-Salem area, founding a wholesale produce business that sold local fruits and vegetables in markets extending to Philadelphia. They were best known for their baked goods, such as sugar cakes (a yeasted, potato bread dough that is covered with brown sugar and cinnamon before baking) and *citron tarts* (tarts with lemon curd filling). Moravians commemorate special occasions with "Love Feasts." On November 17, the founding of North Carolina, the community Love Feast features wine, creamy coffee, and cakes topped with a nut frosting. At Christmas, paper-thin ginger spice cookies and a sweet bread studded with raisins and candied citron, sprinkled with sliced almonds, are specialties.

Oklahoma

Oklahoma started its U.S. history as Indian Territory, lands set aside in the 1820s for the Native American tribes that had been dispossessed of their homes in the Gulf Coast areas. Five major Indian groups lived in the region: Cherokee, Chickasaw, Choctaw, Creek, and Seminole. They were primarily agrarian, growing corn, beans, and squash. They gathered indigenous foods (including acorns, chestnuts, creases, grapes, Jerusalem artichokes, hickory nuts, persimmons, ramp, and sorghum) and hunted small game (see chapter 5 on Native Americans). Today Native Americans make up approximately

8 percent of the total population in the state, over eight times the national average. Traditional foods, such as a Cherokee soup made with hickory nut cream, called *kanuche,* and game dishes, are served mostly at ceremonial occasions. Fry bread and adapted dishes, such as scrambled eggs with spring onions, are more common but have not been accepted into the broader Oklahoma cuisine (Milbauer, 1990).

The earliest whites in Oklahoma were fur trappers, buffalo hunters, railroad workers (including many Mexicans), and ranchers. Although the state was not officially opened up to settlement by whites until 1889, land-hungry homesteaders invaded the state before then (they were called "Sooners"), including German Russians, Italians, and central European immigrants. African Americans, many of whom had been held as slaves by the Indians, purchased land in the region after abolition. Many settlers established small family farms. Most of the plains regions in the state are arid and droughts have occurred regularly; hardship has characterized Oklahoma life.

The cooking of Oklahoma derives more from scarcity of food than from ethnic and regional preferences of the settlers. Wheat became the dominant crop in the state. Rabbit and turnip stew was flavored with a flour gravy, while beef and wheat berries are the primary ingredients of *Oklahoma Stew.* Biscuits were common, and some corn breads were prepared. *Black blizzard cake* (a pound cake whose name refers to the frequent dust storms in the region) was a specialty. The Indians had successfully planted watermelons, which became a popular fruit with all residents, eaten fresh and made into preserves and pickles.

The southern portions of the state are called Little Dixie, and traditional foods of the South are predominant in the fare. A study of foods in the eastern portion of the state found that pork, fried chicken, catfish, biscuits and cream gravy (red-eyed gravy is not as common), corn bread, fried okra, and black-eyed peas were frequent items in local eateries. However, grits and buttermilk were rarely offered (Milbauer, 1990). Other ethnic and regional foods have had less impact over all. Authentic Mexican foods, for example, are uncommon, although a few Tex-Mex items like chili con carne have become very popular. Some dishes with Italian roots are accepted, such as spaghetti and meatballs, particularly around Krebs in the southeast. Sauerkraut, potato soup, and dark breads are evidence of German Russian influence, and central European traditions are maintained at heritage festivals.

■ Oklahoma pioneers of the period lamented the monotony of eating dried apple pie; others wrote that they resorted to using sweetened crackers or potatoes to make pies when fruit was in short supply.

South Carolina

Although South Carolina was part of a British land grant that included North Carolina in the 17th century (see North Carolina), settlement of the state was strikingly different from that of its northern neighbor. The English proprietors thought the fertile soil and warm climate would support Mediterranean fruits and vegetables. Approximately 150 white settlers from England arrived in the area in 1670 and founded Charles Town (now Charleston). They brought ginger, sugar cane, olives, and indigo; only indigo proved successful as a crop, but it was enough to begin a profitable plantation system in the region. Charleston became the primary seaport of the South.

The English were followed by thousands of French Huguenots seeking religious freedom in the 1680s. Most settled in the Charleston area, establishing indigo and later rice and cotton farms. Many who operated plantations in the coastal section also owned townhouses in Charleston. Other Europeans settled the upland Piedmont hills and Appalachian mountain areas of the state, including Germans, Welsh, Swiss, and Scotch Irish. By 1775, half of the white population in South Carolina lived in the central plains and upland regions. Unlike the coastal plains residents, most of these settlers owned small family farms. Like the Piedmont and Appalachian dwellers of North Carolina, these farmers lived on wild game and corn until their garden plots were sufficient to provide other foods.

It is the cooking of coastal South Carolina, especially Charleston, that is unique in the South. Numerous exotic foods arrived by sea, such as chocolate from Central America and bananas from the Caribbean. Unlike the English settlers in other regions of the South, large numbers of those who came to South Carolina had lived first in Barbados and other Caribbean Islands. They brought a taste for tropical flavors and spicy seasonings. Fruit and vegetable pickles

■ The connection between the Caribbean and South Carolina was so strong that, according to historian Peter H. Wood, some legal documents of the 17th century refer to "Carolina in ye West Indies" (Hess, 1992).

were common, for example, mango chutney from India, which was also made with other local fruits and called *Indian pickle.* A French influence was seen in many dishes, particularly elaborate desserts like *Huguenot torte* (a sponge cake with pecans and apples) and *charlotte russe* (a special cylindrical mold lined with ladyfingers, then filled with Bavarian cream and garnished with strawberries and whipped cream).

Seafood is a specialty of the region. Shrimp was especially common in the early years of the state, featured in dishes like shrimp pâté or butter-sautéed shrimp with grits for breakfast and deep-fried shrimp and rice croquettes. Outdoor oyster roasts were the highlight of many festive occasions; they are still favorites today, served with hoppin' John, biscuits, and small sandwiches, such as a crab omelet on slices of bread. *Frogmore stew* is a spicy seafood, sausage, and corn combination similar to gumbo. *She-crab soup,* similar to that of Maryland but garnished with a spoonful of Sherry and a dollop of unsweetened whipped cream, was popular. As in other areas of the coastal South, muddle was prepared from whatever fish were available. A variation unique to South Carolina is *Pinebark stew,* a muddle flavored with bacon and the tiny roots of pine trees.

Most noteworthy are the rice dishes of the region. A variety of rice native to Madagascar was found suitable for the coastal plain climate; thousands of acres of tidal lands were diked and flooded to support the crop. By 1700, it was well established, thanks in part to the skills of slaves from the rice-growing regions of Africa. It was known as *Carolina Gold* (due to its amber color when ripe) and became the foundation of the coastal diet. The rice was traditionally boiled instead of steamed to produce individual fluffy grains that did not stick together. Numerous specialties developed with rice. *Pilau* (also spelled *purlow,* or *pullow*) "is characterized by an elegant simplicity of composition, featuring a single ingredient, that is, in addition to the rice which is the point of the dish" (Hess, 1992, p. 64). Rice is simmered in an aromatic broth (reserved from cooking the second ingredient) until dry, then mixed with the other food. Shrimp pilau and okra pilau are examples. African-influenced hoppin' John, made with black-eyed peas and rice, is also a pilau. Molded rice dishes that are baked until they form a golden crust are called rice pies or rice casseroles. Some include layers of meat or fish. Rice breads, such as *philpy* (cooked rice added to corn bread), and desserts, such as rice pudding and *Carolina snow balls* (a rice dumpling with a whole, peeled and cored apple filling), were common.

Rice production in the region was interrupted by the Revolutionary War and then again during the Civil War. After the abolition of slavery, rice farmers found the crop too labor intensive to use paid workers, and production diminished dramatically. Other crops, such as cotton, became more important in the economy. The Spanish explorers in the 16th century had brought peaches to the Carolinas. At one time, they were so plentiful they were used as hog feed. Today South Carolina is the largest producer of peaches in the South, second only to California in the nation.

■ *Charlotte russe* was created by the great French chef Antoine Carême in honor of the Russian czar Alexander in 1815. It is a favorite special-occasion dessert in South Carolina.

■ *Pepper sherry,* made by infusing bird's eye chiles in sherry, is a popular condiment added to shellfish, soups, and stews for zing.

■ Hoppin' John is served on New Year's Eve in South Carolina and other parts of the South in the homes of both African Americans and whites because eating the rice and black-eyed peas is thought to bring good luck in the upcoming year.

■ *Benne* (sesame) *seeds* were introduced to the South by African slaves. In South Carolina, brown sugar benne wafers and benne brittle (made with sesame seeds instead of nuts) are popular desserts.

Tennessee

At one time, Tennessee was a part of the Virginia frontier. Two waves of settlement occurred, one with families seeking safety from the turmoil of the Revolutionary War, most of whom were of English, Scotch Irish or French Huguenot heritage, and one with Germans, Scots, and Irish who sought farmland in Maryland and Pennsylvania before moving on to available properties in Tennessee. Those who settled in the western regions along the Mississippi River found fertile plains suitable for tobacco and cotton plantations. Horse ranching arose in the middle hill country. Those pioneers who settled in the east existed on subsistence farming in the rugged Blue Ridge and Great Smoky mountains of the Appalachians.

The differences between the western and eastern halves of the state were so pronounced that during the Civil War the east sided with the Union and the west joined the Confederacy. The plantation owners of the west had access to the wealthy markets of the lower Mississippi, such as Natchez and New Orleans. They developed a southern lifestyle typical of plantation life throughout the South, including large, leisurely meals. They are still known for their sumptuous breakfasts, which traditionally feature eggs, tomatoes, potatoes, and cornmeal biscuits with sorghum syrup. During the winter, thick slices of Tennessee country ham, with grits and red-eye gravy made with the ham drippings, may also be served with the meal; in the summer, fried chicken is more common. Tennessee sausages are also a specialty, as is spiced beef (marinated in vinegar, brown sugar, and seasonings, then simmered and sliced thinly). Baked sweet potatoes with butter, as well as greens, are frequent side dishes.

Settlers in the hills and mountains had a diet similar to that of other Appalachian dwellers (see also "Kentucky" and "North Carolina"). Game was the primary meat, including deer, raccoon, opossum, squirrel, and wild turkey. Those who could afford a hog would butcher it in the winter, providing fresh meats, such as ribs barbecued with a tomato-whiskey sauce, and meats for curing as hams, salt pork, and cracklings. Stews were favored. One specialty was *cabbage gumbo,* made without seafood but including country ham and Tennessee smoky sausage in a milk base. As in the wealthier western regions, corn breads, biscuits, and greens were additional core foods. Baked beans were also common. Many farms had at least one apple tree, providing fruit for apple butter and pies. Funnel cakes, undoubtedly introduced by German immigrants, were topped with sorghum syrup. Fried pies were a favorite dessert, and sassafras tea was a popular beverage.

In 1866, Jack Daniel purchased a corn whiskey still and added an extra refinement to the distillation process, using maple wood charcoal to filter the whiskey before aging it in charred oak barrels. This produced a flavor distinct from bourbon, and the liquor became known as Tennessee whiskey.

Texas

Texas is the size of New England, the Mid-Atlantic states, Ohio, and Indiana combined. Although the terrain is usually associated with desert plains, there are also wooded regions, hill country, mountains, and hot, humid lowland areas along the Gulf Coast. It was occupied by Native Americans, claimed by the Spanish and French, ruled by Mexico, and existed as an independent nation before it became part of the United States in 1846. The first American settlement was 300 families in what is now Austin on land granted by the Mexican government. Other Americans were provided land if they were Roman Catholics; thousands from the South (Catholic in name only) entered the state from the east to take advantage of the offer. Germans, Czechs, and Poles emigrated from Europe to central Texas, attracted by later land grants. In the west, small numbers of ranchers and cowboys made their living from raising and driving cattle.

Texas is divided into five culinary regions, each influenced by the available foods, proximity to other ethnic and regional fare, and the preferences of the residents. In the northeast and panhandle areas, immigrants scraped out a living on small family farms, surviving on corn, beans, and native fish and game. When wheat proved a successful crop in the region, corn breads were replaced with biscuits. In the eastern areas, foods of the South dominate. It has always been an affluent region, and meals are often opulent: Dishes frequently feature costly ingredients and tend to be richer (with extra butter, eggs, and cream) than versions from Louisiana, Mississippi, and Arkansas. Corn breads, biscuits, hominy and grits, black-eyed peas, okra, sweet potato pie, bread pudding, and pralines are a few common items. Rice is an important crop, and southern-style rice dishes are popular. Alligator, trout, catfish, and crawfish are available in local rivers. Over 100 million pounds of fish and seafood are produced along the Texas Gulf Coast each year, including shrimp, crab, and oysters (Lee, 1992). *Crab au gratin,* oyster stew, and shrimp casserole are a few specialties. Classic Creole and Cajun dishes, such as jambalaya, étouffée, and gumbo, can also be found.

There are strong German and central European influences in the central regions of the state. Common foods include sausages, ham, sauerbraten, sauerkraut, pumpernickel bread, potato salad, potato dumplings, bierocks, and strudel. Czech kolaches and Polish poppy seed cake are baked for festive occasions. One specialty is "German fried" liver (dredged in cornmeal, cooked in bacon drippings, and served with bacon strips). In the western areas of Texas, beef has always been popular. It is served traditionally as stews and steaks. Chicken-fried steak was created to treat tough cuts—the steak is cut thinly, then pounded with a mallet, coated in flour, and fried. It is traditionally served with a ladleful of gravy made with coffee. Corn bread or tortillas and pinto beans often round out the meal. Few fresh vegetables or fruits were consumed. Today Texas

■ *Moon pies* are a Chattanooga, Tennessee, confection that have become an obsession in the South. They are graham cracker sandwiches with a marshmallow filling covered in chocolate, vanilla, banana, or caramel icing. During the Great Depression, a moon pie and an RC Cola was called a "working-class dessert" because both could be had for a dime.

■ The tiny town of Lynchberg, Tennessee, where the Jack Daniels Distillery is located, has been dry since 1909.

■ Chuck wagon fare was a cooking style all its own, dependent on the skills and whims of the cowboy chefs called "Cookie" or "Miss Sally." Beans, corn bread, sourdough biscuits, and coffee were the staples, but some specialties were created on the trail, including son-of-a-bitch stew (known as *son-of-a-gun stew* in more genteel circles) made with beef organs, including tongue, brain, liver, heart, and kidneys.

■ One specialty of the northeast is rattlesnake, which is chopped into bite-size pieces, marinated in beer and vinegar, coated with cornmeal, and deep-fried.

■ Barbecue and political events go hand in hand in many parts of the country, especially in the South. In 1840 an Alabama plantation owner hosted 4,000 at a celebration barbecue for the election of Martin Van Buren. A German journalist at a Fourth of July barbecue in Little Rock, Arkansas, in 1843 reported that two oxen and two bears, as well as numerous sheep, calves, pigs, deer, turkeys, opossums, raccoons, squirrels, chickens, geese, ducks, partridges, and prairie chickens (grouse) were prepared for the event. President Lyndon B. Johnson was famous for the enormous barbecues held at his Texas ranch.

■ Tex-Mex chili con carne was probably first popularized in the mid-1800s. Chili powder was invented by a German in New Braunfels, Texas in 1902. It usually includes spices like cumin in addition to ground, dried chile peppers.

■ Texas caviar is black-eyed pea salad, seasoned with chiles and garlic and served on tortilla chips.

pasturelands are the leading producer of cattle, sheep, and lambs in the nation.

The most distinctive Texan fare evolved in the south of the state, where Mexican and Spanish influence added their flavors to dishes. Some authentic Mexican foods, such as tortillas, tamales, *chalupas,* salsas, guacamole, and *buñuelos* (see chapter 9), were accepted by white settlers in the region. However, most foods in the area are adapted dishes with Mexican overtones, often referred to as Tex-Mex cuisine. Examples include tamale pie, nachos, and most tacos and enchiladas, which usually feature nontraditional fillings. One regional specialty is *chili con carne,* known in Texas as "a bowl of red," which began as beans, progressed to beans with beef, and is now typically an all-beef stew flavored primarily with hot chili powder. Barbecue is also favored. Unlike barbecue in other regions of the country (e.g., Kansas City), there are two sauces involved in Texas barbecue. The first, called the "mop" or "sop," is used to marinate the meat before cooking and for basting the meat on the spit or grill. (The term *mop* for the sauce basted on barbecued meat may have come from the use of a clean mop to slap the sauce on whole carcasses.) The second sauce is served on the side with the cooked meat. Although barbecued beef is most associated with the state, barbecued goat kid (*cabrito*) is almost as popular in the southern sections.

The unifying element in most of these foods is that they are preferred hot and spicy. In addition to chili powder, chile peppers are used in many dishes. Numerous varieties are added, but worth mention is the *chiletepín* chile, the indigenous precursor to domesticated *pequin* chiles. They are among the hottest of all chiles (they are also called bird's eye peppers, described previously). Chiles often flavor foods in Texas not normally associated with the spice, such as corn bread, and jelly. In addition to chile peppers, other fresh fruits and vegetables are now prevalent in southern Texas due to irrigation. Cantaloupe, pink grapefruit, peaches, sugar cane, and tomatoes are a few examples of specialty crops.

Virginia

Virginia is famous for both its ham and its historic role in the development of American cuisine. Early success with tobacco and the plantation system of agriculture in Virginia established the pattern for the southern lifestyle and led to many food firsts.

The colony at Jamestown, founded in 1607 by English settlers seeking precious metals, was the first white settlement in the state. It began as a communal endeavor, but agricultural efforts were largely unsuccessful. Thousands of settlers died during the early years from starvation and disease. In 1614, land grants were

Barbecue is a traditional Tex-Mex method for preparing food. (© Bob Daemmrich/Stock Boston)

made to individuals and tobacco was planted as a cash crop. Many Virginians living in the coastal Tidewater region of the state quickly made the transformation from subsistence farmers to wealthy, slave-owning plantation proprietors. Those who remained small family farmers migrated west into the hill country and Appalachian mountains, where they made do with limited supplies supplemented by game and wild fruits and vegetables.

Many of the dishes that originated in Virginia were based on Native American ingredients and preparation methods shared with the first settlers. The best known food to evolve is the *Smithfield ham*. The dry-curing process used was learned from the Powhatan Indians, who salt-cured and smoked venison. As early as 1608 the method was being used on wild razorback hogs that had been brought by the Spanish and had since naturalized throughout the South. A few years later, pigs were introduced to the region, and the process was applied to domestic hams. A Smithfield ham is similar to a country ham, but it is made with the shank end of the leg and with the bone in. It is first rubbed with salt, sugar, and pepper for curing, then smoked over hickory, then hung to age. The total processing time is 9 to 20 months. The meat differs from a country ham in that it is saltier, darker in color, and leaner. The flavor is very strong and it is traditionally eaten in very thin slices on beaten biscuits or fried with red-eye gravy and served with fried apples. Another dish adapted from the Indians was *spoon bread,* a corn bread that is enriched with eggs and milk and baked slowly until it forms a crust on the top but remains custardlike underneath.

In addition to ham, the English plantation owners favored the beef and mutton dishes familiar to them from home. Roast beef with Yorkshire pudding and mutton roasted with bacon and herbs are examples. Roasted poultry, including turkey, was common. Seafood from the Chesapeake Bay was also featured, in dishes such as deviled crab, oysters in brown butter, and shad roe sautéed in bacon fat. Soups, such as pumpkin soup or peanut soup, were offered as a first course. Elaborate desserts were the highlight of formal meals, served as a separate course. They included items such as fruitcakes, pound cakes, cheesecakes, *hedgehog* (a molded almond paste filling studded with sliced almonds to look like the English animal), pumpkin pie, and cranberry tarts. Although many plantations had numerous auxiliary buildings, including the kitchen (separate to prevent fires from spreading to the main house), smokehouses, ice houses, cellars, and spring houses (built over a running spring, where perishables like butter could be stored in cold water), the settlers were still dependent on seasonal foods. During the summer, fresh green vegetables and fruits were featured, and more preserved meats were served. In the winter, root vegetables predominated and more fresh meats were available. Virginians today often list Brunswick stew, greens consumed with the pot likker (see chapter 8 on Africans), corn pone, black-eyed peas, pickled watermelon rind, pickled Jerusalem artichokes, and sweet potato pie as traditional specialties.

West Virginia

West Virginia is known as "the Mountain State" because it is located in the heart of the Appalachians. It was originally part of the Virginia territory. Plantation owners in the Tidewater region of Virginia encouraged settlement of the area to act as a buffer from the Native Americans and French in the region. Germans and Scotch Irish from Pennsylvania and New Jersey moved into the state, establishing small farms and orchards in the steep terrain. When Virginia joined the Confederacy during the Civil War, West Virginians voted to form a separate state aligned with the Union. Timber, salt, iron, coal, gas, and oil brought an influx of European immigrants looking for work in the late 19th century. Most were from Ireland, Italy, Germany, Hungary, and Poland.

The rugged landscape, damp climate, and short summers limited most agricultural endeavors. Apples and peaches were successful orchard crops, and small amounts of corn, beans, peanuts, pumpkins, and watermelon were grown in family gardens. Buckwheat was found to be an ideal, fast-growing grain, and buckwheat pancakes traditionally served with applesauce and whole-hog sausage (see Indiana) became popular. Game meats supplemented the small amounts of beef and pork available: Squirrel stew seasoned with bacon, onions, garlic, and thyme was a favorite. In ethnic communities, traditional dishes include European foods. For example, lasagna, *fagiole*

■ Many colonial recipes from Virginia had an everyday version and a special-occasion version with additional eggs, butter, and cream.

■ Spice cookies called *cry babies* were made by Virginia settlers in the Appalachians to appease distressed children. Brunswick stew, *black-pot chicken* (cooked in a Dutch oven), and fried apple pies were other favorites.

■ An elaborate Christmas dish in Virginia was *Yorkshire pie,* made by baking five kinds of boned, seasoned poultry stuffed inside each other in a pastry crust—first a turkey, then a chicken inside the turkey, with pheasant or partridge in the chicken, followed by smaller game birds such as pigeon or dove. The crust was removed, and the multilayered dish was served in slices (Lee, 1992).

■ Thomas Jefferson brought many French specialties to his home in Monticello, such as *boeuf à la daube* (jellied beef) and crêpes. He also brought Italian foods, including pasta, to the United States. His daughter Mary Randolph is credited with introducing macaroni and Parmesan cheese, which became macaroni with Cheddar cheese.

■ The Golden Delicious apple was hybridized in West Virginia by A. H. Mullins in 1890.

■ Social events in early West Virginia centered on church and civic activities. They were called "soups," because when people gathered to meet, they would all bring ingredients to make enough soup to feed the group.

(pasta with beans), *minestra* (vegetable soup), and cannoli are popular in the area around Clarksburg where more than half the population of the county is of Italian descent (Carlson, 1997). Other West Virginia specialties include pumpkin pie; eastern black walnut confections; and huckleberry muffins, cakes, and pies.

The West

Regional Profile

The western United States is the largest region in the nation, encompassing an enormous diversity of lands, from the icy tundra of Alaska to the tropical volcanic islands of Hawaii. The tallest mountains in the country, vast fertile valleys and coastal plains, stretches of scenic desert, and temperate rainforest add to the variety.

It is the history of the open wilderness that links this region. Indigenous peoples adapted their lifestyles to fit each climate and terrain. Pueblo Indians made their homes in the cliffs and cultivated corn, beans, chiles, and squash; the Inuits of Alaska lived in ice igloos and hunted sea mammals and fish for food; and the native Hawaiians enjoyed such fresh abundance that they cooked few dishes (see chapter 5 on Native Americans and chapter 12 on Southeast Asians and Pacific Islanders for more information). The first whites in the West were explorers, trappers, miners, and traders—hardy individuals (mostly men) seeking their fortune. Emigrants came from every direction: the Spanish and Mexicans from Mexico in the south, Russians from the north, Chinese and Japanese from the west, and the numerous pioneers of northern and southern European descent (mostly English, Scottish, Welsh, Danes, Swedes, Slavs, Italians, and Greeks) from the Midwest, looking for new farming, ranching, and fishing opportunities. The West is the most diverse region not only in climate and terrain but also in population.

The West is divided into the Mountain states of Arizona, Colorado, Idaho, Montana, Nevada, New Mexico, Utah, and Wyoming and the Pacific states of Alaska, California, Hawaii, Oregon, and Washington. Approximately 22 percent of Americans reside in the West, and of these, over half live in California. It is a mostly urban population, with a majority of people living in cities or suburbs, ranging from a low of 52 percent in Montana to almost 93 percent in California. It is also a young population, with fewer persons over the age of 65 than the national average; the exceptions are Arizona and Montana, with slightly above average numbers of elders.

Large numbers of many ethnic groups reside in the West (see Figure 15.6). Compared to total U.S. figures, three times as many Asians and Pacific Islanders, more than twice the Latinos, and two times as many Native Americans are part of the western population. African Americans and whites fall below the U.S. average in the region.

Figure 15.6
Demographics in the West

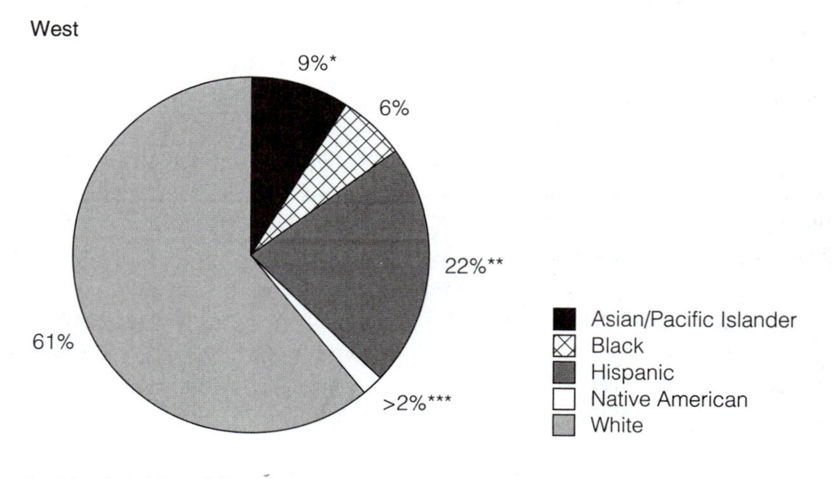

As a portion of each group nationally, 55 percent of all Asians and Pacific Islanders, 47 percent of all Native Americans, and 45 percent of all Latinos live in the West. This includes disproportionate numbers of total Inuit (89 percent), Japanese (72 percent), Aleut (72 percent), Filipinos (68 percent), Salvadorans (62 percent), Chinese (55 percent), Vietnamese (54 percent), Mexicans (52 percent), American Indians (46 percent), and Koreans (44 percent). Among whites residing in western states, there are large numbers with Danish and Spanish ancestry (45 percent each), and Yugoslavian heritage (37 percent, including those from what are currently Croatia and Serbia). In 1991, 31 percent of immigrants to the United States settled in the West, 23 percent in California.

Furthermore, a few individual western states report notable ethnic population figures. Asians and Pacific Islanders account for 63 percent of the population in Hawaii, for example. Latinos are 39 percent of the New Mexico population and 29 percent of the California population. Over 5 percent and 9 percent of the respective populations in Arizona and New Mexico are American Indian, and Native Americans, including Inuits and Aleuts, make up 15 percent of the population in Alaska.

Median income is a little below the U.S. 1997 average in the Mountain states and above the national average in the Pacific states: Alaska boasts the highest median income in the country, $50,829, almost 30 percent above the U.S. average. Yet, despite these figures, poverty levels in the West were approximately 10 percent above the national average. Particularly high levels are found in California and Arizona, and nearly one in every four persons lives below the poverty line in New Mexico, the highest rate in the nation. At the opposite end, low poverty rates are found in Colorado, and Utah and Alaska have the lowest numbers of persons living in poverty in the nation, around 8 percent.

Both the Mountain and Pacific states have above average numbers of high school graduates in their populations, and average and above average numbers respectively of persons over the age of 25 with college degrees. Although only 40 percent of persons living in the West adhere to a Christian faith, the highest percentage of Christians in the nation reside in Utah (80 percent—nearly all are Mormons) and large numbers are also found in New Mexico (58 percent—many are Roman Catholics).

Traditional Fare

The West was largely unknown to whites before the 19th century. Good maps of the region did not exist until the 1870s. Before that time, adventurous trappers and traders made their way into the territory from the Great Plains, where the buffalo was a source of jerky, pemmican, and fat for cooking (see Table 15.7). Miners who followed the strikes in the California Mother Lode, Colorado's Pike's Peak, Montana's Grasshopper Creek, and the Alaska Klondike prepared their own meals, usually pork, beans, and hardtack (tough, dry, unleavened bread or biscuits) three times a day. Some were dependent on the waystations, hotels, and boarding houses that opened to support the rush (Luchetti, 1993). Neighborly hospitality so common in the Midwest and South disappeared in the name of profit; miners were charged the maximum for supplies (e.g., eggs for a half dollar each, potatoes for a dollar a pound, and a box of apples for $500) and a meal would cost about $3 in a local establishment. The farmers and ranchers who later made their way westward survived on game, including rabbit, squirrel, beaver, bear, deer, elk, and moose. Venison, for example, was prepared as steaks, chili con carne, and even as mincemeat for pies. Corn breads and potatoes rounded out the meals. Sourdough breads and biscuits were popular with settlers in the Mountain states, California, and Alaska. Mashed potato or a milk and flour starter was left out to "catch" wild yeast and begin fermentation. Once going, the starter was kept indefinitely, replenished each time a little was used as leavening. Many supplies shipped to the West Coast, such as butter, arrived moldy or rancid.

■ European visitors to the West were appalled by both the poor quality of food and the manners practiced by the inhabitants. They complained of frenzied, rushed meals eaten with knives used like forks and handkerchiefs used like napkins. The "boarding house reach" came about in response to service at the earliest waystations, where a meal was offered at one time only to any and all comers. Diners who hesitated to grab food from the single platter placed in the middle of the large table didn't eat (Luchetti, 1992).

The growth of towns and the success of irrigation increased the food supply in the West. Expensive goods like wines and chocolates became available, and eastern specialties, including Long Island duck and Smithfield ham, were offered at restaurants. Depending on the region, potatoes, corn, apples, wheat, and hops prospered; cattle, dairy cows, and sheep became plentiful. With the arrival of experienced fishermen, more of the Pacific coast seafood was utilized. Salmon, crab, oysters, and

■ One innovative seafood recipe was the *Hangtown fry,* a platter of scrambled eggs with fried oysters, and bacon on the side, created in Placerville, California (known at the time as Hangtown), during the 1860s for a miner who had just struck it rich.

Table 15.7
Western Specialties

Group	Foods	Preparations
PROTEIN FOODS		
Milk/milk products	Milk, cheese (Cheddars like Cougar Gold, Monterey Jack, Tillamook); Basque sheep's milk cheeses	
Meat/poultry/fish/eggs/legumes	Native game, including buffalo, deer, elk, moose, antelope, mountain sheep, mountain goats, bear, *javelina* (wild pig), beaver, rabbit	Game meat steaks, roasts, stews (such as *chili con carne*), hamburger, sausages; beaver tail
	Beef, mutton and lamb, pork	Steaks; beef *enchiladas, tamales, chimichangas, pirozhki; teriyaki;* Indian tacos; *Pueblo pozole;* lamb spit-roasted or roasted with chiles; *chorizo;* luau (pit-cooked) pork
	Clams (e.g., geoducks), crab (Dungeness, king, snow), oysters, shrimp, squid	Clam chowder, Seattle clam hash; *cioppino,* steamed crab, crab cocktails, fried *calamari;* grilled or poached salmon, *lohimuhenno's* (salmon chowder); *sushi, sashimi, teriyaki;* trout grilled with bacon
	Salmon, tuna, halibut, mackerel, sardines, anchovies, *mahi mahi,* bonito, marlin, snapper; freshwater fish, particularly trout	
	Chicken eggs	Hangtown fry
	Dried beans	Chickpeas with lamb, chickpea pudding; Basque beans; lentil soup with lamb, lentil and sausage casserole, white beans cooked with pimento and cheese; split-pea soup
CEREALS/GRAINS	Wheat, corn	Sourdough breads, biscuits, pancakes; *sopapillas;* fry bread; *panocha; capriotada;* whole wheat Mormon bread; *bara brith; malasadas;* Hawaiian bread; tortillas (corn or wheat); *piki;* Asian noodle dishes (e.g., *saimin*) and dough-wrapped foods (egg/spring rolls, *lumpia,* wonton); fortune cookies
FRUITS/VEGETABLES	Apples, apricots, wild and cultivated berries, cactus fruit, cherries, dates, figs, grapes, kiwifruit, lemons, oranges, peaches, pears, pineapple, plums, prunes, sugar cane	Fresh fruit desserts; fruit added to roasts or poultry stuffings; preserves, jellies, wines; cold fruit soups
	Artichokes, avocados, asparagus, broccoli, breadfruit, cauliflower, chile peppers, eggplant, *jicama, nopales,* olives, onions, specialty lettuces (arugala, radicchio, rocket), tomatoes, *tomatillos,* potatoes, taro root, zucchini	Fresh vegetable side dishes; mesclun salads; *guacamole;* Basque potatoes; squash patties; *poi*
ADDITIONAL FOODS		
Seasonings	Chile peppers (especially New Mexico/Anaheim, Jalapeño/Chipolte, Serrano); cinnamon, cilantro, epazote, cumin, garlic, oregano, mint, safflowers (dried petals), *yerba buena;* chocolate; vanilla	Fresh chiles, dried chile powders, smoked chiles, pickled chiles; salsas; red or green chile sauces; mole sauce; fresh, dried, powdered, roasted, pickled garlic
Nuts/seeds	Almonds, hazelnuts, macadamia nuts, pine nuts (piñon seeds), pumpkin seeds	
Beverages	Varietal wines; coffee; tea (camomile, Brigham Young); hot chocolate	Coffee drinks (lattes, etc.); *picón punch*
Fats/oils	Olive oil	
Sweeteners	Sugar from beets, cane	Sugar cane is eaten fresh in Hawaii.

clams were especially popular. Hearty breakfasts of the time featured steak and eggs or ham and eggs, served with hash brown potatoes and buttermilk, buckwheat, or sourdough pancakes.

Many of the immigrants to the West came in search of mining and railroad jobs, including the Chinese and Mexicans. Both of these groups enjoyed highly seasoned foods and promoted the use of chile peppers in the West. Other groups, such as the Italians, Japanese, and some Greeks, became involved in the fishing industry and introduced specialties like seafood *cioppino* (a seafood stew using local fish and shellfish) and *teriyaki.* Many immigrants opened restaurants and markets to serve the needs of the booming towns. German sausages, Italian cannoli, and Chinese stir-fried dishes were all available. Still other immigrants arrived looking for farmland; they planted the fertile Pacific coastal regions and the California Central Valley with temperate fruits and vegetables, such as apples, pears, dates, grapes, plums, prunes, cherries, artichokes, avocados, broccoli, brussels sprouts, lemons, grapefruit, and oranges.

In Alaska and Hawaii, white settlers faced different challenges. Both states became dependent on whaling. In Alaska, seafood was the main food commodity, including salmon, halibut, and several kinds of crab. In Hawaii, pineapple, sugar cane, and other tropical crops were successful exports. Both states were, and still are, reliant on the mainland for the majority of other food supplies.

State Specialties

Alaska

The cooking of Alaska has been hampered in its development by the limited variety of foods. The climate and terrain are by no means uniform throughout the state, with warmer regions found along the coastal panhandle extending south into Canada, around the Yukon River Delta, and the protected Matanuska Valley in the south-central region, but there is little land suited for agriculture.

The Indians, Inuits, and Aleuts of the region lived primarily on seafood and game. Wild berries and roots were harvested in the short summer months (see chapter 5 on Native Americans). The first permanent white settlement in Alaska was on Kodiak Island. The Russians arrived in 1784 to hunt fur seals. They tried to grow wheat to supply the colony, but the grain wasn't hardy enough for the cold climate. The Russians spread southward and established settlements in California to provide goods for the Alaska hunters. They brought *kasha,* a cooked buckwheat porridge, buckwheat *blini,* and soups of fish or cabbage. *Pirogs,* large filled pastries, were made with fish, game, or cabbage. At Easter, they prepared traditional dishes, such as the rich fruitcake called *kulich* (see chapter 7 on Europeans).

In 1896, gold was discovered in the Klondike, starting a rush of prospectors to the territory. Supplies were severely limited, and most of the new settlers lived on little more than flour, bacon, salt pork, lard, and a bit of coffee or tea. Most kept a sourdough culture going to make breads and biscuits. Sourdough specialties included poppy seed potato bread, rye bread with caraway seeds, and whole-wheat bread. Kelp was collected at the coast, and wild flowers were boiled to make syrup. Some miners hunted to supplement their diets, and game, such as deer, caribou, moose, Dall sheep, rabbits, and ptarmigan, was available. Game meats are still popular today, including steaks, roasts, and hamburgers made from moose meat, and caribou meat sausages and Swiss steaks.

Sheep are now raised on the Aleutian Islands, providing lamb for the mainland. Cattle ranches are found on Kodiak Island and in the Delta and Matanuska Valley. Hog farming is on the rise, particularly around Palmer. More important are the dairy operations, providing fresh milk, butter, and cream. Potatoes are the most successful crop, but vegetables like cabbage, cauliflower, and rhubarb also grow in the region. Many of the vegetables attain gigantic proportions in the long daylight hours of summer; for example, cabbages may reach 70 pounds and rhubarb sometimes grows 4 feet tall. Blueberries, gooseberries, lingonberries, raspberries, salmonberries, and strawberries are produced in small amounts. The primary agricultural product of Alaska is seafood, ranking first nationally in quantity and value of the yearly catch. Salmon, herring, and halibut are the most prevalent fish. Shrimp and crab, including Dungeness, Snow (also known as

- Dairying in the West has produced several Cheddar-style cheeses, such as Cougar Gold from Washington, Tillamook from Oregon, and Monterey Jack (a mild, white cheese) from California. More recently, French-style chèvre goat cheeses have become a specialty.

- The Klondike prospectors were so dependent on sour yeast cultures used to make bread that they became known as "Sourdoughs."

- Reindeer herds were imported to Alaska in the late 1800s from Siberia and Scandinavia in the hopes they would become a profitable meat source. Many starved to death when the ranges were overgrazed, and others are thought to have become part of caribou herds. Today some reindeer in the Seward peninsula are raised for meat and to supply reindeer antlers to Asia.

- During the Great Depression, the U.S. government supported the relocation of 200 families from Michigan, Minnesota, and Wisconsin to establish farming in the Matanuska Valley. Many were of German or Scandinavian descent.

Salmon and other seafood are favorites in Alaska and the Pacific Northwest. (© Buddy Mays/CORBIS)

Tanner), and limited King crabs, are trapped during the winter months. Most are frozen or canned for export to the rest of the United States and Japan.

Arizona

The state of Arizona has some of the most scenic terrain in the nation, from the majestic Grand Canyon to the fantastic formations of Monument Valley to the broad plains of the Sonoran Desert. Despite a climate conducive to growing many fruits and vegetables, there is not enough rainfall to support much agriculture without irrigation. Arizona fare reflects the arid conditions and the preferences of the people who settled the region, especially the Native Americans, the Spanish, and the Mexicans.

Some Native Americans who lived in the state prior to European contact, particularly Pueblo peoples, cultivated small amounts of corn, beans, chile peppers, squash, and pumpkins. Pine nuts (piñon seeds) were indigenous foods. Juicy fruits (called *tunas,* or Indian figs) and the young pads (*nopales*) of the saguro and prickly pear cactus were other native items. Small game provided meat in the diet. They prepared stews flavored with chiles and very thin blue corn tortillas called *piki.* Other tribes, such as the Navajo and Apache, were initially hunters who roamed the region in search of big game and wild plants (see chapter 5 on Native Americans for more information). When the Spanish arrived in the 16th century, they introduced many of the foods that became elemental in the cooking of the Indians in both the United States and Mexico, including wheat, hogs, sheep, and cattle. The foods and flavors of the Old World and the New coalesced into a new cuisine. Wheat tortillas became a staple in parts of the American Southwest and in northern Mexico. Hearty beef or lamb stews replaced the mostly vegetarian versions, and fry bread, made with wheat flour and cooked in lard, became common. Dishes that are popular in Arizona with Native American and Spanish or Mexican roots include *menudo* (spicy tripe soup flavored with mint), *pozole* (hominy flavored with pork and chile peppers), *chimichangas* (wheat tortillas wrapped around a beef or chicken and vegetable filling, then deep-fried), beef tamales, quesadillas, burritos and *Indian tacos* (fry bread folded and filled with meats, cheeses, vegetables, and salsa).

Irrigation has provided some agricultural opportunities in the state. Grapefruit, lemons, melons, and figs flourish in the region, as do winter crops of broccoli, cauliflower, potatoes, onions, and carrots. Two vegetables popular in Mexico have found success in Arizona also. *Tomatillos,* a relative of the ground cherry that looks a little like a green tomato in a papery husk, and jicama, a crispy tuber related to the

■ Fresh coriander leaves (also known as *cilantro* or *Chinese parsley*) and cumin were probably introduced to the New World by the Spanish. They were quickly accepted in Mexican cuisine and from there were brought to California with the Spanish missionaries.

■ The Roosevelt Dam on the Salt River in Arizona was completed in 1911. It was the first major irrigation project undertaken by the U.S. government, providing water to the Phoenix region for agriculture and housing.

morning glory, have both been transplanted from Mexico to Arizona. Tomatillos are added to salsa verde and other green sauces, and the sweet pea–flavored jicama frequently provides a crunch in local salads. Cattle ranching and hog farms are common and the Navajo have specialized in raising sheep and goats.

California

California cuisine is a fortuitous fusion of natural resources, agricultural abundance, and a diversity of people. The physical terrain varies between Death Valley (the lowest point in the United States) to Mount Whitney (the highest point in the 48 contiguous states). The climate ranges from cool and moist in the redwood forests of the north to hot and dry in the southern deserts. One range of mountains edges the eastern boundary of the state, and a second runs along the Pacific, dropping down dramatically to the coastline. Seafood is prolific, and the annual catch is second only to Alaska. The state leads the country in agricultural output, however. It supplies over half of the nation's fruits and vegetables and is first in egg and milk production.

The foods of the California Native Americans depended on their location. Coastal tribes survived on clams, mussels, and fish, with local berries and greens. In inland areas, acorns were the foundation of the diet, leached of tannins and processed as a meal or flour (see chapter 5 on Native Americans). The first whites to settle in the state were the Spanish in the 18th century. They built several *presidios* (forts), *pueblos* (small farms), and a series of missions (each one day apart in travel time), with the purpose of protecting their claim to the territory. The Spanish cultivated wheat, olives, grapes, tomatoes, oranges, lemons, pears, apricots, figs, and pomegranates in the lands surrounding the missions and planted native foods including corn, beans, and squash as well. Cattle and hogs were raised. Many of the local Indians were forced into the missions as laborers.

Despite Spanish claims to the contrary, many other European powers sought access to California riches. Russia, in particular, established colonies in northern California to support their fur trade along the Pacific coast and in Alaska. When Mexico gained control of much of the state in the early 19th century, much of the mission territory was redistributed as grants to resident families, who founded wealthy *rancheros* producing mostly beef. At the same time, settlers from throughout the country were making their way across the mountains and founding farms in the fertile valleys of the state. With the discovery of gold in 1848, over 200,000 prospectors and fortune seekers arrived, followed by thousands of Chinese, Italian, and Japanese immigrants. By the beginning of the 20th century, the population had surged to over one and a half million people, many lured by jobs in farming, ranching, fishing, fish processing, meat packing, and the canning of fruits and vegetables. Agriculture continues to draw immigrants from Mexico and Central America looking for migrant farm work. Other recent immigrants who have added to the diversity of the state include Vietnamese, Cambodians, Laotians (particularly Hmong), Koreans, Asian Indians, Ethiopians, Filipinos, and Samoans.

Many California specialties are attributable to the ethnic preferences of the population. The Spanish planted the first vineyards in the state to supply wine for the Catholic sacrament. The Mexicans brought corn tortillas, refried beans, guacamole, and popular filled dishes like tacos and enchiladas. Italians introduced northern Italian favorites, such as polenta and pesto, as well as seafood dishes, such as the tomato-based fish stew made with local fish and Dungeness crab called *cioppino* and fried *calamari* (squid). They were the first to plant artichokes in the region. The Chinese offered authentic southern-style stir-fried dishes, wontons, egg rolls, and adapted dishes like chop suey and fortune cookies. The Japanese added *sukiyaki, teriyaki, tempura, sushi,* and other favorites to the mix. Armenians in the Central Valley brought flavorful lamb skewers, *dolma* (rice- and lamb-stuffed eggplant or grape leaves) and specialty desserts such as baklava. Newer immigrants have popularized hot Thai dishes, Vietnamese *pho,* Filipino *lumpia,* and Indian *Tandoori* cooking and flat breads in the state (see the chapter specific to each ethnic group for more details).

The abundance of fruits and vegetables in the state has led to the creation of "Californian cuisine," dishes that emphasize what is fresh and local. California accounts for nearly the entire U.S. production of avocados, artichokes, garlic (some of which is processed as granules

■ The earliest Spanish settlement in what is now San Francisco was named Yerba Buena (meaning "good herb"), after a prolific native mintlike plant, *Satureja douglasii.* The herb was used by Indians in the region as a flavoring, especially for soups and stews.

■ Although there are undoubtedly other incidents of cannibalism in U.S. history, one of the most famous occurred in 1846–47 when a group of 87 settlers coming from Illinois to California was stranded in the mountains of the Sierra Nevada during the winter months. Almost half the people in the Donner Party perished from hunger, and those who made it survived, in part, through eating the dead.

■ Both San Francisco and Los Angeles claim to be the place where fortune cookies were invented. Sometime between 1907 and 1916 a Chinese chef created the popular treat, probably as a restaurant promotion. The idea of the written fortune may date back to the 12th century, when Chinese insurgents plotting the overthrow of Mongolian rule distributed their plans inside moon cakes.

California produces more than half of the fruits and vegetables consumed in the nation and takes pride in a cuisine based on fresh, local ingredients. (© Paul Barton/The Stock Market)

■ Luther Burbank, a native of Massachusetts, moved to California in 1875 to conduct his experiments on plant breeding. He developed nearly 800 new varieties of fruits, vegetables, and flowers, including the Russet Burbank potato, the plumcot (a cross between plums and apricots), and the Santa Rosa plum.

■ Kiwifruit were first popularized by specialty produce maven Frieda Caplan of Los Angeles. In the 1960s, she imported the fuzzy-skinned, juicy green fruits known as *Chinese gooseberries* from New Zealand. When they met consumer resistance, she promoted them with the catchier name kiwifruit, and a trend was born.

■ By 1853 a settlement with over 500 Chinese fishermen existed at Monterey Bay (later made famous by author John Steinbeck in his 1945 novel *Cannery Row*). They were edged out by Italian fishermen after the turn of the century.

and powder), walnuts, almonds, apricots, nectarines, olives, dates, figs, pomegranates, prunes, and persimmons. It leads the nation in growing lettuce, broccoli, celery, cauliflower, carrots, lima beans, spinach, grapes, lemons, strawberries, plums, peaches, cantaloupes, and honeydew melons. Furthermore, 90 percent of all raisins consumed in the world are produced in California. Specialty fruits, such as kiwifruit and feijoa (a small, green egg-shaped fruit native to Brazil that tastes a little like pineapple and eucalyptus with mint nuances), are now being grown in the state. Innovative dishes such as pasta with chanterelles and grilled duck breasts, goat cheese salad with arugula and radicchio (both created by Wolfgang Puck, who started two well-known Los Angeles restaurants called Spago and Chinois on Main), poached salmon with fresh basil and olive butters, and pears and figs poached in Zinfandel with Cassis cream (by Alice Waters, founder of the restaurant Chez Panisse in Berkeley) are examples of the California approach to cooking. Other trendy items started in California include *mesclun* salads (made with a mixture of baby lettuces) and roasted garlic, which can be added to salads and stews or spread on bread like butter.

Seafood is also popular in the state. During the early 20th century, sardines were the leading catch. As their numbers diminished due to overfishing, other fish, such as tuna, salmon, halibut, mackerel, and anchovies, became the predominant catch. Dungeness crab, traditionally eaten steamed with melted butter and lemon or in crab cocktails, and squid, served deep-fried or over pasta, are popular. Abalone, a large, flat-shelled mollusk that clings to rocks off the California coast, is a specialty. The tough muscle must be pounded into a thin steak to tenderize it before cooking. It has a delicate, sweet flavor and is typically lightly floured and sautéed in butter.

California is probably best known for its wine. The European settlers of the mid-19th century first introduced the superior varietal grapes used in French and German wine making to the region. Today there are many premier wine regions in the state, including Napa Valley, Sonoma, Santa Cruz, Monterey, San Luis Obispo, and Santa Barbara. The Central Valley accounts for most of the grapes cultivated for the bulk wine market. Successful varietals include Barbera, Cabernet Franc, Cabernet Sauvignon, Merlot, Petite Sirah, Pinot Noir, Syrah, and Zinfandel among red wines and Chardonnay, Chenin Blanc, French Colombard, Gewürztraminer, Muscat, Pinot Blanc, Sauvignon Blanc, Semillon, and White (Johannisberg) Riesling among whites. More recently wine makers have been experimenting with other varietals, such as Cinsault, Grenache, Marsanne, and Mourvedre. Sparkling wines, which may be made from red or white grapes, and dessert wines are also specialties of the state.

Colorado

Colorado is the highest state in the nation, with an average elevation of nearly 6,800 feet. The east comprises high plains and Piedmont areas, while the west is noteworthy for two parallel ridges of the Southern Rockies. Although fur trappers and traders were familiar with the mountain region, the first permanent white settlements were established in the fertile San Luis Valley by Spanish and Mexican colonists. A short time later, thousands of prospectors from the Midwest were drawn to the territory when gold was discovered in 1861. Tin miners from Cornwall also arrived. Those who were skilled at farming stayed to homestead on the plains of the east when the gold boom ended and found that wheat was successful in some areas. German and Russian immigrants with beet-farming and sugar-processing experience were attracted to the state when it became known for its sugar beet crop, and Mexicans moved north to work the fields. Chinese immigrants came for jobs building the railroads. A silver boom in the 1890s brought Greeks and Slavs to the area. In more recent years, immigrants from Vietnam, the former states of the Soviet Union, and Guatemala have chosen Colorado as home.

Colorado pioneer fare was similar to that in the rest of the West. Game was a significant food source; beaver tail was a specialty. Bighorn sheep, deer, javelina (wild pig), and buffalo were also prevalent. The region developed a reputation for mountain trout, still a favorite. The Germans brought dishes like sauerbraten, *Rouladen,* and sauerkraut, and the Russians introduced stuffed cabbage, pirozkis, and sliced cucumbers in sour cream. Chile peppers and other spices came with the Mexicans, and dishes like chicken or turkey cooked in *mole* sauce (a rich blend of spices, nuts, and unsweetened chocolate) became popular throughout the state. Legend is that the famous Denver (or Western) sandwich, made with an omelet containing ham, onions, and green peppers between slices of bread, was invented by a Chinese chef making *eggs foo yung* with available ingredients (Jones, 1981).

In some mountain areas and drier plains, cattle and sheep ranching proved the best agricultural use of the land. Today Colorado is the leading producer in the nation of lamb, known as Rocky Mountain lamb.

Hawaii

It is thought that the volcanic islands of Hawaii were first settled by Polynesians from the Marquesa Islands and Tahiti in the 5th century. The foundation of their diet was starchy vegetables, such as taro root (traditionally made into *poi*—boiled, pounded into a paste, and slightly fermented), breadfruit, plantains, cassava, and yams. Seafood and possibly pork and chicken were also eaten (see chapter 12 on Southeast Asians and Pacific Islanders for more information). British explorers discovered the islands in 1778, and missionaries soon followed. The area became a major American port for whaling ships in the 19th century, and Japanese, Chinese, Korean, Filipino, and Asian Indian agricultural workers came to support the developing pineapple and sugar cane industries.

Hawaiian fare today is an amalgamation of traditional native dishes and foods introduced by the many immigrants to the islands. Unlike some areas of the nation, where various foreign contributions have melded into a single cuisine with occasional European or Latino overtones, many dishes in Hawaii maintain their ethnic integrity. Foods from different cultures are commonly served at the same meal, however, representing the state's diverse heritage.

Popular Hawaiian foods include those with Japanese origins, such as teriyaki-grilled meats and fish, *sashimi* (raw, thinly sliced fish), and noodle dishes like *saimin,* an island adaptation of ramen noodles, topped with pork and frequently eaten for lunch. The Chinese brought wok cooking, *dim sum,* long-grain

■ E. & J. Gallo, headquartered in Modesto, California, is the largest wine producer in the world. It controls approximately 30 percent of the national market and over 50 percent of exported U.S. wines.

■ Pioneer cooks in Colorado had difficulties making some baked goods at high altitudes. It took much experimentation to determine that leavening must be reduced and oven temperatures raised to produce moist, finely crumbed cakes.

■ Buffalo is being raised and processed as a specialty meat at several ranches in the West, including Rocky Mountain Natural Meats in Denver.

■ The Hawaiian *pupu* platter with bite-size snacks epitomizes the international nature of food in the state. Typical appetizers include *lomi-lomi* (Hawaiian marinated fish or scallops); *rumaki* (water chestnut–chicken liver–bacon skewers); Chinese chicken wings, spareribs, wontons, and dumplings; roast pork; poi; coconut chips; Portuguese sausage; Korean meatballs; curried fritters; or small *kumaki* skewers of meat, chicken, or fish.

■ Macadamia nuts reputedly have the hardest shell of all nuts to crack (a job traditionally done by placing a board on top of the nut and driving a car over it) and are processed between steel rollers. They are grown on the islands of Hawaii, Maui, Oahu, and Kauai.

■ The Parker Ranch on the island of Hawaii has 60,000 head of cattle and is second only to the King Ranch in Texas in size.

■ Maui onions are considered the sweetest in the world—by Hawaiians, that is. They are the same variety of the Vidalia onions of Georgia.

rice, soybeans, bok choy, lotus root, kumquats, litchi, and ginger to the region. Scottish scones and shortbreads are available, and Portuguese sweet bread is so common it is often called Hawaiian bread. Another Portuguese specialty, *malassadas* (fried doughnuts without a hole), is especially popular, and local variations made with poi or macadamia nuts are novelty items. Filipino fish sauces and *lumpia* (Filipino-style egg rolls), Korean *kimchi* (hot cabbage relish) and spicy beef dishes, and Indian curries are other contributions.

Historically sugar cane is the most important crop in Hawaii, accounting for 20 percent of the raw sugar produced nationally. But soaring costs and environmental concerns have put most sugar cane plantations out of business (only three remain), and Hawaiians are working toward agricultural diversification. Pineapples have also been a significant commodity, including fresh, canned, and juice products. More profitable operations developed in Asia have cut into production on the islands in recent years, however. Hawaii grows most of the world's supply of macadamia nuts and also exports famous kona coffee, which is grown on the western slopes of the island of Hawaii. Cattle ranches are found on the islands of Maui, Hawaii, and the privately owned Niihau; most provide beef for local consumption. Seafood, another Hawaiian specialty, is also mostly fished for Hawaiian markets, although some tuna is canned and exported. In addition to tuna, common food fish include mahi mahi (also called *dorado*), bonito, mackerel, marlin, scad, and both pink and red snapper.

Idaho

Potatoes are synonymous with Idaho. The blight-resistant variety developed in 1872 by horticulturist Luther Burbank (see "California") called the Russet Burbank potato is so common in the state it is now legally known as the "Idaho potato." They are grown primarily in the volcanic soils of the Snake River plain. Ten billion pounds are harvested annually, approximately one-third of national production. Sixty percent of the potatoes from Idaho are frozen, dehydrated, or milled into flour.

■ Potatoes were first grown in Idaho by Henry Harmon Spaulding, a Presbyterian missionary invited by the Nez Percé Indians to bring Christianity to the region. The first commercial crops were grown by Mormons in the area.

■ Trout is farmed in the Magic Valley of Idaho, providing over half of all the trout consumed in the United States.

Idaho was the last state in the country to be explored by whites. The earliest adventurers in the region were trappers and Christian missionaries. Although thousands passed through the state along the Oregon Trail heading west, few settled in the area. Not until the 1860s, when gold was discovered in the north of the state, did tens of thousands of fortune hunters arrive, including numerous Finnish, Welsh, and Chinese miners. The poorest lived on beans, bread, and coffee. Specialties included *lobimuhenno's* (a salmon chowder) brought by the Finns and *bara brith,* a bread studded with currants favored by the Welsh. Trout, salmon, bass, and sturgeon filled the local rivers, and game, such as deer, elk, and buffalo, was available. Recreational fishing and hunting continue today. Venison with huckleberry sauce is a popular dish.

Some miners stayed and took up farming after the gold rush, joined by southern European immigrants in the north of the state. Crops like wheat, oats, barley, hops, peas, lentils, and beans grew well in the dry climate of the north. Farmers in the more southern valleys discovered that, with irrigation, potatoes, corn, broccoli, and peppermint and spearmint (grown for their oils used in flavorings) could be cultivated. Cattle ranchers and sheepherders arrived in the region, particularly Basque immigrants. They introduced sourdough bread and a pencil-thin version of the spicy Spanish sausage *chorizo.* Lamb dishes, including grilled lamb and lamb stew, were a Basque specialty. *Basque potatoes,* still a favorite in the state, is a sliced potato, onion, and bacon casserole. Many dishes in Idaho today feature local ingredients with European nuances, such as split-pea soup, lentil soup with lamb, lentil and sausage casserole, white beans cooked with pimento and cheese, and ham with apple casserole. Apple jelly is added to mayonnaise to make salad dressing. Prunes are used in preserves and desserts, such as prune-whip pie and prune pudding (Lee, 1992).

Montana

Although the name Montana means "mountainous" in Spanish, two-thirds of the state is covered by the Great Plains, with Rocky Mountain ranges found only in the western region. Wheat, oats, barley, sugar beets, potatoes, beans, cherries, and apples are the primary crops. Beef, dairy goods, lamb, and pork are other agricultural commodities.

Native Americans in the state lived mostly in the plains region, surviving on buffalo and wild fruits and nuts. The first whites in the region were fur trappers and Jesuit missionaries. Major settlement did not occur until the gold rush at Grasshopper Creek began in the 1860s. The influx of people to the state provided opportunities in cattle ranching. Longhorns from Texas were driven to Montana, where they were fattened on the abundant prairie grasses before continuing on to distribution via Wyoming. Soon sheepherders of Basque, Russian, Rumanian, or Turkish heritage also made their way to the vast pasturelands. Most survived on the rudimentary meal of the prairie pioneer: salt pork and beans.

Montana specialties reflect the abundant game of the region, especially venison. It is prepared roasted or made into chili con carne or meatballs in a spicy tomato sauce. Elk, moose, and antelope are still plentiful. Wild birds include geese, ducks, pheasant, and grouse. Fish is available in the over 10,000 freshwater lakes and streams, particularly sockeye salmon and many kinds of trout. Native berries are a favorite, especially huckleberries, which are made into breads and pies, and the sour-tasting chokecherries, which are used in pies, cakes, preserves, jellies, and wine. Immigrants to the state have added their foods to the cuisine. Scandinavians who arrived from Minnesota to work in lumbering brought yellow split-pea soup, cold fruit soups, Swedish meatballs, and ham with cherry sauce. Borscht, cheese-filled pastry shells (*vatrousbki*), and cherry desserts were favored by the Russians. The Scots made oatmeal porridge and Mulligatawny stew with mutton, and central Europeans brought stuffed cabbage, dumplings filled with fruit or cheese, and pancakes rolled around cherries.

Nevada

The Nevada landscape is mostly stark, arid desert. Even though early Mormon settlers from Utah were successful in setting up small farms using crop rotation and irrigation, the largest agricultural commodities in Nevada are cattle and sheep; most crops that are grown are used for forage.

It was discovery of gold and silver at the Comstock Lode in 1859 that brought significant numbers of Europeans to the state. Before that time the region was sparsely populated by the Paiute, Shoshone, and Washoe Indians. Mining towns like Virginia City grew out of the dust into bustling urban centers, complete with theaters and operas. Food was in short supply and extremely expensive, yet wealthy prospectors and those who provided services to them supported extravagant restaurants, where internationally known chefs prepared oysters brought from San Francisco over the Sierra mountains as stews, oyster loafs, and stuffings for quail. Champagne imported from France was drunk from crystal goblets.

By 1900, the Comstock Lode was depleted, and many of the mining towns disappeared. Although half the state population of the time had been born outside the United States, few of the immigrants from Asia, Europe, and India who came to work in the mines, on the railroads, or on farms and ranches had the resources to maintain their culinary traditions in the privations of the desert. The exception was the Basques, who came in large numbers to the state, at first as hired shepherds and later as land-owning sheep ranchers. Initially only men came, and then they later brought their families. Women did most of the cooking, providing meals for the ranchers such as sourdough breads and biscuits with sheep's milk cheese and coffee for breakfast, and *Basque beans* (pinto beans with lamb or pork), lamb stews, or Spanish-style potato omelets for the main meal. In the towns, Basques often established hotels, which served as meeting places for Basques doing business in the area and for new immigrants. The hotels became famous for their four- or five-course meals served family-style, and non-Basque visitors often came for the food. Chickpea and meat stews, spit-roasted lamb, spicy chorizo sausages, and even traditional seafood dishes such as *bacalao al pil-pil* (dried salt cod cooked with garlic and olive oil) were offered when the ingredients were available. Red wine, chamomile tea, and *picón punch* (a beverage no longer common in Spain but still available where Basques live in the United States, made from bitter-orange picón liqueur, brandy, grenadine, and soda) were popular drinks.

New Mexico

The state of New Mexico is an arid mix of high desert plateaus, portions of the Great Plains, and

■ In 1923 a cloud of grasshoppers 300 miles long, 100 miles wide, and ½ mile thick descended on Montana, eating every green plant down to its roots.

■ A Montana specialty is *Rocky Mountain oysters,* or "tendergroin," which are calf testicles. They are traditionally served with beans and grilled bread.

■ Deer is ranched commercially in some parts of Nevada, and venison is a specialty.

■ Basques from Nevada may have introduced sourdough bread to San Francisco, where it has become a signature item.

■ The Basque hotels were where the wives of Basque sheepherders went during their pregnancies to escape the isolation of the ranches and to find help with the delivery.

■ Chiles are native to the Americas. There is no consensus on the common names for the hundreds of varieties, nor is there a single heat classification system. Mild New Mexican (also called Anaheim) chiles, hotter jalapeños (sliced, pickled jalapeños are sprinkled on nachos; smoked jalapeños are called *chipotles*), and very hot Serrano chiles are the most popular in New Mexican cuisine. The strings and wreaths of dried red peppers known as *ristras* in New Mexico are made from ripe New Mexican/Anaheim chile peppers.

■ The official state question in New Mexico is "Red or green?" referring, of course, to chile pepper sauce preference.

■ The piñon tree is the state tree of New Mexico.

■ The famous American gourmand and cookbook author James Beard was born in Portland, Oregon, in 1903. His father had come as a youth to the state from Iowa by covered wagon.

■ An adapted New England dish popular in Oregon is *Portland popovers,* essentially Yorkshire pudding batter that is flavored with herbs and baked in ramekins greased with drippings from roast beef or pork (Jones, 1981).

sections of the Rocky Mountains. Although the Spanish had explored the region extensively looking for gold and other minerals, it was Mexican-born settlers who established small farms called *ranchos* and *estancias* along the fertile banks of the Rio Grande River beginning in 1584. Missionaries followed, and Spanish culture pervaded the region. When the Santa Fe Trail opened in the early 19th century, Anglo-Americans (those who spoke English) arrived, adding to the population of Spanish Americans (mostly of Mexican and Native American ancestry who spoke Spanish), and Native Americans. The ethnic heritage of the state is evident in the many Native American celebrations, adobe villages, and numerous examples of Spanish architecture that still exist.

Nowhere is the blending of Native American, Spanish American, and Anglo-American culture more evident than in the cuisine of New Mexico. The Native Americans, mostly Pueblo peoples and nomadic Apache and Navajo tribes (see chapter 5 on Native Americans), and early settlers survived on corn, game like deer and rabbit, beans, squash, tomatoes, and chile peppers. The Mexicans brought beef, mutton, chicken, wheat, rice, chocolate, and numerous spices obtained through Spanish trade with the New World. Traditional dishes that evolved during this period are still common today. Examples include breads (corn tortillas, blue cornmeal piki, and the light, deep-fried wheat flour puffs called *sopaipillas*); soups and stews, such as *Pueblo pozole* (made with dried corn, beef or pork ribs, onions, garlic, and *epazote* or oregano), green chili stew (beef, tomatoes, onions, and a variety of green chile peppers), and garbanzo beans with lamb; and puddings, such as chile-flavored chickpea pudding, *panocha* (similar to Indian pudding from the Northeast, but made with wheat kernels and flavored with cinnamon), and *capriotada* (bread pudding with pine nuts, raisins, and mild cheese, traditionally served with a caramel sauce). Indian roasted lamb with red and green chile peppers, Spanish *arroz con pollo* (updated with chiles), and Mexican *chilaquiles, enchiladas* and *flautas* are other common dishes in the region. Popular desserts include anise cookies called *biscochitos* (a Christmas specialty), sopaipillas served with honey or cinnamon syrup, and *flan.*

Chile peppers, onions, garlic, oregano, *yerba buena, epazote,* safflower blossoms (dried and powdered, reminiscent of saffron), and abundant amounts of mint flavor savory dishes. Sweets depend on chocolate, vanilla, cinnamon, and nutmeg. Toasted pumpkin seeds or pine nuts from the native piñon tree were added to spice blends for sauces, providing a nutty flavor to roasted meats and poultry. Today almonds are used in the same way.

Oregon

Oregon is divided by bands of the Cascade–Sierra Mountains, both physically and climatically. The temperate, wet coastal weather of the west contrasts dramatically with the semi-arid plateaus in the east with their cold winters and warm summers. The coastal region is endowed with a wealth of native foods, including deer, elk, antelope, rabbits, beaver, muskrats, ducks, geese, greens, wild mushrooms, and a multitude of berries. In many eastern regions of the state, near desert conditions exist.

The state was first explored by white trappers seeking beaver pelts. Methodist missionaries who followed had little success in converting the local Native Americans but did encourage settlement of the region. Tens of thousands of pioneers from New England, the South, and the Midwest traveled the Oregon Trail in the middle of the 19th century to participate in the promise of the new frontier. Yankees from the east named their settlements after their hometowns of Portland, Salem, and Albany. Many also survived on the coastal seafood that was familiar in Atlantic fare, particularly clams. Over two dozen varieties (e.g., butter, horse, Japanese littleneck, Manila, razor) live in the sands of the Pacific Northwest. Clam chowder, adapted with whatever local settlers could rustle up, including rice, tomatoes, and cabbage, was common (Jones, 1981). Geoduck (pronounced *gooey-duck*) clams are a large, native bivalve that have necks (siphons) that can stretch several feet. A geoduck may weigh up to 15 pounds, although most are in the 2- to 3-pound range. The body is sliced and pounded into thin steaks (similar to abalone), and the neck is usually minced or ground up for soups or stews. Other seafood specialties of the region include salmon, shrimp, crab, squid, and flounder.

Some settlers started small farms, particularly in the fertile Willamette Valley and the northwestern areas of the state. Fruit, such as pears (mostly Bartlett, with some Anjou, Bosc, Comice, Seckel, and Winter), Hood River apples, Golden Oregon prunes, plums, cherries, and domestic berries (strawberries, raspberries, cranberries, blackberries, and blueberries), proved very successful. Much of the annual fruit harvest is canned or frozen. An Oregon specialty is hazelnuts (also called filberts). The trees were introduced by an immigrant from France in 1858, and today nearly all hazelnuts consumed in the United States are grown in the state. Dairying is an important industry, producing milk, butter, and nationally marketed cheese. In more recent years, Oregon wines have gained a reputation for quality, especially for cool weather varietals like Pinot Gris (a white wine) and Pinot Noir (a red grape used in French-style burgundies). Fruit wines, such as those made from blackberries, currants, cranberries, or peaches, are also favorites in the region.

Fresh seafood, fruits, and vegetables are featured in Oregon cuisine. Grilled and poached salmon is a specialty. Poultry stuffed with prunes is popular, as are various fruit pies, soups, preserves, jams, and jellies. Some ethnic and regional fare traced to the settlers can also be found, such as German foods (schnitzels, sauerbraten, sage sausages, stuffed cabbage, sauerkraut, and strudels), and southern fried chicken served with biscuits and hominy.

Utah

The isolation of Utah attracted the first white settlers to the state. In 1847, an advance party of Mormons led by Brigham Young arrived in the Great Basin area surrounding the Great Salt Lake. He declared it the site where Mormons would found a self-sufficient community free from the persecution suffered in Ohio, Illinois, Missouri, and other states where the members of the Church of Jesus Christ of Latter-Day Saints had lived. That year, another 1,500 settlers joined his group, and by 1860, over 40,000 Mormons were living in the towns of Salt Lake City, Ogden, Bountiful, Provo, and Manti. Most came by wagon train, but nearly 3,000 walked the distance pulling all their possessions in hand trucks.

Most of Utah is scenic desert country, with the exception of the fertile valleys and foothills of the Wasatch Mountains in the northeast. Agriculture was difficult at first, and the early settlers survived on indigenous greens, weeds, and birds such as crows. The Mormons gradually created a sophisticated irrigation system that captured run-off from the mountains and were able to cultivate wheat, potatoes, onions, beans, peas, and a variety of fruit, including apples, apricots, tart cherries, peaches, and pears. Most farms in the state remain small, family-run operations, providing just what is needed for local consumption. Livestock proved a more profitable agricultural pursuit than field crops: Cattle, turkeys, eggs, and sheep became primary products. Swiss Mormon converts started dairying concerns in the Salt Lake City area.

Today approximately 80 percent of the state population are Mormon. Many were of northern European descent, particularly British and Scandinavians. Most had lived first in areas of the Midwest. They brought a preference for hearty foods. Ham, pot roast, roast beef, stews, and fried chicken remain favorite entrees, often served with homemade whole wheat bread or buttermilk biscuits. Spicing is usually mild: *Hamburger bean goulash,* a Utah specialty, is thought to be a denatured version of chili con carne (Lee, 1992). Milk gravy, made with browned flour, pork grease, milk, and seasoned with black pepper, was served with so many foods that cowboys riding through the area dubbed it *Mormon dip.* Potatoes, red cabbage, green beans, and peas are still common side dishes. The Mormons are well known for their love of sweets, and desserts are prominent in the diet. Layer cakes, fruit pies, strawberry shortcake, fruit candies, chocolates, and ice cream are commonly consumed. One notable favorite is *pepparkakor,* a Scandinavian ginger cookie that is often Anglicized as *pepper cookies.* Mormons do not drink alcohol or stimulating beverages like coffee or tea; lemonade and *Brigham Young tea* (sweetened hot water with milk) are traditional beverages.

One immigrant group in Utah that has maintained many of their food traditions are the Roman Catholic Italians who came from Calabria to Carbon County in central Utah. They often grow Mediterranean vegetables and seasonings, such as eggplants, tomatoes, endive, fava beans, fennel, zucchini, garlic, parsley, and

■ *Schnitzel sandwiches,* thin, breaded and fried veal or pork cutlets served between slices of bread, were invented in the United States and introduced to Germany by American soldiers. They are popular in some parts of Oregon (Carlson, 1997).

■ Utah's nickname is the "Beehive State," denoting the industrious agricultural efforts of the first Mormon settlers. The state flower is the sego lily, which some settlers were forced to eat to survive in the first years of settlement.

■ A popular tale in Utah history regards a plague of crickets that threatened the Mormons' first crops in 1848. They prayed to God, and suddenly sea gulls appeared (more than 1,000 miles from the nearest ocean) and ate the insects before they could do harm.

■ Sour cream raisin pie is a 20th-century recipe of undetermined origins that harkens back to the sweetened clabber, milk custards, and dried fruits of early pioneer days. It is popular in Utah.

■ Mormon tea, or Brigham tea, is also the name for a Native American brew common in the arid areas of the Southwest made from *Ephedra nevadenis.*

basil, in home gardens. Everyday fare includes bread sticks, pastas, minestras, salads, and fresh fruits for dessert. Specialties such as goat meats and goat cheeses, variety meats cooked up with eggs in a *frittata* or in a spicy stew, boiled chicken's feet, and deep-fried squash patties made with chopped squash and squash blossoms are traditional favorites (Raspa, 1984).

Washington

The state of Washington has a terrain and history similar to that of Oregon. Like Oregon, the state is divided by the Cascade Mountains into the wet, fertile western coast region and the dry, prairie-like eastern plateaus. Trappers were the first whites to enter the territory, followed in the mid-1800s by other pioneers, mostly farmers seeking land grants, of English, Dutch, German, and Scandinavian descent. Unlike in Oregon, large numbers of African Americans and immigrants from China and Japan came to Washington in search of work on the railroads.

Regardless of ancestry, most settlers lived on cultivated wheat, potatoes and other vegetables, wild game, fish (particularly salmon and halibut), and berries. Few distinctively ethnic flavors are common in Washington fare. Instead, immigrants in the state adapted their cooking to the plentiful food resources of the region: produce and seafood.

In the temperate western areas, corn, asparagus, onions, peas, pears, grapes, strawberries, peaches, raspberries, and plums became important products. Washington leads the nation in the production of both apples and cherries. Apples are especially associated with the state: Grown on the eastern slopes of the Cascades, they require extensive hand labor to thin each cluster of blossoms to a single "king blossom" and to pick any fruit that appears after the initial crop sets. This process produces exceptionally large, well-formed apples. Nearly the entire crop is devoted to Red Delicious and Golden Delicious varieties, although several other types are grown in small amounts, such as Fuji, Gala, Granny Smith, and Winesap. Hops, Walla-Walla onions (a cool weather sweet onion that is a different variety from the sweet onions grown in Georgia and Hawaii), mint, and spearmint are specialty crops. Washington is also known for its white varietal wines, such as Chardonnay, Chenin Blanc, Gewürztraminer, and White Riesling.

Washington has an abundance of seafood. Beginning in the 1800s the five species of salmon in the region (king or chinook, sockeye, dog, humpback, and silver) were sent fresh to West Coast markets or pickled and shipped in barrels to Hawaii, South America, the East Coast, and Europe. Iced fish was also sent to Europe, where it was smoked, then returned to the United States. Washington still leads the nation in total salmon catch and products. Oysters are another specialty. Many species live in the area, but only a few are commercially viable. The small, native Olympia oyster is a favorite, but overharvesting has greatly reduced availability. Pacific oysters, which were imported from Japan in the early 1900s, are the species consumed most often on the West Coast. They are usually named after the location where they are farmed, such as Malaspina and Willapa Bay. European flat oysters and the gigantic geoduck clams (see Oregon) are also harvested. Oysters are typically served on the half shell in the region, although bacon-wrapped broiled oysters, oyster fritters, oyster loaf, and oyster stew are common cooked preparations. Mussels are cultivated at Whidbey Island, and Dungeness crab, named for the town of Dungeness on the Olympic Peninsula, are caught during the winter months.

■ Legend has it that the early settlers of Seattle who came from New England longed for hash but were unable to obtain corned beef. So they substituted clams for the meat, cooking them with bits of bacon, onions, potatoes, and eggs to create a new dish. Today *Seattle clam hash* omits the eggs in favor of more traditional cream (Jones, 1981).

■ Floating oystering communities were established in the Puget Sound area during the late 1800s. Families would rake the mudflats at low tide for prized Olympia oysters. Steamboat merchants would later come by at high tide to purchase burlap bags full of oysters from the gatherers.

■ Coffee drinks such as lattes have become a specialty item associated with Seattle. The city is home to the Starbucks Corporation, which started as a coffee stand at the open-air Pike Place Market in 1971. Founder Harold Schultz almost single-handedly started the trend of U.S. coffee bars (modeled after Italian espresso bars) in 1984. There were nearly 1,400 Starbucks coffee bars throughout the nation by 1997.

Wyoming

The vast grasslands covering much of eastern Wyoming offered grazing for the cattle herds of Texas when pastureland in that state was depleted in the mid-19th century. The fact that it was free added extra incentive for ranchers to move their operations north. By 1886 there were 2 million head in the state, most of which were processed through Cheyenne for East Coast markets. An exceptionally cold winter that year, followed by overgrazing, dramatically reduced the number of cattle in Wyoming. Sheep became a significant commodity, and today cattle and sheep are the main agricultural products. Most crops supply forage for livestock, although sugar cane, winter wheat, barley, beans (especially pinto and Great Northern), and potatoes are cash crops for the region.

Wyoming is best known for its wild fish and game. Some of the most famous names in the history of the West made their living as trappers in the state during the early 1800s, including Jedediah Strong Smith and Kit Carson. Wyoming is amply endowed with big game, including mule deer, white-tail deer, pronghorn antelope, elk, moose, mountain sheep, black bears, and grizzly bears. Game meat is a specialty, especially venison and antelope. Tenderloin, sirloin, and T-bone steaks are cut; the ribs and sirloin tips make roasts (which are sometimes marinated and braised in wine, vinegar, and spices); the brisket, flank, and plate are used for stews or hamburgers; the hams are smoked; and miscellaneous meat is used to make Polish sausage or salami. Game birds include grouse, pheasant, and partridge, and wild turkeys are now being raised for sport. Pheasant roasted with apples or in a pie is a favorite. Many species of trout are found in the mountain streams, and lakes host bass, crappie, and catfish.

The state attracted a diversity of European immigrants, many of whom opened bakeries, confectionery stores, and restaurants. French croissants, Middle Eastern *halvah* (sesame seed candy), German schnitzel, and Chinese wonton soup were reportedly available in southeastern Wyoming as early as 1900 (Lee, 1992).

■ Wyoming was the first state in the nation to grant women the right to vote, own property, and receive equal wages for equal work in 1869. Legend has it that suffragist Ester Morris invited the representatives running for election to tea and "buttered them up" with her cakes and cookies as she persuaded them to support equality issues. A more likely scenario is that Wyoming passed the legislation hoping for publicity and to encourage more women to settle in the territory.

Canadian Regional Specialties

Canada is the second largest nation in the world in area, a diverse land with rainforests, vast prairies, deciduous forests, and sparse tundra regions. There are more lakes and inland waters in Canada than in any other country. It is a nation rich in natural resources, especially abundant freshwater fish and seafood, game, and fertile farmland.

The country is divided into 10 provinces and 3 territories. Just as in the United States, Canadian regional fare reflects indigenous foods and the preferences of the groups who have significantly impacted the nation historically: Native Americans and immigrants from Great Britain and France, with added flavors from smaller populations of immigrant groups from northern Europe and Asia. A recent conference explored Canadian regional fare, and its findings were published in a book titled *Northern Bounty: A Celebration of Canadian Cuisine* (Stewart & Powers, 1995). Although the editors note that there are many nationally known foods, from apple pies and cinnamon buns to lobster and Cheddar cheese, they emphasize that much of Canadian cooking is dependent on local ingredients. They provide an overview of five culinary regions: the Atlantic Provinces, Quebec, Ontario, the Prairies, and British Columbia.

Atlantic Provinces

The cooking of the Atlantic Provinces uses seasonal ingredients prepared as dishes familiar to the Acadians and British immigrants who settled the region, with a little German influence found in Nova Scotia. Seafood is preeminent, including lobster, oysters, clams, and mussels, as well as cod, salmon, haddock, halibut, smelt, and herring. Lobster and corn boils are popular in summer, and "clamburgers" with a little fat rendered from salt pork poured over the top are an Acadian specialty. Planked salmon and *Solomon Gundy* (pickled herring, a Lundberg County favorite) are also common. Malpeque Bay oysters and Prince Edward Island (P.E.I.) mussels are renowned worldwide. A foundation of the diet is many varieties of potatoes, such as Yukon Golds, Red Pontiacs, purple-fleshed potatoes, and Russets grown for their length (to make french fries) or their roundness (for chips). In Nova Scotia, new potatoes are combined with baby peas, beans, and carrots and flavored with butter and cream to make Scottish *hodge podge.* The Acadians make potato dumplings the size of baseballs (*poutine râpée*) that are served as a main course, side dish, or even as dessert with molasses. *Rappie* pie is another Acadian specialty, made from grated potatoes that are squeezed dry, then cooked with enough chicken broth to equal the extracted potato juices, until the potatoes are jellylike. It can be topped with cooked chicken, meat, or clams, flavored with molasses, and covered with bacon or salt pork before it is baked. Fish and chips (sometimes made with lobster) was a tradition in the Atlantic Provinces that is regaining popularity. Other foods of the region include fresh spring shoots of *Salicornia europaea* (crowfoot, glasswort, or samphire green) and fern fronds. Rhubarb, strawberries, saskatoonberries (also called shadbush or juneberries in the United States)

continued

Canadian Regional Specialties—continued

and blueberries are seasonal favorites. Lamb dishes and meat pies are common British foods found in the region (Comfort, 1995; Nightingale, 1995).

Quebec

The cuisine of Quebec is influenced by the French country roots shared by many of the residents. Yet even the most traditional of dishes, such as the meat pie called *tourtière,* vary according to the area where they are prepared. In Charlevoix, *tourtière* is made with cubed meat instead of ground meat, and in Saguena-Lac-St-John, it is more of a layered dish. In the Mauricie, it is made with salmon, and in Quebec City the filling is thickened with oats instead of potatoes, a British contribution.

Specialty foods of the region include leeks, oyster mushrooms, white asparagus, endive, apples (often used to make hard apple cider), and strawberries. Lamb, rabbit, game birds, salmon, and cod are popular. Seventy percent of the world's maple syrup comes from Quebec, and it is used to flavor many products, including *viande boucanée,* salt pork smoked in the maple sugar shack so that it is sweetened by the vapors of the bubbling pots; maple sugar pickles; and even maple sugar wine. Maple sugar pie is a specialty. Dairying is significant in the region. In addition to traditional French cheeses like *Brie,* Quebec produces *oka* (also called Trappist monk cheese), similar to Port Salut; Swiss-style raclette; and several chèvres (goat cheeses). Other common dishes include *cretons* (a pork pâté), *cipaille* (a layered meat pie, sometimes prepared with game), and *fèves au lard* (baked beans) (Armstrong, 1995; Burkhard, 1995).

Ontario

The cooking of Ontario originally emphasized the foods favored in Great Britain. The Scots brought scones, shortbread, and haggis, and the Irish added a love of potatoes. Cookbooks from the late 19th century include many recipes for game, beef, and poultry as entrees and numerous descriptions of sweet and savory pies and puddings. Two ethnic communities have influenced the cuisine of Ontario: the native Cree and Iroquois, and the German Mennonites who came to the Waterloo County from Pennsylvania (Pennsylvania Dutch). Indian foods include corn and bean soups made with venison or pork; bannocks baked in an oven, in a fry pan, or wrapped around a stick over a fire; and many game dishes. Moose muzzle was a specialty. Mennonite foods include those traditional in Pennsylvania, such as smoked hams, sausages, headcheese, dumplings, doughnuts, preserves and relishes, and numerous pies. Wild meat, fowl, and fish are often cooked by the Canadian Mennonites. Fried trout and pot pies made with pigeons or rabbit are examples (Duncan, 1995; Skye, 1995; Staebler, 1995).

Ontario is also known for its plentiful fruits. The temperate climate around the Great Lakes supports numerous varieties of apples, peaches, pears, cherries, berries, grapes, and rhubarb. Specialties include Dutch apple pies topped with streusel, apple or rhubarb coffee cakes, apple dumplings, apples baked in a crust and drenched in caramel sauce, peach crisps, and blueberry jam. Fruit is used to complement meats and poultry, and is added to cabbage salads, mashed rutabagas, and even spicy chili sauces. Wines are produced on the Niagara peninsula, Pelee Island, and the Lake Erie north shore. One unusual dessert wine that has become popular in recent years is *icewine,* similar to the German version known as *Eiswein.* It is made from White Riesling or Vital grapes that are allowed to ripen until they shrivel and the sugars become concentrated. They are harvested after they freeze; the small amounts of the intensely sweet juice that remain are pressed from the grapes and fermented. It is described as having an ambrosial flavor reminiscent of honey, apricots, peaches, oranges, tea, and toffee (Aspler, 1995; Baird, 1995; Ziraldo, 1995).

The Prairies

The Prairie provinces of Canada are red meat country. Buffalo sustained the few Native Americans and white pioneers who made their homes in the region in the early years. The meat was barbecued or dried to make pemmican. The intestines were used for sausages and to wrap a strip of meat that was then grilled over a smoky fire in the dish colorfully called *Crow-Indian-Guts;* baby bison were pit-roasted. As the buffalo herds diminished in the early 19th century, cattle took their place. Chuck wagon and range cooking were specialties. Although coffee, beans and sourdough breads, biscuits, and pancakes were mainstays, fresh beef dishes were the favorites, particularly fried steaks and stews made with onions, potatoes, and canned tomatoes. *Prairie oysters* (breaded, fried beef testicles) were popular. Some of the cowboy terms applied to foods indicated their role in the diet, such as "baked wind pills" (beans), "paperweights" (hot biscuits), and "yesterday, today, and forever" (hash) (Barss, 1995; Schultz, 1995).

The Prairies are also the breadbasket of Canada. The first wheat crops planted by the German Mennonites, Icelanders, Scandinavians, Americans, and British settlers often failed due to poor seed quality, droughts, fires, grasshoppers, and blizzards. But by 1904, Canadian flour won the Gold Medal at the St. Louis World's Fair, and wheat became a significant export crop. Foods of the Prairie farm families often included dishes from immigrant homelands, such as roast beef and Yorkshire pudding, stews, pot roasts, chicken and dumplings, stuffed cabbage, sauerkraut, and perogies (like midwestern bierocks). Biscuits accompanied many meals, including bannocks, *gems* (similar to muffins, cooked in fluted cups), and fried biscuits. During the depression years, bread with gravy and bread puddings filled many stomachs. Traditional baked goods included pound cakes, tea cakes, banana bread, shortbread, chocolate cake, carrot pudding, matrimonial cake (date squares), sour cream raisin pie, cinnamon buns, strawberry shortcake, and ethnic breads like Dutch Kringle bread, French bread, and Ukrainian *paska* (Fraser, 1995).

British Columbia

The West Coast of Canada was settled by British trappers and traders, not farmers. The relatively wealthy populace quickly became reliant on imported foods. As early as 1886, fresh fruit was being shipped to the region from California. Although Chinese, Japanese, and Asian Indians began to immigrate to the West Coast in the 1880s, their ethnic fare was isolated within each of their communities, while most residents favored the mildly seasoned foods of colonial England (Johnson, 1995). The globalization of cooking brought native ingredients and Asian influences to the forefront of British Columbia cuisine. Indigenous greens, such as miner's lettuce and the new shoots of thimbleberries or salmonberries, are lightly dressed as salads; camass bulbs are served as side dishes; and the many wild berries of the region (including thimbleberries, salalberries, salmonberries, huckleberries, saskatoonberries, red elderberries, and soapberries) are used in savory sauces for meats, in fresh desserts, in preserves, and in juices and teas. Sushi and samosas are now available at delicatessens along with the sandwiches and soups. Canadian native foods used in traditional Asian dishes offer new flavors, such as Filipino *sinigang* with salmon instead of pork, Chinese pork and cod braised with garlic, or Japanese geoduck sashimi. Four major varietal grape–growing regions have developed in British Columbia, producing well-respected Chardonnay, White Riesling, Pinot Blanc, and Pinot Noir wines (McWatters, 1995; Turner et al., 1995; Wong, 1995).

Glossary of Ethnic Ingredients

Abalone: Large flat mollusk with finely textured sweet flesh in the broad muscular foot that holds it to rocks (must be pounded before use). It is common in the waters off Asia, California, Mexico, and New Zealand. Available fresh, frozen, canned, and dried.

Abiu: Yellow egg-shaped or round fruit native to the Amazon; popular throughout Brazil and Peru. Translucent white flesh with caramel-like flavor.

Acerola cherries: Caribbean berries resembling cherries that are exceptionally sour.

Achiote: See *Annatto.*

Adzuki bean (aduki, azuki): Very small, dark red bean used primarily in Japanese cooking.

Ahipa: See *Jicama.*

Ajowan (ajwain, carom or lovage seeds): Similar to celery seeds in appearance and to thyme in flavor. Used in Asian Indian and Middle Eastern cooking.

Ajwain: See *Ajowan.*

Akee (ackee, achee): Red fruit with three segments containing large inedible seeds and flesh that looks like scrambled eggs. Nearly all parts toxic. Available canned.

Alligator: Reptile native to rivers and swamps throughout the southern Gulf Coast region, from Florida to Texas. Mild white meat, with texture similar to veal. Tail and other parts eaten.

Almond paste: Arab confection of ground almonds kneaded with sugar or cooked sugar syrup (some brands also contain egg white) used in many European and Middle Eastern desserts. Marzipan is a type of almond paste made with finely ground blanched almonds.

Amaranth (Chinese spinach, pigweed): Leafy, dark green vegetable similar to spinach. The high-protein seeds can be ground into flour and used in baked products, or boiled and eaten as cereal. Popular throughout Asia and Latin America.

Ambarella (golden apple; Jew plum; kadondong; otaheite apple; vi-apple): Small, oval-shaped fruit with very strong flavor native to Polynesia but also found in Southeast Asia and Caribbean. Used unripe for preserves and ripe in desserts.

Amchoor: Dried unripe mango slices or powder, with a sour flavor.

Angelica: Herb with a licorice flavor commonly used in European dishes. Usually available candied.

Annatto (achiote, atchuete): Seeds of the annatto tree used to color foods red or golden yellow. Used in Latin America, India, Spain, and the Philippines. In the United States, annatto is added to some baked goods, Cheddar-style cheeses, ice creams, margarines, and butter for color. May be cooked whole in oil or lard to produce the right hue or used as a ground spice.

Apios: See *Groundnuts.*

Apon seeds (agonbono seeds): Seeds of the wild mango, commonly used in West Africa. Basis of the soup known as agonbono.

Areca nuts: See *Betel.*

Arracacha (Peruvian carrot): Starchy white root of the carrot family used in South America, especially Colombia, Peru, and Venezuela.

Arrowroot: Many varieties of a bland, mealy tuber found in Asia and the Caribbean. When made into a powder, it is used to thicken sauces and stews.

Artichoke: Globelike vegetable member of the thistle family, with multiple edible bracts ("leaves") crowning the undeveloped edible flower (the "heart"). The flavor is slightly sweet. Popular in Middle Eastern and southern European dishes.

Arugula (rocket): Small member of the cabbage family native to the Mediterranean; the peppery leaves are popular in salads throughout Europe.

Asafetida: Dried resin with a pungent odor reminiscent of burnt rubber, which nonetheless imparts a delicate onion-like flavor. It is available as a lump or powder and is commonly used in Asian Indian dishes.

Asian pear (apple pear): Round yellow fruit from Asia with the crispness of an apple and the flavor of a pear.

Avocado (alligator pear): Pear-shaped to round fruit with leathery skin (green to black) and light green buttery flesh. Native to Central America. Three varieties are often available. Eaten mostly as a vegetable, although considered a fruit in some cuisines.

Bacalao (bacalhau, baccala): Cod preserved by drying and salting. Must be soaked, drained, and boiled before use.

Bagoong: See *Fish paste.*

Bagoong-alamang: See *Shrimp paste.*

Bambara groundnut: Legume very similar to peanuts, native to Africa.

Bamboo shoot: Crisp, cream-colored, conical shoot of the bamboo plant. Used fresh (stored in water) or available canned in brine (whole or sliced).

Bangus: See *Milkfish.*

Baobab (monkey bread, lalu powder): Slightly sweet seeds from the large fruit of the native African baobab tree. Used roasted or ground. Pulp of the fruit is also consumed.

Basmati rice: See *Rice.*

Bean curd (Cheong-po, tofu, tobu, tempeh): Custardlike, slightly rubbery white curd with a bland flavor made from soybean milk. Japanese bean curd (called tofu) tends to be softer than Chinese, which is preferred for stir-fried dishes. A chewier version common in Southeast Asia is called *tempeh*. Cheong-po, a Korean bean curd, is made from mung beans.

Beans: See bean type.

Bean sprouts: The young sprouts of mung beans or soybeans popular in Asian cooking (sprouts may also be grown from the tiny seeds of alfalfa, also a legume). The crisp 1- to 2-inch sprouts are eaten fresh or added to stir-fried dishes.

Berbere: Ethiopian spice paste (typically very hot) used to season many foods, usually including allspice, cardamom, cayenne, cinnamon, cloves, coriander, cumin, fenugreek, ginger, nutmeg, and black pepper.

Bergamot orange: Pear-shaped orange with exceptionally tart flesh. Rind used to flavor dishes in the Mediterranean and North Africa; oil extracted from rind flavors Earl Grey tea.

Betel: The heart-shaped leaves of the betel vine (related to black pepper) are used to wrap areca nuts (from the Areca palm; the nuts are usually called betel nuts because of their use with betel leaves) and spices for paan in India. Betel nuts and leaves are chewed together in many Southeast Asian countries and in India to promote digestion. May stain the teeth red.

Bird's nest: Swallows' nests from the cliffs of the South China Sea made from predigested seaweed; used in Chinese soups or sweetened for dessert.

Bitter almond: An almond variety with an especially strong almond flavor, often used to make extracts, syrups, and liqueurs. Grown in the Mediterranean region, bitter almonds are used in European dishes. They contain prussic acid and are toxic when raw (they become edible when cooked) and are unavailable in the United States.

Bitter melon (bitter gourd, foo-gwa): Bumpy-skinned Asian fruit similar in shape to a cucumber that is pale green when ripe. The flesh has melonlike seeds and an acrid taste due to high quinine content (flavor and odor become stronger the longer it ripens).

Bitter orange: See *Seville orange.*

Black bean (turtle bean): Small (less than ½ inch) black bean (dried) used extensively in Central American, South American, and Caribbean cooking.

Black beans, fermented: Black soybeans salted and fermented to produce a piquant condiment. Used in Chinese cooking as a seasoning or combined with garlic, ginger, rice wine, and other ingredients to make black bean sauce.

Black-eyed pea (cow pea): Small legume (technically neither a pea nor a bean), white with a black spot, native to Africa and southern Asia.

Black mushrooms: See *Mushrooms.*

Blood orange: Old variety of orange with deep maroon–colored flesh, sometimes streaked with white. Intense sweet-tart flavor. Common in Spain and North Africa.

Blowfish (fugu, globefish): A popular Japanese specialty, blowfish contain a deadly toxin in the liver and sex organs. After careful cleaning the flesh reportedly has a slight tingle when eaten.

Bok choy (Chinese chard): Vegetable of the cabbage family with long, white leaf stalks and smooth, dark green leaves.

Boxthorn: See *Matrimony vine.*

Breadfruit: Large round tropical fruit with warty green skin and starchy white flesh. It must be cooked and is usually served as a vegetable. Available canned.

Breadroot (Indian breadroot; prairie turnips; *timpsila; tipsin*): Hairy perennial plant (*Psoralea esculenta*) with large brown root eaten by Native Americans of the Plains and adopted by European immigrants who knew it as *pomme de prairie.*

Buckwheat (kasha): Nutty-flavored cereal native to Russia (where it is called kasha), sold as whole seeds (groats) and ground seeds (grits if coarsely ground, flour if finely ground). It is common in Russian and Eastern European cooking.

Buffalo berry: Scarlet berry of the *Sheperdia* genus, so called because it was usually eaten with buffalo meat by the Native Americans.

Bulgur (bulghur, burghul): Nutty-flavored cracked grains of whole wheat that have been precooked with steam. Available in coarse, medium, and fine grades.

Burdock root (gobo): Long thin root with crisp white flesh and thin brown skin with an earthy, sweet flavor. Popular in Asian cooking.

Cactus fruit (cactus pears, *cholla,* Indian figs, *pitaya, sabra,* strawberry pear, *thang long*): Succulent fruit of various cacti popular in numerous nations. Red prickly pear cactus fruit—cactus pears, *cholla,* Indian figs, sabra, tuna—common in Mexico, U.S. Southwest, Central America, Israel, and some Middle Eastern countries, Australia, South Africa, and Italy. Fruit of the organ pipe cactus sold in the United States as strawberry pear or *pitaya.* Fruit of saguaro cactus and apple cactus eaten in desert areas of Mexico and U.S. Southwest. Climbing epiphytic cacti common in South America, Australia, Israel, and Vietnam; one variety called *thang long* or *red pitaya.*

Cactus pads (*nopales, nopalitos*): Paddles of the prickly pear cactus commonly eaten in Mexico, parts of the U.S. Southwest, fresh, cooked, or pickled. Available canned.

Cactus pears: See *Cactus fruit.*

Caimito: See *Star apple.*

Cajú: See *Cashew apple.*

Calabash: Gourdlike fruit of a tropical tree native to the New World.

Calabaza: Round or oblong winter squash with yellow flesh and a flavor similar to pumpkin.

Calamansi: Small sour lime used to flavor foods in Filipino cooking.

Callaloo (cocoyam): Edible leaves of root vegetables, especially amaranth, malanga, and taro. Callaloo is sometimes the name of a dish made from these leaves.

Camass root: Sweet bulb of the camass lily common in the U.S. Pacific Northwest.

Candlenut (kukui nut): Oily tropical nut sold only in roasted form (it is toxic when raw). Popular in Malaysia, Polynesia, and Southeast Asia.

Càng cua: See *Peperomia.*

Cannellini: See *Kidney bean.*

Capers: Small gray-green flower buds from a bush native to the Mediterranean. They are commonly pickled.

Carambola: See *Star fruit.*

Cardoon: Member of the artichoke family that looks like a spiny celery plant, popular in Italian cooking.

Cashew apple (*cajú*): The fleshy "false fruit" attached to the cashew nut. Native to Brazil, it is also eaten in the Caribbean and India.

Casimiroa (white sapote, zapote blanco): Dark green to yellow fruit native to Central America that looks like an Asian pear. Soft, white flesh is eaten fresh or prepared as jellies, ices, milkshakes, and fruit leather.

Cassarep: Caribbean sauce made from the juice of the bitter variety of cassava cooked with raw sugar.

Cassava (cocoyan manioc, yuca): Tropical Latin American tuber (now eaten in most tropical areas of the world) with rough brown skin and mild white flesh. Two types exist: bitter (poisonous unless leached and cooked) and sweet. Flour used in Africa *(gari),* the Caribbean, and Brazil *(farinha).* Cassava starch is used to make the thickening agent tapioca.

Caviar (red caviar, *ikura, tobikko*): Fish roe from a variety of fish eaten worldwide, including sturgeon (technically the only roe that is called caviar), salmon (red caviar, *ikura*), flying fish (*tobikko*), carp, herring, and mullet. Sturgeon caviar graded according to size and quality.

Celeriac (celery root): Gnarled, bulbous root of one type of celery, with brown skin, tan flesh, and nutty flavor.

Cèpes: See *Mushrooms.*

Chanterelles: See *Mushrooms.*

Chayote (christophine, chocho, huisquil, mirliton, vegetable pear): Thin-skinned, green (light or dark), pear-shaped gourd. Native to Mexico, it is now common in Central America, the Caribbean, the southern United States, and parts of Asia.

Cheong-po: See *Bean curd.*

Cherimoya (custard apple): Large, dimpled, light green fruit native to South America. White, creamy, flesh has a flavor reminiscent of strawberries, cherries, and pineapple.

Chicharrónes: Deep-fried pork skin, fried twice to produce puffy strips.

Chickpea (garbanzo bean): Pale yellow round legume popular in Middle Eastern, Spanish, Portuguese, and Latin American cooking. Can be purchased canned or dried.

Chicory: Bitter roasted root of the European chicory plant used as a substitute for coffee. Often added to dark coffee in Creole cooking.

Chile pepper: Although chile peppers, or chiles, are often called *hot peppers,* the fruits are not related to Asian pepper (such as black pepper) but are pods of capsicum plants, native to Central and South America. More than 100 varieties are available, from less than ¼ inch in length to over 8 inches long. Used fresh or dried. The alkaloid *capsaicin,* found mostly in the ribs of the pods, is what makes chile peppers hot. In general, the smaller the chile, the hotter it is. Common types include mild Anaheim (also called *California* or *New Mexico chile*); Cayenne (used mostly dried and powdered as the spice cayenne); dark green, medium hot Jalapeño (often available canned—when smoked are known as *Chipolte*); spicy, rich green Poblano (used fresh, or ripened and dried, called *Ancho*); hot Serrano (small, bright green or red); and very hot Chile de Arbol, Japones, Péquin (tiny berrylike pepper, exceptionally hot, also known as *bird* or *bird's eye* peppers), and Tabasco (small, red chiles, often used dried).

Chili powder: Ground, dried chile peppers, often with added spices such as oregano, cumin, and salt.

Chinese date (dae-chu; jujube): Small Asian fruit (not actually belonging to the date family) usually sold dried. Red dates are the most popular, but black and white are also available.

Chinese parsley: See *Coriander.*

Chitterlings (chitlins): Pork small intestines, prepared by boiling or frying.

Chokecherry: Tart, reddish black native American cherry (*Prunus virginiana*).

Cholla: See *Cactus fruit.*

Chrysanthemum greens ("chop suey" greens; *sook-gat*): Spicy leaves of a variety of chrysanthemum (not the American garden flower), popular in Asian stir-fried dishes, especially in Korea.

Cilantro: See *Coriander.*

Citron: Candied peel of a yellow-green, apple-size citrus fruit. Available crystallized and as preserves.

Citronella: See *Lemon grass.*

Clotted cream (Cornish cream, Devonshire cream): Very thick cream made by allowing cream to separate from milk, then heating it and cooling it so that it ferments slightly. Finally, the cream is skimmed from the milk (although Cornish cream is skimmed before heating and cooling). It is popular in southwest England, where it is spread on bread or used as a topping for desserts.

Cloud (wood) ears: See *Mushrooms.*

Coconut cream: High-fat cream extracted from fresh grated coconut.

Coconut milk: Liquid extracted with water from fresh grated coconut.

Cocoplum: Bland plum with white flesh native to Central America, found in the Caribbean, Central America, and Florida. Eaten fresh or dried.

Cocoyam: See *Callaloo, Cassava.*

Conch: Large, univalve mollusk found in waters off Florida and Caribbean (where it is sometimes called *lambi*). Chewy meat valued for its smoky flavor; can be bitter. Used especially in soups, stews.

Copra: Dried coconut kernels used in the extraction of coconut oil.

Coriander (cilantro, Chinese parsley): Fresh leaves of the coriander plant with a distinctive "soapy" flavor, common in Asian, Middle Eastern, Indian, and Latin American cooking. Seeds used as spice; root used in Thai cooking.

Corn smut (huitlacoche): Fungus (*Ustilaginales*) that grows on corn ears. Prized in Chinese, Mexican and Native American cooking.

Couscous (cuscus, cuzcuz): Small granules of semolina flour used as a grain in African, Italian, Brazilian and Middle Eastern dishes.

Cow pea: See *Black-eyed pea.*

Cracked wheat: Cracked raw kernels of whole wheat used in Middle Eastern cooking.

Crawfish (crawdad, crayfish): Small freshwater crustacean, 4 to 6 inches long, that looks and tastes something like lobster. Found in Europe and the United States (California, Louisiana, Michigan, and the Pacific Northwest). The names *crawfish* and *crayfish* are also applied to the langostino, a saltwater crustacean that lacks large front claws.

Crème fraîche: Slightly thickened, slightly fermented cream popular in France.

Curry leaves (kari): Herb (*Murraya koenigii*) with tangerine overtones used throughout India, Sri Lanka, and in parts of Malaysia. Fresh leaves are briefly fried in ghee, then added to dishes before other seasoning. Not usually a component of curry powder.

Curry powder: The western version of the fresh Asian Indian spice mixture (garam masala) used to flavor curried dishes. Up to twenty spices are ground, then roasted, usually including black pepper, cayenne, cinnamon, coriander, cumin, fenugreek, ginger, cardamom, and turmeric for color.

Custard apple: See *Cherimoya.*

Cuttlefish (inkfish): A mollusk that is similar to squid, but smaller. Available fresh or dried.

Daikon (icicle radish, white radish): Relatively mild white radish common in Asian cooking. The Japanese variety is the largest, often 12 inches long, and is shaped like an icicle. The Chinese variety tends to be smaller.

Dals: Indian term for hulled and split grains, legumes, or seeds. Many types are available, such as lentils or split peas.

Dashi: Japanese stock made from kelp and dried fish (bonita). *Dashi-no-moto* is the dried, powdered, instant mix.

Dilis (daing): Small silvery fish related to anchovies, dried and salted. Used in Filipino dishes.

Dragon's eyes: See *Longan.*

Durian: Cantaloupe-size spiked fruit native to Malaysia, with a strong odor and sweet, creamy flesh.

Eggplant: Pear-shaped to round member of the nightshade family with smooth, thin skin (white or deep purple in color) and spongy, off-white flesh. Native to India, it has a mildly bitter flavor. Two types widely available are the small thin varieties known as Japanese and Chinese eggplant and the more common larger, rounder variety. The Thai type is small, round, white with green stripes.

Egusi: See *Watermelon seeds.*

Elderberries: Small shrubs or trees up to 20 feet. Numerous species found throughout northern hemisphere. In the United States (genus *Sambucus*) produce edible, sweet berries used fresh and in preserves, pies, and wine. Blossoms fried as fritters.

Enoki: See *Mushrooms.*

Epazote (Mexican tea; pigweed): Pungent herb (*Chenopodium ambrosioides*) related to pigweed or goosefoot (and sometimes called by these names). Found in Mexico and parts of the United States. Often added to bean dishes to reduce gas.

Farinha: See *Cassava.*

Fava bean (broad bean, horse bean, Windsor bean): Large green meaty bean sold fresh in the pod. Smaller white or tan fava beans are dried or canned and cannot be used interchangeably with the fresh beans. Common in Italian and Middle Eastern cooking.

Fennel: Light green plant with slightly bulbous end and stalks with feathery, dark green leaves, a little like celery. Used like a root vegetable. Delicate licorice or anise flavor.

Fenugreek: Tan seeds of the fenugreek plant, with a flavor similar to artificial maple flavoring. Essential in the preparation of Asian Indian spice mixtures.

Fiddlehead ferns: Young unfurled fronds a specialty dish of the U.S. Northeast and southeastern Canada. Roots were eaten by Native Americans.

Filé powder: See *Sassafras.*

Fish paste (bagoong, kapi, pa dek, prahoc): Thick fermented paste made from fish, used as a condiment and seasoning in the Philippines and Southeast Asia.

Fish sauce (nam pla, nam prik, nuoc mam, patis, tuk-trey): Thin, salty, brown sauce made from fish fermented for several days. Asian fish sauces vary in taste from mild to very strong, depending on the country and the grade of sauce. Filipino patis is the mildest, Vietnamese nuoc mam is among the most flavorful. Nuoc cham is a sauce made from nuoc mam by the addition of garlic and chile peppers.

Five-spice powder: A pungent Chinese spice mixture of anise, cinnamon, cloves, fennel seeds, and Szechuan pepper.

Fufu: See *Yam.*

Fugu: See *Blowfish.*

Fuzzy melon (hairy melon, mo gwa): Asian squash similar to zucchini with peach fuzz–like covering of the skin.

Gai choy (Chinese green mustard, dai gai choy): Includes several types of mustard family plants grown for their greens. Dark green to reddish-brown leaves are steamed, boiled, or stir-fried.

Garbanzo bean: See *Chickpea.*

Gari: See *Cassava.*

Geoduck: Large (up to 15 pounds) clam native to U.S. Pacific Northwest, with "neck" or "siphon" as long as 3 feet. Neck used in soups, stews; body sliced for steaks.

Ghee: Clarified butter (*usli ghee*) used in India. The term *ghee* is also used for vegetable shortening.

Ginger root: Knobby brown-skinned root with fibrous yellow-white pulp and a tangy flavor. Used sliced or grated in Asian dishes. Also available dried, ground, or candied.

Ginkgo nut: Small pit of the fruit of the ginkgo tree (ancient species related to the pine tree), dried or preserved in brine, common in Japan.

Ginseng: Aromatic forked root with bitter, yellowish flesh, used in some Asian dishes and beverages; best known for therapeutic uses.

Glutinous rice: See *Rice.*

Granadilla: See *Passion fruit.*

Grape leaves: Large leaves of grape vines preserved in brine, common in Middle Eastern cooking.

Gravlax: See *Salmon, cured.*

Greens: Any of numerous cultivated or wild leaves, such as chard, collard greens, creases, cochan (coneflower), dandelion greens, dock, kale, milkweed, mustard greens, pokeweed, purslane, and spinach.

Grits: Coarsely ground grain, especially hominy, which is typically boiled into a thick porridge or fried as a side dish. Served often in the U.S. South.

Ground cherries: Yellow fruit that looks like a tomato, from a bush native to Mexico.

Groundnuts (Apios, Indian potatoes): South American tuber (*Apios americana*) eaten by Native Americans, adopted by European settlers. Different from Africa groundnuts (referring to either peanuts or Bambara groundnuts).

Guanabana: See *Soursop.*

Guarana (Brazilian cocoa): Shrub indigenous to the Amazon. Dried leaves and seeds of the fruit are used to make a stimulating tea (containing caffeine) or mixed with cassava flour to form sun-dried sticks.

Guava: Small sweet fruit with an intense floral aroma, native to Brazil. Skin is yellow-green or yellow, and the grainy flesh ranges from white or yellow to pink and red. Many varieties are available, including strawberry guava and pineapple guava. Guava is popular as jelly, juice, or paste.

Guineps: See *Mamoncilla.*

Head cheese: Loaf of seasoned meat made from the hog's head and sometimes also feet and organs.

Heart of palm: White or light green interior of the palm tree, especially popular in the Philippines. Available canned.

Hickory nuts: Tree indigenous to North America (genus *Carya*), in same family as pecans. Eaten fresh, roasted, or ground into meal or pressed for a cream-like fluid by Native Americans; used in confections in the U.S. South.

Hog peanut: A high-protein underground fruit that grows on the root of the vine *Falcata comosa* in the central and southern United States. The peanut has a leathery shell that can be removed by boiling or soaking. The nut meat can be eaten raw or cooked.

Hoisin sauce: Popular Chinese paste or sauce, reddish brown in color, with a spicy sweet flavor. It is made from fermented soybeans, rice, sugar, garlic, ginger, and other spices.

Hominy (posole, pozole): Lime-soaked hulled corn kernels with the bran and germ removed.

Hot pepper: See *Chile pepper.*

Hot pepper sauce: Sauce made from fermented chile peppers, vinegar, and salt (Tabasco sauce is the best-known U.S. brand).

Huisquil: See *Chayote.*

Icicle radish: See *Daikon.*

Ikura: See *Caviar.*

Indian breadroot: See *Breadroot.*

Indian fig: See *Cactus fruit.*

Indian potato: See *Groundnuts.*

Irish moss: A gelatinous seaweed extract added to milk or rum as a beverage in the Caribbean.

Jaboticaba (*guapura*): Brazilian shrub or small tree with ½- to 1½-inch fruit clustered like grapes. Gelatinous pulp is mild, sweet.

Jackfruit: Large (up to 100 pounds) fruit of the jack tree, related to breadfruit and figs, native to India. Two varieties are widely eaten, one sweeter than the other. Available dried or canned.

Jaggery: Unrefined sugar from the palmyra or sugar palm common in India.

Java plum (rose apple, jambo, jaman): a small sour fruit grown in the Philippines.

Jerusalem artichoke (sunchoke, sunroot): Small nubby-skinned tuber that is the root of a native American sunflower. It is neither from Jerusalem nor related to the artichoke, though the flavor when cooked is similar. It is used raw and cooked.

Jicama (Ahipa; yambean): Legume with medium to large tuber with light brown skin and crisp white flesh, indigenous to Brazil. Used raw in Latin American dishes, it has a sweet, bland flavor, similar to peas or water chestnuts.

Jujube: See *Chinese date.*

Juneberries (saskatoons, serviceberries; shadbush): Red to deep purple berries on large bush native to the Great Plains region of the United States and Canada. White blooms in June associated with shad migratory run on East Coast; favorite of Native Americans.

Juniper berry: Distinctively flavored dark blue berry of the juniper ever-green bush, native to Europe. Used to flavor gin.

Kaffir lime: Aromatic citrus popular in Southeast Asia, especially in Thai cooking. Juice, rind, and leaves used to flavor curries, salad dressings, and sauces.

Kamis: Sour, cucumber-like vegetable native to the Philippines. Used to achieve a sour, cool flavor in Filipino cooking.

Kanpyo (kampyo): Ribbons of dried gourd used mostly for garnishing dishes in Japan.

Kaong: See *Palm nuts.*

Kapi: See *Fish paste.*

Kasha: See *Buckwheat.*

Kava: See *Pepper plant.*

Kadondong: See *Ambarella.*

Kewra (pandanus): Perfumed essence of the screwpine flower used in Asian Indian Indonesian, Malaysian and Thai dishes.

Key lime: Small, tart lime indigenous to the Caribbean, popular in Florida Keys. Known best as primary ingredient in key lime pie.

Kidney bean: Medium-size kidney-shaped bean light to dark red in color (a white variety is popular in Europe, especially Italy, where they are known as *cannellini*). The flavorful beans are common in Europe, Latin America, and the United States.

Kochujang: Korean fermented chile pepper paste added to many dishes.

Kohlrabi: Light green or purple bulbous vegetable that grows above the soil and produces stems bearing leaves on the upper part. A member of the cabbage family, it can be eaten raw or cooked.

Kola nut: Bitter nut of the African kola tree (extracts from this nut were used in the original recipe for Coca-Cola).

Kukui nut: See *Candlenut.*

Kumquat: Small, bright orange, oval fruit with a spicy citrus flavor common in China and Japan. Also available in syrup and candied.

Laverbread: Thick purée of laver (see *Seaweed*) that is baked. Used in sauces and stuffings in Great Britain.

Lemon grass (citronella root): Large, dull green, stiff grass with lemony flavor common in Southeast Asian dishes. Available fresh, dried, or powdered.

Lily buds (golden needles): The dehydrated buds of lily flowers used in the cooking of China.

Lingonberry: Small wild variety of the cranberry found in Canada and northern Europe. Usually available as preserves.

Litchi (lychee): Small Chinese fruit with translucent white flesh and a thin brown hull and single pit. The flavor is grape-like but less sweet. Available fresh and canned. Dried litchis, also called litchi nuts, have different flavor and texture.

Lobster: Ocean-dwelling crustacean valued for its sweet flesh. Two main species consumed in United States. American lobster *(Homarus americanus)* found from Labrador to North Carolina; meat from large claws and tail, premature eggs called "coral," and liver eaten. Spiny lobster (*Panulirus argus* and other species) looks similar to American lobster but is a different animal. Found in warm waters from North Carolina to Brazil; small claws, only tail meat eaten.

Longan (dragon's eyes): Fruit of an Asian Indian tree related to litchis. Used fresh, canned, or dried.

Long bean (Chinese green bean, yard-long bean): Roundish Asian bean, 12 to 30 inches long. Similar in taste to string beans, long beans are more soft and chewy, less juicy and crunchy than string beans.

Long-grain rice: See *Rice.*

Loquat: Slightly fuzzy yellow Asian fruit about 2 inches across, with a tart peach flavor. Available dried and in syrup.

Lotus root (water lily root): Tubular vegetable (holes, as in Swiss cheese, run the length of the root, producing a flower-like pattern when the root is sliced) with brownish skin and crisp, sweet, white flesh. Becomes starchy when overcooked or canned.

Lox: See *Salmon, smoked.*

Luffa (cee gwa, Chinese okra, loofa): Long, thin-skinned Asian vegetable, a member of the cucumber family, with spongy flesh.

Lulo: See *Navanjillo.*

Lupine seeds (*tremecos*): Bitter seeds of a legume used for fodder. Must be leached in water before eating.

Macadamia nut: Round creamy nut native to Australia, now grown in Africa, South America, and Hawaii.

Mahimahi (dolphinfish): A saltwater finfish found off the coasts of Florida and Hawaii.

Mahleb: Middle Eastern spice made from ground black cherry kernels, which impart a fruity flavor to foods.

Malagueta pepper (grains of paradise, guinea pepper): Small West African berries related to cardamom, with a hot, peppery flavor. In Brazil the term refers to a tiny Pequin chile pepper.

Malanga (tannier, yautia): Caribbean tuber with cream-colored, yellow, or pinkish flesh, dark brown skin, and nutty flavor.

Mamey: Medium-size egg-shaped fruit with brown skin and soft flesh ranging in color from orange to yellowish to reddish. It has a flavor similar to pumpkin.

Mammea: South American fruit (*Mammea americana*) with reddish-brown skin and bright yellow flesh that tastes like peaches.

Mamoncilla (guineps; Spanish lime): Small 1- to 2-inch green fruit found in the Caribbean and South America that grow in clusters like grapes but have thicker skin and distinctive sweet, citrusy flesh around a large seed.

Mango: Fruit native to India, yellow to red when ripe, averaging 1 pound in weight. The flesh is pale and sour when the fruit is unripe, bright orange and very sweet when it is ripe. Used unripe for pickles and chutneys, ripe as a fresh fruit.

Manioc: See *Cassava.*

Marzipan: See *Almond paste.*

Masa: Dough used to make tortillas and tamales. Made fresh from dried corn kernels soaked in a lime solution, or from one of two "flours" available: masa harina (tortilla mix made from dehydrated fresh masa) or masa trigo (wheat flour tortilla mix).

Mastic: Resin from the lentisk bush that has a slightly piney flavor, used to flavor Middle Eastern foods. Available in crystal form.

Mate: Plant in holly family native to South America. Dried, powdered leaves, called *yerba,* are brewed to make a stimulating tea (containing caffeine) that is popular in Argentina, Brazil, and Paraguay.

Matrimony vine (boxthorn, wolfberry): Asian vine with culinary and medicinal uses; both leaves and fruit are used in China. Berries are dried, and leaves brewed to make herbal tea.

Mayhaw: Type of hawthorne tree found in U.S. South. Fruit looks like cranberries. Tart apple flavor. Used in preserves, syrups and wines.

Mikan: Japanese citrus related to tangerines and mandarin oranges. Eaten fresh, frozen, and canned in syrup.

Milkfish (bangus): Silvery bony fish with oily flesh popular in Filipino cooking.

Millet: Cereal native to Africa, known for its high-protein, low-gluten content and ability to grow in arid areas. The variety common in Ethiopia is called *teff.*

Mirin: Sweet rice wine used in Japanese dishes.

Miso: Fermented soybean paste, white (mild flavored) or red (strongly flavored), common in Japanese cooking.

Mizuna: Yellow-green, narrow, notched leaves of a mustard family plant common in Japan.

Morels: See *Mushrooms.*

Mullet: Finfish of two families that can be black, gray, or red. The flesh is a mix of dark, oily meat and light, nutty-tasting meat. The texture is firm but tender.

Mung bean: Yellow-fleshed bean with olive or tan skin used in cooking of China, India.

Mushrooms: Fresh or dried fungi used to flavor dishes throughout the world. Common Asian types include *enoki* (tiny yellow mushrooms with roundish caps), *oyster mushrooms* (large, delicately flavored grey-beige caps that grow on trees), *shiitake* (dark brown with wide flat caps, available dried as Chinese black mushrooms), *straw mushrooms* (creamy colored with bell-like caps), and *cloud ears* or *wood ears* (a large flat fungus with ruffled edges, available dried). Popular mushrooms in Europe, available both fresh and dried, include *chanterelles* (a golden mushroom with an inverted cap), *morels* (a delicately flavored mushroom with a dark brown wrinkled cap), and *porcini* or *cèpes* (large brown mushrooms with caps that are spongy underneath; also called *boletus*).

Nam pla: See *Fish sauce.*

Nam prik: See *Fish sauce.*

Nance: Small, yellow tropical fruit similar to cherries with a slightly tart flavor. Two varieties are available.

Napa cabbage (celery cabbage, Chinese cabbage): Bland crunchy vegetable with broad white or light green stalks with ruffled leaves around the edges. Several types are available, similar in taste.

Naranjilla (lulo): Walnut-size, orange-skinned, green-fleshed fruit indigenous to the Americas, used mostly for its juice. Particularly popular in Central America.

Naseberry: See *Zapote.*

Nongus: Fruit of the palmyra palm, grown in India, Indonesia, and Malayasia as a source of sugar.

Nopales (nopalitos): See *Cactus.*

Nopales/nopalitos: See *Cactus pads.*

Nuoc cham: See *Fish sauce.*

Nuoc mam: See *Fish sauce.*

Oca: Tuber of Andean plant (*Oxalis tuberosa*). Resembles a pink potato. Tastes lemony when fresh, sweet after storage. Used in South America prepared like potatoes or eaten fresh.

Okra: Small, green, torpedo-shaped pod with angular sides. A tropical African plant valued for the carbohydrates in it that are sticky and mucilaginous. It is used as a vegetable and to thicken soups and stews.

Olive: Fruit of a tree native to the Mediterranean. Green olives are preserved unripe. Large, soft Kalamata olives are a medium-size, purplish Greek olive. Dark olives (such as Niçoise) are picked in autumn, often cured in salt, with a tannic flavor. Ripe, black olives are smooth-skinned and mild-flavored or wrinkled with a strong tannic flavor.

Olive oil: Extracted from the olive flesh, it is labeled according to percent acidity, from *extra virgin* to *virgin* (or "pure"). U.S. labeling laws restrict the use of the term "virgin" to only olive oil made from the first press; virgin olive oils mixed with refined olive oils to reduce acidity are labeled "pure."

Ostiones: Oyster native to the Caribbean that grows on the roots of mangrove trees.

Otaheite apple: See *Ambarella.*

Oyster mushrooms: See *Mushrooms.*

Oyster sauce: Thick brown Chinese sauce made with soy sauce, oysters, and cornstarch.

Pacaya bud: The bitter flower stalk of the pacaya palm found in Central America. The edible stalk is about 10 inches long and is encased in a tough green skin, which must be removed before cooking.

Pa dek: See *Fish sauce.*

Palillo: Peruvian herb, used dried and powdered to provide a yellowish orange color to foods.

Palmetto cabbage: See *Heart of palm.*

Palm nuts *(Kaong)*: Seeds from palms; pounded into palm butter in West Africa. Also boiled and added to halo-halo mix in Philippines.

Palm oil (dende oil): Oil from the African palm, unique for its red-orange color.

Pandanus: See *Kewra.*

Papaya: Thin-skinned green (under-ripe), yellow, or orange fruit with sweet flesh colored gold to light orange to pink. Mexican (large and round) and Hawaiian (smaller and pear shaped) varieties are available. The shiny round black seeds are edible. Unripe papaya is used in pickles; the ripe fruit is eaten fresh.

Paprika: Powdered red peppers especially popular in Hungarian cooking. Paprika is made from several types of *Capsicum annum,* related to bell and chile peppers. Paprika is usually designated "sweet" or "hot." Spanish paprika, used in Spanish and Middle Eastern dishes, is more flavorful.

Passion fruit (granadilla): Small oval fruit with very sweet gelatinous pulp. Yellow- and purple-skinned varieties are available. Passion fruit is often made into juice.

Patis: See *Fish sauce.*

Pawpaw (tree melon): Light orange fruit that tastes like a cross between a banana and a melon. Native to the Americas, it is approximately 6 inches long.

Peanuts (groundnuts, goobers, monkeynuts): Legume native to South America, introduced to Africa by the Portuguese, then brought to the United States in the 17th century by black slaves. Eaten raw, roasted, or pulverized into peanut butter. Popular in Africa and the United States; used in some Chinese, Southeast Asian, and Asian Indian dishes.

Pejibaye (peach palm): Fruit of a Central American palm, especially popular in Costa Rica.

Peperomia (*càng cua*): Small plant with heart-shaped leaves (*Peperomia pellucida*) found throughout Central and South America, Africa, and Southeast Asia. Used as a culinary herb in Vietnam, and as a medicinal herb in the Philippines, Polynesia, and parts of Latin America.

Pepitas: Pumpkin or squash seeds common in Latin American cooking. May be hulled or unhulled, raw or roasted, salted or unsalted.

Pepper plant (*Piper methysticum*): Leaves of the South Pacific plant used to produce the intoxicating beverage called *kava* or *awa*.

Peppers: Misnamed pods of the capsicum plants native to South and Central America (not actually related to Asian pepper plants, which produce black pepper). Peppers are divided into sweet and hot types (see *Chile pepper*). Sweet peppers include bell peppers (green, red, yellow, and purple), pimentos, and peppers used to make paprika.

Perilla (shiso; beefsteak plant; quen-neep): Aromatic herb with distinctive minty flavor; green or red. Available fresh or pickled. Used mostly as a seasoning or garnish in many Japanese and Korean dishes; sometimes served as a side dish or to wrap rice and other items.

Pigeon pea: Small pea in a hairy pod (a member of the legume family, but not a true pea) common in the cooking of Africa, the Caribbean, and India. Yellow or tan when dried.

Pignoli: See *Pine nut.*

Pigweed: See *Amaranth; Epazote.*

Pili-pili: Hot sauce used in West Africa made with the piri-piri chile pepper and other ingredients such as tomatoes, onions, garlic or horseradish.

Pili nut: Almond-like nut of a tropical tree found in the Philippines.

Pine nut (pignoli, piñon seed): Delicately flavored kernel from any of several species of pine tree. Pine nuts are found in Portugal (most expensive type), China (less costly, with a stronger taste), and the southwestern United States. Common in some Asian, European, Latin American, Middle Eastern, and Native American dishes.

Pink bean (rosada): Small oval meaty bean that is a light tannish pink in color.

Piri-piri: Thin, fiery chile favored in West Africa. Used in sauces and marinades, and is the name of dishes that include some form of the pepper.

Pitanga (Surinam cherry, Brazilian cherry): Small, bright red, ribbed fruit of shrub or small tree native to northeastern South America; found also in the Caribbean and Florida. Thin skin with orange flesh that "melts" in the mouth. Sweet with a slightly bitter bite.

Pitaya/pitahaya/pitajaya: See *Cactus fruit.*

Plantain: Starchy type of banana with a thick skin, which can be green, red, or black. There are many varieties, ranging in size from 3 to 10 inches. The pulp is used as a vegetable and must be cooked. It is similar in taste to squash.

Poi: See *Taro.*

Porcini: See *Mushrooms.*

Posole, Pozole: See *Hominy.*

Prahoc: See *Fish paste.*

Prarie turnips: See *Breadroot.*

Prickly pear: See *Cactus fruit, Cactus pads.*

Pulses: Term used especially in India for edible legume seeds, including peas, beans, lentils, and chickpeas.

Quinoa: Cereal native to the Andes, prepared like rice.

Radicchio: Magenta-colored, slightly bitter member of the chicory family used throughout southern and northern Europe.

Rambutan: Bristly, juicy, orange or bright red fruit used in Southeast Asian cooking; related to the litchi.

Ramp: Strong-flavored indigenous American onion that tastes somewhat like a leek. Both leaves and bulbs are edible.

Red bean: Small, dark red bean native to Mexico and the southwestern United States.

Red caviar: See *Caviar.*

Rice: Grain native to India. More than 2,500 varieties are available worldwide, including basmati rice (small grain with a flavor similar to popcorn, very popular in India and the Middle East); brown rice (unmilled rice with the bran layer intact; can be short-, medium-, or long-grain); glutinous rice (also called *sweet* or *pearl rice;* very short-grain and very sticky when cooked); long-grain rice (white, polished grains that flake when cooked); and short-grain rice (slightly sticky when cooked, popular in Japan).

Rose apple (kopo): small, thin-skinned pink or red fruit native to Southeast Asia with somewhat spongy flesh that has slightly acidic flavor.

Roselle (Florida cranberry; karkadeh; red sorrel; sorrel): Pods of a hibiscus plant relative, common in Africa, the Caribbean, Southeast Asia, Australia, and Florida. Used to make a tart tea popular in Egypt and Senegal and a rum-laced punch in the Caribbean. Also used for chutneys, preserves, and candies.

Sabra: See *Cactus fruit.*

Saewujeot: See *Shrimp paste.*

Saffron: Dried stamens of the crocus flower. It has a delicate, slightly bitter flavor and bright red-orange color. Available as threads or powder.

Salal: Thick-skinned black berries of a native American plant in the heath family (*Gaultheria shallon*).

Salmon, cured: Salmon fillets cured in a mixture of salt, sugar, and dill weed, common in Sweden (where it is known as gravlax), Finland, and Norway.

Salmon, smoked: Raw, tender salmon slices lightly smoked and cured in salt produced in Norway, Nova Scotia, and Scotland. Smoked salmon soaked in a brine solution is called *lox*, a Jewish specialty.

Salmon roe: See *Caviar.*

Salt pork: White fat from the side of the hog, streaked with pork meat, cured in salt.

Saluyot ("okra leaves"): Leaves from Southeast Asian bush with slippery texture when cooked. Added to soups and stews in Filipino cooking.

Sapote: See *Zapote.*

Sapodilla: See *Zapote.*

Saskatoons: See *Juneberries.*

Sassafras (filé powder): Native American herb used to thicken soups and stews.

Sea cucumber (sea slug): Brown or black saltwater mollusk up to 1 foot in length. They lack a shell, but have a leathery skin and look something like smooth, dark cucumbers. Sold dried, they are rehydrated for Chinese dishes, becoming soft and jellylike, with a mild flavor.

Sea urchin roe (uni): Small, delicate eggs of the spiny sea urchin, popular in Japan.

Seaweed (kim): Many types of dried seaweed are used in Chinese, Korean, and Japanese dishes, including *aonoriko* (powdered green seaweed), *kombu* (kelp sheets), *laver* (tissue-thin, dark purple sheets), and *nori* (paper-thin sheets of dark green seaweed). Also popular in the Pacific Islands. See also *Irish moss; Laverbread.*

Serviceberries: See *Juneberries.*

Sesame seeds (benne seeds): Seeds of a plant native to Indonesia. Two types are available: tan colored (white when hulled) and black (slightly bitter). Commonly grown for their oil. Light sesame oil is pressed from raw seeds, dark oil from toasted seeds; the dark oil has a strong taste and is used as a flavoring.

Seville orange (bitter orange; naranja aria, sour orange): Orange with tough skin and dark flesh native to Mediterranean. Inedible raw; juice used in liqueurs (Grand Marnier, Cointreau, Curaçao) and in cooking of the Mediterranean, Caribbean, Central America, and Korea.

Shadbush: See *Juneberries.*

Shallot: Very small bulb covered with a reddish, papery skin, related to onions but with a milder, sweeter flavor.

Shea nut: Nut from the African shea tree, grown for its thick oil, called *shea butter*.

Shiitake mushrooms: See *Mushrooms*.

Shiso: See *Perilla.*

Short-grain rice: See *Rice.*

Shoyu: See *Soy sauce*.

Shrimp paste: Strongly flavored fermented Asian sauce or paste made from small dried shrimp or similar crustaceans. Many types are available (bagoong-alamang is the Filipino variety; saewujeot is the Korean type).

Snail (escargot): Small, edible land snail (a common variety of garden snail, cleansed with a commercial feed). Popular in France.

Snow pea (Chinese pea pod, sugar pea): Flat, edible pod with small, immature peas.

Sorghum (guinea corn): Cereal common to tropical regions of Africa with seeds produced on a stalk. In the Appalachians, Ozarks, and the U.S. South, sorghum is often processed to make sweet syrup.

Sorrel (dock, sourgrass): Small, sour green. See also *Roselle.*

Sour orange: See *Seville orange.*

Soursop (guanabana): Large (often 12 inches long), rough-skinned fruit with cottony, fluffy flesh that can be white, pink, or light orange. Often made into juice or conserves.

Soybean: Small high-protein bean common in Asia. Many varieties of different colors, including black, green, red, and yellow, are available. They are used fresh, dried, and sprouted, most often processed into sauces, condiments, and other products (see *Hoisin sauce, Oyster sauce, Miso, Soy milk, Soy sauce*, and *Bean curd*).

Soy milk: Soybeans that are boiled, pureed, then strained and boiled again to produce a white milklike drink.

Soy sauce (shoyu): Thin, salty, brown sauce made from fermented soybeans. Several types are available. Chinese soy sauces tend to be lighter in flavor than the stronger, darker Japanese shoyu.

Spicebush: Shrub (*Lindera benzoin*) with spicy-smelling bark and leaves; red berries. Used to make Native American teas.

Spiny lobster: See *Lobster.*

Star anise: Eight-armed pods from a plant in the magnolia family, with an anise-like flavor. Native to China.

Star apple (caimito): Purple, apple-size fruit with mild, gelantinous, lavender-colored flesh native to the Caribbean. Seeds form a star around the center.

Star fruit (carambola): Small, deeply ribbed, oval fruit with thin skin shaped like a star when sliced. Green and sour when unripe, yellow and slightly sweet (though still tart) when ripe. Unripe fruit is used in Indian and Chinese dishes. Ripe it is eaten fresh.

Straw mushrooms: See *Mushrooms.*

Strawberry pear: See *Cactus fruit.*

Sumac: Sour red Middle Eastern spice made from the ground berries of a non-toxic variety of the sumac plant.

Sunflowers: Native to the United States (genus *Helianthus*); over 60 varieties. Seeds eaten by Native Americans raw, dried, and powdered (in breads). Unopened flower head can be cooked and eaten like an artichoke. Petals are dried and used like saffron in Southwest.

Sweet peppers: See *Peppers.*

Sweetsop (sugar apple): Sweet, white-fleshed fruit related to the cherimoya and soursop.

Szechwan pepper (fagara): Aromatic berries with a hot flavor popular in some Chinese and Japanese dishes.

Tabasco sauce: See *Hot pepper sauce.*

Tamarind: Tart pulp from the pod of the tamarind bean. Available in the pod, as a paste, in a brick, or as a liquid concentrate.

Tannier (tannia): See *Taro, Malanga.*

Tapioca: See *Cassava.*

Taro (eddo, dasheen, tannier): Starchy underground vegetable similar to cassava with brown hairy skin and white to grayish flesh, common in the Caribbean and Polynesia. In Hawaii the boiled, pounded taro paste called *poi* is a staple in the traditional diet. The large leaves are also eaten (see *Callaloo*).

Tarpon: Large silver fish of the herring family found off the coasts of Mexico and Central America.

Teff: See *Millet.*

Tempeh: See *Bean curd.*

Tepary beans: Small, high-protein bean with wrinkled skin. Grows wild in the southwest United States.

Ti: Tropical plant popular in Polynesia (not related to tea). Ti leaves are used to wrap food packets, and the root is eaten and brewed for a beverage.

Tilapia: Small freshwater fish with sweet, firm, white flesh.

Timpsila: See *Breadroot.*

Tipsin: See *Breadroot.*

Tobikko: See *Caviar.*

Tobu, tofu: See *Bean curd.*

Tomatillo: Small, light green, tomato-like fruit surrounded by a green or tan papery husk, common in Mexico. The flesh is slightly tart and is eaten cooked, usually in sauces and condiments. Available fresh or canned.

Tremecos: See *Lupine seeds.*

Truffle: Black (French) or white (Italian) fungus found underground. Truffles vary from the size of small marbles to as large as tennis balls and are distinctively flavored, similar to a wild mushroom. Available fresh or canned.

Tuk-trey: See *Fish paste.*

Tuna: See *Cactus fruit.*

Turtle: Popular in Carribean, Central America, and U.S. South. Diamondback terrapin (*Malaclemys terrapin*) is the primary ingredient in turtle soups of the Atlantic states. Green turtle (*Chelonia mydas*) is a sea turtle, commonly eaten as steaks or stews. Other turtles eaten occasionally (including eggs) are alligator snapping turtle, common snapping turtle, and loggerhead turtle.

Ugli fruit: Citrus fruit that is a cross between a pommelo and a mandarin orange, with a very bumpy yellow-orange skin and a sweet orange-like flavor. Especially popular in Jamaica.

Uni: See *Sea urchin roe.*

Usli ghee: See *Ghee.*

Verjuice: Juice of unripe lemons used in Middle Eastern fare to give a tang to dishes.

Wasabi: Light green Japanese condiment from root of plant similar to horseradish with a powerful pungency. Available fresh or powdered; green-dyed horseradish often sold as wasabi.

Water chestnut: Aquatic, walnut-size tuber with fibrous brown peel and crunchy, sweet, ivory-colored flesh. Available fresh or canned.

Watermelon seeds: Seeds often eaten in Africa (called *egusi,* toasted, ground or pounded into meal or paste for thickening soups and stews) and in Asia (toasted as a snack; sometimes flavored or dyed red).

Wax gourd: White, oblong fruit of a vine with starchy flesh, common in Southeast Asia.

White bean: Three types of white bean are widely used: cannellini (see *Kidney bean*); Great Northern beans, which are large, soft, and mild tasting; and the smaller, firmer navy beans.

White radish: See *Daikon.*

Wild rice: Seeds of a native American grass.

Winged bean: Edible legume (*Psophocarpus tetragonolobus*) called the "soybean of the tropics." All parts of the plant are consumed, including the shoots, leaves, flowers, pods and seeds, and tuberous root. The pods are large, from 12 to 24 inches long, and feature winglike flanges.

Winter melon (dong gwa, wax melon): Round green-skinned member of the squash family with a waxy white coating and translucent white green or pink flesh. Similar in taste to zucchini, it is used cooked in Chinese dishes.

Wolfberry: See *Matrimony vine.*

Worcestershire sauce: Sauce developed by the British firm of Lea and Perrins including anchovies, garlic, onions, molasses, sugar or corn sweetener, tamarind, and vinegar, among other ingredients.

Yacón (yakon, leafcup): Sweet-tasting root (*Polymnia sonchifolia*) with brown skin and white flesh native to Andes. Eaten throughout South America; in some regions confusingly called *jicama.* See *Jicama.*

Yam (ñame; yampi; cush-cush; mapuey): Tuber with rough brown skin and starchy white flesh (not related to the orange sweet potato called *yam* in the United States). Numerous varieties; may grow quite large, up to 100 pounds. Found in all tropical regions. Yam paste called *fufu* in West Africa.

Yambean: See *Jicama.*

Yautia: See *Malanga.*

Yerba Buena: A variety of mint used in some Native American teas.

Yuca: See *Cassava.*

Yucca (Navajo banana): Spiky-leaved desert plant (*Yucca baccata*) with large, pulpy fruit that ripens in summer. Eaten fresh, boiled, baked, or dried into fruit leather.

Zapote (sapodilla, black sapote, naseberry): Drab-colored fruit of the sapodilla tree (which is the source of chicle used in chewing gum). It has granular, mildly sweet flesh, which can be yellow, red, or black. The zapote is a member of the persimmon family.

Resources

In many ways a book like this poses more questions than it answers. Knowledge of cultural foods is neither balanced nor complete. Many interested readers are undoubtedly asking why there is so little research on adaptations of food habits in the United States or why there are such limited data available on certain cultural groups. As stated in the preface, only the major American cultural groups are presented in this book. Although the authors reviewed many resources, the resulting text is undeniably inadequate in some areas. Thus, the most urgent question is "Where to go from here?"

Classes, seminars, association memberships, research, client interaction, and community involvement are all useful ways to learn more about cultural foods. In the nutrition field, many departments of dietetics and home economics offer courses in food and culture. Culinary schools often offer similar classes. Nutrition, dietetic, and food service organizations occasionally sponsor seminars on general topics in food habits, as well as on culturally specific diets.

Research on foods and food habits is published in a limited number of journals. *American Journal of Clinical Nutrition, American Journal of Public Health, Ecology of Food and Nutrition, Journal of the American Dietetic Association, Journal of Nutrition Education, Journal of Transcultural Nursing,* and *Social Science and Medicine* are a few that frequently feature articles on diet, nutrition, and culture. Food service, hospitality, and restaurant journals sometimes address cultural food issues. Anthropology, folklore, history, home economics, human resources, geography, management, medicine, nursing, psychology, and sociology publications are occasionally good resources as well.

There are numerous books that have contributed to research on foods and food habits. Cookbooks and other popular literature often include anecdotal information of interest. On-line web sites are an additional source of data. Some of the many resources available follow.

Observation and participation in the community, at markets, at festivals and fairs, and at public religious events are also good ways to learn about a population. Community leaders, traditional healers, restaurateurs and grocers can all contribute to cultural knowledge about foods and food habits. Accumulated experience with a minority population can be an important adjunct to printed research.

The authors encourage food professionals to undertake and publish studies on cultural foods to expand the limited information on the topic. We also hope that all professionals share their diversity experiences through associations and journals. In addition, you may contact our web site at www.eatethnic.com to discuss these issues on line. Such research and communication help us to become more culturally competent and thus become more effective health care providers and educators.

General Bibliography

Barer-Stein, T. 1979. *You Eat What You Are.* Toronto: McClelland & Stewart.

Barer-Stein, T. 1999. *You Eat What You Are: People, Culture and Food Traditions.* Willowdale, Canada: Firefly.

Brown, L. K., & Mussell, K. (Eds.). 1984. *Ethnic and Regional Foodways in the United States.* Knoxville: University of Tennessee Press.

Bryant, C. A., Courtney, A., Markesbery, B. A., & Dewalt, K. M. 1985. *The Cultural Feast: An Introduction to Food and Society.* St. Paul, MN: West.

Carlson, B. 1997. *Food Festivals.* Detroit, MI: Visible Ink.

Chevallier, A. 1996. *The Encyclopedia of Medicinal Plants.* New York: DK Publishing.

Claiborne, C. 1985. *Craig Claiborne's The New York Times Food Encyclopedia.* New York: Times Books.

Cole, K. W. 1985. *Minority Organizations: A National Directory.* Garrett Park, MD: Garrett Park.

Counihan, C., & Van Esterik, P. (Eds.). 1997. *Food and Culture: A Reader.* New York: Routledge.

Cunningham, S. 1991. *The Magic of Food: Legends, Lore and Spellwork.* St. Paul, MN: Ilewellyn.

Davidson, A. 1999. *The Oxford Companion to Food.* Oxford: Oxford University Press.

Farb, P., & Armelagos, G. 1980. *Consuming Passions: The Anthropology of Eating.* New York: Washington Square Press.

Fieldhouse, P. 1995. *Food and Nutrition: Customs and Culture* (2nd ed.). London: Chapman & Hall.

Fitzgerald, T. K. (Ed.). 1977. *Nutrition and Anthropology in Action.* Amsterdam: Van Gorcum.

Freedman, R. L. 1981. *Human Food Uses: A Cross-Cultural Comprehensive Annotated Bibliography.* Westport, CT: Greenwood.

Germov, J., & Williams, L. 1999. *The Social Appetite: A Sociology of Food and Nutrition.* South Melbourne, Australia: Oxford University Press.

Graedon, J., and Graedon, T. 1999. *The People's Pharmacy Guide to Home and Herbal Remedies.* New York: St. Martin's.

Harris, M. 1985. *Good to Eat: Riddles of Food and Culture.* New York: Simon & Schuster.

Hopkins, J. 1999. *Strange Foods: Bush Meat, Bats, and Butterflues; An Epicurean Adventure around the World.* North Claredon, VT: Periplus.

Jacobs, J. 1995. *The Eaten Word: The Language of Food, Food in Our Language.* New York: Birch Lane.

Jerome, N. W., Kandel, R. F., & Pelto, G. H. (Eds.). 1980. *Nutritional Anthropology: Contemporary Approaches to Diet and Culture.* Pleasantville, NY: Redgrave.

Jones, E. 1981. *American Food: The Gastronomic Story* (2nd ed.). New York: Vintage.

Lankevich, F. J. 1981. *Ethnic America.* London: Oceana.

Lee, H. G. 1992. *Taste of the States: A Food History of America.* Charlottesville, VA: Howell.

Levenstein, H. 1993. *Paradox of Plenty: A Social History of Eating in Modern America.* New York: Oxford University Press.

Lowenberg, M. E., Todhunter, E. M., Wilson, E. D., Feeney, M. C., Savage, J. R., & Lubawski, J. L. 1979. *Food and People* (3rd ed.). New York: Wiley.

MacClancy, J. 1992. *Consuming Culture: Why You Eat What You Eat.* New York: Holt.

Macdonald, D. 1983. *Resources on Food, Nutrition and Culture.* Toronto: Nutrition Information Service, Ryerson Polytechnic Institute Library.

Mariani, J. F. 1983. *Dictionary of American Food and Drink.* New Haven, CT: Tichnor & Fields.

McClelland, D. A. 1991. *Good as Gold—Foods America Gave the World.* Washington, D.C.: National Museum of History/Smithsonian Institution.

McGee, H. 1984. *On Food and Cooking: Science and Lore of the Kitchen.* New York: Scribner.

Meiselman, H. L. (Ed.). 2000. *Dimensions of the Meal: The Science, Culture, Business, and Art of Eating.* Gaithersburg, MD: Aspen.

Menzel, P. & D'Alusio, F. 1998. *Man Eating Bugs: The Art and Science of Eating Insects.* Berkeley, CA: Ten Speed Press.

Montagne, P. 1977. *The New Larousse Gastronomique.* New York: Crown.

Morgan, L. 1997. *The Ethnic Market Guide: An Ingredient Encyclopedia for Cooks, Travelers, and Lovers of Exotic Food.* New York: Berkley.

National Agricultural Library. 1988. *Diet, Race & Ethnicity in the US: Research and Reference Material.* 1979–1987 (NALBIBL. QB 88–29). Beltsville, MD: Author.

National Research Council. 1945. *The Problem of Changing Food Habits.* Washington, D.C.: National Research Council Bulletin 108.

Newman, J. M. 1993. *Melting Pot: An Annotated Bibliography and Guide to Food and Nutrition Information for Ethnic Groups in America* (2nd ed.). New York: Garland.

Nichter, M., & Nichter, M. 1981. *An Anthropological Approach to Nutrition Education.* Newton, MA: International Nutrition Communication Service.

Powers, J. M., & Stewart, A. (Eds.). 1995. *Our Northern Bounty: A Celebration of Canadian Cuisine.* Toronto: Random House of Canada.

Purnell, L. D., & Paulanka, B. J. 1998. *Transcultural Health Care: A Culturally Competent Approach.* Philadelphia: Davis.

Roberts, C. 1992. *Cultural Perspectives on Food and Nutrition.* Beltsville, MD: National Agricultural Library.

Root, W. 1980. *Food.* New York: Simon & Schuster.

Root, W., & de Rochemont, R. 1976. *Eating in America.* New York: Ecco.

Sanjur, D. 1982. *Social and Cultural Perspective in Nutrition.* Englewood Cliffs, NJ: Prentice Hall.

Schivelbusch, W. 1992. *Taste of Paradise: A Social History of Spices, Stimulants and Intoxicants.* New York: Pantheon.

Schwabe, C. W. 1979. *Unmentionable Cuisine.* Charlottesville: University Press of Virginia.

Shortridge, B. G., & Shortridge, J. R. 1995. Cultural geography of American foodways: An annotated bibliography. *Journal of Cultural Geography, 15,* 79–107.

Shortridge, B. G., & Shortridge, J. R. (Eds.). 1998. *The Taste of American Place.* Lanham, MD: Rowman & Littlefield.

Simoons, F. J. 1994. *Eat Not This Flesh: Food Avoidances from Prehistory to the Present* (2nd ed.). Madison: University of Wisconsin Press.

Tannahill, R. 1989. *Food in History.* New York: Crown.

Thernstrom, S., Orlov, A., & Handlin, O. (Eds.). 1980. *Harvard Encyclopedia of American Ethnic Groups.* Cambridge, MA: The Belknap Press of Harvard University Press.

Trager, J. 1995. *The Food Chronology.* New York: Holt.

USDA/USDHHS. 1986. *Cross-Cultural Counseling: A Guide for Nutrition and Health Counselors.* United States Department of Agriculture FNS-250, 1987–720– 484–1302/60133. Washington, D.C.: U.S. Government Printing Office.

Vissar, M. 1991. *The Rituals of Dinner: The Origins, Evolution, Eccentricities and Meaning of Table Manners.* New York: Grove Weidenfeld.

Von Welanetz, D., & Von Welanetz, P. 1982. *The Von Welanetz Guide to Ethnic Ingredients.* Los Angeles: Tarcher.

Warshaw, H. S., & Blackburn, G. 1995. *The Restaurant Companion: A Guide to Healthier Eating Out.* Chicago: Surrey.

Whit, W. C. 1995. *Food and Society: A Sociological Approach.* Dix Hills, NY: General Hall.

Wilson, C. S. (Ed.). 1979. Food: Custom and nurture. An annotated bibliography on sociocultural and biocultural aspects of nutrition. *Journal of Nutrition Education, 11* (Suppl. 1), 212–261.

Wood, R. C. 1995. *The Sociology of the Meal.* Edinburgh: Edinburgh University Press.

Cookbooks

AFRICAN

Clark, L. (Ed.). 1991. *The Black Reunion Cookbook.* Memphis, TN: Wimmery House.

Copage, E. V. 1991. *Kwanzaa.* New York: Morrow.

DeWitt, D., Wilan, M. J., & Stock, M. J. 1998. *Flavors of Africa Cookbook.* Rocklin, CA: Prima.

Hafner, D. 1993. *A Taste of Africa.* Berkeley, CA: Ten Speed Press.

Harris, J. B. 1998. *The Africa Cookbook: Tastes of a Continent.* New York: Simon & Schuster.

Van der Post, L. (Ed.). 1970. *African Cooking.* New York: Time-Life.

AMERICAN

Adams, M. 1997. *New Recipes from Quilt Country: More Food and Folklore from the Amish and Mennonites.* New York: Clarkson Potter.

Anderson, J. 1997. *The American Century Cookbook.* New York: Clarkson Potter.

Belk, S. 1991. *Around the Southern Table.* New York: Simon & Schuster.

Brown, D. 1968. *American Cooking.* New York: Time-Life.

Brown, D. 1971. *American Cooking: The Northwest.* New York: Time-Life.
Chateau Ste. Michele. 1985. *Taste of Liberty: A Celebration of Our Great Ethnic Cooking.* Woodinville, WA: Documentary Book Publisher.
Claiborne, C. 1987. *Craig Claiborne's Southern Cooking.* New York: Times Books.
Cox. B. 1996. *Spirit of the West: Cooking from Ranch House and Range.* New York: Artisan.
Danforth, R., Feierabend, P., & Chassman, G. 1998. *Culinaria: The United States.* New York: Könemann.
Dedominic, L. M. 1997. *Rocky Mountain Gourmet Cookbook.* Helena, MT: Falcon.
Dupree, 1993. *Southern Memories.* New York: Clarkson Potter.
Feibleman, P. S. 1971. *American Cooking: Creole and Acadian.* New York: Time-Life Books.
Hafner, D. 1997. United Tastes of America: *Recipes and Cultural Origins of Your Favorite Dishes.* New York: Ballantine.
Hibler, J. 1996. *Dungeness Crabs and Blackberry Cobblers: The Northwest Heritage Cookbook.* New York: Knopf.
Jamison, C. A., & Jamison, B. 1995. *The Border Cookbook: Authentic Home Cooking of the American Southwest and Northern Mexico.* Cambridge, MA: Harvard Common Press.
Jones, E. 1981. *American Food: The Gastronomic Story* (2nd ed.). New York: Vintage Books.
Kirlin, K. S., & Kirlin, T. M. 1991. *Smithsonian Folklife Cookbook.* Washington, D.C.: Smithsonian Institution Press.
Landen, R. 1996. *The Food of Paradise: Exploring Hawaii's Culinary Heritage.* Honolulu: University of Hawaii Press.
Leonard, J. N. 1971. *American Cooking: New England.* New York: Time-Life.
Leonard, J. N. 1971. *American Cooking: The Great West.* New York: Time-Life.
Mandel, A. 1996. *Celebrating the Midwestern Table.* New York: Doubleday.
Murphy, M. D., & Richardson, H. W. 1996. *The Boston Globe Cookbook: A Collection of Classic New England Recipes.* Old Saybrook, CN: Globe Pequot.
Page, L. G., & Wigginton, E. 1984. *The Foxfire Book of Appalachian Cookery.* New York: Dutton.
Shenton, J. P., Pellegrini, A. M., Brown, D., & Shender, I. 1971. *American Cooking: The Melting Pot.* New York: Time-Life.
Smith, J. 1990. *The Frugal Gourmet on Our Immigrant Ancestors.* New York: Morrow.
Walter, E. 1971. *American Cooking: Southern Style.* New York: Time-Life.
Weaver, W. W. 1983. *Sauerkraut Yankees: Pennsylvania German Foods and Foodways.* Philadelphia: University of Pennsylvania Press.
Wilson, J. 1971. *American Cooking: The Eastern Heartland.* New York: Time-Life.

ASIAN INDIAN

Bharadwaj, M. 2000. *The Indian Spice Kitchen.* New York: Hippocrene.
Jaffrey, M. 1995. *Madhur Jaffrey's Indian Cooking.* Woodbury, NY: Barron's.
Rau, S. R. 1969. *The Cooking of India.* New York: Time-Life.
Ray, S. 1986. *Indian Regional Cooking.* London: Macdonald.

CARIBBEAN

Creen, L. H. 1991. *A Taste of Cuba.* New York: Dutton.
Harris, J. B. 1991. *Sky Juice and Flying Fish.* New York: Simon & Schuster.
Kaufman, W. I. (Ed.). 1964. *Recipes from the Caribbean and Latin America.* New York: Dell.
Ortiz, E. L. 1986. *The Complete Book of Caribbean Cooking.* New York: Ballantine.
Rivera, O. 1993. *Puerto Rican Cooking in America: Nuyorican and Bodega Recipes.* New York: Four Walls Eight Windows.
Wolfe, L. 1970. *Cooking of the Caribbean Islands.* New York: Time-Life.

CENTRAL AMERICAN

Marks, C. 1985. *False Tongues and Sunday Bread: A Guatemalan and Mayan Cookbook.* New York: Evans.

CENTRAL EUROPEAN AND RUSSIAN

Hazelton, N. S. 1969. *The Cooking of Germany.* New York: Time-Life.
Martin, P. 1981. *The Czech Book: Recipes and Traditions.* Iowa City: Penfield.
Papashvily, H., & Papashvily, G. 1969. *Russian Cooking.* New York: Time-Life.
Visson, L. 1982. *The Complete Russian Cookbook.* Ann Arbor, MI: Ardis.
Wechsberg, J. I. 1968. *The Cooking of Vienna's Empire.* New York: Time-Life.
Won Bremzen, A., & Welchman, I. 1990. *Please to the Table: The Russian Cookbook.* New York: Workingman.
Zermuska, A. 1981. *The Art of Polish Cooking.* Gretna, LA: Pelican.

CHINESE

Chang, W. W. 1973. *An Encyclopedia of Chinese Food and Cooking.* New York: Crown.
Chen, P. K., Chen, T. C., & Tseng, R. 1983. *Everything You Want to Know about Chinese Cooking.* Woodbury, NY: Barron's.
Hahn, E. 1968. *The Cooking of China.* New York: Time-Life.
Horn, K. 1990. *The Taste of China.* New York: Simon & Schuster.
Simonds, N. 1983. *Classic Chinese Cuisine.* Boston: Houghton Mifflin.
Young, G. 1999. *The Wisdom of the Chinese Kitchen: Classic Family Recipes for Celebration and Healing.* New York: Simon & Schuster.

JAPANESE

Omae, K., & Tachibana, Y. 1981. *The Book of Sushi.* Tokyo: Kodansha International USA.
Ortis, E. L., & Endo, M. 1976. *The Complete Book of Japanese Cooking.* New York: Evans.
Steinberg, R. 1969. *The Cooking of Japan.* New York: Time-Life.
Tokimasa, E. C. 1982. *Japanese Foods.* Honolulu: Temmy.

JEWISH

Levy, F. 1991. *International Jewish Cookbook.* New York: Warner.
Nathan, J. 1998. *The Jewish Holiday Kitchen.* New York: Schocken.
Roden, C. 1996. *The Book of Jewish Food: An Odyssey from Samarkand to New York.* New York: Knopf.
Sogolow, R. 1996. *Memories from a Russian Kitchen—From Shtetl to Golden Land.* Santa Barbara, CA: Fithian.

KOREAN

Cho, J. O. 1981. *Home-Style Korean Cooking in Pictures.* New York: Kodansha International USA.

Kwak, J. 1998. *Dok Suni: Recipes from My Mother's Kitchen.* New York: St. Martin's.

Millon, M., & Millon, K. 1991. *Flavours of Korea.* London: Deutsch.

MEXICAN

Bayless, R. 1987. *Authentic Mexican.* New York: Morrow.

Hamman, C. 1998. *Mayan Cooking: Recipes from the Sun Kingdoms of Mexico.* New York: Hippocrene.

Kennedy, D. 1989. *The Cuisines of Mexico.* New York: HarperCollins.

Leonard, J. N. 1970. *Latin American Cooking.* New York: Time-Life.

Long, K. 1992. *Mexican Light Cooking.* New York: Perigee.

Rexach, N. L. 1985. *The Hispanic American Cookbook.* Secaucus, NJ: Stuart.

MIDDLE EASTERN

Abdennour, S. 1998. *Egyptian Cooking: A Practical Guide.* New York: Hippocrene.

Baron, R. 1991. *Flavors of Greece.* New York: Morrow.

Batmanglij, N. K. 2000. *New Food of Life: Ancient Persian and Modern Iranian Cooking and Ceremonies.* Washington, D.C.: Mage.

Corey, H. 1962. *The Art of Syrian Cooking.* Garden City, NY: Doubleday.

Davies, G. 1991. *The Taste of Cyprus.* Lancaster, England: Gazelle.

Day, I. F. 1975. *The Moroccan Cookbook.* New York: Perigee.

Dosti, R. 1982. *Middle Eastern Cooking.* Tucson: HP Books.

Helou, A. 1994. *Lebanese Cuisine.* New York: St. Martin's.

Khalil, N. E. 1980. *Egyptian Cuisine.* Washington, D.C.: Worldwide Graphics.

Najmieh, B. 1986. *Food of Life: A Book of Ancient Persian and Modern Iranian Cooking and Ceremonies.* Washington, D.C.: Mage.

Nickles, H. G. 1969. *Middle Eastern Cooking.* New York: Time-Life.

Rodin, C. 1968. *A Book of Middle Eastern Food.* New York: Vintage.

Uvezian, S. 1974. *The Cuisine of Armenia.* New York: Harper & Row.

Vassilian, H. B. (Ed.). 1991. *Ethnic Cookbooks and Food Marketplace.* Glendale, CA: Armenian Reference Books.

Wolfert, P. 1973. *Couscous and Other Good Food from Morocco.* New York: Harper & Row.

Yianilos, T. K. 1970. *The Complete Greek Cookbook.* New York: Funk & Wagnalls.

MULTICULTURAL

Albya, C. A., & Webb, L. S. 1993. *The Multicultural Cookbook for Students.* Phoenix, AZ: Oryx.

Claiborne, C. 1971. *The New York Times International Cookbook.* New York: Harper & Row.

Foods of the World. (A series of 27 volumes on individual countries or regions). New York: Time-Life.

Gordon, E. 1965. *Cuisines of the Western World.* New York: Golden.

Laas, W. 1967. *Cuisines of the Eastern World.* New York: Golden.

Smith, J. 1992. *The Frugal Gourmet on Our Immigrant Ancestors: Recipes You Should Have Gotten from Your Grandmother.* New York: Avon.

NATIVE AMERICAN

Cox, B., & Jacobs, M. 1991. *Spirit of the Harvest.* New York: Stewart, Tabori & Chang.

Kavasch, E. B. 1995. *Enduring Harvests.* Old Saybrook, CN: Globe Pequot.

Keegan, M. 1977. *Pueblo and Navajo Cookers.* Dobbs Ferry, NY: Earth Books.

Keegan, M. 1987. *Southwest Indian Cookbook.* Weehawken, NJ: Clear Light.

Kimball, Y., & Anderson, J. 1986. *The Art of American Indian Cooking.* Garden City, NY: Doubleday.

Neithammer, C. 1974. *American Indian Food and Lore.* New York: Simon & Schuster.

NORTHERN EUROPEAN

Bailey, A. 1969. *The Cooking of the British Isles.* New York: Time-Life.

Child, J., & Beck, S. 1970. *Mastering the Art of French Cooking* (vol. 2). New York: Knopf.

Child, J. Bertholle, L., & Beck, S. 1961. *Mastering the Art of French Cooking* (vol. 1). New York: Knopf.

Claiborne, C. 1970. *Classic French Cooking.* New York: Time-Life.

Fisher, M. F. K. 1969. *The Cooking of Provincial France.* New York: Time-Life.

FitzGibbon, T. 1983. *Irish Traditional Food.* New York: St. Martin's.

PACIFIC ISLANDER

Bazore, K. 1947. *Hawaiian and Pacific Foods: A Cook Book of Culinary Customs and Recipes Adapted for the American Hostess.* New York: Barrows.

Corum, A. K. 1983. *Ethnic Foods of Hawai'i.* Honolulu: The Bess Press.

Steinberg, R. 1970. *Pacific and Southeast Asian Cooking.* New York: Time-Life.

SCANDINAVIAN

Brown, D. 1968. *The Cooking of Scandinavia.* New York: Time-Life.

Hazelton, N. S. 1987. *Classic Scandinavian Cooking.* New York: Scribner.

Ojakangas, B. 1992. *Scandinavian Feasts.* New York: Stewart, Tabori & Chang.

SOUTH AMERICAN

Harris, J. B. 1992. *Tasting Brazil: Regional Recipes and Reminiscences.* New York: Macmillan.

Rojas-Lombardi, F. 1991. *The Art of South American Cooking.* New York: HarperCollins.

Umaña-Murray, M. 1996. *Three Generations of Chilean Cuisine.* Los Angeles: Lowell House.

SOUTHEAST ASIAN

Alejandro, R. 1982. *The Philippine Cookbook.* New York: Coward-McCann.

Duong, B. 1991. *Simple Art of Vietnamese Cooking.* New York: Prentice Hall.

Gelle, G. G. 1997. *Filipino Cuisine.* Santa Fe, NM: Red Crane.

Law, R. 1990. *The Southeast Asian Cookbook.* New York: Perigee.

Loha-Unchit, K. 1995. *It Rains Fishes: Legends, Traditions, and the Joys of Thai Cooking.* Rohnert Park, CA: Pomegranate.

Sananikone, K. 1999. *Keo's Thai Cuisine.* Berkeley, CA: Ten Speed Press.

Sing, P. 1981. *Traditional Recipes of Laos.* London: Prospect.

Steinberg, R. 1970. *Pacific and Southeast Asian Cooking.* New York: Time-Life.

SOUTHERN EUROPEAN

Anderson, J. 1994. *The Food of Portugal.* New York: Morrow.

Bugialli, G. 1982. *Classic Techniques of Italian Cooking.* New York: Simon & Schuster.

Casas, P. 1983. *The Foods and Wines of Spain.* New York: Knopf.

Casas, P. 1996. *Delisioso! The Regional Cooking of Spain.* New York: Knopf.

Feibleman, P. S. 1969. *The Cooking of Spain and Portugal.* New York: Time-Life.

Field, C. 1997. *Celebrating Italy: The Tastes and Traditions of Italy as Revealed through Its Feasts, Festivals, and Sumptuous Foods.* New York: HarperCollins.

Rogers, A. 1968. *A Basque Story Cookbook.* New York: Scribner.

Root, W. 1968. *The Cooking of Italy.* New York: Time-Life.

PERIODICALS

Dietary Supplement. A new newsletter with comprehensive reviews of current vitamin, mineral, and botanical research and health. Published quarterly by The Dietary Supplement LLC (11905 Bristol Manor Court, MD 20852; phone: 301-881-7008; e-mail: thedietarysupplement@mindspring.com.

Ethnicity and Disease: An International Journal on Population Differences in Disease Patterns. Peer-reviewed research on the ethnic patterns of disease. The official journal of the International Society on Hypertension in Blacks, published three times a year. (PO Box 1897, Lawrence, KS 66044-9968).

Flavor and Fortune: Dedicated to the Art and Science of Chinese Cuisine. Expert articles, reviews, and recipes on Chinese food and cooking. Published quarterly by the Institute for the Advancement of the Science and Art of Chinese Cuisine (PO Box 91, Kings Park, NY 11754; fax: 516-265-9126; Web site: www.flavorandfortune.com).

Fork, Fingers, & Chopsticks: Supporting Food, Nutrition and Health Professionals in Multicultural Settings. Population profiles with traditional/adapted diets, ethnic ingredient information, cultural nutrition issues, and food trends. Published quarterly by Four Winds Food Specialists (PO Box 70015, Sunnyvale, CA 94086; phone: 408-739-8908; Web site: www.eatethnic.com).

Food History News. Food historians tackle subjects in American culinary past in this newsletter. Published quarterly by S. L. Oliver (19061 Main Road, Islesboro, ME 04848).

Journal for the Study of Food and Society. Research emphasizing the social aspects of food and nutrition. Published by the Association for the Study of Food and Society (ASFS). Journal available with membership. (Contact: Jacqueline M. Newman, FNES Department, Queens College, Flushing, NY 11367)

Journal for Transcultural Nursing. Published by the Transcultural Nursing Society with peer-reviewed articles on cultural health practices and intercultural care (Sage Publications, 2455 Teller Road Thousand Oaks, CA 91320; phone: 805-499-9774; Web site: www.tcns.org).

Food and Foodways: Explorations in the History and Culture of Human Nourishment. Research with strong sociology and history perspective. Quarterly peer-reviewed journal (phone: 800-545-8398; fax: 215-750-6343; e-mail: info@gbhap.com; web site: www.gbhap.com/ journals/325/325-top.htm).

On-Line Resources

MULTICULTURAL

Center for International Health Information (http://www.cihi.com): Health data by country.

EatEthnic (http://eatethnic.com): This site is maintained by Four Winds Food Specialists, a partnership of this book's authors. Information on ethnic ingredients and food habits, recipes, holiday food traditions, cultural foods resources, and links to other sites of interest.

Edibilia (http://www.ibmpcug.co.uk/~owls/edibilia.html): This links site has sections on European, African, Sub-Continental (Asian Indian and Bangladesh), Far Eastern, and cuisine of the Americas. Other resources for baking, drinks, etc.

Epicurus (http://www.epicurus.com): Sponsored by a leading food and hospitality industry consultant, this site features a monthly e-zine with short, informative articles on foods, beverages, herbs and spices, interviews with chefs and hoteliers, and breaking food news, as well as tantalizing recipes. Information on some ethnic ingredients and cultural food events.

EthnoMed (http://www.hslib.washington.edu/clinical/ethnomed/index.html): An ethnic medicine guide from the Harborview Medical Center, University of Washington. Provides somewhat limited information on the seven east African and Southeast Asian groups profiled, but everything presented is reviewed by members of the target community for accuracy.

Foodbooks.com (http://www.foodbooks.com/welcome.htm): "Serious books for serious cooks" and an oustanding collection of food history volumes as well.

Food Composition Resource List for Professionals (http://www.nal.usda.gov/fnic/pubs/bibs/gen/97fdcomp. htm): These food composition resources are all available from the National Agriculture Library, including books, U.S. government publications, software databases, and journals. There is also a contact for assistance with specific requests, and a bulletin board.

Mimi's Cyber Kitchen (www.cyber-kitchen.com): A premier links site with thousands of well-organized listings of cooking, food, and recipe sites. Recipe exchange, cookbook reviews, and more.

Multicultural Health Clearinghouse (http://www.omsa.uiuc.edu/clearinghouse/index.html): An exceptional site run by the University of Illinois with information on U.S. ethnic group health needs and links to other cultural health sites.

Nutrition and Food on the Web (http://sfu.ca/~jfremont): Canadian dietitian Jean Fremont has put together a very well-organized links site for both the health professional and the consumer, with an excellent listing of international food and nutrition pages.

Sally's Place (http://www.sallys-place.com): An overview of several international cuisines can be found at this site. Recipes, restaurants, and ingredient sources are listed in some. Another useful page on this site lists professional food organizations, including descriptions and addresses.

SOAR—Searchable Online Archive of Recipes (http://soar.berkeley.edu/recipes/): This site has catalogued over 45,000 recipes from throughout the world. Just enter the name of the dish you are searching for, and a definition appears, with links to actual recipes. Excellent for specific food inquiries.

Spice Pages (http://www.ang.kfunigraz.ac.at/~katzer/engl/index.html: Well-maintained site is run by an Austrian chemist, Gernot Katzer. Over 100 herbs and spices listed, indexed alphabetically, by region, and by part used in cooking; glossary of spice mixtures. Great links.

Transcultural and Multicultural Health Links (http://www.lib.iun.indiana.edu/trannurs.htm): Comprehensive listing of sites for numerous religious, ethnic, and special populations.

The Web of Culture (http://www.webofculture.com): This site is sponsored by a cross-cultural communications consultant. It has information on everything from currencies, languages, gestures, holidays, and religions to embassies, with a small bookstore and a few recipes.

World Food Habits Bibliography (www.ilstu.edu/class/anth273-foodways/ foodbib.html): An outstanding listing of research in the field of food and culture. Search by region or topic (such as eating attitudes, festivals and feasting, and taboos).

World in Your Kitchen (http://www.creative-homeliving.com/World-Kitchen/home.htm): This excellent links site includes international cooking sites, American cooking sites, a spice tour of sites, cookbooks, an interactive chat room, and a culinary tour.

RELIGIOUS

Buddhism and Medical Ethics (www.changesurfer.com/Bud/BudBioEth.html): An introduction to Buddhist perspective on abortion, death and dying, and euthanasia with numerous links.

Catholic Encyclopedia (http://newadvent.org/cathen): This comprehensive resource includes articles on feasting, abstinence, and fasting with detailed histories.

Islamic Culture and the Medical Arts (http://www.nlm.nih.gov/exhibition/islamic_medical/islamic_00html): The role of medicine in the ancient through modern Middle East.

Judaism (http://judaism.miningco.com/culture/judaism/): Extensive, well-organized resource listing of Jewish Web sites, with links including kosher organizations, kosher foods, kosher restaurants, kosher wines, as well as those on religion and holidays. For information on Jewish diseases, see www.jewishdiseases.com.

Orthodox Christian Information Center (http://orthodoxinfo.com/phronema/index.html): Provides extensive guidance on fasting. Click on "Church (Old) Calendar" within the text to get more information on the Orthodox calendar.

ETHNIC

Asian and Pacific Islander Health Network (http://www.apiahf.org): Links to Asian and Pacific Islander health sites.

Black Health Net (http://www.blackhealthnet.com): Health articles on a variety of topics of interest to African Americans.

Indian Health Service (http://www.ihs.gov): Information about health programs and resources; under "other programs," the program statistics provide demographic and health care data.

National Alliance for Hispanic Health (http://www. hispanichealth.org): Health information of interest to Latinos. Catalog of resource materials offers several useful brochures.

BOTANICAL/ALTERNATIVE HEALTH

Alt Diabetes Herbs (www.altdiabetes.com/herbs/herbs/htm): Listing of some herbal remedies used for diabetes and an efficacy ranking system based on the scientific information. A commercial site with extensive advertising.

A Mini-course in Medical Botany (http://www.ars-grin.gov/duke/syllabus/index.html): Excellent source of information on phytochemicals, ethnic plant uses, GRAS botanicals and more, with a link to the search site to obtain data by plant name or active ingredient.

HerbMed (www.amfoundation.org/herbmed.htm): Database maintained by the Alternative Medicine Foundation with information on evidence for activity, warnings, preparations, mixtures, and mechanisms of action.

Herbal Remedies Index (www.vitamins.com/encyclopedia/Index/HERBS.HTM): Lengthy listing of botanical remedies with description, traditional use, and helpful information on activity and contraindications. A commercial site.

National Center for Complementary and Alternative Medicine (http://nccam.nih.gov): This center run by the National Institutes of Health conducts research and disseminates information on complementary and alternative medical practices. Their CAM citation index is especially useful, with over 180,000 listings since 1966; search or browse by disease, method, system, and type of scientific report.

Tropical Plant Database (www.rain-tree.com/plants.htm): Search by common or scientific name, condition, plant properties, or recorded ethnobotanical uses. A commercial site.

USDA-ARS Botanical Databases (http://ars-genome.cornell/botany.html): Databases on ethnobotany, Native American food plants, and plant chemicals.

Wellness Center (http://wellness.UCdavis.edu/safety_info/poison_prevention/take_care_with_plants/index.html): This site, run by University of California, Davis, provides excellent plant toxicity information.

References

Abbotts, J., Williams, R., & Smith, G. D. 1999. Association of medical, physiological, behavioral and socioeconomic factors with elevated mortality in men of Irish heritage in West Scotland. *Journal of Public Health Medicine, 21,* 46–54.

Abraham, N. 1995. Arab Americans. In R. J. Vecoli, J. Galens, A. Sheets, & R. V. Young (Eds.), *Gale encyclopedia of multicultural America.* New York: Gale Research.

Acevedo, M. C. 2000. The role of acculturation in explaining ethnic differences in the prenatal health-risk behaviors, mental health, and parenting beliefs of Mexican-American and European American at-risk women. *Child Abuse and Neglect, 24,* 111–127.

Achaya, K. T. 1994. *Indian food: A historical companion.* Delhi: Oxford University Press.

Ackerman, L. K. 1997. Health problems of refugees. *Journal of the American Board of Family Practice, 10,* 337–348.

Adler, S. R. 1995. Refuge stress and folk belief: Hmong sudden deaths. *Social Science and Medicine, 40,* 1623–1629.

Adrogue, H. J., & Wesson, D. E. 1996. Role of dietary factors in the hypertension of African Americans. *Seminar in Nephrology, 16,* 94–101.

Ahmed, M., & Flatz, G. 1984. Prevalence of primary adult lactose malabsorption in Pakistan. *Human Genetics, 34,* 69–75.

Alejandro, R. 1982. *The Philippine cookbook.* New York: Coward McCann.

Alexander, J. G. 1995. Slovak Americans. In R. J. Vecoli, J. Galens, A. Sheets, & R. V. Young (Eds.), *Gale Encyclopedia of Multicultural America.* New York: Gale Research.

Algert, S. J., Brzezinski, E., & Ellison, T. H. 1998. *Mexican American food practices, customs, and holidays.* Ethnic and Regional Food Practices (Series). Chicago: American Dietetic Association/American Diabetes Association.

Altschiller, D. 1995. Turkish Americans. In R. J. Vecoli, J. Galens, A. Sheets, & R. V. Young (Eds.), *Gale encyclopedia of multicultural America.* New York: Gale Research.

American Dietetic Association. 1993. *Survey of American dietary habits: Executive summary.* Chicago: American Dietetic Association.

American Heart Association. 1999. *2000 heart and stroke statistical update.* Dallas, TX: American Heart Association.

Ammerman, A. S., Kirkley, B. G., Dennis, B., Hohenstein, C., Allison, A., Strecher, V. J., & Bulger, D. 1994. A dietary assessment for individuals with low literacy skills using interactive touch-screen computer technology. *American Journal of Clinical Nutrition, 59* (suppl.), 289S.

Anderson, E. N. 1987. Why is humoral medicine so popular? *Social Science and Medicine, 25,* 331–337.

Anderson, J. 1986. *The food of Portugal.* New York: Morrow.

Anderson, J. 1997. *The American century cookbook.* New York: Clarkson Potter.

Anderson, J. M., Wiggins, S., Rajwani, R., Holbrook, A., Blue, C., & Ng, M. 1995. Living with a chronic illness: Chinese-Canadian and Euro-Canadian women with diabetes—Exploring factors that influence management. *Social Science and Medicine, 41,* 181–195.

Anderson, J. N. 1983. Health and illness in Filipino immigrants. *Western Journal of Medicine, 139,* 811–819.

Andrews, C. 1988. *Catalan cuisine.* New York: Atheneum.

Ang, C. 1999. Tibetan food and beverages. *Flavor & Fortune, 6*(3), 21.

Anonymous. 1957. *The world's great religions.* New York: Time.

Anonymous. 1982. Minorities and high blood pressure. *Dialogues in Hypertension, 4,* 1–8.

Anonymous. 1993. Lead poisoning associated with use of traditional ethnic remedies—California, 1991–1992. *Morbidity and Mortality Weekly Report, 42,* 521–524.

Anonymous. 1995. Deadly coffee: Fungal disease infecting Surui Indians who grow coffee in Brazil. *Discover, 16,* 30.

Armstrong, J. 1995. A taste of Quebec. In J. M. Powers & A. Stewart (Eds.), *Our northern bounty: A celebration of Canadian cuisine.* Toronto: Random House.

Armstrong, J. E. 1990. Dietary practices and concerns of adult urban black men of high socioeconomic status. *Journal of the American Dietetic Association, 90,* 1716–1719.

Aspler, T. 1995. Icewine. In J. M. Powers & A. Stewart (Eds.), *Our northern bounty: A celebration of Canadian cuisine.* Toronto: Random House.

Astin, A. 1998. Why patients use alternative medicine: Results of a national study. *Journal of the American Medical Association, 279,* 1548–1553.

Axelson, M. L. 1986. The impact of culture on food-related behavior. *Annual Review of Nutrition, 6,* 345–363.

Axtell, R. E. 1991. *Gestures: The do's and taboos of body language around the world.* New York: Wiley.

Bachman-Carter, K., Duncan, R. M., & Pelican, S. 1998. *Navajo food practices, customs, and holidays.* Chicago: American Dietetic Association/American Diabetes Association.

Bailey, A. 1969. *The cooking of the British Isles.* New York: Time-Life.

Baird, E. 1995. The future of Ontario's fruit industry. In J. M. Powers & A. Stewart (Eds.), *Our northern bounty: A celebration of Canadian cuisine.* Toronto: Random House.

Ballew, C., & Sugerman, S. B. 1995. High-risk intakes among low-income Mexican women in Chicago, Illinois. *Journal of the American Dietetic Association, 95,* 1409–1413.

Balzer, M. M. 1987. Behind shamanism: Changing voices of Siberian Khanty cosmology and politics. *Social Science and Medicine, 24,* 1085–1093.

Banks, C. G. 1992. "Culture" in culture-bound syndromes: The case of anorexia nervosa. *Social Science and Medicine, 34,* 867–884.

Bankston, C. L. 1995. Cambodian Americans. In R. J. Vecoli, J. Galens, A. Sheets, & R. V. Young (Eds.), *Gale encyclopedia of multicultural America.* New York: Gale Research.

Bankston, C. L. 1995a. Cambodian Americans. In R. J. Vecoli, J. Galens, A. Sheets, & R. V. Young (Eds.), *Gale encyclopedia of multicultural America.* New York: Gale Research.

Bankston, C. L. 1995b. Hmong Americans. In R. J. Vecoli, J. Galens, A. Sheets, & R. V. Young (Eds.), *Gale encyclopedia of multicultural America.* New York: Gale Research.

Bankston, C. L. 1995c. Laotian Americans. In R. J. Vecoli, J. Galens, A. Sheets, & R. V. Young (Eds.), *Gale encyclopedia of multicultural America.* New York: Gale Research.

Bankston, C. L. 1995d. Vietnamese Americans. In R. J. Vecoli, J. Galens, A. Sheets, & R. V. Young (Eds.), *Gale encyclopedia of multicultural America.* New York: Gale Research.

Bar-David, M. L., & Lewinski, Y.-T. 1972. Food. In C. Roth (Ed.), *Encyclopedia Judaica*. Jerusalem: Keter.

Barer-Stein, T. 1981. *You eat what you are: A study of ethnic food traditions.* Toronto: McClelland & Stewart.

Barer-Stein, T. 1999. *You are what you eat.* Willowdale, Canada: Firefly.

Barnes, V. A., Treiber, F. A., Musante, L., Turner, J. R., Davis, H., & Strong, W. B. 2000. Ethnicity and socioeconomic status: Impact on cardiovascular activity at rest and during stress in youth with a family history of hypertension. *Ethnicity & Disease, 10,* 4–16.

Barrett, B. 1995. Ethnomedical interactions: Health and identity on Nicaragua's Atlantic coast. *Social Science and Medicine, 40,* 1611–1621.

Barron, E. R. 1980. French Canadians. In S. Thernstrom, A. Orlov, & O. Handlin (Eds.), *Harvard encyclopedia of American ethnic groups.* Cambridge, MA: Belknap Press of Harvard University Press.

Barss, B. 1995. The chuckwagon tradition in Prairie culture. In J. M. Powers & A. Stewart (Eds.), *Our northern bounty: A celebration of Canadian cuisine.* Toronto: Random House.

Bartholomew, A. M., Young, E. A., & Hazuda, H. P. 1990. Food frequency intakes and sociodemographic factors elderly Mexican Americans and non-Hispanic whites. *Journal of the American Dietetic Association, 90,* 1693–1696.

Bass, M. A., & Wakefield, L. M. 1974. Neutrient intake and food patterns of Indians on Standing Rock Reservation. *Journal of the American Dietetic Association, 64,* 36–41.

Bassford, T. L. 1995. Health status of Hispanic elders. *Ethnogeriatrics, 11,* 25–38.

Batmanglij, N. K. 2000. *New food of life: Ancient Persian and modern Iranian cooking and ceremonies.* Washington, D.C.: Mage.

Bazore, K. 1947. *Hawaiian and Pacific foods: A cook book of culinary customs and recipes adapted for the American hostess.* New York: Barrows.

Bean, L. J. 1976. California Indian shamanism and folk curing. In W. D. Hand (Ed.), *American folk medicine.* Los Angeles: University of California Press.

Becerra, J. E., Hogue, C. J. R., Atrash, H. K., & Perez, N. 1991. Infant mortality among Hispanics: A portrait of heterogeneity. *Journal of the American Medical Association, 265,* 217–221.

Beck, A., 1988. *Love is never enough.* New York: Harper & Row.

Becker, D. M., Yaneck, L. R., Koffman, D. M., & Bonner, Y. C. 1999. Body image preferences among urban African Americans and whites from low income communities. *Ethnicity & Disease, 9,* 377–386.

Beijborn, U. 1980. Swedes. In S. Thernstrom, A. Orlov, & O. Handlin (Eds.), *Harvard encyclopedia of American ethnic groups.* Cambridge, MA: Belknap Press of Harvard University Press.

Bell, R. A., Hillers, V. N., & Thomas, T. A. 1999. Hispanic grandmothers preserve cultural traditions and reduce foodborne illness by conducting safe cheese workshops. *Journal of the American Dietetic Association, 99,* 1114–1116.

Belman, D. 1997. Regional cuisine: Puts American cooking on the map. *Restaurants, USA, 17,* 22–26.

Benkart, P. 1980. Hungarians. In S. Thernstrom, A. Orlov, & O. Handlin (Eds.), *Harvard encyclopedia of American ethnic groups.* Cambridge, MA: Belknap Press of Harvard University Press.

Bentley, M. E. 1988. The household management of childhood diarrhea in rural north India. *Social Science and Medicine, 27,* 75–85.

Bergstrom, R. W., Newell-Morris, L. L., Leonetti, D. L., Shuman, W. P., Wahl, P. W., & Fujimoto, W. Y. 1990. Association of elevated fasting C-peptide level and increased intra-abdominal fat distribution with development of NIDDM in Japanese-American men. *Diabetes, 39,* 104–111.

Bernard, L., Lavallee, C., Gray-Donald, K., & Delisle, H. 1995. Overweight in Cree schoolchildren and adolescents associated with diet, low physical activity, and high television viewing. *Journal of the American Dietetic Association, 95,* 800–802.

Bernardo, S. 1981. *The ethnic almanac.* Garden City, NY: Doubleday.

Berthoff, R. 1980. Welsh. In. S. Thernstrom, A. Orlov, & O. Handlin (Eds.), *Harvard encyclopedia of American ethnic groups.* Cambridge, MA: Belknap Press of Harvard University Press.

Bertron, P., Barnard, N. D., & Mills, M. 1999. Racial bias in federal nutrition policy, Part I: The public health implications of variations in lactase persistence. *Journal of the National Medical Association, 91,*151–157.

Beto, J. A., Sheth, G., & Rewers, P. 1997. Assessing food purchase behavior among low-income black and Hispanic clients using a self-reported shelf inventory. *Journal of the American Dietetic Association, 97,* 69–70.

Bettin, K. J. 1994. A food frequency questionnaire design for a low literacy population. *American Journal of Clinical Nutrition, 59* (suppl.), 289S.

Beyene, Y. 1992. Medical disclosure and refugees: Telling bad news to Ethiopian patients. *Western Journal of Medicine, 157,* 328–332.

Bigelow, B. C. 1995. African Americans. In R. J. Vecoli, J. Galens, A. Sheets, & R. V. Young (Eds.), *Gale encyclopedia of multicultural America.* New York: Gale Research.

Bindra, G. S., & Gibson, R. S. 1986. Iron status of predominantly lacto-ovo vegetarian East Indian immigrants to Canada: A model approach. *American Journal of Clinical Nutrition, 44,* 643–652.

Birchfield, D. L. 1995. Apaches. In R. J. Vecoli, J. Galens, A. Sheets, & R. V. Young (eds.), *Gale encyclopedia of multicultural America.* New York: Gale Research.

Blackhall, L. J., Murphy, S. T., Frank, G., Michel, V., & Azen, S. 1995. Ethnicity and attitudes toward patient autonomy. *Journal of the American Medical Association, 274,* 820–825.

Bladholm, L. 1999. *The Asian grocery store demystified.* Los Angeles: Renaissance.

Blane, D. 1995. Editorial: Social determinants of health—Socioeconomic status, social class, and ethnicity. *American Journal of Public Health, 85,* 903–904.

Blank, R., & Slipp, S. 1994. *Voices of diversity: Real people talk about problems and solutions in a workplace where everyone is not alike.* New York: AMA COM.

Blesch, K. S., Davis, F., & Kamath, S. K. 1999. A comparison of breast and colon cancer rates among native Asian Indians, US immigrant Asian Indians, and whites. *Journal of the American Dietetic Association, 99,* 1275–1277.

Bloch, B. 1983. Nursing care of black patients. In M. S. Orque, B. Bloch, & L. S. A. Monrroy (Eds.), *Ethnic nursing care: A multicultural approach.* St. Louis: Mosby.

Bonham, G. S., & Brock, D. B. 1985. The relationship of diabetes with race, sex, and obesity. *American Journal of Clinical Nutrition, 41,* 776–783.

Borrud, L. G., McPherson, R. S., Nichaman, M. Z., Pillow, P. C., & Newell, G. R. 1989. Development of a food frequency instrument: Ethnic differences in food sources. *Nutrition and Cancer, 12,* 201–211.

Boyd, L. 1979. *British cookery.* Woodstock, NY: Overlook.

Boyle, J. S. 1989. Constructs of health promotion and wellness in a Salvadoran population. *Public Health Nursing, 6,* 129–134.

Boyle, J. S. 1991, April/May. Transcultural nursing care of Central American refugees. *National Student Nurses Association, Inc./Imprint,* 73–77.

Boyle, J. S. 2000. Transcultural nursing: Where do we go from here? *Journal of Transcultural Nursing, 11,* 10–11.

Boyle, J. S., & MacKay, M. C. 1999. Pica: Sorting it out. *Journal of Transcultural Nursing, 10,* 65–67.

Brand, J. C., Snow, B. J., Nabham, G. P., & Traswell, A. S. 1990. Plasma glucose and insulin responses to traditional Pima Indian meals. *American Journal of Clinical Nutrition, 51,* 416–420.

Brandon, E. 1976. Folk medicine in French Louisiana. In W. D. Hand (Ed.), *American folk medicine.* Berkeley: University of California Press.

Brennan, J. 1984. *The cuisines of Asia: Nine great oriental cuisines by technique.* New York: St. Martin's/Marek.

Brissendan, R. 1970. *Joys and subtleties: South East Asian cooking.* New York: Pantheon.

Brod, M., & Heurtin-Roberts, S. 1992. Older Russian émigrés and medical care. *Western Journal of Medicine, 157,* 333–336.

Bronner, Y. L., Gross, S. M., Caulfield, L., Bentley, M. E., Kessler, L., Jensen, J., Weathers, B., & Paige, D. M. 1999. Early introduction of solid foods among urban African-American participants in WIC. *Journal of the American Dietetic Association, 99,* 457–461.

Brookins, G. K. 1993. Culture, ethnicity, and bicultural competence: Implications for children with chronic illness and disability. *Pediatrics, 91,* 1056–1062.

Brown, A. C., & Brenton, B. 1994. Dietary survey of Hopi Native American elementary students. *Journal of the American Dietetic Association, 94,* 517–522.

Brown, D. 1968. *American cooking.* New York: Time-Life.

Brown, D. 1968. *The cooking of Scandinavia.* New York: Time-Life.

Brown, D. 1971. *American cooking: The Northwest.* New York: Time-Life.

Brown, J. D., & Einsiedel, E. F. 1990. Public health campaigns: Mass media strategies. In E. B. Ray & L. Donohew (Eds.), *Communication and health.* Hillsdale, NJ: Erlbaum.

Brown, J. E., Serdula, M., Cairns, K., Godes, J. R., Jacobs, D. R., Elmer, P., & Trowbridge, F. L. 1986. Ethnic group differences in nutritional status of young children from low-income areas of an urban county. *American Journal of Clinical Nutrition, 44,* 938–944.

Brown, L. K., & Mussell, K. 1984. Introduction. In *Ethnic and regional foodways in the United States.* Knoxville: University of Tennessee Press.

Brownell, K. D., Kelman, J. H., Stunkard, A. J. 1983. Treatment of obese children with and without their mothers: Changes in weight and blood pressure. *Pediatrics, 71,* 515–523.

Bruhn, C. M., & Pangborn, R. M. 1971. Food habits of migrant workers in California. *Journal of the American Dietetic Association, 59,* 347–355.

Brverd, B., Kinney, M. B., & Bothwell, E. 1989. Preventing baby bottle tooth decay in American Indian and Native Alaska communities: A model for planning. *Public Health Reports, 104,* 631–640.

Buffington, S. T. 1995a. Cuban Americans. In R. J. Vecoli, J. Galens, A. Sheets, & R. V. Young (Eds.), *Gale encyclopedia of multicultural America.* New York: Gale Research.

Buffington, S. T. 1995b. Dominicans. In R. J. Vecoli, J. Galens, A. Sheets, R. V. Young (Eds.), *Gale encyclopedia of multicultural America.* New York: Gale Research.

Bugialli, G. 1984. *Guiliano Bugialli's foods of Italy.* New York: Stewart, Tabori & Chang.

Bullough, V., & Bullough, B. 1981. The Spanish-speaking minority groups. In *Health care for the other Americans.* New York: Springer.

Burdine, J. N., Chen, M. S., Gottlieb, N. H., Peterson, F. L., & Vacalis, T. D. 1984. The effects of ethnicity, sex, and father's occupation on heart health knowledge and nutrition behavior of school children: The Texas Youth Health Awareness Study. *Journal of School Health, 54,* 87–90.

Burke, C. B., & Raia, S. P. 1995. *Soul and traditional southern food practices, customs, and holidays.* Chicago: American Dietetic Association/American Diabetes Association.

Burke, J. P., Williams, K., Gaskill, S. P., Hazuda, H. P., Haffner, S. M., & Stern, M. P. 1999. Rapid rise in the incidence of type 2 diabetes from 1987 to 1996: Results from the San Antonio Heart Study. *Archives of Internal Medicine, 159,* 1450–1456.

Burkhard, J. 1995. Quebec goes country. In J. M. Powers, & A. Stewart (Eds.), *Our northern bounty: A celebration of Canadian cuisine.* Toronto: Random House.

Burner, O. Y., Cunningham, P., & Hattar, H. S. 1990. Managing a multicultural nurse staff in a multicultural environment. *Journal of Nursing Administration, 20,* 30–34.

Burson, P. J. 1995. Chilean Americans. In R. J. Vecoli, J. Galens, A. Sheets, & R. V. Young (Eds.), *Gale encyclopedia of multicultural America.* New York: Gale Research.

Bursztyn, M., & Raz, I. 1993. Blood pressure, glucose, insulin and lipids of young Ethiopian recent immigrants to Israel and in those resident for 2 years. *Journal of Hypertension, 11,* 455–459.

Byars, D. 1996. Traditional African American foods and African Americans. *Agriculture and Human Values, 13,* 74–78.

Camp, C. 1989. *American foodways.* Little Rock, AR: August House.

Camp, C. 1991. More American than apple pie. In *Good as gold: Foods the Americas gave the world.* Smithsonian Institution Quincentenary Symposium Proceedings. Washington, D.C.: National Museum of History.

Campbell, M. K., Polhamus, B., McClelleand, J. W., Bennett, K., Kalsbeek, W., Coole, D., Jackson, B., & Demark-Wahnefried, W. 1996. Assessing fruit and vegetable consumption in a 5 a Day study targeting rural blacks: The issue of portion size. *Journal of the American Dietetic Association, 96,* 1040–1042.

Campbell, T., & Chang, B. 1981. Health care of the Chinese in America. In G. Henderson & M. Primeaux (Eds.), *Transcultural health care.* Menlo Park, CA: Addison-Wesley.

Capps, L. L. 1994. Change and continuity in the medical culture of the Hmong in Kansas City. *Medical Anthropology Quarterly, 8,* 161–177.

Carlson, B. 1997. *Food festivals.* Detroit: Visible Ink.

Carrese, J. A., & Rhodes, L. A. 1995. Western bioethics on the Navajo reservation: Benefit or harm? *Journal of the American Medical Association, 274,* 826–829.

Carrese, J. A., & Rhodes, L. A. 2000. Bridging cultural differences in medical practice: The case of discussing negative information with Navajo patients. *Journal of General Internal Medicine, 15,* 92–96.

Carstairs, G. M. 1979. *Daru and bhang:* Cultural factors in the choice of intoxicant. In M. Marshall (Ed.), *Beliefs, behaviors, and alcoholic beverages.* Ann Arbor: University of Michigan Press.

Carter-Pokras, O., Pirkle, J., Chavez, G., & Gunter, E. 1990. Blood lead levels of 4–11 year-old Mexican American, Puerto Rican, and Cuban children. *Public Health Reports, 104,* 388–393.

Casas, P. 1983. *The foods and wines of Spain.* New York: Knopf.

Case, K. O., Brahler, C. J., & Heiss, C. 1997. Resting energy expenditures in Asian women measured by indirect calorimetry are lower than expenditures calculated from prediction equations. *Journal of the American Dietetic Association, 97,* 1288–1292.

Cassell, J. A. 1995. Social anthropology and nutrition. A different look at obesity in America. *Journal of the American Dietetic Association, 95,* 424–427.

Cassidy, C. M. 1994. Walk a mile in my shoes: Culturally sensitive food-habit research. *American Journal of Clinical Nutrition, 59* (suppl.), 190S–197S.

Cavelli-Sforza, L. T., Strata, A., Barone, A., & Cucrachi, L. 1987. Primary adult lactose malabsorption in Italy: Regional differences in prevalence and relationship to lactose intolerance and milk consumption. *American Journal of Clinical Nutrition, 45,* 748–754.

Centers for Disease Control and Prevention. 1992. *Chronic disease in minority populations.* Atlanta: Centers for Disease Control.

Centers for Disease Control and Prevention. 1994. Prevalence of selected risk factors for chronic disease by education level in racial/ethnic populations—United States, 1991–1992. *Mortality and Morbidity Weekly Report, 43,* 894–899.

Chakravarty, I. 1972. *Saga of Indian food: A historical and cultural survey.* New Delhi: Sterling.

Chalmers, I., & Glaser, M. 1986. *Great American food almanac.* New York: Harper & Row.

Chance, J. L. D. 1985. *Ma Chance's French Caribbean Creole cooking.* New York: Putnam.

Chang, K. C. (Ed.). 1977. *Food in Chinese culture: Anthropological and historical perspectives.* New Haven, CT: Yale University Press.

Charzewska, J. 1994. Gaps in dietary survey methodology in Eastern Europe. *American Journal of Clinical Nutrition, 59* (suppl.), 157S–160S.

Chase, C. S. 1995. Costa Rican Americans. In R. J. Vecoli, J. Galens, A. Sheets, & R. V. Young (Eds.), *Gale encyclopedia of multicultural America.* New York: Gale Research.

Chau, P., & Lee, H. S. 1987. *Dietary habits, health beliefs, and related food practices of female Chinese elderly.* Master's project, California State University, San Jose.

Chau, P., Lee, H., Tseng, R., & Downes, N. J. 1990. Dietary habits, health beliefs, and food practices of elderly Chinese women. *Journal of the American Dietetic Association, 90,* 579–580.

Chaudry, M. M. 1992. Islamic food laws: Philosophical basis and practical implications. *Food Technology, 46,* 92–93, 104.

Chávez, N., Sha, L., Persky, V., Langenberg, P., & Pestano-Binghay, E. 1994. Effect of length of stay on food group intake in Mexican and Puerto Rican women. *Journal of Nutrition Education, 26,* 79–86.

Chen, P. K., Chen, T. C., & Tseng, R. 1983. *Everything you want to know about Chinese cooking.* Woodbury, NY: Barron's.

Chen-Louie, T. 1983. Nursing care of Chinese American Patients. In M. S. Orque, B. Bloch, & L. S. A. Monrroy (Eds.), *Ethnic nursing care: A multicultural approach.* St. Louis: Mosby.

Chevallier, A. 1996. *The encyclopedia of medicinal plants.* New York: DK Publishing.

Chew, T. 1983. Sodium values of Chinese condiments and their use in sodium-restricted diets. *Journal of the American Dietetic Association, 82,* 397–401.

Chi, I., Lubben, J. E., & Kitano, H. H. L. 1989. Differences in drinking behavior among three Asian-American groups. *Journal of Studies in Alcohol, 50,* 15–23.

Chiapella, A. P., & Feldman, H. I. 1995. Renal failure among male Hispanics in the United States. *American Journal of Public Health, 85,* 1001–1004.

Chin, S. Y. 1992. This, that, and the other: Managing illness in a first-generation Korean-American family. *Western Journal of Medicine, 157,* 305–309.

Chitwood, M. 1999a. Botanical therapies for diabetes. *On the Cutting Edge: Diabetes Care and Education, 20,* 1, 3.

Chitwood, M. 1999b. Commonly used herbal supplements. *On the Cutting Edge: Diabetes Care and Education, 20,* 16–18.

Cho, J. O. 1981. *Home-style Korean cooking in pictures.* New York: Kodansha International.

Choi, E. S. K., McGandy, R. B., Dallah, G. E., Russell, R. M., Jacob, R. A., Schaefer, E. J., & Sadowski, J. A. 1990. The prevalence of cardiovascular risk factors among elderly Chinese Americans. *Archives of Internal Medicine, 150,* 413–418.

Chow, R. T. P., & Krumholtz, S. 1989. Health screening of a RI Cambodian refugee population. *Rhode Island Medical Journal, 72,* 273–277.

Chrisman, N. J., & Kleinman, A. 1980. Health beliefs and practices. In S. Thernstrom, A. Orlov, & O. Handlin (Eds.), *Harvard encyclopedia of American ethnic groups.* Cambridge, MA: The Belknap Press of Harvard University Press.

Christensen, N. K., Sorenson, A. W., Hendricks, D. G., & Munger, R. 1998. Juniper ash as a source of calcium in the Navajo diet. *Journal of the American Dietetic Association, 98,* 333–334.

Chumlea, W. C., Shumei, S. G., Wholihan, K., Cockram, D., Kuczmarski, R. J., & Johnson, C. L. 1998. Stature prediction equations for elderly non-Hispanic white, non-Hispanic black, and Mexican-American persons developed from NHANES III data. *Journal of the American Dietetic Association, 98,* 137–142.

Chyou, P. H., Nomura, A. M. Y., Hankin, J. H., & Stemmermann, G. N. 1990. A case-cohort study of diet and stomach cancer. *Cancer Research, 50,* 7501–7504.

Claiborne, C. 1987. *Southern cooking.* New York: Times Books.

Clark, M. M. 1983. Cultural context of medical practice. *Western Journal of Medicine, 139,* 806–810.

Claudio, V. S. 1994. *Filipino American food practices, customs, and holidays.* Chicago: American Dietetic Association.

Clinton-Davis, L., & Fassil, Y. 1992. Health and social problems of refugees. *Social Science and Medicine, 35,* 507–513.

Coates, R. J., & Monteilh, C. P. 1997. Assessments of food-frequency questionnaires in minority populations. *American Journal of Clinical Nutrition, 65,* 1108S–1115S.

Coates, R. J., Clark, W. S., Eley, J. W., Greenberg, R. S., Huguley, C. M., & Brown, R. L. 1990. Race, nutritional status, and survival from breast cancer. *Journal of the National Cancer Institute, 82,* 1684–1692.

Colahan, C. 1995. Spanish Americans. In R. J. Vecoli, J. Galens, A. Sheets, & R. V. Young (Eds.), *Gale encyclopedia of multicultural America.* New York: Gale Research.

Cole, W. O., & Singh, S. P. 1978. *The Sikhs.* Boston: Routledge & Kegan Paul.

Collins, J. W., & Shay, D. K. 1994. Prevalence of low birth weight among Hispanic infants with United States–born and foreign-born mothers: The effect of urban poverty. *American Journal of Epidemiology, 139,* 184–192.

Colohan, C. 1995. Spanish Americans. In R. J. Vecoli, J. Galens, A. Sheets, & R. V. Young (Eds.), *Gale encyclopedia of multicultural America.* New York: Gale Research.

Comfort, J. 1995. Some good! The Maritime harvest. In J. M. Powers & A. Stewart (Eds.), *Our northern bounty: A celebration of Canadian cuisine.* Toronto: Random House.

Conley, R. J. 1995. Cherokees. In R. J. Vecoli, J. Galens, A. Sheets, & R. V. Young (Eds.), *Gale encyclopedia of multicultural America.* New York: Gale Research.

Constant, J. 1997. Alcohol, ischemic heart disease, and the French paradox. *Clinical Cardiology, 20,* 420–424.

Conzen, K. N. 1980. Germans. In S. Thernstrom, A. Orlov, & O. Handlin (Eds.), *Harvard encyclopedia of American ethnic groups.* Cambridge, MA: Belknap Press of Harvard University Press.

Cooper, C., Campion, G., & Melton, L. J. 1992. Hip fractures in the elderly: A world-wide projection. *Osteoporosis International, 2,* 285–289.

Cortez, C. E. 1980. Mexicans. In S. Thernstrom, A. Orlov, & O. Handlin (Eds.), *Harvard encyclopedia of American ethnic groups.* Cambridge, MA: Belknap Press of Harvard University Press.

Corum, A. K. 1983. *Ethnic foods of Hawai'i.* Honolulu: Bess.

Council on Scientific Affairs. 1991. Hispanic health in the United States. *Journal of the American Medical Association, 265,* 248–251.

Cox, P. 1995. Samoan Americans. In R. J. Vecoli, J. Galens, A. Sheets, & R. V. Young (Eds.), *Gale encyclopedia of multicultural America.* New York: Gale Research.

Crane, N. T., & Green, N. R. 1980. Food habits and food preferences of Vietnamese refugees living in northern Florida. *Journal of the American Dietetic Association, 76,* 591–593.

Cravioto, R., Lockhart, E. E., Anderson, R. K., Miranda, F., & Harris, R. S. 1945. Composition of typical Mexican foods. *Journal of Nutrition, 29,* 317–329.

Crews, D. E. 1994. Obesity and diabetes. In N. W. S. Zane, D. T. Takeuchi, & K. N. J. Young (Eds.), *Confronting critical health issues of Asian and Pacific Islander Americans.* Thousand Oaks, CA: Sage.

Croft, J. B., Keenan, N. L., Sheridan, D. P., Wheeler, F. C., & Speers, M. A. 1995. Waist-to-hip ratio in a biracial population: Measurement, implications, and cautions for cardiovascular disease. *Journal of the American Dietetic Association, 95,* 60–64.

Cronin, A. M., Anders, B., & Moore, M. J. 1996. Latino belief of alleged medical procedure. *Western Journal of Medicine, 164,* 364–365.

Cunningham, S. 1990. *The magic in food: Legends, lore, and spellwork.* St. Paul, MN: Llewellyn.

Curb, J. D., Aluli, N. E., Kautz, J. A., Petrovich, H., Knutsen, S. F., Knutsen, R., O'Conner, H. K., & O'Conner, W. E. 1991. Cardiovascular risk factor levels in ethnic Hawaiians. *American Journal of Public Health, 81,* 164–167.

Cushner, N. P. 1966. Philippines. In *The 1996 Grolier multimedia encyclopedia.* Danbury, CT: Grolier Electronic Publishing.

Cussler, M., & DeGive, M. L. 1952. *'Twixt the cup and the lip.* New York: Twayne.

D'Emilio, F. 1992, December 1. Italians pass on pasta, feast on fat. *San Jose Mercury News,* 8A.

Dagnelie, P. C., & van Stavern, W. A. 1994. Macrobiotic nutrition and child health: Results of a population-based, mixed longitudinal cohort study in the Netherlands. *American Journal of Clinical Nutrition, 59* (suppl.) 1187S–1196S.

Das, D. K., Sato, M., Ray, P. S., Manlik, G., Engelman, R. M., Bertelli, A. A., & Bertelli, A. 1999. Cardioprotection of red wine: Role of polyphenolic antioxidants. *Drugs Experimentation and Clinical Research, 25,* 115–120.

Davis, J. M., Goldenring, J., McChesney, M., & Medina, A. 1982. Pregnancy outcomes of Indochinese refugees, Santa Clara County, California. *American Journal of Public Health, 72,* 742–743.

Davis, M. B. (Ed.), 1994. *Native Americans in the twentieth century: An encyclopedia.* New York: Garland.

De Forge, B. R., Stewart, D. L., De Voe-Weston, M., Graham, L., & Charleston, J. 1998. The relationship between health status and blood pressure in urban African Americans. *Journal of the National Medical Association, 90,* 658–664.

de Lemos, M., & Sunderji, R. 1999, May. Herbal interactions with Warfarin. *Drug & Therapeutics Newsletter.*

de Lorgeril, M. 1999. *The French paradox: What does it tell us?* In Proceedings of the 21st Annual Congress of the ESC, Barcelona, Spain.

de Wit, C. W. 1992. Food–place associations on American product labels. *Geographical Review, 83,* 323–330.

DeBusk, R. M. 1999. Herbal medicines: A primer. *On the Cutting Edge: Diabetes Care and Education, 20,* 4–5.

deGarine, I. 1972. The socio-cultural aspects of nutrition. *Ecology of Food and Nutrition, 1,* 143–163.

deGarine, I., & Pollack, N. J. 1995. *Social aspects of obesity.* Australia: Gordon & Breach.

Deharveng, G., Charrondiere, U. R., Slimani, N., Southgate, D. A. T., & Riboli, E. 1999. Comparison of nutrients in the food composition tables available in the nine European countries participating in EPIC. *European Journal of Clinical Nutrition, 53,* 60–79.

Denny, C. H., & Taylor, T. L. 1999. American Indian and Alaska Native health behavior: Findings from the Behavioral Risk Surveillance System, 1992–1995. *Ethnicity & Disease, 9,* 403–409.

Derecskey, S. 1972. *The Hungarian cookbook.* New York: Harper & Row.

DeSantis, L., & Halberstein, R. 1992. The effects of immigration on the health care system of south Florida. *Human Organization, 51,* 223–234.

Desenclos, J. C. A., & Hahn, R. A. 1992. Years of potential life lost before age 65, by race, Hispanic origin, and sex—United States, 1986–1988. *Morbidity and Mortality Weekly Report, 41,* 13–23.

Devi Dasi, I., & Devi Dasi, S. 1973. *The Hari Krishna cookbook.* Los Angeles: Bhaktivedanta Book Trust.

Devine, C. M., Wolfe, W. S., Frongillo, E. A., & Bisogni, C. A. 1999. Life-course events and experiences: Association with fruit and vegetable consumption in three ethnic groups. *Journal of the American Dietetic Association, 99,* 309–314.

Dewey, K. G., Strode, M. A., & Fitch, Y. R. 1984. Dietary change among migrant and nonmigrant Mexican-American families in Northern California. *Ecology of Food and Nutrition, 14,* 11–24.

Dewey, K. G., Strode, M. A., & Ruiz Fitch, Y. 1984a. Anthropometry of migrant and nonmigrant Mexican-American children and adults in northern California. *Ecology of Food and Nutrition, 14,* 25–35.

Dewey, K. G., Strode, M. A., & Ruiz Fitch, Y. 1984b. Dietary change among migrant and nonmigrant Mexican-American families in northern California. *Ecology of Food and Nutrition, 14,* 11–24.

DeWitt, D., Wilan, M. J., & Stock, M. 1998. *Flavors of Africa cookbook.* Rocklin, CA: Prima.

Diaz, T., Buehler, J. W., Castro, K. G., & Ward, J. W. 1993. AIDS trends among Hispanics in the United States. *American Journal of Public Health, 83,* 504–509.

Dirige, O. V. 1995, February 28. Filipino-American diet and foods. *Asian American Business Journal,* 11–17.

DiSogra, L., Abrams, B., & Hudes, M. 1994. Low prevalence of healthful dietary behaviors in a California agricultural county: Emphasis on white and Mexican-American adults. *Journal of the American Dietetic Association, 94,* 544–546.

Dominguez, L. J., Barbagallo, M., & Sowers, J. R. 1999. Cardiovascular risk factors in South America and the Caribbean. *Ethnicity & Disease, 9,* 468–478.

Donaldson, G. 1980. Scots. In S. Thernstrom, A. Orlov, & O. Handlin (Eds.), *Harvard encyclopedia of American ethnic groups.* Cambridge, MA: Belknap Press of Harvard University Press.

Donohew, L. 1990. Public health campaigns: Individual message strategies and a model. In E. B. Ray & L. Donohew (Eds.), *Communication and health.* Hillsdale, NJ: Erlbaum.

Dosti, R. 1999. *After communism, a cuisine of survival.* Available on-line: www.khao.org/albrecipe.htm.

Dougherty, D., Bankhead, R., Kushner, R., Mirtallo, J., & Winkler, M. 1995. Nutrition care given new importance in JCAHO standards. *Nutrition in Clinical Practice, 10,* 26–31.

Douglas, M. 1966. *Purity and danger: An analysis of concepts of pollution and taboo.* New York: Praeger.

Douglas, M. 1972. Deciphering a meal. *Daedalus, 101,* 61–81.

Douglass, W. A. 1980. Basques. In S. Thernstrom, A. Orlov, & O. Handlin (Eds.), *Harvard encyclopedia of American ethnic groups.* Cambridge, MA: Belknap Press of Harvard University Press.

Dresser, N. 1996. *Multicultural manners.* New York: Wiley.

Dressler, W. W. 1996. Hypertension in the African American community: Social, cultural, and psychological factors. *Seminar in Nephrology, 16,* 71–82.

Dressler, W. W., Mata, A., Chavez, A., & Viteri, F. E. 1987. Arterial blood pressure and individual modernization in a Mexican Community. *Social Science and Medicine, 24,* 679–687.

Drewnowski, A., Henderson, S. A., Shore, A. B., Fischler, C., Preziosi, P., & Hercberg, S. 1996. Diet quality and dietary diversity in France: Implications for the French paradox. *Journal of the American Dietetic Association, 96,* 663–669.

Duke, J. A. 2000. *A mini-course in medical botany.* Available on-line: www.arg-grin.gov/duke/syllabus/index.html.

Duncan, D. 1995. Ontario cooking: Cuisines in transition. In J. M. Powers & A. Stewart (Eds.), *Our northern bounty: A celebration of Canadian cuisine.* Toronto: Random House.

Duyff, R. L., Sanjur, D., & Neslon, H. V. 1975. Food behavior and related factors of Puerto Rican teenagers. *Journal of Nutrition Education, 7,* 99–103.

Easton, S. E., & Ellington, L. 1995. Japanese Americans. In R. J. Vecoli, J. Galens, A. Sheets, & R. V. Young (Eds.), *Gale encyclopedia of multicultural America.* New York: Gale Research.

Ebden, P., Bhatt, A., Carey, O. J., & Harrison, 1988. The bilingual consultation. *Lancet, 13,* 347.

Eckstein, E. F. 1983. Foodways of blacks. In *Menu planning* (3rd ed.). Westport, CT: AVI Publishing.

Eckstein, E. F. 1983a. Foodways of Chicanos. In *Menu planning* (3rd ed.). Westport, CT: AVI Publishing.

Eckstein, E. F. 1983b. Foodways of Puerto Ricans and Cubans living in the United States. In *Menu planning* (3rd ed.). Westport, CT: AVI Publishing.

Eckstein, E. F. 1983. *Menu planning* (3rd ed.). Westport, CT: AVI.

Edmondson, B. 1998. The line between beer and wine. *American Demographics 1,* 8–19.

Edwards, D. M., & Watt, R. G. 1997. Diet and hygiene in the life of Gypsy Travellers in Hertfordshire. *Community Dental Health, 14,* 41–46.

Eisenberg, D. M., Davis, R. B., Ettner, S. L., Appel, S., Wilkey, S., Van Rompay, M., & Kessler, R. C. 1998. Trends in alternative medicine use in the United States, 1990–1997: Results of a follow-up national survey. *Journal of the American Medical Association, 280,* 1569–1575.

Ekoe, J. M. 1988. *Diabetes mellitus: Aspects of the world-wide epidemiology of diabetes mellitus and its long-term complications.* Amsterdam: Elsevier.

Elder, J., Sallis, J. F., Zive, M. M., Hoy, P., McKenzie, T. L., Nader, P. R., & Berry, C. C. 1999. Factors affecting selection of restaurants by Anglo- and Mexican-American families. *Journal of the American Dietetic Association, 99,* 856–857.

Ellis, R., Kelsay, J. L., Reynolds, R. D., Morris, E. R., Moser, P. B., & Frazier, C. W. 1987. Phytate: zinc and phytate × calcium: zinc millimolar ratios in self-selected diets of Americans, Asian Indians, and Nepalese. *Journal of the American Dietetic Association, 87,* 1043–1047.

Elshaw, E. B., Young, E. A., Saunders, M. J., McGurn, W. C., & Lopez, L. C. 1994. Utilizing a 24-hour dietary recall and culturally specific diabetes education in Mexican Americans with diabetes. *Diabetes Educator, 20,* 228–235.

Emboden, W. A. 1976. Plant hypnotics among North American Indians. In W. D. Hand (Ed.), *American folk medicine.* Los Angeles: University of California Press.

Emmons, L. 1986. Food procurement and the nutritional adequacy of diets in low-income families. *Journal of the American Dietetic Association, 86,* 1684–1693.

Emmons, L. 1992. Dietary and purging behavior in black and white high school students. *Journal of the American Dietetic Association, 92,* 306–312.

Erdman, J. W., & Fordyce, E. J. 1989. Soy products and the human diet. *American Journal of Clinical Nutrition, 49,* 725–737.

Erickson, C. J. 1980. English. In S. Thernstrom, A. Orlov, & O. Handlin (Eds.), *Harvard encyclopedia of American ethnic groups.* Cambridge, MA: Belknap Press of Harvard University Press.

Esquivel, G. B., & Keitel, M. A. 1990. Counseling immigrant children in the schools. *Elementary School Guidance & Counseling, 24,* 213–221.

Fabre, J., Baczko, A., Laravoire, P., Dayer, P., & Fox, H. 1983. Arterial hypertension and ethnic factors. *Advances in Nephrology, 12,* 19–39.

Fackelmann, K. A. 1991. The African gene? Searching through history for the roots of black hypertension. *Science News, 140,* 254–255.

Fahm, E. G. 1998, Spring. Nutritional status of black elders: A review of the literature. *Journal of the Family and Consumer Sciences,* 23–27.

Fanelli-Kuczmarski, M., & Woteki, C. E. 1990. Monitoring the nutritional status of the Hispanic population: Selected findings for Mexican Americans, Cubans, and Puerto Ricans. *Nutrition Today, May/June,* 6–11.

Fanelli-Kuczmarski, M., Johnson, C. L., Elias, L., & Najjar, M. F. 1990. Folate status of Mexican American, Cuban, and Puerto Rican women. *American Journal of Clinical Nutrition, 52,* 368–372.

Fannelli-Kuczmarski, M., Kuczmarski, R. J., & Najjar, M. 1995. Food usage among Mexican-American, Cuban, and Puerto Rican adults. *Nutrition Today, 30,* 30–37.

Fang, J., Madnavan, S., & Alderman, M. H. 1996. The association between birthplace and mortality from cardiovascular causes among black and white residents of New York City. *New England Journal of Medicine, 335,* 1545–1551.

Farkas, C. S. 1986. Ethno-specific communication patterns: Implications for nutrition education. *Journal of Nutrition Education, 18,* 99–103.

Fee, C. H. 1998. Increasing cultural competence for effective client counseling: An experimental course. *Journal of Nutrition Education, 30,* 115B–117B.

Feibleman, P. S. 1969. *The cooking of Spain and Portugal.* New York: Time-Life.

Feibleman, P. S. 1971. *American cooking: Creole and Acadian.* New York: Time-Life.

Felitti, V., Perlman, M., & Howard, N. 1996. Iron overload disorders among Hispanics—San Diego, California, 1995. *Mortality and Morbidity Weekly, 45,* 991–993.

Ferguson, C. 1995. Our northern bounty. In J. M. Powers & A. Stewart (Eds.), *Our northern bounty: A celebration of Canadian cuisine.* Toronto: Random House.

Fernandez, N. A., Burgos, J. C., Asenjo, C. F., & Rosa, I. 1971. Nutritional status of the Puerto Rican population: Master sample survey. *American Journal of Clinical Nutrition, 24,* 952–965.

Ferrano, K. F. 1998. Firm believers? Religion, body weight, and well-being. *Review of Religious Research, 39,* 224–244.

Ferro-Luzzi, A., & Branca, F. 1995. Mediterranean diet, Italian-style: Prototype of a healthy diet. *American Journal of Clinical Nutrition, 61* (suppl.), 1338S–1354S.

Fewster, W. J., Bostian, L. R., & Powers, R. D. 1973. Measuring the connotative meanings of foods. *Home Economics Research Journal, 2,* 44–53.

Field, M., & Field, F. 1970. *A quintet of cuisines.* New York: Time-Life.

Fisher, M. F. K. 1969. *The cooking of provincial France.* New York: Time-Life.

Fitzgerald, T. K. 1979. Southern folks' eating habits ain't what they used to be . . . if they ever were. *Nutrition Today, 14,* 16–21.

FitzGibbon, T. 1968. *A taste of Ireland: Irish traditional food.* New York: Avenal.

Fitzpatrick, J. P. 1980. Puerto Ricans. In S. Thernstrom, A. Orlov, & O. Handlin (Eds.), *Harvard encyclopedia of American ethnic groups.* Cambridge, MA: Belknap Press of Harvard University Press.

Fitzpatrick-Nietschmann, J. 1983. Pacific Islanders—Migration and health. *Western Journal of Medicine, 139,* 848–853.

Flegal, K. M., Ezzati, T. M., Harris, M. I., Haynes, S. G., Juarez, R. Z., Knowler, W. C., Perez-Stable, E. J., & Stern, M. P. 1991. Prevalence of diabetes in Mexican Americans, Cubans, and Puerto Ricans from the Hispanic Health and Nutrition Examination Survey, 1982–84. *Diabetes Care, 14* (suppl. 3), 628–638.

Fong, A. K. H. 1991. Educating clients: Successful communication creates change. *Journal of the American Dietetic Association, 91,* 289–290.

Fordyce, J. K., Christakis, G., Kafatos, A., Duncan, R., & Cassidy, J. 1983. Adipose tissue fatty acid composition of adolescents in a U.S.–Greece cross-cultural study of coronary heart disease risk factors. *Journal of Chronic Diseases, 36,* 481–486.

Forsythe, H. E., & Gage, B. 1994. Use of a multicultural food-frequency questionnaire with pregnant and lactating women. *American Journal of Clinical Nutrition, 59* (suppl.), 203S–206S.

Fraser, M. 1995. Canada's breadbasket: Decades of change. In J. M. Powers & A. Stewart (Eds.), *Our northern bounty: A celebration of Canadian cuisine.* Toronto: Random House.

Freedman, A. D. 1996. Medicine, traditional. In *The 1996 Grolier multimedia encyclopedia.* Danbury, CT: Grolier Electronic Publishing.

Freeze, K. J. 1980. Czechs. In S. Thernstrom, A. Orlov, & O. Handlin (Eds.), *Harvard encyclopedia of American ethnic groups.* Cambridge, MA: Belknap Press of Harvard University Press.

Freidenberg, J., Mulvihill, M., & Caraballo, L. R. 1993. From ethnology to survey: Some methodological issues in research on health seeking in east Harlem. *Human Organization, 52,* 151–161.

Freimer, N., Echenberg, D., & Kretchmer, N. 1983. Cultural variation—Nutritional and clinical implications. *Western Journal of Medicine, 139,* 928–933.

Fuchs, M., & Bashshur, R. 1975. Use of traditional Indian medicine among urban Native Americans. *Medical Care, 13,* 915–927.

Furnham, A., & Alibhai, N. 1983. Cross-cultural differences in the perception of female body shapes. *Psychological Medicine, 13,* 829–837.

Gabriel, A., Gabriel, K. R., & Lawrence, R. A. 1986. Cultural values and biomedical knowledge: Choices in infant feeding. *Social Science and Medicine, 23,* 501–509.

Galanis, D. J., McGarvey, S. T., Quested, C., Sio, B., & Afele-Fam'amuli, S. 1999. Dietary intake of modernizing Samoans: Implications for risk of cardiovascular disease. *Journal of the American Dietetic Association, 99,* 184–190.

Gann, P., Nghiem, L., & Warner, S. 1989. Pregnancy characteristics and outcomes of Cambodian refugees. *American Journal of Public Health, 79,* 1251–1257.

Gardenswartz, L., & Rowe, A. 1988, July. Why diversity matters. *HR Focus,* S1–S3.

Gardner, W. E. 1994. Mortality. In N. W. S. Zane, D. T. Takeuchi, & K. N. J. Young (Eds.), *Confronting critical health issues of Asian and Pacific Islander Americans.* Thousand Oaks, CA: Sage.

Gardner, W. E. 1972. The differential effects of race, education and experience in helping. *Journal of Clinical Psychology, 28,* 87–89.

Garrett, J. T. 1990. Indian health: Values, beliefs, and practices. In M. S. Harper (Ed.), *Minority aging: Essential curricula content for selected health and allied health professions.* Health Resources and Services Administration, Department of Health and Human Services, DHHS Publication No. HRS (P-DV-90-4). Washington, D.C.: U.S. Government Printing Office.

Garrett, L., 1994. *The coming plague: Newly emerging diseases in a world out of balance.* New York: Farrar, Strauss & Giroux.

Geiger, H. J. 1996. Race and health care: An American dilemma. *New England Journal of Medicine, 335,* 815–816.

Geissler, E. M. 1998. *Pocket guide to cultural assessment.* St. Louis: Mosby.

Geissler, E. M. 1998. *Pocket guide to cultural assessment* (2nd ed.). St. Louis: Mosby.

Gelle, G. G. 1997. *Filipino cuisine: Recipes from the islands.* Santa Fe, NM: Red Crane.

Geronimus, A. T., Bound, J., Waidmann, T. A., Hillemeier, M. M., & Burns, P. B. 1996. Excess mortality among blacks and whites in the United States. *New England Journal of Medicine, 335,* 1552–1558.

Ghaemi-Ahmadi, S. 1992. Attitudes toward breast-feeding and infant feeding among Iranian, Afghan, and Southeast Asian immigrant women in the United States: Implications for health and nutrition education. *Journal of the American Dietetic Association, 92,* 354–355.

Gilbert, S. E., & Gay, G. 1989. Improving the success in school of poor black children. In B. J. Robinson Shade (Ed.), *Culture, style, and the educative process.* Springfield, IL: Thomas.

Gillespie, A. K. 1984. A wilderness in the megalopolis: Foodways in the Pine Barrens of New Jersey. In *Ethnic and regional foodways in the United States.* Knoxville: University of Tennessee Press.

Gillis, M. 1995. Iranian Americans. In R. J. Vecoli, J. Galens, A. Sheets, & R. V. Young (Eds.), *Gale encyclopedia of multicultural America.* New York: Gale Research.

Gillum, R. F. 1996. Epidemiology of hypertension in African American women. *American Heart Journal, 131,* 385–395.

Gilman, S. C., Justice, J., Saepharn, K., & Charles, G. 1992. Use of traditional and modern health services by Laotian refugees. *Western Journal of Medicine, 157,* 310–315.

Gladney, V. M. 1966. *Food practices of Mexican Americans in Los Angeles County.* Los Angeles: Los Angeles County Health Department.

Gladney, V. M. 1972. *Food practices of some black Americans in Los Angeles County.* Los Angeles: County of Los Angeles Department of Health Services.

Gleason, N. A. 1995. A new approach to disordered eating—Using an electronic bulletin board to confront social pressure on body image. *Journal of American College Health, 44,* 78–80.

Gobetz, E. 1995. Slovenian Americans. In R. J. Vecoli, J. Galens, A. Sheets, & R. V. Young (Eds.), *Gale encyclopedia of multicultural America.* New York: Gale Research.

Godel, J. C., & Hart, A. G. 1984. Northern infant syndrome: A deficiency state? *Canadian Medical Association Journal, 131,* 199–204.

Gofton, L. 1986. The rules of the table: Sociological factors influencing food choice. In C. Ritson, L. Gofton, & J. McKenzie (Eds.), *The Food Consumer.* New York: Wiley.

Gold, S. J. 1992. Mental health and illness in Vietnamese refugees. *Western Journal of Medicine, 157,* 290–294.

Goldberg, D. L., Novotny, R., Kieffer, E., Mor, J., & Thiele, M. 1995. Complementary feeding and ethnicity of infants in Hawaii. *Journal of the American Dietetic Association, 95,* 1029–1031.

Goldberg, J., Rudd, R. E., & Deitz, W. 1999. Using 3 data sources and methods to shape a nutrition campaign. *Journal of the American Dietetic Association, 99,* 717–722.

Good, M. J. D. 1980. Of blood and babies: The relationship of popular Islamic physiology to fertility. *Social Science & Medicine, 14B,* 147–156.

Goody, J. 1982. The high and low: Culinary culture in Asia and Europe. In *Cooking, cuisine, and class.* Cambridge: Cambridge University Press.

Gordeuk, V., Mukiibi, J., Hasstedt, S. J., Samowitz, W., Edwards, C. Q., West, G., Ndambire, S., Emmanuel, J., Nikanza, N., Chapanduka, Z., Randall, M., Boone, P., Romano, P., Martell, R. W., Yamashita, T., Effler, P., & Brittenham, G. 1992. Iron overload in Africa. *New England Journal of Medicine, 326,* 95–100.

Gordon, A. M. 1982. Nutritional status of Cuban refugees: A field study on the health and nutrition of refugees processed at Opa-Locka, Florida. *Americans Journal of Clinical Nutrition, 35,* 582–590.

Gornick, M. E., Eggers, P. W., Reilly, T. W., Mentnech, R. M., Fitterman, L. K., Kucken, L. E., & Vladeck, B. C. 1996. Effects of race and income of mortality and use of services among Medicare beneficiaries. *New England Journal of Medicine, 335,* 791–799.

Gostin, L. O. 1995. Informed consent, cultural sensitivity, and respect for persons. *Journal of the American Medical Association, 274,* 844–845.

Gould-Martin, K., & Ngin, C. 1981. Chinese Americans. In A. Harwood (Ed.), *Ethnicity and medical care.* Cambridge, MA: Harvard University Press.

Graedon, J., & Graedon, T. 1999. *The people's pharmacy guide to home and herbal remedies.* New York: St. Martin's.

Graham, J. S. 1976. The role of the *curandero* in the Mexican American folk medicine system in West Texas. In W. D. Hand (Ed.), *American folk medicine.* Berkeley: University of California Press.

Granger, B. H. 1976. Some aspects of folk medicine among Spanish-speaking people in southern Arizona. In W. D. Hand (Ed.), *American folk medicine.* Los Angeles: University of California Press.

Granquist, M. A. 1995. Swedish Americans. In R. J. Vecoli, J. Galens, A. Sheets, & R. V. Young (Eds.), *Gale Encyclopedia of Multicultural America.* New York: Gale Research.

Grant, M. D., Rudberg, M. A., & Brody, J. A. 1998. Gastrostomy placement and mortality among hospitalized Medicare beneficiaries. *Journal of the American Medical Association, 279,* 1973–1976.

Greeley, A. M. 1972. *That most distressful nation: The taming of the American Irish.* Chicago: Quadrangle.

Green, D. 1995. Puerto Ricans. In R. J. Vecoli, J. Galens, A. Sheets, & R. V. Young (Eds.), *Gale encyclopedia of multicultural America.* New York: Gale Research.

Greenberg, B. 1989. *How to run a traditional Jewish household.* Northvale, NJ: Aronson.

Greene, V. 1980. Poles. In S. Thernstrom, A. Orlov, & O. Handlin (Eds.), *Harvard encyclopedia of American ethnic groups.* Cambridge, MA: Belknap Press of Harvard University Press.

Greenlund, K. J., Liu, K., Dyer, A. R., Kiefe, C. I., Burke, G. L., & Yunis, C. 1996. Body mass index in young adults: Associations with parental body size and education in the CARDIA study. *American Journal of Public Health, 86,* 480–485.

Grim, C. E., Wilson, T. W., Nicholson, G. D., Hassell, T. A., Fraser, H. S., Grim, C. M., & Wilson, D. M. 1990. Blood pressure in blacks: Twin studies in Barbados. *Hypertension, 15,* 803–809.

Grivetti, L. E., & Paquette, B. 1978. Nontraditional ethnic food choices among first-generation Chinese in California. *Journal of Nutrition Education, 10,* 109–112.

Gudykunst, W. B. 1994. *Bridging differences: Effective intergroup communication* (2nd ed.). Thousand Oaks, CA: Sage.

Gudykunst, W. B., & Nishida, T. 1994. *Bridging Japanese/North American differences.* Thousand Oaks, CA: Sage.

Guendelman, S. 1995. *High risk outcomes: The health paradox of Latina mothers and infants.* Chicano/Latinos Policy Project (CLPP) Working Paper 1 (4). Berkeley: Institute for the Study of Social Change at the University of California at Berkeley.

Guendelman, S., & Abrams, B. 1995. Dietary intake among Mexican-American women: Generational differences and a comparison with white non-Hispanic women. *American Journal of Public Health, 85,* 20–25.

Gugliotti, G. 2000, March 19. Herbal products take a human toll. *Washington Post.* IFIC. 1999. *Functional foods: Attitudinal research.* Washington, D.C.: International Food Information Council.

Gumaa, K. A., Mustafa, K. Y., Mahmoud, N. A., & Gader, A. M. 1978. The effect of fasting in Ramadan: Serum uric acid and lipid concentration. *British Journal of Nutrition, 40,* 573–581.

Gupta, R. 1987. *Caribbean prenatal nutrition project* (extended abstract). Dorchester, MA: Codman Square Health Center.

Gupta, S. P. 1976. Changes in food habits of Asian Indians in the United States: A case study. *Sociology and Social Research, 60,* 87–99.

Gutierrez, C. P. 1984. The social and symbolic uses of ethnic/regional foodways: Cajuns and crawfish in South Louisiana. In *Ethnic and regional foodways in the United States.* Knoxville: University of Tennessee Press.

Haffner, L. 1992. Translation is not enough: Interpreting in a medical setting. *Western Journal of Medicine, 157,* 255–259.

Haffner, S. M., Hazuda, H. P., Mitchell, B. D., Patterson, J. K., & Stern, M. P. 1991. Increased incidence of Type II diabetes mellitus in Mexican Americans. *Diabetes Care, 14,* 102–108.

Hafner, D. 1993. *A taste of Africa.* Berkeley, CA: Ten Speed Press.

Hagar, J. M., & Rahimtoola, S. H. 1991. Chagas' heart disease in the United States. *New England Journal of Medicine, 325,* 763–768.

Hahn, E. 1968. *The cooking of China.* New York: Time-Life.

Hahn, H. W. L. 1994. Hepatitis B. In N. W. S. Zane, D. T. Takeuchi, & K. N. J. Young (Eds.), *Confronting critical health issues of Asian and Pacific Islander Americans.* Thousand Oaks, CA: Sage.

Hahn, R. A. 1992. The state of federal health statistics on racial and ethnic groups. *Journal of the American Medical Association, 267,* 232–268.

Haider, S. Q., & Wheeler, M. 1979. Nutritive intake of black and Hispanic mothers in a Brooklyn ghetto. *Journal of the American Dietetic Association, 75,* 670–673.

Halderson, K. 1998. *Alaska native food practices, customs, and holidays.* Chicago: American Dietetic Association/American Diabetes Association.

Hall, E. T. 1977. *Beyond culture.* Garden City, NY: Anchor/Doubleday.

Hall, E. T. 1979. Learning the Arab's silent language. *Psychology Today, August,* 45–54.

Hall, E. T., & Hall, M. R. 1990. *Understanding cultural differences: Germans, French, and Americans.* Yarmouth, ME: Intercultural Press.

Hall, R. B. 1996a. North Korea. In *The 1996 Grolier multimedia encyclopedia.* Danbury, CT: Grolier Electronic Publishing.

Hall, R. B. 1996b. South Korea. In *The 1996 Grolier multimedia encyclopedia.* Danbury, CT: Grolier Electronic Publishing.

Hamdallah, M., & Bhatia, A. J. 1995. Prevalence of sickle-cell trait in USA adolescents of Central American origin. *Lancet, 346,* 707–708.

Hampl, J. S., Holland, K. A., Marple, J. T., Hutchins, M. R., & Brockman, K. K. 1997. Acute hemolysis related to consumption of fava beans: A case study and medical nutrition therapy approach. *Journal of the American Dietetic Association, 97,* 182–183.

Hancock, I. 1991. Romani foodways: Gypsy culinary culture. *The World and I,* 666–677.

Hand, W. D. 1980. *Magical medicine.* Berkeley: University of California Press.

Hanft, S. 1995. English Americans. In R. J. Vecoli, J. Galens, A. Sheets, & R. V. Young (Eds.), *Gale encyclopedia of multicultural America.* New York: Gale Research.

Hankin, J. H., & Wilkins, L. R. 1994. Development and validation of dietary assessment methods for culturally diverse populations. *American Journal of Clinical Nutrition, 59* (suppl.), 198S–200S.

Hann, R. S. 1994. Parasitic infections. In N. W. S. Zane, D. T. Takeuchi, K. N. J. Young (Eds.), *Confronting critical health issues of Asian and Pacific Islander Americans.* Thousand Oaks, CA: Sage.

Hann, R. S. 1994b. Tuberculosis. In N. W. S. Zane, D. T. Takeuchi, & K. N. J. Young (Eds.), *Confronting critical health issues of Asian and Pacific Islander Americans.* Thousand Oaks, CA: Sage.

Hanna, J. M., Pelletier, D. L., & Brown, V. J. 1986. The diet and nutrition of contemporary Samoans. In P. T. Baker, J. M. Hanna, & T. S. Baker (Eds.), *The changing Samoans: Behavior and health in transition.* New York: Oxford University Press.

Harnack, L., Story, M., & Holy Rock, B. 1999. Diet and physical activity patterns of Lakota Indian adults. *Journal of the American Dietetic Association, 99,* 829–835.

Harnack, L., Story, M., Martinson, B., Neumark-Sztainer, D., & Stang, J. 1998. Guess who's cooking? The role of men in meal planning, shopping, and preparation in U.S. families. *Journal of the American Dietetic Association, 98,* 995–1000.

Harris, J. B. 1992. *Tasting Brazil: Regional recipes and reminiscences.* New York: Macmillan.

Harris, J. B. 1998. *The Africa cookbook: Tastes of a continent.* New York: Simon & Schuster.

Harris, K. M. 1999. The health status and risk behaviors of adolescents in immigrant families. In D. J. Hernandez (Ed.), *Children of immigrants: Health, adjustment, and public assistance.* Washington, D.C.: National Academy Press.

Harris, M. 1972. *Cows, pigs, wars, and witches.* New York: Random House.

Harris, M. 1985. *Good to eat: Riddles of food and culture.* New York: Simon & Schuster.

Harris, P. R., & Moran, R. T. 1987. *Managing cultural differences* (2nd ed.). Houston, TX: Gulf.

Harrison, E., Adjei, A., Ameho, C., Yamamoto, S., & Kono, S. 1998. The effect of soybean protein on bone loss in a rat model of postmenopausal osteoporosis. *Journal of Nutritional Science and Vitaminology, 44,* 257–268.

Hart, D. V. 1981. Bisayan Filipino and Malayan folk medicine. In G. Henderson & M. Primeaux (Eds.), *Transcultural health care.* Menlo Park, CA: Addison-Wesley.

Hart, D. V. 1996. Guam. In *The 1996 Grolier multimedia encyclopedia.* Danbury, CT: Grolier Electronic Publishing.

Harvey-Berino, J., Hood, V., Rourke, J., Terrance, T., Dorwaldt, A., & Secker-Walker, R. 1997. Food preferences predict eating behavior of very young Mohawk children. *Journal of the American Dietetic Association, 97,* 750–753.

Harwood, A. 1981. *Ethnicity and medical care.* Cambridge, MA: Harvard University Press.

Harwood, A. 1981. Mainland Puerto Ricans. In A. Harwood (Ed.), *Ethnicity and medical care.* Cambridge, MA: Harvard University Press.

Hashizume, S., & Takano, J. 1983. Nursing care of Japanese-American patients. In M. S. Orque, B. Bloch, & L. S. A. Monrroy (Eds.), *Ethnic nursing care: A multicultural approach.* St. Louis: Mosby.

Hazelton, N. S. 1969. *The cooking of Germany.* New York: Time-Life.

Hazelton, N. S. 1987. *Classic Scandinavian cooking.* New York: Scribner's.

Healy, J. R. 1996. Hawaii. In *The 1996 Grolier multimedia encyclopedia.* Danbury, CT: Grolier Electronic Publishing.

Heberle, M. O. 1985. *Polish cooking.* Tucson, AZ: HP Books.

Heimlich, E. 1995a. Acadian Americans. In R. J. Vecoli, J. Galens, A. Sheets, & R. V. Young (Eds.), *Gale encyclopedia of multicultural America.* New York: Gale Research.

Heimlich, E. 1995b. Gypsy Americans. In R. J. Vecoli, J. Galens, A. Sheets, & R. V. Young (Eds.), *Gale encyclopedia of multicultural America.* New York: Gale Research.

Helman, C. G. 1990. *Culture, health and illness: An introduction for health professionals* (2nd ed.). London: Wright.

Henderson, G., & Primeaux, M. 1981. Health care. In G. Henderson & M. Primeaux (Eds.), *Transcultural health care.* Menlo Park, CA: Addison-Wesley.

Henley, J. 2000, January 16. French town decrees local fare for students. *San Jose Mercury News,* 5AA.

Henry, J. P., & Cassel, J. C. 1969. Psychosocial factors in essential hypertension: Recent epidemiologic and animal experimental evidence. *American Journal of Epidemiology, 90,* 171–200.

Herter, G. L., & Herter, B. E. 1960. *Bull cook and authentic recipes and practices.* Waseca, MN: Herter's.

Hertzler, A. A., Wenkam, N., & Standal, B. 1982, Classifying cultural food habits and meanings. *Journal of the American Dietetic Association, 80,* 421–425.

Hess, K. 1992. *The Carolina rice kitchen.* Columbia: University of South Carolina Press.

Heymann, T. D., Bhupulan, A., Zureikat, N. E., Bomanji, J., Drinkwater, C., Giles, P., & Murray-Lyon, I. M. 1995. Khat chewing delays gastric emptying of a semi-solid meal. *Alimentary Pharmacological Therapy, 9,* 81–83.

Hickling, F. W., & Griffith, E. E. 1994. Clinical perspectives on the Rastafari movement. *Hospital and Community Psychiatry, 45,* 49–53.

Higgins, C., Laredo, R., Stollar, C., & Warshaw, H. S. 1998. *Jewish food practices, customs, and holidays.* Chicago: American Dietetic Association/American Diabetes Association.

Higonnet, P. L. R. 1980. French. In S. Thernstrom, A. Orlov, & O. Handlin (Eds.), *Harvard encyclopedia of American ethnic groups.* Cambridge, MA: Belknap Press of Harvard University Press.

Hillstrom, L. C. 1995. French Americans. In R. J. Vecoli, J. Galens, A. Sheets, & R. V. Young (Eds.), *Gale encyclopedia of multicultural America.* New York: Gale Research.

Hines, D. 1959. *Adventures in good eating.* Ithaca, NY: Duncan Hines Institute.

Hitti, P. K. 1961. *The Near East history.* New York: Van Nostrand.

Ho, G. P., Nolan, F. L., & Dodds, M. L. 1966. Adaptation to American dietary patterns by students from oriental countries. *Journal of Home Economics, 58,* 277–280.

Hoang, G. N., & Erickson, R. V. 1982. Guidelines for providing medical care to Southeast Asian refugees. *Journal of the American Medical Association, 248,* 710–714.

Hodes, R. M. 1997. Cross-cultural medicine and diverse health beliefs: Ethiopians abroad. *Western Journal of Medicine, 166,* 29–36.

Hoerr, S. L., Nelson, R. A., & Lohman, T. R. 1992. Discrepancies among predictors of desirable weight for black and white obese adolescent girls. *Journal of the American Dietetic Association, 92,* 450–453.

Hoglund, A. W. 1980. Finns. In S. Thernstrom, A. Orlov, & O. Handlin (Eds.), Cambridge, MA: Belknap Press of Harvard University Press.

Holt, T. 1980. Afro-Americans. In S. Thernstrom, A. Orlov, & O. Handlin (Eds.), *Harvard encyclopedia of ethnic groups.* Cambridge, MA: Belknap Press of the Harvard University Press.

Hong, M. 1995. Guatemalan Americans. In R. J. Vecoli, J. Galens, A. Sheets, & R. V. Young (Eds.), *Gale encyclopedia of multicultural America.* New York: Gale Research.

Horner, R. D., Lackey, C. J., Kolasa, K., & Warren, K. 1991. Pica practices of pregnant women. *Journal of the American Dietetic Association, 91,* 34–38.

Hosking, R. 1996. *A dictionary of Japanese food: Ingredients and culture.* Rutland, VT: Tuttle.

Hosokawa, B. 1969. *Nisei: The quiet American.* New York: Morrow.

Hostetler, J. A. 1976. Folk medicine and sympathy healing among the Amish. In W. D. Hand (Ed.), *American folk medicine.* Berkeley: University of California Press.

Howard, A. 1980. Hawaiians. In S. Thernstrom, A. Orlov, & O. Handlin (Eds.), *Harvard encyclopedia of American ethnic groups.* Cambridge, MA: Belknap Press of Harvard University Press.

Howard, A. 1986. Samoan coping behavior. In P. T. Baker, J. M. Hanna, & T. S. Baker (Eds.), *The changing Samoans: Behavior and health in transition.* New York: Oxford University Press.

Howard, B. V., Bogardus, C., Ravussin, E., Foley, J. E., Lillioja, S., Mott, D. M., Bennett, P. H., & Knowler, W. C. 1991. Studies of the etiology of obesity in Pima Indians. *American Journal of Clinical Nutrition, 53,* 1577S–1585S.

Howell, W. 1982. *The emphatic communicator.* Belmont, CA: Wadsworth.

Hughes, L. O. 1990. Insulin, Indian origin and ischaemic heart disease. *International Journal of Cardiology, 26,* 1–4.

Hull, D. 1979. Migration, adaptation, and illness: A review. *Social Science and Medicine, 131,* 25–36.

Hunter, J. M. 1973. Geophagy in Africa and the United States. *Geographical Review, 63,* 170–195.

Hussaini, M. M. 1993. *Islamic dietary concepts and practices.* Bedford Park, IL: Islamic Food and Nutrition Council of America.

Hutchens, A. R. 1973. *Indian herbalogy of North America.* Boston: Shambala.

Hysa, K., & Hysa, R. J. 1998. *The best of Albanian cooking: Favorite family recipes.* New York: Hippocrene.

Ichiro, H., Fujio, I., Tsuneya, W., & Keiichi, Y. 1972. *Japanese religion.* Palo Alto, CA: Kodansha International.

Ifkovic, E. 1995. Croatian Americans. In R. J. Vecoli, J. Galens, A. Sheets, & R. V. Young (Eds.), *Gale encyclopedia of multicultural America.* New York: Gale Research.

Ikeda, J. P. 1999. *Hmong American food practices, customs, and holidays.* Chicago: American Dietetic Association/American Diabetes Association.

Ikeda, J. P., Ceja, D. R., Glass, R. S., Harwood, J. O., Lucke, K. A., & Sutherlin, J. M. 1991. Food habits of the Hmong living in central California. *Journal of Nutrition Education, 23,* 168–175.

Ikeda, J. P., Murphy, S., Mitchell, A., Flynn, N., Mason, I. J., Lizer, A., & Lamp, C. 1998. Dietary quality of Native American women in rural California. *Journal of the American Dietetic Association, 98,* 812–813.

Ikeda, J., & Gonzales, M. M. 1986. *Food habits of Hispanics newly immigrated to the San Francisco Bay area* (abstract). Oakland: California Dietetic Association Annual Meeting.

Ikeda, J., Dugan, S., Feldman, N., & Mitchell, R. 1993. Native Americans in California surveyed on diets, nutrition needs. *California Agriculture, 47,* 8–10.

Illich, I. 1976. *Medical nemesis.* New York: Random House.

Immink, M. D. C., Sanjur, D., & Burgos, M. 1983. Nutritional consequences of U.S. migration patterns among Puerto Rican women. *Ecology of Food and Nutrition, 13,* 139–148.

Ingham, J. M. 1970. On Mexican folk medicine. *American Anthropologist, 72,* 76–87.

Isern, T. D. 1990. Bierocks. In J. Hoy & T. Isern (Eds.), *Plains folk: The romance of the landscape* (vol. 2). Norman: University of Oklahoma Press.

Ismail, A. I., & Szpunar, S. M. 1990. The prevalence of total tooth loss, dental caries, and periodontal disease among Mexican Americans, Cuban Americans, and Puerto Ricans: Findings from HHANES 1982–1984. *American Journal of Public Health, 80* (suppl.), 66–70.

Iwu, M. M. 1993. *Handbook of African medicinal plants.* Boca Raton, FL: CRC Press.

Jackson, B. 1976. The other kind of doctor: Conjure and magic in black American folk medicine. In W. D. Hand (Ed.), *American folk medicine.* Berkeley: University of California Press.

Jackson, J. J. 1981. Urban black Americans. In A. Harwood (Ed.), *Ethnicity and medical care.* Cambridge, MA: Harvard University Press.

Jackson, M. Y. 1986. Nutrition in American Indian health: Past, present, and future. *Journal of the American Dietetic Association, 86,* 1561–1565.

Jackson, M. Y. 1993. Height, weight and body mass index of American Indian schoolchildren 1990–1991. *Journal of the American Dietetic Association, 93,* 1136–1140.

Jackson, M. Y., & Broussard, B. A. 1987. Cultural challenges in nutrition education among American Indians. *Diabetes Educator, 13,* 47–50.

Jaffrey, M. 1975. *Invitation to Indian cooking.* New York: Vintage.

Janes, C. R. 1990. *Migration, social change, and health: A Samoan community in California.* Stanford: Stanford University Press.

Jefferson, A. W. 1995. Brazilian Americans. In R. J. Vecoli, J. Galens, A. Sheets, & R. V. Young (Eds.), *Gale encyclopedia of multicultural America.* New York: Gale Research.

Jeffery, R. W. 1991. Population perspectives on the prevention and treatment of obesity in minority populations. *American Journal of Clinical Nutrition, 53,* 1621S–1624S.

Jelliffe, D. B. 1967. Parallel food classifications in developing and industrialized countries. *American Journal of Clinical Nutrition, 20,* 279–281.

Jelliffe, D. B., & Bennett, F. J. 1961. Cultural and anthropological factors in infant and maternal nutrition. *Proceedings of the Fifth International Congress of Nutrition, 20,* 185–188.

Jenkins, C. N. H., & Kagawa-Singer, M. 1994. Cancer. In N. W. S. Zane, D. T. Takeuchi, & K. N. J. Young (Eds.), *Confronting critical health issues of Asian and Pacific Islander Americans.* Thousand Oaks, CA: Sage.

Jensen, J. M. 1980. East Indians. In S. Thernstrom, A. Orlov, & O. Handlin (Eds.), *Harvard encyclopedia of American ethnic groups.* Cambridge, MA: Belknap Press of Harvard University Press.

Jerome, N. W. 1969. Northern urbanization and food consumption of Southern-born Negroes. *American Journal of Clinical Nutrition, 22,* 1667–1669.

Jerome, N. W. 1997. Culture-specific strategies for capturing local dietary intake patterns. *American Journal of Clinical Nutrition, 65,* 1166S–1167S.

Johnson, E. 1995. Historical influences on West Coast cooking. In J. M. Powers & A. Stewart (Eds.), *Our northern bounty: A celebration of Canadian cuisine.* Toronto: Random House.

Johnson, P. B. 1997. Alcohol-use-related problems in Puerto Rican and Irish-American males. *Substance Use and Misuse, 32,* 169–179.

Jones, E. 1981. *American food: The gastronomic story* (2nd ed.). New York: Vintage.

Jones, M. A. 1980. Scotch-Irish. In S. Thernstrom, A. Orlov, & O. Handlin (Eds.), *Harvard encyclopedia of American ethnic groups.* Cambridge, MA: Belknap Press of Harvard University Press.

Jones, S. 1995a. Austrian Americans. In R. J. Vecoli, J. Galens, A. Sheets, & R. V. Young (Eds.), *Gale Encyclopedia of multicultural American.* New York: Gale Research.

Jones, S. 1995b. Polish Americans. In R. J. Vecoli, J. Galens, A. Sheets, & R. V. Young (Eds.), *Gale encyclopedia of multicultural America.* New York: Gale Research.

Joos, S. K. 1980. Diet, obesity, and diabetes mellitus among the Florida Seminole Indians. *Florida Science, 43,* 148–152.

Josling, T., & Ritson, C. 1986. Food and the nation. In C. Ritson, L. Gofton, & J. McKenzie (Eds.), *The Food Consumer.* New York: Wiley.

Jouris, D. 1994. *All over the map: An extraordinary atlas of the United States.* Berkeley, CA: Ten Speed Press.

Juarez, C. A. B. 1990. Professional practice at a Native American community. *Contemporary Dialysis and Nephrology, 11,* 23–28.

Jurgens, J. 1995. Albanian Americans. In R. J. Vecoli, J. Galens, A. Sheets, & R. V. Young (Eds.), *Gale encyclopedia of multicultural America.* New York: Gale Research.

Jurgens, J. 1995. Greek Americans. In R. J. Vecoli, J. Galens, A. Sheets, & R. V. Young (Eds.), *Gale encyclopedia of multicultural America.* New York: Gale Research.

Kakarala, M., Song, W., Hoerr, S., & Bond, J. 1994. Computerized database of Indian foods. *American Journal of Clinical Nutrition, 59* (Suppl.), 291S.

Kalŏik, S. 1984. Ethnic foodways in America: Symbol and the performance of identity. In *Ethnic and regional foodways in the United States.* Knoxville: University of Tennessee Press.

Kanders, B. S., Ullmann-Joy, P., Foreyt, J. P., Helmsfield, S. B., Heber, D., Elashoff, R. M., Ashley, J. M., Reeves, R. S., & Blackburn, G. L. 1994. The Black American Lifestyle Intervention (BALI): The design of a weight-loss program for workingclass African-American women. *Journal of the American Dietetic Association, 94,* 310–312.

Kannan, S., Carruth, B. R., & Skinner, J. 1999. Cultural influences on infant feeding beliefs of mothers. *Journal of the American Dietetic Association, 99,* 88–90.

Kant, A. K., Block, G., Schatzkin, A., Ziegler, R. G., & Nestle, M. 1991. Dietary diversity in the U.S. population NHANES II, 1976–1980. *Journal of the American Dietetic Association, 91,* 34–38.

Kao, G. D., & Devine, P. 2000. Use of complementary health practices by prostate carcinoma patients undergoing radiation therapy. *Cancer, 88,* 615–619.

Karim, N., Bloch, D. S., Falciglia, G., & Murthy, L. 1986. Modifications of food consumption patterns reported by people from India, living in Cincinnati, Ohio. *Ecology of Food and Nutrition, 19,* 11–18.

Katone-Apte, J. 1975. Dietary aspects of acculturation: Meals, feasts, and fasts in a minority community in South Asia. In M. L. Arnott (Ed.), *Gastronomy: The anthropology of food habits.* The Hague: Mouton.

Kavanagh, K. H., & Kennedy, P. H. 1992. *Promoting cultural diversity: Strategies for health professionals.* Newbury Park, CA: Sage.

Kavasch, B. 1977. *Native harvests.* New York: Random House.

Kavasch, E. B. 1995. *Enduring harvests.* Old Saybrook, CT: Globe Pequot.

Kavasch, E. B., & Baar, K. 1999. *American Indian healing arts.* New York: Bantam.

Kawano, Y., Matsuoka, H., Takishita, S., & Omae, T. 1998. Effects of magnesium supplementation in hypertensive patients: Assessment by office, home, and ambulatory blood pressures. *Hypertension, 32,* 260–265.

Kay, M. A. 1977. Health and illness in a Mexican American Barrio. In E. H. Spicer (Ed.), *Ethnic medicine in the Southwest.* Tucson: University of Arizona Press.

Keegan, M. 1977. *Pueblo and Navajo cookery.* Dobbs Ferry, NY: Earth Books.

Kelly, A. A. 1985. Hippocrates of Cos: The founder of western medicine. In S. Bliss, E. Bauman, L. Piper, A. I. Brint, & P. A. Wright (Eds.), *The new holistic health handbook.* Lexington, MA: Green.

Kelly, J. L. 1983. Loco moco: A folk dish in the making. *Social Process in Hawaii, 30,* 59–64.

Kemnitzer, L. S. 1996. Folk medicine. In *The 1996 Grolier multimedia encyclopedia.* Danbury, CT: Grolier Electronic Publishing.

Khanna, G. 1986. *Herbal remedies: A handbook of folk medicine* (4th ed.). New Delhi: Tarang Paperbacks.

Khare, R. S. 1976. *The Hindu hearth and home.* Durham, NC: Carolina Academic Press.

Kiefer, C. W., Kim, S., Choi, K., Kim, L., Kim, B. L., Shon, S., & Kim, T. 1985. Adjustment problems of Korean American elderly. *Gerontologist, 25,* 477–482.

Kilara, A., & Iya, K. K. 1992. Food and dietary practices of the Hindu. *Food Technology, 46,* 94–102, 104.

Kim, H. C. 1980. Koreans. In S. Thernstrom, A. Orlov, & O. Handlin (Eds.), *Harvard encyclopedia of American ethnic groups.* Cambridge, MA: Belknap Press of Harvard University Press.

Kim, K. K., Kohrs, M. B., Twork, R., & Grier, M. R. 1984. Dietary calcium intakes of elderly Korean Americans. *Journal of the American Dietetic Association, 84,* 164–169.

Kim, K. K., Yu, E. S., Liu, W. T., Kim, J., & Kohrs, M. B. 1993. Nutritional status of Chinese-, Korean-, and Japanese-American elderly. *Journal of the American Dietetic Association, 93,* 1416–1422.

Kimball, Y., & Anderson, J. 1986. *The art of American Indian cooking.* New York: Simon & Schuster.

Kinloch, P. 1985. *Talking health but doing sickness: Studies in Samoan Health.* Wellington, New Zealand: Victoria University Press.

Kinsella, M. 1983. *An Irish Farmhouse Cookbook.* Belfast: Appletree Press.

Kirlin, K. S., & Kirlin, T. M. 1991. *Smithsonian folklife cookbook.* Washington, D.C.: Smithsonian Institution Press.

Kitano, H. H. 1976. *Japanese-Americans: The evolution of a subculture* (2nd ed.). Englewood Cliffs, NJ: Prentice Hall.

Kittler, P. G., Sucher, K., & Tseng, R. Y. 1986. Cultural foods education: An exploratory study of dietitians and Plan IV programs in California. *Journal of the American Dietetic Association, 86,* 1705–1708.

Klatsky, A. L., & Armstrong, M. A. 1991. Cardiovascular risk factors among Asian Americans living in northern California. *American Journal of Public Health, 81,* 1423–1428.

Kleinman, A. 1980. *Patients and healers in the context of culture.* Berkeley: University of California Press.

Kleinman, A., Eisenberg, L., & Good, B. 1978. Culture, illness, and care: Clinical lessons from anthropologic and cross-culture research. *Annals of Internal Medicine, 88,* 251–258.

Kleinman, J. C., & Kessel, S. S. 1987. Racial differences in low birthweight: Trends and risk factors. *New England Journal of Medicine, 317,* 749–754.

Klessig, J. 1992. The effects of values and culture on life-support decisions. *Western Journal of Medicine, 157,* 316–322.

Knapp, J. A., Haffner, S. M., Young, E. A., Hazusa, H. P., Gardner, L., & Stern, M. P. 1985. Dietary intakes of essential nutrients among Mexican-Americans and Anglo-Americans: The San Antonio heart study. *American Journal of Clinical Nutrition, 42,* 307–316.

Knapp, R. B., & Houghton, M. D. 1999. Breast-feeding practices of WIC participants from the former USSR. *Journal of the American Dietetic Association, 99,* 1269–1271.

Koh, F. M. 1985. *Creative Korean cooking.* Minneapolis: East-West Press.

Kokinos, M., & Dewey, K. G. 1986. Infant feeding practices of migrant Mexican-American families in northern California. *Ecology of Food and Nutrition, 18,* 209–220.

Kollipara, U. K., & Brittin, H. C. 1996. Increased iron content of some Indian foods due to cookware. *Journal of the American Dietetic Association, 96,* 508–510.

Koo, L. C. 1984. The use of food to treat and prevent disease in Chinese culture. *Social Science and Medicine, 18,* 757–766.

Kramer, B. J. 1992. Health and aging in urban American Indians. *Western Journal of Medicine, 157,* 281–285.

Kraybill, D. B. 1995. Amish Americans. In R. J. Vecoli, J. Galens, A. Sheets, & R. V. Young (Eds), *Gale encyclopedia of multicultural America.* New York: Gale Research.

Kreps, G. L. 1990. Communication and health education. In E. B. Ray & L. Donohew (Eds.), *Communication and health.* Hillsdale, NJ: Erlbaum.

Kreps, G. L., & Kunimoto, E. N. 1994. *Effective communication in multicultural health care settings.* Thousand Oaks, CA: Sage.

Krieger, N., & Sidney, S. 1997. Prevalence and health implications of anti-gay discrimination: a study of black and white women and men in the CARDIA cohort: Coronary Artery Risk Development in Young Adults. *International Journal of Health Service, 27,* 157–176.

Kromhout, D., Keys, A., Aravanis, C., Buzina, R., Fidanza, F., Giampaoli, S., Jansen, A., Menotti, A., Nedeljkovic, S., Pekkarinen, M., Simic, B. S., & Toshima, H. 1989. Food consumption patterns in the 1960s in seven countries. *American Journal of Clinical Nutrition, 49,* 889–894.

Kuhnlein, H. V. 1995. Benefits and risks of traditional food for indigenous peoples: Focus on dietary intakes of Arctic men. *Canadian Journal of Physiological Pharmacology, 73,* 765–771.

Kuhnlein, H. V., Calloway, D. H., & Harland, B. F. 1979. Composition of traditional Hopi foods. *Journal of the American Dietetic Association, 75,* 37–41.

Kuhnlein, H. V., Soueida, R., & Receveur, O. 1996. Dietary nutrient profiles of Canadian Baffin Island Inuit differ by source, season, and age. *Journal of the American Dietetic Association, 96,* 155–162.

Kulesza, W., Rywik, S., Balaz, V., Budlovsky, I., & Marczuk, A. 1984. Prevalence of ischaemic heart disease risk factors in male population aged 45–54 years in Warsaw and Bratislava. *Corpus Vasorum Antiquorum, 26,* 61–71.

Kumanyika, S. 1987. Obesity in black women. *Epidemiological Reviews, 9,* 31–50.

Kumanyika, S. K., & Morssink, C. B. 1997. Cultural appropriateness of weight management programs. In *Overweight and weight management: The health professional's guide to understanding and practice.* Gaithersburg, MD: Aspen.

Kumanyika, S., & Helitzer, D. L. 1985. Nutritional status and dietary pattern of racial minorities in the United States. In *Report of the Secretary's Task Force on Black and Minority Health* (Vol. II). Washington, D.C.: U.S. Department of Health and Human Services.

Kumanyika, S., Wilson, J. F., & Guilford-Davenport, M. 1993. Weight-related attitudes and behaviors of black women. *Journal of the American Dietetic Association, 93,* 416–422.

Kunitz, S. J., & Levy, J. E. 1981. Navajos. In A. Harwood (Ed.), *Ethnicity and medical care.* Cambridge, MA: Harvard University Press.

Kurson, K. 1995. Palestinian Americans. In R. J. Vecoli, J. Galens, A. Sheets, & R. V. Young (Eds.), *Gale encyclopedia of multicultural America.* New York: Gale Research.

Kushi, L. H., Lew, R. A., Stare, F. J., Curtis, R. E., Lozy, M., Bourke, G., Daly, L., Graham, I., Hickey, N., Mulcahy, R., & Kevaney, J. 1985. Diet and 20-year mortality from coronary heart disease: The Ireland–Boston diet–heart study. *New England Journal of Medicine, 312,* 811–818.

Kuster, A. E., & Fong, C. M. 1993. Further psychometric evaluation of the Spanish language health-promoting lifestyle profile. *Nursing Research, 42,* 266–269.

Kwak, J. 1998. *Dok Suni: Recipes from my mother's Korean kitchen.* New York: St. Martin's.

La Farge, O. 1956. *A pictorial history of the American Indian.* New York: Crown.

Lackey, C. J. 1982. Pica—Pregnancy's etiological mystery. In *Alternative dietary practices and nutritional abuses in pregnancy.* Washington, D.C.: National Academy Press.

Laguerre, M. S. 1981. Haitian Americans. In A. Harwood (Ed.), *Ethnicity and medical care.* Cambridge, MA: Harvard University Press.

Lalbachan, P. 1994. *The complete Caribbean cookbook.* Rutland, VT: Tuttle.

Landale, N. S., Oropesa, R. S., & Gorman, B. K. 1999. Immigration and infant health: Birth outcomes of immigrant and native-born women. In D. J. Hernandez (Ed.), *Children of immigrants: Health, adjustment, and public assistance.* Washington, D.C.: National Academy Press.

Lands, E. M., Hamazaki, T., Yamazaki, K., Okuyama, H., Sakai, K., Goto, Y., & Hubbard, V. S. 1990. Changing dietary patterns. *American Journal of Clinical Nutrition, 51,* 991–993.

Lang, G. C. 1985. Diabetes and health care in a Sioux community. *Human Organization, 44,* 251–260.

Lannon, J. 1986. How people choose food: The role of advertising and packagings. In C. Ritson, L. Gofton, & J. McKenzie (Eds.), *The Food Consumer.* New York: Wiley.

Larson-Presswalla, J. 1994. Insights into Eastern health care: Some transcultural nursing perspectives. *Journal of Transcultural Nursing, 5,* 21–24.

Latourette, K. S. 1964. *The Chinese: Their history and culture* (4th ed.). New York: Macmillan.

Lau, G., Kee, M. M., & Ng, A. 1998. *Chinese American food practices, customs, and holidays.* Chicago: American Dietetic Association/The American Diabetes Association.

Lauderdale, D. S., Jacobsen, S. J., Furner, S. E., Levy, P. S., Brody, J. A., & Goldberg, J. 1997. Hip fracture incidence among elderly Asian-American populations. *American Journal of Epidemiology, 146,* 502–509.

Law, M., & Wald, N. 1999. Why heart disease mortality is low in France: The time lag explanation. *British Medical Journal, 318,* 1471–1480.

Lazarou, J., Pomeranz, B. H., & Corey, P. N. 1998. Incidence of adverse drug reactions in hospitalized patients: A meta-analysis of prospective studies. *Journal of the American Medical Association, 279,* 1200–1205.

Lazebnik, N., Kuhnert, B. R., & Kuhnert, P. M. 1989. The effect of race on serum ferritin during parturition. *Journal of the American College of Nutrition, 8,* 591–596.

LeBreton, M. M. 1980. Acadians. In S. Thernstrom, A. Orlov, & O. Handlin (Eds.), *Harvard encyclopedia of American ethnic groups.* Cambridge, MA: Belknap Press of Harvard University Press.

Lee, C. 1965. *Chinatown, U.S.A.* Garden City, NY: Doubleday.

Lee, E. 1988. Cultural factors in working with Southeast Asian refugee adolescents. *Journal of Adolescence, 11,* 167–179.

Lee, H. G. 1992. *Taste of the states: A food history of America.* Charlottesville, VA: Howell.

Lee, R. H. 1960. *The Chinese in the United States of America.* Hong Kong: Hong Kong University Press.

Lee, S. K., Sobel, J., & Frongillo, E. A. 1996a. Acculturation and dietary practices among Korean Americans. *Journal of the American Dietetic Association, 99,* 1084–1089.

Lee, S. K., Sobel, J., & Frongillo, E. A. 1999b. Acculturation, food consumption, and diet-related factors in Korean Americans. *Journal of Nutrition Education, 31,* 321–330.

Leininger, M. 1991. Becoming aware of types of health practitioners and cultural imposition. *Journal of Transcultural Nursing, 2,* 36.

Leininger, M. M. 1991. *Culture care diversity and universality: A theory of nursing.* New York: National League for Nursing Press.

Leistner, C. G. 1990. Letter to the editor. *Journal of the American Dietetic Association, 90,* 1658–1659.

Leistner, C. G. 1996. *Cajun and Creole food practices, customs, and holidays.* Chicago: American Dietetic Association/American Diabetes Association.

Leonard, A. R., Jang, V. L., Foerster, S., Igra, A., Ransom, B., & Lambert, C. B. 1981. *Dietary practices, ethnicity, and hypertension: Preliminary results of the 1979 California hypertension survey.* Sacramento: California Department of Health Services.

Leonard, J. N. 1968. *Latin American cooking.* New York: Time-Life.

Leonard, J. N. 1970. *American cooking: New England.* New York: Time-Life.

Leonard, J. N. 1971. *American cooking: New England.* New York: Time-Life.

Leonard, J. N. 1971. *American cooking: The great West.* New York: Time-Life.

Leonelli, L. 1987. *We eat what we are: Food use patterns of Hmong and Mien in Sacramento, California.* Master's thesis, California State University, Sacramento.

Levin, J. S., & Vanderpool, H. Y. 1987. Is frequent religious attendance *really* conducive to better health? Toward an epidemiology of religion. *Social Science and Medicine, 24,* 589–600.

Levi-Straus, C. 1969. *The raw and the cooked.* New York: Harper & Row.

Lewis, G. H. 1989. The Maine lobster as regional icon: Competing images over time and social class. *Food and Foodways, 3,* 303–316.

Lewis, J. S., & Glaspy, M. F. 1975. Food habits and nutrient intakes of Filipino women in Los Angeles. *Journal of the American Dietetic Association, 67,* 122–125.

Li, D. K., Ni, H., Schwartz, S. M., & Daling, J. R. 1990. Secular change in birthweight among Southeast Asian immigrants to the United States. *American Journal of Public Health, 80,* 685–688.

Lillie-Blanton, M., & Laveist, T. 1996. Race/ethnicity, the social environment, and health. *Social Science and Medicine, 43,* 83–91.

Lin-Fu, J. S. 1988. Population characteristics and health care needs of Asian Pacific Americans. *Public Health Reports, 103,* 18–27.

Ling, S., King, J., & Leung, V. 1975. Diet, growth, and cultural food habits in Chinese American infants. *American Journal of Chinese Medicine, 31,* 125–132.

Lipson, J. G. 1992. The health and adjustments of Iranian immigrants. *Western Journal of Nursing Research, 14,* 10–29.

Lipson, J. G., & Hafizi, H. 1998. Iranians. In L. Purnell, & B. J. Paulanka (Eds.), *Transcultural nursing care.* Philadelphia: Davis.

Lipson, J. G., & Meleis, A. I. 1983. Issues in health care of Middle Eastern patients. *Western Journal of Medicine, 139,* 854–861.

Liu, H. Y., Lu, Y. F., & Chen, W. J. 1995. Predictive equations for basal metabolic rate in Chinese adults: A cross-validation study. *Journal of the American Dietetic Association, 95,* 1403–1408.

Lloyd, T. C. 1981. The Cincinnati chili culinary complex. *Western Folklore, 40,* 28–40.

Lock, M. 1990. Rationalization of Japanese herbal medication: The hegemony of orchestrated pluralism. *Human Organization, 49,* 41–47.

Lockwood, Y. R., & Lockwood, W. G. 1991. Pasties in Michigan's Upper Peninsula. In S. Stern & J. A. Cicala (Eds.), *Creative ethnicity: Symbols and strategies of contemporary ethnic life.* Logan: Utah State University Press.

Looker, A. C., Johnson, C. L., Woteki, C. E., Yetley, E. A., & Underwood, B. A. 1988. Ethnic and racial differences in serum vitamin A levels of children aged 4–11 years. *American Journal of Clinical Nutrition, 47,* 247–252.

Looker, A. C., Johnston, C. C., Wahner, H. W., & Dunn, W. L. 1995. Prevalence of low femoral bone density in older U.S. women from NHANESIII. *Journal of Bone Mineral Research, 10,* 796–802.

Looker, A. C., Loria, C. M., Carroll, M. D., McDowell, M. A., & Johnson, C. L. 1993. Calcium intakes of Mexican Americans, Cubans, Puerto Ricans, non-Hispanic whites, and non-Hispanic blacks in the United States. *Journal of the American Dietetic Association, 93,* 1274–1279.

Loria, C. M., McDowell, M. A., Johnson, C. L., & Woteki, C. E. 1991. Nutrient data for Mexican-American foods: Are current data adequate? *Journal of the American Dietetic Association, 91,* 919–922.

Loria, C., Arroyo, D., & Briefel, R. 1994. Cultural biases influencing dietary interviews with Mexican Americans: The Hanes experience. *American Journal of Clinical Nutrition, 59* (suppl.), 290S.

Lovoll, O. S. 1995. Norwegian Americans. In R. J. Vecoli, J. Galens, A. Sheets, & R. V. Young (Eds.), *Gale encyclopedia of multicultural America.* New York: Gale Research.

Lowenberg, M. E. 1970. Socio-cultural basis of food habits. *Food Technology, 24,* 27–32.

Lowenberg, M. E., Todhunter, E. N., Wilson, E. D., Savage, J. R., & Lubawski, J. L. 1979. *Food and people* (3rd ed.). New York: Wiley.

Luchetti, C. 1993. *Home on the range.* New York: Villard.

Ludman, E. K., & Newman, J. M. 1984. Yin and yang in the health-related food practices of three Chinese groups. *Journal of Nutrition Education, 16,* 3–5.

Ludman, E. K., Kang, K. J., & Lynn, L. L. 1992. Food beliefs and diets of pregnant Korean-American women. *Journal of the American Dietetic Association, 92,* 1519–1520.

Ludman, E. K., Newman, J. M., & Lynn, L. L. 1989. Blood-building foods in contemporary Chinese populations. *Journal of the American Dietetic Association, 89,* 1122–1124.

Luna, L. 1994. Care and cultural context of Lebanese Muslim immigrants: Using Leininger's theory. *Journal of Transcultural Nursing, 5,* 12–20.

Lusky, A., Lubin, F., Barell, V., Kaplan, G., Layani, V., & Wiener, M. 2000. Body mass index in 17-year-old Israeli males of different ethnic backgrounds: National or ethnic-specific references? *International Journal of Obesity and Related Metabolic Disorders, 24,* 88–92.

Lynch, J. 1990. Organ donation & approaching the African American family. *Contemporary Dialysis & Nephrology, 11,* 21–22.

Ma, Y., Murthy, V., Roderer, G., Monsalve, M. V., Clarke, L. A., Normand, T., Julien, P., Gagne, C., Lambert, M., Davignon, J., Lupien, F. J., Brunzell, J., & Hayden, M. R. 1991. A mutation in the human lipoprotein lipase gene as the most common cause of familial chylomicronemia in French Canadians. *New England Journal of Medicine, 324,* 1761–1766.

MacClancy, J. 1992. *Consuming culture.* New York: Holt.

Maduro, R. 1983. *Curanderismo* and Latin view of disease and curing. *Western Journal of Medicine, 139,* 868–874.

Magana, A., & Clark, N. M. 1995. Examining a paradox: does religiosity contribute to positive birth outcomes in Mexican American populations? *Health Education Quarterly, 22,* 96–109.

Magocsi, P. R. 1980. Russians. In S. Thernstrom, A. Orlov, & O. Handlin (Eds.), *Harvard encyclopedia of American ethnic groups.* Cambridge, MA: Belknap Press of Harvard University Press.

Magocsi, P. R. 1995. Russian Americans. In R. J. Vecoli, J. Galens, A. Sheets, & R. V. Young (Eds.), *Gale encyclopedia of multicultural America.* New York: Gale Research.

Mahaffey, K. R., Gartside, P. S., & Glueck, C. J. 1986. Blood lead levels and dietary calcium intake in 1- to 11-year-old children: The second National Health and Nutrition Examination Survey, 1976–1980. *Pediatrics, 78,* 257–262.

Mandel, A. 1996. *Celebrating the midwestern table.* New York: Doubleday.

Mani, U. V., Bhatt, S., Mehta, N. C., Pradhan, S. N., Shah, V., & Mani, I. 1990. Glycemic index of traditional Indian carbohydrate foods. *Journal of the American College of Nutrition, 9,* 573–577.

Mar, M. H., & Zeisel, S. H. 1999. Betaine in wine: Answer to the French paradox? *Medical Hypotheses, 53,* 383–385.

Maretzki, T. W. 1987. The *kur* in West Germany as an interface between naturopathic and allopathic ideologies. *Social Science and Medicine, 24,* 1061–1068.

Mariani, J. F. 1983. *The dictionary of American food and drink.* New Haven, CT: Ticknor & Fields.

Markides, K. S., & Coriel, J. 1986. The health of Hispanics in the southwestern United States: An epidemiological paradox. *Public Health Reports, 101,* 253–263.

Marks, C. 1985. *False tongues and Sunday bread: A Guatemalan and Mayan cookbook.* New York: Evans.

Marks, C., & Kim, M. 1993. *The Korean kitchen.* San Francisco: Chronicle.

Marks, G., Garcia, M., & Solis, J. M. 1990. Health risk behaviors of Hispanics in the United States: Findings from HHANES, 1982–84. *American Journal of Public Health, 80* (suppl.), 20–26.

Marks, G., Solis, J., Richardson, J. L., Collins, L. M., Birba, L., & Hisserich, J. C. 1987. Health behavior of elderly Hispanic women: Does cultural assimilation make a difference? *American Journal of Public Health, 77,* 1315–1319.

Marmot, M. G., & Syme, S. L. 1976. Acculturation and coronary heart disease in Japanese Americans. *American Journal of Epidemiology, 104,* 225–247.

Marsh, C. E. 1984. *From Black Muslims to Muslims: The transition from separatism to Islam, 1930–1980.* Metuchen, NJ: Scarecrow.

Marshall, J. A., Lopez, T. K., Shetterly, S. M., Morgenstern, N. E., Baer, K., Swenson, C., Baron, A., Baxter, J., & Hamman, R. 1999. Indicators of nutritional risk in a rural elderly Hispanic and non-Hispanic white population: San Luis Valley Health and Aging Study. *Journal of the American Dietetic Association, 99,* 315–322.

Marti, J. E. 1995. *The alternative health & medicine encyclopedia.* New York: Gale Research.

Martin, P. 1981. *The Czech book: Recipes and traditions.* Iowa City: Penfield.

Martinez, G. A., & Krieger, F. W. 1985. 1984 milk-feeding patterns in the United States. *Pediatrics, 76,* 1004–1008.

Martinez, N. C. 1993. Diabetes and minority populations: Focus on Mexican Americans. *Nursing Clinics of North America, 28,* 87–95.

Martins, Y., Pelchat, M. L., & Pliner, P. 1997. "Try it; it's good for you": Effects of taste and nutrition information on willingness to try novel foods. *Appetite, 28,* 89–102.

Masi, R. 1989. Multicultural medicine: Fad or forgotten concept? *Canadian Medical Association Journal, 140,* 1086–1087.

Massara, E. B., & Stunkard, A. J. 1979. A method of quantifying cultural ideals of beauty and the obese. *International Journal of Obesity, 3,* 149–152.

Matson, M. 1994. *Food in Missouri.* Columbia: University of Missouri Press.

Maxwell, W. 1995. Honduran Americans. In R. J. Vecoli, J. Galens, A. Sheets, & R. V. Young (Eds.), *Gale encyclopedia of multicultural America.* New York: Gale Research.

May, K. M. 1992. Middle-Eastern immigrant parents' social networks and help-seeking for child health care. *Journal of Advanced Nursing, 17,* 905–912.

Mayeno, L., & Hirota, S. M. 1994. Access to health care. In N. W. S. Zane, D. T. Takeuchi, & K. N. J. Young (Eds.), *Confronting critical health issues of Asian and Pacific Islander Americans.* Thousand Oaks, CA: Sage.

Mbiti, J. S. 1970. *African religions and philosophy.* Garden City, NY: Doubleday Anchor.

McBean, A. M., & Gornick, M. E. 1994. Differences in rates of procedures performed in hospitals for Medicare beneficiaries. *Health Care Financial Review, 15,* 77–90.

McClelland, D. A. 1992. *Good as gold: Foods the Americas gave the world.* Washington, D.C.: Smithsonian.

McComber, D. R., & Postel, R. T. 1992. The role of ethnic foods in the food and nutrition curriculum. *Journal of Home Economics, 84,* 52–54, 59.

McCord, C., & Freeman, H. P. 1990. Excess mortality in Harlem. *New England Journal of Medicine, 322,* 173–177.

McGarvey, S. T. 1991. Obesity in Samoans and a perspective on its etiology in Polynesians. *American Journal of Clinical Nutrition, 53,* 1586S–1594S.

McGee, H. 1984. *On food and cooking: The science and lore of the kitchen.* New York: Scribner's.

McKeigue, P. M., Shah, B., & Marmot, M. G. 1991. Relation of central obesity and insulin resistance with high diabetes prevalence and cardiovascular risk in South Asians. *Lancet, 337,* 382–386.

McMurray, M. P., Ceriqueira, M. T., Conner, S. L., & Conner, W. E. 1991. Changes in lipid and lipoprotein levels and body weight in Tarahumara Indians after consumption of an affluent diet. *New England Journal of Medicine, 325,* 1704–1708.

McWatters, H. 1995. British Columbia wines take on the world. In J. M. Powers & A. Stewart (Eds.), *Our northern bounty: A celebration of Canadian cuisine.* Toronto: Random House.

Meftuh, A. B., Tapsoba, L. P., & Lamounier, J. A. 1991. Breastfeeding practices in Ethiopian women in southern California. *Indian Journal of Pediatrics, 58,* 349–356.

Meleis, A. I. 1981. The Arab-American in the health care system. *American Journal of Nursing, 81,* 1180–1183.

Meleis, A. I., Lipson, J. G., & Paul, S. M. 1992. Ethnicity and health among five Middle Eastern immigrant groups. *Nursing Research, 42,* 98–103.

Melendy, H. B. 1995. Filipino Americans. In R. J. Vecoli, J. Galens, A. Sheets, & R. V. Young (Eds.), *Gale encyclopedia of multicultural America.* New York: Gale Research.

Melville, M. B. 1985. Salvadorans and Guatemalans. In D. W. Haines (Ed.), *Refugees in the United States.* Westport, CT: Greenwood.

Mendes, H. 1971. *African heritage cookbook.* New York: Macmillan.

Mendoza, F. S., & Dixon, L. B. 1999. Hispanic health and nutritional status. In D. J. Hernandez (Ed.), *Children of immigrants: Health, adjustment, and public assistance.* Washington, D.C.: National Academy Press.

Mermelstein, N. H. 1992. Seeds of Change: The Smithsonian Institution's Columbus Quincentenary Exhibition. *Food Technology, 46,* 86–89.

Messer, E. 1981. Hot/cold classification: Theoretical and practical implications of a Mexican study. *Social Science and Medicine, 158,* 133–145.

Messina, M., & Messina, V. 2000. Soyfoods, soybean isoflavones, and bone health: A brief overview. *Journal of Renal Nutrition, 10,* 63–68.

Mikhail, B. I. 1994. Hispanic mothers' beliefs and practices regarding selected children's health problems. *Western Journal of Nursing Research, 16,* 623–638.

Milbauer, J. A. 1990. The geography of food in Eastern Oklahoma: A small restaurant study. *North American Culture, 6,* 37–52.

Miller, M. E. 1996. Mayans. In *The 1996 Grolier multimedia encyclopedia.* Danbury, CT: Grolier Electronic Publishing.

Miller, R. 1996. Central America. In *The 1996 Grolier multimedia encyclopedia.* Danbury, CT: Grolier Electronic Publishing.

Miralles, M. A. 1989. *A matter of life and death: Health-seeking behavior of Guatemalan refugees in south Florida.* New York: AMS Press.

Mo, B. 1992. Modesty, sexuality, and breast health in Chinese-American women. *Western Journal of Medicine, 157,* 260–264.

Mogelonsky, M. 1995. *Everybody eats.* Ithaca, NY: American Demographic Books.

Mogelonsky, M. 1996. *Who's buying food and drink.* Ithaca, NY: New Strategist Publications.

Mogelonsky, M. 1998. Watching in tongues. *American Demographics.*

Molesky, J. 1986. Pathology of Central America refugees. *Migration World, XIV,* 19–23.

Molina, C. D., Molina, M. M., & Molina, J. M. 1988. Intestinal parasites in Southeast Asian refugees two years after immigration. *Western Journal of Medicine, 149,* 422–425.

Molinari, C. 1995. Czech Americans. In R. J. Vecoli, J. Galens, A. Sheets, & R. V. Young (Eds.), *Gale encyclopedia of multicultural America.* New York: Gale Research.

Molony, D. 1998. *The American Association of Oriental Medicine's complete guide to Chinese herbal medicine.* New York: Berkley.

Monrroy, L. S. 1983. Nursing care of Raza/Latina patients. In M. S. Orque, B. Bloch, & L. S. A. Monrroy (Eds.), *Ethnic nursing care: A multicultural approach.* St. Louis: Mosby.

Monsen, E. R. 1992. Respecting diversity helps American eat right. *Journal of the American Dietetic Association, 92,* 282.

Montepio, S. N. 1986–87. Folk medicine in the Filipino American experience. *Amerasia Journal, 13,* 151–162.

Moran R. T., & Harris, P. R. 1981. *Managing cultural synergy.* Houston, TX: Gulf.

Morrison, T., Conaway, W. A., & Borden, G. A. 1994. *Kiss, bow, or shake hands: How to do business in sixty countries.* Holbrook, MA: Adams.

Mossavar-Rahmani, Y., Pelto, G. H., Ferris, A. M., & Allen, L. H. 1996. Determinants of body size perceptions and dieting behavior in a multiethnic group of hospital staff women. *Journal of the American Dietetic Association, 96,* 252–256.

Muecke, M. A. 1983. In search of healers—Southeast Asian refugees in the American health care system. *Western Journal of Medicine, 139,* 835–840.

Muecke, M. A. 1983. Caring for Southeast Asian refugee patients in the USA. *American Journal of Public Health, 73,* 431–438.

Muecke, M. A. 1983a. Caring for Southeast Asian patients in the USA. *American Journal of Public Health, 73,* 431–438.

Muecke, M. A. 1983b. In search of healers—Southeast Asian refugees in the American health care system. *Western Journal of Medicine, 139,* 835–840.

Muecke, M. A. 1992. New paradigms for refugee health problems. *Social Science and Medicine, 35,* 515–523.

Mumford, J. 1995. Ecuadoran Americans. In R. J. Vecoli, J. Galens, A. Sheets, & R. V. Young (Eds.), *Gale encyclopedia of multicultural America.* New York: Gale Research.

Mumford, J. 1995. Salvadoran Americans. In R. J. Vecoli, J. Galens, A. Sheets, & R. V. Young (Eds.), *Gale encyclopedia of multicultural America.* New York: Gale Research.

Munch, P. 1980. Norwegians. In S. Thernstrom, A. Orlov, & O. Handlin (Eds.), *Harvard encyclopedia of American ethnic groups.* Cambridge, MA: Belknap Press of Harvard University Press.

Murphy, N. J., Schraer, C. D., Thiele, M. C., Boyko, E. J., Bulkow, L. R., Doty, B. J., & Lanier, A. P. 1995. Dietary change and obesity associated with glucose intolerance in Alaska natives. *Journal of the American Dietetic Association, 95,* 676–682.

Murphy, S. P., Castillo, R. O., Martorell, R., & Mendoza, F. S. 1990. An evaluation of food group intakes by Mexican-American children. *Journal of the American Dietetic Association, 90,* 388–393.

Murray, C. J. L. 1990. Mortality among black men. *New England Journal of Medicine, 322,* 205–206.

Murray, R. H., & Rubel, A. J. 1992. Physicians and healers—Unwitting partners in health care. *New England Journal of Medicine, 326,* 61–64.

Murrell, S. 1995. Jamaican Americans. In R. J. Vecoli, J. Galens, A. Sheets, & R. V. Young (Eds.), *Gale encyclopedia of multicultural America.* New York: Gale Research.

Nabhan, G. P. 1997. *Gathering the desert.* Tucson: University of Arizona Press.

Naff, A. 1980. Arabs. In S. Thernstrom, A. Orlov, & O. Handlin (Eds.), *Harvard encyclopedia of American ethnic groups.* Cambridge, MA: Belknap Press of Harvard University Press.

Nash, A. 1995. Korean Americans. In R. J. Vecoli, J. Galens, A. Sheets, & R. V. Young (Eds.), *Gale encyclopedia of multicultural America.* New York: Gale Research.

National Heart, Lung, and Blood Institute. 2000, May 17. NHLBI study shows large blood pressure benefit from reduced dietary sodium. *NIH News Release.*

National Public Radio, Henry J. Kaiser Family Foundation, and Kennedy School of Government. 1999, February. *Survey of Americans on dietary supplements.*

National Research Council. 1945. (Cited by M. E. Lowenberg.) Sociocultural basis of food habits. *Food Technology, 24,* 27–32.

National Research Council. 1975. *Herbal pharmacology in the People's Republic of China.* Washington, D.C.: National Academy of Sciences.

Natow, A. B., Heslin, J., & Raven, B. C. 1975. Integrating the Jewish dietary laws into a dietetic program. *Journal of the American Dietetic Association, 67,* 13–16.

Ndidi Uche Griffin, F. 1994. Perceptions of African American women regarding health care. *Journal of Cultural Diversity, 1,* 32–35.

Nelli, H. S. 1980. Italians. In S. Thernstrom, A. Orlov, & O. Handlin (Eds.), *Harvard encyclopedia of American ethnic groups.* Cambridge, MA: Belknap Press of Harvard University Press.

Nelli, H. S. 1983. *From immigrants to ethnics: The Italian American.* Oxford: Oxford University Press.

Nelson, M. S., & Javanovic, L. 1987. Pregnancy, diabetes, and Jewish dietary law: The challenge for the pregnant diabetic women who keep kosher. *Journal of the American Dietetic Association, 87,* 1054–1058.

Netland, P. A., & Brownstein, H. 1985. Anthropometric measurements for Asian and Caucasian elderly. *Journal of the American Dietetic Association, 85,* 221–223.

Neuendorf, K. A. 1990. Health images in the mass media. In E. B. Ray & L. Donohew (Eds.), *Communication and health: systems and applications.* Hillsdale, NJ: Erlbaum.

Neumark-Sztainer, D., Story, M., Resnick, M. D., & Blum, R. W. 1997. Psychosocial concerns and weight control behaviors among overweight and nonoverweight Native American adolescents. *Journal of the American Dietetic Association, 97,* 598–604.

Newman, J. M. 1984. *The melting pot.* New York: Garland.

Newman, J. M. 1998a. China: Transformations of its cuisine, a prelude to understanding its people. *Journal for the Study of Food and Society 2,* 5–6.

Newman, J. M. 1998b. Chinese ingredients: Both usual and unusual. In J. M. Powers (Ed.), *From Cathay to Canada: Chinese cuisine in transition.* Willowdale, Canada: Ontario Historical Society.

Newman, J. M. 1999a. Fujian, the province and its foods. *Flavor & Fortune, 6*(2), 13, 20.

Newman, J. M. 1999b. Tibet and Tibetan foods. *Flavor & Fortune, 6*(4), 7–8, 12.

Newman, J. M. 2000a. Chinese meals. In H. L. Meiselman (Ed.), *Dimensions of the meal: The science, culture, business, and art of eating.* Gaithersburg, MD: Aspen.

Newman, J. M. 2000b. Mongolians and their cuisine. *Flavor & Fortune, 7*(1), 9–10, 24.

Newman, V., Norcross, W., & McDonald, R. 1991. Nutrient intake of low-income Southeast Asian pregnant women. *Journal of the American Dietetic Association, 91,* 793–799.

Nguyen, M. D. 1985. Culture shock: A review of Vietnamese culture and its concepts of health and disease. *Western Journal of Medicine, 142,* 409–412.

Nickles, H. G. *Middle Eastern cooking.* New York: Time-Life.

Nielsen, M., & Peterson, P. L. 1995. Danish Americans. In R. J. Vecoli, J. Galens, A. Sheets, & R. V. Young (Eds.), *Gale encyclopedia of multicultural America.* New York: Gale Research.

Niethammer, C. 1974. *American Indian food and lore.* New York: Macmillan.

Nightingale, M. 1995. The enduring influences on today's East Coast cooking. In J. M. Powers & A. Stewart (Eds.), *Our northern bounty: A celebration of Canadian cuisine.* Toronto: Random House.

NIH Osteoporosis and Related Bone Diseases—National Resource Center. 2000. Available on-line: http:www.osteo.org/osteo.html.

Nomani, M. Z. A., Hallak, M. H., & Siddiqui, I. P. 1990. Effects of Ramadan fasting on plasma uric acid and body weight in healthy men. *Journal of the American Dietetic Association, 90,* 1435–1436.

Norden, E. E. 1995. Portuguese Americans. In R. J. Vecoli, J. Galens, A. Sheets, & R. V. Young (Eds.), *Gale encyclopedia of multicultural America.* New York: Gale Research.

Novas, H., & Silva, R. 1997. *Latin American cooking in the U.S.A.* New York: Knopf.

Novello, A. C., Wise, P. H., & Kleinman, D. V. 1991. Hispanic health: Time for data, time for action. *Journal of the American Medical Association, 265,* 253–254.

Nurko, C., Aponte-Merced, L., Bradley, E. L., & Fox, L. 1998. Dental caries prevalence and dental health care of Mexican American workers' children. *ASDC Journal of Dentistry in Children, 65,* 65–72.

O'Connell, B. 1999. Herbal supplements in diabetes: One dietitian's perspective. *On the Cutting Edge: Diabetes Care and Education, 20,* 11–16.

Ojakangas, B. 1992. *Scandinavian feasts.* New York: Stewart, Tabori & Chang.

Olsen, M. E. 1978. *The process of social organization: Power in social systems* (2nd ed.). New York: Holt, Rinehart, & Winston.

Orimo, H., Hashimoto, T., Sakata, K., Yoshimura, N., Suzuki, T., & Hosoi, T. 2000. Trends in the incidence of hip fracture in Japan, 1987–1997: The third nationwide survey. *Journal of Bone Mineral Metabolism, 18,* 126–131.

Orlov, A., & Ueda, R. 1980. Central and South Americans. In S. Thernstrom, A. Orlov, & O. Handlin (Eds.), *Harvard encyclopedia of American ethnic groups.* Cambridge: MA: Belknap Press of Harvard University Press.

Ornish, D. 1990. *Dr. Dean Ornish's program for reversing heart disease.* New York: Random House.

Orque, M. S. 1983a. Nursing care of Filipino American patients. In M. S. Orque, B. Bloch, & L. S. A. Monrroy (Eds.), *Ethnic nursing care: A multicultural approach.* St. Louis: Mosby.

Orque, M. S. 1983b. Nursing care of South Vietnamese patients. In M. S. Orque, B. Bloch, & L. S. A. Monrroy (Eds.), *Ethnic nursing care: A multicultural approach.* St. Louis: Mosby.

Ortiz, E. L. 1979. *The book of Latin American cooking.* New York: Vintage.

Ots, T. 1990. The angry liver, the anxious heart and the melancholy spleen: The phenomenology of perceptions in Chinese culture. *Culture, Medicine, and Psychiatry, 14,* 21–58.

Packard, D. P., & McWilliams, M. 1993, May–June. Cultural foods heritage of Middle Eastern immigrants. *Nutrition Today,* 6–12.

Packard, D. P., & McWilliams, M. 1994. Tips on counseling Middle Eastern clients. *Journal of the American Dietetic Association, 94,* 1254.

Packel, J. 1995. Peruvian Americans. In R. J. Vecoli, J. Galens, A. Sheets, & R. V. Young (Eds.), *Gale encyclopedia of multicultural America.* New York: Gale Research.

Page, L. G., & Wigginton, E. 1984. *The Foxfire book of Appalachian cookery.* New York: Dutton.

Palinkas, L. A. 1995. Health under stress: Asian and Central American refugees and those left behind. *Social Science and Medicine, 40,* 1591–1596.

Palinkas, L. A., & Pickwell, S. M. 1995. Acculturation as a risk factor for chronic disease among Cambodian refugees in the United States. *Social Science and Medicine, 40,* 1643–1653.

Pan, Y. L., Dixon, Z., Himburg, S., & Huffman, F. 1999. Asian students change their eating patterns after living in the United States. *Journal of the American Dietetic Association, 99,* 54–57.

Pang, K. Y. 1989. The practice of traditional Korean medicine in Washington, DC. *Social Science and Medicine, 28,* 875–884.

Pang, K. Y. C. 1991. *Korean elderly women in America: Everyday life, health, and illness.* New York: AMS Press.

Pang, K. Y. C. 1994. Understanding depression among elderly Korean immigrants through their folk illnesses. *Medical Anthropology Quarterly, 8,* 209–216.

Pangborn, R. M., & Bruhn, C. M. 1971, Winter. Concepts of food habits of "other" ethnic groups. *Journal of Nutrition Education,* 106–110.

Pap, L. 1981. *The Portuguese-American.* Boston: Twayne.

Papashvily, H., & Papashvily, G. 1969. *Russian cooking.* New York: Time-Life.

Pareo-Tubbeh, S. L., Romero, L. J., Baumbartner, R. N., Garry, P. J., Lindeman, R. D., & Koehler, K. M. 1999. Comparison of energy and nutrient sources of elderly Hispanics and non-Hispanic whites in New Mexico. *Journal of the American Dietetic Association, 99,* 572–582.

Parker, A. 1980. Austrians. In S. Thernstrom, A. Orlov, & O. Handlin (Eds.), *Harvard encyclopedia of American ethnic groups.* Cambridge, MA: Belknap Press of Harvard University Press.

Parker, J. G. 1994. The lived experience of Native Americans with diabetes within a transcultural nursing experience. *Journal of Transcultural Nursing, 6,* 5–11.

Parodi, P. W. 1997. The French paradox unmasked: The role of folate. *Medical Hypotheses, 49,* 313–318.

Pasquali, E. A. 1985. The impact of acculturation on the eating habits of elderly immigrants: A Cuban example. *Journal of Nutrition for the elderly, 5,* 27–36.

Pasquali, E. A. 1994. Santeria. *Journal of Holistic Nursing, 12,* 380–390.

Passariello, P. 1990. Anomalies, analogies and sacred profanities: Mary Douglas on food and culture 1957–1989. *Food and Foodways, 4,* 53–71.

Passim, H., & Bennett, J. W. 1943. Social process and dietary change: In *The Problem of Changing Food Habits* (p. 108). Washington, D.C.: National Research Council Bulletin.

Patel, C., & Nicol, A. 1997. Adaptation of African-American cultural and food preferences in end-stage renal disease diets. *Advances in Renal Replacement Therapy, 4,* 30–39.

Patel, S., Unwin, N., Bhopal, R., White, M., Harland, J., Ayis, A. S., Watson, W., & Alberti, K. G. 1999. A comparison of proxy measures of abdominal obesity in Chinese, European, and South Asian adults. *Diabetic Medicine, 16,* 853–860.

Pavri, T. 1995a. Asian Indian Americans. In R. J. Vecoli, J. Galens, A. Sheets, & R. V. Young (Eds.), *Gale encyclopedia of multicultural America.* New York: Gale Research.

Pavri, T. 1995b. Pakistani Americans. In R. J. Vecoli, J. Galens, A. Sheets, & R. V. Young (Eds.), *Gale encyclopedia of multicultural America.* New York: Gale Research.

Peck, R. E., Marks, J. S., Dibley, M. J., Lee, S., & Trowbridge, F. L. 1987. Birth weight and subsequent growth among Navajo children. *Public Health Reports, 102,* 500–507.

Pelican, S., Batchelor, B., Belshaw, J., Osborn, W., Pearce, J., Przekurat, C., Schumacher, P., & Strauss, K. 1994. Nutrition services for alcohol/substance abuse clients: Indian Health Services tribal survey provides insights. *Journal of the American Dietetic Association, 94,* 835–836.

Pelto, G. H. 1981. Anthropological contributions to nutrition education research. *Journal of Nutrition Education, 13* (suppl.), S2–S8.

Pelto, G. H., Pelto, P. J., & Lung'aho, M. S. 1981. "Modern" and "traditional" food use in West Finland: An example of quantitative pattern analysis. *Nutrition Research, 1,* 63–71.

Perez, L. 1980. Cubans. In S. Thernstrom, A. Orlov, & O. Handlin (Eds.), *Harvard encyclopedia of American ethnic groups.* Cambridge, MA: Belknap Press of Harvard University Press.

Perl, L. 1965. *Red-flannel hash and shoo-fly pie.* Cleveland: World.

Perl, L. 1967. *Rice, spice, and bitter oranges: Mediterranean food festivals.* New York: World.

Perry, C. 1999, June 2. Tequila: The fusion hootch. *Los Angeles Times.*

Peterson, J., & Peterson, D. 1995. *Eat smart in Brazil.* Madison, WI: Gingko.

Peterson, J., & Peterson, D. 1996. *Eat smart in Turkey.* Madison, WI: Gingko.

Plawecki, H. 1992. Cultural considerations. *Journal of Holistic Nursing, 10,* 4–5.

Plawecki, H. M., Sanchez, T. R., & Plawecki, J. A. 1994. Cultural aspects of caring for Navajo Indian clients. *Journal of Holistic Nursing, 12,* 291–306.

Pobocik, R. S., & Shovic, A. C. 1996. Development of a nutrition exchange booklet for Guam using culturally accepted foods. *Journal of the American Dietetic Association, 96,* 285–287.

Poe, D. M. 1986. *Profile of Portuguese elderly nutrition participants: Demographic characteristics, nutrition knowledge and practices.* Master's thesis, MGH Institute of Health Professions, Boston, MA.

Pongstaporn, W., & Bunyaratavej, A. 1999. Hematological parameters, ferritin and vitamin B_{12} in vegetarians. *Journal of the Medical Association of Thailand, 82,* 304–311.

Potvin, L., Desrosiers, S., Trifonopoulos, M., Leduc, N., Rivard, M., Macaulay, A. C., & Paradis, G. 1999. Anthropometric characteristics of Mohawk children aged 6–11 years: A population perspective. *Journal of the American Dietetic Association, 99,* 955–961.

Pouesi, D. 1994. *An illustrated history of Samoans in California.* Carson, CA: KIN Publications.

Pousada, L. 1995. Hispanic-American elders: Implications for health-care providers. *Ethnogeriatrics, 11,* 39–52.

Pozzetta, G. 1995. Italian Americans. In R. J. Vecoli, J. Galens, A. Sheets, & R. V. Young (Eds.), *Gale encyclopedia of multicultural America.* New York: Gale Research.

Pribila, B. A., Hertzler, S. R., Martin, B. R., Weaver, C. M., & Savaiano, D. A. 2000. Improved lactose digestion and intolerance among African-American adolescent girls fed a dairy-rich diet. *Journal of the American Dietetic Association, 100,* 524–528.

Primeaux, M., & Henderson, G. 1981. American Indian patient care. In G. Henderson & M. Primeaux (Eds.), *Transcultural health care.* Menlo Park, CA: Addison-Wesley.

Rabin, S. 1994. How to sell across cultures. *American Demographics, 16,* 56–57.

Ragucci, A. T. 1981. Italian Americans. In A. Harwood (Ed.), *Ethnicity and medical care.* Cambridge, MA: Harvard University Press.

Raheja, B. S. 1987. Diabetes-associated complications and traditional Indian diet. *Diabetes Care, 10,* 382–383.

Rairdan, B., & Higgs, Z. R. 1992. When your patient is a Hmong refugee. *American Journal of Nursing, March,* 52–55.

Raloff, J. 1998. Strong bones: A sodium connection? *Science News, 153,* 303.

Ramakrishna, J., & Weiss, M. G. 1992. Health, illness, and immigration. East Indians in the United States. *Western Journal of Medicine, 157,* 265–270.

Randall, T. 1991. Key to organ donation may be cultural awareness. *Journal of the American Medical Association, 265,* 176–177.

Randall-David, E. 1989. *Strategies for working with culturally diverse communities and clients.* Bethesda, MD: Association for the Care of Children's Health.

Rapple, B. A. 1995. Irish Americans. In R. J. Vecoli, J. Galens, A. Sheets, & R. V. Young (Eds.), *Gale encyclopedia of multicultural America.* New York: Gale Research.

Raspa, R. 1984. Exotic foods among Italian-Americans in Mormon Utah: Food as nostalgic enactment of identity. In *Ethnic and regional foodways in the United States.* Knoxville: University of Tennessee Press.

Ratner, M. 1995. Thai Americans. In R. J. Vecoli, J. Galens, A. Sheets, & R. V. Young (Eds.), *Gale encyclopedia of multicultural America.* New York: Gale Research.

Rau, S. R. 1969. *The cooking of India.* New York: Time-Life.

Raymond, C. A. 1986. Biology, culture, dietary changes conspire to increase incidence of obesity. *Journal of the American Medical Association, 256,* 2157–2158.

Reddy, M. A. (Ed.). 1993. *Statistical record of native North Americans.* Washington, D.C.: Gale Research.

Reeler, A. 1990. Injections: A fatal attraction. *Social Science and Medicine, 31,* 1119–1125.

Regenstein, J. M., & Regenstein, C. E. 1992. The kosher food market in the 1990s—A legal view. *Food Technology, 46,* 122–124, 126.

Reischauer, E. O. 1974. *Japan: The story of a nation.* New York: Knopf.

Reiter, A., Boylan, L. M., Driskell, J., & Moak, S. 1987. Vitamins B_{12} and folate intakes and plasma levels of black adolescent females. *Journal of the American Dietetic Association, 87,* 1065–1067.

Renaud, S., & Gueguen, R. 1998. The French paradox and wine drinking. *Novartis Foundation Symposium, 216,* 208–217.

Retchin, S. M., Wells, J. A., Valleron, A. J., & Albrecht, G. L. 1992. Health behavior changes in the United States, the United Kingdom and France. *Journal of General Internal Medicine, 7,* 615–622.

Reuters. 1999. Minority women at high risk for osteoporosis. Available on-line: http:www.allheath.com/conditions/bone/news/0,4800,57_120895,00.html.

Rimm, E. B., & Ellison, R. C. 1995. Alcohol in the Mediterranean diet. *American Journal of Clinical Nutrition, 61* (suppl.), 1378S–1382S.

Rippley, L. J. 1980. Germans from Russia. In S. Thernstrom, A. Orlov, & O. Handlin (Eds.), *Harvard encyclopedia of American ethnic groups.* Cambridge, MA: Belknap Press of Harvard University Press.

Rippley, L. J. 1995. German Americans. In R. J. Vecoli, J. Galens, A. Sheets, & R. V. Young (Eds.), *Gale encyclopedia of multicultural America.* New York: Gale Research.

Risser, A., & Mazur, L. J. 1995. Use of folk remedies in a Hispanic population. *Archives of Pediatric and Adolescent Medicine, 149,* 978–981.

Robinson, R. H., & Johnson, W. L. 1982. *The Buddhist religion.* Belmont, CA: Wadsworth.

Robison, J. I., Hoerr, S. L., Petersmarck, K. A., & Anderson, J. V. 1995. Redefining success in obesity intervention: The new paradigm. *Journal of the American Dietetic Association, 95,* 422–423,

Rodale Press, 1995. *The Prevention Index: A report card on the nation's health, 1995 summary report.* Emmaus, PA: Rodale.

Roden, C. 1974. *A book of Middle Eastern food.* New York: Vintage.

Roden, C. 1987. *Mediterranean cookery.* New York: Knopf.

Rodriguez, J. 1995. Argentinean Americans. In R. J. Vecoli, J. Galens, A. Sheets, & R. V. Young (Eds.), *Gale encyclopedia of multicultural America.* New York: Gale Research.

Rogers, A. 1968. *A Basque story cook book.* New York: Scribner's.

Rogers, M. R. 1980. Portuguese. In S. Thernstrom, A. Orlov, & O. Handlin (Eds.), *Harvard encyclopedia of American ethnic groups.* Cambridge, MA: Belknap Press of Harvard University Press.

Rojas-Lombardi, F. 1991. *The art of South American cooking.* New York: HarperCollins.

Romero-Gwynn, E., Gwynn, D., Grivetti, L., McDonald, R., Stanford, G., Turner, B., West, E., & Williamson, E. 1993. Dietary acculturation among Latinos of Mexican descent. *Nutrition Today, 28,* 6–12.

Romero-Gwynne, E. 1989. Breast-feeding pattern among Indochinese immigrants in northern California. *American Journal of Diseases of Children, 143,* 804–808.

Romieu, I., Hernandez-Avila, M., Rivera, J. A., Ruel, M. T., & Parra, S. 1997. Dietary studies in countries experiencing a health transition: Mexico and Central America. *American Journal of Clinical Nutrition, 65,* 1159S–1165S.

Root, W. 1966. *Foods of France.* New York: Vintage.

Root, W., & DeRochemont, R. 1976. *Eating in America.* New York: Ecco.

Root, W., & deRochemont, R. 1976. *Eating in America: A history.* New York: Ecco.

Rosanhoff, A., & Calloway, D. H. 1982. Calcium source in Indochinese immigrants. *New England Journal of Medicine, 306,* 239–240.

Ross, P. D., Norimatsu, H., Davis, J. W., Yano, K., Wasnich, R. D., Fujiwara, S., Hosod, Y., & Melton, L. J. 1991. A comparison of hip fracture incidence among native Japanese, Japanese Americans, and American Caucasians. *American Journal of Epidemiology, 133,* 801–809.

Rotberg, R. I. 1996. African history. In *The 1996 Grolier multimedia encyclopedia.* Danbury, CT: Grolier Electronic Publishing.

Roth, C. 1972. Dietary habits. In C. Roth (Ed.), *Encyclopedia Judaica.* Jerusalem: Keter.

Rozin, P. 1991. The promise of plenty: Food in American history. In *Good as gold: Foods the Americas gave the world.* Smithsonian Institution Quincentenary Symposium Proceedings. Washington, D.C.: National Museum of History.

Rozin, P. 1996. The socio-cultural context of eating and food choice. In H. L. Meiselman & H. J. H. MacFie (Eds.), *Food choice, acceptance and consumption.* London: Blackie Academic & Professional.

Rozin, P. 1997. The socio-cultural context of eating and food choice. In H. L. Meiselman, & H. J. H. MacFie (Eds.), *Food choice, acceptance, and consumption.* London: Blackie Academic & Professional.

Rozin, P., Fallon, A., & Augustoni-Ziskind, M. 1985. The child's conception of food: The development of contamination sensitivity to "disgusting" substances. *Developmental Psychology, 21,* 1075–1079.

Ruben, B. D. 1990. The health caregiver–patient relationship: Pathology, etiology, treatment. In E. B. Ray & L. Donohew (Eds.), *Communication and health.* Hillsdale, NJ: Erlbaum.

Rumbaut, R. G., & Weeks, J. R. 1989. Infant health among Indochinese refugees: Patterns of infant mortality, birthweight and prenatal care in comparative perspective. *Research in the Sociology of Health Care, 8,* 137–196.

Sadella, E., & Burroughs, J. 1981, October. Profiles in eating: Sexy vegetarians and other diet-based stereotypes. *Psychology Today,* 51–57.

Sakr, A. H. 1971. Dietary regulations and food habits of Muslims. *Journal of the American Dietetic Association, 58,* 123–126.

Sakr, A. H. 1975. Fasting in Islam. *Journal of the American Dietetic Association, 67,* 17–21.

Salazar-Schettino, P. M. 1983. Customs which predispose to Chagas' disease and cysticercosis in Mexico. *American Journal of Tropical Medicine and Hygiene, 32,* 1179–1180.

Salmond, S. W. 1999. Managed care: Pitfalls for cultural bias. *Journal of Transcultural Nursing, 10,* 295–296.

Samet, J. M., Coultas, D. B., Howard, C. A., Skipper, B. J., & Hanis, C. L. 1988. Diabetes, gallbladder disease, obesity, and hypertension among Hispanics in New Mexico. *American Journal of Epidemiology, 128,* 1302–1311.

Sanjur, D. 1982. *Social and cultural perspectives in nutrition.* Englewood Cliffs, NJ: Prentice Hall.

Sanjur, D. 1995. *Hispanic foodways, nutrition, and health.* Boston: Allyn & Bacon.

Saracino, J., & Michael, P. 1996. Positive steps toward a multicultural association. *Journal of the American Dietetic Association, 96,* 1242–1244.

Sargent, J. D., Stukel, T. A. Dalton, M. A., Freeman, J., & Brown, M. J. 1996. Iron deficiency in Massachusetts: Socioeconomic and demographic risk factors among children. *American Journal of Public Health, 86,* 544–550.

Sarma, D. S. 1953. The nature and history of Hinduism. In K. W. Morgon (Ed.), *The religion of the Hindus.* New York: Ronald.

Sawyers, J. E., & Eaton, L. 1992. Gastric cancer in the Korean-American: Cultural implications. *Oncology Nursing Forum, 19,* 619–623.

Scarborough, J. 1998. *The origins of cultural differences and their impact on management.* Westport, CT: Quorum.

Schaefer, L. J., & Kumanyika, S. K. 1985. Maternal variables related to potentially high sodium infant feeding practices. *Journal of the American Dietetic Association, 85,* 433–438.

Schaffer, D. M., Velie, E. M., Shaw, G. M., & Todoroff, K. P. 1998. Energy and nutrient intakes and health practices of Latinas and white non-Latinas in the 3 months before pregnancy. *Journal of the American Dietetic Association, 98,* 876–884.

Schelbert, L. 1995. Swiss Americans. In R. J. Vecoli, J. Galens, A. Sheets, & R. V. Young (eds.), *Gale encyclopedia of multicultural America.* New York: Gale Research.

Schilling, B., & Brannon, E. 1986. *Cross-cultural counseling: A guide for nutrition and health counselors.* FNS #250. Alexandria, VA: U.S. Department of Agriculture/U.S. Department of Health and Human Services.

Schindler, R., & Schindler, G. 1970. *Hawaii Kai cookbook.* New York: Hearthside.

Scholl, T. O., Hediger, M. L., Schall, J. I., Khoo, C. S., & Fischer, R. L. 1996. Dietary and serum folate: Their influence on the outcome of pregnancy. *American Journal of Clinical Nutrition, 63,* 520–525.

Schroll, K., Moreiras-Varela, O., Schlettwein-Gsell, D., Decarli, B., de Groot, L., & van Staveren, W. 1997. Cross-cultural variations and changes in food-group intake among elderly women in Europe: Results from the Survey in Europe on Nutrition and the Elderly: a Concerted Action (SENECA). *American Journal of Clinical Nutrition, 65,* 1282S–1289S.

Schultz, J. 1995. Development of the Prairie palate: The red-meat eaters. In J. M. Powers & A. Stewart (Eds.), *Our northern bounty: A celebration of Canadian cuisine.* Toronto: Random House.

Schultz, J. D., Spindler, A. A., & Josephson, R. V. 1994. Diet and acculturation in Chinese women. *Journal of Nutrition Education, 26,* 266–272.

Schwab, T., Meyer, J., & Merrell, R. 1994. Measuring attitudes and health beliefs among Mexican Americans with diabetes. *Diabetes Educator, 20,* 221–227.

Scourby, A. 1984. *The Greek Americans.* Boston: Twayne.

Scribner, R., & Dwyer, J. H. 1989. Acculturation and low birthweight among Latinos in the Hispanic HANES. *American Journal of Public Health, 79,* 1263–1267.

Seakins, J. M., Elliott, R. B., Quested, C. M., & Mataumua, A. 1987. Lactose malabsorption in Polynesian and white children in the southwest Pacific studied by breath hydrogen technique. *British Medical Journal, 295,* 876–878.

Segal, M. 1992, January–February. Native food preparation fosters botulism. *FDA Consumer,* 23–27.

Serdula, M. K., Cairns, K. A., Williamson, D. F., Fuller, M., & Brown, J. E. 1991. Correlates of breast-feeding in low-income population of whites, blacks, and Southeast Asians. *Journal of the American Dietetic Association, 91,* 41–45.

Shad, J. A., Chinn, C. G., & Brann, O. S. 1999. Acute hepatitis after ingestion of herbs. *Southern Medical Journal, 92,* 1095–1097.

Sheikh, A. A., & Sheikh, K. S. (Eds.). 1989. *Eastern and Western approaches to healing: Ancient wisdom and modern knowledge.* New York: Wiley.

Sheikh, A. A., Kunsendorf, R. G., & Sheikh, K. S. 1989. Healing images: From ancient wisdom to modern science. In A. A. Sheikh & K. S. Sheikh (Eds.), *Eastern and Western approaches to healing: Ancient wisdom and modern knowledge.* New York: Wiley.

Shenton, J. P., Pellegrini, A. M., Brown, D., & Shenker, I. 1971. *American cooking: The melting pot.* New York: Time-Life.

Shepard, R., & Raats, M. M. 1996. Attitudes and beliefs in food habits. In H. L. Meiselman and H. J. H. MacFie (Eds.), *Food choice, acceptance and consumption.* London: Blackie Academic & Professional.

Sherwood, N. E., Harnack, L., & Story, M. 2000. Weight-loss practices, nutrition-beliefs, and weight-loss program preferences of urban American Indian women. *Journal of the American Dietetic Association, 100,* 442–446.

Shintani, T. T., Hughes, C. K., Beckham, S., & O'Conner, H. K. 1991. Obesity and cardiovascular risk intervention through the ad libitum feeding of traditional Hawaiian diet. *American Journal of Clinical Nutrition, 53,* 1647S–1651S.

Shore, B. 1980. Pacific Islanders. In S. Thernstrom, A. Orlov, & O. Handlin (Eds.), *Harvard encyclopedia of American ethnic groups.* Cambridge, MA: Belknap Press of Harvard University Press.

Shortridge, B. G., & Shortridge, J. R. 1998. Introduction: Food and American culture. In B. G. Shortridge & J. R. Shortridge (Eds.), *The taste of American Place.* Lanham, MD: Rowman & Littlefield.

Shovic, A. C. 1994. Development of a Samoan nutrition exchange list using culturally accepted foods. *Journal of the American Dietetic Association, 94,* 541–543.

Siemering, A. 1999. Class-conscious fusion. *Trend/Wire, 6*(50), 2.

Singer, E. A. 1984. Conversion through foodways acculturation: The meaning of eating in an American Hindu sect. In L. K. Brown, & K. Mussels (Eds.), *Ethnic and regional foodways in the United States.* Knoxville: University of Tennessee Press.

Singh, G. K., & Yu, S. M. 1995. Infant mortality in the United States: Trends, differentials, and projections, 1950 through 2010. *American Journal of Public Health, 85,* 957–964.

Siris, E., Miller, P., Barrett-Conner, E., Abbott, T., Sherwood, L., & Berger, M. Design of NORA, the National Osteoporosis Risk Assessment Program: A longitudinal US registry of postmenopausal women. *Osteoporosis International, 8* (suppl. 1), S62–S69.

Sisti, V. V. 1998. *Leith's Latin-American cooking.* Philadelphia: Running Press.

Skaer, T. L., Robison, L. M., Sclar, D. A., & Harding, G. H. 1996. Utilization of *curanderos* among foreign born Mexican American women attending migrant health clinics. *Journal of Cultural Diversity, 3,* 29–34.

Skardel, D. B. 1974. *The divided heart.* Lincoln: University of Nebraska Press.

Skardel, D. B. 1980. Danes. In S. Thernstrom, A. Orlov, & O. Handlin (Eds.), *Harvard encyclopedia of American ethnic groups.* Cambridge, MA: Belknap Press of Harvard University Press.

Skye, B. 1995. Traditional Cree and Iroquois foods. In J. M. Powers & A. Stewart (Eds.), *Our northern bounty: A celebration of Canadian cuisine.* Toronto: Random House.

Slemenda, C. W., Christian, J. C., Hui, S., Fitzgerald, J., & Johnston, C. C. 1991. No evidence for an effect of lactase deficiency on bone mass in pre- or post-menopausal women. *Journal of Bone Mineral Research, 6,* 1367–1371.

Sloan, R. P., Bagiella, E., & Powell, T. 1999. Religion, spirituality, and medicine. *Lancet, 353,* 664–667.

Smagula, S. 1995. Nicaraguan Americans. In R. J. Vecoli, J. Galens, A. Sheets, & R. V. Young (Eds.), *Gale encyclopedia of multicultural America.* New York: Gale Research.

Smith, C. J., Nelson, R. G., Hardy, S. A., Manahan, E. M., Bennett, P. H., & Knowler, W. C. 1996. Survey of the diet of Pima Indians using quantitative food frequency assessment and 24-hour recall. *Journal of the American Dietetic Association, 96,* 778–784.

Smith, L. 1979. Mexican-American view of Anglo medical and dietetic practices. *Journal of the American Dietetic Association, 74,* 463–464.

Smith, T. 1998. *Analysis of the 1990–96 General Opinion Social Surveys.* N.p.: National Opinion Research Center.

Snaden, J. W. 1996. Tonga. In *The 1996 Grolier multimedia encyclopedia.* Danbury, CT: Grolier Electronic Publishing.

Snow, L. F. 1983. Traditional health beliefs and practices among lower-class black Americans. *Western Journal of Medicine, 139,* 820–828.

Snyder, P. 1981. Ethnicity and folk healing in Honolulu, Hawaii. *Social Science and Medicine, 15B,* 125–132.

Snyder, P. 1984. Health service implications of folk healing among older Asian Americans and Hawaiians in Honolulu. *Gerontologist, 24,* 471–476.

Sobal, J. 1999. Social change and foodways. In *Proceedings of the Cultural and Historical Aspects of Food Symposium.* Corvallis: Oregon State University.

Sobel, J., & Stunkard, A. J. 1989. Socioeconomic status and obesity: A review of the literature. *Psychological Bulletin, 105,* 260–275.

Sodowsky, G. R., & Carey, J. C. 1988. Relationships between acculturation-related demographics and cultural attitudes of an Asian-Indian immigrant group. *Journal of Multicultural Counseling and Development, 16,* 117–136.

Sofowara, A. 1982. *Medicinal plants and traditional medicine in Africa.* Chichester, England: Wiley.

Solis, J. M., Marks, G., Garcia, M., & Shelton, D. 1990. Acculturation, access to care, and use of preventative services by Hispanics: Findings from HHANES 1982–84. *American Journal of Public Health, 80* (suppl.), 11–19.

Solomon, J., & Solomon, O. P. 1979. *Cracklin bread and asfidity.* Tuscaloosa: University of Alabama Press.

Sorlie, P. D., Backlund, E., & Keller, J. B. 1995. U.S. Mortality by economic, demographic, and social characteristics: The National Longitudinal Study. *American Journal of Public Health, 85,* 949–956.

Spector, R. E. 1991. *Cultural diversity in health and illness* (3rd ed.). Norwalk, CT: Appleton & Lange.

Spector, R. E. 1996. *Cultural diversity in health and illness* (4th ed.). Norwalk, CT: Appleton & Lange.

Spence, H. 1960. *The story of America's religions.* New York: Holt, Rinehart & Winston.

Spicer, E. H. 1980. American Indians. In S. Thernstrom, A. Orlov, & O. Handlin (Eds.), *Harvard encyclopedia of American ethnic groups.* Cambridge, MA: Belknap Press of Harvard University Press.

Spindler, A. A., & Schultz, J. D. 1996. Comparison of dietary variety and ethnic food consumption among Chinese, Chinese American, and white women. *Journal of Agriculture and Human Values, 131,* 64–73.

Spring, M. A., Ross, P. J., Etkin, N. L., & Deinhard, A. S. 1995. Sociocultural factors in the use of prenatal care by Hmong women, Minneapolis. *American Journal of Public Health, 85,* 1015–1017.

Stadler, J. 1980. *Genetic component of celiac-sprue.* Paper presented at the annual Midwestern Celiac-Sprue Association Conference, Des Moines, IA.

Staebler, E. 1995. Faith, family, and food: The old-order Mennonites in Waterloo County. In J. M. Powers & A. Stewart (Eds.), *Our northern bounty: A celebration of Canadian cuisine.* Toronto: Random House.

Stavig, G. R., Igra, A., & Leonard, A. R. 1988. Hypertension and related health issues among Asians and Pacific Islanders in California. *Public Health Reports, 103,* 28–37.

Stavish, S. 1998. OTC herbal preparations have mixed safety record. *Skin & Allergy News, 29,* 41.

Steinberg, R. 1969. *The cooking of Japan.* New York: Time-Life.

Steinberg, R. 1970. *Pacific and Southeast Asian cooking.* New York: Time-Life Books.

Stephenson, P. H. 1995. Vietnamese refugees in Victoria BC: An overview of immigrant and refugee health care in a medium-sized Canadian urban centre. *Social Science and Medicine, 40,* 1631–1642.

Stern, M. P., Gaskell, S. P., Hazuda, H. P., Gardner, L. I., & Haffner, S. M. 1983. Does obesity explain excess prevalence of diabetes among Mexican Americans? Results of the San Antonio Heart Study. *Diabetologia, 24,* 272–277.

Stevanovic, B. 1995. Serbian Americans. In. R. J. Vecoli, J. Galens, A. Sheets, & R. V. Young (Eds.), *Gale encyclopedia of multicultural America.* New York: Gale Research.

Stevens, J., Kumanyika, S. K., & Keill, J. E. 1994. Attitudes towards body size and dieting: Differences between elderly black and white women. *American Journal of Public Health, 84,* 1322–1325.

Stewart, A., & Powers, J. M. 1995. Introduction. In J. M. Powers & A. Stewart (Eds.), *Our northern bounty: A celebration of Canadian cuisine.* Toronto: Random House.

Steyn, K., Fourie, J., Lombard, C., Katzenellbogen, J., Bourne, L., & Jooste, P. 1996. Hypertension in the black community of the Cape Peninsula, South Africa. *East African Medical Journal, 73,* 758–763.

Stolarik, M. M. 1980. Slovaks. In S. Thernstrom, A. Orlov, & O. Handlin (Eds.), *Harvard encyclopedia of American ethnic groups.* Cambridge: MA: Belknap Press of Harvard University Press.

Storey, M., Bass, M. A., & Wakefield, L. M. 1986a. Food preferences of Cherokee teenagers in Cherokee, North Carolina. *Ecology of Food and Nutrition, 19,* 51–59.

Storey, M., Tompkins, R. A., Bass, M. A., & Wakefield, L. M. 1986b. Anthropometric measurements and dietary intakes of Cherokee Indian teenagers in North Carolina. *Journal of the American Dietetic Association, 86,* 1555–1560.

Story, M., & Harris, L. J. 1988. Food preferences, beliefs, and practices of Southeast Asian refugee adolescents. *Journal of School Health, 58,* 273–276.

Story, M., & Harris, L. J. 1989. Food habits and dietary change of Southeast Asian refugee families living in the United States. *Journal of the American Dietetic Association, 89,* 800–803.

Story, M., Strauss, K., Zephier, E., & Broussard, B. 1998. Nutritional concerns in American Indian and Alaska Native children: Transitions and future directions. *Journal of the American Dietetic Association, 98,* 170–176.

Storz, D. 1998. Filipino culture. *News and Views: The Newsletter of the San Jose Peninsula Dietetic Association, 29,* 7.

Stowers, S. L. 1992. Development of a culturally appropriate food guide for pregnant Caribbean immigrants in the United States. *Journal of the American Dietetic Association, 92,* 331–336.

Sturner, P. 1995. Colombian Americans. In R. J. Vecoli, J. Galens, A. Sheets, & R. V. Young (Eds.), *Gale encyclopedia of multicultural America.* New York: Gale Research.

Suarez, F. L., Adshead, J., Furne, J. K., & Levitt, M. D. 1998. Lactose maldigestion is not an impediment to the intake of 1500 mg calcium daily as dairy products. *American Journal of Clinical Nutrition, 68,* 1118–1122.

Sucher, K. P., & Kittler, P. G. 1991. Nutrition isn't colorblind. *Journal of the American Dietetic Association, 91,* 297–298.

Sue, D. W., & Sue, D. 1990. *Counseling the culturally different: Theory and practice* (2nd ed.). New York: Wiley.

Suganuma, E. K., Alexander, G. R., Baruffi, G., & Gilden, S. R. 1988. Infant feeding practices in Hawaii. *Hawaii Medical Journal, 47,* 112–119.

Sutherland, A. 1992. Gypsies and health care. *Western Journal of Medicine, 157,* 276–280.

Svetkey, L. P., Simons-Morton, D., Vollmer, W. M., Appel, L. J., Conlin, P. R., Ryan, D. H., Ard, J., & Kennedy, B. M. 1999. Effects of dietary patterns on blood pressure: Subgroup analysis of the Dietary Approaches to Stop Hypertension (DASH) randomized clinical trial. *Archives of Internal Medicine, 159,* 285–293.

Swartz, L. 1987. Illness negotiation: The case of eating disorders. *Social Science and Medicine, 24,* 613–618.

Swinburn, B. A., Boyce, V. L., Bergman, R. N., Howard, B. V., & Bogardus, C. 1991. Deterioration in carbohydrate metabolism and lipoprotein changes induced by modern, high fat diet in Pima Indians and Caucasians. *Journal of Clinical Endocrinology and Metabolism, 73,* 156–165.

Szathmary, E. J. E., Ritenbaugh, C., & Goodby, C. S. M. 1987. Dietary change and plasma glucose levels in an Amerindian population undergoing cultural transition. *Social Science and Medicine, 24,* 791–804.

Takooshian, H. 1995. Armenian Americans. In R. J. Vecoli, J. Galens, A. Sheets, & R. V. Young (Eds.), *Gale encyclopedia of multicultural America.* New York: Gale Research, Inc.

Tamir, A., & Cachola, S. 1994. Hypertension and other cardiovascular risk factors. In N. W. S. Zane, D. T. Takeuchi, and K. N. J. Young (Eds.), *Confronting critical health issues of Asian and Pacific Islander Americans.* Thousand Oaks, CA: Sage.

Tate, D. J. M. 1979. *The making of modern South East Asia* (Vols. I & II). Oxford: Oxford University Press.

Taylor, T. W. 1972. The states and their Indian citizens. In S. Thernstrom, A. Orlov, & O. Handlin (Eds.), *The Harvard encyclopedia of American ethnic groups* (1980). Cambridge, Mass.: Belknap Press of Harvard University Press.

Teeguarden, R. 1984. *Chinese tonic herbs.* New York: Japan Publications.

Terry, R. D. 1994. Needed: A new appreciation of culture and food behavior. *Journal of the American Dietetic Association, 94,* 501–503.

Terry, R. D., & Bass, M. A. 1984. Food practices in an eastern Cherokee township. *Ecology of Food and Nutrition, 14,* 63–70.

Teufel, N. I., & Dufour, D. L. 1990. Patterns of food use and nutrient intake of obese and non-obese Hualapai Indian women in Arizona. *Journal of the American Dietetic Association, 90,* 1229–1235.

The Henry J. Kaiser Family Foundation. 1999. *Key facts: Race, ethnicity, & medical care.* Menlo Park, CA: Author.

Thomas, D. N. 1981. Black American patient care. In G. Henderson & M. Primeaux (Eds.), *Transcultural health care.* Menlo Park, CA: Addison-Wesley.

Thomas, I., Gupta, S., Sempos, C., & Cooper, R. 1986. Serum lipids of Indian physicians living in the U.S. compared to U.S.-born physicians. *Atherosclerosis, 61,* 99–106.

Thomas, J. D. 1985. Gypsies and American medical care. *Annals of Internal Medicine, 102,* 842–845.

Thomas, J. D. 1987. Disease, lifestyle, and consanguinity in 58 American Gypsies. *Lancet, 2:* 377–379.

Thompson, T. L. 1990. Patient health care: Issues in interpersonal communication. In E. B. Ray & L. Donohew (Eds.), *Communication and health.* Hillsdale, NJ: Erlbaum.

Tirtha, S. S. S. 1998. *The Ayurvedic encyclopedia: Natural secrets to healing, prevention, and longevity.* Bayville, NY: Ayurvedic Holistic Center Press.

Tong, A. 1986. Food habits of Vietnamese immigrants. *Family Economic Review, 2,* 28–30.

Trager, J. 1995. *The food chronology.* New York: Holt.

Trevino, F. M., Moyer, M. E., Valdez, R. B., & Stroup-Benham, C. A. 1991. Health insurance coverage and utilization of health services by Mexican Americans, mainland Puerto Ricans, and Cuban Americans. *Journal of the American Medical Association, 265,* 233–237.

Trifonopoulos, M., Kuhnlein, H. V., & Receveur, O. 1998. Analysis of 24-hour recalls of 164 fourth- to sixth-grade Mohawk children in Kahnawake. *Journal of the American Dietetic Association, 98,* 814–816.

Trillin, C. 1974. *American fried.* New York: Doubleday.

Trostler, N. 1997. Health risks of immigration: The Yemenite and Ethiopian cases in Israel. *Biomedicine and Pharmacotherapy, 51,* 352–359.

Trotter, R. T. 1991. A survey of four illnesses and their relationship to intracultural variation in a Mexican-American community. *American Anthropologist, 93,* 115–125.

Tsunehara, C. H., Leonetti, D. L., & Fujimoto, W. Y. 1990. Diet of second-generation Japanese-American men with and without non-insulin-dependent diabetes. *American Journal of Clinical Nutrition, 52,* 731–738.

Tumulty, P. 1970. What is a clinician and what does he do? *New England Journal of Medicine, 283,* 20–24.

Turner, N. J., Philip, S., & Turner, R. D. 1995. Traditional native plant foods in contemporary cuisine in British Columbia. In J. M. Powers & A. Stewart (Eds.), *Our northern bounty: A celebration of Canadian cuisine.* Toronto: Random House.

U.S. Department of Agriculture, Food and Nutrition Service. 1980. *Southeast Asian American food habits.* Publication No. FNS-225. Washington, D.C.: U.S. Department of Agriculture.

U.S. Department of Health and Human Services. 1997. Asian Americans and Pacific Islanders. *Prevention Report, 12,* insert.

Umaña-Murray, M. 1996. *Three generations of Chilean cuisine.* Los Angeles: Lowell House.

Unaeze, F. E., & Perrin, R. E. 1995. Haitians. In R. J. Vecoli, J. Galens, A. Sheets, & R. V. Young (Eds.), *Gale encyclopedia of multicultural America.* New York: Gale Research.

Usunier, J. C. 1993. *International marketing.* New York: Prentice Hall.

US–USSR Steering Committee for Problem Area I. 1984. The pathogenesis of atherosclerosis. Nutrient intake and its association with high-density lipoprotein and low-density lipoprotein cholesterol in selected U.S.–U.S.S.R. subpopulations. *American Journal of Clinical Nutrition, 39,* 942–952.

Valassi, K. 1962. Food habits of Greek Americans. *American Journal of Clinical Nutrition, 11,* 240–248.

Valle, V. M., & Valle, M. L. 1995. *Recipe of memory: Five generations of Mexican cuisine.* New York: New Press.

Van der Post, L. 1970. *African cooking.* New York: Time-Life Books.

Van der Post, L. 1978. *First catch your eland.* New York: Morrow.

van Straalen, A. 1986. *The book of holidays around the world.* New York: Dutton.

Várdy, S. B., & Szendrey, T. 1995. Hungarian Americans. In R. J. Vecoli, J. Galens, A. Sheets, & R. V. Young (Eds.), *Gale encyclopedia of multicultural America.* New York: Gale Research.

Varela, G., Moreiras-Varela, O., & Blazquez, M. J. 1985. Urbanization, nutritive status, and food habits in the Spanish populations. *Bibliotheca Nutritio et Dieta, 36,* 55–71.

Vaughan, L. A., Benyshek, D. C., & Martin, J. F. 1997. Food acquisition habits, nutrient intakes, and anthropometric data of Havasupai adults. *Journal of the American Dietetic Association, 97,* 1275–1282.

Visson, L. 1982. *The complete Russian cookbook.* Ann Arbor, MI: Ardis.

Voeks, R. A. 1997. *Sacred leaves of Candomblé: African magic, medicine, and religion in Brazil.* Austin: University of Texas Press.

Vogel, V. J. 1976. American Indian foods used in medicine. In W. D. Hand (Ed.), *American folk medicine.* Los Angeles: University of California Press.

Vogel, V. J. 1981. American Indian medicine. In G. Henderson & M. Primeaux (Eds.), *Transcultural health care.* Menlo Park, CA: Addison-Wesley.

Wallendorf, M., & Reilly, M. D. 1983. Ethnic migration, assimilation, and consumption. *Journal of Consumer Research, 10,* 292–302.

Walsh, R., & McCarthy, J. 1995. *Traveling Jamaica with knife, fork & spoon.* Freedom, CA: Crossing.

Walter, E. 1971. *American cooking: Southern style.* New York: Time-Life.

Walter, E. 1971. The South's great gift of soul food. In *American cooking: Southern style.* New York: Time-Life.

Wang, L. L. 1995. Chinese Americans. In R. J. Vecoli, J. Galens, A. Sheets, & R. V. Young (Eds.), *Gale encyclopedia of multicultural America.* New York: Gale Research.

Wang, M. C., & Sabry, Z. I. 1994. A culturally-sensitive method for the assessment of diet in a group of American Chinese children. *American Journal of Clinical Nutrition, 59* (suppl.), 290S.

Wansiuk, B. & Sangerman, C. 2000. *The taste of comfort.* American Demographics (July).

Wargelin, M. 1995. Finnish Americans. In R. J. Vecoli, J. Galens, A. Sheets, & R. V. Young (Eds.), *Gale encyclopedia of multicultural America.* New York: Gale Research.

Waterlow, J. C. 1989. Diet of the classical period of Greece and Rome. *European Journal of Clinical Nutrition, 43* (suppl. 2), 3–12.

Waxler-Morrison, N., Anderson, J., & Richardson, E. 1990. *Cross cultural caring: A handbook for health professionals in western Canada.* Vancouver: University of British Columbia Press.

Weaver, H. N. 1999. Transcultural nursing with Native Americans: Critical knowledge, skills, and attitudes. *Transcultural Nursing, 10,* 197–202.

Weaver, W. W. 1982. *Sauerkraut Yankees: Pennsylvania-German foods and foodways.* Philadelphia: University of Philadelphia Press.

Wechsberg, J. 1968. *The cooking of Vienna's empire.* New York: Time-Life.

Wechsberg, J. 1974. *The cooking of Vienna's empire.* Alexandria, VA: Time-Life.

Wein, E. E. 1995. Evaluating food use by Canadian aboriginal peoples. *Canadian Journal of Physiological Pharmacology, 73,* 759–764.

Wenkan, N. S., & Wolff, R. J. 1970. A half century of changing food habits among Japanese in Hawaii. *Journal of the American Dietetic Association, 57,* 29–32.

Werstman, V. F. 1995. Romanian Americans. In R. J. Vecoli, J. Galens, A. Sheets, & R. V. Young, (Eds.), *Gale encyclopedia of multicultural America.* New York: Gale Research.

Westermeyer, J., Neider, J., & Vang, T. F. 1984. Acculturation and mental health: A study of Hmong refugees at 1.5 and 3.5 years postmigration. *Social Science and Medicine, 18,* 87–93.

Wetzel, M. S., Eisenberg, D. M., & Kaptchuk, T. J. 1998. Courses involving complementary and alternative medicine in US medical schools. *Journal of the American Medical Association, 280,* 784–787.

Wheeler, M., & Haider, S. Q. 1979. Buying and food preparation patterns of ghetto blacks and Hispanics in Brooklyn. *Journal of the American Dietetic Association, 75,* 560–563.

White, S. L., & Maloney, S. K. 1990. Promoting healthy diets and active lives to hard-to-reach groups: Market research study. *Public Health Reports, 105,* 224–231.

Whitney, E. N., & Rolfes, S. R. (1993). *Understanding nutrition* (6th ed.). St. Paul, MN: West.

Whittemore, A. S., Wu-Williams, A. J., Lee, M., Shu, Z., Gallagher, R. P., Deng-ao, J., Lun, Z., Xianghui, W., Kun, C., Jung, D., Teh, C. Z., Chengde, L., Yao, X. J., Paffenbarger, R. S., & Henderson, B. E. 1990. Diet, physical activity, and colorectal cancer among Chinese in North America and China. *Journal of the National Cancer Institute, 82,* 915–926.

Will, J. C., & Casper, M. 1996. The contribution of diabetes to early deaths from ischemic heart disease: U.S. gender and racial comparisons. *American Journal of Public Health, 86,* 576–579.

Willet, W. C., Sacks, F., Trichopoulou, A., Drescher, G., Ferro-Luzzi, A. Helsing, E., & Trichopoulos, D. 1995. Mediterranean diet pyramid: A cultural model for healthy eating. *American Journal of Clinical Nutrition, 61* (suppl.), 1402S–1406S.

William, P. T., Fortman, S. P., Terry, R. B., Garay, S. C., Vranizan, K. M., Ellsworth, N., & Wood, P. D. 1987. Association of dietary fat, regional adiposity and blood pressure in men. *Journal of the American Medical Association, 257,* 3251–3256.

Wilson, C. S. 1985. Nutritionally beneficial cultural practices. *World Review of Nutrition and Diet, 45,* 68–96.

Wilson, D. K., Bayer, L., Krishnamoorthy, J. S., Amprey-Thornhill, G., Nicholson, S. C., & Sica, D. A. 1999. The prevalence of salt sensitivity in an African-American adolescent population. *Ethnicity & Disease, 9,* 350–358.

Wilson, D. K., Kliewer, W., Plybon, L., & Sica, D. A. 2000. Socioeconomic status and blood pressure reactivity in healthy black adolescents. *Hypertension, 35,* 496–500.

Wilson, E. G. 1971. *A West African cookbook.* New York: Avon.

Wilson, J. 1971. *American cooking: The Eastern Heartland.* New York: Time-Life.

Wilson, T. W. 1986. History of salt supplies in West Africa and blood pressures today. *Lancet, 1,* 784–786.

Wilson, T. W., & Grim, C. E. 1991. Biohistory of slavery and blood pressure differences in blacks today: A hypothesis. *Hypertension, 17,* I122–I128.

Wilson, T. W., Hollifield, L. R., & Grim, C. E. 1991. Systolic blood pressure levels in black populations in sub-Saharan Africa, the West Indies, and the United States: A meta-analysis. *Hypertension, 18,* I87–I91.

Wilson, U. M. 1983. Nursing care of American Indian patients. In M. S. Orque, B. Bloch, & L. S. A. Monrroy (Eds.), *Ethnic nursing care: A multicultural approach.* St. Louis: Mosby.

Windham, C. T., Wyse, B. W., & Hansen, R. G. 1983. Nutrient density in diets in the USDA Nationwide Food Consumption Survey, 1977–78: Impact of socioeconomic status on dietary density. *Journal of the American Dietetic Association, 82,* 28–34.

Winters, E., & Swartz, M. 1995. Hawaiians. In R. J. Vecoli, J. Galens, A. Sheets, & R. V. Young (Eds.), *Gale encyclopedia of multicultural America.* New York: Gale Research.

Winzelberg, A. J., Eppstein, D., Eldredge, K. L., Wilfrey, D., Dasmahapatra, R., Dev, P., & Taylor, C. B. 2000. Effectiveness on an Internet-based program for reducing risk factors for eating disorders. *Journal of Consulting Clinical Psychology, 68,* 346–350.

Wolfe, L. 1970. *The cooking of the Caribbean Islands.* New York: Time-Life.

Wolfe, W. S., & Sanjur, D. 1988. Contemporary diet and body weight of Navajo women receiving food assistance: An ethnographic and nutritional investigation. *Journal of the American Dietetic Association, 88,* 822–827.

Wolfe, W. S., Weber, C. W., & Arviso, K. D. 1985. Use and nutrient composition of traditional Navajo foods. *Ecology of Food and Nutrition, 17,* 323–326.

Woloshin, S., Bickell, N. A. Schwartz, L. M., Gany, F., & Welch, H. G. 1995. Language barriers in medicine in the United States. *Journal of the American Medical Association, 273,* 724–728.

Wong, S. 1995. Asian flavors. In J. M. Powers & A. Stewart (Eds.), *Our northern bounty: A celebration of Canadian cuisine.* Toronto: Random House of Canada.

Wood, J. B. 1989. Communicating with older adults in health care settings: Cultural and ethnic considerations. *Educational Gerontology, 15,* 351–362.

Wood, R. C. 1995. *The sociology of the meal.* Edinburgh: Edinburgh University Press.

Woolf, N., Conti, K. M., Johnson, C., Martinez, V., McCloud, J., & Zephier, E. M. 1999. *Northern Plains Indian food practices, customs, and holidays.* Chicago: American Dietetic Association/American Diabetes Association.

Wray, L. A. 1992. Health policy and ethnic diversity in older Americans: Dissonance or harmony? *Western Journal of Medicine, 157,* 357–361.

Wright, A. L., Clark, C., & Bauer, M. 1993. Maternal employment and infant feeding practices among the Navajo. *Medical Anthropology Quarterly, 7,* 260–280.

Wyant, K. W., & Meiselman, H. L. 1984. Sex and race differences in food preferences of military personnel. *Journal of American Dietetic Association, 84,* 169–175.

Wyman, C. 1993. *I'm a Spam fan.* Stamford, CT: Longmeadow.

Yagalla, M. V., Hoerr, S. L., Song, W. O., Enas, E., & Garg, A. 1996. Relationship of diet, abdominal obesity, and physical activity to plasma lipoprotein levels in Asian Indian physicians residing in the United States. *Journal of the American Dietetic Association, 96,* 257–261.

Yardley, M. 1970. *Hawaii cooks.* Tokyo: Tuttle.

Yetley, E. A., Yetley, M. J., & Aguirre, B. 1981. Family role structure and food-related roles in Mexican-American families. *Journal of Nutrition Education, 13* (suppl. 1), S96–S101.

Yianilos, T. K. 1970. *The complete Greek cookbook.* New York: Funk & Wagnalls.

Yip, R., Li, Z., & Chong, W. H. 1991. Race and birth weight: The Chinese example. *Pediatrics, 87,* 688–693.

Yoder, D. 1976. Hohman and Romanus: Origins and diffusion of Pennsylvania German powwow manual. In W. D. Hand (Ed.), *American folk medicine.* Berkeley: University of California Press.

Yom, M. S., Gordon, B. H. J., & Sucher, K. P. 1995. Korean dietary habits and health beliefs in the San Francisco Bay area. *Journal of the American Dietetic Association, 95* (suppl.), A-98.

Zane, N. W. S., & Kim, J. H. 1994. Substance use and abuse. In N. W. S. Zane, D. T. Takeuchi, & K. N. J. Young (Eds.), *Confronting critical health issues of Asian and Pacific Islander Americans.* Thousand Oaks, CA: Sage.

Zapata, J., & Shippee-Rice, R. 1999. The use of folk healing and healers by six Latinos living in New England: A preliminary study. *Journal of Transcultural Nursing, 10,* 136–142.

Zelinsky, W. 1973. *The cultural geography of the United States.* Englewood Cliffs, NJ: Prentice Hall.

Zeranska, A. 1989. *The art of Polish cooking.* Gretna, LA: Pelican.

Zhou, Y. D., & Britten, H. C. 1994. Increased iron content of some Chinese foods due to cooking in steel woks. *Journal of the American Dietetic Association, 94,* 1153–1154.

Ziegler, V. S., Sucher, K. P., & Downes, N. J. 1989. Southeast Asian renal exchange list. *Journal of the American Dietetic Association, 89,* 85–92.

Zillikens, M. C., & Conway, J. M. 1990. Anthropometry in blacks: Applicability of generalized skinfold equations and differences in fat patterning between blacks and whites. *American Journal of Clinical Nutrition, 52,* 45–51.

Ziraldo, D. 1995. Founding the Vintners Quality Alliance (VQA). In J. M. Powers & A. Stewart (Eds.), *Our northern bounty: A celebration of Canadian cuisine.* Toronto: Random House.

Zive, M. M., Taras, H. L., Broyles, S. L., & Frank-Spohrer, G. C. 1995. Vitamin and mineral intakes of Anglo-American and Mexican-American preschoolers. *Journal of the American Dietetic Association, 95,* 329–335

Index

D

E

O

P